# Extruding Plastics

# Extruding Plastics

## *A practical processing handbook*

D. V. Rosato

*Plastics Institute of America*
*Rhode Island School of Design*
*Chatham, MA, 02633*
*USA*

KLUWER ACADEMIC PUBLISHERS
BOSTON/DORDRECHT/LONDON

**Distributors for North, Central and South America:**
Kluwer Academic Publishers
101 Philip Drive
Assinippi Park
Norwell, Massachusetts 02061 USA
Telephone (781) 871-6600
Fax (781) 871-6528
E-Mail <kluwer@wkap.com>

**Distributors for all other countries:**
Kluwer Academic Publishers Group
Distribution Centre
Post Office Box 322
3300 AH Dordrecht, THE NETHERLANDS
Telephone 31 78 6392 392
Fax 31 78 6546 474
E-Mail services@wkap.nl>

Electronic Services <http://www.wkap.nl>

A catalogue record for this book is available from
the British Library

ISBN 0-412-82810-3



*Printed on acid-free paper.*

Printed in the United States of America

# Contents


**Preface** **xiii**

**1 The complete extrusion process** **1**
Overview 1
Extruder type and construction 6
Extruder operation 10
Extrusion coating 13
Other lines 14
Processing plastic 19
Extruder heating and cooling profile 22
Auxiliary equipment 24
Plant operation 28
Plant safety 32
Acceptable risk 39
Energy conservation 40
Processor 42
Custom 43
Captive 43
Proprietary 43
Predicting performance 43
Scale-up 44
Myths and fact 44
People and productivity 46
Processing and patience 47
Troubleshooting 47
History 50

**2 Extrusion machine and line** **54**
Features of extruder and production line 54
Overview 54

Single-screw extruder 56
Twin-screw extruder 60
Extruder components 65
Alignment 67
Borescoping 69
Drive system 70
Barrel 75
Barrel and feed unit 93
Auxiliary equipment 98
Operating extruder and complete line 116
Planning a line 118
Start-up and shut-down 121
Temperature/pressure and output 125
Maintenance 129
Troubleshooting 130
Training 142
Coextrusion 144
Melt flow characteristics 146
Types of construction 150
Selecting equipment 152
Safety 158

**3 Processing plastic material 160**
Overview 160
Plastic type 162
Properties 168
Heat profile 173
Melt flow and rheology 174
Melt flow and defect 178
Thermal properties 180
Purging 187
Recycling 188
Drying 190

**4 Plasticizing screw process 199**
Overview 199
Screw design 208
Output 210
Screw torque 212
Screw cooling 213
Performance 214
Mixing devices 216
Barrier screw 219
Melting action 220
Venting 221
Process performance 225

| | |
|---|---|
| Wear | 226 |
| Screw inspection | 226 |
| **5 Die design and performance** | **228** |
| Overview | 228 |
| Melt behavior | 241 |
| Die land | 242 |
| Slit | 246 |
| Rod | 246 |
| Tear drop | 246 |
| Temperature | 246 |
| Manifold and die | 250 |
| Coextrusion dies | 256 |
| Special dies | 266 |
| Die design | 270 |
| Construction materials | 273 |
| Maintenance | 275 |
| Troubleshooting | 276 |
| **6 Process control and computer** | **283** |
| Overview | 283 |
| Trade-off | 287 |
| Sensors | 290 |
| Accuracy | 296 |
| Integration | 301 |
| Intelligent processing | 304 |
| **7 Blown film** | **305** |
| Overview | 305 |
| Plastic materials | 312 |
| Blown tube characteristics | 315 |
| Start-up | 320 |
| Orientation | 323 |
| Process optimization | 324 |
| Line control | 325 |
| Output rate | 326 |
| Die | 334 |
| Multi-layer or coextrusion | 335 |
| Film randomization | 338 |
| Troubleshooting | 348 |
| **8 Flat film** | **349** |
| Overview | 349 |
| Other type film | 356 |
| Flat or blown film | 357 |
| Plastic materials | 358 |

Flat film characteristics 361
Start-up 362
Shut-down 362
Orientation 363
Process control 367
Line control 369
Output rate 369
Die 370
Troubleshooting 372

**9 Sheet** **376**
Overview 376
Plastic materials 383
Processing ABS 383
Processing PET 386
Orientation 387
Process optimization 388
Line components 388
Roll stack 388
Coextruded or laminated sheets 396
Pull roll 400
Air knife 400
Trim and slitting 400
Antistatic bath 401
Process control 402
Cut-off 403
Winder 406
Die 409
Applications 411
Troubleshooting 412

**10 Calendering** **418**
Overview 418
Calendering or extrusion 420
Calendering operation 422
Surface finishing 427
Plastic materials, 430
Fluxing and feeding 432
Heat sensitivity 433
Contamination 434
Recycling 434
Orientation 435
Processing optimization 435
Roll 435

Control 437
Application 437
Safety 438
Costing 438
Troubleshooting 440

**11 Coating and lamination 441**
Overview 441
Substrate 447
Plastic materials 448
Processing characteristics 449
Shut-down 456
Process optimization 457
Gauge variation 457
Neck-in 457
Substrate 459
Adhesion 460
Printing 462
Block and slip surface quality 463
Control 463
Dimensional stability 464
Heat sealing 464
Die 464
Applications 465
Safety 465
Troubleshooting 466

**12 Wire and cable 469**
Overview 469
Plastic materials 471
Processing characteristics 474
Other coating lines 478
Start-up 483
Process optimization 485
Die 486
Troubleshooting 493

**13 Pipe and tube 494**
Overview 494
Plastic materials 501
Processing characteristics 503
Process optimization 514
Orientation 518
Die 520

Applications 524
Cost 525
Troubleshooting 527

**14 Profiles** **528**
Overview 528
Plastic materials 532
Processing characteristics 533
Coated profile substrate 535
Process optimization 536
Cooling profile 536
Window frame precision fast line 537
Rod 539
Robotic profile 540
Mechanical strength 540
Applications 541
Die 542
Troubleshooting 551

**15 Blow molding** **552**
Overview 552
Plastic materials 554
Coextrusion 556
Processing characteristics 556
Extrusion blow molding 556
Injection blow molding 562
Stretch blow molding 570
Process optimization 574
Die/mold/tool 577
Applications 584
Cost 584
Troubleshooting 592

**16 Fiber and filament** **593**
Overview 593
Plastic materials 595
Processing characteristics 596
Melt spinning 597
Dry spinning 598
Wet spinning 598
Other processes 599
Fiber twist 599
Filtration 600
Nonwoven 601

Spinneret 602
Process optimization 603
Theoretical versus actual value 604
Applications 606
Troubleshooting 606

**17 Compounding** **608**
Overview 608
Plastic materials 609
Blend 615
Processing characteristics 616
Batch compounding 616
Continuous compounding 617
Feeder and blender 618
Reactive compounding 620
Pellet 622
Dicer 627
Type of mixer 628
Process optimization 628
Control 629
Die 629
Troubleshooting 629

**18 Other processes** **631**
Introduction 631
Injection molding 631
Noncontinuous extruder 638
Continuous extruder 640
Mold 641
Injection–compression molding 642
Injection/blow molding with orientation 643
Continuous molding with extruder 645
Velcro strip Ferris wheel molding 645
Carousal molding 650
Thermoforming 650
Scrapless forming 661
Postforming 661
Netting 663
Foam 663

**19 Testing and quality control** **673**
Introduction 673
Quality assurance 674
Testing 675

Type of test 677
Density and specific gravity 677
Melt index 680
Melt rheometer 682
Quality control 686
Statistical process control 687
Monitoring process variables 688
Assessing statistical practice 689
Quality system regulation 689
Extrusion variables 690
Plastic material 691
Melt flow 692
Process control 693
Product 695
Processing intelligence 696
Judgment and experience 696
Problem and solution 697

**20 Summary** **699**
Introduction 699
Success by design 700
Cost 700
Estimating product cost 705
Direct and indirect cost 706
Technology cost modeling 707
Processing improvement 711
Process types 712
Plant control 719
Energy 720
Plastic and energy 721
People 725
Processing and patience 725
Plastics growth with extrusion 725

**Appendices** **732**
Metric conversion charts 732
Standard metric symbols 735
Mathematical symbols and abbreviations 735
Greek alphabet 736
Properties of water 736
Angle conversions 736

**References** **737**

**Index** **755**

# Preface

Worldwide, extrusion lines successfully process more plastics into products than other processes by consuming at least 36 wt% of all plastics. They continue to find practical solutions for new products and/or problems to meet new product performances.

This book, with its practical industry reviews, is a unique handbook (the first of its kind) that covers over a thousand of the potential combinations of basic variables or problems with solutions that can occur from up-stream to down-stream equipment. Guidelines are provided for maximizing processing efficiency and operating at the lowest possible cost. It has been prepared with an awareness that its usefulness will depend greatly upon its simplicity and provision of essential information.

It should be useful to: (1) those already extruding and desiring to obtain additional information for their line and/or provide a means of reviewing other lines that can provide their line with operating improvements; (2) those processing or extruding plastics for the first time; (3) those considering going into another extrusion process; (4) those desiring additional information about employing the design of various products more efficiently, with respect to both performance and cost; (5) those contemplating entering the business of extrusion; (6) those in new venture groups, materials development, and/or market development; (7) those in disciplines such as nonplastics manufacturers, engineers, designers, quality control, financial, and management; and (8) those requiring a textbook on extrusion in trade schools and high schools or colleges.

Only a few theoretical concepts are included. Persons dealing with practicalities will find the theoretical explanations enlightening and understandable. Theorists will gain insight into the practical limitations of equipment, plastics, and people. There are various excellent books that provide detailed theoretical analysis; they are listed in the Reference section. However, this review of the industry's operating experiences presented with 387 figures and 128 tables will help make extrusion lines operate more efficiently and expand their capabilities. These experiences establish empirical analyses that, in turn, develop theoretical concepts.

The book's practical approach provides for the novice and experienced personnel a description of extrusion advances and trends. This leads to 'streamlining' processing lines by anticipating and understanding problems that could occur, what causes them, how to eliminate them, and/or how to take corrective action. The content is arranged to provide a natural progression from simple to complex situations. It correlates materials, equipment, process controls, product requirements, and people. Thus, one learns how to avoid 'preventable' down-time and/or unacceptable products. A step-by-step approach is used to simplify and understand practical fundamentals that interrelate processes with plastics-to-products.

This comprehensive book explains in a clear, concise, simple format the different extrusion processes that produce many different commodity and engineered plastic products. Included are: (1) flexible-to-rigid-to-foam-to-decorative constructions; (2) both simple and complex shapes; (3) processing virgin and/or recycled plastics; (4) procedures for start-ups and shut-downs; (5) postforming in-line, troubleshooting, and so on. A target has been to open up new possibilities in the art of extruding, with the goal of approaching zero defect processing.

It is essential to re-emphasize what is reviewed in the text, namely that all data and information presented on equipment and plastics are to be used as guides. Obtain the latest, most complete, information from suppliers.

Information is derived from the author's experience as well as personal contacts with industry worldwide, including many industry people, and industry sources, as is evident by the references listed at the end of this book. Every possible care has been taken to ensure that the content of this book is correct. While the information contained is believed to be true and accurate, no one, including the editor, contributors, sources, and publisher, can accept any legal responsibility for any errors, omissions, operating risks/damages, or other factors. All information is provided in good faith but without legal responsibility. This book will serve to help people understand the process of extrusion and that includes maintaining the highest degree of safety and avoiding injury to personnel and damage to equipment.

Information contained in this book may be covered by US and worldwide patents. No authorization to utilize these patents is given or implied; they are discussed for information only. Disclosures are neither a license to operate nor a recommendation to infringe any patent. No attempt has been made to refer to patents by number, title, or ownership.

D. V. Rosato
Chatham, MA., USA
December, 1997

# 1

# The complete extrusion process

## OVERVIEW

The extrusion processes offer the advantages of complete versatile plastic processing techniques unsurpassed in economic importance by any other process. Worldwide, extruder lines are the largest converters of plastics and can be considered the most important production machinery in the plastic industry. Commercially, extrusion lines are targeted to give advantages with regard to operating cost (output per hour). The two main reasons that make them attractive to the processors and markets are their almost unlimited range of applications and their continuous production capabilities to meet new market challenges [1]. There are also batch or noncontinuous processing of plastics (Chapter 18) that include injection blow molding (Chapter 15), injection molding (Chapter 18), and the major market of compounding plastic materials (Chapter 17).

This book provides practical information that affects processing performances when changing individual variables during extrusion as well as on the important up-stream and down-stream equipment. The variables all relate to the common factors of temperatures, pressures, and times which, in turn, relate to the manufacturing output rates and costs. Each chapter in this book contains important information on different variables and the behavior of plastics during processing. Detailed information in one chapter that applies elsewhere will not be repeated in subsequent chapters. Thus, to gain the maximum benefit from this book, review or examine all chapters. The table of contents and index provide helpful cross references. Advantages and disadvantages as well as troubleshooting guides are provided throughout this book. The information presented comes from many worldwide industry sources, individuals and companies that include those listed in the Reference section at the end of this book.

All processes fit into an overall scheme that requires the interaction and

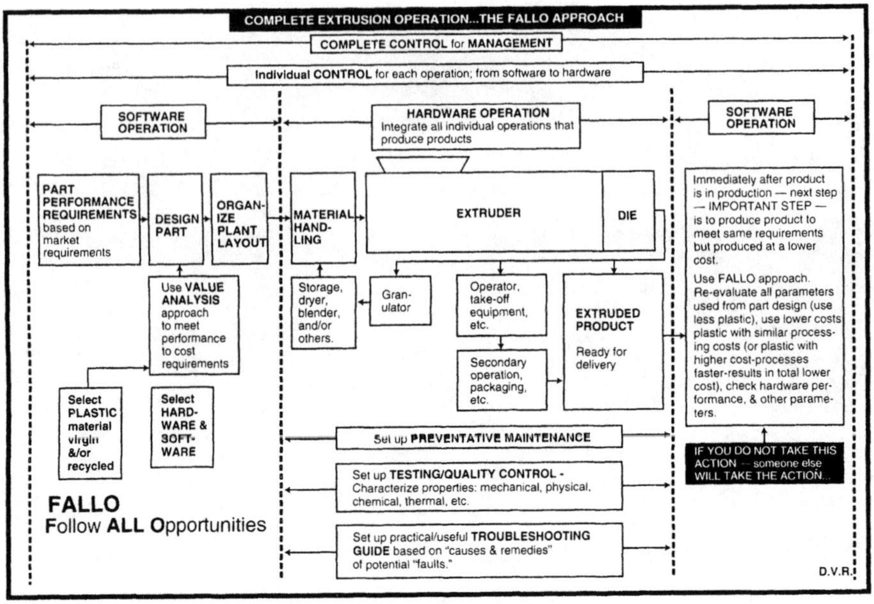


**Figure 1.1** The FALLO approach.

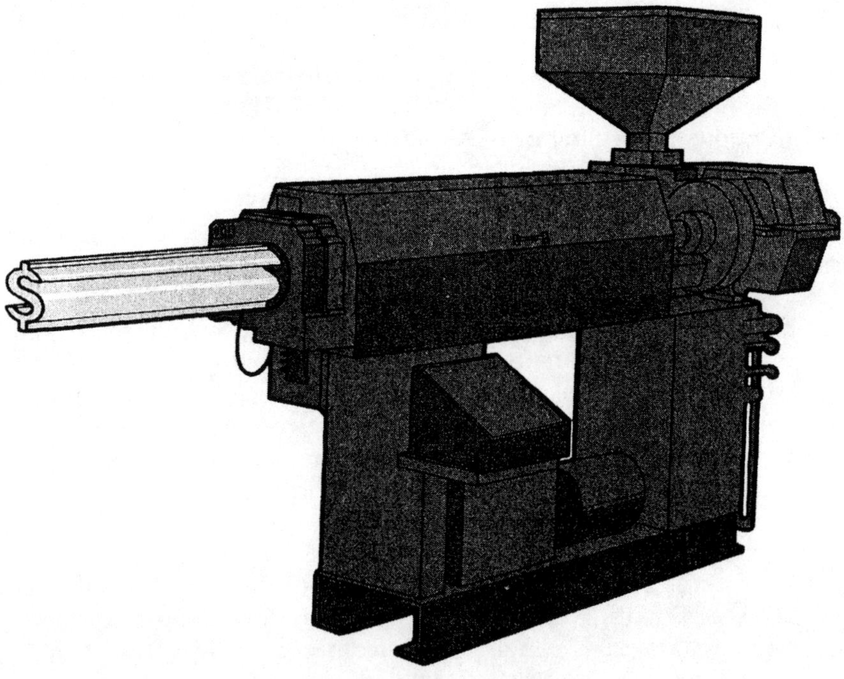

**Figure 1.2** Extruders provide products used worldwide with profits.

proper control of different operations. An example is shown in Fig. 1.1 where the block diagram pertains to any process system. The FALLO (Follow ALL Opportunities ) approach makes one aware that many steps are involved in processing and all must be properly understood and coordinated [1–6]. Basically the FALLO approach consists of: (1) designing a product to meet performance and manufacturing requirements at the lowest cost; (2) specifying the proper plastic material(s) that meet product performance requirements after being processed; (3) specifying the complete equipment line by (a) designing the die 'around' the product, (b) putting the 'proper performing' extruder 'around' the die, (c) setting up auxiliary equipment (up-stream to down-stream) to 'match' the operation of the complete line, and (d) setting up the required 'complete controls' (such as testing, quality control, troubleshooting, maintenance, data recording, etc.) to produce 'zero defects;' and (4) purchasing and properly warehousing plastic materials. Using this type of approach leads to maximizing the product's profitability (Fig. 1.2).

Plastics, predominantly thermoplastics (TPs) (Chapter 3), are usually

**Figure 1.3** Double 'H' Plastics Co. with 14 complete pipe/profile lines with Welex extruders and Gonair/Gatto coolers.

extruded using a single-screw where a solid material is fed through the extruder's plasticator (barrel/screw). The extruder forms a homogeneous plastic melt and forces it through a die orifice that relates to the shape of the product's cross section. The formed TP melt (extrudate) is cooled as it is being drawn away from the die exit through down-stream equipment (auxiliary equipment includes up-stream and down-stream equipment). Products produced include films, sheets, profiles, pipes, tubes, rods, wire/cable coverings, coatings, filaments, blown shapes, and others as reviewed throughout this book.

Figures 1.3 and 1.4 are examples of extrusion lines. Figure 1.5 shows a Conair Group completely automated system from railcar to the finished products with: (1) railcar unloading plastics into silos and bins with high volume conveyers via inventory control systems; (2) plastics conveying by compressed and/or vacuum lines with dust collectors, loading controls, etc.; (3) high volume blending and feeding with different units such

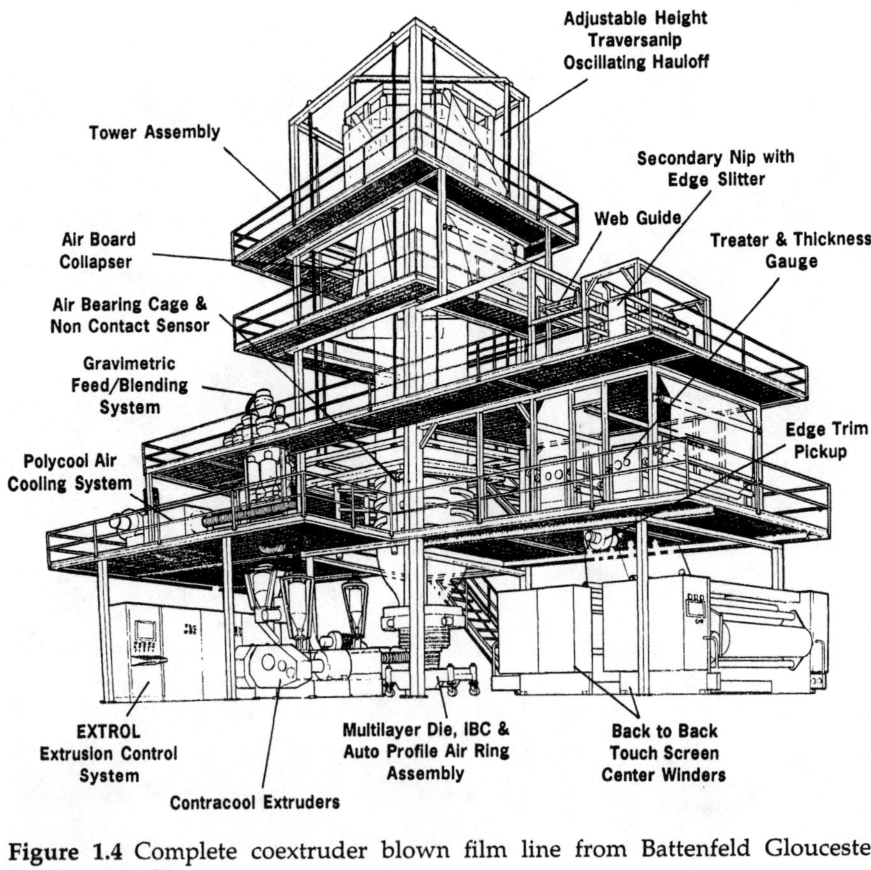


**Figure 1.4** Complete coextruder blown film line from Battenfeld Gloucester Engineering Co.

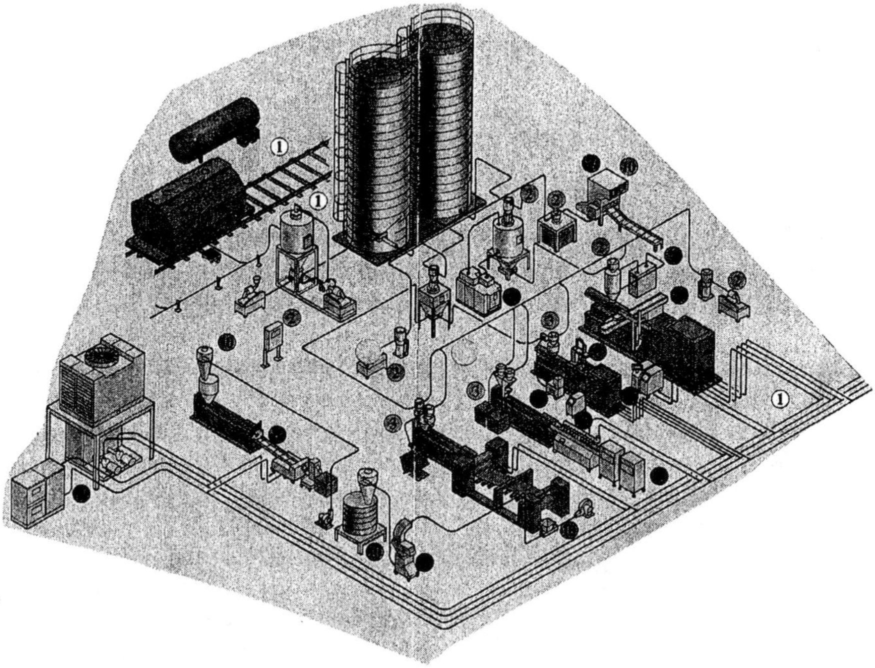

**Figure 1.5** Example of the Conair Group's schematic of auxiliary equipment and production support systems.

as auto-weight batch and/or continuous blenders, auto-color blenders, precise metering at-the-hopper throat, etc.; (4) different types of plastics driers; (5) heat transfer systems to meet different requirements on-line; (6) strand pelletizers to meet requirements such as underwater and water-ring types; (7) granulators to handle different requirements such as beside-the-press and/or central operation, sheet/film/pipe, etc., shredders, and robot feed; (8) robots for handling parts; (9) dies, sizing/cooling tanks, pullers, cutters, saws, coilers, conveyors, etc., for fabricating profile/pipe/tubing; and (10) reclaim systems providing different capabilities of dry and wet separators, fluff refeeding, scrap washing, fines collector, etc.

Worldwide total plastics consumption is at least 150 million tons (340 billion lb) with about 90% TPs and 10% thermosets (TSs). Use by the basic processes to produce products is estimated at 36 wt% by extruders, 32% injection molding, 10% blow molding, 6% calendering, 5% coating, 3% compression molding, and others at 8%. When analyzing processes that includes producing products and compounding all types of plastics for use in all the basic processes, at least 65 wt% is estimated to go through extruders. It is estimated that in the USA there are 18000 extruders, 80000 injection molding machines, and 6000 blow molding machines, producing about 25% of the world's plastic products.

## EXTRUDER TYPE AND CONSTRUCTION

There are many different extruder types that have been designed (some patented) for over a century to produce the wide variety of plastics and products. The action between the screws and barrels basically provide shearing (heating) action of the plastic [391]. Extruders can be classified as: (1) continuous with single-screws (single and multi-stage) or multiscrews (twin-screw, etc.); (2) continuous disk or drum, that use viscous drag melt actions (disk pack, drum, etc.) or elastic melt actions (screwless, etc.); and (3) discontinuous, that use ram actions (very low viscosity TPs, TSs and RPs (reinforced plastics), plastics and rubbers/elastomers) and reciprocating actions (injection molding, etc.) [222, 272, 278, 308].

The widely used are the single-screw extruders and twin-screw extruders (Fig. 1.6). In turn, each of these types have different designs to meet specific fabricating requirements based on technical and practical approaches. Most of them have a plasticator (screw/barrel) in the horizontal position. To meet certain production and cost requirements, there are a few designs positioned vertically.

There are the usual and more popular single-screw types using conventional designs with uniform diameters of the screw and barrel. Examples include extruders having decreasing screw channel volume, continuous variable speed, pressure control, and venting (devolatilization) system. Special designs use conical or parabolic screws for special mixing and

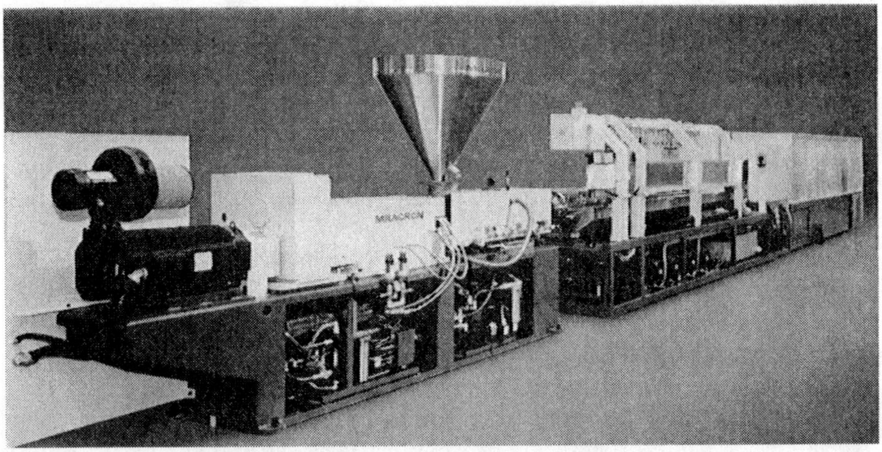

**Figure 1.6** Cincinnati Milacron twin-screw window and custom profile system with vacuum calibration table.

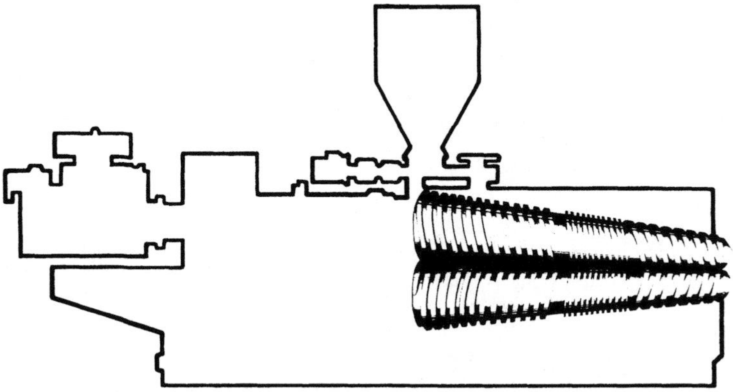

**Figure 1.7** Extruder with tapered twin-screws.

kneading effects. They can include eccentric cores, variable pitch superimposed flights of different pitch, kneading rotors, fitted core rings, periodic axial movement, etc. Barrels may have internal threads, telescopic screw shapes, feeding devices, etc. (Chapters 2 and 4).

There are also the popular common twin-screw extruders [in the family of multiscrew extruders that include tapered screws (Fig. 1.7)] with at least one feed port through a hopper, a discharge port to which a die, and process controls such as temperature, pressure, screw rotation (rpm), melt output rate, etc., are attached. For all types of extruders targeted to deliver a high-quality melt at the end of the screw, the plasticating or melting process must be completed prior to reaching the end of the screws (Chapters 2–4).

There are screwless extruders, principally for research and development (R&D), of various designs, such as using rotating drums, slit/stepped plates/disks, and/or conical rotors with eccentric barrels. They can have varying gradual clearance between barrel and rotor or plates to plasticize (melt) the TP by frictional heat. Designs to increase their output include using a screw extended from the center of the rotor or a doctor blade transferring melt through a slit die opening. Different designs of this isothermal system are targeted to reduce processing energy consumption, reduce and provide more uniform shear to improve properties, reduce residence time, etc.

Because of the many different products as well as plastics processed, one cannot provide a specific output rate (a very important parameter) for a machine unless a dedicated extrusion line is used for a product using a specific plastic; ranges are provided. As an example, output capacities for large sheet lines can now range up to at least 4500 kg/h (10 000 lb/h) and can average well over 454–1360 kg/h (1000–3000 lb/h); thickness can be regulated to less than 1% in all directions. The considerations regarding optimum output rate of machines are dependent on the characteristics of the extruder and type of plastic being processed. Extruders available have a wide range of performance or technical capabilities that in turn relate to the cost of the machine. A similar situation exists with auxiliary equipment. Usually the only common characteristic is the screw diameter and some times that comparison is questionable.

Important in machine specifications and requirements are factors such as screw design, type controls, drive power, and relationship of output rate with its operating cost (Chapters 2 and 20). Cost includes the usual costs of the machine, electric power, labor, plant overhead, maintenance, downtime, and so on. Note that the greater output capacity of a specific and more expensive machine meeting performance requirements may justify its purchase cost because it produces more profit. Greater capacity is obtained without additional labor costs, production costs, and so on, provided the machine is properly operating.

In choosing the size of an extruder and its driving mechanism, considerations to be evaluated include: (1) the range of screw speed screw needed; (2) the requirement that screw speed should be infinitely variable, and may require certain speed levels or several ranges; (3) maximum power required from the drive based on the plastic to be processed; (4) relationship required between the screw speed and the torque on the screw shaft; and (5) whether the machine to be used for a single or multiple products. The drives consist of motor or belt drives, which are linked to the screws through systems such as double reduction gearboxes. Belt drives are used for the smaller machines. The drives simplify the transformation of high motor speeds into lower speeds and high torque required to operate and control the screws.

Consideration is given to the constructional elements of the drive in the direction of flow power, i.e., the transmission of torque to the screw shaft and the axial thrust bearing. Torque can be transferred from the gearbox shaft by means of a key or slot. Usually the better method is via a coupling sleeve with longitudinal splines which require a push-fit onto the screw shaft. The different arrangements between gearboxes and screw shafts each have their advantages and disadvantages.

The feed box and barrel of the extruder form the casing for the screw. The screw must fit closely into the inner wall of its barrel so that the material feeding, pumping, and plasticizing actions are accomplished efficiently. The clearance between screw and barrel varies depending on the plastic to be processed and capability of the extruder. The essential factor in the pumping action during the extrusion process is the interaction between the rotating flights of the screw and the stationary barrel wall. If the plastic is to be conveyed at all, its frictional build-up must be low at the screw surface but high at the barrel wall. If this action does not occur, the plastic may rotate with the screw and not move into the axial direction to exit the die. It is usually an advantage to have automatic feeding with controlled fill levels weightwise. With this action, relatively small hopper capacities are sufficient.

Different feeding devices are available to provide specific plastic material handling systems appropriate to the extruder being used. The melt flow through the die under pressure transmits the usual throttling effects of the die to the movement of the melt in the screw threads. These effects are complicated and influence output rates. Reduction in outputs caused by counter-pressure can be calculated with some degree of accuracy prior to the actual processing. However, what always makes 'calculating' more efficient is to develop experience. The amount of plastic conveyed by the screw in free discharge is about half the channel volume per revolution, because the melt adheres to both the screw and barrel walls, causing its average speed to be midway between the speeds of the two surfaces. This drag flow (quantity) depends on the dimensions of the screw and the rotational speed.

When the melt is extruded under pressure through the die, its output rate is reduced by an amount which is called back flow or pressure flow. Quantitatively this loss can be calculated as if the melt were flowing backwards through the channel under the influence of the pressure at the screw tip. Basically this pressure flow depends on the dimensions of the channel as well as the pressure and viscosity of the melt. Screw speed does not directly influence the flow, however, indirectly it does affect pressure flow since altering melt viscosity via speed changes results in pressure changes.

There is a leakage flow over the screw flights from one thread to the next in the direction of the pressure gradient that reduces effective output rate. This flow loss can usually be discarded when the clearances between screw flights and barrel wall is small. With worn screws, particularly when processing highly fluid melts and dies with high pressure resistance, the leakage flow contribution becomes significant with the result of increased operating costs per output rate. This action may also effect product performance, altering dimensions and properties, for example. Thus, screw replacement or refurbishing occurs; the approach to be used is based on factors such as screw cost, delivery time, and life expectancy.

Small clearances between the screw flights and the barrel wall are important to prevent backflow of melt and possible surging in extruder output. It is usually suggested that radial clearances of 0.1 mm (0.004 in) are maintained up to 60 mm (2 in) diameter. As the diameter increases, so do the clearances such as 0.13 mm (0.005 in) for 120 mm (4.5 in) diameter. To aid in the plastic flow an undercut (clearance increased) for a short distance can be used at the feed end of the screw of about 0.25 mm (0.010 in) and/or an undercut at its tip of 0.50 mm (0.020 in).

When developing theoretical analysis of the extrusion process, as with other processes, one has to start with certain basic and proven theoretical assumptions targeted to meet specific requirements (Chapters 3–5). This approach continues to be used in expanding the capability and increase performances of extruders so that practical understanding is applied to the hardware and software. Experienced personnel continue to provide better understanding of the basic assumptions followed with theoretical analyses. With this type action extruders continue to be upgraded from those previously built. Practical analysis followed with theory continue to show the way to build and operate machines with greater efficiency and reduce cost to operate (improved melt control, lower energy consumption, reduced maintenance, etc.).

Theoretical boundary conditions are established empirically in the region between the feed and melt zones. They are difficult to evaluate. Basically experimental studies are conducted so that quantitative values can be correlated. In any study it is important, for example, not to consider the output rate as an isolated unit without taking into account the screw length and channel depths, etc. Extrusion research with its mathematical laws of thermodynamics and rheology continue to extend the practical use of extrusion lines [3, 125, 149–154, 163, 187, 206, 210, 212, 238, 248, 269, 348–353, 370, 410].

## EXTRUDER OPERATION

Generally, the material being fed flows by gravity (usually controlled weightwise) from the feed hopper down into the throat of the extruder barrel. Special measures are taken and devices used for materials that do not flow easily or can cause hang-ups (bridging or solidification resulting in plastic not flowing through the hopper). As the screw turns in the heated barrel, plastic falls down into its channel. Frictional forces develop in the plastic during plasticizing so that the melt moves forward toward the die [113, 167, 187, 370].

The initial action is where the plastic is in a solid state with its temperature below its melting point. As the temperature of the plastic rises over its melting point and starts its plasticizing action, a melt film will form on the barrel surface. This change-over can occur at any location downstream in the plasticator, although generally not prior to the start of the compression (transition) zone. The actual location depends on the plastics rheological behavior, machine geometry which is general fixed, and operating conditions. The plastic's solid state reduces as it moves downstream to where it is a complete melt by the end of the transition zone (Chapter 4). Thereafter the metering zone pumps the melt to the die. In turn, the melted extrudate exiting the die is shaped and proceeds through downstream auxiliary equipment as it cools. The die exerts a resistance to flow so that a melt pressure in the plasticator is required to force the melt through the die.

Melt pressure in the barrel can vary from a low of about 3 MPa (500 psi) to a high of at least 34.5 MPa (5000 psi) depending on die/product requirements. An average pressure ranges from 14–20 MPa (2000–3000 psi). The die head pressure is determined by the die orifice shape and dimensions, melt temperature, rheological properties of the melt, and melt flow rate through the die. In turn, the extrudate exiting the die is shaped and proceeds through the downstream equipment as it cools (Table 1.1).

Machine operation can take place in three stages. The first stage requires operating the extruder for warm-up with operational settings of up-stream and down-stream equipment. The next stage involves setting the required processing conditions to meet product requirements at the lowest cost. The final stage is devoted to fine-tuning and problem solving the complete line. A successful operation requires close attention to many details, such as the melt quality, temperature profile adequate to melt but which does not degrade the plastic, production of a minimum of scrap, and procedures for start-up and shut-down that will not (or minimize) degrade the plastic. Figure 1.8 shows examples of temperature profiles for different plastics.

Within various types of the family of plastics [polyethylene (PE), polyviny chloride (PVC), polypropylene (PP), or others], each type can have different profiles. Experience shows how to set the profile and/or obtain preliminary information from the material supplier. Degrading or oxidizing certain plastics is a potential hazard that occurs particularly when the extruder is subject to frequent shut-downs. In this respect, the shut-down period is even more critical than the start-up period as discussed throughout this book (troubleshooting, etc.).

The following start-up procedure provides a general review since each line has its specific requirements. Throughout this book, different start-up

**Table 1.1** Examples of extruder diehead pressures

| | *Melt pressure at the die* | |
|---|---|---|
| *Extruded shape* | *psi* | *MPa* |
| Film, blown | 1000–5000 | 6.9–34.5 |
| Film, cast | 200–1500 | 1.4–10.4 |
| Sheet | 200–1500 | 1.5–10.4 |
| Pipe | 400–1500 | 2.8–10.4 |
| Wire coating | 1000–5000 | 6.9–34.5 |
| Filament | 1000–3000 | 6.9–20.7 |

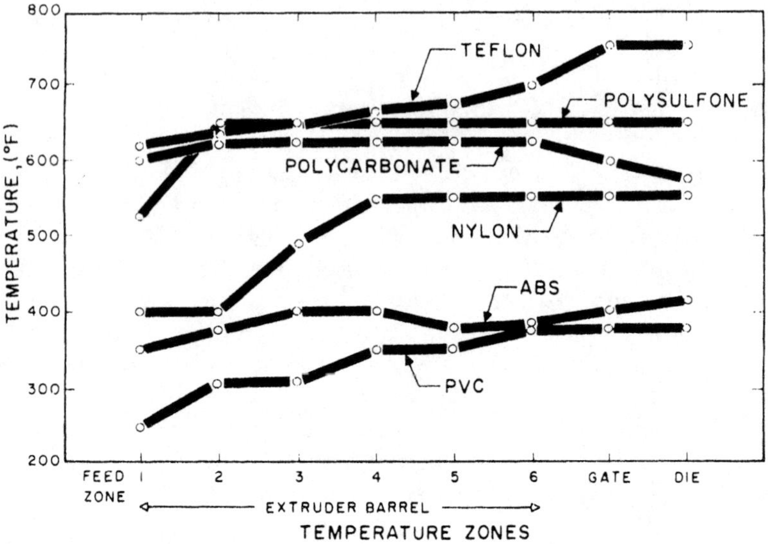


**Figure 1.8** Temperature profiles of different plastics going through an extruder.

procedures will be reviewed. A plant operational manual should contain the procedure and should incorporate experience, machine (extruder and auxiliary equipment) manufacturer instructions, and/or plastic materials instructions. It is important that an authorized operator(s) receive proper in-house and/or industry training in operating the complete line. From start-up through shut-down, dangerous situations can occur (as with other equipment such as the automobile). For the proper trained person, the problems can be eliminated. As an example, with an over-heated machine the melt may decompose and exit/split from the die; nobody should be standing in front of the die and the hopper lid should be firmly in place.

When threading-up, also called stringing (leading the extrudate from the die through the down-stream equipment), the operator is in close contact with machinery that can cause hazardous conditions if not properly handled. To save time and simplify the usual threading, use a remaining extrudate, string/rope, etc., that can be attached to the new hot extrudate and in turn follow the down-stream equipment. Where roll nips exist, they are usually left open so that any joints (hot to cold extrudate bond) will easily pass through.

To avoid accidents, do not let more personnel than required stand around the fast-moving, hot operating line. Become familiar with the control panel that is usually located at a strategic location near or on the extruder. If control panels exist in different locations that handle the separate down-stream equipment, become familiar with them. But probably, during equipment set-up, the important control panels were strategically located so that the set-up person, or operator, can easily operate the line. Of course a central process control panel is desirable to simplify the operation. Settings for all the required line operation may all be programmed for start-up going from slow speed to full line speed. The operator has to check the settings versus operation capabilities and evaluate that the best performance-to-cost exists (Chapters 6 and 20).

Set all barrel, die, and down-stream equipment temperature controllers at the desired start-up temperatures. Never start the extruder cold since cold plastic can do a lot of damage; it may snap the screw, split the barrel, or break parts in the extruder power pack/head. If required and available, turn on hopper cooling water jacket gradually to avoid plastic bridging in the hopper. Wait until the die temperature reaches the required level prior to starting the screw rotation. If the die was removed during shut-down, tighten it to the adapter when both have become hot, before the extrusion run has been started. Establish a bolt-tightening pattern, such as tightening opposite sides of the bolt circle alternately. Set the screw speed at minimum and later run it up to the specified requirement.

Flush out (purge) remainder of last run if plastic remained in the extruder. This action may not be required for certain materials. Purge until a clear extrudate exits the die. This action could take up to at least 15–20 min. If necessary, readjust the temperature settings while purging. Certain dies may require their lands to be cleaned; do this with proper tools and protective gloves and face shield.

It may be necessary to purge a heat sensitive plastic with another more heat stable compound since it will withstand subsequent reheating. Special compounds may be required; with certain materials all that is needed is low-density polyethylene (LDPE). These purging agents are usually introduced into the hopper when the screw has been pumped dry or free of the more heat unstable plastic. If the melt exiting the die looks discolored, increase the frequency of purging.

All temperature controls should be programmed so that overshooting does not occur and that the heating-up time should also be reasonably short in time. This action can be accomplished by using appropriate temperature controllers. Once the machine is at the set temperatures, then it should be allowed to balance out or equilibrate and balance out within a narrow temperature range. Check temperature of the extrudate using a hand-held temperature probe or the equivalent.

## Extrusion coating

During certain operations, such as coating, the extruder may be required to be moved over the coating line without the extrudate exiting. You may be required to place two deckles (metal inserts used to close off die opening ends) in the die ends to obtain roughly the required coating width (Chapter 5). Run the screw speed down to the minimum. Stop the extruder. This stoppage should not be longer than a few minutes, preferably one minute or less – just enough time to carry out the following steps fast. Move the extruder into the final coating position. String the substrate to be coated into the nip (a pinch is formed between two rolls); nip is in the open position. Start the chill roll. Close the nip. The pressure roll is now turning, moving along with the revolving chill roll. Push the start controls for all the other driven rolls except the windup; keep the speed low. All is set to go to start the extruder screw again (Chapter 11).

Take the substrate from the backside of the chill roll and wind it over the idler rolls to the windup. Start the windup roll and wind the substrate around it. Simultaneously with these preceding steps, bring the screw and line speed up to running speeds. This is done by one operator while another threads the substrate through the rolls. In the mean time, the temperatures have gone up. Check them and wait until they have reached the desired values. Also, if required, adjust the deckles to the final desired width.

Simultaneously, on the windup station, adjust the windup tension so that the take-off meets with the specification requirement. This is done by sight until a firm roll is obtained and/or by tension adjustment control. Set timers and if required, set slitters to the desired width using the proper controls. Run trim through the trim disposal (granulating) system to ensure all is operating properly. Now the line is running so raise the line speed to the desired target. Flag the roll by inserting a piece of paper, etc, into one side of the roll as it is being wound. The flag describes the job which begins from the flag point. It can carry a job number, etc.

## Other lines

For lay flat film and sheet (75% film/25% sheet of 14 billion 16 market) check that the die is clean and level. With clean die lips, consider applying a very thin coating of silicone grease to the lips to aid melt flow. For all lines when at operating speeds, the weight and dimensions of the product become important parameters. Haul off speed and screw speed can be increased to speed up production. However, the increase in screw speed can cause undesirable such as an increase in weight and increase in melt temperature. Other changes can be made however when making changes a logical approach has to be used. Changes should be made one at a time (allowing each change to occur timewise) and the effect of each change noted or else one could end up with lots of scrap.

Start-up differs with the type of product to be extruded, but there are some common aspects as reviewed above and throughout this book. The procedure will differ if a clean–empty extruder is used or one which contains plastic and is reheated. It is necessary to wait until the barrel and die are at the correct operating heat before startup based on plastic being used. It is common practice to purge some plastics through the extruder before stringing up the extrudate. Purge may simply be permitted to fall on the floor. Since melt is very hot and sticky, prevent contacting with your skin and clothing.

Actual start-up is generally at the rate the operator can handle. It may be very slow compared to the targeted operating speed. Usually, once a reasonable product is being made, take-off speed can be increased. All speed control equipment must be done simultaneously either manual or preferably by computer process control (Chapter 6).

The proper heating procedure produces a flowable melt through the extruder and eliminates leaving solid, unmelted plugs. Plugs would allow gas to form from plastic decomposition that in turn would build up pressure. When pressure becomes to high, or when the plug becomes loose, the plug would blow out through the die; followed by molten, decomposed plastic possibly causing the melt stream to be interrupted and requiring the operator to rethread the line. If the plug is first melted, pressure cannot develop behind it. Gas may form but it can escape as bubbles without serious pressure build-up.

During shut-down it is common to run the extruder to an empty condition, however for certain plastics, it becomes beneficial to keep the extruder full; experience or material suppliers recommendations determine the approach. A logical and well planned shut-down procedure should be prepared with an empty or filled plasticator. Procedures should also be prepared for temporary stops. These procedures ensure to save the cost of unnecessary complete shut-down or a complete machine cleanup especially when only the temporary stop was required. With an empty condition there is no start-up with a cold plastic, a condition that could overload the extruder motor. However, some operations such as those for conventional LDPE film lines prefer to shut down with the screw full of plastics. This procedure prevents air from entering and oxidizing the plastic.

Because PVC readily decomposes with heat, to ensure that this material is completely removed from the extruder at shut-down, a LDPE is often used as a purging material. This low melt plastic can remain in the barrel since it softens readily on start-up. With this type of start-up it is preferable to raise barrel temperatures slightly above normal operating temperatures, to be reduced latter. The higher temperature ensures that unmelted plastic will not produce excessive torque in the screw.

Usually shut-down is very simple. The flow of material from the hopper to the feed throat is stopped, but the screw rotation is continued until nothing exits the die. Then the screw is stopped. This action is called pumping the screw. At the same time that feed is stopped, the electrical power to the barrel and die heaters may be turned off. Full cooling occurs in the barrel.

When very high barrel temperatures are used, the previously described shut-down procedure may not be good enough. It may be necessary to continue passing plastic through the extruder while bringing the barrel temperature down. Simply pumping the barrel dry would allow the plastic remaining to be oxidized and crosslinked by air at the very high temperature. Leave the die heaters on. After reducing the barrel temperature down and approaching the plastic melt temperature, pump the plastic through the machine while it continues to cool. Do not allow the melt pressure to exceed the maximum allowable barrel pressure that may be 45 MPa (6500 psi). After the plastic is relatively cool (at high pressure) close the base of the feed hopper. Follow by purging the cylinder clean by pumping the screw dry so that no plastic exists in the die. Then reduce the screw rotation to where it stops. Turn off the die heat and make sure that the barrel cooling is at the maximum setting. Clean die lips with copper pad and apply a light coat of silicone grease (or equivalent) if the silicone does not interfere with the products performance such as secondary decorating or bonding operations. After machine cools, turn off the water and the main electrical power switch. Check that the correct procedures were followed before complete shut-down of the machine or before changing to another material.

Leaving the extruder in the full condition initially follows the procedure just described for an empty condition. After closing the slide at the base of the hopper, reduce the screw rotation until it just stops; no purging occurs. After the screw stops, turn off the barrel and die heaters as well as barrel cooling. As above, follow with the cleaning, etc.

The usual procedure for temporary stoppage is to purge the plasticator. Pass the material through the machine at minimum screw speed. Do not allow melt to build up on or around any part of the die. The die may require cleaning and coating.

Cleaning the extruder is important. In fact, keeping the machine as clean as possible at all times is of vital importance. Cleaning is a nonproductive necessity. It could involve very long time periods during which time the line is not in production. Set up a procedure that when the line is 'down' consider cleaning. At that time also consider preventative maintenance for the extruder as well as all auxiliary equipment up-stream and down-stream.

A major cleanup involves the screw, barrel, breaker plate with screen, adapter, and die (Fig. 1.9). The following steps generally occur. Let the extruder run with plastic at a low temperature such as 330°C (450°F) without further feeding until you can see the screw when you uncover the

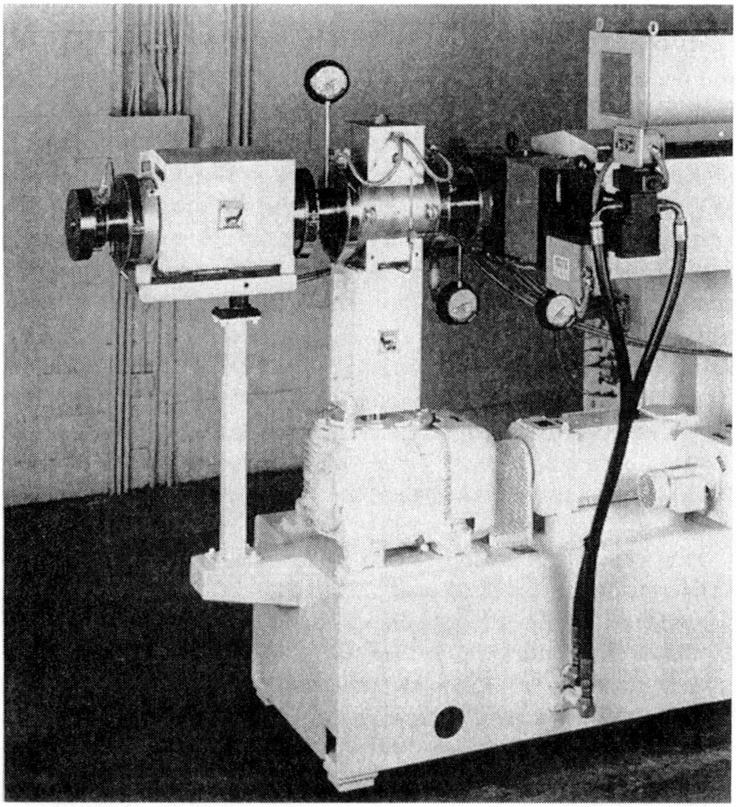

**Figure 1.9** Front end of Welex extruder showing in-line screen changer, gear pump, static mixer, and downstream adapter.

hopper. Turn off all electricity and water. While the machine is still hot, remove any equipment in-line with the screw that could include a gear pump. Remove the die which may require heavy duty lifting equipment such as an overhead crane, lift truck, or portable chain hoist. Remove the adapter plate with screen. Push the screw forward, usually from the back end, and remove it from the barrel. The parts are now ready to be cleaned or may need replacements.

The cleaning process basically involves a copper or brass bladed scraper to remove most of the plastic adhering to any of the metal parts. Follow this scraping by cleaning with copper or brass wool. Sprays are available, such as silicone grease spray, for assistance in removing tightly adhering melt. Then coat the screw with a protective coating such as silicone grease. Always clean the barrel when cleaning the screw. A simple procedure is run a brass brush at the end of a long 'handle' through the barrel. Use another similar arm for lubrication with silicone grease. Various techniques are used to support parts being cleaned. In the case of the screw it may be beneficial to start cleaning the screw as you remove it from the barrel [2].

Plastic may be burned safely by using a flaming torch, furnace, hot salt bath, etc., being careful not to damage any of the metal surfaces being cleaned. Furnaces can collect exhaust material so that the surrounding air is not contaminated. For the other methods and for safety, the cleaning should be done inside a vented hood behind a protective window, or equivalent operation. Cleaning/burning a breaker plate may take 15 or more minutes, so consider using a spare breaker plate to eliminate loss in production time. Any material such as colorant or carbon residues can be easily removed by using compressed air and/or brushing away with a soft brush. Lubricate the seat of the clean adapter and breaker plate, also the end of the screw if the screw has not been removed for cleaning. The face of the adapter as well as that of the barrel will be clean and need no cleaning if they were fitted accurately during the previous run.

After long periods of continuous operation, such as with polyethylene film, a slow build-up layer of oxidized plastic will occur particularly on the inside of the barrel wall and screw. This degraded material will finally begin to flake off. This action, in turn, will result in the appearance of film defects such as gels or yellow-brown oxidized particles. These defects cause the film to have a poor appearance and, especially in very thin films, can actually cause holes or tear-off of the film as it comes out of the extruder die.

The danger point has been reached at which the manual screen pack has little or no effect in holding back the impurities. The result is that a very thorough clean-up is immediately required. The time period between thorough cleanings can best be determined by experience. If the extruder is continuously in use for long periods, every two months might be a good time interval to consider. Regardless, a definite periodic cleaning schedule should be established. Whatever product is being fabricated, surface appearance visually or by instrumentation is the best indicator for your cleaning schedule.

Usually the best time to clean a die is when it is not extremely hot and after the heaters and sensor, such as temperature and pressure transducers, have been disconnected. Remove accessible plastic with the usual tools; consider even a wooden scraper. Disassemble the die and clean all parts. As an example, with a coating die scrape the manifold, fix and adjust jaws, and ensure die lands are free of any plastic. Follow with polishing the curved surfaces of the manifold (carefully) with a very fine grit polishing cloth, and hone its flat surface first with a less fine cloth and then with a fine grit polishing cloth to remove any imperfections such as scratches or mars.

A guide to cleaning die lands is as follows: (1) remove die jaws while they are hot but below 200°C (390°F) with actual temperature based on the plastic being processed; (2) remove the die lands from the jaws if they are separate; (3) scrape excess plastic from the jaws and lands with copper, brass; and/or wood devices to fit the surface being scraped followed with a final action using copper or brass wool; (4) polish lands using a very fine grit cleaning cloth such as a 2400-grit lapping compound on a soft damp cloth or tissue; and (5) apply a light coat of silicone grease or spray (or equivalent) to prevent molten plastic, such as PE, from sticking to the lands during start-up.

Chrome plated dies must be handled with extreme care. Honing will remove the chrome plating so this procedure is not recommended. If the chrome cannot be made smooth by cleaning and/or polishing, they should be sent for replating to an appropriate operation; possibly the original die manufacturer. Improper chrome plating is very damaging to the successful running of a line.

Follow by re-assembling the die carefully in the reverse procedure that was used to disassemble. As an example, ensure the die is level. Tighten the collar halves with equal pressure on both sides using a prescribed pattern. Bolts are tightened on opposite sides of the bolt circle alternately. This procedure prevents cocking the parts by tightening one side more than the other. Where required, set the opening between the die lands to the desired dimensions using appropriate measurement tools that will not damage the die. Final opening for certain dies is made after heated on the extruder.

Precautionary procedures should be set up to minimize the build-up of oxidized plastic on die lands. As an example, a brass shim or blade which is thinner than the die opening can be inserted between the lands and gently pushed along its length to remove contamination. It should never be pushed in and out of one place on the die. This procedure may be required daily.

## PROCESSING PLASTIC

The target after the extruder has its start-up and running is to ensure that the melt temperature, pressure, and output rate (time) are consistent within a processing 'window' of operation (Chapter 6). These factors directly effect the final product (Fig. 1.10). The time period for the melt exiting is influenced by such factors as the speed of both the extrusion screw and of the haul-off equipment. When one factor varies, the other factors are effected. The usual TPs have rather wide operating windows for these three factors. However, it is best to determine their 'ideal' settings in order to obtain maximum performance/cost efficiency by using an EAD and EVD (Extrusion Area Diagram or Extrusion Volume

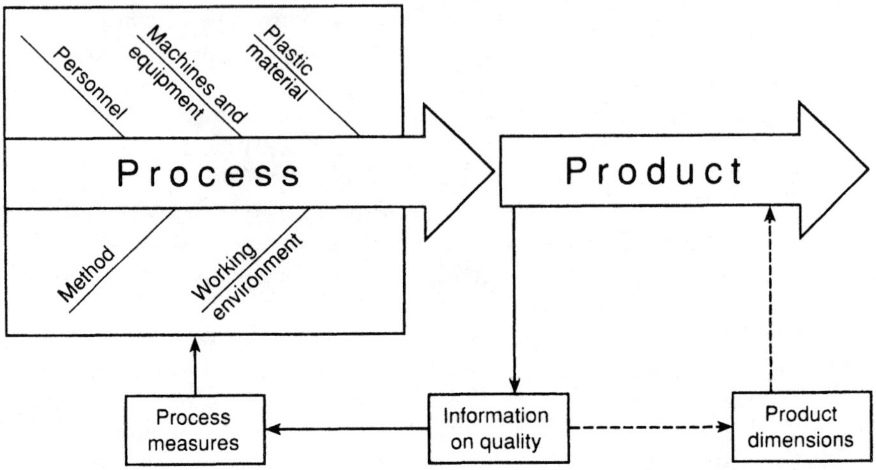


**Figure 1.10** Example of a simplified fish bone diagram applicable to extrusion.

Diagram) approach; see Molding Area Diagram and Molding Volume Diagram in Ref. 2.

An extruder is a combination of screw conveyor and compressor. As a conveyor, it mechanical acts as 'screwing' itself backward. Since the screw itself stays in one place, the plastic does the moving and is forced out through the die. Since the plastic is viscous and develops frictional resistance to flow, power is required to turn the screw. As a compressor, the screw moves solid plastic (pellets, etc.) initially and gradually the plastic melts. In the solid stage, it is denser, especially the crystalline plastic; with this plastic air spaces exist with the result that they have a low bulk density (weight per volume). The screw compresses the solid plastic together. With this pressure action, the air usually takes the least path of resistance and is pushed back out of the feed opening. If any air remains in the melt, it could be released when the melt exits the die with no damage to the product or it may remain in the product as a contamination [3, 206, 209, 232, 349, 370, 397].

Care should be taken to prevent conditions that promote surface condensation of moisture on the plastic and moisture absorption by additives such as pigments in color concentrates (Chapter 3 reviews the importance of properly drying plastics). Processors must avoid contamination from other plastics, dust, fines, paper clips, nuts/bolts, and so on. Special care should be taken in cleaning silos, bins, feed hoppers, blenders, scrap

**Figure 1.11** Example of a Welex extruder with crammer feeder to handle low bulk density plastics that otherwise are difficult to handle; overview and close-up.

Welex

Welex

granulators, and other material handling equipment. Plastic silos, containers, and hoppers should be kept covered to prevent contamination. Follow established procedures (or set them) so that start-up will emphasize the importance of a clean environment.

In an extrusion line, the reference base with which the results are compared is usually based on degree of process control. In turn, two factors – plastic material and extruder (and line) capabilities – directly influence control (Chapters 3 and 6), including controlling rate of feeding plastics (Fig. 1.11).

## EXTRUDER HEATING AND COOLING PROFILE

The temperatures required along a barrel, adapter, and the die depend largely on the specific extrusion operation being conducted with its screw design, plastic used (Tables 1.2 and 1.3), and available process control. The thermal condition of the plastic is essentially determined, for a given material, by screw geometry (Chapter 4) with its rotational speed and the total restriction or pressure existing in the die. The electrical heaters are normally placed along the barrel grouped in separate and adjoining zones. Small machines usually have two to four zones. Larger machines have five to ten zones. In most cases, each zone is controlled independently.

Most of the drive motor energy supplied to the extruder is used to heat and melt the plastic to be extruded. In order to bring the melt back to the solid state, a large fraction of heat must be removed from the extrudate within a short distance from the die. Shape and appearance of the hot molten extrudate have to be precisely maintained until sufficient cooling and setting have taken place. Output can increase for a given shape resulting in higher lineal take-off speed. This action can reduce the available cooling time near the die which is essential for proper shape molding. Based on this combination of facts, highly developed cooling methods and take-off procedures are indispensable for high-speed extrusion production.

Heat to soften the plastic is supplied in two ways: by external barrel heating and internal frictional forces brought about on the plastic due to the action of the metal screw in the metal barrel. The amount of such frictional heat supplied is appreciable. In many extrusion operations it represents most of the total heat supplied to the plastic. Electricity, fluid such as steam, or hot oil can be used. Electrical heating is generally preferred because it is the most convenient, responds rapidly, easiest to adjust, easy to clean, requires a minimum of maintenance, covers a much larger temperature range, and is generally the least expensive in terms of initial investments.

Accurate control of the barrel temperatures is essential because the

**Table 1.2** Examples of extrusion operations for different products in LDPE: all temperatures in °C

| *Product* | *HP* | *Barrel temperature* | | *Die temperature* | *Pressure (psi)* | *Material temperature* |
|---|---|---|---|---|---|---|
| | | *Rear* | *Front* | | | |
| Pipe | 40 | 150 | 160 | 165 | 1500 | 165 |
| Tubular film | 40 | 150 | 160 | 165 | 1500 | 165 |
| Flat film | 40 | 200 | 240 | 250 | 1000 | 250 |
| Coating | 100 | 250 | 315 | 325 | 1000 | 320 |
| Wire coating | 50 | 220 | 240 | 240 | 3000 | 240 |
| Contours | 20 | 175 | 200 | 205 | 1500 | 200 |

**Table 1.3** Examples of melt temperatures in an extruder

| *Plastic* | *Temperature (°C)* |
|---|---|
| Cellulose acetate | 160–205 |
| Cellulose acetate butyrate | 160–205 |
| Cellulose propionate | 160–205 |
| Ethylcellulose | 190–230 |
| FEP fluorocarbon | 380–400 |
| Ionomer | 230–290 |
| Nylon-6 | 250–270 |
| Polycarbonate | 290–300 |
| Polyethylene, low-density | 230–260 |
| Poly(ethylene terephthalate) | 285 |
| Poly(methyl methacrylate) | 180–210 |
| Polypropylene | 270–300 |
| Polystyrene | 240 |
| Polyurethane elastomer | 165–220 |
| Poly(vinyl chloride) | |
| Rigid | 160–165 |
| Nonrigid | 190–205 |
| Copolyester | 200–250 |
| Chlorotrifluoroethylene copolymers | 260–320 |
| Vinylidene chloride – vinyl chloride copolymers | 175–190 |

viscosity of plastics can easily change with temperature fluctuations. The hopper is usually water cooled to prevent the plastic from melting prematurely and bridging or sticking to the hopper throat before reaching the screw eliminating probable the complete shut down of a line.

It is possible by sacrificing machine productivity to do considerable cooling in an extruder. The melt temperature can be kept low. The approach is a combination of minimizing viscous working of the melt with its accompanying heat generation and transferring heat out of the melt by way of metal surfaces in contact with it. Heat generation for a given production rate increases with machine size; however, it goes down very fast as screw speed is reduced. Heat transfer is primarily a matter of how much melt-to-metal surface can be provided. This action increases rapidly with machine size. What this means is that you can accomplish cooling if you are willing to go to a large, very slowly turning machine that generates little viscous heat and offers a large heat transfer area. This extruder may be one or even two sizes larger than the machine that would normally provide an equal output. This desired freedom of melt temperature control is at the price of machine productivity. The drive would cost less than the drive of a standard machine of this larger size. For certain plastics and products, the larger machine approach may be required.

In choosing the size of the extruder and its driving mechanism, it is important to consult with a plastics technologist who is usually the material supplier. You will be informed with some certainty of the processing requirements of the TP to be used. This is a situation very different from that encountered in the operation or design of other machines such as water pumps, machine tools, etc. The reason being that water and metal rarely undergo such complicated changes of (rheological) condition as a TP (Chapter 3). Factors to consider with a TP include: (1) the operating range of screw; (2) whether speed will be infinitely variable or is it sufficient to establish a range of speed levels; (3) with several speed ranges, which TP will best suit the product requirements; (4) the maximum power required from the drive; (5) the required relationship between the screw speed and the torque at the screw shaft; and (6) the power losses in the drive mechanism. When determining any machine requirements, it is important to know and explain whether the extruder will be used for a single product (shape and dimension) or whether many different products will be manufactured.

## AUXILIARY EQUIPMENT

As stated earlier, understanding and control of TP melt flow through an extruder is important. However just as important is understanding the up-stream and down-stream auxiliary equipment. The complete production line has to properly interface as if it were only 'one piece of equipment'. Latter in this chapter and throughout this book, information will be presented on the up-stream equipment that specifically concerns material preparation and handling. It is important that the material reaching the extruder hopper meet the tightest quality control and moves to the extruder screw at a consistent rate.

Down-stream of the die different types of takeoff equipment are used to meet different product requirements. They include sizing fixtures, cooling tanks, coiling devices, line speed controls (mechanical, electrical, and/or sonic), tension controls (capstans, etc.), pulling devices (caterpillar, pinch rolls, opposite moving belts, etc.), cutters in-line, collecting equipment (troughs, tanks, stackers, etc.), postforming, fast takeoff rolls, and so on.

After exiting the extruder the TP hot melt is to be cooled. Target is to cool the extrudate below its melting point ($T_m$) or glass transition temperature ($T_g$) to impart dimensional stability (Chapter 3). This cooling action can often be done by immersing in a tank of water, by a cold water and/or air spray, over cooling rolls, or their combinations. This control is achieved even though plastics basically are poor conductors of heat with poor thermal stability and have high specific heats.

Many different types of auxiliary equipment (AE) and supporting secondary operations (SO) can be used. They can maximize overall processing productivity and efficiency with the result of reducing fabricated product cost. Their proper selection, use, and maintenance are as important as that of the basic extruder; they can cost more than the base machine and can cause more down time for the line if they do not properly interface, causing problems with cooling rate, speed of production, etc. Requirements for the individual equipment include: ease of operation, control accuracy, and maintenance; safety devices; energy requirements; life expectancy; and compatibility with other equipment.

All in-line equipment has to be properly interfaced or else the line operates inefficiently and more costly. Much of the equipment in the past did not properly interface operations. A major problem was that plants wanted to use auxiliary equipment already in the plant but really did not provide the performance required. A set of rules can be developed that govern the communication and transfer of data between extruders and auxiliary equipment. This communication protocol provides a set of rules governing communication or transfer of data between computer hardware and/or software. When related to plastic processing equipment, communication includes reference to exchange of process control, meeting standards, following production schedules, etc.

It is important to first determine production requirements. The highest or lowest cost equipment does not necessarily equal the best performance. As an example, cooling installations in certain lines are not properly engineered for the line speed. The result generally is a higher energy cost that eliminates any cost reduction occurring due to increased line speeds. Throughout this book developing electrical energy efficiency in the plant is reviewed. With auxiliary equipment, major changes are continually occurring to reduce energy costs, such as in drying and preheating units. More use of sophisticated controls are providing processors with improved diagnostics and temperature controls to significantly aid operating equipment at the lowest cost. As an example, more integrated electronic control on plastic dryers have been regulating the dew point of air flow and reducing the amount of time heaters operate.

An example of important design and use of equipment is the cooling and take-off units. As the plastic melt leaves the die, it could be supported by a shaping fixture/sizing plate to retain the desired shape through the initial period when the material cools. For rod and tubing, a metal cooling sleeve or a set of steel or brass sizing plates can be used. The sizing plates and/or tubes with the required down sizing shape can be external and are usually used to hold the plastic as it passes through the water. Depending on shape, other means may be used such as rollers, adjustable fingers or blocks of material to direct its movement, disks, etc. Basically, cooling is provided by a water trough, water cascade, air, or their combinations. The simplest method is to draw the extrudate directly into a long water bath and keep it submerged so that the water can provide direct and complete cooling. The water bath can be sectionalized so that the water temperature progressively is reduced from boiling to cold depending on material and shape.

There are exceptions for certain rigid plastics that can be cooled in air. The air cooled plastics include various types of polystyrene (PS), poly(methylmethacrylate) (PMMA), unplasticized polyvinyl chloride (UPVC), and cellulose actetate (CA); with these materials, water cooling tends to set up internal stresses and gives a poor surface appearance. Air cooling follows much of the same approach as water cooling except that streams of pressure controlled air are blown against the extrudate. Sizing and other devices are used to hold and/or guide the plastic. Cryogenic cooling is used in certain lines for extruding pipe, blown film, blow molding, profiles, etc., resulting in faster line speeds.

Another method for making hollow sections is to use different forms of a vacuum box. In this system the extrudate is directed through a box (tank) partly filled with water. At the entrance end there would be a sizing unit (rings, plates, tubes, etc.). When the lid of the box is closed, a vacuum is induced inside the box. The air inside the plastic section can push the wall of the extrudate out against the sizing device and thus hold the section rather firmly while it is cooled and set. The water acts as a lubricating agent between the plastic and sizing device.

A setup worker takes the extrudate and pulls it from the die through the cooling and sizing units to the takeoff drives. These drives are usually the caterpillar type with product-shaped treads. By changing the speed of the take-off unit, it is possible to control the dimension of the contour. A faster take-off will produce greater draw down. Pipe, tube, and contours/ profiles are either cut into lengths or coiled. There are TPs that present few problems in extrusion and can be produced into complex contours. The difficult TPs to process are limited to simple profiles.

Another example is in-line cutting equipment. When this equipment can be used, it can eliminate very costly secondary operations, improving product quality and profitability. There are applications that require the ends of an extruded product to be cut at an angle rather than square. Although most products have square cuts on both ends, it is not unusual for one end to be square and the other angled. The simplest method of achieving the same angle on both ends is to machine the bore of a bushing or cutting die at the required angle. The product, which is moving perpendicular to the knife and through the bushing, must be angled bore by a feed tube.

This method can only be used in limited cases where the angle does not exceed 30°, the product is semirigid, and the coefficient of friction is low. Soft products with high coefficient of friction tend to jam, due to the change in feed direction. With this method, it is necessary to adjust the bushing holder away from the cutter so that the product can move smoothly from the belt centerline to the feed tube. When the job of cutting angles is complete, the bushing can be removed, the bushing holder realigned, the regular bushing inserted, and the cutter returned to normal cutting service (square cuts).

A different technique is used when the cut angle begins to exceed 30°, the material is more rigid, or the coefficient of friction is higher. This approach does require the purchase of additional hardware, but this added hardware will minimize feed problems associated with the first method and allow a wider range of products to be cut. Here the product is kept in-line with the feed device and the downstream transport. First, bushings are machined so that the faces form the required angle. The bore is kept in-line, but the cutter is repositioned. The cutter must be securely mounted to the table top in order to stay in-line. The knife guard and bushing holder are custom built for each application.

There are also requirements for two different cut-end angles. Examples are ends cut one square and the other angled or at two different angles. Mitered cuts such as those required on window products are representative of such applications. Such parts can be reproduced by coupling cutting heads that have been precisely aligned to a special type bushing. The most versatile way to cut the same angle on each end from 0° to 45° is to use on-demand cutters mounted on a special table with a pivoting plate. The cutter can be pivoted to the desired angle and locked in place.

There are constraining elements in operating extrusion line auxiliary equipment that can limit their operations. They range from the capability of the plastics as well as their uniform rate of feeding the extruder to line speeds. As an example, a major constraint in any high speed mono or coextruded/multilayer film (blown, cast, etc.) lines is the winder speed capability. Generally surfaces winders are best at accommodating large diameter finished rolls; up to 165 cm (65 in) diameter. Center-driven turret winders are generally better suited for small diameter rolls with multiple cores on a shaft often described as a 3- or 5-up film system; up to 60 cm (24 in) diameter.

There have been winders with average operating rates of 150–230 m/min (500–750 ft/min). The higher speed lines operate up to at least 300 m/min (1000 ft/min). There are 3-layer cast film lines with feedblocks and dies that operate up to 550 m/min (1800 ft/min); these lines handle up to 320 cm (126 in) face width, 5-up production, and include roll-handling/recore system. Each line has annual nameplate capacity of 9 million kg (20 million lb)

The different winders are available with increased torque capacities that offer tapeless automatic roll transfer. They also provide digital drive contact pressure control capability that enables processors to adjust lay-on pressure to control a roll's hardness, etc. The dual-lip versus single-lip air rings used in blown film lines can boast line output about 25% with film gauge quality remaining steady.

Extrusion coating lines are available that can deliver up to at least $90\,g/m^2$ ($126 \times 10^{-6}\,lb/in^2$) of plastic. Average layer thickness using a single pass over a substrate is at 0.0035 mm (1.4 mil). Usually coating lines have required multiple passes to achieve this level of coating. This thickness using applicable plastics meets or exceeds Fedral Drug Administration (FDA) requirements for oxygen/moisture barrier properties in medical packaging.

## PLANT OPERATION

Because of the great variety of extrusion operations, it has always been impossible to give a universally applicable rule for plant operation. Planning and designing will differ according to the type and magnitude of the production, space available, availability of labor, and the marketing conditions going from local to worldwide. Proper planning is required to setup and operate the production line that starts with availability of receiving and handling plastic material to the product's proper packaging that interrelates materials, process, and products (Fig. 1.10).

Material handling is important. It involves the use of equipment to unload railroad cars and/or trucks to ensure fast transfer of material to proper storage of materials in silos, etc. (Fig. 1.12). Plastic conveying systems from storage outside to an inside location should take into

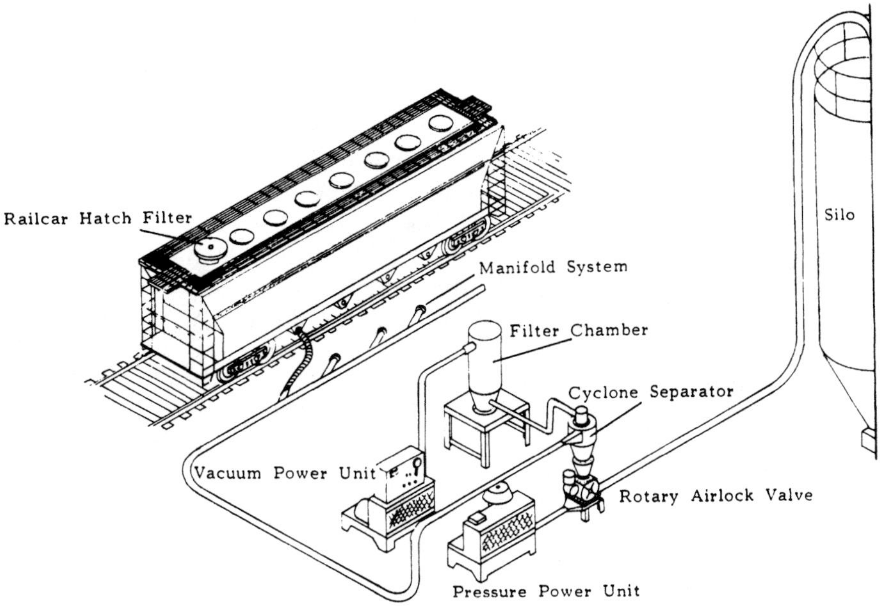


**Figure 1.12** Example of a method for unloading a railcar of plastics.

account temperature variations so that plastics reaches the hopper at the required temperature. In moving the plastic, take advantage of gravity and where possible make a direct feed to the hopper. Locate transfer material handling pipe lines correctly to eliminate hang ups and reduce the cost of moving the plastic. Properly install or locate dryers, preheaters, blenders, etc., to take advantage of using them most efficiently (Fig. 1.5) [2].

Utilities and financial considerations form an important combination if we are to operate most efficiently and at the lowest cost. As discussed above, it is important to locate material handling pipe or conduit lines and auxiliary equipment at the most convenient heights and locations. If the building has an extremely high ceiling, keep the pipe lines as low as possible to significantly reduce the energy required for moving the plastic. Keep connecting lines, such as electric lines, away from areas or trenches where water could collect. When purchasing any equipment, primary and auxiliary, examine what maintenance is needed and most important whether you understand the instructions. The task is to set up preventative maintenance procedures.

The basic consideration, when choosing an extruder, is to determine the type of production for which the machine is to be used and the desired output per hour. The extruder capacity, $Q_e$ in pounds per hour, can be calculated from the barrel diameter, $D_b$ in inches, by the empirical relation,

$$Q_e = 16(D_b)^{2.2}.$$

Extruder output and capacity is estimated by different theoretical methods. These estimates take into account different analyses. As an example, most of the energy needed to melt the TP comes from the mechanical working, and the barrel heaters serve mainly to insulate the plastic. At an efficiency from drive to screw of about 80%, the capacity $Q_c$ can be approximately related to the horsepower (hp) supplied. Heat capacity, temperature rise from feed to extrudate, and other factors of the plastic are taken into consideration so that other equations can be developed. These equations are not exact because the heat of melting and other thermal effects are not exact or have been ignored. However, they do provide a guide to the output rate. The real value will be known when the extruder is operational where the precise output is determined based on machine settings.

In extrusion, it can be quite useful to be able to predict the change in output resulting from a change in temperature. The relationship between output and temperature change is

$$Q(\%) = 100 \times (b/n) \times T.$$

The change in output $Q$ is a percentage, the temperature change is in °C (or °F) and temperature coefficient $b$ is in $°C^{-1}$ (or $°F^{-1}$). The factor $n$ is the power law index of the plastic index (Table 2.14). This power law index describes how the viscosity of a plastic melt changes with shear action. The temperature coefficient $b$ describes how the viscosity of a plastic melt changes with temperature. This is a general description of the effect of temperature. The equation is more specific to the die. It is necessary to study the interaction of the die and the extruder to predict the overall change in output from a temperature increase.

A custom extruder has to supply a variety of shapes in a variety of compounded plastic materials. They frequently are in short runs. These machine requirements, such as output rate, are different from those of long run producers who use the machine for one (dedicated line) or a few specific products. In either case, the output capacity of an extruder does not necessarily determine the possible hourly production, since the extruder is only part of the extrusion system or line. The other frequently more important phase is the ability to take-off and cool the melt at the desired production rate, while holding the specified shape within tolerances. The appearance of the product is also dependent upon cooling rate and take-off speed. For short-run production, the cost of specialized cooling and take-off equipment is usually limited and stays in relationship to the quantity of the product to be produced. For long runs, the highly engineered and mechanized cooling and take-off machinery pays for itself very rapidly, even though its cost can be considerably higher than that of the extruder. These installations permit taking full advantage of the higher extruder output rates.

Logical reasoning and technical capability has to be applied to the planning. As an example, for a given project a successful operation can take advantage of the output rate of a 25 hp, 60 mm (2.34 in) extruder. If production increase of 300% is desired and a larger extruder with a 75 hp, 120 mm (4.68 in) extruder is installed, it will meet three times more provided the take-off is capable of shaping and cooling at a 3-times faster lineal speed. The speed is often the main limiting factor of extrusion production. Cooling (as well as heating) rates for the different plastics vary. There is no direct relationship of cooling rate with thickness; cooling rate does not increase or decrease linearly with wall thickness. Thus, a section twice as thick as another requires more than twice the cooling rate. This differential can be drastically changed using plastics with fillers such as high loading of short fiber glass available in different forms [43]. But the usual sea-saw will exist where a gain will balanced with loss and in this example with glass one has the extruder wear and tear problem creeping into the operation.

With the conventional TPs being processed, the extruded products with dimensionally identical cross sections made from different plastics should have very different take-off speeds or production rates. Thinner parts can be more easily cooled and permit higher lineal speeds. Take-off rates in weight per hour are determined by wall thickness and total surface area exposed to cooling. To reduce take-off speed in a long production run and an extruder with higher output, processors may use multiple die heads. The economics of using a large extruder with multiple head versus several smaller and take-off trains can be compared. A consideration would be the fact that with two lines if one went down, at least the other would be running. However, costs would be in favor of the single system.

There tends to be a positive correlation between the quality of the products offered by a company with fulfillment of explicit requirements or implicit customer expectations and its profit margin. Studies indicate that the ROI (return on investment) as a yardstick for a company's profit depends not only on the market share but above all on product quality. Hence, the notion 'quality first, because profit is its logical consequence.' Generally, the customer is only in a position to assess a few of the quality features at the instant of purchase. So purchase is and would remain a matter of trust. An endeavor to improve the market share calls for strengthening this trust. Clearly there is a close connection between quality, cost-effective production, and the plant operation.

A plant operation is a system made up to meet the complete extrusion

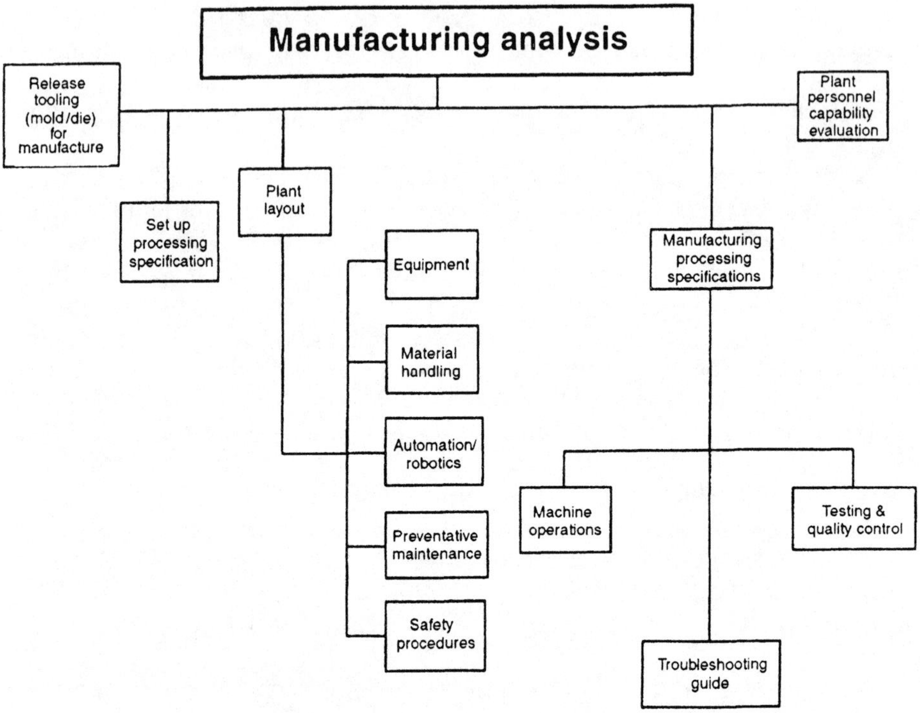


**Figure 1.13** Flow diagram for manufacturing analysis.

operation (Figs. 1.1 and 1.13) so that products meet their performance objectives at the lowest cost. The plant's maximum productivity can only be gained if the whole system works effectively and efficiently. Also the system must be responsive to change (new material, new controls, product design change, personnel, etc.) and must evolve and improve with time. Across all plant systems, there is the one basic common denominator that is collection and analyzing data. Planning for manufacturing automation must focus on data as the key to systems integration. Be aware that bringing a new high-technology processing equipment into a poorly managed plant environment will only guarantee that it will suffer the same delays and mishandling as the ones in place.

## PLANT SAFETY

All processing equipment have procedures to operate and meet safety requirements; they are available from equipment suppliers who also help to understand how to handle plastics. They include safe start-ups, location of safety devices, etc. (Table 1.4). Processing plastics usually generate a lot of force and heat. They are built to run safely but they must be treated with understanding and respect.

Most plastics will decompose if heated too long a time and/or at too high a temperature. When decomposition occurs there may be a hazard. If gas is trapped in the plasticator very high pressures may develop. Equipment such as the plasticator and die may break and explode with the hot decomposed melt. Burns and wounds on personnel may result. Rather than damage the metallic barrel, hot melt can shoot out of the die or hopper with explosive force. These reactions very rarely occur; they can be prevented by using the correct operating procedures for handling equipment and processing conditions for the plastic. Part of the responsibility rests with the operator to follow directions and to be alert for

**Table 1.4** Examples of equipment safety check list that pertain to an extruder sheet line

| *Potential hazard (causes)* | *Safe practice checks* |
|---|---|
| Tooling change | |
| Burns (hot tooling and polymer) | Wear gloves when handling hot tooling. |
| Electrical shock (extruder pipes, wiring) | Follow electrical lockout procedures when working on tooling. |
| Preparation | |
| Slipping/falling (materials on floor) | Check housekeeping.<br>Check for any leaks. |
| Startup | |
| Projectiles and burns (excessive head pressure) | Check zone temperatures.<br>Monitor head pressure (max. 4000 psi). |
| All hazards (malfunctioning E-stop) | Check at emergency stops. |
| Cuts/lacerations (slitter blades) | Keep slitter blades in up position when threading sheet. |
| Operation | |
| Pinch points/burns (heated rollers) | Keep hands clear of moving rollers. |
| Burns (hot sheeting) | Wear gloves when handling hot sheeting. |
| Cuts/lacerations (utility knife) | Wear gloves when cutting samples. |
| Shutdown | |
| Fire, flooding, etc. (equipment left on) | Make sure all utilities and auxiliary equipment are off. |
| Cleanup | |
| Slipping/falling (materials on floor) | Check housekeeping.<br>Check for any leaks. |
| Burns (hot polymer) | Wear gloves when handling hot drool. |

unusual operating conditions. Either overheating caused by a faulty controller or a freeze-off from a burned out heater can be a hazard.

There is another potential danger on startup. If all the plastic between the screw and the melt path exit is not melted, a frozen plastic plug could form. This condition could generate a barrel pressure in excess of 70 MPa (10 000 psi). This pressure build-up can be detected using a pressure sensor(s) in the barrel. Precautions to be used to eliminate potential problems include: (1) do not degrade plastic; (2) be sure that the plastic in the system is melted before starting/turning the screw; (3) never be next to unprotected areas in front of the extruder die opening during startup; (4) remember that pressure could be building in the equipment and you will not know it unless pressure sensor(s) are used; (5) become familiar with the operation of equipment and material prior to startup via instruction manuals, supervisor training, and so on; and (6) above all, use common sense. Table 1.5 provides a safety guide for operators [397].

It could be a practice to plug a vented barrel and use it the same way as a solid barrel (Chapter 4). It is rare but the internal pressure can exceed the strength limit of the bolts retaining the plug and in turn the plug would be released violently from the barrel. To prevent this potential hazard a number of safety precautions are taken, such as to ensure retaining bolts with more than enough strength are used, rotate barrel downward or away from the operator, use a pressure sensor at the head of the extruder to provide a preliminary warning at a minimum safety pressure value followed with shut-off of the machine at higher pressures (if practical or otherwise alert all in the plant), check correct installation of shear pins and/or rupture disk, and ensure that the machine is heated adequately and temperature controls are in place at the forward barrel end. Shear pins and/or rupture disks are installed in extruders to release pressure of 70 MPa (10 000 psi); added safety exist when set at 50 MPa (7500 psi).

In the various extrusion lines, pairs of rolls are situated in different positions (calender, sheet line, coating line, etc.). There is a V-shaped gap between a pair of rolls where incoming or down stream material is 'nipped' and drawn between the rolls. These nip rolls can be used to pull plastic melt from the die exit, control thickness of sheet, etc. Care has to be exercised by workers when setting up these type lines so that fingers, etc., are protected. Different safety measures are used. An example is reviewed in Chapter 10 on calendering where a double-action interlock safety bar at the bite of the rolls is described.

When running a coextrusion line, make provision to adequately alert the operators if one of the extruder drives trips out (stops operating). This action will prevent extrudate from back flowing in the stopped machine and creating a blockage in the feed section and potential hazard.

**Table 1.5** Safety guide for extruder operators

| *Problem* | *Hazard* | *Solution* |
|---|---|---|
| Feeding, esp. from bags or drums | Fire risk<br>Trip hazard | Avoid spillage on to heaters or floor<br>Clean up with vacuum, not brush or compressed air |
| Feeding with powders | Fire/explosion risk | Avoid spillage<br>Earthing against static electricity<br>No smoking |
| | Possible toxicity | Wear dust mask |
| Clearing bridging or blockage in feed | Injury and machine damage if caught in revolving screw | Don't remove safety/magnetic grid<br>Don't lean over hopper<br>Use only long strip of polymer, never metal |
| Excess pressure giving leakage of molten polymer from joints | Fire if contacting heaters | Stop/reduce speed |
| | Injury if bolts break | Wear goggles and gloves |
| | Burns from hot polymer from die[a] | Never lean or stand in front of die, even when screw stopped[a] |
| Accidental contact with metal parts or heaters[b] | Burns | Avoid contact<br>Wear gloves and footwear<br>Post warning notices 'HOT' or 'LIVE ELECTRICS' if leaving machine |
| Contact with molten polymer | Skin burns[c]<br>Eye injury | Avoid contact<br>Wear goggles, protective gloves and footwear<br>*Isolate* motors when working on extruder, die or haul-off |
| Contact with live electrics, e.g., exposed terminals, wet cables | Electric shock<br>Consequential injuries | Wear gloves<br>Keep cables off floor<br>*Isolate* heaters when covers removed |
| Mechanical movement or falling components | Trapping, esp. fingers | *Isolate* drive motors |
| | Crushing, esp. feet | Keep guards in place whenever possible<br>Avoid in-running 'nips' and close clearances<br>Wear protective gloves and footwear |

[a] Molten polymer may be suddenly and violently ejected from the die after several minutes due to decomposition and gas formation at high temperature.
[b] May be much hotter than set temperatures.
[c] Avoid instinct to pull off polymer, which will cause an open wound. Though very painful, plunge in cold water and shake about until cold; then get medical attention *immediately*.

All equipment has (better have) electrical, mechanical, and/or hydraulic safety devices. The movable barriers on certain equipment allows the operator access with safety. As an example, a safety device on electrically operated equipment ensures that, upon opening a panel, gate, or other protective device, the electrical circuit is broken thereby not permitting the equipment to operate.

Procedures are set up for the steps to be taken in proper lock-out of machine's operation. As an example, properly lock-out of a machine's electrical circuit before starting repairs protects the maintenance worker from accidental start ups which could cause severe injury. The following steps for the proper lock-out procedure can be used as guide: (1) shut down all possible switches at the point of operation, then physically 'open' the main disconnect switch; (2) snap your own lock on the locking device – an ordinary padlock can be used for most electrical lockouts; (3) check the lock-out device and safety interlock to make sure that the switch cannot be operated; (4) place a name tag on the shank of the lock to indicate that the machine has been locked out; (5) notify the supervisor when repair work has been completed – only a supervisor should give the go ahead to remove your lock; and (6) take off the name tag and remove the lock.

Safety interlocks are safety devices designed to ensure that equipment will not operate until certain precautions have been taken. As an example, guarding of safety circuits are used in programmable controllers (PCs). The purpose is to establish a procedure for the guarding of OEM (original equipment manufacturer) supplied safety circuits on machines equipped with programmable logic controllers. It is well know that programmable controllers offer substantial freedom and flexibility in the design and modification of logic circuitry. It is very important and imperative that OEM supplied circuitry incorporated for the protection of the machine operators not be subject to modification or deletion by the user. For this reason, programmed safety circuitry must be guarded against access by the end processor to prevent inadvertent or intentional safety circuit alterations.

Two methods are suggested as a means of accomplishing protection of safety circuits. One approach is external guarding that can be accomplished by supplying all hard-wired safety circuits external to the PC in addition to the internal programmed circuits. Thus, the modification of the PC program or failure of the memory cannot compromise the operation of the safety circuits supplied by the OEM. The other approach is internal guarding of safety circuits that could be supplied such that the safety circuit addresses are confined to a nonprogrammable portion of the memory. Since this memory cannot be accessed by the user, modification to these circuits is not possible.

Certain design criteria should be followed when using programmable controllers. Applications of PC require careful consideration of their advantages, disadvantages, and limitations. External guarding of OEM supplied safety circuits should be done in addition to the internally programmed safety logic. PCs are used as a cost effective means of manipulating inputs while offering an extended degree of reliability in the control system.

The following considerations should be observed when utilizing programmable logic controllers: (1) standard practices and codes normally observed in the design of electro-mechanical systems should also be used in the circuit design for PCs; (2) documentation of logic programs and records must be retained by the equipment manufacturer for future reference and use; and (3) the following 'sounds and engineering practices' should be observed when using PCs with (a) the motor starting interlock device should not be subject to a failure of the PC, (b) all pilot devices should be wired from a common bus and fed through a master relay contact, (c) the AC power supply to the PC should give the programmable logic controller sufficient time to stabilize all voltages prior to the motor(s) being started, (d) all inputs from the PC should be correctly fused with fast blow rectifier fusing, (e) the AC feed all output modules should also originate off the incoming bus so that the output solenoids cannot be energized until the master relay contact has been closed, (f) all double solenoids should be interlocked with each other in the logic program, and (g) solenoid suppression may be added as required by PC manufacturer.

In addition to process safety, there is also industry safety that includes protection of employees and other personnel from the hazards of being struck against, struck by, caught in, falling, slipping, over exertion, contact with hot surface or temperature extremes, contact with electric current, and inhalation, absorption, or ingestion. Fire protection should be included. These hazards are usually confined within the site boundary and do not affect the public.

Very important in the development of safe work practices and a safe work environment are the development of a safety organization and its continuing activities. It should include the highest levels of management to demonstrate that safety has a priority equal to that of production, quality efficiency, cost control, and employee relations. Only in this way can responsibility and authority be accepted or shared throughout the operation to develop and maintain attitudes essential to safe work practices and conditions.

Other important aspects are inspection of fabrication equipment, dies and molds, audits of work practices, guards on moving equipment, training, accident investigations, record keeping, personal hygiene, personal protective equipment, fire prevention, noise control, and control of thermal radiation, ionization radiation, and chemical exposures. Process safety includes the safe transportation, handling, and storage of materials (plastics, chemicals, solvents, and so on), the control of extruders and other processing equipment, the protection of employees and the public from fires, explosions, releases of toxic materials, and protection of property from the effects of such accidents.

To protect operating personnel from recognized hazards, standards have been prepared to assign certain responsibilities to machine manufacturers, remanufacturers, modifiers, and employers. These standards are periodically updated; prepared by the American National Standards Institute (ANSI) with Society of Plastics Industry (SPI) participation. All processing equipment has standard procedures to operate and meet safety requirements. They are available from equipment manufacturers/suppliers and help to understand how to handle the equipment with plastics. They include safe start-ups, location of safety devices, and so on.

Figure 1.14 is an example of where accidents occur in all types of manufacturing plants that is updated by the National Safety Council, Chicago in their annual publication on safety. Safety manuals and standards are available from different sources such as SPI (Society of Plastics

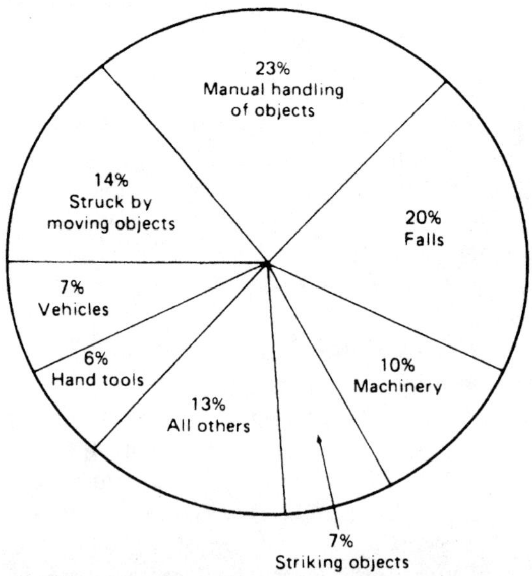


**Figure 1.14** General statistics on where accidents occur in all types of manufacturing plants, including all plastics plants.

Industry), ANSI (American National Safety Institute), etc., and the literature. The SPI Machinery Division continually releases new guidelines and updating old ones on safe construction, care, and use of machines that includes keeping plants clean and continually check safety devices on a planned schedule.

## ACCEPTABLE RISK

There are many risks that people are subjected to in the plant, at home, and elsewhere that can cause harm, health problems, or death. Precautions should be taken and enforced based on what is practical, logical, and useful. However, those involved in laws and regulations, as well as the public and, particularly, the news media should recognize there is acceptable risk.

This is the concept that has developed in connection with toxic substances, food additives, air and water pollution, fire and related environmental concerns, and so on. It can be defined as a level of risk at which a seriously adverse result is highly unlikely to occur but it cannot be proven whether or not there is 100% safety. In these cases, it means living with reasonable assurance of safety and acceptable uncertainty. This concept will always exist such as the use of automobiles, aircrafts, boats, lawnmowers, food, medicine, water, air, and so on. Practically all elements around us encompass some level of uncertainty.

The goal is to approach perfection in a zero-risk society. Basically, no product is without risk; failure to recognize this factor may put excessive emphasis on achieving an important goal while drawing precious resources away from product development and approval. The target or goal should be to attain a proper balance between risk and benefit using realistic factors and not the 'public–political panic' approach. People are exposed to many risks. Some pose a greater threat than others. The following data concerns the probability over a lifetime of premature death per 100 000 people in USA: 290 as the result of being hit by a car while being a pedestrian, 200 as the result of tobacco smoke, 75 from diagnostic X-ray exposure, 75 while bicycling, 16 when passengers in a car, 7 in Miami/New Orleans from drinking water, 3 when hit by lightning, 3 in a hurricane, and 2 in fire.

Around 1995 a young intern at the FDA made some amusing calculations. If the FDA permitted the packaging of Coca Cola in acrylic barrier plastic bottles, and if you drank 37 000 gallons of coke per day for a lifetime, you would have a 10% risk of getting cancer. Since normal people have a 25% risk of getting cancer, reducing the risk to 10% was a real plus for the acrylic barrier plastic bottles. So why have the unrealistic public not have a law enacted that the public should drink lots of coke?

## ENERGY CONSERVATION

When examining energy consumed or lost, the equipment used in the complete production line as well as the plastic is involved. Every extruder can be pictured thermodynamically as an open system with a complex working medium, the medium having interim phase transitions, particularly the transition from the plastic solid to the melted or fluid condition. There is a continuing effort by equipment manufacturers, particularly extruder manufacturers, to reduce energy consumption as well as energy loss. Figure 1.15 is an example of where energy/heat losses from an extruder occur.

Like the output capacity, the energy efficiency of an extruder is dependent on the torque available on the screw, actually screw rotational speed, heat control, and material being processed. Unfortunately, costly energy losses can occur, ranging from 3 to 20% and due to various factors, with the major loss occurring in the drive system. The power for screw rotation is supplied by a variable speed motor drive system, and is transmitted through a gear reduction unit, a coupler, and a thrust bearing. Gear reducers impart the final speed and torque to the screw. Most gear reducers use double-reaction helical or herringbone gears for ruggedness

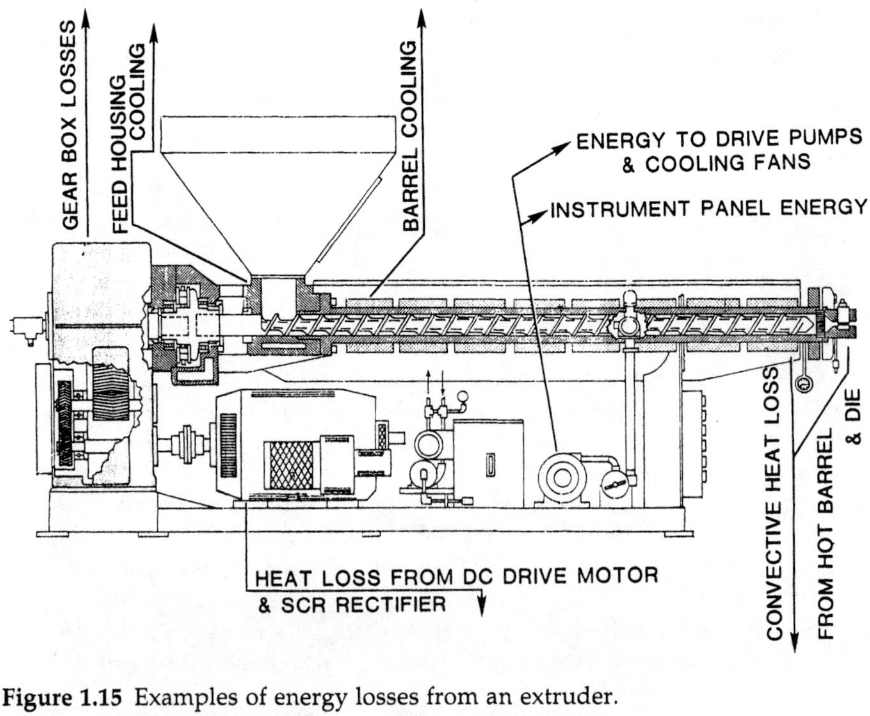


**Figure 1.15** Examples of energy losses from an extruder.

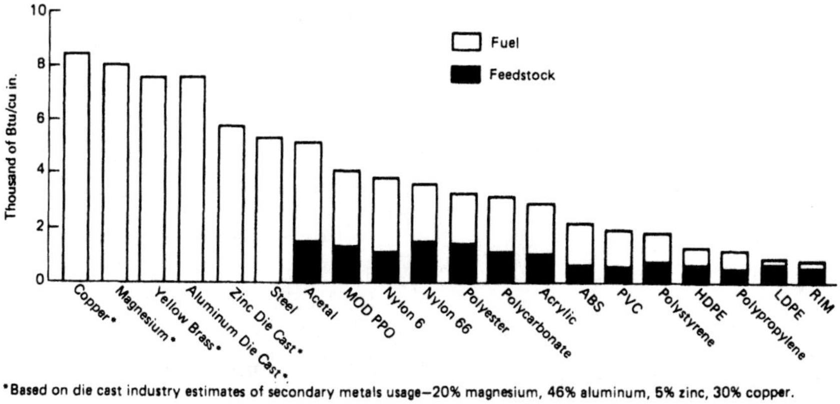


**Figure 1.16** Energy requirements for different materials.

and to hold down noise levels within acceptable limits. Worm and pinion gear combinations have been used on smaller extruders. The efficiency of the power transmission gear with the worm has a maximum of about 85%, that of the helical gear reaches 95%, and the herringbone reaches 97%.

In comparison with other materials, plastics have the lowest specific energy requirement for their manufacture (Fig. 1.16), fabricating products, and lowest recycling energy consumption. Energy conservation can be considered in the extrusion as well as the overall plastic industry from several viewpoints that go from producing plastic to processing to recycling products. In practically all evaluations, utilization of plastics saves or reduces energy requirements. In automobiles, aircraft, and other means of transportation, the increasing use of plastics reduces weight and fuel consumption; an example is replacement of the small whiskey bottles aboard aircraft with extruded blow molded plastic bottles rather than glass, resulting in lots of fuel energy saved. Using plastics in construction saves energy via insulation gains. Also, just shipping the lighter weight plastic products save energy.

Increasing use of plastics highlights the fact that they are made from petroleum, a non-renewable resource. However, most of the plastics familiar to consumers are manufactured from ethylene. During the distillation of petroleum , the ethylene fraction comes off as a byproduct which use to be burned for its energy value in the refinery processes. In the past, this waste gas became the building block for most of today's plastics. The plastics industry consumes roughly 3–4% of total annual petroleum consumption by the USA.

This use is more than offset by the savings that plastics create. An old study conducted for the British Plastics Federation determined that: (1)

plastics in the average car cuts fuel consumption by nearly 5% and contributes to improved safety performance; (2) plastic packaging has helped reduce food spoilage in the industrialized world to around 2%; developing companies have a spoilage rate 30–50%; (3) plastics neither rots nor rust (although there are degradable plastics) and applications continue to increase – witness the increased use of extruded plastic lumber; and (4) best of all plastics can be recycled many different ways.

Plastics are suitable for thermal reclamation (recycling) because their energy content or heat value is mostly higher than that of coal and other materials. Often energy content is similiar to that of heating oil. Energy consumption is in the order of about 2MJ/kg (2–2.5MJ/l), and the highest recoverable energy content of about 42MJ/kg. Some recycling comparisons are as follows: (1) paper requires about 6.7MJ/kg – as a general rule about twice as much paper is needed compared to plastics; (2) in glass production, if one uses about 10% of recycled glass, this only reduces the energy consumption of the process by 2% – thus recycled glass requires about 8MJ/kg but the comparative figure is higher when considered in relation to each product; need about 10–20 times as much material compared with plastics; (3) scrap steel and tin-plate is about 6MJ/kg; and (4) aluminum requires about 60MJ/kg. For this reason it is rational to reuse plastic wastes that are not suitable for material or chemical reclamation thermally as well as those in recycling that do not have a cost advantage. Especially in waste incineration plants, energy is obtained in the form of community heat and electric power. Most plastics can be incinerated without problems or with a low, controllable and confining emission of pollutants. For certain plastics, such as halogenated PVC, special available systems can be used to permit incineration.

In a cradle-to-grave analysis, conducted by Franklin Associates, that compared blow molded plastic bottles to aluminum and glass for the delivery of 1000gal (3.8mm$^3$) of soft drink, the study found that plastic out-performed aluminum and glass in all categories. The study compared the materials on the basis of energy consumption, air emissions, waterborne wastes, and solid wastes produced.

## PROCESSOR

This is part of the plastics industry where advances that are continually occurring in materials and machinery are combined to continually expand and create the finished products that most industries and the consumer public buys. Products range from the triviality of drinking straws to the lifesaving glory of the firefighters' protective suit and so on. There are three types of processor: custom, captive, and proprietary.

### Custom

These are operations that in the metalworking field might be known as job shops. They process plastics into products or components used in other industries. For example, a manufacturer of refrigerators could retain a custom processor to extrude sheets and thermoform the sheets into inner door liners. Custom processors typically have a close relationship with the companies for whom they work. They may be involved (to varying degrees) in the design of the product and the die, they may have a voice in material selection, and in general they assume a responsibility for the work they turn out. There is a subgroup in custom processing known as 'contract' fabricators. They have little involvement in the business of their customers. In effect, they just sell machine time.

### Captive

These are in-house operations of companies who have acquired plastics processing equipment to make products they need for the product they manufacture. For example, a refrigerator manufacturer may acquire a thermoforming machine to produce inner door liners. Or they may acquire a complete extrusion line to produce sheet and also acquire a thermoformer to produce the completed inner door liners. Generally speaking, these manufacturers will install a captive operation when their component requirements are large enough to make it economical (or a secret product/process exists). Some manufacturers who run their own plastics fabricating lines will nevertheless place a portion of their requirements with outside vendors to keep their own capital investment down, to avoid internal single-source supply, to maintain contact with the 'outside world' and the pricing intelligence it provides, and so on. The vendor could be a custom processor or a captive operation for their requirements.

### Proprietary

These are operations where the processor makes a product for sale directly to the public or to different companies. They usually have their own tradename.

## PREDICTING PERFORMANCE

Avoiding product failures can depend, in part, on the ability to predict the performance of all types of plastic materials and their shapes. With available time, the usual approach of product laboratory and/or field testing provides useful and reliable performance data. Engineers and designers continue to develop sophisticated computer methods for calculating stresses in complex structures while using different materials. The computational methods have replaced the oversimplified models of material behavior formerly relied on. However, for new and very complex product structures that are being designed to significantly reduce the volume of materials used and, in turn the product cost, computer analysis is conducted on prototypes already fabricated and undergoing testing. This computer approach can result in early and comprehensive analysis of the effects of conditions such as temperature, loading rate, environment, and material defects on non-structural and/or structural reliability. This information is supported by stress–strain behavior collected in actual material evaluations [3].

When combined with the use of computers, the finite-element analysis (FEA) method has greatly enhanced the capability of the structural analyst to calculate displacement and strain–stress values in complicated structures subjected to arbitrary loading conditions. In its fundamental form, the FEA technique is limited to static, linear elastic analysis. However, there are advanced FEA computer programs that can treat highly nonlinear dynamic problems efficiently. Important features of these programs include their ability to handle sliding interfaces between contacting bodies and the ability to model elastic–plastic material properties. These program features have made possible the analysis of impact problems which in the past had to be handled with very approximate techniques. FEA has made these analyses much more precise, providing better direction in locating high stress areas. Final verification of load-carrying capability may require actual testing of the fabricated product based on computational analysis.

### Scale-up

A limited amount of development has been conducted investigating scale-up of extruders [348–353, 370, 376]. Experience continues to be the main approach since no direct relationships exist; many variables have to be considered. Scale-up techniques have been used where shear rate is keep constant, Newtonian flow characteristics are used, and so on (Chapters 3–6).

## MYTHS AND FACTS

Understanding the extrusion process and plastic behavior during fabrication continues to improve and expand our capability to produce products that meet performance requirements and reduce costs. In the mean time there are factual and mythical problems. The types of problems vary. As an example, there can be a processing problem where machine settings do not properly match the required processing conditions; this is a fact. Unfortunately there are times that myths develop due to factors that range from competition to ignorance. Competition does evolve 'at times' among companies, environmentalists, politicians, and so on. To believe we know something when we have an opinion is a myth (lie). A revealing exercise is to say with every opinion, 'I do not know but in my opinion . . .' Any kind of pretence is a myth that can result in people saying 'plastics are bad and uncontrollable' [347].

An example of a myth is that an extruder can make a homogeneous blend from unhomogeneous concentrates or master batches. The fact is that the continuous extrusion process is a 'plug flow' phenomena in which minimal flight-to-flight mixing occurs with single and twin screw systems. If a concentrated feeder is adding inconsistent concentrations, the extruder will mirror these inconsistencies at the discharge. This action is true even if mixing action by the screw is used such as incorporating mixing pins, Maddox mixer (Chapter 4), melt pump, and/or static mixer. If the poor distribution is easily noticeable or measurable, no amount of post blending can evenly distribute this error [187, T. J. Chauvin].

In order to understand potential problems versus solutions (and eliminate myths), it is important to consider the relationships of machine/equipment capabilities, plastics processing variables, and product performances. A distinction has to be made between machine conditions and processing variables. As an example, machine conditions include operating temperatures and back pressures, screw rotation speed, die temperature, etc. Processing variables are more specific such as melt conditions in the plasticator and die, melt flow rate versus temperature, etc.

Computers permeate different areas of extrusion. This technology is implied or directly involved in product design, machine capability requirements, die design, machine processing requirements, quality control, cost analysis, etc. Computers are not a panacea but they have their place. Most important is the person(s) involved in their use with proper knowledge in using the hardware and software in order for the computer to properly and efficiently operate. They are another tool and guide for the user; not a mythical concept that they perform the task of eliminating all problems; in fact they could setup problems.

Recognize that what comes naturally to us in conceptualizing a processing line or design is a real extension for a computer. It can be said that the computer needs people (fact) since computers do not have the capabilities of translating our thought processes into computational and cognitive models that will make intelligent systems such as computer-aided design (CAD), computer-aided manufacturing (CAM), etc. Outside the realm of science fiction or myths, computers are not particularly innovative. They are most adept at making decisions when logical relations can be expressed in algorithm format. This means the human activity or mind cannot be modeled and defined by some mathematical method (fact).

The computer supports rather routine tasks of embodiment and detailed operation rather than the human creative activities of conceptual operation. The computer can make things better (fact), but usually not easier, for example as in product design that has to relate to processing the plastic. A computer makes it easier if one is knowledgeable about the computer's capability in specific areas of interest, such as machine settings, product design, die design, etc. By using the computer tools properly, the results are a much higher level of processing that will result in no myths.

Successful designed products require the combination of various factors that includes sound judgment and knowledge of processing. Until the designer becomes familiar with processing, a fabricator must be taken into the designer's confidence early in development and consulted frequently. It is particularly important during the early design phase when working with conditions such as shapes and sizes. There are certain features that have to be kept in mind to avoid degradation of plastic properties. Such features may be called property detractors or constrains. Most of them are responsible for the unwanted internal stresses that can reduce the available stress for load bearing purposes.

Since myths develop about the 'world of extrusion and plastics' that they are not perfect processes and materials, so what is new (in life)? They are facts but also the fact exists that they are controllable. Of course this statement can be made for all processes, materials (steel, wood, etc.), as well as anything around us. What continually occurs with extrusion (and all the others) are the improvements as time passes with the important fact that the knowledge of people are extended.

## PEOPLE AND PRODUCTIVITY

The recipe for productivity includes a list of ingredients. They include research and development, new technologies, updated equipment, and automated systems with modern facilities. Each of these have certain limitations depending on factors such as production quantities and product performance requirements interrelated to cost. However, the one ingredient that ties the recipe together is people; none of the above factors has much impact without the properly trained people. Operations optimization requires a blend of labor, management, materials science, control engineering, and mechanical design. Cross train for success and spend time on the shop floor.

An example of important people are the shop foremen and supervisors who are usually former line workers with experience and know-how of setting up and running extrusion lines. They represent key people in a company by eliminating or at least reducing down-time. It is important to have these people pass their knowledge to the line workers that are involved in specific areas of production. Suggestions on improving people's capabilities and knowledge are: (1) maintain a good, but disciplined, relationship of all people; (2) impress upon the foremen and supervisors that they can have a good future and that their position is not the end of the line; (3) conduct in-house training sessions; and (4) in small plants, where in-house training is not feasible timewise, encourage your foremen and others to seek more education [187, D. Biklen].

## PROCESSING AND PATIENCE

When making processing changes, allow enough time to achieve a steady state in the complete extrusion line before collecting data. It may be important to change one processing parameter at a time. For example, with one change, such as extruder screw speed, temperature zone setting, cooling roll speed, blown film internal air pressure, or another parameter, allow four time constants to achieve a steady state prior to collecting data.

## TROUBLESHOOTING

Guides are provided in most of the following chapters so all the different extrusion processes have their specific troubleshooting approaches; also maintenance guides are provided. However, the 'problems-to-solutions' are interchangeable so that reviewing them all can be more helpful than just reviewing one set of a troubleshooting guide [1, 3, 187, 190].

A simplified approach to troubleshooting is to develop a checklist that incorporates the rules of a problem-to-solving procedure such as: (1) have a plan and keep updating it based on the experienced gained in operating the equipment; (2) watch the processing conditions; (3) change only one condition at a time; (4) allow sufficient time for each change and keep some kind of a log of the action, with results, that are occurring; (5) check housekeeping, storage areas, granulators, personnel clothing and behavior; and (6) narrow the range of areas in which the problem belongs, e.g. material storage and handling, extruder, die, cooling tank, puller, specific control, part design, environment (humidity, ventilation location and direction of forced air, dust, etc.), people, and management. Table 1.6 provides a simplified troubleshooting guide [397].

When a start-up for a line is initially conducted, the experienced operator will set the operation in all its different modes of operation going from not enough (or too little) to more than is required to fabricate acceptable products. Different objectives are examined, such as using the minimum amount of plastic, increase output rate, and so on. This procedure

**Table 1.6** Guide to troubleshooting

| *Problem* | *Condition* | *Cause* | *Solution* |
|---|---|---|---|
| Heating interrupted | Temperature falling | (a) Control action to avoid overshoot, time-proportioning or interaction with adjacent zones | Observe if heating resumed in say 30 s, otherwise check (b), etc. |
| | | (b) Fail-safe control, indicating broken thermocouple or connection | Check thermocouple continuity |
| Temperature continues to rise | Temperature above upper control limit | (a) Thermocouple slipped out or not in contact with heater/extruder | Replace, ensuring good thermal contact |
| | | (b) Thermocouple incorrectly connected | Check connections and temperature indication, e.g. in hot water or body heat |
| Output suddenly falls | Steady screw speed | (a) Feed hopper empty | See Barrel and feed unit section in this chapter. Investigate blocked feed line |
| | | (b) Bridging in hopper or into screw | Stop screw and clear with non-metallic rod, e.g. strip of product. Check adequate feed pocket cooling. Check air entrapment esp. with powders |
| | | (c) Polymer melting in feed section | Reduce first barrel zone temperature<br>Increase feed pocket cooling<br>Cool feed end (only) of screw |
| | | (d) Surging | See Table 2.18. |
| Polymer leakage from joints or fittings | Excessive pressure | (a) Blocked die or screen | Immediately reduce screw speed.<br>Check die and adaptor temperature<br>Shut down and clear |

**Table 1.6** *Continued*

| *Problem* | *Condition* | *Cause* | *Solution* |
|---|---|---|---|
| | | (b) Failed die heater or control | Immediately reduce screw speed<br>Check thermocouple<br>Shut down and replace heater |
| | | (c) Incompletely melted polymer, esp. at start-up or after feed interruption | Stop screw<br>Wait to heat up and restart *slowly* |
| Screw stops or speed drops | Excessive torque | (a) Low barrel temperature | Reduce speed setting and check heaters |
| | | (b) Excessive pressure | As for 'Leakage' above |
| | | (c) Excess feed, e.g. of strips | Stop or reduce speed setting<br>Increase temperature of first barrel zones |
| | | (d) Excessive speed | Reduce speed and/or increase barrel temperatures |
| Lumpy or unmelted polymer at die | Incomplete melting, esp. of powders | (a) Insufficient shearing/heating in barrel | See Tables 2.15 and 2.17. |
| | | (b) Excessive screw speed | Reduce screw speed *or* improve melting. See Table 2.16. |
| | | (c) Feed contamination, e.g. with hard-grade polymer | Reduce screw speed<br>Check feed material<br>If necessary, increase barrel temperatures |
| Foaming or bubbles at die (distinguish from later contraction bubbles due to rapid cooling), surface blisters | Steady running or start-up | (a) Moist polymer | Stop feed and restart with dried polymer |
| | | (b) Excess temperature causing decomposition | Stop or reduce screw speed. See Table 2.15. |
| | | (c) Air entrapment in feed | Raise first barrel zone temperature (to reduce unmolten length)<br>Fit vent at rear of feed opening |

follows the FALLO approach (Fig. 1.1). By going through the complete steps to set up the line, one gets exposed to many, perhaps all, of the problems that the line will encounter during any time of its operation. One should record all the problems encountered and how they were corrected. This type input can include data that can be included in a computer control system and/or an easy understood troubleshooting guide for the specific line to produce acceptable products at the lowest cost. Include personnel involved and specifically what equipment was used from up-stream, through the extruder, to down-stream equipment. This type of action would set up an operation targeted for what can be called total control.

## HISTORY

Without going back to the Roman times and those prior to the Romans, who had some type of extrusion device, let us start during 1797 with officially recorded information. At that time Joseph Bramah of England constructed a hand-operated piston press (ram action) for the manufacture of seamless lead pipes, which may have been the first documented 'extruder.' Other non-ferrous metals were later processed in similar presses that included wood pulp, paste-like foodstuff such as macaroni, building materials such as tiles and ceramics, soap, and so on. Material was forced out through an appropriate die to form a rod, pipe, sheet, etc.

The use of continuously operating ram extruders for a natural TP substance, gutta percha initially and latter TS rubber, for cable coverings occurred in England, Germany, and USA. Gutta percha is a milky juice from Malaysian tress. It is hard but not brittle, acid resistant, waterproof, and an excellent insulator. It was first observed and documented in 1650 when a visiting Englishman observed natives molding knife handles with the material. In 1845 Bewley and Brooman received the first patent that only used manual operation to ram extrude tubes, pipes, and other products. For half a century, if there was a useful or decorative fabricated part (extruded, molded, etc.) in any home it was made of gutta percha. About 1855 the ram machine was converted to a mechanical drive.

In 1879 Englishman Matthew Gray patented an extruder with feed roller. A mill roll attached and located above the material entry into the screw provided ease of feeding, ensure an even build-up of pressure, and allowed ventilation and partial dehydration of the gutta percha. The feed rolls were heated by steam or hot water and the screw casing was heated by hot water. Trapped air could escape through an opening in the material inlet region. A relevant patent also described processing rubber; hot rubber strip stock feeding started.

In 1882 Willoughby Smith (England) introduced a new type of extrusion based on the Gray patent but using a gear pump instead of the ram action. This type design has been used ever since in covering cables and other products however eventually the gutta percha was replaced by polyethylene, etc. Machine builders produced mechanically driven screw extruders prior to 1879; they included John Royle (USA) and Francis Shaw and Iddon (England).

A drawing dated 1873, owned by Phoenix Gummiwerke AG, Hamburg-Harburg, Germany, is probably the earliest design for an extruder. It used a double-threaded screw for a tube machine extruding 82 mm (3.5 mm) diameter with 5–7 diameters in length. A large demand for these type extruders developed and by 1900 about 600 TS rubber extruders were operating in different countries producing different products. By 1900 cellulose acetate and casein were added to the materials being extruded by the Bell laboratories (USA). By the mid-1900s the use of extruders, both ram and screw, increased significantly.

One of the first screw extruders, which appears to have been specially designed for processing TPs, was produced by P. Troester (Germany) in 1935 which was fitted with combining steam and electrical heating, as well as a considerable longer screw than that use with TS rubber extruding.

In 1936 American Bruce H. Maddock (Union Carbide) started research on single screw extrusion at which time the process was principally used in processing TS rubber. Maddock made many contributions over decades that included: (1) technology for coating wire with extruded TP insulation, including polyethylene (PE), vinyl, and fluoropolymers; (2) establishment of a systematic body of knowledge regarding how productivity and product quality are affected by temperature, pressure, melt flow, mixing, and other process variables; (3) the 'push-out' or 'screw-freeze' technique for analyzing the extrusion process by stopping the machine, rapidly cooling the melt containing color tracers, removing the solid filled screw from the barrel, and unwrapping the plastic helix, which serves as a specimen for studying process variables along the extruder length and in the cross section directions; and (4) developing (US 3486192 patent free) the addition to a screw of a barrier mixing section, called the Maddock Mixer or the Union Carbide Mixing Head, which improves product quality while increasing output.

In 1936 Horst Heidrich in Germany built a machine for the direct use of granulating or chopped plastics; it only used electrical heating. In 1937–8 Francis Shaw (now USA) produced a comparable machine but with a longer screw using electrical or oil heating. By 1939 Troester built a completely electrically heated plastics extruder which for the first time was fitted with compressed air cooling for removing surplus heat. The machine had variable speed drives and automatic temperature controls. This Troester machine basically resembles those currently used.

The first departure from single screw extrusion dates from the early 1930s when Roberto Colombo and Carl Pasquetti of Italy developed a twin screw extruder for TPs. During 1935 kneading pumps were developed mainly for rubber and ceramics by I. G. Farbenindustrie AG and by Leistritz and Burghauser of Germany. An older example of twin screw extruders was in 1869 by Fellows and Bate (England) to manufacturer sausage.

The technology of TP film extrusion originated in 1933. Patents then granted to the Norddeutschen Seekabelwerke AG (Germany) related to special polymerization products, particularly polystyrene, and the 'Styroflex' method based on it was originally applied only to PS. A wider field of application was initiated and in developed in USA only when PE became marketable during 1939–1945.

Credit for the original success of vented TP sheet extrusion is given to Ernest Bernhardt (DuPont, USA) during 1957 when he published a paper describing a vented two-stage screw for running acrylic sheet without drying. Bernhardt's screw drew vacuum on the decomposition zone via a hole in the flight which communicated with the longitudinal bore of the screw. A rotary union at the rear of the screw was connected to a vacuum pump. Later, at Prodex, American Frank R. Nissel studied the shortcoming of this screw where it had to be removed and unplugged every time the vent hole flooded. The approach, at that time, was to combine the vent hole in the cylinder and operate at a precisely controlled head pressure to make the vented screw run with precision and remove the plugging problem. In addition, the Bruce Maddock valve extrusion principal was used to further simplify its operation.

From this beginning, advances in plastics have been followed by advances in extrusion equipment particularly since the 1930s. With all this action, there has been extensive work in rheology and thermodynamics in better understanding the process of extruding plastics and other materials. The results continue to increase production rates, increasing amount of control over the extrusion variables, and/or reducing costs. Major developments continue to be in the refinement of existing extrusion methods. In turn the developments and particularly the literature on this subject proliferate [315, 370].

The growth of the extrusion business includes the important requirement of the market/people acceptance of the many plastic products that total in the billions of dollars. They include major markets for plastic bags, pipes, medical devices, and the many that are reviewed in this book. There are also thousand of specialty examples that now exist and have existed during the past century.

An example of a specialty application occurred within months of the opening in 1965 when the Houston-Texas Astrodome's grass field had simply died. In the following year the natural grass was replaced with over 11 600 $m^2$ (124 000 $ft^2$) of new Astro Turf fabric made from Nylon 6.6 extruded fibers (Chapter 12) containing additives such as pigments and ultraviolet (UV) inhibitors (Chapter 3). This fabric was secured by more than 5 km (3 mi) of zippers that used noncontinuous extruder/injection molding machines; they included the use of continuous molding systems (Chapter 18).

Another example is a product to replace the 'ice' in ice skating rinks. It is a sandwich construction of high density polyethylene (HDPE) extruded sheet attached to a wood core. The sandwich sections are 0.8 × 1.5 m (2.5 × 5 ft) with overall thickness at about 3.2 cm (1.25 in). The most well-known skating rink using this 'Super Ice' system is Madison Square Garden; they have been using this structure since 1985. It is reported to have 94% of the glide factor of wet ice when it has been properly maintained with a 'glide' solution. The surface is nor slippery to foot traffic and can be used for 'in-line skating' without the glide solution that is used for ice skating. The surface shaves just like ice with each side of the sandwich construction lasting about five years.

# 2

# Extrusion machine and line

## FEATURES OF EXTRUDER AND PRODUCTION LINE

### Overview

The single-screw and multi-screw types have their differences (Fig. 2.1, Table 2.1). Each has its benefits, depending on the plastic being processed and the products to be fabricated. At times their benefits can overlap, so that either type could be used. In this case, the type to be used would depend on cost factors, such as cost to produce a quality product, cost of equipment, and cost of maintenance.

Similar extruders from different manufacturers, and even those from the same manufacturer processing the same TP, will often require different operating settings to produce similar products. This is true even when screw designs are 'identical.' The reasons for the differences include factors such as variability of plastic (most important), barrel nonuniform internal dimensions, control sensor locations, variable or limited available heater wattage, coolant flow rates, etc. To obtain a consistent performance for the same material from one extruder to another, one has to know the variables that exist in setting up the machine controls on both machines. Good quality extrusions require: (1) up-stream equipment delivering properly controlled TP to the hopper; (2) homogeneity by the extruder in terms of the melt heat profile and mix, with accurate and sustained flow rates; (3) a good die design; and (4) accurately controlled down-stream equipment for cooling and handling the product [158, 187, 235, 279, 303, 370].

It is interesting to review patents and literature from the mid-1980s because actions and designs being taken, thereafter, basically expanded many of those initiated by the original pioneers in the extrusion business; see the historical review in Chapter 1. Since this period, plasticators have had conical smear heads, kneading section of the screw where a tapered

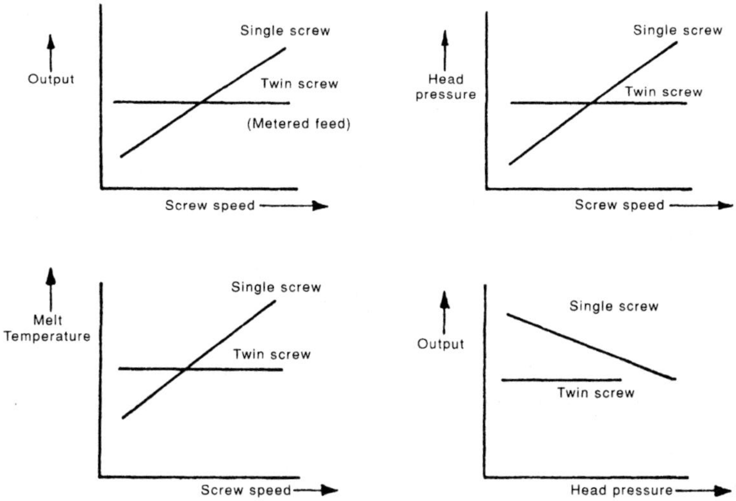


**Figure 2.1** Effect of single- and twin-screw machine variables.

**Table 2.1** Comparison of single- and twin-screw extruders

| | *Single-screw* | *Twin-screw* |
|---|---|---|
| Flow type | Drag | Near positive |
| Residence time and distribution | Medium/wide | Low/narrow (useful for reaction) |
| Effect of back pressure on output | Reduces output | Slight/moderate effect on output |
| Shear in channel | High (useful for stable polymers) | Low (useful for PVC) |
| Overall mixing | Poor/medium | Good (useful for compounding) |
| Power absorption and heat generation | High (may be adiabatic | Low (mainly conductive heating) |
| Maximum screw speed | High (output limited by melting, stability, etc.) | Medium (limits output) |
| Thrust capacity | High | Low (limits pressure) |
| Mechanical construction | Robust, simple | Complicated |
| First cost | Moderate | High |

cone formed with the barrel wall using annular slots with grooves and channels, controlled feeding rates, and even adiabatically operating extruders. An adiabatic extruder is a machine that theoretically operates 'on its own heat' after the extruder has been heated sufficiently by conventional means to melt the plastic. This type of machine is also called an autothermal or autogeneous extruder.

**Single-Screw Extruder**

Features of this machine are shown in Fig. 2.2. This schematic from Welex shows: (1) drive motor (from 20–2000 hp infinitely variable speed drives directly coupled to reducer for maximum efficiency all deigned to save floor space); (2) high efficiency gears to process all plastics (heavy duty, heat treated helical or herringbone gears equipped with shaft-driven oil pumps and oil cooler); (3) long-lasting thrust bearing (with life expectancy well in excess of 30 years' continuous operation); (4) large rectangular standard feed opening [round with lining, optional, for use with crammer feeders (Fig. 1.11)]; (5) cast-in heater/cooler elements (heat quickly, quickly, and last a long time); (6) cast-in stainless steel cooling tubes run parallel with heating elements (closed-loop, non-ferrous distilled water is automatically adjusted via microprocessor-based temperature controllers providing uniform, efficient cooling); (7) high-performance screws with bimetallic lined cylinder designed for the plastic requirement (long life at all temperatures and cored for cooling); (8) prepiped and prewired (ready for single power drop installation); (9) guards fully insulated (one-piece, hinged, no loose parts, no disassembly needed for access); (10) heavy fabricated steel single-unit base (preassembled so all parts are in place ready to be used); (11) when required, patented two-stage venting which eliminate predrying for most plastics (vents can be plugged in minutes); (12) screen changer (optional) for continuous operation without shut down (hinged swing-bolt gate standard); (13) gear pump (optional) to ensure absolute volumetric output stability; (14) static mixer (optional) to provide thermal and viscosity homogeneity; and (15) die (optional) (designed to produce single or multi-layer sheet without modification; also strand dies, etc.).

The essential parameter in the extruder's pumping process is the interaction between the rotating flights of the screw and the stationary barrel wall. For the plastic material to be conveyed, its friction must be low at the screw surface but high at the barrel wall. If this basic criterion is not met, the plastic will probably rotate with the screw and not move in the axial/output direction.

In the output zone, both screw and barrel surfaces are usually covered with the melt, and external forces between the melt and the screw channel walls has no influence except when processing extremely high viscosity materials such as rigid PVC and UHMWPE. The flow of the melt in the output section is affected by the coefficient of internal friction (viscosity) particularly when the die offers a high resistance to the flow of the melt [237, 238, 391, 394].

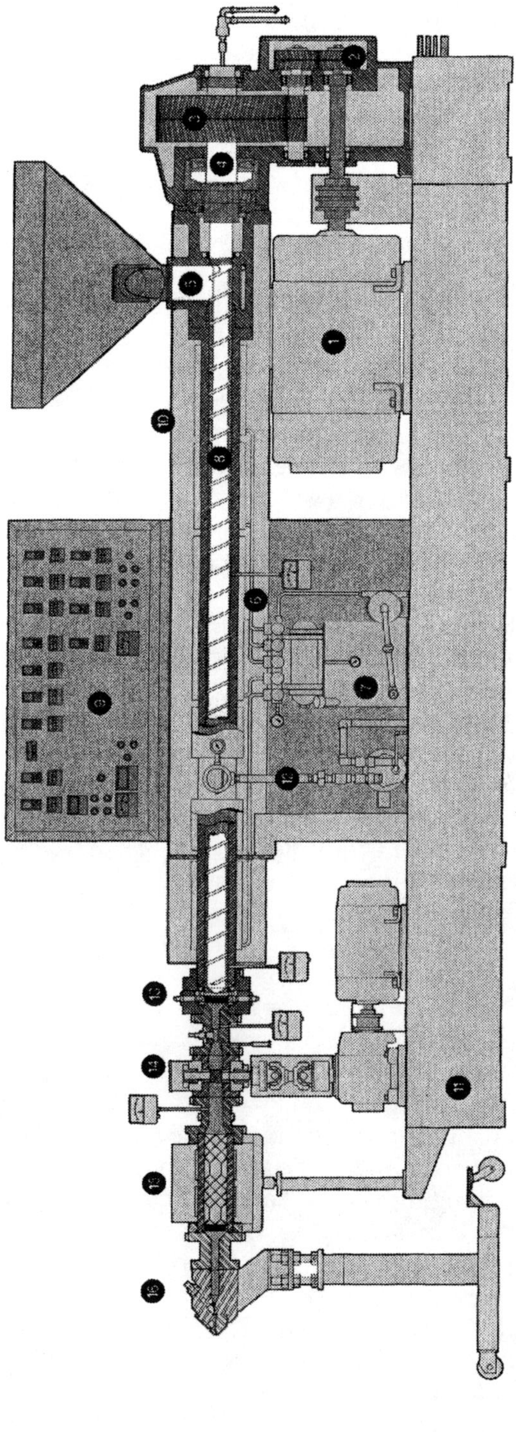

**Figure 2.2** Example of a Welex extruder schematic.

Control of temperature is a very important processing factor in all extruders, but more important with the larger machines because more metal exists. It can be especially critical in the case of reactive extrusion where the extent of reaction and side reactions will be extremely temperature sensitive.

Understanding the heat transfer basic key elements in extruders has been extensively reviewed in the literature [2, 237, 348, 397, 448]. As an example, in order to estimate the plastic overall heat transfer capability, equations can be used based on an extruder size [162]. The overall heat transfer coefficient ($U_i$) based on the barrel's internal diameter (ID) is an indication of the ability to move heat into and out of the process. Its optimization is less critical when the requirement is to add heat to the process as the driving force. It can usually be easily increased by boosting the temperature of the electrical or oil heating source. In the case of electrical heaters, there should be high-temperature cut-outs in place to prevent burnout or meltdown especially when using cast aluminum heaters. In the case of oil heat, prevent the heat transfer fluid from degrading. This can be done by choosing a construction material that maximizes thermal conductivity, the plastic film coefficient, and the jacket side heat transfer coefficient. In the case of lower conductivity materials, minimizing the barrel thickness will help.

This summary shows that: (1) for a low interior heat transfer coefficient, heat transfer is limited on the plastic side of the process; (2) for higher coefficients, heat transport limitations through the barrel become more significant and the negative influence on overall heat transport of using more corrosion resistance materials of construction occurs; and (3) even for the maximum coefficient, the plastic film ($h_i$) presents a significant barrier. This shows the importance of estimating $h_i$ on evaluating the ability of the extrusion to control temperature through external heating/cooling means. The experimental measurements have shown $h_i$ to be typical in the range of 40–90 BTU/ft$^2$ hr °F. Factors that affect $h_i$ are, aside from the plastic thermal conductivity, screw outside diameter (OD) to barrel ID clearance, and rpm. As the barrel-screw clearance decreases, the $h_i$ increases and thereby improving $U_i$. The effect of rpm is to increase $h_i$ due to the effect of more frequent renewal of the plastic film on the barrel.

Two opposing factors control the pumping capability of the machine. The screw, if the feed zone operates correctly, builds up a pressure gradient in the material filling the screw channel. A pressure gradient is generated by the feed section as well as the transition (compression), and metering (output) sections. This gradient, particularly its point of maximum pressure, depends on factors, such as screw rotation speed, barrel temperature profile, type of material, screw dimensions/design, and flow resistance due to the die. With screw deeply cut feed zone and a die with very little resistance, the maximum pressure usually occurs in the last third of the screw length. This action is called free discharge or unthrottled flow. The output zone is over-run by the previous zones. When the output zone is made deeper, the pumping of the earlier zones is less throttled and the maximum pressure shifts towards the exit end of the screw [370].

With the pressure being zero at both the screw entrance and die exit, the pressure in the melt always acts in two opposing directions. The melt flow through the die under pressure transmits the throttling effect of the die to the movement of the melt in the screw threads. These effects are complicated, but the reduction in output caused by the counter-pressure can be calculated with some degree of accuracy. The amount of plastic conveyed by the screw in free discharge is about half the channel volume per revolution, because the melt adheres to both the screw and barrel walls, causing its average speed to be midway between speeds of the two surfaces. The quantity, which is called the drag flow, depends only on the dimensions of the screw and its rotational speed [397].

With the melt extruded under pressure through the die, the output is reduced by an amount which is called pressure flow or back flow. Quantitatively the output loss can be calculated as if the melt were flowing backwards through the screw channel under the influence of the pressure at the screw tip. This pressure action on flow depends only on the dimensions of the channel and on the pressure and viscosity of the melt; it does not depend on the screw speed, although indirectly the speed does affect the pressure flow by altering the viscosity and in turn the pressure.

Another pressure loss is the leakage flow where melt flows over the screw flights from one thread to the next in the direction of the pressure gradient. This flow loss is usually disregarded when the clearance between screw flight and barrel wall is small. With worn screws, highly fluid melts and dies of high resistance, the leakage flow becomes significant.

Good pumping consistency on extruders is necessary in most applications to allow good product consistency. As reviewed, numerous other factors influence the ability of the extruder to deliver the melt at good pumping stability. They include: (1) feed material bulk density; (2) friction characteristics of the feed (lubricated, etc.); (3) feed material temperature (long-term effect); (4) feed section opening and hopper design; (5) screw feed area design; (6) screw transition/meter section design or barrier/meter section designs (Chapter 4); (7) screw tip pressure level; (8) operating screw speed range; (9) barrel temperature profile; (10) motor rpm stability; (11) screw/barrel/feed section wear; and (12) others. All these factors and others reviewed in this book should be investigated when instability, such as surging and material variation, occurs.

Since down-stream from the die other factors are also candidates for product variation, the extruder should first be singled out as the problem through pressure variation checks near the screw tip in the die adapter. If more than 3% total pressure variation is occurring, some attention to the extruder is warranted [187, E. Steward].

**Twin-screw extruder**

With the development of extrusion techniques for newer TP materials, it was found that some plastics with or without additives required higher pressures and needed higher temperatures than were required with rubber. There was also the tendency for the material to rotate with the screw. The result was degraded plastics. The peculiar consistency of some plastics interfered with the feeding and pumping process. The problem magnified with bulky materials, also certain types of emulsion PVC and HDPE, as well as loosely chopped PE film or sticky pastes such as PVC plastisols.

During the early 1930s, the twin and multi-screw extruders were developed to correct the problems with the single-screw extruder. The conveyance and flow processes of multi-screw extruders are very different from those in the single-screw extruder. The main characteristic of multi-screw extruders include: (1) their high conveying capacity at low speed; (2) positive and controlled pumping rate over a wide range of temperatures and coefficients of frictions; (3) low frictional (if any) heat generation which permits low heat operation; (4) low contact time in the extruder; (5) relatively low motor-power requirements self-cleaning action with high degree of mixing; and (6) very important, positive pumping ability which is independent of the friction of the plastic against the screw and barrel which is not reduced by back flow. Even though the back flow does not exist, their flow phenomena are more complicated and therefore far more difficult to treat theoretically than single-screw flow. Result has been that the machine designer has to rely mainly on experience [107, 146, 151, 187, 204, 234, 242, 256,276, 281, 297, 312, 323, 379, 395, 411].

Similar to the single-screw extruder, the multi-screw extruder, including the more commonly used twin-screw, has advantages and disadvantages. Figure 2.3 shows the different designs used with the twin-screw extruders. The market for counter-rotating twin-screws (TSs) is dominated by the cylindrical screws (parallel TS) and the extruder fitted with conical screws (Fig. 1.7). The type of design to be used will depend on performance requirements for a specific material to produce a specific product (however single-screw types dominate the industry). With the multi-screws, very exact metered feeding is necessary for certain materials otherwise output performance will vary. With overfeeding, there is a possibility of overloading the drive or bearings of the machine, particularly with counter-rotating screw designs. For mixing and homogenizing plastics, the absence of pressure flow is usually a disadvantage. Disadvantages also include their increased initial cost due to their more compli-

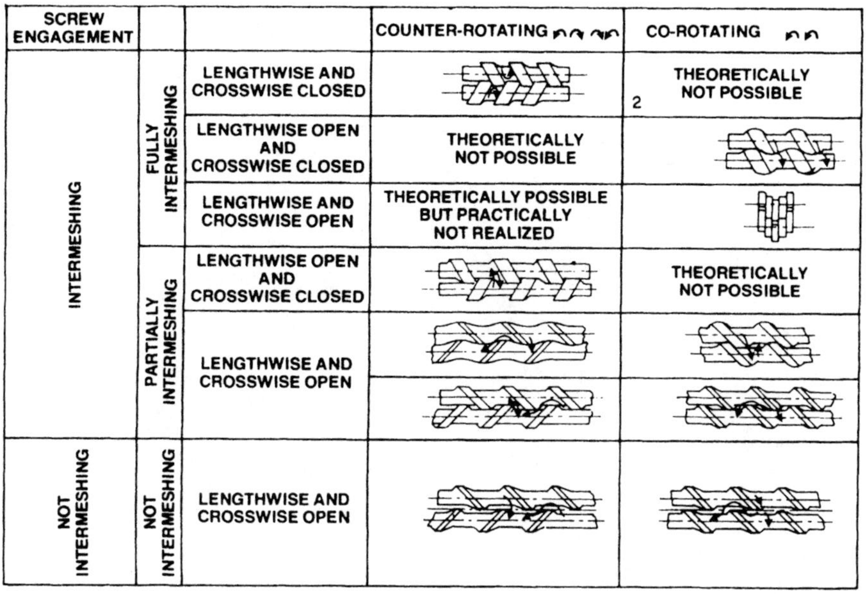


**Figure 2.3** Example of different twin-screw mechanisms.

cated construction as well as their maintenance and potential difficulty in heating. However, there are many applications where these disadvantages are outweighed by the performance required, such as in compounding (Chapter 17).

Twin-screws with intermeshing counter-rotating screws are principally used for compounding, including situations where volatile must be removed during extrusion. Twin-screws have found a substantial market in difficult compounding and devolatilization processes. To provide specialized compounding and mixing, particularly in the laboratory, different mixing techniques are required, such as using interchangeable screw sections on a splined shaft (Fig. 2.4).

Certain plastics require very gradual heating using external heaters and sometimes require specific mixing effects not available with single-screws. With these machines, the terms external and internal are not preferred. They are replaced by the terms extensive and intensive to avoid confusion with internal mixers. The multi-screw extensive mixing action is characterized by the fact that the plastic constantly changes in flow direction. Each particle moves sometimes on the surface and other times in the body of the mass of plastic. The surface of the mass is constantly renewed and after a certain time a more or less statistical distribution of all components is achieved. Typical of these extensive mixers are agitators and most of the kneading machines.

**Figure 2.4** A Werner & Pfleider screw assembly offers specialized processing combinations.

The intermeshing twin-screw extruders known as kneading pumps made possible continuous and enforced pumping, due to the effect of the intermeshing screws. The gear pump is a special case of the twin-screw extruder. Original machine designs encountered different operational problems when high pressures and temperatures were required for processing certain TPs. Severe construction problems occurred due to overloading the screws and mechanical stresses on both the barrel and screw; bending occurred due to the separating forces, wear was caused due to different stresses on the thrust bearings, etc. These early problems were resolved by the 1960s; in contrast, early problems with single-screw extruders were overlooked.

The intensive mixing action of multi-screw extruders is highlighted by the fact that adjacent layers of plastic have different speeds, i.e. velocity gradients or shear rates. In addition to the actual mixing effect, due to displacement of layers with respect to each other, pigment particles or ungelled particles of the plastic (fish eyes) are broken down by its limited friction action, especially in highly viscous plastics.

There are extruder types that combine the intensive and extensive mixing effects in one machine such as in the single-screw extruder and a Banbury mixer. The shearing of the plastic in the shallow screw channels produces intensive mixing and the transverse with pressure flows provide extensive mixing. With multi-screw extruders, this extensive mixing effect is missing. As reported decades ago, the mixing action can be seen using a small amount of color pigment that enters the hopper with the plastic. This color shot is reproduced in the extrudate of the multi-screw extruder as a slightly extended and relatively sharply defined region. The intensive mixing effect of the multi-screw machine is usually more pronounced than that of the single-screw machine. The shear rate between the flights of the intermeshing screws varies inversely with the separation of the flights and increases with the screw speed. This mixing action has an important effect on the conversion of drive power into heat and the design of the drive unit with respect to power and required speed.

In the past, multi-screw extruders had a virtual monopoly when certain plastics could not be processed through a single-screw extruder. The type plastics included very low bulk densities (below $0.3\,g/cm^3$), finely granulated emulsion–polymerized PVC, liquid or paste materials such as PVC plastisols, powdered LDPE, PE scrap reduced to flakes, and those with extremely low granular friction (0.2–0.3 mm) such as crystalline PS. With design changes, particularly in feeding systems and screws, many of these materials became processable in single-screw extruders. They produced pipes and profiles after developments occurred with materials such as powdered dry-blends of suspension polymerized PVCs as well as certain emulsion PVCs after preblending in a heated high-speed mixer. Other methods of handling certain plastics involved combinations of single-with twin-screw extruders. They included a short tapered auxiliary screw fitted either beside or above the main screw with flights of both screws intermeshing.

An interesting feature of nonintermeshing twins is the possibility of running the two screws at different speeds, thus creating frictional relationships between them, which in some cases can be exploited for the rapid melting of powders. In some twins, one screw is significantly shorter than the other. This design is used for plastics that may be adequately conveyed by a single screw, once in the form of a melt; they are difficult to feed into screw flights because of low bulk density or very low coefficient of friction of the solid against the surrounding walls. Thus, after melting by the twins, the melt moves through the single screw [107, 152, 146, 204, 289, 332, 335, 411, 449].

When twin-screw extruders are being studied for the difficult process of removing solvents, monomers, water, etc., from a reactor product, pilot scale testing is definitely in order. These tests are necessary to determine the screw design and barrel arrangement required. Special care must be exercised in the extrapolation process. Pilot scale problems can be much greater in the productions size equipment and should not be ignored when designing the screws and barrels.

With rigid PVC, when using powder plastics, the twin-screw has advantages particularly with changing materials or mixtures. Most of the commercial machines on the market and in use today are twin screw extruders and are widely used in the manufacture of rigid PVC products that include pipe, window profiles, and siding directly (Fig. 1.6) [187, L. F. Sansone].

With twin-screw types, the area of condensed phase plastic reaction engineering continues to expand as a preferred approach for obtaining desired product properties and for eliminating the source of solvent emissions. The roles of distributive and dispersive mixing in these machines are very useful. Different factors affect mixing based on screw configurations and extruder designs. The nature of distributive mixing for partially filled channels in the counter-rotating, non-intermeshing (NITSE) and co-rotating, fully-intermeshing (COTSE) twin-screw extruders have provided useful information.

Distributive mixing is the commingling of two 'fluids' so that the scale of fluid separation reduces to scale where another process (diffusion or a chemical reaction) can occur. The mixing is in a laminar flow regime which is characteristic of neat plastics. It is distinguished by the deformation of the fluid interfaces as a result of the applied shear strain. Distributive mixing relates the amount of interfacial area growth to the fluid strain rate, as distinguished from dispersive mixing, which is a function of the magnitude of the stress. The latter accomplishes droplet and agglomerate breakup; the former is the distribution of those components.

When reviewing specifically dewatering of plastics, an industry method for removing liquids from solids in a twin-screw extruder was patented by Werner & Pfleiderer Corp. in 1992–3. The use of mechanical dewatering in the polymerization process by substituting a twin-screw for a thermal dryer results in meeting requirements and a reduction in long term energy costs. The system's use includes chopped and washed PE film, engineering plastics, chopped and washed EPS (expandable PS) food packaging, and coagulated styrenic latex.

In principle, the dewatering in twin-screw extruders is based on counteraction of three forces which together form a pinch point where the material is compacted and the water is extracted. Wet plastic is conveyed downstream toward a restrictive element, which exerts a force in the opposite direction, causing plastic compression and water extraction. The third force, exerted by the screws of the drainage port, is acting transverse to the axis of opposing, preventing the compressed material with its high viscosity from moving out of the pitch zone. However, the extracted water is permitted to leave the zone through the drainage port.

An example of this capability is in the dewatering chopped EPS packaging food scrap. Output rates achieved in several industrial lines were 3500 pph of wet material that contained up to 60% water, with a dry plastic discharge rate of 1700 pph. Hot water is discharged from the first drainage port. The remaining water is taken out through atmospheric and vacuum vents. Foreign matter in the form of paper and foil requires a screen pack changer which creates 3.5–4.1 MPa (500–600 psi) back pressure. The melt is then stranded and pelletized [423].

Now available are higher output twin screw compounders using oversized motors providing screw speeds of 1000–1200 rpm. These increased torque motors permit taking plastic into the screw flights resulting in further raising their output rates. With barrels using the standard L/D (length/diameter) of 28–40, they practically double output rates than machines of similar size. As an example with a conventional twin screw operating at 400 rpm its output is 454 kg/h (1000 lb/h). With just a 30% higher torque and operating at the same rpm, it extrudes 590 kg/h (1300 lb/h) or provides a 30% increase. At a screw speed of 460 rpm output is raised another 15% or to 680 kg/h (1500 lb/h). This is a 50% increase over the conventional unit at 400 rpm.

Other gains include shorter plastic residence time (Chapter 3) and the fact that the processing window is narrower. Compounding extruders are typically starve-fed (to be reviewed latter in this chapter) but with these high torque units more material can be fed into the screws. This action helps to keep the melt temperature down at the higher speeds. Note that machines specifically optimized for high speed (just like any other machine designed for a specific operation) may not perform well at slower speeds or may require different screw designs, and some materials do not lend themselves to their mixing/shearing action. They have found to be useful in processing materials, such as preparing engineering plastics, alloying/blending, and compounds that are filled and reinforced such as chopped glass fibers used extensively in injection molding (Chapter 18) [2], compression molding [43], etc.

## EXTRUDER COMPONENTS

A schematic diagram and view of a single-screw extruder are shown in Figs. 2.2 and 2.5. This section reviews extruder components. Information on screws as well as vented barrels and dies are reviewed in Chapters 4 and 5. Extruder manufacturers provide machines with options since not all processors require or use certain components such as automatic screen changers, static mixers, melt pumps, and control cabinets such as free standing or base mounted, with either discrete or microprocessor control.

Figure 2.6 shows that extruders can be built to provide different output

**Figure 2.5** Example of Welex extruder with vent hole, screen changer, gear pump, static mixer, and sheet die.

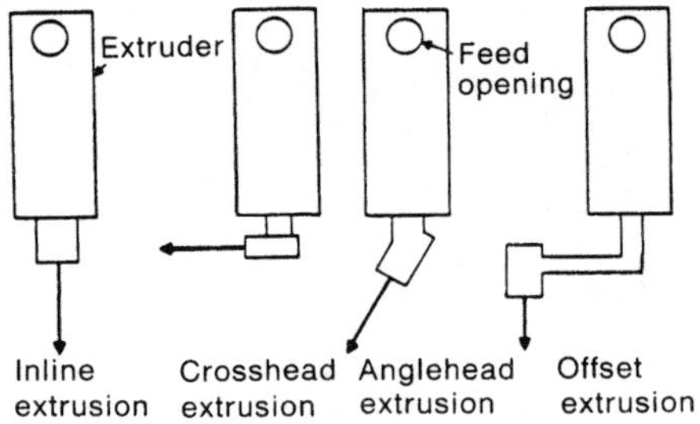


**Figure 2.6** Extruder output directions.

directions to meet different line requirements. The 'in-line' is principally used and takes advantage of the preferred laminar melt flow direction without 'bending.' The crosshead permits continuous coating of substrates such as wire.

## Alignment

Without proper machine installation, the precision alignment built into extruders (and other in-line equipment) is lost when not properly supported on all its mounting points. Installation involves factors such as ground support stability, precise alignment of equipment, uniform support, and effective control of vibration. Instructions should be in the manufacturer's manual, if not do not buy the equipment. The following paragraphs give a procedure for alignment.

Installation and alignment has to be done with extreme accuracy. Extruders larger than 60 mm (2.5 in) and/or screws with longer than 30:1 L/D (length to diameter) should definitely be aligned during installation as well as after barrel, feedblock, transmission repairs, or any significant mechanical interference machine problem occurs. This can be done in most cases with optical or laser equipment when the system is at room temperature. Assuming proper alignment occurs at room temperature and significant movement occurs during heat up or during operation, the causes of movement must be reconciled to prevent excessive wear or even failure of components [187, J. Frankland].

The extruder requires attention during installation. Its components must be held in alignment with each other despite torsion twisting, bending loads, thermal gradients, and the pull of attached auxiliary equipment. Its base is usually of modular steel design constructed to provide support and ensure alignment to all vital points of the extruder.

The prime objective is to keep the screw and barrel centerlines coincident with the production line height requirement. Installation is a multistep procedure that consists of building a foundation, setting and leveling the machine supports, and aligning the machine components to each other. A good foundation provides a level bed for the machine and does not move or settle during operation. The contractor laying the foundation should be familiar with the underlying soil conditions and meet the machinery recommended requirements.

Depending on the soil and other ground conditions, a bed of reinforced concrete is put down with sufficient thickness and strength so that it can distribute both static and dynamic loads to the ground without settling. To ensure proper load distribution, a steel sole plate is usually set in the concrete. The machine is attached to the foundation using anchor bolts, jack screws, shims, and grout. These mounts are carefully adjusted to permit accurate leveling and absorption of any shock. The anchor bolts are part of the foundation. A washer or other device attached at the bottom is needed to prevent the bolts from unscrewing during tightening of the supports.

Once it is determined that the supports are level, alignment of the various components to each other can start. One component of the assembly should be the starting point. Usually it is the gearbox. All other components are aligned to it. If the extruder is not shipped in one piece or it was aligned and assembled on a base structure at the factory, it should be rechecked after site installation.

To align the drive motor to a gearbox, the coupling must be opened. The coupling allows a small amount of axial 'play' as well as radial and angular misalignment. A dial indicator is usually mounted on one coupling half to check the axial and radial runout of the other half. Jack screws are placed on the motor base to move it back and forth as well as from side to side. Shims are placed under the motor footings to get the proper elevation. After the motor has been checked electrically, the coupling halves are bolted together.

The barrel assembly, or individual barrels, are attached to machined surfaces on the gear box. A screw jack can be placed under the barrel to make it level. The barrel bore should be swept by a dial indicator attached to the gearbox drive shaft to check the horizontal and vertical positions. This action is followed by bolting barrels together (for multi-barrels) using a procedure that tightens the bolts in sequence to prevent flange bending and with the proper torque to prevent flange bending. This step is critical for good barrel alignment. Because the barrels are heated, allowance must be made for the steel's expansion. Growth in the axial direction is significant and mechanisms such as slides must be provided under the barrel supports to allow movement (Fig. 2.7). Most machines have at least one support forward of the transmission. The barrel must slide through the support(s) very freely or the machine alignment will be compromised. The support points should be smooth and lubricated with graphite or other high temperature lubricant.

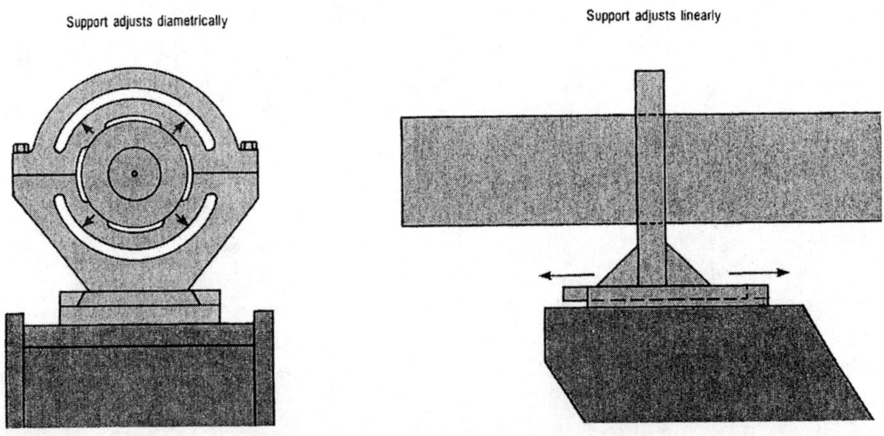


**Figure 2.7** Example of barrel alignment.

The final step is to put grout around the machine base and foundation, removing all air gaps. Grout is an insert, high strength, nonshrinking cement filler of plastic consistency. After the grout hardens, the extruder is ready to operate.

Incorrect alignment, whether at the time of installation or as a consequence of structural settlement, vibration, or machine age, is often the source of product inconsistency. The usual problem is due to the lack of awareness of the vital importance of proper alignment in controlling the operation and maintenance costs. If the extension of the life of the equipment is added, the impact on the 'bottom line' or profit becomes even more important.

If the barrel tends to wobble when the screw is rotating, it is a symptom of a misalignment. It could be a deficient alignment between the transmission, feedblock, or barrel. This should be investigated immediately as it can result in a high wear situation, screw breakage, heater creep, etc. Correcting problems requires someone who is thoroughly experienced in extruder alignment and construction. Larger machines will require borescoping.

**Borescoping**

A problem is the alignment of a barrel with the screw. Their clearance can range from 0.05 to 0.20 mm (small to large diameter) on any side of the screw. Alignment using a scope, or borescoping, is not guaranteed to correct all problems. Other areas must also be examined. Regardless, aligning with a scope provides a much better opportunity of extruding better products with less down-time, less scrap, and extending the life of the barrel/screw; otherwise it will be costly. Most machines can be adjusted in a day at very little cost. The machine will have at least a 25% life extension. As an example, if a barrel costs $12 000 and screw at $9000 with a total of $21 000, extending their life only 25% saves over $5250, plus more production.

With an alignment scope, one can tell exactly the shape of barrel; it should be straight but in could be curved or even S-shaped. Keep a record so that future checks can be used for comparison.

With a machine that has been operating, a clean barrel is required prior to alignment. An approach is to take the barrel temperature down to 93°C (200°F), open the front of the barrel and slowly pour PVC pellets or regrind into the feed section. Run at low speed, observing the amperage load and being sure not to overload the machine. Continue this process until the plastic exits looking like what entered. Result will be a clean barrel and screw. If plastic remains a complete clean-out with screw removal will be required.

After cleaning, measure the complete ID of the barrel by some type of device such as an electronic measuring device. If the barrel is not straight, it may go up–down, left–right, or both. You may be able to correct this situation by just adjusting the barrel supports or using shims under the barrel supports, feed sections, or gearbox. Just setting the output end is not sufficient; also set the feed end and center sections.

In adjusting the barrel, allow a slight droop at the exit end such as 0.25 mm (0.010 in) for a 120 mm (4.5 in) extruder. Check with the machine builder as to the exact amount. The reason for this action is due to the heat that will only cause the barrel to go up since it is bolted to the floor and cannot go down [187, M. J. Stinson].

## Drive system

The drive turns the screw at the required rotating speed with a capability to provide a constant speed or a degree of undesirable speed fluctuations resulting with throughput of extrudate fluctuating. The drive also provides the required amount of torque to the shank of the screw. Various drive systems are used to meet performance requirements at the lowest cost. Plastics with their low to high force and output requirements dictate the torque needed in the machine.

The power to turn the screw comes from the drive system with an often very large motor. Power requirements are usually 5–10 lb/h/motor hp. Thus, an extruder processing 45 kg/h (100 lb/h) output requires 10–20 hp motor. The speed of electric motors is usually much too fast, such as 1700 rpm for direct connection to the screw. Common screw speeds range from 20 to 200 rpm. For this reason, gear reducer is used between the motor and screw. Reduction can also be sheaves with pulleys and belts connecting the motor and gearbox.

There are many choices and options available for drive motors to meet the different processing requirements (Figs. 2.8–2.10). The selection of the optimum drive is fairly straightforward, though, DC drives have a long and successful history and are still a good choice for most applications, offering low cost and familiarity to service people. AC variable frequency drives have become more popular, they are available from many suppliers, and offer low motor maintenance. In either case, most applications do not require the ultra precision feedback controls. DC drives offer 0.1% regulation at a reasonable cost. However, the open loop, flux vector (FV) drives offer the required precision without a costly encoder [187, W. Kramer].

The drive motor technology offers four viable options that are the standard DC, brushless DC, variable frequency AC, and flux vector AC (Table 2.2). Most applications can be satisfied with the conventional variable frequency AC drives using a tachometer or encoder feedback. The FV drives are appropriate alternates to brushless DC drives in medical and

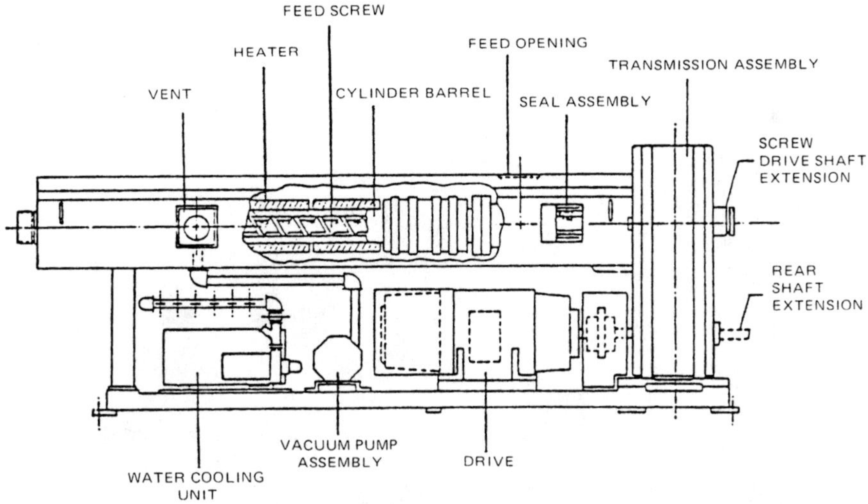


**Figure 2.8** Example of a direct-coupled extruder.

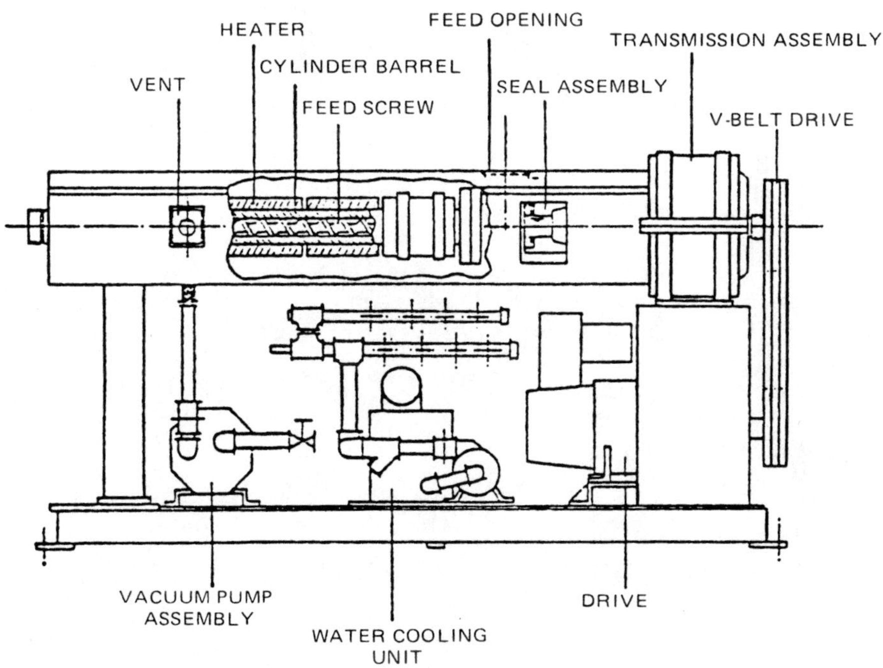


**Figure 2.9** Example of belt-driven extruder.

**Figure 2.10** Belt-driven Welex extruder with belt guard opened.

**Table 2.2** Comparison of drive motors

| | *Conventional DC with Tach Feedback* | *Brushless DC* | *Variable Frequency AC with Tach Feedback* | *Flux Vector AC* |
|---|---|---|---|---|
| Power range | 1/4–2000 hp* | 1/4–600 hp | 1/4–1000 hp | 1/4–300 hp |
| Speed regulation | 1.0–0.01%* | 0.5–0.01% | 0.5–0.03% | 0.5–0.01% |
| Constant torque | 20:1 | 100:1 | 40:1 | 1000:1 |
| Breakaway torque | 1.5 | 1.25 | 0.6 | 1.5 |
| Power factor | 20–85% | 0.98 | 0.98 | 0.98 |

other precision applications. FV controlled variable frequency AC drives have been made possible by developments in solid-state power switching devices and control processors.

A FV drive is a variable frequency AC drive with the ability to control both the magnetizing current and torque producing current through vector calculations performed in microprocessor memory. FV drives achieve torque and speed control with performance superior to that of an advanced DC drive while using much more reliable and economical brushless AC motors. Their 100:1 speed range with 0.01% regulation is standard with constant torque characteristics, even at zero speed. AC drives operate at very favorable power factors throughout the speed and load range, providing an efficiency advantage over DC types, and they do not require isolation transformers. Operating advantages exist with an extreme quiet operation when using a high carrier frequency of over 10 kHz. With an oversized frame, the FV drive does not require a separate cooling fan which is usually a major source of noise.

There is a strong trend to increase horsepower and output for all size extruder. Horsepower requirements suggested for various extruder sizes are 60 mm (2.34 in) with 40–60 hp and 120 mm (4.68 in) with 150–250 hp. With horsepower sufficient for very high screw speeds of up to 200 rpm, the capability of a 60 mm (2.34 in) extruder may be increased from 150 to 300 lb/h (330 to 660 kg/h).

Drives normally provide automatic current limitation to prevent screw breakage as a result of excessive torque. Whatever type drive is used, it is essential that some form of overload safety device is incorporated in the drive system. Additional protection should be provided by means of a rupture disk installed in the zone between the extruder screw and breaker plate/screen pack, or a pressure transducer with high pressure cut-out interlock to the drive.

Calculating DC drive power can be done easily even if the extruder panel only gives screw speed and motor current. Use the relationship

$$\text{Power}\left(\text{watts}\right) = \text{Current}\left(\text{amperes}\right) \times \text{Voltage}\left(\text{volts}\right)$$

and the conversion

$$\text{Horsepower} = \text{Kilowatts divided by 746.}$$

The most difficult reading to obtain is the drive armature voltage which is proportional to speed, and may be estimated from motor name plate data where

$$\text{Arm voltage} = \text{Screw speed} \times \left(\text{Maximum armature voltage}/\text{Maximum screw speed}\right).$$

Make sure to check the calibration of your screw speed and current readouts using

$$\text{Horsepower} = \left(\text{Amperes}/746\right) \times \left(\text{Speed}/\text{Maximum speed}\right) \times \text{Maximum voltage.}$$

Motors deliver approximately constant torque over their speed range. For more power, the motor should be run near to its top-rated speed. To match optimum motor speed with maximum screw speed, a gearbox with a reduction ratio from 20:1 to 30:1 are usually used. The power transmission capability of the gearbox is matched to the maximum motor power. Speed reducers are constant-torque-rated devices because hp is directly proportional to speed. As an example, a transmission for 100 hp at 100 rpm screw speed is only good for 50 hp at 50 rpm screw speed. The ultimate limitation in the rating of speed reducers which can usefully be used on an extruder is the strength of the screw. The maximum torque which a screw can safely transmit continuously can readily be calculated (Chapter 4).

If herringbone style gears are used, usually in a two-stage process, efficiency is high – in the order of 98%. Regular maintenance of these gearboxes involving checks of oil levels and use of the correct grade of oil are essential. A belt drive with some speed reduction can also be incorporated in the drive transmission. This provides some safety against sudden overloads but power losses can be as high as 10%. All belts and pulleys must be guarded. A change of pulleys can be useful, if the extruder is used for another application whose power requirements differ.

When a single-screw extruder gearbox reduces motor speed to screw speed, it experiences generated gear forces as well as screw and pressure thrusting, and possible twisting and bending screw forces on the drive shaft. The reducer is designed and sized accordingly using industry standards.

A twin-screw gearbox, in addition to the above, distributes torque to both shafts in proportion to the relative rotational resistance of each screw. Also, unlike a single-screw gearbox, the close spacing of the two shafts restricts the amount of allowable torque that can be transmitted to the screws. Numerous gear train design approaches are used to compensate for this limitation.

The extrusion process emphasis is on power requirements, but it is the torque, or resistance to rotation, that determines the design and limitations of the gearboxes and other drive components. Meshing gears generate equal and opposite forces at the point of contact. These forces include a tangential force with a radial separating force which acts toward the center of each gear. With a helical gear, axial forces may add to or subtract from the screw end pressure thrust on the output shafts [187, T. Crouch].

Thrust bearings in the drive system, an essential part of most extruders, have a significant effect on reliability, capacity, and performance. Pressure developed at the end of the screw during processing acts to force the screw back into the gearbox. This force on the screw is the diehead pressure times the screw's cross-sectional area. With larger screws, the force will increase to the screw diameter squared. The thrust bearing resists this force. Rolling contact bearings are the usual type of bearing used. An essential part of each bearing is a lubricating film of oil without which a bearing will be damaged in a very short period of time. With shaft misalignment, an uneven load on the bearing rollers may drastically shorten bearing life.

A serious problem encountered by roller thrust bearings is skidding which results from the loss of traction between rollers and raceways. Skidding can be caused by load surging or axial vibration of the screw. Contact is intermittent. A solution of the problem is to eliminate end play. However, because a bearing is mechanically very stiff 34.5–69 $\times 10^3$ MPa (5–10 $\times 10^6$ psi), simply removing play from the bearing assembly by shimming or machining can result in heavy loads and considerable heat generation. A better method is to use disk springs in conjunction with reverse thrust bearings. These have a much lower and controlled spring rate. This also smoothes out axial vibration which occur in extruder screw [187, T. P. Harrington].

## Barrel

The barrel is a cylinder that houses a screw. It provides the bearing surface where shear is imparted to the plastic granules. Heating and cooling media are housed around it to keep the barrel (melt) at the desired temperatures. Extruder sizes are classified by the barrel's ID (inside diameter) and overall length. It is common practice to refer to the L/D ratio which is the barrel length to diameter ratio. For low output, such as filament or profile extrusion, 40–60 mm (1.6–2.3 in) diameter extruders are normally used, whereas for sheet, 120 and 150 mm (4.7 and 5.8 in) diameter screws are more common.

The trend has been toward the use of longer barrel lengths. In the past, L/D ratios of 24/1 were the rule. Now 30/1 and longer are frequently specified. These longer barrels permit more melt inventory time; consequently, a higher output rate is possible without sacrificing the plasticization rate or reducing the homogenization of the melt. To take full advantage of the increased output rates of longer barrels, a more powerful drive unit should be used. Typically, 150 hp motors are used on 120 mm (3.2 in), 24/1 L/D extruders. For a 30/1 L/D, 200 hp are frequently specified. Smaller 60 mm (1.6 in) machines have drive motors of up to 50 hp.

There are many options for barrel construction with most designed to withstand up to 69 MPa (10 000 psi) internal pressures; higher pressure units are manufactured for certain processes. (The SPI Machinery Component Division Bulletin *Recommended Guidelines For Single Barrels* provides working tolerances that produce effective performance with economy of manufacture.) The choice of materials may be determined by the need for corrosion and/or wear protection, cost, repair, or their combinations. They can be made from a solid piece of metal. The most common material is carbon steel with a nitride interior. Nitriding gives a very thin hardened layer of only about 0.3 mm (0.01 in) which results in a limited life span especially when processing abrasive filled compounds. For large extruders repair can be achieved by inlaying weld material. Smaller machines can be repaired by machining out the opening diameter and inserting a liner.

Solid barrels can also be fabricated from stainless steel with a hardened interior. However, the hardening process often lessens the corrosion protection aspects of stainless steel. Other negative aspects to solid stainless steel are poor heat transfer and cost. Therefore, this is usually an option for smaller machines of less than 60 mm (2.3 in). With castings applied to the interior in place of hardening, a superior wear and/or corrosion protection occurs but at a higher cost.

Lined barrels are used where repair or high-performance alloys are considerations. The liners are thin walled and inserted into an opening in the larger load-bearing barrel. Liners can be made from various hardened carbon steel, stainless steel, nickel based alloys, etc. They can be used with a bimetallic coating. With liners, heat transfer can become a problem if air gaps between liner and barrel materials exist.

Bimetallic coatings are used when wear and corrosion are of concern. They can be applied by traditional welding techniques for large barrels. Centrifugal casting is the most common technique for both single and twin bore barrels. A wide variety of bimetallic coatings are used. The choice of coating that should be specified depends on the plastic to be processed. For non-corrosive processes with up to about 30–40 wt% of filler, the iron-based alloy should be considered. Highly corrosive systems using materials such as PVC and fluoropolymers can benefit from the Ni/Co family of alloys. Processes compounding high levels of abrasive fillers should use the family of alloys with tungsten [187, W. M. Davis].

Table 2.3 lists the recommendations for single bore barrels prepared by the SPI Machinery Component Manufacturers Division. It is a guide to manufacturers and processors. These recommendations have been developed over many years and provide working tolerances that produce effective performance with economy of manufacture.

### *Heating and cooling*

There are three principal methods of barrel heating: electrical, fluid, and steam. Electrical heating is the most common because of its advantages. It can cover a much larger temperature range. It is clean, relatively inexpensive, and efficient. The heaters are generally placed along the barrel,

**Table 2.3** Recommended SPI single barrel dimensional guidelines

**Lengths, depths and widths**

| | *English units* | | *Metric units* | |
|---|---|---|---|---|
| *Dimension description* | *Range (in)* | *Tolerance (in)* | *Range (mm)* | *Tolerance (mm)* |
| I. Most linear dimensions | To 12 | ±0.010 | To 300 | ±0.25 |
| | 12–60 | ±0.030 | 300–1500 | ±0.75 |
| | 60–120 | ±0.045 | 1500–3000 | ±1.00 |
| | 120–200 | ±0.060 | 3000–5000 | ±1.50 |
| | over 200 | ±0.090 | over 5000 | ±2.25 |
| II. Counterbore depths or pilot lengths | All sizes | ±0.005 | All sizes | ±0.125 |

**Diameters**

| | *English units* | | |
|---|---|---|---|
| *Dimension description* | *Range (in)* | | *Tolerance (in)* |
| I. Bores (Including counterbores) | To 1½ dia. | (<60 long) | +0.001–0.000 |
| | 1½–3 dia. | (>60 long) | +0.002–0.000 |
| | 3–5.5 dia. | (<120 long) | +0.002–0.000 |
| | 3–5.5 dia. | (>120 long) | +0.0025–0.000 |
| | 5½–8 dia. | (<180 long) | +0.0025–0.000 |
| | 5½–8 dia. | (>180 long) | +0.003–0.000 |
| | 8–12.5 dia. | (<180 long) | +0.003–0.000 |
| | 8–12.5 dia. | (>180 long) | +0.004–0.000 |

| | *Metric units* | | |
|---|---|---|---|
| | *Range (mm)* | | *Tolerance (mm)* |
| | 0–38 dia. | (<1500 long) | +0.025–0.000 |
| | 3–75 dia. | (<1500 long) | +0.050–0.000 |
| | 25–75 dia. | (>1500 long) | +0.038–0.000 |
| | 75–140 dia. | (<3000 long) | +0.051–0.000 |
| | 75–140 dia. | (>3000 long) | +0.063–0.000 |
| | 140–200 dia. | (<4500 long) | +0.063–0.000 |
| | 140–200 dia. | (>4500 long) | +0.076–0.000 |
| | 200–315 dia. | (<4500 long) | +0.076–0.000 |
| | 200–315 dia. | (>4500 long) | +0.102–0.000 |
| II. Outside diameters | All sizes and lengths | ±0.005 in | ±0.102 mm |
| III. Pilot diameters | All sizes | ±0.001 in | ±0.038 mm |

**Table 2.3** *Continued*

**Concentricity of diameters**

Concentricity between all inside and outside diameters should be held within 0.002 in TIR or 0.051 mm TIR within one bore diameter of each end. Concentric diameter dimensions further inside of one bore diameter from the ends should be avoided and/or specially tolerated when required.

**Straightness**

Straightness is generally specified for the bore of the barrel and is measured by TIR.

| | | | |
|---|---|---|---|
| 1–3 in | bore diameter | (≤24 L/D) | 0.008 in TIR |
| 1–3 in | bore diameter | (≥24 L/D) | 0.010 in TIR |
| 3–5.5 in | bore diameter | (≤24 L/D) | 0.010 in TIR |
| 3–5.5 in | bore diameter | (≥24 L/D) | 0.012 in TIR |
| 5.5–8 in | bore diameter | (≤24 L/D) | 0.012 in TIR |
| 5.5–8 in | bore diameter | (≥24 L/D) | 0.015 in TIR |
| 8–12.5 in | bore diameter | (≤24 L/D) | 0.014 in TIR |
| 8–12.5 in | bore diameter | (≥24 L/D) | 0.016 in TIR |
| 25–75 mm | bore diameter | (≤24 L/D) | 0.203 mm TIR |
| 25–75 mm | bore diameter | (>24 L/D) | 0.254 mm TIR |
| 75–140 mm | bore diameter | (≤24 L/D) | 0.254 mm TIR |
| 75–140 mm | bore diameter | (>24 L/D) | 0.305 mm TIR |
| 140–200 mm | bore diameter | (≤24 L/D) | 0.305 mm TIR |
| 140–200 mm | bore diameter | (>24 L/D) | 0.380 mm TIR |
| 200–315 mm | bore diameter | (≤24 L/D) | 0.355 mm TIR |
| 200–315 mm | bore diameter | (>24 L/D) | 0.406 mm TIR |

To prevent short 'kinks' to which the screw could not conform, any two successive measurements taken less than twice the bore diameter apart should have no more than 0.5 the total allowable TIR.

**Bar test**

A second way to check both straightness and bore size is by using a precision ground test bar.

**Finish**

The following surface finishes apply unless specified otherwise.

| | |
|---|---|
| Outer diameters | 125 RMS |
| Counterbores | 32 RMS |
| Pilots | 32 RMS |
| Cylinder bores | 8–32 RMS |
| Feed ports | 125 RMS |
| *Vent ports | 32 RMS |

*Required within 0.25 in of cylinder bore.

The bore shall be free from visual surface defects in the bore over its entire length.

**Table 2.3** *Continued*

**Parallelism and perpendicularity**
Parallel surfaces can be determined by TIR and all dimensions meeting the concentricity recommendations are acceptable.

Perpendicular surfaces to the cylinder bore can be tested by placing the barrel on a surface plate and indicating with an adjustable height table indicator or with a precision square. All mating surfaces should be perpendicular as follows:

| | |
|---|---|
| Pilots and counterbores | 0.0015 in TIR |
| Flanges | 0.001 in TIR per 10 in of dia. |
| Pilots and counterbores | 0.038 mm TIR |
| Flanges | 0.038 mm TIR per 250 mm of dia. |

**Threads**
The variation in threads used is too broad to be addressed in these recommendations. It is suggested that whenever thread selection is possible that either ANSI or ISO standards are adhered to for ease of measurement and compatibility.

**Flange attachments**
Barrel flanges may be affixed in accordance with standard procedures depending on the resultant thrust from internal pressure and other design factors. These procedures apply for flanges at either end of the barrel.

**Venting**
The following warning label should be affixed to a vented extruder or injection barrel supplied without a venting plug.

*Warning.* This vented injection/extrusion barrel was designed for operation with the vent *OPEN ONLY.*

Operation of this machine with the vent plugged or otherwise closed off may result in serious injury to persons in the vicinity. Check with your supervisor if this plug is intentionally or unintentionally plugged.

**Downsizing injection barrels**
When downsizing (reducing the barrel/screw bore size and resultant shot capacity) an injection molding machine must have the pressure capability reduced or be redesigned for the new resultant pressure. This is necessary to reduce the screw thrust pressure in proportion to the reduction in area of the bore to prevent the barrel from being over-pressurized. All units that are downsized should be equipped with a warning label on the barrel where it can be easily seen.

**Barrel construction**

I. Bimetallic centrifugally cast and hot isostatically pressed barrels

   The inlay thickness is to be 1.6 mm (0.06 in) nominal, with a 3.2 mm (0.125 in) maximum and 0.8 mm (0.03 in) minimum.

**Table 2.3** *Continued*

Hardness standards are those of the manufacturer and are based on the arithmetic means of five hardness measurements taken at random points along the bore. The hardness values vary greatly with the type of inlay. The inlay should have a 100% metallurgical bond at all points determined by ultrasonic techniques.

II. Nitrided barrels or other single metal barrels

Depending on the material used in the outer tube and the bore hardening method, both the hardness and depth of hardness will vary. As a result the hardness standards are those of the manufacturer.

III. Sleeved barrels (new)

All sleeves should be inserted by either shrink fit or press fit with no less than the interference requried to maintain the stress levels applying to all barrels. In all cases a minimum interference of 0.025 mm (0.001 in) diametral shrunk per inch of diameter should be used.

Hardness standards are those of the manufacturer and vary with the material and hardening method selected for the sleeve.

IV. Sleeved barrels (relined)

The liner material should have a coefficient of thermal expansion within ±30% of the base (outer tube) material from room temperature through to maximum expected operating temperature.

All internal sleeves should blend to any remaining bore with no visible lip and with no more than the following mismatch in TIR.

| | | | |
|---|---|---|---|
| To 2 in | bore dia. 0.001 in | To 50 mm | bore dia. 0.025 mm |
| 2–3.5 in | bore dia. 0.002 in | 50–90 mm | bore dia. 0.051 mm |
| >3.5 in | bore dia. 0.003 in | >90 mm | bore dia. 0.076 mm |

In the case of injection barrels, relining of the discharge end should extend a minimum of 50 mm (2 in) past the maximum backward travel of the rear edge of the check ring on the nonreturn valve.

grouped in zones. Each zone is usually controlled independently, so the desired temperature profile can be maintained along the barrel (Fig. 2.11).

Fluid heating, such as the use of heated oil, allows an even temperature over the entire heat-transfer area, avoiding local overheating. If the same heat-transfer fluid is used for cooling, an even reduction in temperature can be achieved. The maximum operating temperature of most fluids is relatively low for processing TPs, generally below 250°C (482°F). With its even temperature, the required fluid heating is desirable with TS plastics so that no accidental overheating occurs to chemically react and solidify in the barrel. However to date, very little TS is used in extrusion.

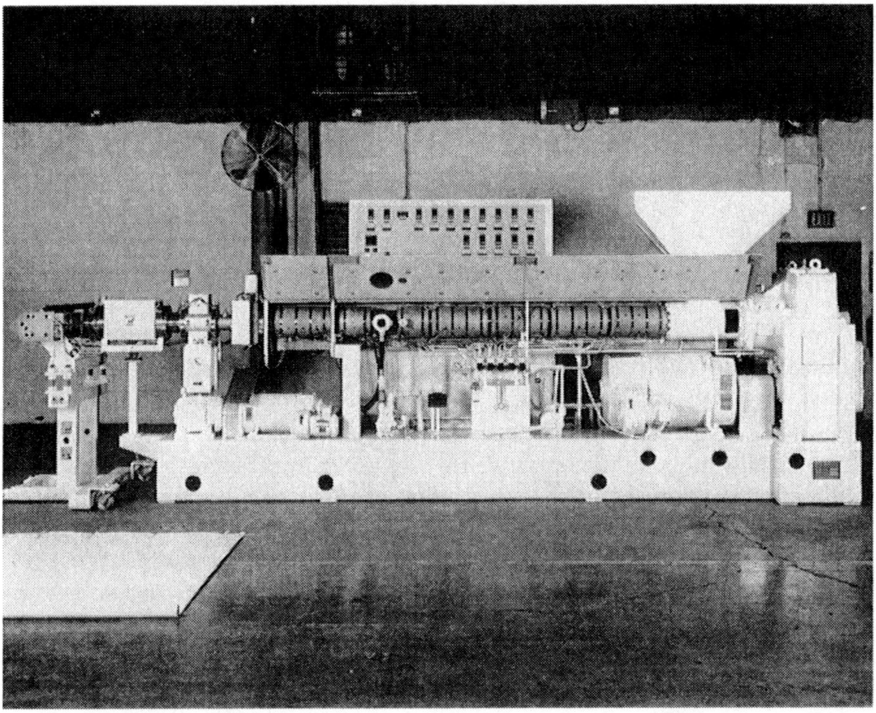

**Figure 2.11** Welex sheet line extruder with barrel cover guard in the open position revealing heater/cooler elements.

Steam heating was used in the past, particularly when processing natural rubber. Now it is rarely used. Steam is a good heat-transfer fluid because of its high specific heat capacity, but it is difficult to get steam to the temperatures required for TP processing of 200°C (392°F) and greater.

The cooling of barrels is an important aspect. The target is to minimize any cooling and, where practical, to eliminate it. In a sense, cooling is a waste of money. Any amount of cooling reduces the energy efficiency of the process, because cooling directly translates into lost energy; it contributes to machine's power requirement. If an extruder requires a substantial amount of cooling, when compared to other machines, it is usually a strong indication of improper process control, improper screw design, excessive L/D, and/or incorrect choice of extruder (single- versus twin-screw).

Cooling is usually required with forced-air blowers mounted underneath the barrel. The external surface of the heaters or the spacers between the heaters is often made with cooling ribs to increase the heat-transfer area (ribbed surfaces will have a larger area than flat surface), which significantly increases cooling efficiency. Forced air is not required with small diameter extruders because their barrel surface area is rather large compared to the channel/rib volume, providing a relatively large amount of radiant heat losses.

Fluid cooling is used when substantial or intensive cooling is required. Air cooling is rather gentle because its heat-transfer rates are rather small compared to water cooling. But it does have the advantage in that, when the air cooling is turned on, the change in temperature occurs gradually. Water cooling produces rapid and steep change; a requirement in certain operations. This faster action requires much more accurate control and is more difficult to handle without proper control equipment.

The larger barrels are often liquid cooled, using cored channels to circulate the cooling medium because they require intense cooling action (feed-throats also use water cooling). However if not properly controlled, problems could develop. If the water temperature exceeds its boiling point, evaporation can occur. The water system is an effective way to extract heat, but can cause a sudden increase in cooling rate resulting in a nonlinear control problem; resulting in more difficulty to regulate temperature. Nevertheless, the water cooling approach is used very successfully with adequate installation in the extruder and adequate control and startup procedures.

*Adapter system*

The adapter systems refers to everything located between the screw and the die. It may include a screen pack, static mixer, and coextrusion adapter (Figs. 1.9, 2.12, and 2.13). This adapter system must be of streamlined design. Flow channels should not contain sudden changes in cross section surface interruptions (caused by mismatched assembly joints or damage, etc.) or other 'dead spots.' Areas of stagnation can give rise to localized plastic degradation and subsequent release of particles of degraded plastic into the melt stream.

Adequately sized heaters must be provided for the adapter, since it is generally a large piece of metal. It is especially important to control the temperature of the adapter and head separately, since they usually differ greatly in size and energy requirements. The die, where it extends beyond the extruder head, should also have its own thermocouple and temperature controller.

It is important to recognize that the flow through the center of an adapter is generally at a higher temperature than at the wall. Plastic is typically a poor conductor of temperature so cooling at the wall will not significantly affect the center of the melt stream. Consequently, the

**Figure 2.12** A 4-bolt swing gate die clamping system at the front end of a Welex extruder.

adapter temperature(s) should be set at, or even slightly above, the actual melt temperature. If an adapter zone is set below the actual melt temperature, it can induce or exacerbate a temperature gradient. Temperature gradients often lead to undesired flow variations and/or create gauge problems.

While valves are not ordinarily required, they are preferred over screen packs for control of back pressure in some processing operations with certain heat sensitive plastics. Valving can also be used in certain vented extruders and in some extruders where screw design is not optimum. The design should be streamlined to prevent excessive holdup of the melt.

**Figure 2.13** Single bolt quick die clamping front end of a Welex extruder with pressure safety bolt.

*Screen pack*

Melt from the screw is usually forced through a breaker plate with a screen pack (Figs. 1.9 and 2.14). Extra heat develops when melt goes through the screens, so some heat-sensitive plastics cannot use a screen pack. The function of a screen pack is to reduce rotary motion of the melt, removing large unmelted particles, and other contaminants. This situation can be related to improper screw design, contaminated feedstock, poor control of regrind, overheating melt, and/or so on. Sometimes screen packs are used to control the operating pressure of extruders (Table 2.4). However, there are advantages in processing with matched and controlled back pressure, operating within the required melt pressure and temperature, as this can facilitate mixing, effectively balancing out melt heat problems.

In operation, the screen pack is backed up by a breaker plate that has a number of passages, usually many round holes ranging from 3.2 to 4.8 mm (0.125 to 0.187 in) in diameter. One side of the plate is recessed to accommodate round disks of the usual wire cloth, which make up the

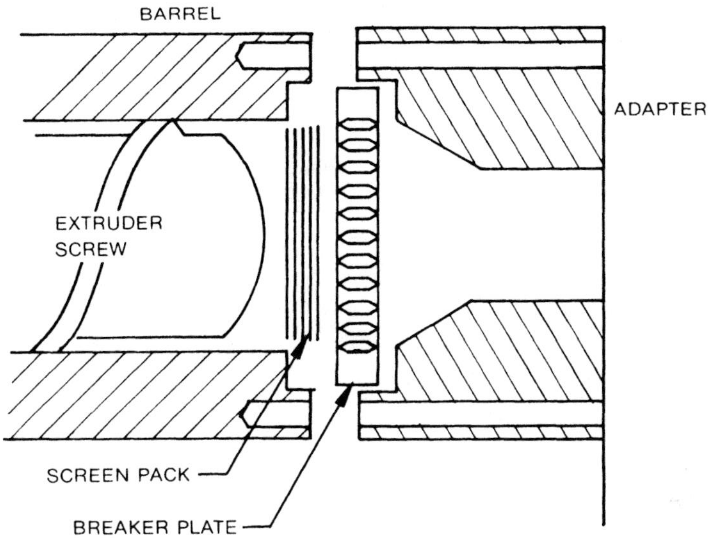


**Figure 2.14** Basic layout of a screen pack.

**Table 2.4** Means of increasing output and improving product quality

| *Machine conditions* | *To increase output* | *To improve end product quality* |
|---|---|---|
| Barrel pressure through valving (back pressure) | Decrease | Increase |
| Screen pack (Screen mesh or number of screens) | Decrease | Increase |
| Screw temperature | Increase | Decrease |
| Screw speed | Increase | |

screen pack (Tables 2.5 and 2.6). When using fine screens in a pack, use a coarser screen as the first upstream screen to catch large dirt particles and to also support the following finer screens. To help prevent screen blow-out, back up fine screens using progressively coarser screens downstream. Pressure controls should be used on both sides of the breaker plate to ensure the pressure on the melt remains within the required limits.

Table 2.7 lists the particle sizes corresponding to mesh sizes. Mesh size indicates the finest US Standard Sieve through which more than 95% of the plastic will pass. The mesh number is determined by the number of screen openings per linear inch (or 2.54 cm). Therefore, particle size increases with decreasing mesh size.

**Table 2.5** Screens used before the breaker plate to filter out contaminants in the melt[a]

| Contaminant | Metal fibers | Wire mesh | | Sintered powder |
|---|---|---|---|---|
| | | *Square weave* | *Dutch twill*[b] | |
| Gel captured | 5 | 1 | 2 | 3 |
| Contaminant capacity | 6 | 2 | 3 | 3 |
| Permeability | 4 | 4 | 1 | 2 |

[a] Range is from poorest (1) to best (6). Multiple screens are normally used; example screen pack has 20 mesh against breaker plate, followed by 40, 60, and 100 mesh (coarsest mesh has lowest mesh number).
[b] Woven in parallel diagonal lines.

Based on the processing requirements, the screen changers may be manual to highly sophisticated. Manual systems are used for limited runs or infrequent changes. The packs are usually mounted outside the extruder between the head clamp and the die; they can be changed via mechanical or hydraulic devices. Continuous screen changers also are used. The more sophisticated the system, the higher costs; however, continuous screen changers are used since they provide processing cost advantages.

The commonly used square weave mesh has poor filtering performance but good permeability. If filtering is really important, another filter media should be used. Metal fibers stand out in their ability to capture gels and hold contaminants (Table 2.8) [352]. Gel problems are particularly severe in small gauge products such as fibers and films [349]. It is particularly in these applications that metal fiber filters have been used. If the plastic is heavily contaminated, the screen will clog quickly. If the screens have to be replaced frequently, an automatic screen changer should be considered. In these devices, pressure drop across the screen is monitored continuously. If the pressure drop exceeds a certain value, a hydraulic piston is usually used to move the breaker plate with the screen pack out of the way; simultaneously, a breaker plate with clean screens are moved into position. These units are called slide-plate screen changers.

Screen changers are generally classified as manual, slide-plate (discontinuous), continuous flow and constant pressure, and backflush. Each have performance variations to meet different requirements. Manual changers usually require shut-down of the extruder to change the filter.

A common type is a manual shifted side-plate changer; following shutdown and depressurization, the change is made within seconds. Shifting a slide-plate while the line is running may introduce air into the melt,

**Table 2.6** Screen scale sieves classification, adopted by the US Bureau of Standards

| *Mesh* | *Diameter of wire, decimal of an inch* | *For closer sizing sieves from 0.0015 in to 3.000 ratio* $\sqrt[4]{2}$ *or 1.189* |
|---|---|---|
| $2\frac{1}{2}$ | 0.099 | 0.312 |
| 3 | 0.070 | 0.263 |
| $3\frac{1}{2}$ | 0.065 | 0.221 |
| 4 | 0.065 | 0.185 |
| 5 | 0.044 | 0.156 |
| 6 | 0.036 | 0.131 |
| 7 | 0.0328 | 0.110 |
| 8 | 0.032 | 0.093 |
| 9 | 0.033 | 0.078 |
| 10 | 0.035 | 0.065 |
| 12 | 0.028 | 0.055 |
| 14 | 0.025 | 0.046 |
| 16 | 0.0235 | 0.0390 |
| 20 | 0.0172 | 0.0328 |
| 24 | 0.0141 | 0.0276 |
| 28 | 0.0125 | 0.0232 |
| 32 | 0.0118 | 0.0195 |
| 35 | 0.0122 | 0.0134 |
| 42 | 0.0100 | 0.0168 |
| 48 | 0.0092 | 0.0116 |
| 60 | 0.0070 | 0.0097 |
| 65 | 0.0072 | 0.0082 |
| 80 | 0.0056 | 0.0069 |
| 100 | 0.0042 | 0.0058 |
| 115 | 0.0038 | 0.0049 |
| 150 | 0.0026 | 0.0041 |
| 170 | 0.0024 | 0.0035 |
| 200 | 0.0021 | 0.0029 |
| 250 | 0.0016 | 0.0024 |
| 270 | 0.0016 | 0.0021 |
| 352 | 0.0014 | 0.0017 |
| 400 | 0.001 | 0.0015 |

**Table 2.7** Guide to converting mesh size into particle size

| *Mesh* | *10* | *16* | *20* | *35* | *50* | *120* |
|---|---|---|---|---|---|---|
| mil | 78.7 | 46.9 | 33.1 | 19.7 | 11.7 | 4.9 |
| micron | 2000 | 1190 | 840 | 500 | 297 | 125 |

**Table 2.8** Comparison of different filter media

| | *Wire mesh square weave* | *Wire mesh Dutch twill* | *Sintered powder* | *Metal fibers* |
|---|---|---|---|---|
| Gel capture | Poor | Fair | Good | Very good |
| Contaminant capacity | Fair | Good | Fair | Very good |
| Permeability | Very good | Poor | Fair | Good |

causing a momentary disruption of the process. An array of continuous screen changer technologies have developed to eliminate this problem. There are the continuous flow and constant pressure types. No air is introduced with continuous flow designs, but the melt pressure upstream of the breaker plate will still rise during the screen change. However, the variation in die pressure may be sufficiently minor as to have no major effect on a number of processes such as blown and cast film. Technologies include a design that splits the flow through two filters in a single slide-plate. Another involves changers that split the flow through single or dual filters in two slide plates or sliding bolts, or even a rotating wheel. Others offer prefilled filters that create momentary disruption as they are hydraulically placed into the melt.

Constant pressure machines provide an absolutely minimal change in die pressure while continuing to filter impurities. They have been applied on virtually every type process. They are well suited for pipe, profiles, thin-gauge film, and foam production.

Recycling is exerting pressure upon screen changers. To meet this situation on changing frequency, screen use, and operator intervention, backflush screen changers are being used. Most rely on down-stream melt pressure that forces clean melt to flush contamination off the screen pack in the screen changer, in an off-line position. Some backflush changers use a piston pump to meter and inject clean plastic up-stream against a small portion of the on-line screen pack, which has been isolated from plastic flow. Backflush screen changers can extend the time between screen changes by a factor of 10–100.

There are also filtration devices that use laser-drilled drums in place of typical screen packs. The holes frequently number 600 000 to over 3 000 000 and are equivalent in filtration to screens of 100–150 mesh. Contaminants trapped on the surface of the holes are wiped away (removed) by mechanical arms.

When adding an automatic screen changer to a new or existing extruder, the screw needs to extend into the screen changer body. This extension is needed to eliminate the existence of additional plastic inventory, which would require additional soak time during startups. The screw should be extended to within 6–9.5 mm (0.25–0.38 in) of the breaker plate. The amount of extension can be supplied by the screen changer manufacturer.

Screw manufacturer can make a modification to an existing screw by making a threaded extension, or to have the additional length added as an integral part of a new screw. The new extension should be either non-flighted or, preferably, flighted with the extended portion of flights undersize from the main flight diameter by 0.4–0.5 mm (0.015–0.020 in) per side. The reason for the non-flighted or undersizing of the flights in the extended area is that the entrance bore of the screen changer is not hardened like the barrel liner. Otherwise, the flights will gall in the bore of the screen changer [187, T. Womer].

Available are cleaning devices for the screen changers (and other extruder components, such as dies) that operate economically and safely removing contaminated plastics. The routine techniques used include blow torches, hot plates, hand working, and scraping. Personnel have to be careful not to damage expensive tooling by spot annealing, mechanical abuse, etc. There are also other commercial cleaning systems used, such as aluminum oxide beds (fluidized beds), salt baths, hot air ovens, and vacuum pyrolysis. As an example, the Beringer's automatic jet vacuum pyrolysis cleaner utilizes heat and vacuum to remove the plastic. Most of the plastic is melted and trapped. Remaining plastic is vaporized and appropriately collected in a secondary trap.

### *Gear pump*

Gear pumps, also called melt or metering pumps, have been standard equipment for over a half century in textile fiber production. During the 1980s they established themselves in all kinds of extrusion lines. They consist of a pump, a drive for the pump, and pump controls located between the screen pack (or screw) and die. Two counter-rotating gears transport a melt from the pump inlet (extruder output) to the pump discharge outlet (Fig. 2.15). Gear rotation creates a suction that draws the melt into a gap between one tooth and the next. This continuation action from tooth to tooth develops a surface drag that resists flow, so some inlet pressure is required to fill the cavity. This pump device is basically a closely intermeshing, counter-rotating, twin-screw extruder. However, since gear pumps are solely used to generate pressure, they are seldom called an extruder, even though they are extruders.

The inlet pressure requirements vary with material viscosity, pump speed, and mixing requirements. These pressures are usually less than 69 MPa (1000 psi) but cannot go below specified pressures such as 21 MPa (300 psi). An extruder specifically designed for use with a pump only has to 'mix,' with no need to operate at high pressures to move the melt. It

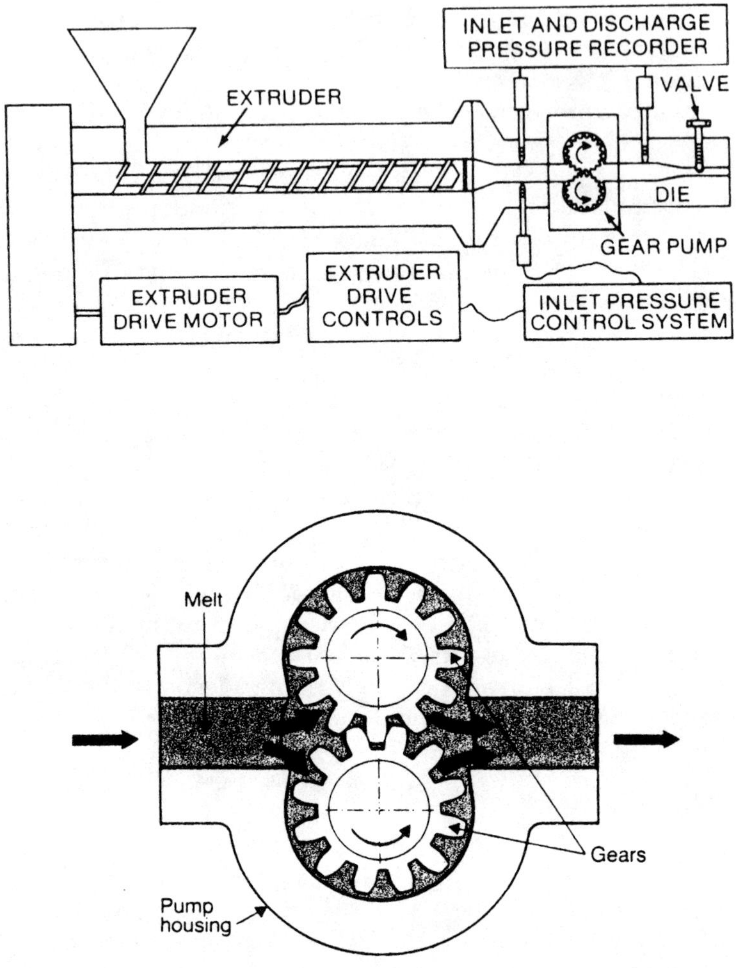


**Figure 2.15** Example of gear pump in an extruder.

only has to generate the low pump-inlet pressure, thus it can deliver melt at a lower than usual heat, hence requiring less energy and often yielding a higher output rate. This positive displacement gear device pumps the melt at a constant rate. It delivers the melt to the die with a very high metering accuracy and efficiency (Fig. 1.9). It is common to have pressure differentials as high 276 MPa (4000 psi) between pump inlet and discharge [240].

The available pump's volumetric efficiency varies from 85–98%. Some melt is deliberately routed across the pump to provide lubrication, some slips past the gears. An incomplete fill on the inlet side will show up as a fast change in output and pressure at the exit. The extended loss of inlet pressure can damage the pump by allowing it to run dry. Overpressurization at the inlet, caused by the extruder's sudden surge, will at least change the melt conditions and, in extreme cases, can be dangerous to both equipment and operator. The problem can be eliminated by close-loop pressure controls at its inlet and exit. To prevent overfeeding and overpressure, the metering screw's metering section should have a barrel clearance larger than normal.

Melt pumps are most appropriate when the characteristics of the screw and die combine to give a relatively poor pumping performance. This can happen when the die pressures are low but more often occurs when they are extremely high, or when the melt viscosity is extremely low. When pumps are used to increase production rate by reducing the extruder head pressure without a corresponding increase in the screw speed, it often increases the solids content of the extrudate, creating an inferior product. This problem often necessitates additional filtration, which serves to increase pressure and may counteract many of the benefits expected from the pump, as well as increasing cost investments further.

Depending on the screw design, the extruder often creates pulses (surges), causing the production rate to fluctuate. Some products can seldom tolerate even minor fluctuations, and a pump can often assist in removing these minor product nonuniformities. A pump can generally provide output uniformity of ±0.5% or better. Products include films [down to 19 mm (0.75 mil) thickness], precision medical tubing, HIPS film with 1600 kg/h (3500 lb/h), fiber-optic sheathing, fibers, PET magnetic tape, PE cable jacketing (wt/length variation) reduced from 14 to 2.7%, and so on.

Pumps are very helpful to sheet extruders who also do in-house thermoforming, as they often run up 50 wt% regrind mixes. Normally having a variable particle size, this mix promotes surging and up to 2% gauge variation. Pumps practically eliminate the problem and make cross-web gauge adjustments much easier. Pumps are recommended in: (1) most two-stage vented barrels where output has been a problem, such as ABS sheet; (2) extrusions with extremely critical tolerances, such as CATV cable, where slight cyclic variations can cause severe electrical problems; (3) coextrusion where precise metering of layers is necessary and low pressure differentials in the pump provide fairly linear outputs; and (4) twin-screw extruders where pumps permit long wear life of bearings and other components, thus helping to reduce their high operating costs.

Beside improving gauge uniformity, a pump can contribute to protect quality by reducing the plastic's heat history. This heat reduction can help blown film lines, particularly those running high viscosity melts such as LLDPE and heat sensitive melts such as PVC. Heat drops of at least 11–

17°C (20–30°F) will occur. In PS foam sheet lines, a cooling of 6–8°C (10–15°F) occurs in the second extruder as well as 60% reduction in gauge variation by relief of back pressure. One must be aware that all melts require a minimum heat and back pressure for effective processing.

Pumps cannot develop pressure without imparting some energy or heat. The heat increase of the melt depends on its viscosity and the pressure differential between inlet and outlet (or P). The rise can be 3°C (5°F) at low viscosity and below P, and up to 17°C (30°F) when both these factors are higher. By lowering the melt heat in the extruder, there is practically no heat increase in the pump when P is low. The result is a more stable process and a higher output rate. This approach can produce precision profiles with a 50% closer tolerance and boost output rates by 40%. Better control of PVC melt heat could increase the output up to 100%. As an example, the output of totally unstabilized, clear PVC blown film meat wrap went from 270 to over 450 kg/h (600 to 1000 lb/h) with the use of a gear pump.

With pump use, potential energy savings amount to 10–20%. Pumps are 50–75% energy efficient, whereas single-screw extruders are about 5–20% efficient.

Although they can eliminate or significantly improve many processing problems, gear pumps cannot be considered a panacea. However, they are worth examining since they could boost productivity and profits very significantly. Their major gains tend to be in: (1) melt stability; (2) temperature reduction in the melt; and (3) increased throughput with tighter tolerances for dimensions and weights. They can cause problems when the plastic contains abrasive additives (small clearances make the gear pump very susceptible to wear) and when the plastic is susceptible to degradation (they are nor self-cleaning, combined with high temperatures, this will cause degraded plastic to be pumped).

*Static mixer*

Most extrusion operations involve either the processing of regrinds with virgin plastics and/or additives such as color. Uniform mixing of the feed mix is important to achieve acceptable product properties but also more difficult as output rates increase. Head pressure and stock temperature must be held constant in order to maintain extrudate control.

More extruders are being equipped with a static mixer to overcome some of the problems just reviewed (Fig. 1.9). These mixers are located at the end of the screw before the screen changer and/or die. If a gear pump is used, the static mixer is located between the screw and gear pump. Static mixers contain a series of passive elements placed in a flow channel (Fig. 2.16).

These elements cause the plastic compound to subdivide and recom-

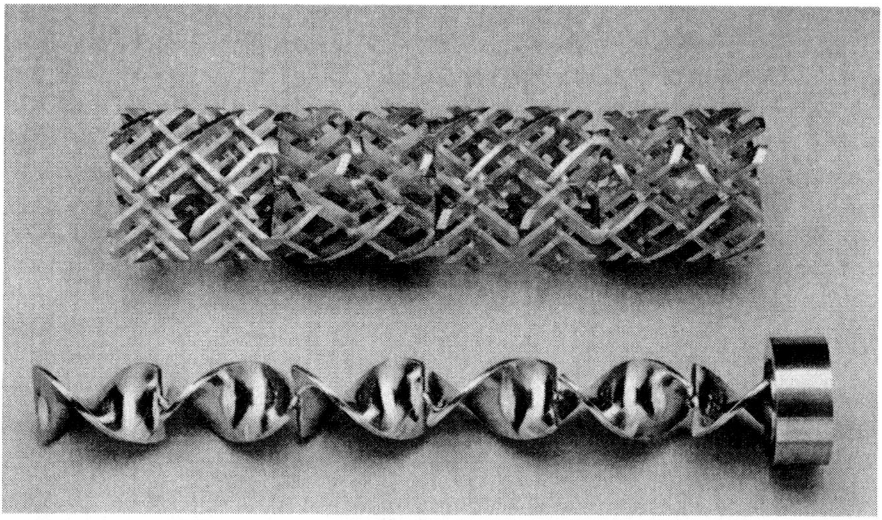

**Figure 2.16** Two different styles of static mixers.

bine in order to increase the homogeneity of the melt. There are no moving parts and only a small increase in the screw energy is needed to overcome the resistance of the mechanical baffles.

The installation of a static mixer increases the effective L/D of the extruder and usually results in some increase of the melt temperature and head pressure. The overall result is a more homogeneous melt and a more stable extrusion process with less output surge. When using static mixers, sufficient heat-up time should be provided with startups of the extruders. To prevent excessive pressures and damage to the extruder during startups, the screw's rpm should be kept low until plastic flows uniformly from the die.

**Barrel and feed unit**

When selecting an extruder, the size and shape of the feed throat is very important (Fig. 2.17). Feed throat size and shape can have a significant affect on output and stability in an extruder. In general, the smaller the hole, the more adverse the effect. Sometimes, smaller feed holes can be compensated for by screw designs, but more often the feed hole geometry must be modified [2, 4, 157, 239].

Output rates have been observed to vary as much as 25% with the only variable being the feed throat geometry. Round feed throats are sufficient for 100% pellet feed but when 20% or more regrind is added with the virgin feed rate is reduced; a rectangular or oblong opening will improve

**Figure 2.17** Detail view of Welex extruder feed section with guards removed.

the feeding characteristics (Fig. 2.18). An elongated opening also helps in eliminating potential bridging problems in the throat. There are materials, such as flakes, regrind and others, that have poor flow so they present their own problems that need to be resolved.

The feed throat is the section in an extruder barrel, where plastic is directed into the screw channel. It is fitted around the first few flights of the screw. Some extruders do not have a separate feed throat. In these machines, the feed throat is an integral part of the barrel, even though it may not be the best design approach.

If a plastic feeding problem exists, an approach to consider is installing grooves in the feed area Barrel grooves have been around for a long time and produce excellent results when processing materials that have problems moving into the screw channel. Some grooving systems require special metal carbide insertable feed sections, special high-capacity cooling in the grooved area, and/or specialized screw designs. Prior to considering grooving, determine if a feeding or melting problem exists.

As an example, the use of grooved feed extrusion for blown and flat film as well as sheet provides many benefit with different plastics particularly HMWHDPEs and certain PPs. They include a lower and controlled melt

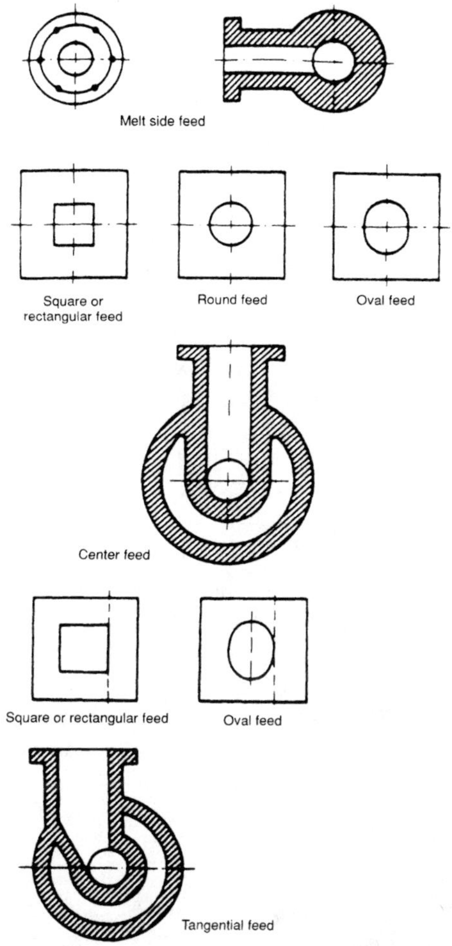


**Figure 2.18** Examples of different feed position openings.

temperature, increased extrusion rate which remains nearly insensitive to head pressure, and lower residence times. The process becomes feed controlled, making the output rate nearly independent of back pressure. As screw is increased over its range, the specific rate changes only a few percent. Plastic passing down the grooves is compressed to very high pressures; with all the pumping force already generated, the balance of the screw can be dedicated to mixing without being compromised as only a pump.

With a screw designed to accept the high flow from the grooved feed, the specific feed rate is always higher, often almost double that of a smooth bore screw of the same diameter. This has the advantage of allowing plastics to be processed with considerably shorter residence times; thus, for heat sensitive plastics, less time at elevated temperatures.

The extruder's performance is not compromised by adding lubricants to the plastic. As an example, adding a plastic processing aid to HMWHDPE for reasons of improved processability and elimination of melt fracture has the added advantage of higher specific throughput. In theory, this occurs because the processing aid selectively lowers the coefficient of friction between screw and plastic. However, the important action of the grooves prevents this friction action from happening between the screw and barrel. The result is that the axial friction is reduced further and the circumferential friction remains unchanged. Contrary to most smooth bore extrusion, there is little change in the melt temperature with increased screw speed. This feature benefits film blowing of plastics which exhibit low melt strength such as LLDPE [12].

Placement of holes in the barrel for sensors must be carefully determined to ensure the sensors operate most efficiently and no damage occurs to the barrels. There can be vent holes to be used with two-stage extruders providing an escape for moisture, air, and volatile to escape when plasticizing certain plastics (Chapter 3).

The feed-throat casting is generally water-cooled to prevent an early temperature rise of the plastics. A good starting point is to have the temperature about 43–49°C (110–120°F), or 'warm to the touch' to help ensuring that a stable feed is developed. If the temperature rises too high, it may cause the plastic to adhere (stick) to the surface of the feed opening, causing a material-conveying problem to the screw. The overheated plastic solidifies at the base of the hopper or above the barrel bore causing bridging whereby material no longer can enter the screw. The problem can also develop on the screw, with plastic sticking to it, restricting forward movement of material.

Over cooling the hopper can have a negative effect on extruder performance. The reason for this action is due to its heat sink effect that pulls most of the heat from the feed zone of the barrel. The idea of hopper block cooling is primarily to prevent sticking or bridging in that area. Thus, it should not be run colder than necessary. Always control water flow in the throat cooling systems from the outlet side to prevent steam flashing and to minimize air pockets

Where the feed-throat casting connects with the barrel, a thermal barrier is usually included to prevent barrel heat from escaping through the feed throat. This action is not possible in a barrel with integral feed opening.

The usual target is to feed plastic at a controlled weight rate rather than by volume. The key advantage of gravimetric or loss-in-weight technology lies in its ability to meter material with pinpoint accuracy despite variations in bulk density. They basically consist of the feeder (including discharge device), scale, and control unit. They can control weight per length, screw speed (weight per time), or both, depending on the process and whether the end product is sold according to weight or length. There are both batch and continuous weighing units.

Although gravity flow may be suitable for bulky materials, additional devices are used to ensure steady flow. A vibrating pad may be attached to the hopper to dislodge any bridges or bonds that may want to form. Different designed stirrers are used to mix materials, to provide an even distribution, or to prevent separation. The stirrers can be used to wipe material from the hopper wall if the bulky material tends to stick to it. Crammer feeders are used with certain bulky materials when other devices do not have the capability to provide the proper feeding action (Fig. 1.11).

Air entrapment can occur with certain plastics, particularly those having low bulk density. If air cannot escape through the feed hopper (due to less pressure at hopper than exit end), it will be carried with the plastic melt through the die and problems can develop. One way to eliminate air entrapment is to use a vacuum feed hopper, but this is rarely practical. Problems can develop over loading the hopper without losing vacuum. Double-hopper vacuum systems are used where plastic is loaded into a top hopper where air is removed before the plastic is dumped into the main hopper. This system has merit when used very carefully but it can encounter problems such as leaks. Another method is the use of a vented barrel (Chapter 4).

To maintain the maximum and most consistent feeding, it is necessary to exercise care when changing hopper dimensions, feed throat openings, or adding any intermediate sections (side feeders, magnet packs, adapters, etc.). In summarizing what should be considered when re-engineering the solids delivery system, consider the following information: (1) minimum taper for hoppers is 60° included angle for general purpose use with some plastics requiring a smaller angle/steeper sides; (2) be sure the system is streamlined with no ledges, projections, or rough surfaces; (3) avoid, as much as possible, changes in shape such as round to square, because each change of shape causes a restriction to flow; and (4) the absolute minimum cross section in any solid's flow channel should be the cross section of the barrel bore with a preference of about 1.5 times the cross section of the barrel bore.

These rules can apply to solids flow as much as to liquid flow. That is, anything that has entry problems, is not streamlined, has shape changes, or restrictive flow area will result in excessive pressure drops. Unless there is a minimum pressure of the solid's mass at the entry to the screw, the screw channel will not fully fill. This is particularly true as the screw speed increases and obviously varies with the characteristics of the solids.

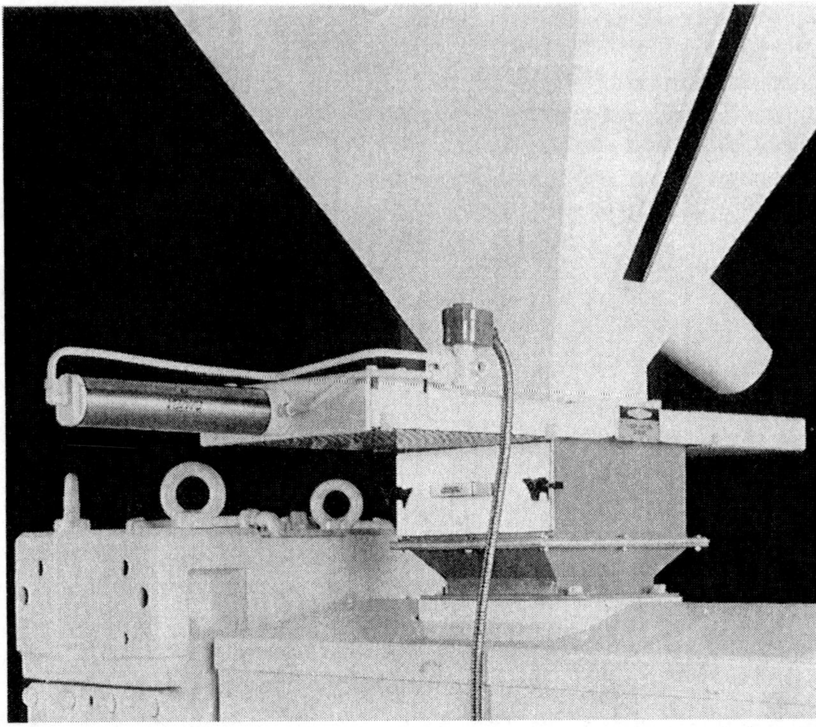

**Figure 2.19** Base of Welex feed hopper with pneumatic sliding shut-off and a drawer style hopper magnet.

The hopper can be fitted with devices to perform different functions. As an example, a hopper can be fitted with a hinged or tightly fitted sliding cover (Fig. 2.19) and a magnetic screen for protection against moisture pick-up and metal ingress, respectively. It is usually advisable to install a hopper drier, especially when processing certain materials such as regrind and colors. It can be of value in limiting material handling, as well as removing moisture.

**Auxiliary equipment**

Many different types of auxiliary equipment are used upstream through downstream Table 2.9 lists a few of the equipment [2, 4, 165, 344]. The equipment is used to maximize overall processing productivity and efficiency, and reduce manufacturing cost. Their total cost is usually more than the extruder. Proper selection, use, and maintenance are as important as the selection of the extruder.

**Table 2.9** Auxiliary equipment used with plastic processes

| | |
|---|---|
| Adhesive applicator | Process control for individual or all equipment |
| Bonding | Pulverizing/grinding |
| Cutting | Recycling system |
| Die cutting | Robotic handling |
| Dryer | Router |
| Dust-recovery | Saw |
| Freezer/cooler | Screen changer |
| Granulator | Screw/barrel backup |
| Heater | Sensor/monitor control |
| Knitting | Software |
| Leak detector | Solvent recovery |
| Material handling | Testing/instrumentation |
| Metal treating | Trimming |
| Metering/feeding material | Vacuum debulking |
| Mold extractor | Vacuum storage |
| Mold heat/chiller control | Water-jet cutting |
| Oven | Welding |
| Pelletizer/dicer | Others |
| Printing/marking | |

All in-line equipment has to be properly interfaced so it operates efficiently. Much of the equipment used in the past did not properly interface, so operations were rather less efficient than they could have been. Rules have been developed that govern the communication and transfer of data along the complete line. The processor must determine what is needed based on what has to be accomplished, what controls are required, ease of operation and maintenance, safety devices, energy requirements, compatibility with existing equipment, and so on.

This section will review a few of the common equipment that are used by many of the lines. Each of the processes (blown film, sheet, tube, etc.) to be reviewed will contain some information applicable to their specific lines. There is an extrusion line with limited use called the 'free extruder' which just takes the extrudate and uses no special equipment. All it requires is to cool the extrudate (usually air, water tank, or their combination) and no special forming device, controlling fixture, sizing device, and so on.

New generation equipment that are continually produced are meaningful contributions to plant productivity via greater reliability of operation, and pin point equipment control so that products are more reliable and profitable. So it is important to keep up to date. As an example, fluid chillers and temperature control systems are major beneficiaries of the computer revolution. They become more energy-efficient, reliable, and cost-effective. The application of microprocessor- and computer-compatible controls that can communicate with the extruder results in pin point control of the line.

*Communication protocol*

Within the plastic processing SPI communication protocol cell, there are two basic types of auxiliary devices: (1) those that require minimum configuration or minimum data (chillers, dryers, loaders, etc.); and (2) those that require large amounts of configuration data or provide large amounts of process data (gauging/sensors, robots, die controllers, cutters, etc.). A set of rules govern communication or transfer of data between computer hardware and/or software. When related to plastic processing equipment, communication includes reference to exchange of process controls, meeting standards, and following production schedules; these activities permit an interchange of actions between all in-line equipment.

The information that is required to monitor and configure a manufacturing operation is distributed among the various auxiliary equipment. This information is transferred to the central control. Communication interfaces and communication protocols have been developed to allow this information to be exchanged. Successful communication requires a durable interface and a versatile protocol. The interface is mechanical and electrically durable. Mechanical durability is achieved by suitable hardware attachments and cable strain relief. Electrical durability is achieved by suitable transceiver circuitry. Circuitry meeting the requirements of ANSI RS-485 is ideal for half- and full-duplex protocols. It is designed for use where more than one device may 'talk' at a time. Less durable circuitry, such as RS-232 and RS-422, may fail if more than one device 'talks.'

*Roll*

There are many different types of rolls to handle films, sheets, and coating lines. They include winders, dancer rolls, spreader rolls, textured rolls, engraved rolls, and cooling rolls. All have the common feature that they are required to be extremely precise in all their measurements, surface conditions that include commercial grade mirror finishes, center line operations, bearings and all ancillary mounted on journals, and rotating speed [245, 290, 344].

The controls and uniformity of speed operations in many operations are very critical. They can weight at least 6.8 tonnes (15 000 lb) with diameters at least up to 150 cm (60 in) and widths at least up to 100 m (30 ft). Some rolls require roundness and surface finishes to be within 0.00127 mm (0.00005 in) TIR (total indicator reading). Many winders offer sophisticated features and are highly automated, but some are designed to answer the need for simplicity, versatility, and economy. There are surface winders with gap-winding ability for processing tacky films such as EVAs and the metallocene plastics.

The winders roll a continuous film or sheet into certain weight (kg or lb) or diameters on spools or reels. Material speeds are up to at least 670 m/min (2200 ft/min) in cast film lines; at least 330 m (1000 ft/min) in blown film lines. Blown film lines may require use of reverse winding systems to allow coextruded films to be wound with a particular material as the inside or outside layer. Throughputs are over 1000 kg/h (2200 lb/h). Transfers from one roll to another can take less than a second.

The major trouble spot is material tension control. There is a proportional relationship between winding tension and lay-on-roll forces. As an example, wrinkles may be the result of an alignment problem. Winding shafts are potential problem areas were bumps and valleys develop.

There are various tension control techniques available (Chapter 6). The proper selection involves decisions on how to produce the tension, how to sense the tension, and how to control the tension. The tension system selection process depends primarily on the data sensitive to the application. For instance, if the material has a very low tension requirement and if exact control is required, then perhaps using a magnetic particle brake with an electrical transducer roll with appropriate electronic control is best. However, if the material is on large diameter rolls and moves at slow speed, then a roll size follower system can be used effectively.

There are manual adjustment systems, such as the dancer roll, canvas drag brake, various pony brakes, and pneumatically operated brakes. The most expensive is the regenerative drive systems. The transducer rolls and dancer rolls is a close second. These systems are usually required in high web speed applications where accurate tension control of expensive and/or sensitive material is paramount. With roll windup systems, different roll or reel-change systems are used to keep the lines running at their constant high speeds.

Dancer rolls can be used as a tension sensing device in film, sheet, and coating (wire, film, etc.) lines. They provide an even controlled rate of material movement. Type roll can have an influence on the rolls performance. As an example, chrome plated steel casting drums would seem to be very durable dancers. If used in the absence of a nip roll, they should last many years. However, these rolls are in fact very soft due to the annealing which good rolls receive for stress relieving the steel. A situation can occur where a casting drum has been coupled with a steel chill roll to nip polish a cast film web. The casting drum was imprinted by hard plastic edges or die drips. This action occurs because the compressive stresses in a solid plastic passing through the nip roll will exceed the yield strength of the soft steel drum surface. Higher line speeds make the problem worse. In order to prevent this damage, the roll must be hardened [187, E M. Mount III].

Winding strain can occur. It is the phenomena of a wound roll of film turning into a hard rock corrugated nightmare in a few days. This action is caused by several factors: (1) Trapped air, as the roll is being wound makes a roll feel soft. Static charges helps trap air. Lay-on rolls help to squeeze air out but can also create other problems. The rapid escape of air can produce telescoping. (2) Tension creates a compression load which will squeeze out the very thin film of air, crush underlayers, and crush cores. Tension also tends to even out some of the wrinkles and irregularities. (3) Room temperature recoverable strains are residual processing strains that will release themselves at room temperature to produce a stress and/or shrinkage. Techniques are available for predicting the level of room temperature recoverable strains. (4) Crystallization of crystalline plastics also produces shrinkage of a magnitude generally 0.5–2%. Crystals take less space and thus, as the crystal structure goes to completion, shrinkage occurs. As an example, PE film is commonly loosely wound. It is permitted to shrink for about one to two days, slit, and rewound [187, P. Limbach].

Roll/drum cooling systems are used. They range from a very inexpensive, rather poor surface nonuniform temperature control to the usually more desirable, expensive, and very uniform surface temperature control. Recognize that what you pay for is what you get. When sizing chillers, be sure to include the heat generated by the pumps. For example, a 20 hp water pump can require up to 7 tonne (8000 lb) of additional chilling capacity to remove the heat generated by the pump.

One type of film spreader or stretching roller is the grooved metal idler roll. The grooved roller has opposing, etched spiral grooves which start at the roll's center and spiral toward the ends. As the roll turns, air flows and follows the direction of the grooves along the metal surface. Because of the grooves, flow is generated and moves from the center of the roll outward. This action forces any web wrinkles out toward the ends of the roll.

The major advantage of these rolls is that they are free turning and existing idler rolls can be easily modified. However, as the web's processing speeds increase, these rollers gradually loose their effectiveness. Because the roll has a smooth surface, it has a low coefficient of friction. Subsequently, as the line speeds increase, so does the air flow, causing the web to slide over without making complete contact with the web. If the roller does not make contact with the web, it does not rotate. So its ability to remove wrinkles is effected. The surface of this type roller also has other limitations for certain applications. For films metallized or printed and coated webs, the roller's hard surface can produce hard undesirable marks on the substrate as it passes the grooves.

Another type of spreader roller is the expander roller. It originally consisted of metal or wooden slats with internal elastic bands connected to angled end plates. The connection of the bands to the pitched end plates caused the bands to expand in the center section as it rotated. There was a pulling apart of the slats to provide a spreading action. There are several variations that include some with internal mechanisms. The next generation eliminated the metal and wood slats with more substantial elastic cords. However, they were still attached to the pitched end plates. With these expander rollers, the substrate enters at a point where the cords are relaxed and then exits at the point of maximum cord expansion to achieve web spreading. The amount of spreading action can be adjusted by changing the angle of the roller's end plates. The direct elastic cord and substrate contact provide better coefficient of friction and are less abrasive than the metal or wooden slats.

However, while these rollers offer adjustable spreading action, they eventually lose their effectiveness because the rubber cords do not fully recover to their original state after continuous stretching. Another limitation is that at high film speed, air enters between the elastic bands and is trapped under the web. This causes the web to float over the roller surface negating its function. There can also be web markings. One of the latter generation operates on the same principle, but features a stretchable one-piece rubber sleeve supported by a series of brushes. As the roll rotates, the entire roller sleeve, as opposed to individual cords, expands and contracts to provide spreading action. The amount of spreading is controlled by the two factors of the wrap or angle at which the web enters onto the roller and the angular displacement of the end caps. Notable advancements in this expandable sleeve roller include a smooth, continuous surface which does not produce marking or allow air to enter under the web. However, the stretching of the rubber still causes the roller to eventually wear over time.

When the melt leaves the die and enters roll nips, it is soft enough to take the finish of the rolls it contacts. Thus, in addition to smooth and highly glossy finishes, textured or grain rolls can be used. They can impart a mirror image. They can give both functional and aesthetic qualities to the film or sheet. There are as many different grains as the imagination can conjure up.

Various methods are used to render inert plastics, such as polyolefins, more receptive to inks, adhesives, or various decorations. They include flame treatments, chemicals, additives in the plastics, and corona discharge. The electrical corona discharge treatment is used on webs. The flow of electrical energy from a conductor to the surrounding air or gas occurs. The phenomenon occurs when the voltage is high enough (over 5000 V) to cause partial ionization of the surrounding gas. The discharge is characterized by a pale violet glow, a hissing noise, and the odor of ozone.

The treatment takes place only on the exposed surface to a depth of only 1 nm (10 Å). Proper handling of the web is required to eliminate problems such as air entrapment and wrinkles on the reverse side.

*Orientation*

Orientation consists of a controlled system of stretching plastic molecules to improve their strength (Table 2.10), stiffness optical, electrical, and other properties with the result that improved product cost occurs. Depending on the properties of a specific plastic, the stretch ratio may vary from 2.5:1 to as high as 10:1. Some specialty films may have an even higher stretch ratio.

Used for almost a century, orientation became prominent during the 1930s for stretching fibers up to 10 times. Latter it was adapted principally to films and other products (Table 2.11) taking advantage of its benefits. Practically all TPs can undergo orientation, although certain types find it particularly advantageous (PET, PP, PVC, PE, PS, PVDC, PVA, PC, etc.). The largest market for plastics worldwide, consuming about 20 wt% of total , is oriented plastics.

A nylon 6/6 fiber, which is an unoriented glassy plastic, has a modulus

**Table 2.10** Effects of orientation on PP films

| | Stretch (%) | | | | |
|---|---|---|---|---|---|
| *Property* | *None* | *200* | *400* | *600* | *900* |
| Tensile strength, psi | 5600 | 8400 | 14000 | 22000 | 23000 |
| (MPa) | (38.6) | (58.0) | (96.6) | (152.0) | (159.0) |
| Elongation at break (%) | 500 | 250 | 115 | 40 | 40 |

| *Property*[a] | *As cast* | *Uniaxial orientation* | *Balanced orientation* |
|---|---|---|---|
| Tensile strength, psi (MPa) | | | |
| MD | 5700 (39.3) | 8000 (55.2) | 26000 (180) |
| TD | 3200 (22.1) | 40000 (276) | 22000 (152) |
| Modulus of elasticity, psi (MPa) | | | |
| MD | 96000 (660) | 150000 (1030) | 340000 (2350) |
| TD | 98000 (680) | 400000 (2760) | 330000 (2280) |
| Elongation at break (%) | | | |
| MD | 425 | 300 | 80 |
| TD | 300 | 40 | 65 |

[a] MD = machine direction, TD = transverse direction and direction of uniaxial orientation.

**Table 2.11** Example of orientation used to fabricate different types of TP film tapes

| *Ranges of application* | *Demands made* | *Rate of stretching* | *Thermoplastic* |
|---|---|---|---|
| Carpet basic weave | Low shrinkage | 1:7 | PP |
| | High strength | 1:5 | PETP |
| | Temperature stability | | |
| | Specific splicing tendency | | |
| | Matt surface | | |
| Tarpaulins | High strength | 1:7 | PP |
| | | | PE |
| Sacks | High strength | 1:7 | PP |
| | High friction value | | PE |
| | Specific elongation | | |
| | Weather resistant | | |
| Ropes | High tensile strength | 1:9 to 1:11 | PP |
| | Specific elongation | (15) | |
| | Good tendency to splicing | | |
| Twine | High tensile strength | 1:9 to 1:11 | PP |
| | High knotting strength | | PP/PE |
| Separating weave | High strength | 1:7 | PP |
| Filter weave | Low shrinkage | 1:7 | PP |
| | Abrasion resistant | 1:5 | PETP |
| Reinforcing weave | Low shrinkage | 1:7 | PP |
| | Specific elongation | 1:5 | PETP |
| | Temperature resistance | | |
| Tapestry and home textiles | UV resistance | 1:7 | PE |
| | Low static charge | | |
| | Uniform coloration | | |
| | Textile-type handle | | |
| Outdoor carpets | Low shrinkage | 1:7 | PP |
| | Wear resistance | 1:5 | PETP |
| | Weather resistance | | |
| | Elastic recovery | | |
| | Uniform coloration | | |
| | Defined splicing | | |
| Decorative tapes | Effective surface | 1:6 | PP with blowing agent |
| | Low specific gravity | | |
| Knitted tapes, sacks, and other packagings, seed and harvest protective nets | High knotting strength | 1:6.5 | PP |
| | Low splicing tendency | | PE |
| | Suppleness | | |
| | UV resistance | | |
| Packaging tapes | High strength | 1:9 | PP |
| | Low splicing tendency | 1:7 | PETP |
| Fleeces | Fiber properties | 1:7 | PP and blends |

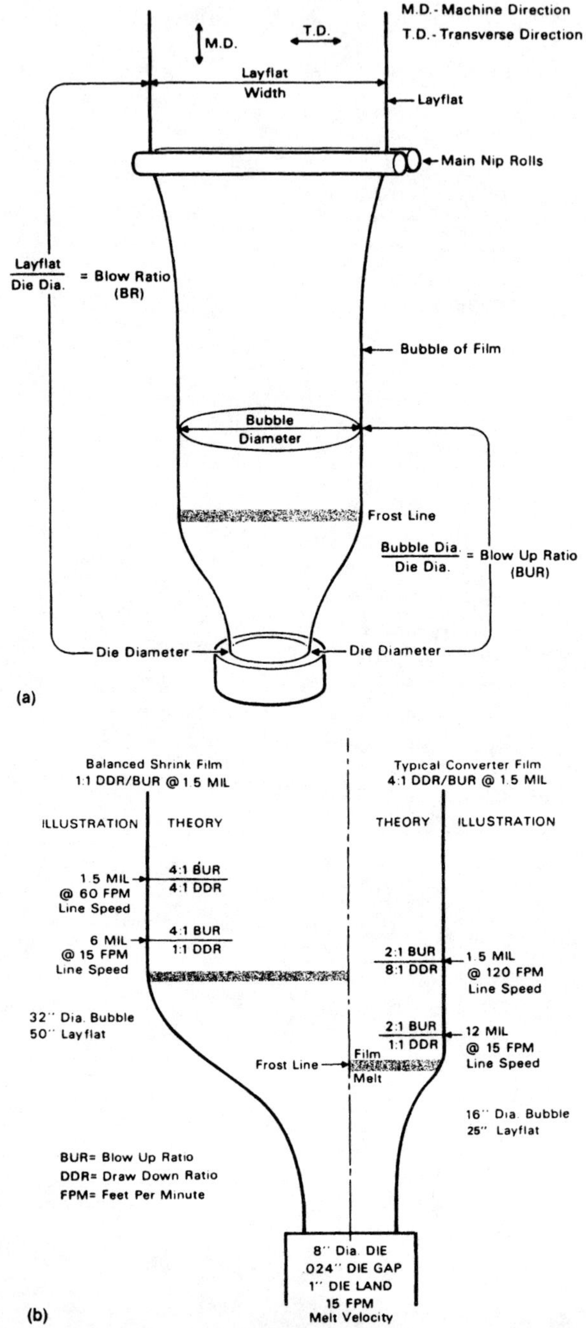


**Figure 2.20** Blown film: (a) terminology, (b) orientation.

of elasticity ($E$) of about 2000 MPa (300 000 psi) (this development occurred during the 1930s). Above its glass transition temperature ($T_g$) its modulus of elasticity drops. Once it is extended and has its molecules oriented in the direction of stress, large stresses occur with $E$ (and other properties) increasing. The next step is to cool the nylon below its $T_g$ without removing the stress, retaining its molecular orientation. Result is its $E$ in the oriented/tension direction increases to $15–20 \times 10^3$ MPa ($2–3 \times 10^6$ psi); almost a 20-times increase occurs.

In extrusion the most important orienting processes are for blown film, flat film/sheet, and blow molding. During blown film processing (Chapter 7), the blowup ratio determines the degree of circumferential orientation, and the pull rate of the bubble determines longitudinal orientation (Fig. 2.20). The optimum stretching heat for amorphous plastic (PVC, etc.) is usually just above its $T_g$; for crystalline type (PET, PE, etc.) it is below its $T_g$ (Table 2.12).

In orienting flat film or sheet the processor uses a tentering frame (takeoff from textile weaving) which is enclosed in a heat-controlled oven; a very gentle and accurate airflow is used (Fig. 2.21). The frame has continuous speed control and diverging tracks with holding clamps. As the clamps move apart at prescribed diverging angles, the hot plastic is stretched in the transverse direction, producing a single orientation (O).

To obtain bidirectional orientation (BO) an approach includes the use of in-line heat-controlled rolls located between the extruder (or roll of material) and tenter frame (Fig. 2.22).

The rotation of each succeeding roll is increased, based on the desired longitudinal stretch properties. Each line has interrelating control operations, as well as specific operations to simplify operations. The extruder is followed by a cooling system (chill roll contact, air and/or water). This annealing action can be a duplication of the heated rolls with only about two rolls used.

**Table 2.12** Examples of orienting conditions for plastics

| Plastic | Melting temperature, $T_m$ (°C) | Glass transition temperature, $T_g$ (°C) | Modulus of elasticity (GPa): Biaxially oriented film, Longitudinal | Biaxially oriented film, Transverse | Unoriented film | Oriented fiber, Melt spun | Oriented fiber, Gel spun | Density (g cm$^{-3}$) |
|---|---|---|---|---|---|---|---|---|
| PVC | – | 70 to 90 | 2.5–2.7 | 3–3.5 | 2.2 | 5.5 | – | 1.35 |
| HDPE | 138 | –70 to –110 | 3–4 | 3–4 | 1.2 | 5 | 170 | 0.96 |
| PP | 134 | –5 to –20 | 3–4 | 2–3 | 0.9 | 5 | 18 | 0.90 |
| PA | 260 | 50 to 75 | 2–2.5 | 2–2.5 | 0.5 | 4.5 | 19 | 1.13 |
| PET | 250 | 70 to 110 | 4–5.7 | 4.5–8.5 | 1.5 | 15 | 28 | 1.35 |

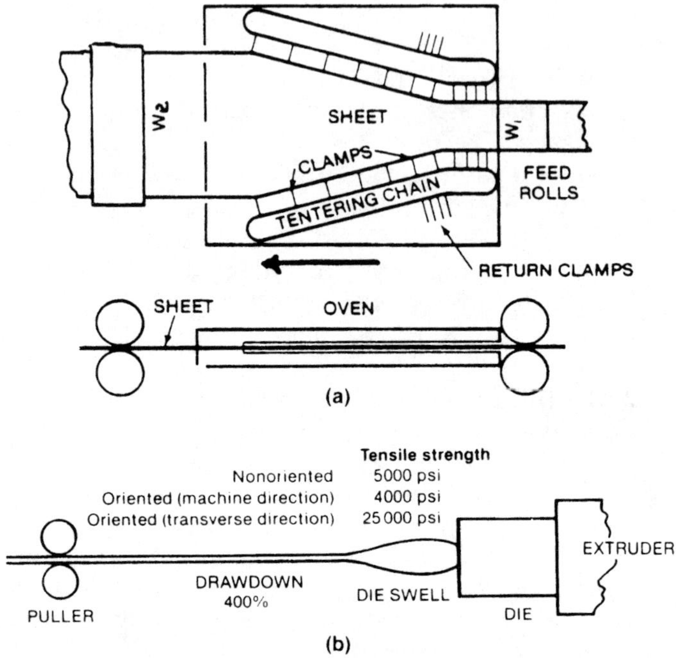


**Figure 2.21** Use of tenter frame to biorient film or sheet: (a) example or ratio of puller-roll to feeder-roll is 4/1 (ratio of W2 to W1); (b) drawdown phenomenon with die swell to produce orientation in the longitudinal direction.

Depending on the plastic being processed, equipment systems can vary. As an example, the number of heated rolls for PS (OPS) with an output of 1800 kg/h (4000 lb/h) can be as few as four. With a PP (OPP) system, producing the same output rate may need 12 or 14 heated rolls.

Depending on the plastic used, the flat/casting film equipment varies. For instance, a highly crystalline plastic such as PP is quenched rapidly on a single drum rotating in a water bath. Variations of this approach include the use of cooling rolls instead of the water bath. Casting machines for accommodating extruder outputs between 1400 and 3200 kg/h (3000 and 7000 lb/h) call for rolls exceeding 1.5 m (5 ft) in diameter, with cooling water flow rates in the area of 4 $m^3$/min (1000 gal/min). PS casting film equipment typically uses a three-roll vertical stack arrangement for the cooling or casting rolls (Chapter 9). With this amorphous plastic, the cooling is more gradual, therefore requiring a different technique from that of PP (Marshall & Williams Co.).

Tapes can be uniaxially oriented using a series of heated and speed controlled rollers. An example for a fast, in-line tape operation is shown in Fig. 2.23. Another product is an oriented solid rod. The rod (billet) is

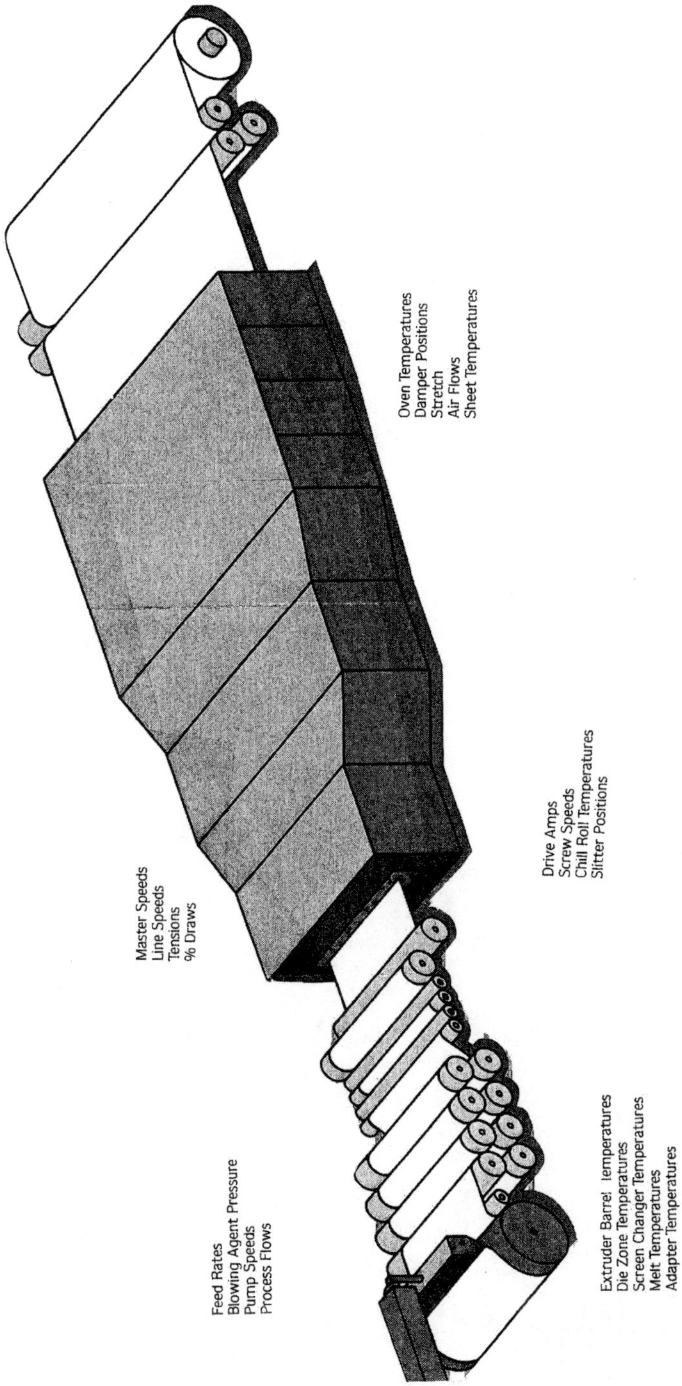


**Figure 2.22** Schematic of biorienting film (or sheet) up to at least 6 m (20).

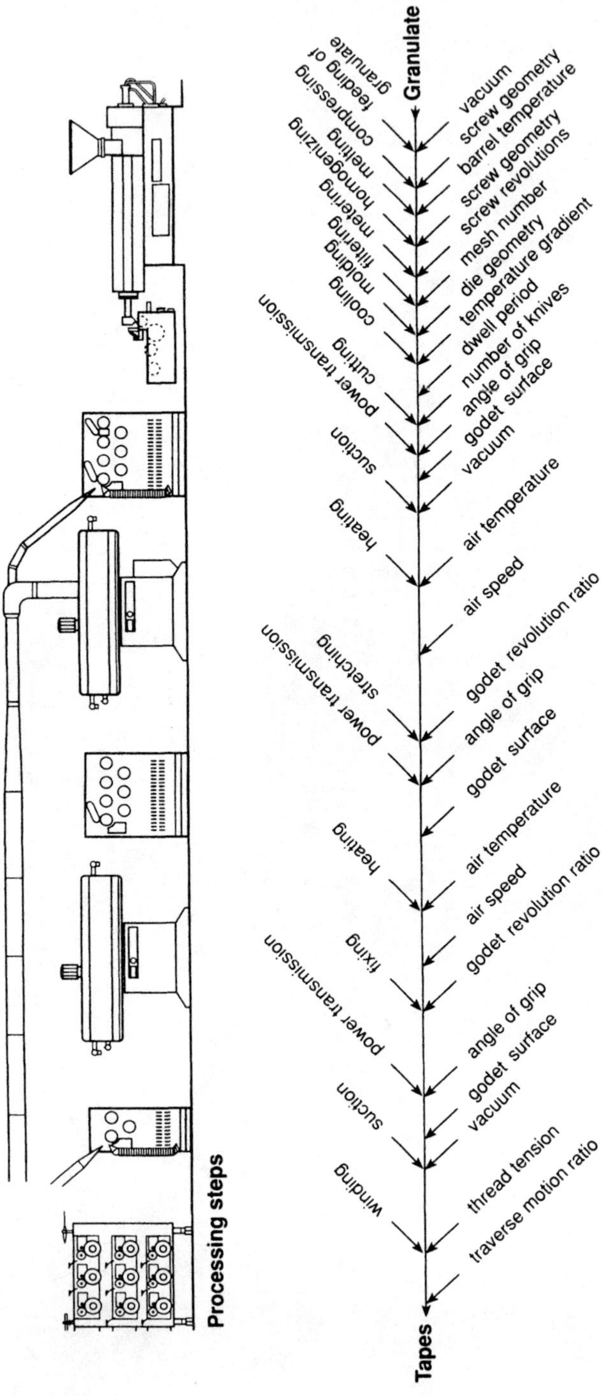


**Figure 2.23** Extruder chill roll process line used to produce uniaxial oriented film tape.

directed through a reducing conical die, usually heated. This technique is a take-off used in the metal industry with rods, wires, etc. The amount of orientation is based on the deformation ratio used (diameter before and after going through the die), travel speed, and temperature. The compressing (squeezing) action orients the plastic molecules.

Some general rules for orienting by stretching are: (1) lowest temperature will give the greatest orientation (tensile strength, etc.); (2) highest rate of stretching will give the greatest orientation at a given temperature and percent stretch; (3) highest percent stretch will give the greatest orientation at a given temperature and rate of stretching; and (4) greatest quench rate will preserve the most orientation under any stretching condition. With orientation, the thickness is reduced and the surface enlarged. If film is longitudinally stretched, its thickness and width are reduced in the same ratio. If lateral/transverse contraction is prevented, stretching reduces the thickness only [287, 389].

For plastics whose $T_g$ is above their decomposition temperature, orientation can be accomplished by swelling them temporarily with liquids to lower $T_g$ of the total mass, particularly in solution processing. As an example, cellulose viscous film and fiber can be drawn during solution spinning, while they still contain enough fugitive solvent to plasticize them and lower their $T_g$ below decomposition temperature. Final removal of the plasticizing solvent makes the orientation permanent.

Shrink film made by orientation is a very popular product; the film shrinks around a product when heat is applied. Heat applied depends on plastic used; temperatures required are about 65–150°C (150–300°F). Uses of shrink film includes labels, stacking products to be shipped, tamper-evident neck bands, combination label/neck bands, reinforcing glass bottles, and devices for promotional packaging (Table 2.13). The big market with these dry bands, in contrast to cellulose wet bands, are made principally from PVC or vinyl copolymers. They can be made of flat or tubular/blown film.

The degree to which the band shrinks is determined when the film is stretched. To obtain a band that will eventually shrink 60%, the film must be stretched 60%. In effect, stretching the film program's memory into the plastic. That memory is recalled by applying heat when the band is to be shrunk, returning the plastic to its original dimensions.

When the plastic is quenched (very quick cooling) after orientation, the molecules are 'frozen' in the oriented position; however, they still 'remember' their original position. Upon reheating to the orienting temperature, the plastic returns to its original shape. To obtain permanently oriented plastic, that will not shrink, instead of quenching the oriented plastic is exposed to its softening temperature and allowed to cool. This temperature heat-set change destroys the 'memory' eliminating shrinkage.

**Table 2.13** Performance of extruded shrink films

| *Film type* | *Advantage* | *Possible problem* |
|---|---|---|
| Polyethylene (low density) | Strong heat seals<br>Low-temperature shrink<br>Medium shrink force for broad application<br>Lowest cost | Narrow shrink temperature range<br>Low stiffness<br>Poorer opticals<br>Sealing wire contamination |
| Polypropylene | Good optical appearance<br>High stiffness<br>High shrink force<br>No heat-sealing fumes<br>Good durability | High shrink temperature<br>High shrink force, not suitable for delicate or fragile products<br>Brittle seals<br>High sealing temperature |
| Copolymers | Strong heat seals<br>Good optical appearance<br>High shrink force<br>No heat-sealing vapors | High shrink force, not suitable for fragile products<br>Higher shrink temperature<br>Higher heat-seal temperature<br>Lower film slip may give machine problems |
| Polyvinyl chloride | Lowest shrink temperature range<br>Wide shrink temperature range<br>Excellent optical appearance<br>Controlled stiffness by plasticizer-content control<br>Lowest shrink force for wrapping fragile products | Weakest heat seals<br>Least durable after plasticizer loss<br>Toxic and corrosive gas emission from heat sealing, good ventilation required<br>Durability problem at low temperature<br>Low shrink force inhibits use as a multiple-unit bundling film<br>Low film slip causes machine wrapping difficulties |
| Multi-layer coextrusions | Excellent optical appearance<br>Good machineability<br>Low shrink temperature | In coextruded films, one ply compensates for the deficiencies of the other, so they are superior films with no significant performance shortcomings; the wide variability in layer composition and number of layers makes performance analysis difficult |

The cost of these processes are fairly expensive and initially increases the cost per unit weight of the product. However, the yield increases considerably and the quality improves greatly. Thus with increased performance of stretched products, develops major cost advantages.

*Material handling*

In most processes, from small to large production runs, the cost of the plastics used compared to the total cost of production may be 30 to at least 60%, so it is important to handle plastic with care and to eliminate unnecessary production problems and waste. Any plastic handling system has the three basic subsystems of unloading, storage, and transfer. With small quantity users of commodity or expensive engineering plastics, containers such as bags and gaylords have to meet these three objectives.

For the large-scale operations, railcars can be unloaded in many different ways. Some suppliers simply recommend inexpensive, back-to-back, flexible-hose assemblies to unload them. To save time at the loading dock, one should consider approaches such as a stationary pipe manifold arrangement along the length of the rail siding with connections about every 4.5 m (25 ft). Thus transport is achieved with one short-flex hose from car to car instead of by handling as much as 30 m (100 ft) of heavy-flex hose. Labor and unloading time will be saved and the pressure drops inherent in long, multiple-connection, flex-hose runs will be avoided [258, 443].

Easy methods of unloading include the use of a vacuum pump/dust collector which can be located in the silo skirt. The pump induces a vacuum in the line, drawing plastic from the railcar into a vacuum loader. When the loader fills, the pump stops, and the plastic dumps into the silo. This on/off batching effect keeps transfer rates relatively fast, typically 2700–3200 kg/h (6000–7000 lb/h). If the manifold pickup is far from the silo, the transfer rate will drop.

If unloading requires high speeds or transfer over long distances, consider a push/pull system: plastic is pulled from the railcar by negative pressure, then pushed by positive pressure to its final location. Some plants use a one-blower system, but this limits transfer capacities, requires more work-hours, and can actually degrade plastic through excessive heat transfer within the air stream. One-blower systems can also lead to line blockages, as they lack purging capabilities.

Another way to unload railcars and transfer plastic to silos is to use a two-blower system. One blower handles car unloading through negative pressure, the other transfers plastic to silos using positive pressure. Splitting up the unloading and transfer plastic between two blowers permits peak efficiency in each operation. This system times purging to virtually eliminate any line blockages. With this arrangement, the pressure blower continues operating to ensure complete transfers in silos, while the vacuum remains out of the transfer loop.

Silos and other containers provide more than just a place to stow away plastics until it is ready for use; silos can protect plastics from environmental damage caused by excess moisture, atmospheric pollutants, and solar radiation. Moreover, they should require a minimum amount of maintenance; they should not leak, nor should they introduce rust or other contaminants.

As with other process equipment, suppliers of dryers, blenders, and metering equipment are increasing the feeding accuracy of the equipment by the incorporation of microprocessor-based controllers that can be easily interlinked with the computer facilities of the plant. To mix components easily and economically, processors use blenders mounted on machine hoppers. In this operation, which offers great flexibility, precisely predetermined proportions of each ingredient flow into a mixing chamber, which will discharge a very accurate mix. Feeders meter by volume or weight.

In volumetric blending, variable speed metering augers feed multiple components into the mixing chamber. A microprocessor provides accurate, repeatable programming and closed-loop control over the variable speed of each metering auger.

Gravimetric blending improves accuracy and requires less operator involvement in calibration, particularly in running processes where great accuracy is needed. Metering by weight eliminates overfeeding of expensive additives. The principle of this type of feeding with metered weight control is well established; to an accuracy of at least 0.25–0.50% for ingredient and blend ratio setting points at 2 (two standard deviations). Gravimetric metering is advantageous in coextrusion.

*Drier*

Numerous types of equipment are used to remove moisture or water from plastics before they are processed. This section reviews equipment; see Chapter 3 regarding the importance and understanding of drying conventional plastics and hygroscopic plastics in order to obtain maximum performance and/or no degradation. There is a great variety of choices available for drying, such as belt/conveyor, centrifugal, convection, flash, fluid-bed, freeze, rotary drum/tray, vacuum, screw, etc. systems, each with their different capabilities and costs. (Vented barrel drying is reviewed in Chapter 4.) The important parameters to consider are the amount and flow rate of plastic that has to be handled and amount of heat required. There is a relationship between the heated air temperature flowing and its volume to the degree of evaporation.

The residence time for drying to the desired residual moisture content varies with the material and on the initial moisture content. It can range from 2–3 h for PC up to 5–6 h for PA. A guide to compare different dryers is to use the following concepts: (1) a drier with the highest air-density rating is best but a small difference between the rating will not be decisive; (2) a drier with the higher impinging air velocity will be better if the air speed at the beginning can be reduced without affecting the supply volume; (3) a drier with the longest dwell time is better; and (4) an important factor is the volume of air making contact with the plastic. The greater the air volume, the greater the evaporation rate.

### *Granulating/recycling*

Scrap and product rejects in the production of different products can be granulated and recycled. If possible, the target is to eliminate or at least reduce 'scrap' because it has already cost money with time to go through the process. Different types of granulators are available from many different sources. Selection depends on factors such as the type of plastic used, product thickness, shape, etc. With heat-sensitive plastics (or even others), the granulator has to be of the type not to cause overheating during the cutting actions. Thick plastics may require a series of different granulators so that incremental reduction occurs, eliminating over heating.

An easy 'cutting' action is required to granulate with minimum frictional heat. General-purpose types have definite limitations. Some mixes of recycled with virgin TPs may have properties in products with very little loss and may only require a slight change in the processing setup.

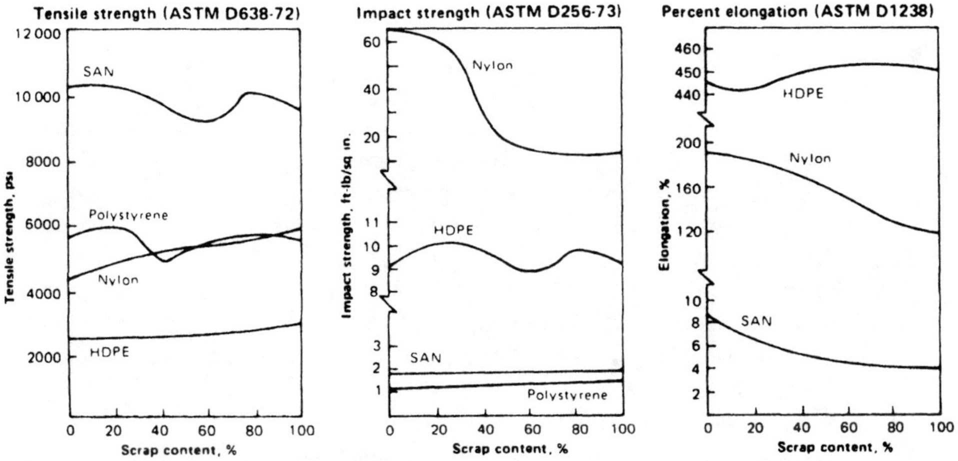


**Figure 2.24** Shows how regrind levels affect mechanical properties of certain plastics once through the machine and blended with the same type virgin plastics.

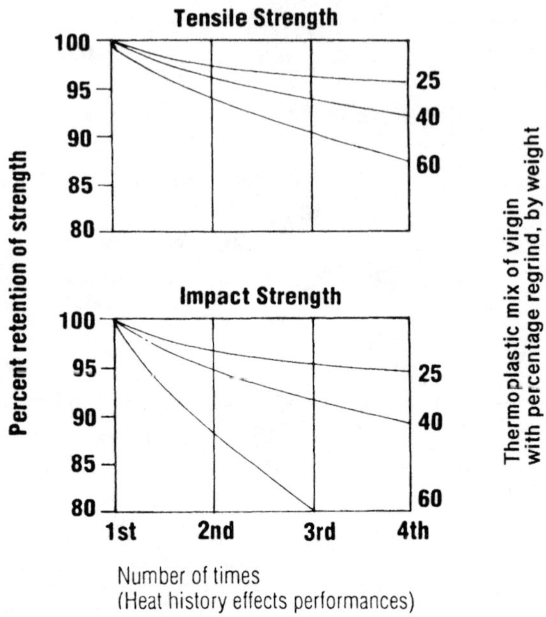


**Figure 2.25** Examples of potential effects on the performance of regrind mixed with virgin plastics.

However, others could have significant changes in performance (Figs. 2.24 and 2.25) and processing conditions because recycled plastic is denser and usually has a variable size.

*Secondary operation*

After extruding in the basic functional processing machines, secondary operations may be required to produce the final product. These operations can occur in-line or off-line. They include any one or a combination of the following: annealing (to relieve or remove residual stresses and strains), postcuring (to improve performance), joining and assembling (adhesive, ultrasonic welding, vibration welding, heat welding, etc.), and decorating/printing. The type of operation to be used depends on the type of plastic used. As an example with decorating or bonding certain plastics can be easily handled while others require special surface treatments to produce acceptable products [1–4].

## OPERATING EXTRUDER AND COMPLETE LINE

An overview of what happens in an extruder can be summarized as follows:

1. Screws for single-screw extruders are usually right-hand thread, and turn counter-clockwise when looking from the rear or power drive mechanism. Their rotation appears as if to unscrew itself backward out of the material. The machine's thrust bearing prevents this backward action. Twin-screws may turn in either direction, or both, but the principle is the same.
2. Most of the heat to melt the plastics comes from the motor drive system, not the heaters. The exceptions are very small extruders, slow-moving twin-screws, processing high temperature plastics, and when using a coating line.
3. Motor speeds are generally reduced from 10:1 to 30:1 to obtain reasonable screw speeds, which are lowest for very small machines because they need residence time and very large machines that have to avoid excessive shear rates as the screw diameter increases.
4. The major cooling action in the extruder is due to the plastic material entering it. The plastic absorbs the heat from the motor and heaters so that it can melt. When the feed is stopped, the systems equilibrates. The result with this stop is usually that the front end transfers heat to the rear feed end of the screw. In turn, there is danger of material sintering, bridging, and/or sticking to the screw root.
5. Pellets must stick to the barrel and slip on the screw root in its feed zone, and must stick to each other as much as possible. This action permits maximum conveying through that zone. More of this action is not always better, though, as some screws can 'bite off' more than their front ends can 'chew.'
6. Material is by far the biggest component of the manufacturing cost (30–60%). Therefore, almost anything that saves material can be justified; material management is an important criteria to follow even at the expense of equipment development.
7. Energy is a very small proportion of manufacturing cost, as an extruder is basically an energy-efficient machine when compared to other processes, such as injection molding, etc. Moreover, excess energy would overheat the material and make it unextrudable. However, it appears that there is always room to reduce energy losses in the extruder and complete lines.
8. Pressure at the screw tip is important, as it relates to thrust bearing wear, mixing efficiency, personal safety, screen contamination, melt temperature, and so on. This pressure is the cumulative demand of the head, from the screens to the die lips, and is not something generated independently by the extruder.
9. Production rate is the drag flow displacement of the last screw flights of the extruder, less the effect of resistance (pressure demand of the head), plus the effect of overbite at the feed end. Leakage over the

screw flights may also have an effect, sometimes in a positive direction if there is a pressure peak along the barrel.

10. Effective shear rates are around 100–500 rev/s in single screw extruders, as well as in most die lips, compared to values of from 1–10 in a typical ASTM melt index (MI) test. Proper comparison of materials demand at least two viscosities (not a ratio) with measurement or extrapolation into the practical 100–500 range [187, A. L. Griff].
11. In the numerical simulation of an extrusion process, it is essential to use the physical data that is as accurate as possible before one can evaluate the usefulness of the program.

## Planning a line

A processing line is a complete fabricating operation that goes from upstream material storage and handling equipment, through the extruder, and down-stream to produce and transport the finished product. Within the line are many factors to ensure fabricating products to meet requirements at the lowest cost using the FALLO approach (Fig. 1.1). The line includes inspection, quality control, packaging, and delivery to a destination such as a warehouse bins or transportation vehicles.

Many different types of auxiliary equipment (AE) and secondary operations (SO) are utilized with extruders to maximize overall productivity, efficiency, and reduce fabricating cost. Their proper selection, use, and maintenance are as important as the selection of the extruder. The complete up-stream and down-stream auxiliary equipment could cost much more than the extruder.

The processor must determine what is needed based on what the equipment has to accomplish going from material storage/handling to handling the finished product, what controls are required, the ease of operations and maintenance, safety devices, energy requirements, and desired compatibility with existing equipment. Equipment is available from many sources that provide many different functions to permit extruding products that meet performance requirements at the lowest cost. The highest- or lowest-cost equipment does not necessarily provide the best performance. As an example, chillers for barrel heat control or blow molding mold temperature control in many applications are not properly engineered to meeting their requirements. The result is generally a higher energy cost that could eliminate any cost reduction occurring based on original cost investment.

With knowledge of what is available and using common sense provides a means of planning an extrusion line. If the new line is intended to increase production for an existing product, you will be familiar with the requirements that the line must meet. The following example concerns setting up a new product line [117]. Different factors are to be considered, beginning with the targeted output rate. Next is to determine information that can be highlighted that include the following important factors: (1) desired minimum output; (2) maximum linear speed; (3) direction of line; (4) space (width, height, length) required for the line to operate and maintain; (6) centerline height; (7) lighting; and (8) floor drains and piping requirements. These factors as well as others definitely require at least some familiarity of extruders and auxiliary equipment and applying common sense with logic.

Extruders have different output ratings for different plastics. It is important, for example, to make certain the take-away equipment can handle the full extruder output. As an example, a change was made on a line producing 250–750 cm (100–300 ft) coils of PE pipe. It was decided to change from a 120 to 90 mm (4.5 to 3.5 in) extruder because the take-away equipment could not handle the speed of the pipe. A footage counter activated the pipe cutter. Because the linear speed of the pipe would have been too fast for the operator to attach the cut pipe to the rotating coiler (winder), the 90 mm (3.5 in) machine was a practical decision. With the change, in order to meet the new cooling requirements, ethylene glycol mix was used for cooling in place of just water.

It is important to consider the direction of extrusion in respect to the plant layout. Unless outdoor plastic silos are used, ample space must be provided close to the extruder, preferably to the rear of the machine. The finished product is usually moved to a place in the storage area.

Overall width of the line is important. A definite requirement is to provide ample walking space on each side or around the line. Space is required for setting controls that exist along the line as well as the space required by maintenance people with their hardware and instrumentation. Recognize that certain equipment may have to be moved in case of repair or replacement and there must be sufficient space to move it away from the line. You may need space for a forklift truck in moving all types of equipment in and out of the line. An overhead crane or its equivalent may be required. Height should be sufficient to accommodate the hopper and any other device needed above the hopper. You may have to put a hole in the ceiling which, in turn, may require you to provide some type of heat/cold insulation so the material handling will operate at maximum efficiency.

The length of the planned line may have to include accommodating room as a backup for unanticipated auxiliary equipment. An example is a line extruding vinyl house siding with holes drilled in the siding for future assembly. The drilling operation was initially performed away from the line. Fortunately when the company later acquired a special machine to perform the drilling operation in line, space was available for this improvement that reduce cost for the complete line operation.

Another example is in a plant where two lines for about a year produced MDPE pipe. After the lines were setup and operating, a major material supplier developed and made commercially available a new plastic. It was a HMWHDPE that was reported to be superior to any other plastic on the market for pipe extrusion. The pipe plant carefully evaluated the new material and determined that the claims were true except it had one drawback. It was impossible to extrude using their existing line. With a working pressure of 40 MPa (6000 psi), on a single die head, it was very difficult to see the melt move in the extruder.

To make this a practical operation using the better performance-to-cost material, six dies were positioned around the extruder head. Each die had a gear pump. Because the linear speed was so slow, specially designed takeoff coilers were designed and manufactured. After many weeks of debugging, this new setup became practical and operational. As the plant personnel working on this project explained, by their experience and good fortune (and the usual luck), there was sufficient space to handle this special line operation. What helped was the elimination of the water cooling troughs. Cooling was accomplished by water cooling the die cores.

Analyze what is required to move the finished product from the end of the line quickly to a destination that could be the warehouse or the shipping dock. The target is to avoid cluttering the end of the line which could result in damaged products, develop poor house keeping, and invite accidents.

The centerline height for the line can set up complications, particularly when a product line requires additional operations in the line; the relevant height is from the centerline of the extruder to the ground. As an example, the typical height for a 60 mm (2.4 in) extruder is 100 cm (42 in). Most line equipment can be adjusted to this height. However, when selecting a new machine for special products, it is best to specify height that can accommodate unusual down-stream operations. For example, a company was coating steel wire through a crosshead die. Following this were three large and heavy wire braiding machines having a centerline height of 60 cm (24 in). To raise these machines to 107 cm (42 in) height was not practical. Rather than angle the coated wire down to a 60 cm (24 in) height, the extruder was designed to a 60 cm (24 in) centerline height.

When reviewing the subject of lighting along the extrusion line, it is important to recognize that people require proper lighting to do their work efficiently. This is particularly true where controls exist, inspectors operate, etc. Also consider the potential of additional operations that may be required.

There is a tendency to have water collect on the floor along cooling troughs. So floor drains and piping becomes an important consideration when setting up the line since the proper approach will eliminate or minimize personnel slipping and may prevent a serious accident if a worker contacts ungrounded or even grounded electrical equipment. On a new floor, screen covered trenches with drains should be considered to catch spilled water. Though there may be small amounts of water, it is advisable to provide a wood platform along the cooling tank area. Also evaluate the method to be used in piping water or coolant as well as its recirculation path from the tank so that problems do not develop when maintenance is required.

In addition to considering the above factors, other factors will be important depending on your specific requirements and perhaps the existing plant layout. Such factors include material storage and handling that would probably include drying and preheating, recycling area that may be isolated from the line, electrical power requirements, clean room requirements, safety measures, and so on.

**Start-up and shut-down**

Machine operation takes place in three stages. The first stage covers the running of a machine and its peripheral equipment. The next involves setting processing conditions to a prescribed number of parameters for a specific plastic, with a specific die in a specific processing line, to meet product performance requirements. Figure 2.26 relates to blown film settings. The final stage is devoted to problem solving and fine tuning of the complete line (Fig. 1.1) that leads to meeting performance requirements at the lowest cost. A successful operation requires close attention to many details, such as quality and flow of feed material(s), a heat profile adequate to melt but not degrade the plastic. Processors must also become familiar with a troubleshooting guide.

Prior to start-up one must check certain machine conditions.

1. Unless the same plastic is ready in the machine from a previous run, the entire machine should be cleaned and/or purged, including the hopper, barrel, breaker plate, die, and downstream equipment. If a plastic was left in the barrel for a while, with heat off, the processor must determine if the material is subject to shrink and could have caused moisture entrapment from the surrounding area, producing contamination that would require cleanup (this situation could also be a source of corrosion in/on the barrel/screw).
2. One must check heater bands and electrical connections, handling electrical connections very carefully.
3. Check thermocouples, pressure transducers, and their connections very carefully.
4. Be sure the flow path through the extruder is not blocked.
5. Have a bucket or drum, half filled with water, to catch extrudate

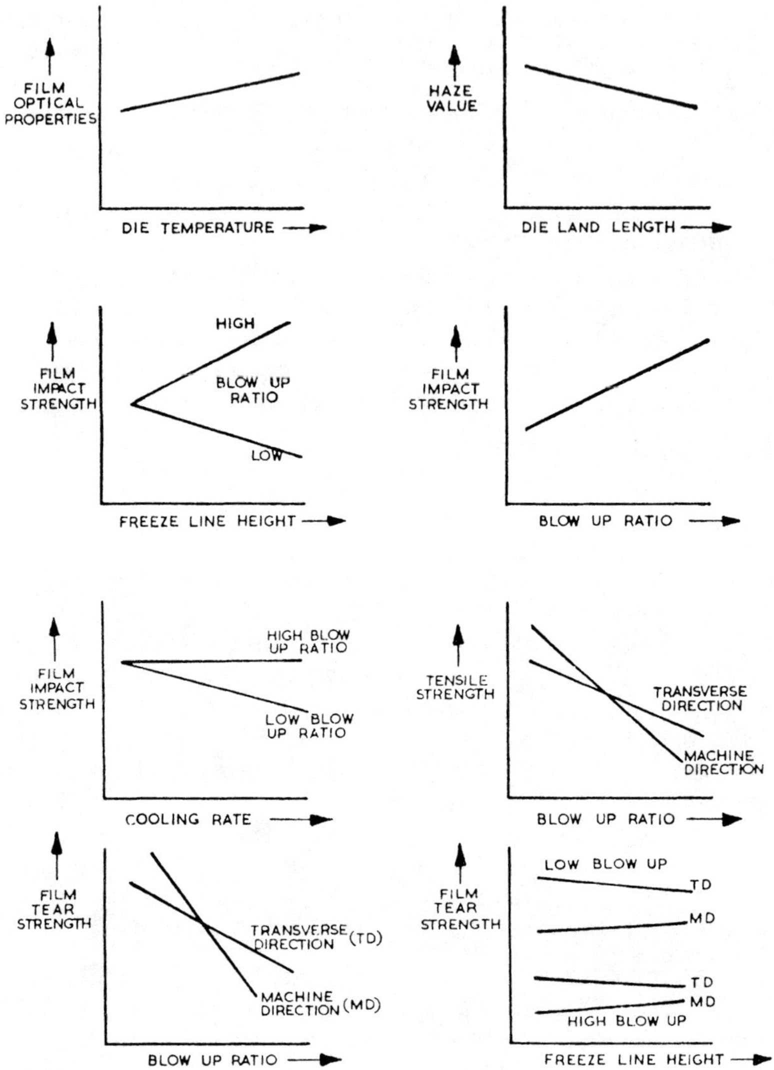


**Figure 2.26** Example of how blown film extruder settings can effect the properties of plastics.

whenever purging or initial processing of plastics that has contaminated gaseous by-products.

6. Inspect all machine ventilation systems to ensure adequate airflow.
7. Check operating manual of the machine for other startup checks and requirements that have to be met such as motor load (amperage) readings.

Details for operating the machine are based on what the plastic being processed requires, such as temperature settings, screw rpm, etc. available from the material supplier and/or experience. Start-up procedures involves certain precautions.

1. Starting with the front and rear zones (die end and feed section), one should set heat controllers slightly above the plastic melting point and turn on the heaters. Heat-up should be gradual from the ends to the center of the barrel to prevent pressure buildup from possible melt degradation.
2. Increase all heaters gradually, checking for deviations that might indicate burned-out or run-away heaters by slightly raising and lowering the controller set point to check if power goes on and off.
3. After the controllers show that all heaters are slightly above the melt point, adjust to the desired operating heats (based on experience and/or plastic manufacturer's recommendation), checking to ensure that any heat increase is gradual, particularly in the front/crosshead.
4. The time required to reach temperature equilibrium may be 30–120 min, depending on the size of the extruder. Overshooting is usually observed with on/off controllers.
5. Hot melts can behave many different ways, so no one should stand in front of the extruder during start-up, and one should never look into the feed hopper because of the potential for blow-back due to previous melt degrading, and so on.
6. After set heats have been reached, one puts the plastic in the hopper and starts the screw at a low speed such as 2–5 rpm; some plastics, such as nylon, may require 10–20 rpm.
7. The processor should observe the amperage required to turn the screw, stop the screw if the amperage is too high, and wait a few minutes before restart.
8. In working with a melt requiring high pressure, the extruder barrel pressure should not exceed 7 MPa (1000 psi) during the start-up period.
9. One should let the machine run for a few minutes, and purge until a good quality extrudate is obtained visually (experience teaches what it should look like; a certain size and amount of bubbles or fumes may be optimum for a particular melt, based on one's experience after setting up all controls). If plastic was left in the extruder, a longer purging time may be required to remove any slightly degraded plastic.
10. For uniform output, all of the plastic needs to be melted before it enters the screws metering zone (Chapter 4), which needs to be run

full. When time permits, after running for a while, the processor should stop the machine, let it start cooling, and remove the screw to evaluate how the plastic performed from the start of feeding to the end of metering. Thus one can see if the melt is progressive and can relate it to screw and product performances.

11. Turn up the screw to the required rpm, checking to see that maximum pressure and amperage are not exceeded.
12. Adjust the die with the controls it contains, if required, at the desired running speed. Once the extruder is running at maximum performance, set up controls for takeoff/downstream equipment, which may require more precision settings, if required.
13. Extrudate can start its tract from the die by 'threading' through the downstream equipment to its haul-off.
14. One may get into a balancing act of interrelating extruder and downstream equipment. Extruder screw speeds and haul-off rates may then be increased. Downstream equipment is adjusted to meet their maximum operating performance, such as having the vacuum tank water operate with its proper level and vacuum applied.
15. The extruder can be 'fine-tuned' to obtain the final required settings for meeting the desired output rate and product size.

Shut-down procedures vary slightly, depending on whether or not the machine is to be cleaned out or just stopped briefly. If the same type of plastic is to be run again, cleanout is generally not required. The goal is to avoid degradation by reducing exposure of the plastic to high heat. If cleanout is required, because a different type plastic is to be processed, it is necessary to disassemble the machine at a heat high enough to allow cleanout before material solidifies.

A procedure for shut-down without cleanout is as follows.

1. Empty the hopper as well as possible.
2. If coating take the action required to remove the substrate such as with wire coating remove the wire from the crosshead die; in coating paper, film, etc., the extruder is usually on a track and can be withdrawn from the substrate; etc.
3. Reduce all heat settings to the melt heat.
4. Reduce the screw speed to 2–5 rpm, purging the plastic, if required, into a water bucket or drum prior to reducing the melt heat.
5. When the screw appears to be empty, stop the screw and shut off the heaters and the main power switch (however, steps 3 and 4 must be completed before the melt drops significantly or a premature shut-down will occur with plastic remaining in the barrel).
6. If a screen pack with breaker plate is used, disconnect the crosshead (or die) from the extruder and remove the breaker plate and screen. If necessary, appropriate action is taken to clean them.

For clean-out of the extruder at shut-down, the first three steps are the same steps 1, 2, and 4 in the preceding paragraph. Then the procedure is as follows.

4. When the extruder appears to be empty, stop the screw.
5. Shut off and disconnect the crosshead or die) heaters. Reduce other heaters to about 170–330°C (400–625°F), depending on the plastic's temperature at the melt point.
6. Disassemble the crosshead and clean it while still hot. Remove the die, and gear pump if used, and remove as much plastic as possible by scraping with a copper spatula or brushing with a copper wire brush. Remove all heaters, thermocouples, pressure transducers, and so on. Consider using an exhaust duct system above the disassembly and cleaning area, even if the plastic is not a contaminating type; this procedure keeps the area clean and safe.
7. Push the screw out gradually while cleaning with a copper wire brush and copper wool. Care should be exercised if a torch is used to burn and remove plastic; tempered steel, such as Hastelloy, may be altered and the screw distorted or weakened as well as subjected to excessive wear, corrosion, or even failure (broken).
8. After screw removal, continue the cleaning, if necessary.
9. Turn off heaters and the main power switch.

Final cleaning of parts, particularly disassembled parts, is best done manually, or much better, in ventilated burnout ovens, if available, operating at about 540°C (1000°F) for about 90 min. For certain parts with certain plastics, the useful life could be shortened by corrosion; check with the part manufacture. After burnout, remove any grit that is present with a soft, clean cloth. If water is used, air-blast to dry. With precision machined parts, water cleaning could be damaging because the potential of corrosion when certain metals are used.

## Temperature/pressure and output

With a change in barrel temperature, one is able to predict the change in output. The following information pertains to the extruder even though it relates more to the die. It is necessary to study the interaction of the die and extruder to predict the overall change in output from a temperature increase [349]. The relationship between output and temperature change is: $Q = 100 \times (b/n) \times T$ where $Q$ is output in percent, $T$ is the temperature change in °C (or °F), $b$ is the temperature coefficient in $°C^{-1}$ (or $°F^{-1}$), and $n$ is the power law index of the plastic melt (Table 2.14). This power law index describes how viscosity of a melt changes with shear rate. The temperature coefficient $b$ describes how the viscosity of a melt changes with temperature.

**Table 2.14** Power law index for different plastics

| *Plastic* | *Power law index*, n | *Temperature coefficient*, b ($°C^{-1}$) |
|---|---|---|
| PS | 0.30 | 0.08 |
| PVC | 0.30 | 0.20 |
| PMMA | 0.25 | 0.20 |
| SAN | 0.30 | 0.20 |
| ABS | 0.25 | 0.20 |
| PC | 0.70 | 0.05 |
| LDPE | 0.35 | 0.03 |
| LLDPE | 0.60 | 0.02 |
| HDPE | 0.50 | 0.02 |
| PP | 0.35 | 0.02 |
| PA-6 | 0.70 | 0.02 |
| PA-6.6 | 0.75 | 0.03 |
| PET | 0.60 | 0.03 |
| PBT | 0.60 | 0.03 |
| PVF2 | 0.38 | 0.03 |
| FEP | 0.60 | 0.04 |

As an example, for a melt with a $b$ of 0.02°$C^{-1}$ and an $n$ of 0.50, a 1°C temperature change will cause a 4% change in output. For a PVC or PMMA with a high $b$ of 20°C, the change will be more dramatic causing an 80% change in output. Amorphous plastics tend to have high values of $b$. Thus, it is obvious that good temperature control is required to avoid output fluctuations particularly when processing amorphous plastics [187, C. Rauwendaal].

The relationship between temperature and pressure change is $Q = P/n$ where output $Q$ and pressure $P$ are in percent, and $n$ is the power loss index of the plastic melt (Table 2.14). As an example, for PP a 2% variation in pressure causes a 5.7% variation in output. For a shear thinning plastic melt, the output change will always be larger than the pressure change. Understanding this equation helps one to realize the importance of maintaining a uniform head pressure, if constant flow rate is important [187, C. Rauwendaal].

*Heat profile*

A typical heat profile that is adequate at low screw speed may be inadequate when speeds are increased, because the greater rate of melt flow

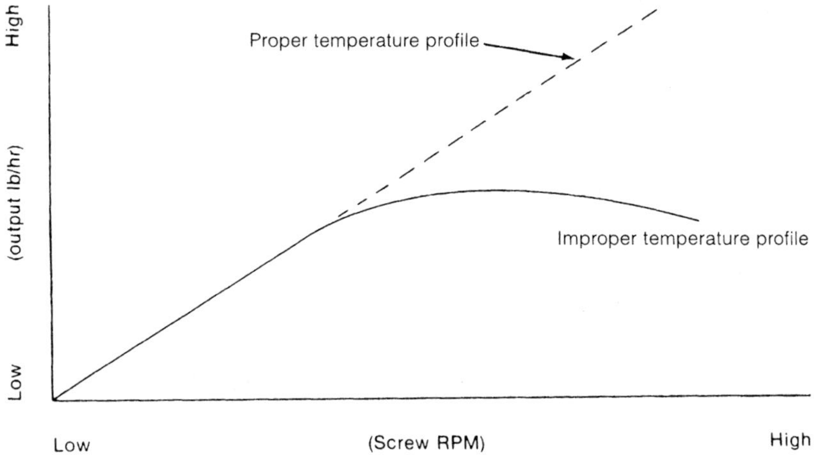


**Figure 2.27** Plastics entering the barrel throat at a higher temperature will increase extruder output with an increase in rpm.

through the extruder. The output may not increase at a linear rate with increasing speed unless the heat profile increases (Fig. 2.27). With a proper heat profile, a linear relationship between speed and output can be maintained through a wider operating range. However, too high a heat at low speed is likely to cause melting all the way back to the barrel throat, causing plastic bridging and degradation.

The intent of barrel feed zone temperatures is to find those conditions where the plastic sticks to the barrel's ID surface so advancing flite can scrape it off and forward it downstream. If the zone is too cold, the plastic will not stick and simply turn around with the screw starving the melt conveying zone. When too hot a zone exists, the plastic can melt prematurely, creating an annular melt film which can act as a lubricant also causing the screw to be starved. Different approaches can optimizing zone settings such as having the rear zone at 30°C (50°F) below the plastic melt point leaving the other zones at their normal settings [187, T. Black]. Figure 2.28 shows how preheating plastic in a hopper can aid processing.

*Water/air cooling*

There are advantages and potential disadvantages of barrel water cooling. The advantages can be summarized as: (1) it provides maximum cooling available and temperature difference available for driving force (most barrels will allow cold water use, when needed); (2) it does not dump heat removed into extruder environment and add to room heat load; (3) it can

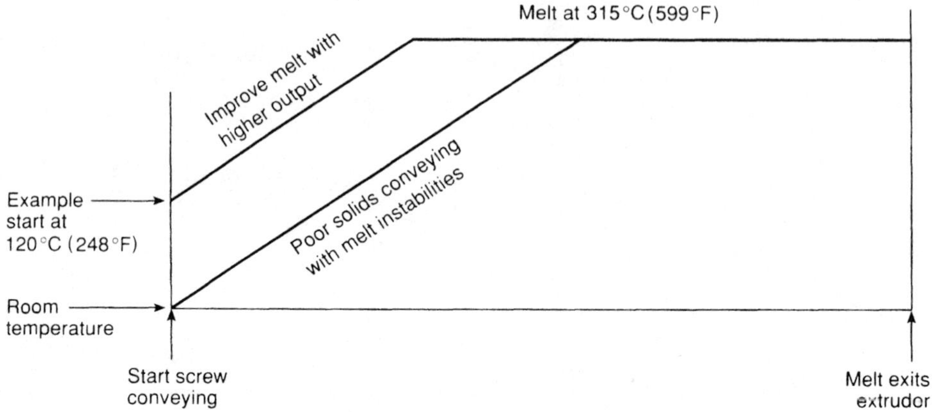


**Figure 2.28** Improving plastic and/or machine performance by preheating plastics entering the extruder throat.

be timed pulsed for control; and (4) it can latter be modified for greater cooling via reduced fluid temperature. Disadvantages are: (1) it requires a recirculation water system or zero solids once through water; (2) it can be difficult to control due to the high heat flux available; and (3) it can cause safety problems, as a plug-up can result in high-pressure steam generation and unexpected release.

All disadvantages can be overcome by installing a well-designed system which incorporates the following features. The recirculating water cooling media can be made up from plant steam condensate or zero solids water can be purchased. The feed lines to each cooling zone should have a calibrated water flow limiting valve and a high-pressure steam rupture disk properly vented. The best control system provides time pulsing via a proper controller and a solenoid valve in the water cooling line to each zone. Routine maintenance should be directed at chemical cleaning the barrel water flow passages, as slow buildup of water solids will occur if steam generation is allowed by the water flow [187, F. Hoffman].

Controlling cooling is important and the methods available provide different functions. Ideally, it would be better to cool using true proportioning. Practicality almost dictates that cooling be on–off because if true proportioning were used, cooling would have to be applied uniformly around the barrel. This is not the usual case. When heating, heater bands do apply heat uniformly.

However, on–off heaters do not provide for smooth barrel heating. Too much heat will result in more difficulty in handling the melt. Proportioning will smooth out the barrel heating. Air cooling is gentle by not 'shocking' the barrel. This keeps the plastic viscosity constant, resulting in easier handling and cooling of the extrudate. Thus, current proportioning heaters work well and do not impart excessive heating, resulting in a fairly uniform plastic viscosity. On–off air cooling is good; it is gentle and does not shock the barrel as it occurs with water. So before you blame the heating and cooling, take a close look at the screw and/or die design [187, D. Biklin].

## Maintenance

Procedures to be used in maintaining extruders are reviewed throughout this book. A few comments are provided here, such as after changing the configuration of the screws of a twin-screw machine, before reinstalling, roll them flat on a flat surface (together). This procedure will ensure that configuration and element orientation are identical on both shafts [104].

When starting up a new extrusion setup without plastic in the barrel, start the screw rotation at a slow speed such as about 5 rpm. Gradually look into the air gap between the feed throat and throat housing and make sure the screw is turning. Screws have been installed without having their key in place, or the key has fallen out during installation. Also make sure that anti-seize material is applied to the drive hub, to help installation and removal. Also if the key is left out and the drive quill is turning and the screw is not, the screw will not gall to the drive quill.

On shut-down of a blown or cast film line, put a 1% antioxidant concentrate into the material system just prior to pulling the switches. This is particularly effective in protecting coextrusion systems from cross linked gels caused by heat aging during final cooling and heating during startup.

As a general statement, it can be stated that periodic checkups and regular maintenance should become a habit (at least using your maintenance manuals from the equipment supplier). Include factors such as: (1) watch out for plastic, water, and oil leaks; (2) change oil in gear box and transmission (if oil-driven) at least every six months if operating on one shift or three to four months if operating on more shifts; (3) lubricate required parts; (4) check band heaters (for tightness), thermocouples, pressure transducers, etc.; (5) check control circuits (electrical, hydraulic, mechanical, etc.); (6) check die opening with a device such as a feeler gauge; (7) routinely check the line for alignment, level, and parallelism, particularly windup equipment; (8) keep all rolls clean and rust-free; (9) check safety devices on all equipment; (10) keep shop clean; (11) keep spare parts in stock based on experience and/or suppliers; (12) set up sessions for personnel to 'repeat' instructions on safety equipment procedures and prevent hazards harmful to personnel, equipment, and product; and (13) others. Know the signs of trouble, such as unusual noise, before they occur.

It is important to set up preventive maintenance procedures on all equipment. Equipment is built to operate if properly maintenance is used. Processors should make it a habit to perform regular checkups and maintenance work based on equipment suppliers' recommendations and one's practical experience.

**Troubleshooting**

Troubleshooting guides are reviewed throughout this book. This section provides additional information that includes some common operating problems and possible solutions. Tables 2.15–2.17 describe common problems and solutions. With equipment, materials, and products, troubleshooting guides are set up (usually required) to take fast, corrective action when products do not meet performance requirements. This problem-solving approach fits into the overall fabricating–design interface as summarized in the FALLO approach (Fig. 1.1).

Two important requirements for efficient troubleshooting are good instrumentation and good understanding of the extrusion process. Instrumentation is important in process control but is absolutely essential in troubleshooting. Without good instrumentation, it tends to be a useless approach at best, no matter how well one understands the process. Lack of instrumentation can prove very costly if it causes a certain problem to remain unsolved for even a short time.

A simplified approach to troubleshooting is to develop a checklist that includes the basic rules of problem solving such as: (1) have a plan and keep updating it based on experience gained; (2) watch the processing conditions; (3) change only one condition at a time; (4) allow sufficient time for each change to occur; (5) check housekeeping, storage granulators, etc.; and (6) narrow the range in which a problem belongs such as material, extruder, die, etc. To aid processors remote troubleshooting has become available from various equipment suppliers (and much more of this will be occurring). Via telephone links or PC modems (with even wireless in the future) from your controller to a central service computer, a specialist at the supplier's service computer can immediately check out the equipment and have corrections made directly or provide the diagnostic information to the processor.

When setting up troubleshooting guides, it is important to recognize that one common cause of fluctuations in extruder conditions and product properties are caused from variations in properties of plastic being processed. Causes include changes in melt viscosity and elasticity due to changes in molecular weight (MW), MW distribution, heat history, etc., (Chapter 3). To minimize this problem material suppliers may be able to 'minimize' variations. In-house monitoring equipment (melt viscosity, swell ratio, shear rate, etc.) can aid the processor in relating material

**Table 2.15** Troubleshooting guide: common extrusion problems and how to solve them

| *Problem* | *Causes* | *Solution* |
|---|---|---|
| | **General considerations** | |
| Surging | Resin bridging in hopper | Eliminate bridging |
| | Incorrect melt temperature | Correct melt temperature |
| | Improper screw design | Check design |
| | Rear barrel temperature too low or too high | Increase or decrease rear temperature |
| | Low back pressure | Increase screen pack |
| | Improper metering length | Use proper screw design |
| Gels (Contaminants that look like small specks or bubbles) | Melt temperature too high | Lower melt temperature |
| | Not enough progression in screw | Use new screw |
| | Bad resin | Check resin quality |
| | Melt temperature too high | Check melt temperature |
| Melt fracture (Rough surface finish. Also called 'sharkskin') | Melt temperature too low | Increase melt temperature |
| | Die gaps too narrow | Heat die lips |
| | | Increase die gaps |
| | | Use processing aids |
| Bad color | Color concentrate incompatible with resin | Ensure melt index of concentrate base material is close to melt index of resin |
| Bubbles | Wet material | Dry thoroughly |
| | Overheating | Decrease temperature; check thermocouples |
| | Shallow metering section | Use proper compression-ratio screw |
| Overheating | Improper screw design | Use lower-compression screw |
| | Restriction to flow | Check die for restrictions |
| | Barrel temperature too low | Increase temperature |
| Die lines | Scratched die | Refinish die surface |
| | Contamination | Clean head and die |
| | Cold polymer | Check for dead spots in head; adjust barrel and head temperature to prevent freezing |
| Flow lines | Overheated material | Decrease temperature |
| | Poor mixing | Use correct screw design |
| | Contamination | Clean system |
| | Improper temperature profile | Adjust profile |

**Table 2.15** *Continued*

| *Problem* | *Causes* | *Solution* |
|---|---|---|
| | **Blown film** | |
| Wrinkles | Dirty collapsing frame | Clean frame |
| | Too much web tension | Adjust tension |
| | Improperly designed air ring | Use new air ring |
| | Gauge variations | See gauge variations |
| | Insufficient cooling | Use refrigerated air<br>Increase flow of air<br>Reduce output |
| | Misalignment between nip rolls and die | Check alignment |
| Folds, creases | Excessive stretching between nip and roller | Reduce winding speed |
| | Nip assembly drive not constant | Adjust or replace drive |
| Blocking | Inadequate cooling | Use better cooling method |
| | Excessive winding tension | Adjust tension |
| | Excessive pressure on nip rolls | Adjust pressure |
| | Bad resin | Check resin |
| Port lines | Melt temperature too low | Increase melt temperature |
| | Die too cold or too hot in relation to melt temperature | Adjust die temperature |
| Splitting | Excessive orientation in machine direction | Increase blowup ratio |
| | Die lines | See die lines |
| | Degraded resin | Reduce melt temperature |
| | Poor resin choice | Ensure resin is suitable |
| Die lines | Nick on die lip | Change die |
| | Dirty die | Clean die |
| | Inadequate purging | Increase purging time between resin changes |
| Gauge variations (machine direction) | Surging | Check temperature<br>Check hopper for bridging |
| | Inconsistent take-up speed | Check take-up speeds |
| Gauge variations (transverse direction) | Nonuniform die gap | Adjust gap<br>Center air ring on gap |

**Table 2.15** *Continued*

| *Problem* | *Causes* | *Solution* |
|---|---|---|
| Printing problems | Insufficient treatment<br>Additives interfering with ink | Use properly treated film<br>Use resins with no interfering additives<br>Erratic treatment<br>Reduce slip levels to about 600 ppm for water-based inks |
| | **Sheet**[a] | |
| Poor gauge uniformity | Melt flow is not stable | Use gear pump to stabilize flow |
| Viscosity not stable | Poor mixing | Use static mixer |
| Streaks | Contaminated system | Clean hopper<br>Check screw and die; clean if necessary |
| Total discoloration | Excessive regrind | Check amount of regrind used |
| Discontinuous lines | Too much moisture | Increase resin drying<br>Use hot regrind |
| | **Pipe and tubing**[b] | |
| Poor output | Improper die or screw design | Ensure die and screw are designed for desired output |
| ID blisters | Insufficient vacuum<br>Excessive moisture<br>Gases entrapped<br>Water inside pipe | Increase vacuum<br>Maintain normal percentage of moisture in compound<br>Reduce temperature<br>Stop water access |
| ID burn streaks | Mandrel heat too high<br>Stock temperature too high | Check mandrel heat<br>Reduce temperature slowly |
| ID grooves | Mandrel is coated with material | Clean mandrel |
| ID wavy surface | Screw clearance set improperly<br>Puller drive slipping | Adjust clearance<br>Adjust or replace puller drive |
| OD burn streaks | Material hung up on die<br>Temperatures too high | Clean die<br>Reduce temperatures slowly |
| OD uneven circumference | Too much air pressure on puller<br>Insufficient air pressure | Reduce air pressure<br>Check air pressure and all connections |

**Table 2.15** *Continued*

| *Problem* | *Causes* | *Solution* |
|---|---|---|
| OD discolored | Stabilizer level too low | Check stabilizer level |
| OD pock marks | Air bubbles adhering to pipe in flotation tank | Install wiper in tank |
| | Improper adjustment of spray rings that surround water tank | Readjust spray rings |
| OD oversized | Air supply too high | Adjust air supply |
| | Insufficient water supply | Increase water supply |
| | Pipe hot when measured | Allow pipe to cool before measuring |
| Wall too thick | Misadjusted die bushing | Adjust die bushing to achieve uniform thickness |
| | Wrong die setup | Use correct setup |

[a] Most of the problems covered under blown film are also relevant to sheet extrusion.
[b] These data pertain to extrusion lines using water-filled vacuum-sizing tanks.

variations to processing variations (Chapter 19). There can also be incorrect material blending in storage and/or feeding, variation in size, shape, or surface friction of solid raw materials particularly with recycled material, feed rate insufficient (hopper and/or screw), and so forth.

If a barrel zone on the barrel constantly overrides or requires too much cooling to maintain a set point, it may be that the melting is being concentrated in that section. This can either exist because of screw design or an improper barrel heat profile. A simple and hopeful solution is to increase the melting prior to the 'hot zone.'

For those unfamiliar with it, the barrel/screw combination is a very complex heat transfer system. To understand something that seems as simple as a zone override can require a complete analysis of the system. Just a few of the factors that can cause a zone override are screw design, barrel mass, thermocouple placement, cooling jacket fit, barrel/screw wear, head pressure, overall temperature profile, defective temperature controllers, and inadequate cooling. Before assuming that zone override is strictly a screw design problem, you should analyze the system as a complete heat transfer mechanism. Although the screw is responsible for most of the heat transfer, it cannot control the heat distribution.

A pyrometer with a surface contact probe is essential to troubleshooting. The contact probe can be used to check for heater burnout, heat flow,

**Table 2.16** Troubleshooting guide for screen changers

| *Problem* | *Solution* |
|---|---|
| **Seal leakage** | |
| Head pressure above maximum set pressure | Reduce head pressure |
| Excessive overhung load | Support downstream die adapter, etc. |
| Seals improperly installed | Remove, clean, and reinstall |
| Excessively scored seals or slide plate | Correct cause of damage (see below) or replace |
| Disk spring relaxed | Replace |
| Initial pressure setting of upstream body too low | See procedure for setup in manual |
| **Excessively scored seals** | |
| Maximum set pressure rating too high | See procedure for setting U/S body |
| Wire strands protruding above face of slide plate | Ensure correct screen size and installation |
| Screens tearing out or folding on shift | Utilize positive screen retention |
| Tramp metal caught by screens | Utilize hopper magnet or other metal-catching device |
| **No slide-plate movement** | |
| Guards not closed | Close guards |
| Pressure ready light not on (w/opt. solenoid valve) | Check hydraulic power unit |
| Limit switches or interlock wiring faulty | Repair or replace |
| Direction control valve not operating | Repair or replace |
| **Erratic or slow slide-plate movement** | |
| Air in system | Purge air |
| Low precharge in accumulator | Recharge accumulator |
| Oil viscosity too high | Replace with oil of correct viscosity |
| Pressure drop across breaker plate too high | Reduce pressure drop |
| Polymer viscosity too high | Decrease viscosity |
| Initial pressure setting of upstream body too high | Procedure for setting U/S body |
| Leakage past the direction control valve or changer cylinder | Repair or replace |
| **No pressure from hydraulic power unit** | |
| Motor not running | Check disconnects, motor overloads, motor |
| Unload solenoid valve not closed | Replace |
| Pump/shaft damaged | Replace pump |
| **Low pressure, motor stops** | |
| Faulty pressure switch | Replace switch |

**Table 2.16** *Continued*

| *Problem* | *Solution* |
|---|---|
| **Low pressure, motor continues to run** | |
| Low fluid level | Fill reservoir |
| Suction strainer blocked | Clean or replace |
| Oil overheated | See below |
| Air entrained in fluid | Allow bubbles to subside; use antifoam |
| Leakage past direction control valve | Repair or replace |
| Leakage past screen-changer cylinder | Repair or replace |
| Pump damaged/worn | Replace pump |
| **Normal pressure, does not stop** | |
| Faulty pressure switch | Replace switch |
| **Rapid pump/motor cycling** | |
| Faulty pressure switch | Replace switch |
| Leakage past check valve | Replace pump |
| Pump damaged or worn | Replace pump |
| Leaking past direction control valve or screen-changer cylinder | Repair or replace |
| **Oil overheating** | |
| Rapid screen-changer cycling | Shift less frequently |
| Restricted air movement or high temperature around reservoir | Assure free movement of air around hydraulic unit |
| Fluid bypass through system causing excessive pump-up time or rapid pump cycling | See above |

and temperature balance. It can check melt thermocouple accuracy and also the actual melt temperature as it exits the die. Die exit temperature is often much higher than the melt probe temperature which is usually indicated near the screw discharge and is influenced by adapter metal temperature. No plant should be without a pyrometer in working condition.

Trying to measure the melt temperature of an extrudate with a room temperature pyrometer probe will often give a false reading because when the cold probe is inserted, it becomes sheathed with the plastic that has been cooled by the probe. A more effective method is by using what some call the 15/15 (30/30) method. One simply raises the temperature of the probe about 15°C (30°F) above the melt temperature and then keep the probe surrounded with hot melt for 30 s. The easiest way to preheat the probe is to place the probe on, near, or in a hole in the die. By preheating

**Table 2.17** Troubleshooting coloring problems

| *Problem* | *Cause* | *Solution* |
|---|---|---|
| Streaks, contaminants | Dirty die plate | Clean |
| | Clogged screen pack | Clean or replace |
| | Inadequate purge from prior run | Increase purge times<br>Run lighter colors first |
| | Hangup in feed system | Flush out completely |
| | Dirty belt feeder | Clean with vacuum system |
| Color off-specification | Erratic feeding | Check and adjust for flow characteristics of material<br>Check if feeder type is appropriate |
| | Air currents affecting feed | Put wind screen around feeder |
| | Leaks in feed system | Check valves, etc. |
| | 'Blowback' from equipment up-stream or down-stream, causing over feeding | Make sure feeder is adequately supported |
| Poor color dispersion | Not enough shear stress | Increase screw rpm if not running at full throttle. If already at full capacity, increase cooling to raise viscosity. Reduce rates to increase specific energy |
| | Worn screw element | Check and replace<br>Use abrasion-resistant coating on screw |
| | Improper screw design (with singles) | Change to more 'intense' mixing screw |
| | Poor pigment quality | Call supplier |
| Bleedover from prior run | See streaks, contaminants | See streaks, contaminants |
| Agglomerates | Dirty screen pack (with singles) | Clean |
| | Damp material-storage area | Use climate-controlled storage area |
| | Moisture in mixer | Avoid running too long, too hot |
| | Wrong type of feeder | If running abrasives, use more gentle single-screw feeder |
| | Not enough shear stress | See above |
| | Selection of resin type | Use powder instead of pellets<br>Powder melts easier and requires less grinding and compression |
| Color fluctuations | Bridging in feeder or extruder hopper | Check. Consider running pre-blend |
| | Segregation of power-pellet mixture | Switch to all-powder formulation |

above the anticipated temperature, just prior to inserting it into the melt, then it requires the probe to actually be cooled by the melt. The lowest temperature reached will be the stock temperature. It also helps to move the probe around in the melt to have the probe more quickly reach a state of equilibrium. To be more accurate, repeat the procedure.

If the barrel on an extruder tends to wobble when the screw is rotating, as previously reviewed in this chapter, it is a symptom of a misalignment. It could be an alignment deficiency between the transmission, feed block, and/or barrel. This should be investigated immediately as it can result in a high wear situation, screw breakage, and/or heater creep.

To avoid liquid injection hardware becoming plugged with plastics, consider using a spring loaded pin type nozzle. The spring loading allows you to set the pressure so that it is higher than the pressure inside the extruder barrel, thus keeping the port clean and open.

The first indication of a loss to an extruder is often a drop in motor amperage. The ammeters rarely have an alarm. Another early indication is zone one barrel temperature which will under- and over-shoot the set point depending on whether the process is heating or cooling. Temperature controllers do usually have alarms. However, operators should be trained to look at motor amperage and check the hopper immediately upon seeing a zone one temperature alarm.

When an extrusion line is used to run much higher viscosity plastics than it was designed, the drive can be overloaded from excessive torque created by the higher shear stress on the screw. Several modifications to the system, such as a larger drive motor, a shallower screw, and/or changing the drive system by changing the belt with sheaves or gearbox ratio, will correct the drive overload. The latter is preferable, especially if there are belts and sheaves, and as long as the gearbox is strong enough to run at the higher torque and shaft overhung load. All these engineering numbers can be calculated. If the reduced top speed from the retiming procedure interferes with the required output of the lower viscosity plastic, field regulation can be added to the drive to keep the higher torque capability and allow higher top speed.

When an empty hopper is not the cause of machine failure, plastic might have stopped flowing through the feed throat because of screw bridging. An overheated feed throat, or start-up followed with a long delay, could build up sticky plastics and stop flow in the hopper throat. Plastics can also stick to the screw at the feed throat or just forward from it. When this happens, plastic just turns around with the screw, effectively sealing off the screw channel from moving plastic forward. As a result, the screw is said to be 'bridged' and stops feeding the screw. The common solution is to use a 'proper rod' to break up the sticky plastic or to push it down through the hopper.

*Air entrapment*

Air entrapment is a common problem. It is caused by air being taken in with plastics from the hopper. Compression of the solid plastic in the screw feed section will normally force air out of the solid melt bed back through the hopper; least path of pressure resistance. When this action does not occur, it moves forward with the melt until it exits the die. These air pockets can cause problems such as inclusions or surface imperfections.

There are solutions to this problem. The more popular and simple approach is to change the screw temperature in the solids conveying zone to achieve a more positive compacting. Often an increase of the first barrel section reduces the entrapment, however, a lower temperature sometimes gives an improvement. In any case, the temperature in the solids conveying zone are important parameters. The next step is to increase diehead pressure to alter the screw pressure profile to achieve a more rapid compaction. Another possible solution is to starve feed the extruder, but this will probably reduce output and requires additional hardware such as an accurate feeding device.

If the problem remains, a change in plastic particle size or shape could help. Other options include a vacuum hopper system (as previously reviewed), a grooved barrel section (rapid compaction occurs), reduce friction on the screw by applying a coating on the screw, and increased compression ratio of the screw (Chapter 4). Note that the problem may not be air entrapment. It could be volatile from the plastic, plastic degrading, additives or surface agents, and/or moisture. By determining which problem exists, it may be easier to resolve the problem of 'air.' Vented barrels may be the solution.

*Surging*

This is a condition of an unsteady-state extrusion operation. The screw is not delivering a constant melt flow. It is accompanied by fluctuations in the motor drive ammeter or load meter. Table 2.18 provides a guide to surging problems with solutions.

Single screws seldom have a perfectly matched feeding and pumping rate. If they do, it is only at one screw speed and/or set of operating conditions. Thus, it is necessary to have sufficient head pressure that the screw's metering section is filled for at least several turns, or output may be unstable. This type of surging (instability) is usually characterized by a steady, rhythmic surge that defies correction with operating conditions. It is very prevalent in two-stage screws where the second stage has been designed for excessive capacity vs. the first stage. With this type of surge, addition of screens, restrictive breaker plate, control valve, or other means

**Table 2.18** Guide to cause of surging with solution

| *Cause of surging* | *Solution* |
|---|---|
| Drive variation | Have drive serviced |
| Plugging of screw channel with solids | (a) screw design – evaluate<br>(b) raise temperatures<br>(c) raise polymer temperature (preheat)<br>(d) screw/barrel wear – have either or repair<br>(e) excessive fillers, i.e., inadequate % of polymer to allow for fluxing |
| Irregular feed | (a) screw design – evaluate<br>(b) throat restriction – remove or redesign<br>(c) bridging – agitate, vibrate, redesign screw<br>(d) slippery additives – eliminate or recompound into masterbatch<br>(e) sticky additives – precompound or add downstream |
| Metering section too deep – running partially filled as in two stage screw | (a) increase head pressure<br>(b) screw design |
| Incorrect barrel temperature profile | Determine optimum temperature profile |
| Improper screw design | (a) preheat the material<br>(b) starve feed the extruder<br>(c) reduce or increase the screw speed<br>(d) increase die back pressure with adjustable valves or screen packs<br>(e) cool or heat screen, if possible<br>(f) change the screw design tension |
| Surge in feed (for starve fed machines) | Check the volumetric or gravimetric flow of feeder |
| Inconsistent feed rate | Check the feeders |
| Melt point (in extruder) instability | (a) redesign the melting configuration<br>(b) modify temperature profile for heat input |
| Poor mixing of inhomogeneous materials | Improve distributive mixing of screw design |
| Poor feeding due to slip/stick in feed section for polypropylene | Increase screw cooling and lower zone 1 temperature as low as possible |
| 99% of all surges are caused in the feed-transition area. Reasons: | |
| Lack of temperature control | Increase watt density of heaters in 1st and 2nd zones, check controllers |
| Feed section or transition too short | Lengthen by recutting screw |

**Table 2.18** *Continued*

| *Cause of surging* | *Solution* |
|---|---|
| Feed section too deep | New screw |
| Bad mix of feed stocks from hopper | Add blending/mixing equipment |
| Uneven heating of predried materials | Add insulation to drying hopper |
| Starved screw due to feed section over-heating | Screw cooling |
| Varying bulk density | Troubleshoot scrap (fluff) addition system to regulate bulk density |
| Screw is solids feed limited – too shallow, low bulk density | Deepen feed section add shallow grooves to cast iron feed section bore |
| Screw periodically fouls and releases | Add screw cooling to feed zone only (short bore) |
| In compounding operation, surging is exaggerated by pellets of irregular shape and size | (a) pumping/pressurizing extruder is running too fast<br>(b) material is too hot at the metering zone – cooling the screw<br>(c) die is too large, the rate/hour/hole is too low, reduce the number of holes |

of building head pressure will improve stability. Even though one may not completely eliminate the instability with increased pressure, there will be a significant reduction or the surge is probably due to another cause [187, J. Frankland]. As reviewed above, the gear pump provides a solution.

*Gels*

Gels appear particularly in sheet and film as small, local imperfections of many different forms. Since their appearance is often similar, regardless of source, they are often mistakenly treated as a single problem. In fact, they have many sources and several different solutions. As an example poorly homogenized melt can cause some feed pellets to maintain their integrity through the process and appear as imperfections in the product. Process adjustment such as higher heats, finer screens, etc., probably solves the problem.

Gels can, and do sometimes, come with plastic. Particles sintered during its manufacture, as well as impurities within the plastic, will float within the melt and not loose their identity. This results in gels where process changes will rarely help. The extrusion process itself can create gels by allowing some material to stay on the hot side for too long. Plastic that reaches zero or close to zero forward velocity can cause partially degrade, become incompatible with the plastic flowing around it, break loose, and appear as gels. To correct this situation, examine streamlining the system, compound lubricity and elasticity, and metal temperatures as they relate to stock temperature.

Contaminants introduced in the plant can turn up as gels. The solution to this problem lies in correcting materials handling problems and cleanup. Chasing gels with process conditions is rarely the answer [187, D. R. Hopkins].

*Coextrusion instability*

Flow instabilities, such as interfacial instability, melt fracture, or layer nonuniformity, are a common problem that could cause quality problems and/or limit production rates in coextrusion. There are several options to correct these problems, but the idea is to select the best option(s) and minimize other possible adverse effects. Instabilities include: (1) extruder instability/melt nonuniformity where pressure and temperature vary with time and location; (2) steady-state instability/layer nonuniformity where plastic viscosities mismatch and/or poor die design or assembly; and (3) nonsteady-state instability/elastic instability where melt fracture exists and interfacial instability (zig-zag, wave, die swell, etc.) [187, T. I. Butler].

These problems have been reviewed regarding corrective action to take. In solving these coextrusion instability problems, no predictive theory exists for what are usually related to complicated rheological interactions. What is required is what processor's use; namely practical empirical experience with logic.

## Training

In-house training by qualified personnel is used in plants. However there are very capable outside sources that include colleges and organizations, such as the Plastics Institute of America (PIA) in Lowell, MA, that have specialized extrusion training programs.

Another example is the Techware Designs, subsidiary of Spirex Corp. with a computer software package called the 'Extruder's Technician' (ET). They provide services for installation and use as well as software upgrades. It is ideal for the processor as well as engineers, sales people, purchasing agents, managers, and anyone involved in extrusion. Like Techware's injection molding software program, it is customized in explaining specific extrusion processes that include blown and cast film, sheet, coating with wire and cable, pipe and tube, profile, and so on. It includes cost estimating capabilities, engineering calculations that can easily be understood and used, reference guides, troubleshooting information with problems and solutions, and so on. The ET user interface is built around a set of icons at the top of the computer screen. Clicking one of these summons drop-down menus and sub-menus from which users activate the software's features.

Users manually enter most of their extrusion data by filling in clearly marked input data. At the same time, ET's designers kept data entry requirements to a minimum in two ways. First, information computed in one section, such as extrudate cross-sectional area, can flow directly into another section where it is needed for a related calculation, such as product cost estimation. Second, if you are not sure what data to enter in a field, clicking on an adjacent button calls up a menu of typical entries for the field. You simply pick one that best matches your process or material. For example, one menu supplies melt density values which you will not find on a typical plastic data sheet.

ET's calculation programs let users quickly answer a number of specific extrusion questions, including those related to pressure drop at the end of the die and production rate. Other calculation programs automatically determine extrudate volumes, areas, lengths, and weights. ET also provides a host of generic engineering calculations covering mechanical, electrical, and statistical problems. Conversions can be made from Imperial (English) units to metric ones and vice versa.

The screw output estimator stands out as a particular time saver. This program estimates the extrusion rate for a given screw, plastic, and process conditions. It can forecast the throughput of a worn screw and also gives the bayback by replacing the worn screw.

For estimating product costs by length and/or weight, users input details are on the plastic, product, and machine on which it will be produced. Specify a desired profit, and the program will recommend a cost for the job. Among the cost inputs, ET asks for the product dimensions, weight, and output rates all of which can flow in from the package's of other calculation programs. The pricing information also takes into account factors such as: (1) extruder related costs such as machine-hour rates, set-up expenses, etc.; (2) material costs for one or more plastics (also for coextrusion); and (3) secondary costs such as those associated with tooling, upstream and downstream equipment, and packaging.

An extensive reference section is included on generic guides to electrical, mechanical, and chemical information; screw and barrel information, such as alloy compatibilities, rebuilding, and hard facing material choices, plastic data such as viscosities, drying conditions, etc., and so on. This section also lets users create their own documents. For instance, ET's set-up sheets will store commonly used processing parameters by job number. Users can also build procedure manuals for start-ups, shut-downs, and process monitoring tasks.

Troubleshooting features for specific extrusion processes match with processing problems with cause and solutions. On-line help comes in two forms. One is an on-line manual to the entire ET package. The other is more specific context such as drop-down 'windows' that define the variables used in a calculation.

## COEXTRUSION

Coextrusion provides multiple molten layers usually using one (in turn provides two separate layers) or the usual of two or more extruders with melts going through one die that are bonded together (Fig. 2.29). This technique permits the use of melt to bond various plastics or using an

**Figure 2.29** Large coextrusion sheet system (hoppers to die) with Frank Nissel, president of Welex Inc. (Blue Bell, PA 19422) and a leader in developing advanced coextrusion lines.

**Table 2.19** Examples of compatibility between plastics for coextrusion

| | *LDPE* | *HDPE* | *PP* | *Ionomer* | *Nylon* | *EVA* |
|---|---|---|---|---|---|---|
| LDPE | 3 | 3 | 2 | 3 | 1 | 3 |
| HDPE | 3 | 3 | 2 | 3 | 1 | 3 |
| PP | 2 | 2 | 3 | 2 | 1 | 3 |
| Ionomer | 3 | 3 | 2 | 3 | 3 | 3 |
| Nylon | 1 | 1 | 1 | 3 | 3 | 1 |
| EVA | 3 | 3 | 3 | 3 | 1 | 3 |

[a] 1 = layers easy to separate, 2 = layers can be separated with moderate effort, 3 = layers difficult to separate.

'adhesive' tie-layer between those that are difficult to bond (Table 2.19). This technique is an economical competitor to conventional laminating processes by virtue of reducing materials-handling costs, and machine time costs (Chapter 11). Pin holing is also reduced even when it uses one extruder and divides the melt into at least a two-layer structure [121, 173, 205, 336].

Other gains include elimination or reduction of delamination, air entrapment, and provide an excellent way to integrate/entrap recycled contaminated plastic with one side or both sides using virgin plastic depending on application. All kinds of advantages have developed in all types of markets. A novel example is coextruding film that produces startling iridescent effects simply by taking advantage of some basic optical principles. Years ago, this Dow Chemical process fabricated 0.001 mm (0.5 mil) film that contained 115 alternating layers of two commercially available plastics (PE and PP). The number of layers for this same thickness has more than doubled since. A special rotating type die provides the end results.

A processor desiring to enter the field once had little choice of equipment but expanding markets has led to a proliferation of equipment, with more emerging. There are many types of coextruded structures that range from two to at least seven layers. Common structures include: PP/filled PP/PP used for many food trays and containers; oriented PP (OPP) products generally produced in a copolymer/homopolymer/copolymer configuration; PP/adhesive/HIPS often used for fruit juice containers; and PP-virgin/PP-reclaim/adhesive/EVOH/adhesive/PP-reclaim/PP-virgin for rigid food containers. Note that PP will normally only bond to other olefins without a tie-layer; a tie-layer is generally required with other plastics [270, 273, 286, 434].

Choosing an adhesive layer is by no means a simple operation. There are many different types, each with specific capabilities. EVAs form the bulk being used. Proper choice can improve performance, such as increasing melt strength and bubble stability in blown film. High melt strength can also help in cast film used in thermoforming or a coating. Good melt draw is required to run higher take-up speeds and/or thinner structures without causing flow distribution or edge-weave problems.

**Melt flow characteristics**

Flow instabilities, such as interfacial instability, melt fracture, or layer nonuniformity, can become problems that could cause quality problems with the coextruded product. There are several options to correct these problems. The key to success is to select the best option or combination of options to eliminate the problem and minimize other possible adverse effects. Flow instabilities can cause limited production rates. Appropriate operating conditions of the coextrusion process combined with the proper plastic selections can successfully be used to solve the problems.

There are several types of flow instabilities that can cause nonuniformity to occur. They include: (1) extruder instability (melt uniformity) where pressure varies with time and/or temperature varies with time and location; (2) steady-state instability (layer uniformity) when plastic viscosity is mismatched and/or poor die design or assembly exists; and (3) nonsteady-state stability (elastic instability) due to melt fracture and/or interfacial instability of zig-zag, wave, and/or die swell.

Extruder surging is measured by variations in output rate with time. Surging is normally also associated with variations of pressure in the adapter. Good pressure variation for a single-screw extruder should be less than 1% (in the adapter plate). Gravimetric feed control will assure that the average layer ratio targets are achieved.

The steady-state instability means the plastic flows will reach some equilibrium condition, but the melt film does not have a uniform distribution of layers. The nonuniform layer distribution can be caused either by a mismatch of plastic viscosity or by the die design. The importance of viscosity in maintaining layer uniformity is important. The higher viscosity will tend to flow to the path of least resistance; the lower viscosity shaping can correct a nonuniform layer distribution in flat dies. In blown film dies, cleaning and precise alignment of the mandrels is required.

The nonsteady-state flow instabilities are generally believed to be the result of a dynamic response of the elastic components of the plastic viscosity within the environment of the extrusion process. These elastic instabilities are found to occur in the region of the highest shear stress.

They can be described as folds or waves of plastic flow at an interface which results in a small periodic thickness variation in the plastic layer. This situation will increase the roughness between the interface boundary.

Melt fracture is instability at the plastic/metal interface on the melt film surface. It is the folding of the plastic flow on the surface of the film. Many theories on the nature of melt fracture have been proposed with most agreeing that the elastic nature of the plastic is responsible for the instability. Melt fracture most often is reported to occur in the entrance of the die land area, but some studies have seen melt fracture occurring at the exit of the die as well.

A key attribute relating to melt fracture is the sudden change in shear stress as in a contraction to high shear stress at the entrance of the die land or the sudden expansion to non-strained flow at the exit of the die land. Birefringence measurements have confirmed that these are the locations of large stress concentrations in the melt flow. Using a critical shear stress, melt fracture has been reported for LLDPE to occur from 1.0 to $2.3 \times 10^6$ dyne/cm$^2$ (15 to 33 psi).

Melt fracture is reduced by lowering the critical shear stress at the wall. Ways of reducing melt fracture are: (1) viscosity in the skin layer could be reduced; (2) total output rate could be reduced; (3) die gap could be increased; and (4) processing aids can be used to reduce the shear stress at the die surface.

The elastic interfacial instability, sometimes called zig-zag, is an instability at a plastic/plastic interface and has been observed to occur in the die land. At low output rates, these instabilities are nearly undetectable; they do not interfere with the coextruded film. If a large amplitude wave form develops at the interface location, the melt flow can cause intermixing at the interface boundary. When there are significant differences in the refractive index of plastics, interfacial instability can more readily be seen.

A wave pattern, sometimes called W instability, occurs at the plastic/plastic interface and most often in asymmetrical structures. Wave instability has been found to occur where melt flows converge with large differences in velocities between plastic streams. Die swell cause some instabilities. It occurs most frequently when one of the plastics has little die swell, while the adjacent plastic has high die swell. The instability initiates at the die exit.

With nonsteady-state instabilities, the target is to reduce the shear stress at the interface. This is accomplished in the coextruded structure either at the plastic/plastic interface or the plastic/metal interface. The basic steps to take are: (1) lower output rates (not an economical approach); (2) increase skin layer ratio (highest shear stress is at the die wall, increasing the skin layer ratio shifts the plastic/plastic interface toward the center reducing shear stress on the interface); (3) reduce viscosity of adjacent layer (raise temperature to lower viscosity or use a plastic with a lower viscosity, higher melt index, with result that the stress shifts from the wall); and (4) increase die gap (result is a change in the geometry of the flow area, reducing the shear stress at the wall). Recognize that 'viscosity matching' is not always the best approach. There will be occasions to deliberately mismatch viscosities.

These methods can be used one at a time or in combination to solve the different instability problems. Although no complete predictive theory exists for these complicated rheological interactions, these guidelines with some good empirical experience can help eliminate the coextrusion problems [187, T. I. Butler].

In another study on wave pattern instability, test results did show the critical deviation (skew) required to induce the wave instability was found to be feedblock dependent (Chapter 5). It was shown that, in some cases, the wave instability can be initiated in the feedblock. Performance was evaluated in the context of feedblock geometry and greater sheet symmetry was attainable when the feedblock had a small expansion ratio and two-fold symmetry.

The geometry of the die has been shown to significantly influence whether or not the extruded sheet will exhibit wave instability. Die expansion ratio, channel length, and symmetry showed a consistent trend with die performance. More severe skews were attainable when the die had a small expansion ratio, a short channel length, and a two-fold symmetry. These results were unaffected by the initiation point of the instability; relative die performance was shown to be independent of the extrusion line and plastic structure.

For all cases where data was obtained, it was shown that structures with a large extension viscosity ratio or with a large extension viscosity skin are more susceptible to the wave instability. Two layer structures with large extension viscosity exhibited wave instability. Since extension viscosity is only important in expansion or contraction melt flows, the feedblock, die, and plastic are consistent [343].

Coextruding identical viscoelastic plastics through channels of various geometries can cause the layers or strands to rearrange. This layer movement may be caused by secondary flows that can occur in different channel geometries due to the elastic characteristics of the plastic. The results have many implications for the coextrusion of viscoelastic plastics. Since the layer and strand movements observed occurred without any differences in viscosity between the layers, it implies that layer arrangement may occur in coextrusion in which the plastic viscosities are well matched.

Since the strands continue to arrange as the plastic flows down the channel, results also imply that this geometric effect becomes more pronounced in sheet dies as they are scaled up to larger widths. The layer arrangements observed in a three plastic (PS, PE, and PC) study appear to follow the trend of increasing layer arrangement with increasing melt flow characteristics. In this study, the PS showed the most layer rearrangement, followed by the PE, and very little movement occurred with the PC [173].

Coextruding plastics with different processing behaviors, such as PC and PP, is accomplished by developing a degree of melt flow matching. PC adds heat resistance, strength, and clarity to PP-based coextrusions, but processing know-how is needed. An approach is to adjusting melt temperatures to achieve a processing working window that meets product performance. PC is very viscous with its viscosity being 12000 poise at typical extrusion melt temperatures and shear rates. In contrast, commodity plastics commonly used as structural materials in packaging, such as 5 MI (melt index) for PP (9000 poise), LDPE (1400 poise), and PS (1700 poise), are considerably thinner.

Since PP, PE, and PS are shear sensitive, where viscosity decreases with increasing shear rate, the actual processing viscosity can vary significantly with volumetric flow rate or die geometry. The PC's high viscosity would appear to be a problem when coextruding typical partners such as PP, PET, EVOH, PE, and other based tie layers through conventional feedblocks. Fortunately, the viscosity of PC is highly temperature sensitive, declining steeply as melt temperature rises. The plastic is also relatively insensitive to changes in shear rate.

These qualities give the processor a window of melt temperature to bring the PC's viscosity within a desired 3:1 ratio to other components in the structure that is generally recommended with conventional coextrusion feedblocks. Exceeding this ratio results in poor layer distribution, edge flow of the thinner plastic, and significant flow instabilities with typical monolayer coathanger dies. There are limits to how far the melt temperature can be varied. PC exhibits excellent thermal stability and, dependent on residence time, can reach temperatures as high as 343°C (650°F) without significant degradation. For normal production, a maximum melt temperature of 304°C (580°F) is recommended.

At the other extreme, as the melt temperature is lowered the viscosity increases dramatically, requiring high input screw torque from the extruder drive. The size of the drive motor and the torque ratings of the screw and gearbox may limit how cool the melt can be run. In raising the melt temperature of PC to lower its viscosity, the processor must avoid burning the other components in the coextrusion. The others usually have much lower melt temperatures. This is particularly true with EVOH, commonly used as a gas barrier. In typical flat-die feedblock coextrusion, the component layers are combined in the feedblock and remain together through the adapter and die. Heat conduction will try to establish an equilibrium temperature level somewhere between the hottest and the coldest component in feed melt temperatures.

The key to avoiding degradation of the more sensitive plastics, such as EVOH, by the more stable and hotter PC is to bring both melt streams into the feedblock as cold as viscosity matching and homogenization of the melt allows. With this action, it is also important to minimize the intimate contact time in the die. Accomplishing the first part is a function of shear heat input and barrel zone temperature profiles, while the second part is a function of melt flow rate and die design.

The problems encountered when coextruding high-temperature, high-viscosity plastics, such as PC, with a low temperature plastic can be overcome through careful process design and applying processing experiences. Although this may require running extruders at temperatures and speeds that would be considered not logical for monolayer structures, the end result is the successful combination of plastics into unique structures/products that fill the growing market [270].

It has been shown numerically and experimentally that coextruding identical viscoelastic plastics through channels of various geometries can cause the layers to rearrange. As reviewed above, this layer movement is caused by secondary melt flows that occur in different channel geometries as a result of the elastic characteristics of the plastic.

Finite-element simulations can be used to predict secondary melt flows in channels of arbitrary cross-sections. The choice of viscoelastic flow model is critical to the ability to predict secondary flow patterns. The flow model must have the capability to predict a nonzero second normal stress difference. The range of shear rates over which the rheological data are fitted is also important in determining the accuracy of the output of the simulation [173].

## Types of construction

Coextrusion products with many different combinations of plastics are being used extensively in every major sheet and film application area; it is big business worldwide. Products range from garbage bags to sophisticated multi-layer barrier food packaging products meeting exceptional performance requirements and economic requirements. Garbage bag costs have been reduced via coextrusion; good product properties and a down gauged product were achieved and continue to be achieved into the future. The same results are true for the more exotic barrier layer products which go into microwaveable shelf-stable containers. With their four to at least eleven layers, the necessary properties are obtained at lower costs than can be achieved by any other means in a plastic container. These quality products which sell well at the lowest cost are the driving force [251, 313, 440].

Shelf-stable food packaging by definition requires packaging which will maintain its contents with imperceptible changes for long periods of time [313]. An example of commercialization and progress in the coextrusion markets is the development of the dairy container. In 1967 a customer of Frank Nissel (Welex Inc.) wanted a glossy high-impact PS sheet to make thermoformed yogurt cups as pretty as their injection molded cups . . . and that market started [314].

The large degree of interest in improving products via coextrusion is leading to developments in all fields which are important to successful coextrusion. Coextrusion feedblocks and blown film dies are more versatile and easier to adjust and maintain, specialized interply adhesives are proliferating, and new plastics continue to be developed which are aimed at coextrusion applications, especially barrier types of plastics. Included is a special PC developed to be extruded at a low enough melt temperature to allow successful coextrusion with PVDC.

An interesting and special coextrude system is a coextruded film and sheet that utilizes the superior barrier and heat resistance properties of thermoplastic liquid crystal polymers (LCP) for use in commercial packaging applications. Superex Polymer Inc./Foster-Miller Inc. (Waltham, MA) has a process to overcome the major drawback for LCPs which is that the plastic tends to orient itself only in the direction of flow, creating weaknesses in transverse directions. The technology is a three-layer composed of an LCP layer, a proprietary thermoplastic tie layer, and a PET layer. The tie layer permits LCPs to be coextruded with other materials.

The techniques involve downward blown film extrusion using a special annular die in which concentric inner and outer mandrels rotate in opposite directions. Shear produced by this counter rotation produces some orientation, and the film is further oriented by stretching after it leaves the die (as in conventional blown film systems).

As reported earlier, though typically considered a high-priced ($10/lb) engineering plastic, the Superex process for coextruding the LCP film enables it to be competitive with traditional barrier plastics, such as EVOH and PVDC, on a cost/performance basis. There may be a special niche in the packaging industry where stringent requirements for oxygen and moisture exist. Superex also provides the ability to down gauge existing packaging film and sheet, due to the LCP's superior gas barrier performance [202].

There are equipment developments which make coextrusion more practical. Examples include the microprocessor controllers which can control five or more extruder and gear pumps which can be incorporated in the process control to aid layer thickness control and uniformity. There are gauge controllers that can measure individual layer thicknesses as well as total gauge in certain cases. To summarize equipment developments, there is a lot of activity toward improving and optimizing these systems that will make them more practical, stable, and economical.

The advance in high rate coextrusion equipment of at least 660 kg/h (3000 lb/h) has made the plastic much more the dominate cost factor. All other costs such as depreciation, labor, and utilities, may only represent 20% or less of the final product cost. That means small reductions in plastic/pound of product can mean major savings which can rapidly pay-off the extra equipment costs of a coextrusion line [187, C. Finch].

There are always obstacles that exist with coextrusion. With coextrusion, some unique problems may be encountered which do not exist in monolayer extrusion. As reviewed, the major unique problems are layer nonuniformity, melt interfacial instability, and the scrap recycle of dissimilar plastics (trim, etc.). Fortunately with the help of material suppliers, equipment manufacturers and particularly the processors themselves, problems are solved. The challenges exist but so do the growth markets with all types of capabilities with coextrusion. Highlighting a few, they can include changing the surface of film and sheet (gloss, color, etc.), functionality (antistatic, UV resistance, etc.), barrier properties (oxygen, moisture, carbon dioxide, etc.), decrease cost ('burying' low cost plastics, recycled plastics, etc.), and so on.

Sheet and film fabricators who have not examined coextrusion should study the opportunities and challenges it presents. It has its many advantages of performance and cost, but, on the other side of the coin, if approached improperly, it will give nothing but headaches. Coextruded products are replacing metal cans and glass jars as well creating completely new packaging designs. Behind all this action is development work that started over 30 years ago. Some applications have taken a long time and others a very short time to appear. For example, the simple two-layer GPPS/HIPS structures for dairy containers took less than a year to become fully commercial.

## SELECTING EQUIPMENT

When the opportunity exists, purchasing new equipment normally makes sense. A guide to requirements for the different extrusion processing equipment are given in Tables 2.20 and 2.21 [397]. This information shows the different requirements of processes to produce the different commercial products.

An option might be that you want someone's line or parts that you know is operating and in 'excellent' condition that meet your needs.

**Table 2.20** Examples of conditions for different processes

| *Process* | *Screw diameter (relative)* | *Screw length L/D ratio* | *Screw type* | *Metering depth* | *Screw speed (relative)* | *Melt temperature* | *Degree of control* *Speed* | *Temperature* | *Mechanical power* | *Comments* |
|---|---|---|---|---|---|---|---|---|---|---|
| Rod/section and profiles | High | 20/24 PE, PP<br>15/20 PVC<br>20/24 Nylon | T<br>T<br>S | Shallow | Low | Medium | ×× | ×× | ××× | Output relative to haul-off speed, for dimensions. Mimimum elastic memory. Medium temperature for rigidity while cooling. Add screens to promote mixing. |
| Tube | High | 20/24 PE, PP<br>15/20 PVC<br>20/24 Nylon | T<br>T<br>S | Medium | Medium | Medium except die | ×× | ×× | ×× | Minimum elastic memory. High die temperature for gloss. Medium temperature for drawdown of small tubes. |
| Wire covering/sheathing | Low | 20/24 PE, PP<br>15/20 PVC<br>20/24 Nylon | T<br>T<br>S | Shallow | High | High | ×× | × | ×× | High temperature for good drawing and avoiding melt fracture. High pressure for thin insulation. Output relative to haul-off speed for thickness |

**Table 2.20** *Continued*

| *Process* | *Screw diameter (relative)* | *Screw length L/D ratio* | *Screw type* | *Metering depth* | *Screw speed (relative)* | *Melt temperature* | *Degree of control* | | | *Comments* |
|---|---|---|---|---|---|---|---|---|---|---|
| | | | | | | | *Speed* | *Temperature* | *Mechanical power* | |
| Sheet/ laminating (including continuous vacuum forming) | High | 20/24 PE, PP<br>15/20 PVC<br>20/24 Nylon | T<br>T<br>S | Medium | Medium | Medium | ×× | ×× | ×× | Medium speed is compromise for high output and uniformity. Steady output to avoid ripples. High output for thick sheet. Uniform temperature for die distribution. Medium temperature for drawdown. |
| Flat film | Medium | 24/28 PE, PP<br>20 PVC | T<br>T | Shallow | Medium | High | ××× | ××× | ××× | High temperature for drawdown. High pressure for die. Medium speed for uniform elastic memory. Uniform output for thickness. |

| | | | | | | | | | | | |
|---|---|---|---|---|---|---|---|---|---|---|---|
| Paper coating | Medium | 24/28 | PE, PP | T | Shallow | Medium | High | ××× | ××× | × | Output relative to drum/haul-off speed for thickness. Medium output for rapid cooling. Very high temperature for flow and surface finish. |
| Tapes | Medium | 20/24 | PE, PP | T | Medium | Medium | Medium | ××× | ××× | ×× | Medium temperature for crystallization and drawdown. Uniform output for dimensions and crystallization. Uniform temperature for dimensions. |
| Filament/fibres | Medium | 20/24 | PE, PP | T | Shallow | Medium | High | ××× | ××× | × | High temperature for drawing. |
| Tubular film | Medium | 20/24 | PE, PP | T | Shallow | Medium | Medium | ×× | ××× | ××× | High pressure for die. Medium temperature for drawing and blowing. Medium speed for uniform elastic memory. |
| Foam | Medium | 20/24 | | T | Medium | Low | Low | × | ×× | ×× | Temperature depends on blowing agent. Low shear heating for decomposition. Medium pressure and mixing for uniform distribution and pore size. |

**Table 2.20** *Continued*

| *Process* | *Screw diameter (relative)* | *Screw length L/D ratio* | | *Screw type* | *Metering depth* | *Screw speed (relative)* | *Melt temperature* | *Degree of control: Speed* | *Degree of control: Temperature* | *Degree of control: Mechanical power* | *Comments* |
|---|---|---|---|---|---|---|---|---|---|---|---|
| Compounding/ blending | Medium/ high | 24/28 | PE, PP | T | Deep | Medium | Low | × | ×× | ×× | Deep screw at high pressure for mixing. Large diameter and medium speed for high output. Low temperature for mixing and heat history. |
| | | 20 | PVC | T | | | | | | | |
| | | 20/24 | POM | T | | | | | | | |
| | | | PMMA | S | | | | | | | |
| | | | Nylon | S | | | | | | | |
| Degrading | Medium | 24/28 | PP | T | Shallow | High | High | × | ××× | ×× | |
| Blow moulding | High | 24/28 | PE, PP | T | Medium | Medium | Low | ×× | ×× | ××× | No haul-off. Melt temperature and swell ratio compromised with output by control, e.g. of speed and diameter. |
| Injection moulding | High | 24/28 | PE, PP | T | Medium/ shallow | Medium | Medium/ high | × | ×× | ×× | No haul-off. Melt temperature control important. Back pressures low in practice. |
| | | 20 | PVC | | | | | | | | |
| | | 20/24 | PS | | | | | | | | |

Notes: The screw type and dimensions are largely dependent on the polymer and temperatures. Screw diameter and length are a compromise with capital cost. Screw speed is a compromise between output and quality. The length/diameter ratio and channel depth depend largely on the polymer used. 'Temperature' under 'Control' refers to set temperatures in the melt pumping section. Mechanical power input is controlled mainly by temperatures in the melting zone. Screw cooling is excluded since it is always very sensitive, requiring precise control.

Suggested *running* limits are:

| Speed: | | Temperature: | | Mechanical power: | |
|---|---|---|---|---|---|
| × | ±1% | × | ±2°C | × | ±3% |
| ×× | ±0.3% | ×× | ±0.5°C | ×× | ±1% |
| ××× | ±0.1% | ××× | ±0.1°C | ××× | ±0.4% |

**Table 2.21** Comparison of process requirements

| *Process* | *Uniform/constant* | | | | | | | *Maximum* (×) *or minimum* (✓) | | | | | | | |
|---|---|---|---|---|---|---|---|---|---|---|---|---|---|---|---|
| | *Output, Q* | *Pressure, P* | *Melt temperature,* $T_m$ | *Mixing* | *Residence time,* $t_{res}$ | *Specific power, E/Q* | *Elastic memory* | *Output, Q* | *Pressure, P* | *Melt temperature,* $T_m$ | *Mixing* | *Residence time,* $t_{res}$ | *Specific power, E/Q* | *Shear history* | *Elastic memory* |
| Rod, section, profile | | × | | | | | × | | | | | | | | ✓ |
| Tube, hollows | | × | | | | | | ×[a] | ✓ | | | | | | ✓ |
| Wire covering, sheathing | | | | × | | | × | | × | × | | | | | |
| Sheet, laminating | | × | × | | | | | × | | | | | | | ✓ |
| Flat film[b] | × | | × | × | | | × | | × | | × | | | | |
| Paper coating | × | | × | | | | | | × | × | × | | | | |
| Tape | × | | × | × | | | | | | | | | | | |
| Fibers, filament, netting | × | | × | | | | | | | × | × | | | | |
| Tubular film | × | | × | × | | | × | | × | | × | | | | |
| Foam | | × | × | × | × | | | | | ✓ | × | | | | |
| Compounding[c] | | | × | × | | ×[d] | | × | | ✓ | × | ×[e] | | × | |
| Blending, degrading | | | × | × | × | × | | × | | ×[f] | × | × | × | × | |
| Degassing, drying | | | | | | | | × | | ✓ | × | × | ✓ | | |
| Filtering | | × | | | | | | × | | ✓ | | | ✓ | | |
| Scrap recovery | | | | × | | | | × | | ✓ | × | | ✓ | ✓ | |
| Dewatering | | | | | | | | × | | ✓ | | | ✓ | ✓ | |
| Polymerizing | × | | × | × | × | | | | | | × | × | ✓ | ✓ | |
| Blow molding | × | | × | | | | × | × | | ✓ | | | | | ✓ |
| Injection molding | | | × | × | | | | × | | | × | | | | |

[a] Large pipes.
[b] Incl. multilayer and laminating.
[c] Incl. pumping of additives.
[d] And shear.
[e] Usually.
[f] Degrading (×), Blending (✓).

Another option is upgrading or rebuilding your equipment to expand your line's capability.

In addition it is easy to purchase all types of new equipment to meet all industry requirements; if it has not been built and you have the requirements, it will be designed and built for you. This action is typical of a progressive industry locally and worldwide.

Not unlike the used car market, there is a whole industry devoted to the resale of pre-owned (used) equipment for any part of the lines. For some extrusion plants, the decision to obtain equipment comes along quickly based on many different reasons, such as new or anticipated new business. One of the problems in getting new equipment sometimes is the time required to receive it in your plant. With used machinery, it is possible to purchase everything from one vendor with immediate delivery.

With used equipment, there is a compromise, particularly if one is not familiar with the equipment. But even those who are familiar with the equipment have the same problem as the buyer of a used car. But like buying a used car, one has to deal with responsible people; at least those you determine you can trust. Contracts of one type or another can be helpful if you have no time schedule to meet.

Developing meaningful quotations from new or used equipment suppliers obviously is important. What makes it interesting in buying used equipment is the potential that its cost could be about 25% of new equipment. But, one might say it is as risky as getting married in today's 'market.'

## SAFETY

Processing equipment has standard procedures to operate and has to meet safety requirements. The procedures and requirements are available from equipment suppliers. Safety information and standards are available from various sources that include the Society of Plastics Industry (SPI) and American National Standards Institute (ANSI).

For the past century, we have observed increasing effort on the part of equipment manufacturers to improve safety, and fabricating plants. Examples of safety features are many and differ for the different equipment in the lines. Safety interlocks ensure that equipment will not operate until certain precautions have been taken. Safety machine lockout procedures are set up for action to be taken in proper lockout of the machine's operation, such as electrical and mechanical circuits. There are preloaded pressure bolts around dies, pressure rupture disks on barrels, turret winder emergency stops, coextrusion line alarm if one extruder stops, and so on.

The operating environment is continuously upgraded to reduce sound and noise in the operating areas. Equipment arrives with a seemingly infinite number of yellow warning labels, plates, and notices. At times it is difficult to locate equipment specifications due to the large number of warning literature and signs on the machines. Safety gates and guards are well defined. All this action is required since many hazardous conditions, such as hot melt, plugged vent, overheating plastics, fingers caught between rolls, and so on, exist.

As reviewed in this chapter's section on communication protocol, programmable controller safety is used. Purpose is to establish protection of the safety circuits so that someone can not accidentally delete the circuits [4].

# 3

# Processing plastic material

## OVERVIEW

To obtain the best processing melts for any plastics, one starts with the plastic manufacturer's recommended heat profile and/or one's own experience. There are different starting points for the various types of plastics. Time and effort on start-up make it possible to achieve maximum efficiency of performance versus cost for the processed plastics. By the application of logic, the information gained can be stored and applied to future set-ups.

The terms plastic, polymer, and resin are usually taken as synonymous, but there are technical differences. A polymer is a pure unadulterated material that is usually taken as the 'family' name for a group of materials (PE, PVC, etc.); it is a NEAT material (Nothing Else Added To). Pure polymers are seldom used, so the terms plastic or resin are used identifying polymers with additives, etc. Resin tends to be used with TS plastics.

In order to understand the melt process mechanism of plastics it is first necessary to consider: (1) the basic special properties of plastics (specifically TPs since they represent practically all the plastics extruded); and (2) the typical physical changes which they undergo during processing. There most important property is the reversible change of plasticity with temperature, i.e., their **rheological** behavior or the deformation and flow of material. A plastic is more or less rigid at room temperature, becomes elastic, and then viscous or 'plastic' as temperature increases. At too high a temperature, the plastics become damaged through degradation (depolymerization) or decomposition (breakdown of the monomer or oxidation). According to type of plastic, the transitions between physical states of the plastics are defined.

From the approximaately room temperature of a TP entering the hopper, the exit temperature of the TP melt lies somewhere above its

**Table 3.1** Melt processing temperature for TPs

| *Material* | *Processing temperature rate* °C | °F |
|---|---|---|
| ABS | 180–240 | 356–464 |
| Acetal | 185–225 | 365–437 |
| Acrylic | 180–250 | 356–482 |
| Nylon | 260–290 | 500–554 |
| Polycarbonate | 280–310 | 536–590 |
| LDPE | 160–240 | 320–464 |
| HDPE | 200–280 | 392–536 |
| Polypropylene | 200–300 | 392–572 |
| Polystyrene | 180–260 | 356–500 |
| PVC, rigid | 160–180 | 320–365 |

softening and below its degradation temperatures. The TP in the screw moves through all the intermediate melt processing stages from ambient to flow temperatures (Table. 3.1). The most important of these stages is its viscoelastic behavior, which is characterized by certain relaxation phenomena. The effects of melt properties on local flow velocity are complicated but definable when based on specific processing conditions, such as temperature, time, and pressure. The time dependence of flow behavior has to be considered. Under certain conditions, the flow behavior also depends on melt pressure via the apparent viscosity. When the melt (extrudate) exits the die orifice, the melt pressure has a major effect on the swelling of the extrudate which is observed at the exit point.

If the path length or residual time of the melt in the die is too short (actually, it is always too short) for its potential energy to be completely converted into heat within the die, the last stage of relaxation does not occur until the melt leaves the die's orifice. This action usually causes controllable die-swell based on their rheological behaviors (Chapter 5). Another characteristic of plastics is their poor thermal conductivity that influences their melting and cooling.

This chapter describes important characteristics of plastics that influence processing and relate to performance of extruded products. Information presented pertains to practically all plastics behavior. However, there are exceptions, such as where a few plastics almost have no swelling at the die exit or, in fact, based on how the machine operates, the melt shrinks rather than swells [3]. If important exceptions exist, they will be reviewed.

## PLASTIC TYPE

Plastics are usually obtained in the form of granules, powders, flakes, and pellets; sometimes in liquid form. They are classified as thermoplastic (TP) or thermoset (TS), where TPs represent at least 90 wt% of all plastics used. TPs become soft and extrudable on heating without undergoing significant chemical or performance changes. On cooling, they harden (Fig. 3.1). This cycle can be repeated. TS plastics are rigid and undergo crosslinking or chemical change during processing and after curing cannot normally be resoftened or reprocessed. There are TP and TS elastomers (rubber-like).

Crosslinking is the principal difference between TSs and TPs. During

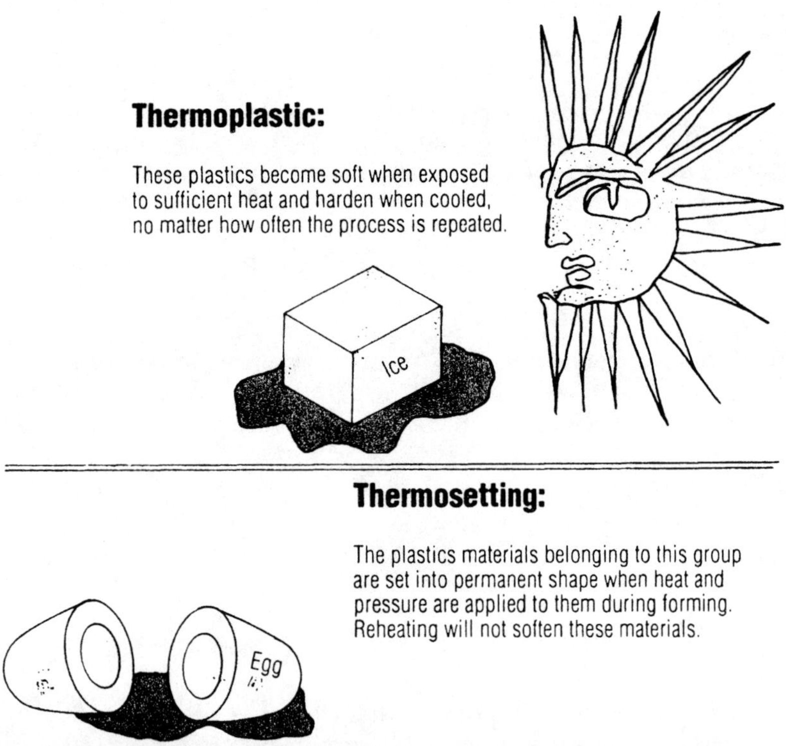


(a)

**Figure 3.1** (a) Characteristics of thermoplastics (TPs) and thermosets (TSs); (b) Melting characteristics of TPs and TSs.

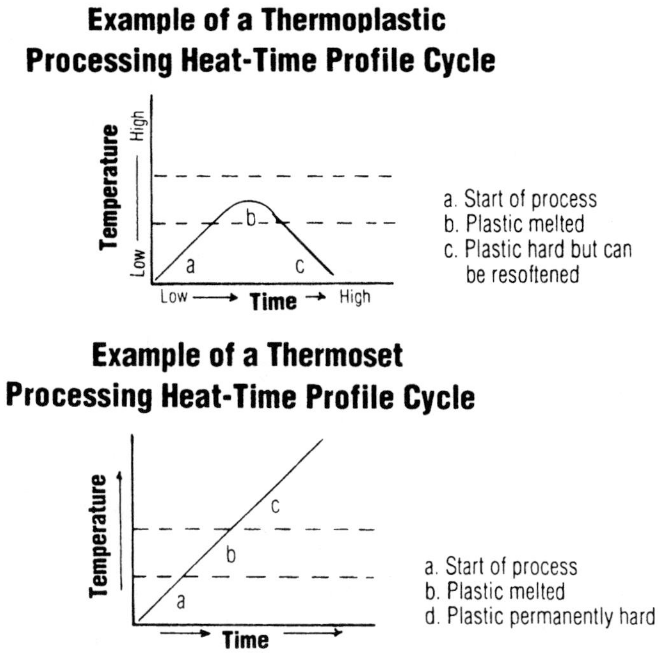


(b)

**Figure 3.1** *Continued*

curing or hardening of TSs, the crosslinks are formed between adjacent molecules producing a complex, interconnected network that can be related to its viscosity and performance. These crosslinks prevent the slippage of individual chains, thus preventing plastic flow under the addition of heat. If excessive heat is applied, degradation rather than melting will occur.

Certain TPs can be converted to TSs providing improved properties. They can be crosslinked by different processes such as chemicals and irradiations. PE is a popular plastic that is crosslinked; its abbreviation is XLPE.

About 200 basic types are commercially recognized (Table 3.2) with less than 20 that are commonly used. About two-thirds of all plastics consumed are LDPE, HDPE, PP, PVC, and PS. The relatively new generation high-performance metallocene and plastomer plastics are predicted to be in this group (Table 3.3) [156]. The basic types, with their many modifications of different additives, fillers, reinforcement, grafting, alloying, etc., [1, 3] provide different processing capabilities and product performances. Worldwide there are about 17000 different processable plastics that include many extrudable types.

**Table 3.2** Types of plastics

| | |
|---|---|
| Acetal (POM) | Liquid crystal polymer (LCP) |
| Acrylics | Aromatic copolyester (TP polyester) |
| Polyacrylonitrile (PAN) | Melamine formaldehyde (MF) |
| Polymethylmethacrylate (PMMA) | Nylon (Polyamide) (PA) |
| Acrylonitrile butadiene styrene (ABS) | Parylene |
| Acrylonitrile/methylmethacrylate (AMMA) | Phenolic |
| | Phenol formaldehyde (PF) |
| Acrylonitrile/styrene/acrylate (ASA) | Phenoxy |
| Alkyd | Polyallomer |
| Allyl diglycol carbonate (CR-39) | Polyamide (nylon) (PA) |
| Allyls | Polyamide-imide (PAI) |
| Diallyl isophthalate (DAIP) | Polyarylethers |
| Diallyl phthalate (DAP) | Polyaryletherketone (PAEK) |
| Aminos | Polyaryl sulfone (PAS) |
| Melamine formaldehyde (MF) | Polyarylate (PAR) |
| Urea formaldehyde (UF) | Polybenzimidazole (PBI) |
| Carboxymethyl cellulose (CMC) | Polycarbonate (PC) |
| Casein (CS) | Polyesters |
| Cellulosics | Aromatic polyester (TS polyester) |
| Cellulose acetate (CA) | Thermoplastic polyesters |
| Cellulose acetate butyrate (CAB) | Crystallized PET (CPET) |
| Cellulose acetate propionate (CAP) | Polybutylene terephthalate (PBT) |
| Cellulose nitrate (CN) | Polyethylene terephthalate (PET) |
| Cellulose plastics, general (CE) | Polyethylene terephthalate, saturated (PETG) |
| Cellulose propionate (CP) | Unsaturated polyester (TS polyester) |
| Cellulose triacetate (CTA) | Polyetherketone (PEK) |
| Ethyl cellulose (EC) | Polyetheretherketone (PEEK) |
| Chlorinated polyether | Polyetherimide (PEI) |
| Cresol formaldehyde (CF) | Polyimide (PI) |
| Epoxy (EP) | Thermoplastic PI |
| Ethylene vinyl acetate (EVA) | Thermoset PI |
| Ethylene vinyl alcohol (EVOH) | Polymethylmethacrylate (acrylic) (PMMA) |
| Fluorocarbons | Polymethylpentene |
| Ethylene-tetrafluoroethylene copolymer (ETFE) | Polyolefins (PO) |
| Fluorinated ethylene propylene (FEP) | Chlorinated PE (CPE) |
| Polymonochlorotrifluoroethylene (CTFE) | Cross-liked PE (XLPE) |
| Polytetrafluoroethylene (PTFE) | Cyclic olefin copolymer (COC) |
| Polyvinyl fluoride (PVF) | High-density PE (HDPE) |
| Polyvinylidene fluoride (PVDF) | Ionomer |
| Furan | Linear LDPE (LLDPE) |
| Furan formaldehyde (FF) | Low-density PE (LDPE) |
| Ionomer | Polyallomer |
| Ketone | Polybutylene (PB) |

**Table 3.2** *Continued*

| | |
|---|---|
| Polyethylene (PE) | Styrene butadiene (SB) |
| Polypropylene (PP) | Syndiotactic polystyrene (SPS) |
| Ultrahigh molecular weight PE (UHMWPE) | Sulfones |
| | Polyether sulfone (PES) |
| Polyoxymethylene (POM) | Polyphenyl sulfone (PPS) |
| Polyphenylene ether (PPE) | Polysulfone (PSU) |
| Polyphenylene oxide (PPO) | Urea formaldehyde (UF) |
| Polyphenylene sulfide (PPS) | Vinyls |
| Polyurethane (PUR) | Ethylene vinyl acetate (EVAC) |
| Silicone (SI) | Chlorinated PVC (CPVC) |
| Styrenes | Polyvinyl acetate (PVAC) |
| Acrylonitrile butadiene styrene (ABS) | Polyvinyl butyrate (PVB) |
| | Polyvinyl chloride (PVC) |
| General-purpose PS (GPPS) | Polyvinylidene chloride (PVDC) |
| High-impact PS (HIPS) | Polyvinylidene fluoride (PVF) |
| Polystyrene (PS) | Unplasticized rigid PVC (UPVC) |
| Styrene acrylonitrile (SAN) | Etc. . . . |

With different metallocene based plastics their abbreviations are preceded with 'm', such as mLDPE, mPP, etc.

In addition to broad categories of TPs and TSs, TPs can be further classified in terms of their structure (morphology) as either crystalline (PE, PP, PA, PET, POM, etc.), amorphous (PS, PC, PVC, ABS, PMMA, etc.), or liquid crystalline polymer. There are also those in the oriented stage. As reviewed in Chapter 2, 'oriented' describes molecules that are stretched into a rather linear configuration and lie fairly parallel to each other. These characteristics relate to plastic's molecular structure and processability. Crystalline types tend to have their molecules arranged in a regular repeating structure. Technically they are semicrystalline since typically less than 80% of their content are crystalline; the remainder is amorphous. They tend to pack into neat orderly arrangement of high density, sharp melting point ($T_m$), and directional properties. The amorphous types have molecular chains in random coil configurations (such as water-boiled spaghetti). These plastics have a more distinct and strong glass transition temperature ($T_g$). Since they do not contain crystals to scatter light, they are transparent. Liquid crystal polymer (LCP) types are a unique class of TPs with molecules that are stiff, rod-like structures organized in large parallel arrays. They have low melt viscosity and are more easily processed (Table 3.4).

The processability characteristics of plastics relate to their morphology i.e., factors such as molecular arrangement structure, molecular weight (MW), and molecular weight distribution (MWD). MW refers to the

**Table 3.3** New-advanced plastic for extrusion

| *Plastic* | *Density* | *Application* | *Target replacement* |
|---|---|---|---|
| mPE plastomers | 0.900–0.915 g/cc. | Coextrusion, heat-seal layers | EVA resins with 9% VA content and ULDPE/VLDPE |
| mPE elastomers | 0.854–0.900 g/cc. | Stretch-cling films | Ethylene methyl-acrylate |
| mLLDPE | 0.916–0.925 g/cc. | Stretch films, industrial sacks, shopping bags, form/fill/seal applications, frozen food packaging | C4 LLDPE, LDPE/LLDPE blends, EVA, ULDPE, ionomer layers |
| mMDPE | 0.932 g/cc. | Good-optics food packaging films, personal care, agriculture | LDPE |
| High-performance, high-clarity LLDPE | 0.909–0.912 g/cc. | Liner, stretch film, lamination, highly transparent food packaging | HAO C8-LLDPE, LDPE/LLDPE blends |
| Ethylene-propylene resins, random heterophasic PP copolymer | 0.890 g/cc. | Coextruded form/fill/seal film, hot fill, heavy duty, medica land personal care films, blends with other PP or PE grades | Soft PVC, nonpolymer packaging, LDPE, LLDPE |

**Table 3.4** General properties of crystalline, amorphous, and liquid crystalline plastics

| *Property* | *Crystalline* | *Amorphous* | *Liquid crystalline* |
|---|---|---|---|
| Specific gravity | Higher | Lower | Higher |
| Tensile strength | Higher | Lower | Highest |
| Tensile modulus | Higher | Lower | Highest |
| Ductility, elongation | Lower | Higher | Lowest |
| Resistance to creep | Higher | Lower | High |
| Max. usage temperature | Higher | Lower | High |
| Shrinkage and warpage | Higher | Lower | Lowest |
| Flow | Higher | Lower | Highest |
| Chemical resistance | Higher | Lower | Highest |

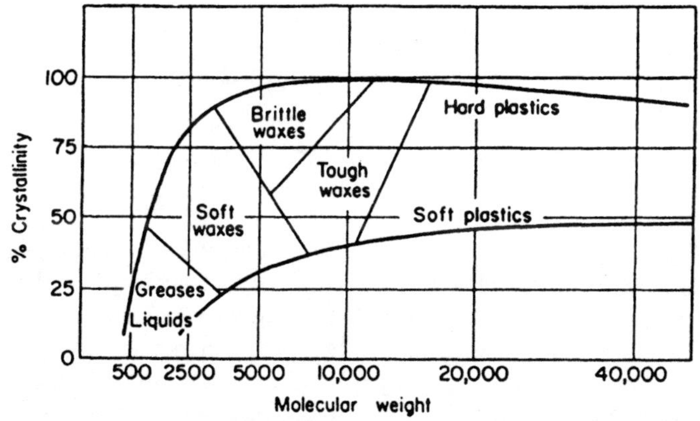


**Figure 3.2** Effect of crystallinity and molecular weight upon physical properties of PE.

molecules' weight and size; they are identified as average MW for a plastic (Fig. 3.2). Put simply, processsability is easy with low MW, whereas properties of finished products increase with high MW. The MWD is important for the processor. Basically the MWD gives a general picture of the ratio of large, medium, and small molecular chains. It is called narrow if the plastic is made up of chains close to the average length; wide identifies a wide variety of lengths (Fig. 3.3). A narrow MWD enhances the performance of plastic products.

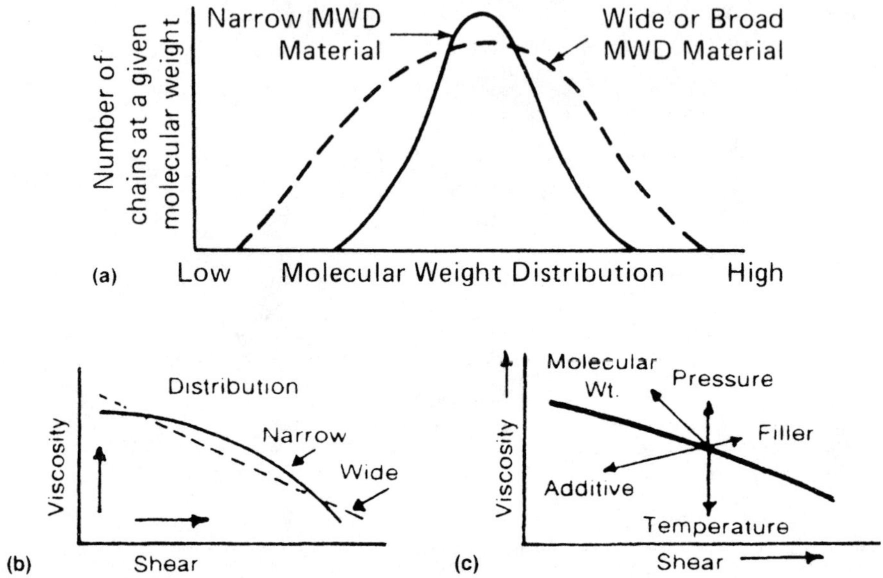


**Figure 3.3** (a) Molecular weight distribution (MWD); (b) viscosity versus shear rate as related to MWD; (c) factors influencing viscosity.

## PROPERTIES

Plastics commercially provide more types than all other materials (steel, wood, etc.) put together. Many variations are available, providing a wide range of properties (Fig. 3.4). See Table 16.2 on the future properties. Unfortunately no one plastic has all the 'best' properties.

Processing requirements also vary depending on the plastic to be used (Table 3.5). The trend is that for the higher product performance requirements, process controls are more exact using smaller 'windows' of operation.

Almost 90 wt% of all TPs are classified as 'commodity plastics' and the others are 'engineering plastics'. The engineering types (PA, PC, POM, etc.) are characterized by improved performances such as higher properties (mechanical, heat resistance, etc.) than the commodity (PE, PP, PVC, etc.) so they are more expensive and usually require tighter process controls.

Plastics, like any other material, are never perfect. They have advantages as well as disadvantages both for meeting processing and product performances. This situation is not a problem as long as one properly evaluates the situation. Some plastics process with ease and others can develop problems if not properly processed within tight control settings. This situation occurs when processing TPs with certain additives

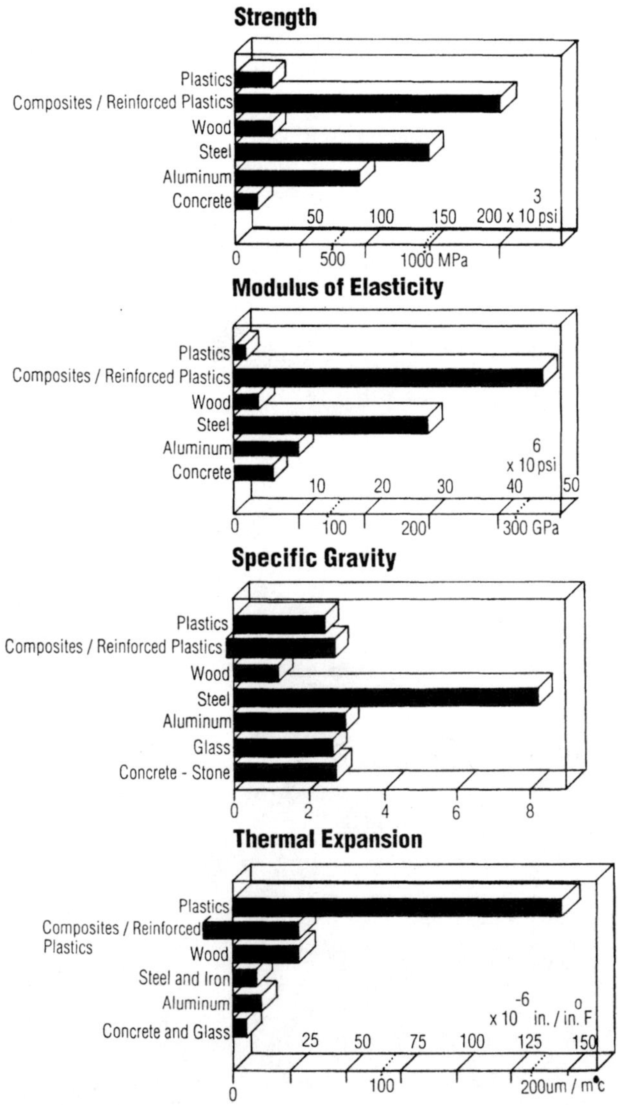


**Figure 3.4** General comparison of different materials.

such as certain colorants, heat stabilizers, etc. Plastics decompose at certain temperatures (Table 3.6) and, if higher temperatures are required, adding certain fillers/reinforcements can raise the decomposition temperature.

Utilizing the laws of physics, chemistry, and mechanics, theoretical values of materials properties have been determined for different

**Table 3.5** Guide for extrusion and injection molding machine settings[a]

| *Resin data*[b] | *Specific gravity* ($g\,cm^{-3}$) | *Density* ($lb\,ft^{-3}$) | *Specific volume* ($in^3\,lb^{-1}$) | *Specific volume* ($cm^3\,g^{-1}$) | *Extrusion temperature* (°F) | *Injection temperature* (°F) | *Linear mold shrinkage* | *Specific heat* ($BTU\,lb^{-1}\,°F^{-1}$) | *Water absorption in 24h* (%) | *Maximum water content allowable for molding* |
|---|---|---|---|---|---|---|---|---|---|---|
| ABS, extrusion | 1.02 | 64.0 | 27.0 | 0.980 | 435 | | 0.005 | 0.34 | 0.25 | |
| ABS, injection | 1.05 | 65.0 | 26.0 | 0.952 | | 500 | 0.005 | 0.40 | 0.40 | 0.20 |
| Acetal, injection | 1.41 | 88.0 | 19.7 | 0.709 | | 390 | 0.020 | 0.35 | 0.25 | |
| Acrylic, extrusion | 1.19 | 74.3 | 23.3 | 0.839 | 375 | | 0.004 | 0.35 | 0.30 | |
| Acrylic, injection | 1.16 | 72.0 | 24.1 | 0.868 | | 450 | 0.005 | 0.35 | 0.20 | 0.08 |
| CAB | 1.20 | 74.6 | 23.1 | 0.833 | 380 | 440 | 0.004 | 0.35 | 1.50 | 0.15 |
| Cellulose acetate, extrusion | 1.28 | 80.2 | 21.6 | 0.781 | 380 | | 0.005 | 0.40 | 2.50 | |
| Cellulose acetate, injection | 1.26 | 79.0 | 21.9 | 0.794 | | 450 | 0.005 | 0.36 | 2.40 | 0.20 |
| Cellulose proprionate, extrusion | 1.22 | 76.1 | 22.7 | 0.821 | 380 | | 0.004 | 0.40 | 1.70 | |
| Cellulose proprionate, injection | 1.22 | 75.5 | 22.9 | 0.828 | | 425 | 0.004 | 0.40 | 2.00 | 0.25 |
| CTFE | 2.11 | 134.0 | 13.1 | 0.473 | | 550 | 0.008 | 0.22 | 0.01 | |
| FEP | 2.11 | 134.0 | 12.9 | 0.465 | 600 | 600 | 0.010 | 0.28 | <0.01 | |
| Ionomer, extrusion | 0.95 | 59.6 | 29.0 | 1.050 | 500 | | 0.007 | 0.54 | 0.07 | |
| Ionomer, injection | 0.95 | 59.1 | 29.2 | 1.060 | | 420 | 0.007 | 0.54 | 0.20 | |
| Nylon-6 | 1.13 | 70.5 | 24.5 | 0.886 | 520 | 550 | 0.013 | 0.40 | 1.60 | 0.15 |

| | | | | | | | | | | |
|---|---|---|---|---|---|---|---|---|---|---|
| Nylon-6,6 | 1.14 | 71.2 | 24.3 | 0.878 | 510 | 510 | 0.015 | 0.40 | 1.50 | 0.15 |
| Nylon-6,10 | 1.08 | 67.4 | 25.6 | 0.927 | | 450 | 0.011 | 0.40 | 0.40 | 0.15 |
| Nylon-6,12 | 1.07 | 66.8 | 25.9 | 0.935 | 475 | 500 | 0.011 | 0.40 | 0.40 | 0.20 |
| Nylon-11 | 1.04 | 64.9 | 26.6 | 0.962 | 460 | 450 | 0.005 | 0.47 | 0.30 | 0.10 |
| Nylon-12 | 1.02 | 63.7 | 27.1 | 0.980 | 450 | 445 | 0.003 | | 0.25 | 0.10 |
| Phenylene oxide based | 1.08 | 67.5 | 25.6 | 0.926 | 480 | 525 | 0.006 | 0.32 | 0.07 | |
| Polyallomer | 0.90 | 56.2 | 30.7 | 1.110 | 405 | 405 | 0.015 | 0.50 | 0.01 | |
| Polyarylene ether | 1.06 | 66.2 | 30.7 | 0.940 | 460 | 535 | 0.006 | | 0.10 | |
| Polycarbonate | 1.20 | 74.9 | 23.1 | 0.832 | 550 | 575 | 0.006 | 0.30 | 0.20 | 0.02 |
| Polyester PBT | 1.34 | 83.6 | 20.7 | 0.746 | | 460 | 0.020 | | 0.08 | 0.04 |
| Polyester PET | 1.31 | 8.18 | 21.1 | 0.746 | 480 | 490 | 0.002 | 0.40 | 0.10 | 0.005 |
| HD polyethylene, extrusion | 0.96 | 59.9 | 28.8 | 1.040 | 410 | | 0.025 | 0.55 | <0.01 | |
| HD polyethylene, injection | 0.95 | 59.3 | 29.1 | 1.050 | | 480 | 0.025 | 0.55 | <0.01 | |
| HD polyethylene, blow molding | 0.95 | 56.9 | 28.8 | 1.040 | 410 | | 0.025 | 0.55 | <0.01 | |
| LD polyethylene, film | 0.92 | 57.44 | 30.1 | 1.090 | 350 | | 0.032 | 0.55 | <0.01 | |
| LD polyethylene, injection | 0.92 | 57.4 | 30.1 | 1.090 | | 400 | 0.032 | 0.55 | <0.01 | |
| LD polyethylene, wire | 0.92 | 57.4 | 30.1 | 1.090 | 400 | | 0.025 | 0.55 | <0.01 | |
| LD polyethylene, ext. coating | 0.92 | 57.1 | 30.0 | 1.090 | 600 | | 0.025 | 0.55 | <0.01 | |
| LLD polyethylene, extrusion | 0.92 | 57.4 | 30.1 | 1.087 | 500 | | | | | |
| LLD polyethylene, injection | 0.93 | 58.0 | 29.8 | 1.075 | | 425 | | | | |
| Polypropylene, extrusion | 0.91 | 56.8 | 30.4 | 1.100 | 450 | | 0.005 | 0.50 | 0.03 | |
| Polypropylene, injection | 0.90 | 56.2 | 30.7 | 1.110 | | 490 | 0.018 | 0.50 | <0.01 | |
| Polystyrene, impact sheet | 1.04 | 64.9 | 26.6 | 0.963 | 450 | | 0.005 | 0.34 | 0.10 | |
| Polystyrene, gp crystal | 1.05 | 65.5 | 26.2 | 0.943 | 410 | 425 | 0.004 | 0.32 | 0.03 | |
| Polystyrene, injection impact | 1.04 | 64.9 | 26.6 | 0.968 | | 440 | 0.006 | 0.34 | 0.10 | |
| Polysulfone | 1.25 | 77.4 | 22.3 | 0.807 | 650 | 680 | 0.007 | 0.28 | 0.30 | 0.05 |
| Polyurethane | 1.20 | 74.9 | 23.1 | 0.834 | 400 | 400 | 0.020 | 0.40 | 0.10 | 0.03 |
| PVC, rigid profiles | 1.39 | 86.6 | 19.9 | 0.720 | 365 | | 0.025 | 0.25 | 0.02 | |
| PVC, pipe | 1.44 | 87.5 | 19.7 | 0.714 | 380 | | 0.025 | 0.25 | 0.10 | |

**Table 3.5** *Continued*

| *Resin data*[b] | *Specific gravity* ($g\,cm^{-3}$) | *Density* ($lb\,ft^{-3}$) | *Specific volume* ($in^3\,lb^{-1}$) | *Specific volume* ($cm^3\,g^{-1}$) | *Extrusion temperature* (°F) | *Injection temperature* (°F) | *Linear mold shrinkage* | *Specific heat* ($BTU\,lb^{-1}\,°F^{-1}$) | *Water absorption in 24 h* (%) | *Maximum water content allowable for molding* |
|---|---|---|---|---|---|---|---|---|---|---|
| PVC, rigid injection | 1.29 | 83.6 | 21.0 | 0.756 | | 380 | 0.025 | 0.25 | 0.10 | 0.07 |
| PVC, flexible wire | 1.37 | 85.5 | 20.2 | 0.731 | 365 | | 0.025 | | | |
| PVC, flexible extruded shapes | 1.23 | 76.8 | 22.5 | 0.814 | 350 | | 0.025 | | | |
| PVC, flexible injection | 1.29 | 80.5 | 21.4 | 0.776 | | 300 | 0.025 | | | |
| PTFE | 2.16 | 134.8 | 12.9 | 0.464 | | | | 0.25 | <0.01 | |
| SAN | 1.08 | 67.4 | 25.6 | 0.927 | 420 | 470 | 0.005 | 0.31 | 0.03 | 0.02 |
| TFE | 1.70 | 106.1 | 16.3 | 0.589 | | 610 | 0.040 | 0.46 | 0.01 | |
| Urethane elastomers | 0.83 | 51.6 | 33.5 | 1.210 | 390 | 400 | 0.001 | 0.46 | 0.07 | 0.03 |

[a]Specific information on all machine settings and plastic properties is initially acquired by using the resin supplier's data sheet on the particular compound or resin to be used.
[b]These are strictly typical average values for a resin class; consult your resin supplier for values and more accurate information.

**Table 3.6** Temperature decomposition ($T_d$) ranges for various plastics

| *Type of material* | $T_d$ (°F) |
|---|---|
| PP | 610–750 |
| PC | 645–825 |
| PVC | 390–570 |
| PS | 570–750 |
| PMMA | 355–535 |
| ABS | 480–750 |
| PA | 570–750 |
| PET | 535–610 |
| Fluoropolymer | 930–1020 |

Note: Adding certain fillers/reinforcements can raise decomposition temperatures.

materials. These are compared to actual values in Table 16.2; these data were developed during 1944. For steel, aluminum, and glass, the theoretical and actual experimental values are practically the same, whereas for PE, PP, and PA (also others), they are very different. The properties of plastic materials have the potential of reaching values that are far superior to those of all other materials (steel, etc.). With the passing of time, the properties of plastic materials have significantly improved.

## HEAT PROFILE

General heat-time profiles are shown in Fig. 3.1(b) and Table 3.5. Amorphous material usually requires a fairly low initial heat in the screw plasticator; its purpose is to preheat material but not melt it in the screw's feed section prior to entering the compression zone (Chapter 4). Crystalline material requires higher initial heating to ensure that it melts prior to reaching the compression zone. Careful implementation of these procedures produces the best melts, which in turn produce the best products.

Process control is extremely important for obtaining good performance properties, particularly tolerance (or shrinkage) control – sometimes it is the most influential factor. The dimensional accuracy of the product relates to the process settings, the accuracy of the die, and the process controls. Adequate process control and its associated instrumentation are essential for product quality control (Chapter 6). Sometimes the goal is precise adherence to a control point or simply to maintain the temperature within a comparatively narrow range.

Economical production requires that tolerances are not specified more tightly than necessary, however after a production target is met, one should process to 'tighter' tolerances if possible, resulting in more profit by using less plastic and time during cooling. Many plastics change dimensions after being processed, principally because their molecular orientations or molecules are not relaxed. To ease or eliminate the problem, one can change the process set-up, for example by including annealing or by extending the in-line cooling time so that the plastic is 'stress relieved.'

The process cycle of heating and cooling TPs can be repeated indefinitely using granulating scrap, defective products, and so on. During this cycle, plastics develop a time-to-heat history or 'residence time'. With only limited repetition of the cycle for certain TPs, their processing and product performance requirements are not significantly affected. However, some TPs can significantly lose certain behaviors requiring changes in the machine settings.

## MELT FLOW AND RHEOLOGY

Rheology is the science that deals with the deformation and flow of matter under various conditions, such as a plastic's melt flow. Even though the rheology of TPs is complex, it is manageable. These plastics combine the properties of an ideal viscous liquid (pure shear deformations) with those of an ideal elastic solid (pure elastic deformation). Plastics are therefore said to be 'viscoelastic'. Their viscous behaviors relate to their processing capabilities.

There are two different types of flow; in viscous flow, the energy causing deformation is dissipated and in elastic flow that energy is stored. The combination produces viscoelastic plastics. Viscosity is a plastic's resistance to viscous deformation (flow). The resistance to elastic deformation is the modulus of elasticity ($E$); its range for a melt is 1000–7000 kPa (145–1000 psi) where it is called the rubbery range [1].

As a melt is subjected to a fixed stress (or strain), the deformation–time curve will show an initial rapid deformation followed by a continuous flow. Figure 3.5 relates elasticity and strain with: (1) basic deformation versus time curve; (2) stress–strain deformation versus time (the creep effect); (3) stress–strain deformation versus time (the stress-relaxation effect); (4) material exhibiting elasticity; and (5) material exhibiting plasticity. The relative importance of elasticity (deformation) and viscosity (flow) depends on the time scale of the deformation. For a short time, elasticity dominates; over a long time, the flow becomes purely viscous. This behavior influences processes. When a product is annealed (stress relieved using heat), it will change shape or swelling occurs with post-extrusion. Deformation contributes significantly to process flow defects.

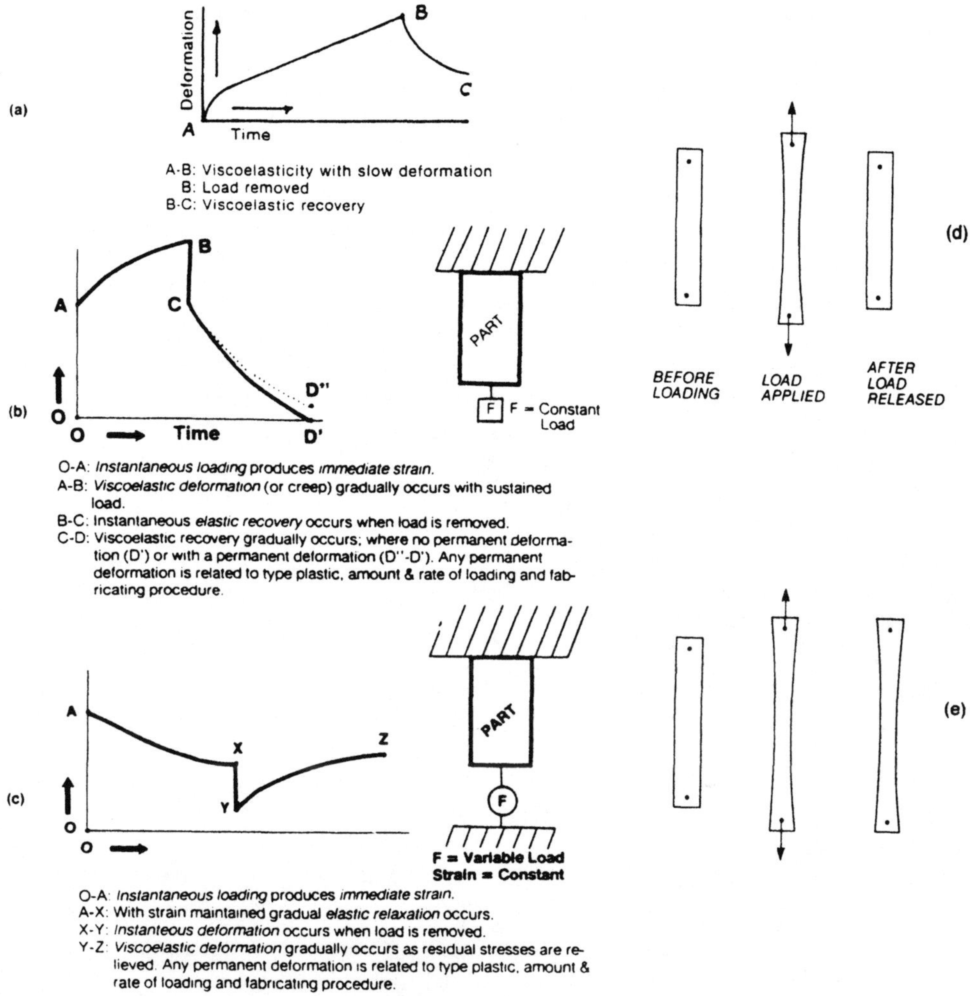


**Figure 3.5** Elasticity and strain behaviors.

Melts with small deformation have proportional stress–strain behaviors [3]. As the stress on the melt is increased, the recoverable strain tends to reach a limiting value. It is the high-stress range, near the elastic limit, that processes operate.

MW, temperature, and pressure have little effect on elasticity; the main controlling factor is MWD. Practical elasticity phenomena often exhibit little connection with the actual value of the modulus and the viscosity.

Although the modulus is influenced only slightly by MW and temperature, these parameters have a great effect on viscosity and can therefore alter the balance of a process.

In addition, there are also two modes in which deformation is produced – simple shear and simple tension. The actual action during melting tends to be theoretically complex (but controllable) with this shear–tension combination. Together with engineering design, deformation determines the screw's pumping efficiency and controls the relationship between pressure and output rate through a die.

When a melt moves in a direction parallel to a fixed surface (screw/barrel or die wall), it is subject to a shearing force. As the screw rotational speed increases, so does the shear rate with potential advantages or disadvantages. Advantages of an increase are a less viscous melt and easier flow. This shear-thinning action is required to 'move' plastic. When water, a Newtonian liquid, is in an open-ended pipe, pressure can be applied to move it; doubling the water pressure results in flow rate of the water to double. Water does not have a shear-thinning action. However, in a similar situation using plastic (a nonNewtonian liquid), if the pressure is doubled, the melt flow may increase from 2–15 times depending on the plastic used (Fig. 3.6).

As an example, LLDPE with its low shear-thinning action, experiences a very low rate increase, which can explain why it can cause more processing problems compared to other PEs. The higher flow melts include PVC and PS.

A disadvantage with the higher shear rates is that too high a heat increase may occur, potentially causing problems in cooling as well as degradation and discoloration. A high shear can lead to a rough product surface (melt fracture, etc.). For each plastic and every processing condi-

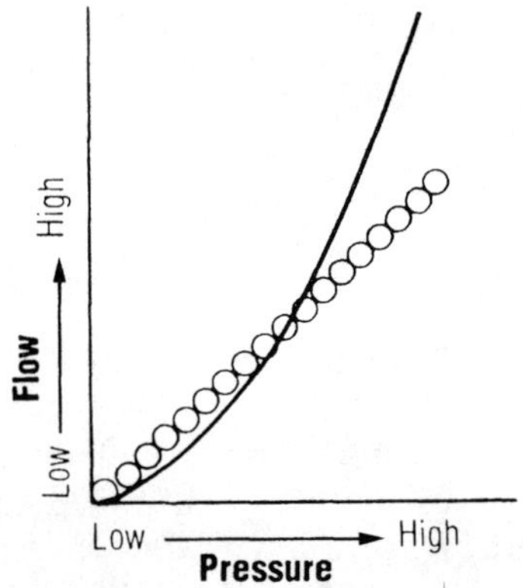


**Figure 3.6** Rheology and flow properties of plastics (solid line) and water (circles).

tion, there is a maximum shear rate beyond which such problems develop.

Shear in the screw channel is equal to $\mu DN/60h$, where $D$ = average barrel inside diameter, $N$ = screw rpm, and $h$ = average screw channel depth. This equation does not include the melt slippage between the barrel wall and screw surfaces, but the shear rate obtained is still useful for purposes of comparison. A 60 mm (2.5 in) screw with a 3.6 mm (0.140 in) channel rotating at 100 rpm produces a shear rate of $93.5\,s^{-1}$. This value is approximately the desired value in most extrusion processes with $100\,s^{-1}$ generally being the target.

The same equation can be use to determine the shear rate of the slippage between barrel and screw. With a new barrel, which usually has a small clearance of about 0.127 mm (0.005 in), a very high shear rate of about $2618\,s^{-1}$ can exist. With this small clearance, only a small amount of melt is subject to the higher heat, so that any overheating is overcome by the mass of melt it encounters (melt mixing action occurs). As the screw wears, more melt flows through enlarged clearances, and the shear rate is lower. The effect of wear on overheating is usually very small and is not the main reason why the complete melt overheats.

Shear rates can also be determined in melt flow through dies. The equations applicable to the different die shapes seldom account for slippage of melt on the die surfaces, but they can be used to compare the processability of melts and to control melt flow. The shear for a rod is $4Q/Rh^2$, for a long slit it is $6Q/wh^2$, and an annulus die it is $6Q/wh^2$ where $Q$ = volumetric flow rate, $R$ = radius, $w$ = width, and $h$ = die gap.

Shear stresses and shear rates vary depending on type of plastic processed. Figure 3.7 shows the flow conditions for a fluoropolymer [Teflon, 179] at a given constant melt and die temperature. Initially with a slow output rate, the plastic degrades as a result of excessive residence time. The next region shows where normal operation occurs. In the next region, the plastic goes from smooth to a rough-fractured melt. By increasing its flow rate further, the plastic goes through a 'critical shear rate' where again it becomes smooth. With the final higher flow rate, it becomes rough again. This situation, with its distinct flow regions, is not typical of many plastics; it shows what occurs during a certain plastic melt flow.

Fluoropolymers are considerably more viscous at normal processing temperatures than other TPs. Both melt viscosity and melt flow number (MFN, the amount of plastic in grams passing through a given die orifice under constant pressure and temperature per 10 min) gives the effect of temperature and time. An increase in MFN of up to 10% during processing is normal for Teflon. The change should be kept below 25% if properties are not to be changed significantly.

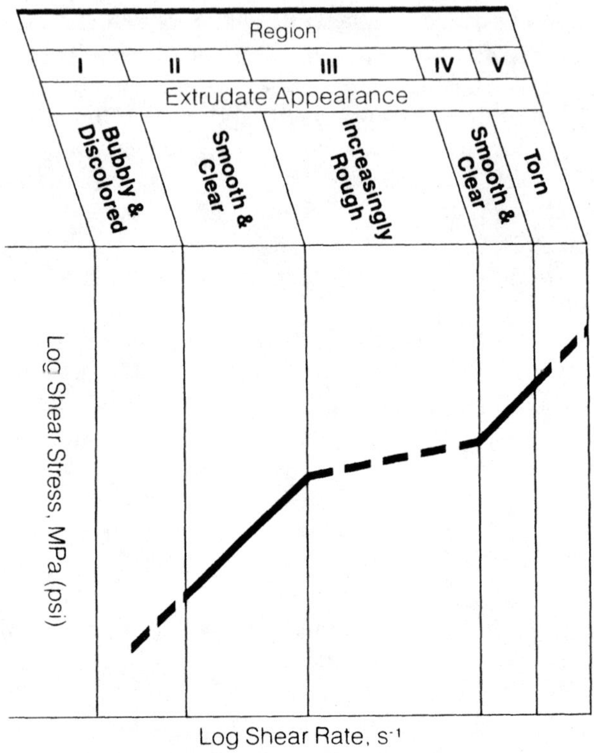


**Figure 3.7** Shear stress versus shear rate.

## MELT FLOW AND DEFECT

In any deformation occurs there are local stress concentrations. Should the **viscosity** (resistance of flow) increase with stress, the deformation at the stress concentration will be less rapid than in the surrounding plastic; the stress concentration will be smooth and the deformation will be stable. However, when the viscosity decreases with increased stress, any stress concentration will cause catastrophic failure of the plastic. Viscosity is the ratio of shearing stress to the rate of shear of a fluid (Fig. 3.8).

The viscosity of plastics varies inversely with their temperature, assuming constant shear stress. It is usually taken to mean Newtonian viscosity in which case the viscosity is constant. In nonNewtonian (plastic melt) behavior, the viscosity varies (Fig. 3.8). Such viscosity is often called apparent viscosity at the corresponding shear rates [3, 4].

Unfortunately, undesirable melt flow defects develop during extrusion. The cause can be analyzed. As an example, melt fracture (MF) is when any melt flow through a die is in an unstable condition of flow. Instead of a

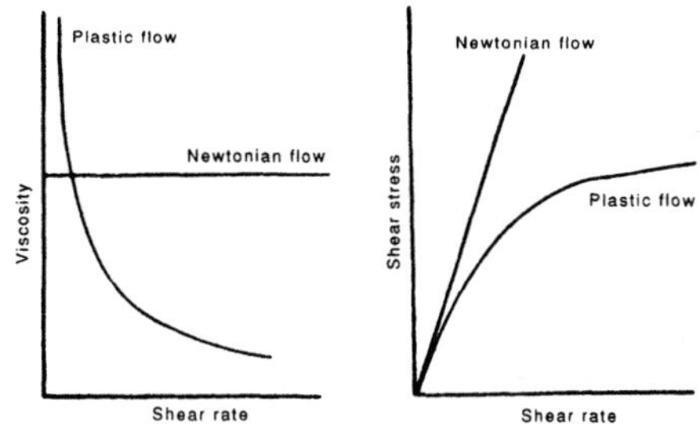


**Figure 3.8** Viscosity behavior of plastics and other materials.

steady, streamlined flow in and out of the die, the flow is distorted. It appears as a rough surface on the extrudate. As an example, in tubing the roughness first appears on the inside surface unless the die's inter mandrel is heated. This roughness is a tearing of the melt caused by excessively high shear stress. Extruding too fast is the most common cause. MF occurs with flow above its critical shear rate. This rate is temperature dependent. Increasing output without MF can be obtained by increasing the melt and die temperatures. However, this approach is limited to melt temperatures below those that would cause plastic degradation. If required to reduce or eliminate this defect, the entry to the die is tapered or streamlined. This defect is also called elastic turbulence, alligatoring, or herringbone.

During flow through the die, the melt next to the die tends not to move, whereas that in the center flows rapidly. When the melt leaves the die, its flow profile is abruptly changed to a uniform velocity. This change requires a rapid acceleration of the surface layer, resulting in a high local stress. If this stress exceeds a critical value, the surface breaks, giving a rough appearance (sharkskin) The worst condition occurs in plastic with a high modulus and high viscosity, or in high MW plastics of narrow MWD at low temperatures and high extrusion rates. The addition of die-lip heating, locally reducing the viscosity, is effective in reducing sharkskin.

Nonplastication of melt uneven stress distribution occurs with subsequent lumpiness. The product could look ugly or have a fine matte finish. With a wide MWD, there could be a lack of gloss.

Many plastics contain small quantities of material that boils at processing temperatures, or they may be contaminated with moisture/water. These volatiles can cause bubbles, a scarred surface, and/or other defects. Removial of volatiles is reviewed in Chapter 4.

The transition from room temperature to a high processing temperature may decrease plastic's density up to 25%. Cooling causes possible shrinkage (up to 3%) and may cause surface distortions or voiding with internal frozen strains. As reviewed in other chapters, this condition can be eliminated or reduced by minimizing elevated temperatures and cooling rates.

Processing operations are usually carried out under highly nonisothermal conditions in which the melt passes through a rather complicated temperature history with solidification. Depending on the process and product, the cooling rates can be as high as five hundred to thousands of degrees per minute. During the initial stage of cooling, the surface layer cools so fast that the temperature difference between the surface layer and the core attains a maximal value. Then the core cools faster than the surface layer until uniform temperature is attained. During the initial stage of cooling, thc contraction of the surface layer is greater than the contraction of the core. This action introduces tensile stresses at the surface and compressive stresses in the core [3]. Later the core contracts more than the surface, leading to tensile stresses in the core and compressive stresses in the surface.

Stresses that arise during rapid nonhomogeneous cooling of the melt through the glass transition temperature or melting point are called thermal, cooling, or residual stresses; stresses due to a change in temperature. In this example the thermal stresses are introduced into a plastic that was originally stress-free and externally not constrained.

Thermal stresses also arise in externally constrained plastic products subjected to a change in temperature (heating or cooling) even without the occurrence of a solidification or phase transition. Thermal stresses that arise in products are developed from a melt which is not originally stress-free. They occur in the presence of unrelaxed flow stresses. Accordingly, development of residual stresses in products during processing is due to the coupled effect of flow stresses and of thermal stresses arising from a temperature gradient and an external constraint at the boundary layer. Thus, residual stresses identifies a system of stresses that are in effect locked into a product, even without external forces acting on it.

It is well known that the presence of residual stresses greatly affects the mechanical properties and performance characteristics of products. Understanding the factors governing their development is of great importance to the design of products. It is important to be able to measure and predict the development of residual stress during processing.

## THERMAL PROPERTIES

In order to select plastics that will maintain acceptable mechanical and other performance requirements, one must be aware of both the normal and extreme operating environments to which a product will be subjected, including the environment (conditions) during processing to fabricate the product. Processes are influenced by factors such as the thermal characteristics of plastics, such as melt temperature ($T_m$), glass transition temperature ($T_g$), thermal conductivity, thermal diffusivity, heat capacity, coefficient of linear thermal expansion, and decomposition temperature ($T_d$). Tables 3.7 and 3.8 provide this type of information.

All these properties relate to the selection of the optimum processing conditions. There is a maximum processing temperature, or to be more precise a maximum time-to-temperature ratio, for all plastics before they initiate decompostion or complete destruction. Thermal properties also change according to the types of additives in plastics.

Practically all TPs are processed as a melt at elevated temperatures, necessitating the transfer of substantial quantities of heat, first to the plastic and then to return the extrudate to a solidified state at room temperature. Because of the low thermal conductivities and thermal diffusivities of plastics in general, the process tends to be slow and expensive. The heaters and heat-exchange equipment must therefore be carefully evaluated and designed to ensure efficient and effective heat transfer.

Other more subtle heating and cooling effects may also play an important role. For example, when a mass of crystallizable plastic is cooled, the inner regions of the mass tend to cool more slowly than the surface because of the slow overall heat-transfer process. This cooling-induced morphology may be beneficial or detrimental, depending on the use of the product. Thus, an intimate relationship exists between thermal properties, energy requirements, dimensional considerations and in-line equipment design and fabrication technique. Ultimately, process optimization rests on these relationships.

The melt temperature ($T_m$) occurs at a relatively sharp point for crystalline plastics. The amorphous plastics do not have a true $T_m$; they start melting as soon as the heat cycle begins. In reality, there is no single $T_m$ point, but a range. It is often taken at the peak of the DSC (differential scanning calorimeter) thermal analysis test equipment [3].

The melt temperature $T_m$ depends on the processing pressure and time at heat, particularly during a slow temperature change for relatively thick melts. Also if $T_m$ is too low, the melt's viscosity is high so that more power is required to process the plastic. Degradation can occur if the viscosity is too high.

The glass transition temperature ($T_g$) is the point below which plastic behaves like glass; it is brittle but very strong and rigid. Above this temperature it is neither as strong or rigid as glass, but neither is it brittle. At $T_g$ the plastic's volume or length increases as shown in Figs. 3.9 and 3.10. As shown, TPs have a more definite $T_g$.

When melting plastic in the extruder, the temperature range is between $T_g$ and $T_m$; it is usually easier to process at the higher temperature. However, a temperature just above $T_g$ reduces heating and cooling

**Table 3.7** Thermal properties of TPs and other materials

| Plastic and (morphology)[a] | | Density $g\,cm^{-3}$ ($lb\,ft^{-3}$) | Melt temperature ($T_m$) °C (°F) | Glass transition temperature ($T_g$) °C (°F) | Thermal conductivity $10^{-4}\,cal\,s^{-1}\,cm^{-1}\,°C^{-1}$ ($BTU\,lb^{-1}\,°F^{-1}$) | Heat capacity $cal\,g^{-1}\,°C^{-1}$ ($BTU\,lb^{-1}\,°F^{-1}$) | Thermal diffusivity $10^{-4}\,cm^2\,s^{-1}$ ($10^{-3}\,ft^2\,h^{-1}$) | Thermal expansion $10^{-6}\,°C^{-1}$ ($10^{-6}\,°F^{-1}$) |
|---|---|---|---|---|---|---|---|---|
| PP | (C) | 0.9 (56) | 168 (334) | 5 (41) | 2.8 (0.068) | 0.9 (0.004) | 3.5 (1.36) | 81 (45) |
| HDPE | (C) | 0.96 (60) | 134 (273) | −110 (−166) | 12 (0.290) | 0.9 (0.004) | 13.9 (5.4) | 59 (33) |
| PTFE | (C) | 2.2 (137) | 330 (626) | −115 (−175) | 6 (0.145) | 0.3 (0.001) | 9.1 (3.53) | 70 (39) |
| PA | (C) | 1.13 (71) | 260 (500) | 50 (122) | 5.8 (0.140) | 0.075 (0.003) | 6.8 (2.64) | 80 (44) |
| PET | (C) | 1.35 (84) | 250 (490) | 70 (158) | 3.6 (0.087) | 0.45 (0.002) | 5.9 (2.29) | 65 (36) |
| ABS | (A) | 1.05 (66) | 105 (221) | 102 (215) | 3 (0.073) | 0.5 (0.002) | 3.8 (1.47) | 60 (33) |
| PS | (A) | 1.05 (66) | 100 (212) | 90 (194) | 3 (0.073) | 0.5 (0.002) | 5.7 (2.2) | 50 (28) |
| PMMA | (A) | 1.20 (75) | 95 (203) | 100 (212) | 6 (0.145) | 0.56 (0.002) | 8.9 (3.45) | 50 (28) |
| PC | (A) | 1.20 (75) | 266 (510) | 150 (300) | 4.7 (0.114) | 0.5 (0.002) | 7.8 (3.0) | 68 (38) |
| PVC | (A) | 1.35 (84) | 199 (390) | 90 (194) | 5 (0.121) | 0.6 (0.002) | 6.2 (2.4) | 50 (128) |
| Aluminum | | 2.68 (167) | 1000 | | 3000 (72.5) | 0.23 | 4900 (1900) | 19 (10.6) |
| Copper/bronze | | 8.8 (549) | 1800 | | 4500 (109) | 0.09 | 5700 (2200) | 18 (10) |
| Steel | | 7.9 (493) | 2750 | | 800 (21.3) | 0.11 | 1000 (338) | 11 (6.1) |
| Maple wood | | 0.45 (28.1) | 400 (burns) | | 3 (0.073) | 0.25 | 27 (10.5) | 60 (33) |
| Zinc alloy | | 6.7 (418) | 800 | | 2500 (60.4) | 0.10 | 3700 (1430) | 27 (15) |

[a] C = crystalline resin, A = amorphous resin.

**Table 3.8** General properties of a few plastics

| *Polymer* | *Shear modulus G' (GPa)* | *Young's modulus E' (GPa)* | *Tensile stress* $\sigma_{UT}$ *(MPa)* | *Tensile strain* $\varepsilon_{UT}$ *(%)* | *Poisson's ratio* $\mu'$ | *Specific heat C* $(J kg^{-1} K^{-1})$ | *Thermal conductivity* $\lambda$ *at 2 K* $(mW m^{-1} K^{-1})$ | *Integral thermal expansion* $\Delta L/L$ *(%)* |
|---|---|---|---|---|---|---|---|---|
| Epoxy resin | 2.6 | 8.1 | 179 | 2.1 | 0.37 | 1.6 | 70 | −1.2 |
| Thermoplastic polymers | | | | | | | | |
| HDPE | 3.5 | 9.7 | 175 | 4.0 | 0.30 | 1.0 | 26 | −1.8 to −2.2 |
| PTFE | 2.8 | 8.5 | 81 | 1.2 | | | | −1.6 |
| PS | 2.1 | 5.0 | 68 | 2.1 | 0.35 | 4–6 | 25 | −1.5 |
| PSU | 1.9 | 6.0 | 150 | 3.0 | 0.37 | | | −1.1 |
| PVC | 2.6 | 7.9 | 110 | | | | 20 | |
| PC | 2.2 | 6.0 | 170 | 3.5 | 0.39 | | 25 | −1.4 |
| POM | 5.0 | 13.0 | 170 | 1.5 | 0.32 | 1.1 | 30 | −1.5 |
| Polyamide | 2.55 | 7.5 | 190 | 2.2 | 0.30 | 1.6 | | −0.8 |
| Polyimide | 2.35 | 6.6 | | | 0.39 | | 40 | −0.8 |

requirements. A 'happy' medium is determined based on the capability of the interface of the plastic extruder. After the melt exits the die orifice, the target is to be below $T_g$, so that it starts its solidification.

A plastic's thermal properties, particularly its $T_g$, influences its processability in many different ways. The selection of a plastic should take these thermal properties into account. As an example, more expensive plastic could cost less to process because of its shorter procesing time and because it uses less energy, and so on.

As can be seen from Table 3.7, the value of $T_g$ for a particular plastic is not necessarily a low temperature, which helps explain some of the differences observed in plastics. For example, because at room temperature PS and PMMA are below their $T_g$, one observes them in their glassy state. But at room temperature, natural rubber is above its $T_g$ [$T_g$ = −75°C (−130°F); $T_m$ = 30°C (86°F)], with the result that it is very flexible. When cooled below $T_g$, natural rubber becomes very hard and brittle.

The different plastics have different thermal stablities. Fluoropolymers have exceptional thermal stability compared to most plastics. This is due to the extremely strong carbon-to-carbon and carbon-to-fluorine bonds. Such stability is not only advantageous in extruded products, but is essential during their melt processing.

Dimensional stability is an important thermal property for the majority of plastics. For most plastics, the main determinant of dimensional stability is their $T_g$. Only with highly crystalline plastics is $T_g$ not a limitation.

Thermal conductivity (TC) is the rate at which a material will conduct heat energy along its length or through its thickness. As explained throughout this book, TC has an important effect on heating and cooling the extrudate which tends to be an insulator. This condition requires special understanding and operating equipment to maximize heating and cooling efficiencies. The TC of plastics depends on several variables, but it is possible to ascertain the two principal variables, temperature and molecular orientation (MO). In fact, MO may vary within an extruded product, producing a variation in thermal conductivity. It is important for the product designer and processor to recognize such a situation. Certain products require skill to estimate a part's performance under steady-state heat flow.

The heat capacity or specific heat of a unit mass of material is the amount of energy required to raise its temperature by 1°C. Heat capacity and specific heat can be measured either at constant pressure or constant volume. For amorphous plastics, heat capacity and specific heat increases with temperature in an approximately linear fashion below and above its $T_g$, but step-like change occurs near $T_g$. No such stepping occurs with crystalline plastics.

Whereas heat capacity is a measure of energy, thermal diffusivity is a

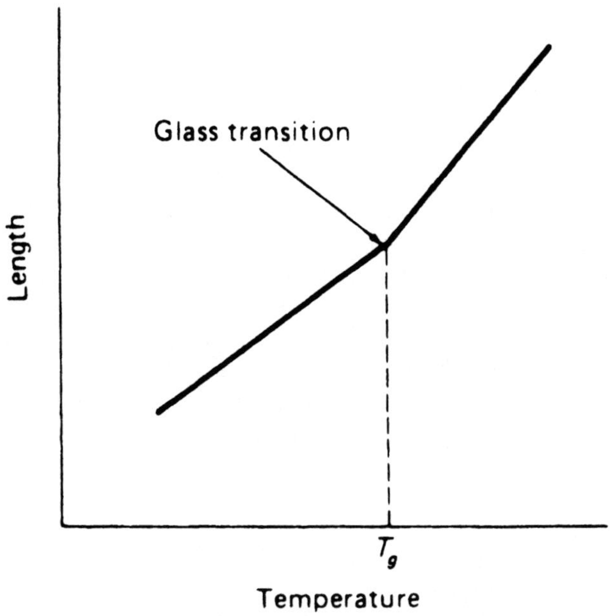


**Figure 3.9** Example of $T_g$ on length (or volume) of plastics.

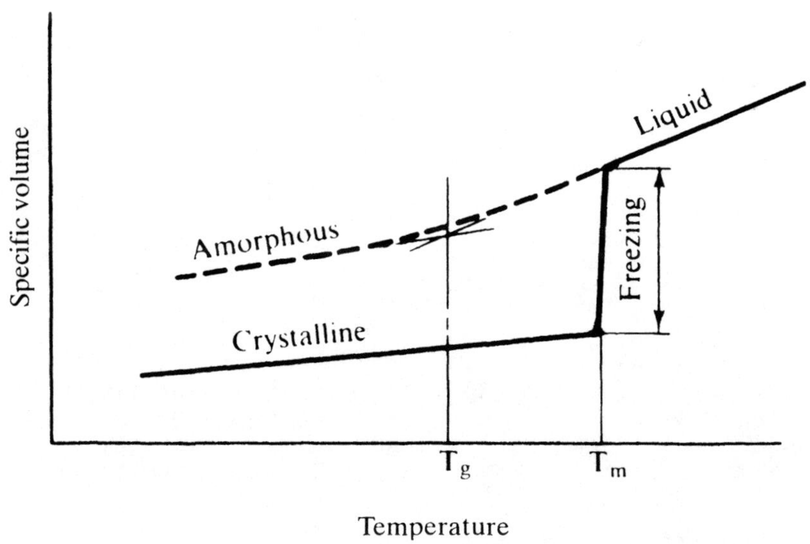


**Figure 3.10** Solidification during processing of glassy/amorphous and crystalline TPs.

measure of the rate at which energy is transmitted through a given plastic. It relates directly to processability. In contrast, metals have thermal diffusivity values hundreds of times larger than those of plastics. Thermal diffusivity determines a plastic rate of energy change with time. This function depends on thermal conductivity, specific heat at a constant pressure, and density, all of which vary with temperature; nevertheless, thermal diffusivity is relatively constant.

Plastics degradation is a deleterious change in characteristics, such as the chemical structure, physical and mechanical properties, and/or appearance of plastic. A degraded appearance usually means discoloration. Degradation can occur during processing (Table 3.6). Factors that determine the rate of degradation are: (1) residence time; (2) stock (melt) temperature and distribution of stock temperature; (3) deformation rate and deformation rate distribution; (4) presence of oxygen or other degradation-promoting additive; and (5) presence of antioxidants and other stabilizers.

**Table 3.9** Guideline for purging agents

| *Material to be purged* | *Recommended purging agent* |
|---|---|
| Polyolefins | HDPF |
| Polystyrene | Cast acrylic |
| PVC | Polystyrene, general-purpose, ABS, cast acrylic |
| ABS | Cast acrylic, polystyrene |
| Nylon | Polystyrene, low melt index HDPE, cast acrylic |
| PBT polyester | Next material to be run |
| PET polyester | Polystyrene, low melt index HDPE, cast acrylic |
| Polycarbonate | Cast acrylic or polycarbonate regrind; follow with polycarbonate regrind; do not purge with ABS or nylon |
| Acetal | Polystyrene; avoid any contact with PVC |
| Engineering resins | Polystyrene, low melt index, HDPE, cast acrylic |
| Fluoropolymers | Cast acrylic, followed by polyethylene |
| Polyphenylene sulfide | Cast acrylic, followed by polyethylene |
| Polysulfone | Reground polycarbonate, extrusion-grade PP |
| Polysulfone/ABS | Reground polycarbonate, extrusion-grade PP |
| PPO | General-purpose polystyrene, cast acrylic |
| Thermoset polyester | Material of similar composition without catalyst |
| Filled and reinforced materials | Cast acrylic |
| Flame-retardant compounds | Immediate purging with natural, non-flame-retardant resin, mixed with 1% sodium stearate |

**Table 3.10** Guidelines for plastic changes

| *Material in machine* | *Material changing to* | *Mix with rapid purge and soak* | *Temperature-bridging material* | *Follow with* |
|---|---|---|---|---|
| ABS | PP | ABS | – | PP |
| ABS | SAN | SAN | – | SAN |
| ABS | Polysulfone | ABS | PE | Polysulfone |
| ABS | PC | ABS | PE | PC |
| ABS | PBT | ABS | PE | PBT |
| Acetal | PC | Acetal | PE | PC |
| Acetal | Any material | PE | – | New material |
| Acrylic | PP | Acrylic | – | PP |
| Acrylic | Nylon | Acrylic | – | Nylon |
| TPE | Any material | PE | – | New material |
| Nylon | PC | PC | – | PC |
| Nylon | PVC | Nylon | PE | PVC |
| PBT | ABS | PBT | PE | ABS |
| PC | Acrylic | PC | – | Acrylic |
| PC | ABS | PC | PE | ABS |
| PC | PVC | PC | PE | PVC |
| PE | Ryton | PE | PE | Ryton |
| PE | PP | PP | – | PP |
| PE | PE | PE | – | PE |
| PE | PS | PS | – | PS |
| PETG | Polysulfone | PETG | – | Polysulfone |
| Polysulfone | ABS | Polysulfone | PE | ABS |
| Polysulfone | ABS | Cracked acrylic | – | ABS |
| PP | ABS | ABS | – | ABS |
| PP | Acrylic | Acrylic | – | Acrylic |
| PP | PE | PE | – | PE |
| PP | PP | PP | – | PP |
| PS | PP | PP | – | PP |
| PVC | Any material | LLDPE or HDPE | – | New material |
| PVC | PVC | LLDPE or HDPE | – | PVC |
| PPS | PE | PPS | PE | PE |
| SAN | Acrylic | Acrylic | – | Acrylic |
| SAN | PP | SAN | – | SAN |

## PURGING

Purging has always been a necessary 'evil' that consumes substantial nonproductive amounts of plastics, labor, and machine time. It is sometimes necessary to run hundreds of pounds of plastic to clean out the last traces of a dark color before changing to a lighter one. Sometimes there is no choice but to pull the screw for a thorough cleaning [187, R. Gregory, K. Powell, W. Foerster, T. Womer].

Although there are few generally accepted rules on purging agents to use (Table 3.9) and how to purge, the following tips should be considered: (1) try to follow less viscous with more viscous plastics; (2) try to follow a lighter color with a darker color plastic; (3) maintain equipment by using preventative maintenance; (4) keep the materials handling equipment clean; and (5) use an intermediate plastic to bridge the temperature gap such as that encountered in going from acetal to nylon (Table 3.10).

Ground/cracked cast acrylic and PE-based materials generally are the main purging agents. Others are used for certain plastics and machines (Table 3.9). Cast acrylic, which does not melt completely, is suitable for virtually any plastic. PE-based compounds containing abrasive and release agents have been used to purge the 'softer' plastics, such as other olefins, styrenes, and certain PVCs. These type purging agents function by mechanically pushing and scouring residue out of the extruders. About 18 g/mm (5–10 lb/in) of the extruder screw diameter is a guide as to what is needed. As reviewed in Chapter 2, special conditions and preparations are required (removing dies, screen packs, etc.).

## RECYCLING

Most processing plants have been reclaiming reprocessable TP materials (Fig. 1.5) such as blown film trim (Fig. 3.11), scrap, flash, rejected products,

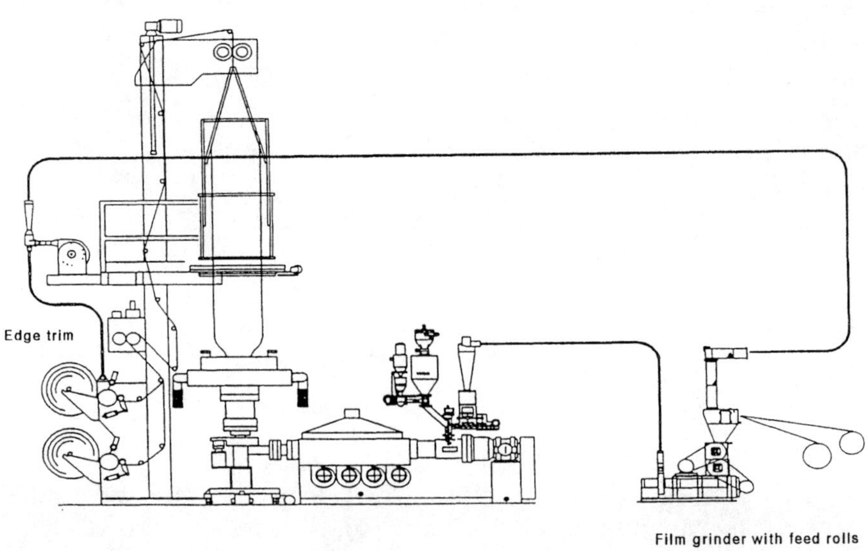


**Figure 3.11** Standard TE-TFH scrap reclaim system.

and so on. If possible the goal is to significantly reduce or eliminate 'trim, scrap, etc.' because it has already cost money and time to go through the process; granulating just adds more money and time. Also it usually requires resetting the process to handle blending it with virgin plastics that does not have uniform particle sizes, shapes, and properties.

Since scrap can be a mixture ranging from fine dust to large irregular chunks of different shapes, thicknesses, etc., it is important to use a granulator that provides the most uniformity and the least damage to the scrap. The principal damage is caused by overheating during the cutting action of the granulator. A granulator that handles soft plastics will not work well when granulating hard plastic. One that handles thin plastics is

**Figure 3.12** Goodyear had this two-piece suit and matching tie made from recycled 2-liter PET beverage bottles in 1978.

not the proper type to handle thick plastics, and so on. Keeping the scrap clean before and after granulating is a requirement.

Granulated plastics may or may not be reformulated by the addition of stabilizers, pigments, plasticizers, fillers, reinforcements, and other additives. Certain blends, particularly general purpose commodity plastics, can be blended with additives to improve their processability and/or product performances. This type of action is considered to improve granulated material that has lost too much of their properties (Figs. 2.24 and 2.25).

Figure 3.12 is an example of Goodyear's two-piece suit and matching tie made from extruded fibers that used recycled two-liter PET stretched injection blow molded bottles (Chapter 15). This suit was shown to the public during 1978; in 1980 it was donated to the Ripley's Believe It or Not Museum, Dells, WI. The recycling process shredded the bottles into small flakes for reprocessing.

## DRYING

All plastics, to some degree, are influenced by the amount of moisture or water they contain before processing. With minimal amounts of moisture in many plastics, mechanical (Fig. 3.13), physical, electrical, aesthetic, and other properties may be affected, or may be of no consequence. However, there are certain plastics that, when compounded with certain additives such as color, could have devastating results. There are also hygroscopic plastics that require special drying equipment.

It is important to recognize that moisture is a major problem which has continuously influenced degradation during processing. Even those

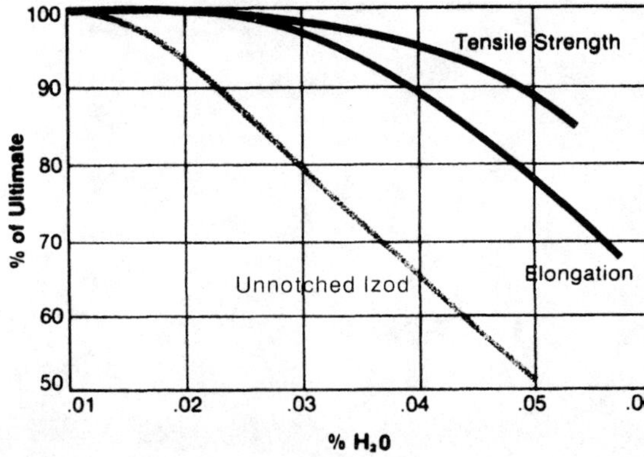


**Figure 3.13** Moisture affects the mechanical properties of hygroscopic PET plastic during processing.

which are not generally affected by moisture can only tolerate a certain amount. Day-to-night moisture contamination can be a source of problems if not adequately eliminated; otherwise it has an accumulative effect.

During the drying process (Fig. 3.14) at ambient temperature and 50% relative humidity, the vapor pressure of water outside a plastic is greater than within. Moisture migrates into the plastic, increasing its moisture content until a state of equilibrium exists inside and outside the plastic.

But conditions are very different inside a drying hopper (etc.) with controlled environment (Fig. 3.15). At a temperature of 170°C (350°F) and −40° (−40°C) dew point, the vapor pressure of the water inside the plastic is much greater than the vapor pressure of the water in the surrounding area, so moisture migrates out of the plastic and into the surrounding air stream, where it is carried away to the desiccant bed of the drier.

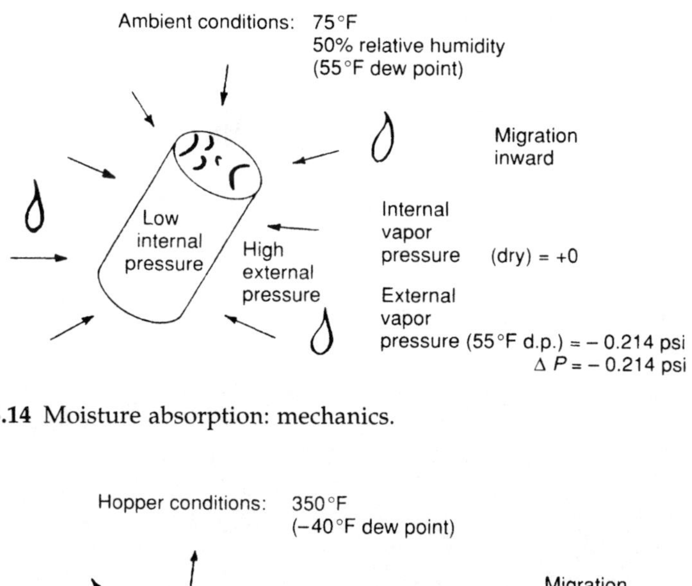


**Figure 3.14** Moisture absorption: mechanics.

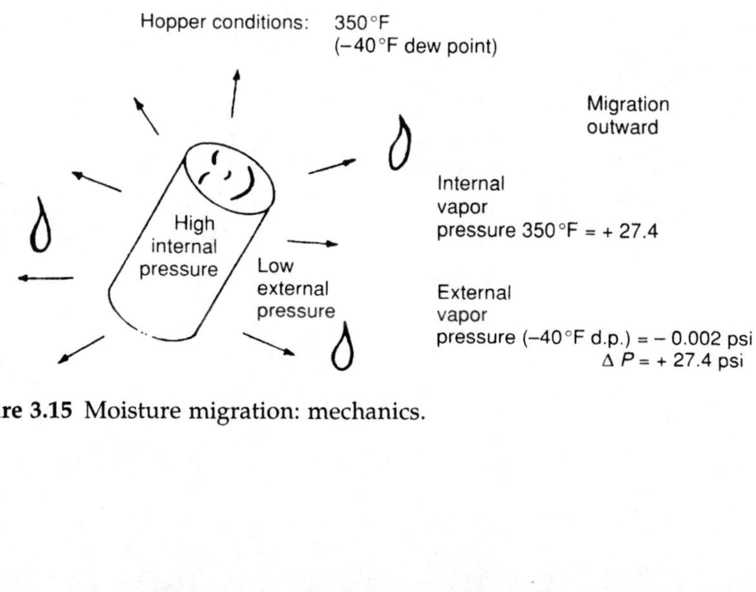


**Figure 3.15** Moisture migration: mechanics.

Before drying can begin, a wet material must be heated to such a temperature that the vapor pressure of the liquid content exceeds the partial pressure of the corresponding vapor in the surrounding atmosphere. The effect of the atmospheric vapor content on the rate of the drier as well as the effect of the material temperature is conveniently studied by different devices such as the construction of a psychometric chart (Fig. 3.16). It plots moisture content dry-bulb temperature, wet-bulb or saturation temperature, and enthalpy at saturation.

First, one determines from the material supplier and/or experience the plastic's moisture content limit. Next, one determines which procedure

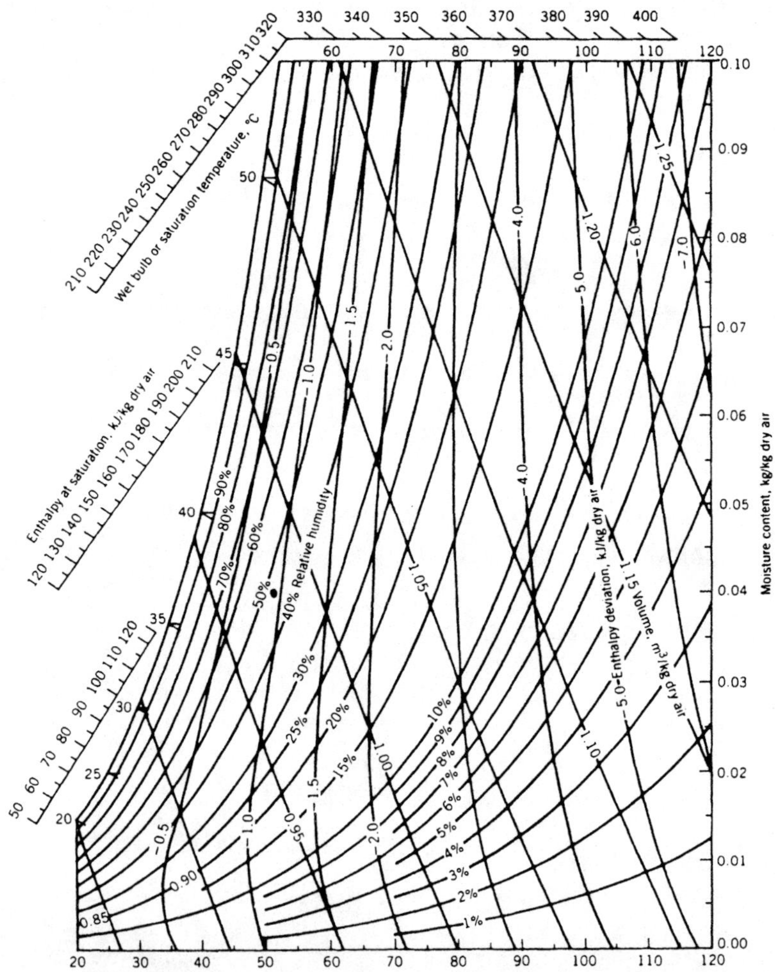


**Figure 3.16** Psychrometric chart: air-water vapor at 1 atm (101 kPa).

will be used in determining water content, such as weighing, drying, and/or reweighing. This procedure has definite limitations. Fast automatic analyzers, suitable for use with a wide variety of plastic systems, are available that provide quick and accurate data for obtaining the in-plant moisture control of plastics.

Drying or keeping moisture content at designated low levels is important, particularly for hygroscopic types. They have to be dry prior to processing. Usually the moisture content is 0.02 wt%. In practice, a drying heat 30°C (86°F) below the softening heat has proved successful in preventing caking of the plastic in a drier. Drying time varies in the range of 2–4 h, depending on moisture content. As a rule of thumb, the drying air should have a dew point of −34°C (−30°F) and the capability of being heated up to 121°C (250°F). It takes about $0.03\,m^3/min$ ($1\,ft^3/min$) of plastic processed when using a desiccant drier. The pressure drop through the bed should be less than 1 mm $H_2O$ per mm of bed height.

Simple tray driers or mechanical convection, hot-air driers, while adequate for certain plastics, are simply incapable of removing enough water for the proper processing of hygroscopic plastics, particularly during periods of high humidity.

Hygroscopic plastics are commonly passed through dehumidifying hopper driers before entering a screw plasticator. However, except where extremely expensive protective measures are taken, the drying may be inadequate, or the moisture regained may be too rapid to avoid product defects unless barrel venting is provided (Chapter 4). To ensure proper drying for 'delicate' parts, such as lenses and compact disks, the combination of drying the plastics and using vented extruders provides a double check. However, just using vented extruders can be suitable.

Plastic usage for a given process should be measured so as to determine

**Table 3.11** Drying equipment for coaters

| *Heat transfer* | *Web handling* |
|---|---|
| Convection dryers | |
| Parallel airflow | Idler-supported dryers |
| Impingement air | Conveyor dryers |
| Airfoil | Catenary dryers |
| Through dryers | U-type dryers |
| Infra-red radiation dryers | Arch dryers |
| Near infrared (electric) | Tenter frame dryers |
| Far infrared (electric or gas) | Floater dryers |
| Conduction dryers | |
| Hot-roll dryers | |

**Table 3.12** Moisture troubleshooting guide

| *Problem* | *Cause* | *Solution* |
|---|---|---|
| Silver streaks, splay | Wet material due to improper drying, high percent regrind, over wet virgin resin | Follow resin manufacturer's drying instructions and dryer manufacturer's operating and maintenance instructions |
| | | Use desiccant dryer |
| Brown streaks/burning | Contamination | Purge barrel/screw and clean dryer/auxiliary equipment |
| | | Check resin |
| | Overheating of material | Check molding equipment settings and controls |
| | | Check resin manufacturer's instructions about processing temperatures |
| Bubbles | High moisture content | Check each step of drying process |
| | | Check dried resins exposure to air |
| | Trapped air | Force air out of feed vent |
| | | Increase screw speed and/or back pressure |
| Brittle parts | Wet resin or overdried resin | Check drying instructions and conditions |
| | Molded-in stresses | Increase melt temperature/reduce injection pressure |
| | | Review part design |
| | Poor part design | Review design for notches and other stress concentrators |
| Flash | Wet material | Check drying procedures |
| | Insufficient clamp tonnage | Use larger machine |
| | Excessive vent depth | Change mold design |
| | High injection pressure | Decrease injection pressure |
| | Damaged mold | Repair damage |
| | Misaligned platen | Realign platen |

| | | |
|---|---|---|
| | Material temperature too high | Decrease material temperature by lowering cylinder temperature, decrease screw speed/lower back pressure (screw machine) |
| | Cycle time too long | Decrease overall cycle time |
| | Plunger pushing forward too long | Decrease plunger forward time |
| Material in drying hopper caking or meltdown occurring | Process temperature set too high | Check resin data sheet for meltdown temperature |
| | | Make sure operators know correct process temperature set point |
| Dew point reading too high | Dirty process/auxiliary filter(s) | Clean or replace filters[a] |
| | Desiccant saturated | Dry cycle machine for several complete cycles. (This is common with equipment which is not operated on a continual basis) |
| | Material residence time in hopper too short | Replace with larger hopper[b] |
| | Return air temperature too high | Add 'after-cooler' to return air line |
| | Heaters burned out | Replace |
| | Bad heater thermostat or thermocouple | Replace |
| | Cycle timer malfunctioning | Adjust or replace |
| | Air control valves not seating properly | Adjust |
| | Contaminated or wornout desiccant | Replace[c] |

**Table 3.12** *Continued*

| | | |
|---|---|---|
| | Incorrect blower rotation | Check and correct rotation |
| | Regeneration heating elements inoperative | Check electrical connections; replace elements if needed |
| | Desiccant assembly not transferring | If valve system, check and repair valve/drive assembly<br>If rotational system, adjust drive-assembly. Check electrical connections on motor and replace motor if needed |
| | Moist room air leading into dry process air | Check hopper lid, all hose connections, hoses and filters. Tighten, replace, repair as needed |
| | Dew point meter incorrect | Check meter and recalibrate |
| Dew point cycling from high to low | Electrical malfunctions<br>Desiccant bed(s) contaminated | Check electrical connections on heaters/controller. Repair/ replace[c] |
| Process air temperature too high | Incorrect temperature setting | Reset for correct temperature |
| | Thermocouple not properly located | Secure thermocouple probe into coupling at inlet of hopper |
| | Electrical malfunctions | Check electrical connections and replace if necessary<br>Insulate hopper and hopper inlet air line |
| Excessive changeover temperature | Insufficient reactivation airflow | See 'Insufficient airflow' |
| | Malfunctioning cycle time | Adjust or replace |
| | Blades of blower wheel dirty | Clean |
| Process air temperature too low | Incorrect temperature setting on controller | Reset for correct temperature |
| | Controller malfunctioning | Check electrical connections. Replace/repair if needed |
| | Process heating elements | Check electrical connections<br>Replace/repair if needed |

| | | |
|---|---|---|
| | Thermostat malfunction | Replace or repair |
| | Voltage differentials | Check supply voltage |
| | Inadequate airflow | Check/clean filters, check blower rotation and correct, check and repair airflow meter |
| | Hose connections incorrect | Check connections. Delivery hose should enter hopper at bottom |
| | Inadequate insulation | Insulate hopper and hopper inlet air line |
| | Dryer inadequate for required temperatures | Replace with high temperature dryer |
| Insufficient airflow (Dew point reading could be good but resin is still wet) | Process or auxiliary filter(s) blocked | Clean or replace[a] |
| | Blower rotation incorrect | Check manufacturer's electrical instructions, and change blower rotation |
| | Air ducts blocked | Remove obstruction |
| | Airflow meter incorrect | Disengage line exiting dryer repair if needed |
| | Desiccant bed contaminated | Replace desiccant[c] |
| | Tightly packed material in hopper | Increase hopper size or drain hopper and refill |
| Heater burn out | Excessive vibration | Relocate dryer, reduce vibration |
| | High voltage condition | Reduce voltage, relocate dryer, or use heaters rated for actual voltage |
| | Malfunction in heater thermostat | Adjust or replace |
| | Blades on blower wheel dirty | Clean |

[a] An inexpensive pressure-differential switch, common option for almost every brand of dehumidifying dryer, will signal when a filter is restricting airflow.
[b] Since drying systems tend to be designed for a specific material, different materials may need longer residence times or higher drying temperatures.
[c] Plastic dust contaminants, because of their flash point, can ignite during regeneration of the desiccant bed causing a fire inside the dehumidifier.

how much plastic should be loaded into the hopper. Usually the hopper should hold enough dried plastic for 0.5–1 h of production. This action is taken so as to prevent storage in the hopper for any length of time eliminating potential moisture contamination from the surrounding atmospheric area. Care should be taken to ensure that hygroscopic plastics are in an unheated hopper for no more than 0.5–1 h, or as specified by the material supplier (and/or experience).

Table 3.11 provides information on drying equipment for coaters. Table 3.12 provides a troubleshooting guide based on what can happen with improper drying.

# 4

# Plasticizing screw process

## OVERVIEW

A very important component of any extruder is the plasticator with its screw and barrel. Figure 4.1 highlights the screw's three processing sections (zones) of feed, transition (or compression), and metering. In extruding plastic, if the proper screw design is not used, products may not meet their performance and cost requirements. The hard steel shaft screw has helical flights which rotates within a barrel to mechanically process and advance the plastic. There are general purpose and dedicated screws used. Type of screw used is dependent on the plastic material to be processed [233, 262, 333, 392].

This review primarily concerns single-screw extruders. The single-screw extruder basically consists of a screw, barrel, drive mechanism, plastic feed arrangement, screen pack, die, and controls (Chapter 2). Table 4.1 provides simplified examples of different screw designs based on product and type plastic used. All screw diameters used in this comparison are 90 mm (3.5 in).

Since practically all plastics processed are TPs, this review concerns the processing of TPs. When extruding TS, the screw is usually limited in design, having a compression ratio (feed to metering depth ratio) of 1 to eliminate possible overheating the TS (Fig. 4.2). The TS cannot be permitted to overheat in the barrel; if overheated, it will solidify (Chapter 3). If it does, the screw must be removed from the barrel and the solidified TS removed from the screw.

The features common to all plastics screw extruders are screw(s) with matching barrel(s) are at least one feed port, and one discharge port to which a die is fitted. The essential factor in their 'pumping' process is the interaction between the rotating flights of the screw and the stationary barrel wall. If the plastic is to be mixed and conveyed at all, its friction must be low at the screw surface but high at the barrel wall. If this basic

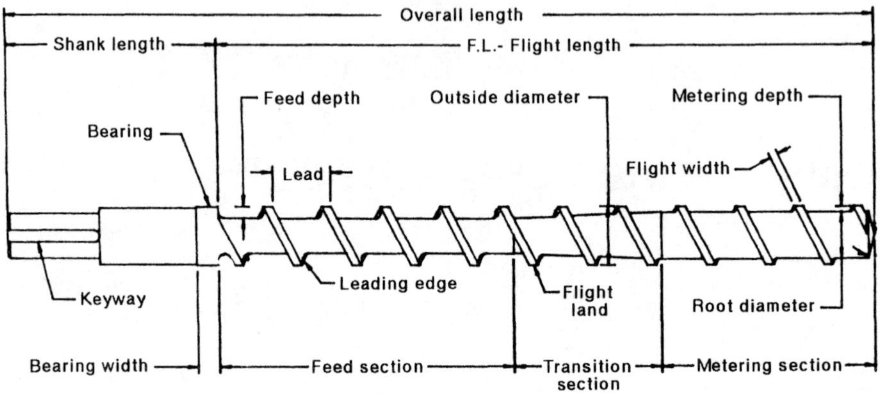


**Figure 4.1** Extrusion screw nomenclature with usual 17.6° flight helical angle.

**Table 4.1** Examples of different screws for different products

| | | *Depth (mm)* | | | *Length (diameters)* | | | | |
|---|---|---|---|---|---|---|---|---|---|
| *Purpose* | *Polymer* | *Feed* | *Transition*[a] | *Meter* | *Feed* | *Transition* | *Meter* | *Total* | *Compression ratio*[b] |
| High-output compounding | LDPE | 15 | 15–6.5T | 6.5 | 6 | 12 | 6 | 24 | 2.1:1 |
| General purpose | LDPE, HDPE, PS | 15 | 15–5.5T | 5.5 | 4 | 10 | 6 | 20 | 2.4:1 |
| Film | LDPE | 15 | 15–3.0S | 3.0 | 10 | 11 | 9 | 20 | 4.3:1 |
| High-output compounding | PP powder | – | 15–7.9T | 7.9 | – | 18 | 2 | 20 | 1.8:1 |
| General purpose | PP | – | 15–6.0T | 6.0 | – | 16 | – | 16 | 2.2:1 |
| General purpose | PP | – | 12.7–5.6T | 5.6 | – | 10.9 | 9.1 | 20 | 2.1:1 |
| General purpose | POM | 15 | 15–5.0T | 5.0 | 4 | 12 | 4 | 20 | 2.7:1 |
| General purpose | PMMA | 15 | 15–4.0S | 4.0 | 14 | 1 | 5 | 20 | 3.3:1 |
| Compounding | Nylon 66 | – | 11.1–9.6T/S | 2.8 | – | $12 + \frac{1}{2}$ | 7.5 | 20 | 3.6:1 |
| Film | PET | 15 | 15–3.0S | 3.0 | 14 | 1 | 9 | 24 | 4.3:1 |
| Film | UPVC | 15 | 15–7.0T | 7.0 | 4 | 12 | 4 | 20 | 1.9:1 |
| Low-output compounding | UPVC | 15 | 15–9.0T | 9.0 | 4 | 7 | 4 | 15 | 1.5:1 |

[a]T = tapered; S = stepped.
[b]Cross-section basis.

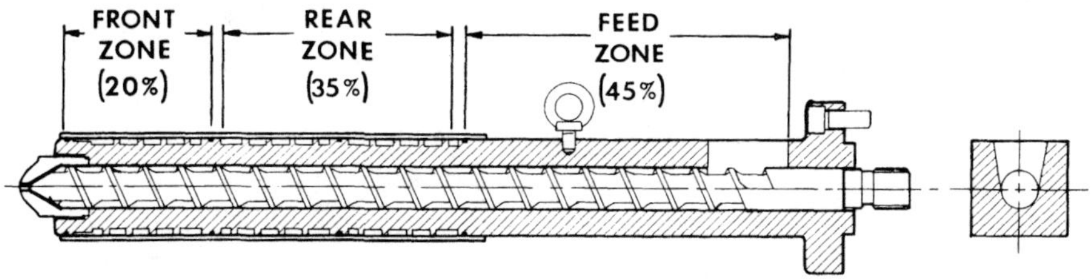


**Figure 4.2** Thermoset plastic screw with a compression ratio (CR) of 1.

criterion is not met, the material may rotate with the screw without moving at all in the axial direction and out through the die. The clearance between the screw and barrel is usually extremely small.

The general-purpose single-screws are designed to suit as wide a range of plastics as possible. They will not be the ideal answer for the extrusion of specific plastics. As an example, a screw designed for a semicrystalline (usually called crystalline) plastic must provide initially at least a greater heat input than an amorphous thermoplastic screw (Chapter 3). Thus, when a specific material is going to be used for a long run, it becomes economically very beneficial to use a dedicated screw. The design of a screw is determined by the melt flow or rheological characteristics of the plastics; plastic material dictates screw design.

Although an all-encompassing (or universal) design was once employed by many (prior to about 1967), today's screws are generally customized for a particular process. The specific screw type involves factors such as the type of plastic being processed, its temperature limitations, the degree of mixing required, amount of pressure required to move the plastic, and form of extrudate. Current designs are configured for such specific needs as high dispersion, high distribution (mixing), high to low temperatures, and good homogeneity.

Another design factor involves removal of unwanted vapors during processing (which may be facilitated by use of vents at various points in the barrel) and the need for sufficient pressure to move the plastic through a die. Pressure requirements usually run from 100 MPa (1500 psi) for a compounding or reclaim material to 275–350 MPa (4000–5000 psi) for a blown film line.

Extruders are usually designated by their barrel diameter and screw L/D. Diameters are 120–150 mm (4.5–6 in). The L/Ds range from 6–48 with the popular types having values of 24–36 L/D (24 being the most popular). The English standard sizes for screws are 0.75, 1, 1.5, 2, 2.5, 3.5, 4.5, 6, 8, 10, 12, 14, 16, 18, and 24 inches; metric standards are 20, 25, 30, 35, 40, 50, 60, 120, 150, 200, 250, 300, 350, 400, 450, 500, and 600 mm. (The English and metric sizes do not directly relate mathematically.) The longer

L/Ds of 40 and higher are used to meet special requirements such as in vented extruders requiring extraction of moisture and volatiles from the plastic melt [234, 307, 318, 319, 421].

Spacing between screw and barrel vary depending on screw size. The 64 mm (2.5 in) screw size has clearance of about 0.05 mm (0.002 in), the 76 mm (3 in) screw has clearance of about 0.10 mm (0.004 in), the 114 mm (4.5 in) screw has clearance of about 0.13 mm (0.005 in), and the 152 mm (6 in) screw has clearance of about 0.18 mm (0.007 in). To help move the plastic at the feed end there is usually an undercut of about 0.25 mm (0.010 in). At the screw tip end, an undercut could be about 0.50 mm (0.20 in) to help move the melt forward.

In Table 4.2 recommendations for single-screws are provided as a guide to manufacturers and processors. These standards provide working tolerances that produce effective performance with economy of manufacture prepared by the SPI's Machinery Component Manufacturers Division.

Rate of output (throughput), or the speed at which plastic is moved

**Table 4.2** Single screw dimensions/tolerances standards prepared by SPI

**Lengths, depths and widths**

| Dimension description | *English units* Range (in) | Tolerance (in) | *Metric units* Range (mm) | Tolerance (mm) |
|---|---|---|---|---|
| I. Most linear dimensions | To 12 | ±0.010 | To 300 | ±0.25 |
| | 12–60 | ±0.030 | 300–1500 | ±0.75 |
| | 60–120 | ±0.045 | 1500–3000 | ±1.00 |
| | 120–200 | ±0.060 | 3000–5000 | ±1.50 |
| | over 200 | ±0.090 | Over 5000 | ±2.25 |
| II. Flight width | To 0.500 | ±0.015 | To 12 | ±0.38 |
| | 0.500–1.000 | ±0.020 | 12–25 | ±0.50 |
| | over 1.000 | ±0.030 | Over 25 | ±0.75 |
| III. Channel depth | To 0.100 | ±0.003 | To 2.5 | ±0.08 |
| | 0.100–0.500 | ±0.007 | 2.5–13.0 | ±0.18 |
| | Over 0.500 | ±0.012 | Over 13.0 | ±0.30 |
| IV. Barrier flight depth | To 6.0 in dia. | ±0.002 in | | ±0.05 mm |
| | Over 6.0 in dia. | ±0.003 in | | ±0.075 mm |
| V. Section length | All sizes | ±1/8 turn | ±1/8 turn | |
| VI. Keyways | All sizes | Depth ±0.005 in | Depth ±0.13 mm | |
| VII. Hollowbores | All sizes | Length ±0.030 in | Length ±0.76 mm | |
| VIII. Thread pilots | | | | |
| *Injection | All sizes | Length ±0.005 in | Length ±0.13 mm | |
| Extrusion | All sizes | Length ±0.015 in | Length ±4 mm | |

*This should particularly be observed for screws fitted with nonreturn valves.

**Table 4.2** *Continued*

**Diameters**

| Dimension description | English units Range (in) | English units Tolerance (in) | Metric units Range (mm) | Metric units Tolerance (mm) |
|---|---|---|---|---|
| I. Flight and bearing (hub) diameters | To 6.0 | +0.000–0.002 | To 150 | +0.00–0.05 |
| | Over 6.0 | +0.000–0.004 | Over 150 | +0.00–0.10 |
| II. Shank diameters (drive fit) | All sizes | +0.0–0.002 | All sizes | +0–0.50 |
| III. Hollowbores | To 3 | ±0.020 | To 75 | ±0.5 |
| | 3–6 | ±0.040 | 75–150 | ±1.0 |
| | Over 6 | ±0.060 | Over 150 | ±1.5 |
| IV. Thread pilots | | | | |
| *Injection | All sizes | dia. +0.001–0.000 | Dia. +0.025–0.000 | |
| Extrusion | All sizes | dia. +0.002–0.000 | Dia. +0.050–0.000 | |

*This should particularly be observed for screws fitted with back-flow valves.

**Concentricity – outside diameters**

| English units Length (in) | English units T.I.R. (in) | Metric units Length (mm) | Metric units T.I.R. (mm) |
|---|---|---|---|
| To 100 | 0.004 | To 2500 | 0.100 |
| 100–200 | 0.006 | 2500–5000 | 0.150 |
| 200–300 | 0.010 | 5000–7600 | 0.250 |
| Over 300 | 0.015 | Over 7600 | 0.400 |

Concentricity of outside diameters to include all except root diameter of flighted sections.

Concentricity of diameters in flighted sections cannot be accurately measured due to the interrupting effect of the flight. Flight depth variations taken from a true OD are used as a measure of concentricity in this area.

**Concentricity – inside diameters**

| | | | | |
|---|---|---|---|---|
| I. Thread pilots (non-return registers) | All sizes | 0.0005 in TIR | All sizes | 0.013 mm TIR |
| | Measured to the flight OD | | | |
| II. Thread pilots, threads, and hollowbores | All sizes | 0.001 in TIR | All sizes | 0.025 mm TIR |

Inside diameters are measured concentric to all inside diameters and/or the nearest outside flight diameter.

**Table 4.2** *Continued*

**Flight radii**
Unless otherwise specified, the root radius will not be less than 0.5 of the flight depth up to a 1 in or 25 mm radius.

When specified, the following radius tolerances should be observed: ±0.030 in, ±0.75 mm

**Parallelism and Perpendicularity**
All flights, unless otherwise specified, shall be perpendicular to the screw axis from the root radius to the OD on both sides.

Other surfaces perpendicular to the screw axis can be tested by use of a surface plate and an adjustable height table indicator or a precision square. The following tolerances should be observed:

| | | |
|---|---|---|
| All sizes | ±0.001 in/in | ±0.0025 mm/mm |
| OR | | |
| All sizes | 0.006° | 0.006° |

Parallel surfaces can be determined by TIR or by using a surface plate and an adjustable height gauge. All dimensions meeting the concentricity and/or flight depth standards are acceptable.

**Surface finish**

| | | | | |
|---|---|---|---|---|
| I. | Unplated surfaces | All surfaces | 32 μ | 0.80 μm |
| II. | Chrome plated surfaces | Channel | 16 μ | 0.40 μm |
| | | OD | 32 μ | 0.80 μm |
| | | Bearing/shank | 32 μ | 0.80 μm |

**Screw threads**
The variation in threads used is too broad to be addressed by these recommendations. It is suggested that whenever thread selection is possible that either the ANSI or ISO standards are adhered to for ease of measurement and compatibility.

**Hard surfacing**
All hard surfacing materials should be specified as to alloy, width of weld, and depth of weld.

**Measurement temperature**
All measurements are to be taken at room temperature of 22 ± 11°C or 72 ± 20°F.

through the extruder, has been continuously pushed higher as a result of design advances in equipment and materials. Output rates generally range from a few kilograms to over 5 tonnes per hour on single-screw machines. Twin-screw extruder diameters range from 14–300 mm (0.55–12 in). Their output rates range from a few kilograms to at least 30 tonnes per hour.

Table 4.3 provides a guide to single-screw output rates. Output deviation averages 10–15% based on different processing settings. A rough

**Table 4.3** Guide to extruder output rates in lb/h

| | *Screw diameter (in)* | | | | | |
|---|---|---|---|---|---|---|
| *Material* | *1½* | *2½* | *3½* | *4½* | *6* | *8* |
| ABS | 280 | 400 | 825 | 1350 | 2270 | 4100 |
| Acrylic | 320 | 470 | 900 | 1500 | 2700 | 4750 |
| PC | 210 | 320 | 680 | 1025 | 1850 | 3200 |
| PP | 280 | 400 | 825 | 1350 | 2270 | 4100 |
| HIPS | 340 | 560 | 1100 | 1800 | 3250 | 5750 |
| PVC, flexible | 300 | 450 | 900 | 1500 | 2700 | 4750 |
| PVC, rigid | 180 | 250 | 500 | 800 | 1450 | 2300 |
| LDPE | 310 | 525 | 1050 | 1750 | 3000 | 5500 |
| LLDPE | 200 | 300 | 600 | 1000 | 1800 | 3260 |
| HDPE | 215 | 325 | 725 | 1175 | 2150 | 3750 |

estimate for output rate (OR) in lb/h can be calculated by using the barrel's ID in inches and using the following equation: $OR = 16\,ID^2$ ($lb/h \times 0.4536 = kg/h$).

With nearly all machines, only the cylinder temperature is directly controlled. The actual heat of the melt, within the screw and as it exits, can vary considerably depending on the efficiency of the screw used and the method of operation. Factors affecting melt heat include: the time plastic remains in the plasticator (residence time); the internal surface heating area of the cylinder and the screw per volume of plastic being heated; the thermal conductivity of the cylinder, screw, and plastic (Chapter 3); the heat differential between the cylinder and the melt; and the amount of melt turbulence in the cylinder. In designing the screw, a balance must be maintained between the need to provide adequate time for heat exposure and the need to maximize output most economically.

The constantly turning screw augers the plastic through the heated barrel, where it is heated to a proper temperature profile and blended into a homogeneous melt. The rotation causes forward transport, and is the major contributor to heating the plastic once the initial barrel heat start-up occurs. Before the melt can leave the barrel, it usually passes through a breaker plate and screen pack. This device builds up back pressure in the barrel, filters out contaminants, and tends to convert the turbulent, rotating melt flow into a laminar flow. The melt continues through the die and extruded into the desired shape with the aid of the down-stream equipment [392]. An explanation of the melting action is given in Fig. 4.3, where:

**Figure 4.3** Melt model for standard screw.

1. The feed section initiates solids conveying. This is enhanced by sliding with low friction on the screw and high friction on the barrel. In this section, there is also some material compacting and a little heating of the plastic.
2. At the beginning of the transition, the plastic is further heated and more compression occurs. The solid plastic is forced against the barrel causing a sliding action. This frictional heat creates a thin film of melted plastic on the inner barrel surface.
3. As the plastic proceeds down the transition, there is more melting and more compression. Usually most of the melting takes place in the transition zone. Here the plastic is divided into three parts: a compacted solid bed, a melt film along the barrel surface, and a melt pool. The melt pool is formed as the melt film is collected by the advancing flight. Most of the melting continues to be the result of sliding friction of the solids bed against the heated barrel. This is a rapid and efficient melting action, similar to the melting of an ice cube by pushing it against a hot grinding wheel.
4. The channel depth continues to decrease as plastic progresses down the transition zone. Melting continues and the width/volume of the solids bed decreases, while the width/volume of the melt pool increases. Unfortunately, as the channel gets shallower, shear rate increases. Now the already melted plastic continues to be heated. With too high a heating, undesirable conditions can develop, such as degrading the plastic.
5. Continuing downstream through the plasticator, the solids bed breaks up; the unmelted plastics are distributed throughout the channel like ice cubes in water. The efficient melting by friction of the solids bed against the barrel tends to stop. Now only less efficient melting occurs where overheating in the melt continues in the shallow metering zone. Within this zone the complete melting action should occur.
6. Plastic continues down the shallow metering section to its exit from the plasticator (screw/barrel). There is a possibility that unmelted plastics or the melt has nonuniform temperature and viscosities. This situation of a nonuniform melt usually results in poor product performances, color mixing, and so on. Improved mixing can be obtained by reducing the screw's channel depth; however, overheating and less output occurs. As reviewed in Chapter 2, method of feeding plastics from the hopper can help (or resolve) this situation. The constant depth meter-

ing section is not considered a good mixer when improved mixing is required in this section. This is because smooth laminar flow patterns are established causing the different portions of melt to continue to move in a fairly constant pattern so that no mixing action occurs to eliminate the problem. Screw design plays an important part to eliminate the problem.

In general, heat transfer problems have led screw designers to concentrate on turning screws into more efficient heat transfer devices. As a result, the internal design and performance of screws vary considerably to accommodate the different plastics that are used. Most machines are single-screw, constant pitch, and metering-type screws to handle the majority of plastics.

In contrast to conventional single-screw extruders, the multiple-screw extruders involve the use of the usual two or four screws (Chapter 2). The more popular is the twin-screws with barrels placed side by side. The screw profiles are designed to meet different operations along the screw length for feeding, melting, mixing, venting, die pressurization, etc. Both co- and counter-rotating intermeshing machines have certain general screw geometries in common. Conveying screws are used to transport material, characterized by pitch and conveying direction of forward or reverse. Screw pitch is used to control the residence time distribution and degree of fill in the channels. Conveying screws can have single, double, or triple flights.

Nonintermeshing counter-rotating extruders vary the screw channel depth to effect changes in volume rather than by varying screw pitch or the number of screw flights. Thus, residence time and degree of fill are controlled by selection of screw depth.

Reverse pitch screw elements are used to create back pressure. The degree of fill increases to 100% upstream of the reverse pitch screw; the preceding forward conveying screws must overcome this resistance. The resulting pressure drop is used to creat a melt seal. Reverse pitch screws have the significant impact of increasing residence time distribution and shear impact.

Kneading elements are used for melting, mixing, dispersing, and homogenizing. The kneading elements are comprised of individual kneading disks of various widths, which are offset from each other. The kneading disks can be staggered for forward conveying, reverse conveying, or zero conveying; many have one, two, or three lobes. The level of shear input and type of mixing action are controlled by selecting a disk width, offset angle, number of kneading disks, and direction of conveying.

Various other types of screw geometry's utilize interrupted or cut screw flights. The working principle is the same as for kneading elements;

conveying efficiency is reduced to the benefit of back flow for mixing. The degree to which the screw flight is interrupted directly influences the degree of mixing.

## SCREW DESIGN

Screw design technology is considered to be empirical and/or secretive. The production rate of single-screw extruders is often limited by the melting capacity. The melting capacity of the screw depends on the plastic properties, the processing conditions, and the particular geometry of the screw. Once the melting capacity is predicted, the screw can be designed to match the melting capacity. Scientific approaches to screw designs based on an analytical melting model can be used [152].

The rotating geometric-helical flighted screw mechanically plasticizes, with the help of heat and pressure at a controlled flow rate, and advances a melt to be pumped through a die. Plastic in the screw channel is subject to changes during operation. Each operation of the screw subjects the plastic to different thermal and shear situations. Consequently, the plasticizing process becomes rather complex. However, it is controllable and repeatable within the limits of the equipment and material capabilities [152, 169, 234, 348, 391, 392].

A fixed screw speed, screw pitch, and channel depth determines output. A deep channel screw is much more sensitive to pressure changes than a shallow screw. In the lower pressure range, a deep channel will provide more output; however, the reverse is true at high pressures. Shallower channels tend to give better mixing and flow patterns. Thus, although the screw is usually a simple looking device, it accomplishes many different operations with its three basic sections or zones: (1) solids conveying or feeding; (2) transition or compressing and pressurizing melt; and (3) metering the final mix or melt refinement with its pressure – temperature stabilization.

Figure 4.4 is an example of a general mixing action. Its flow of plastic in the screw is based on an open discharge (A) and/or blocked discharge (B). The schematic drawing of Fig. 4.5 depicts the melting of plastic in a screw where in area A the plastic is melted by heat conduction and area B the plastic is melted by shearing. Area C represents a partially melted plastic. Area D is unmelted plastic that is usually referred to as the solid bed of plastic material.

In the melting zone, the solid bed of plastic is usually pushed against the trailing edge of the flight with a thin melt film between the solid bed and the barrel. From an analysis of the melting process, it can be determined that melting of most plastics (particularly commodity plastics) is improved by using small flight clearance, large helix angle, narrow flight width, barrier and multiple flights, and internal screw temperature con-

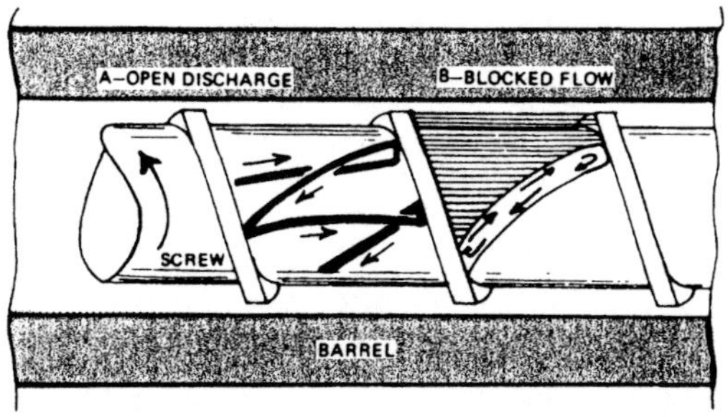


**Figure 4.4** General mixing action and flow of plastic in a screw based on open discharge (a) and/or blocked discharge (a).

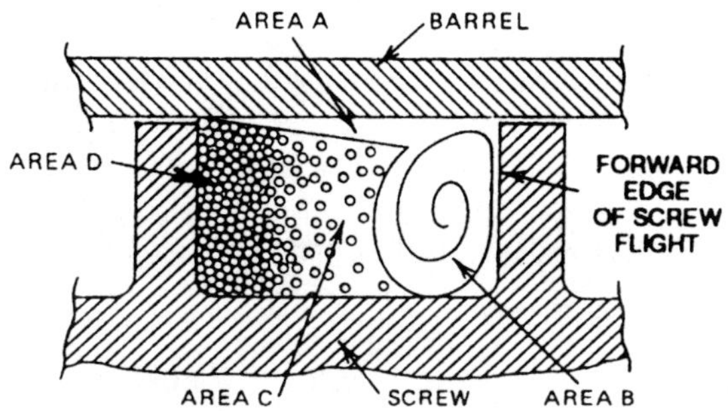


**Figure 4.5** Schematic drawing of plastic melting in a screw.

trol. Limitations can exist due to melt behavior, particularly for plastics that are heat sensitive.

Hypothetical data on screws are given in Table 4.4 for a short screw. It provides examples of variations on the same L/D screw processing different plastics. Advantages of a short L/D screw include: (1) less residence times in the barrel so heat-sensitive plastics are kept at melt heat for a shorter time, lessening the chance of degradation; (2) the design occupies less space; (3) the screw requires less torque, reducing the screw strength and horsepower requirement; and (4) the screw requires a smaller investment cost initially and for replacement. Long L/D screw advantages include: (1) the screw can be designed for a greater output or recovery rate provided that sufficient extruder torque is available; (2) the screw can be designed for more uniform output and greater mixing; (3) the screw will pump at higher pressures and give greater melting with less shear; and (4) the screw will provide more conductive heat from the barrel.

The compression ratio (CR) relates to the compression that occurs on

**Table 4.4** Hypothetical screw designs for general types of plastic

| *Dimensions (in.)* | *Rigid PVC* | *Impact polystyrene* | *Low-density polyethylene* | *High-density polyethylene* | *Nylon* | *Cellulose acetate/butyrate* |
|---|---|---|---|---|---|---|
| Diameter | $4\frac{1}{2}$ | $4\frac{1}{2}$ | $4\frac{1}{2}$ | $4\frac{1}{2}$ | $4\frac{1}{2}$ | $4\frac{1}{2}$ |
| Total length | 90 | 90 | 90 | 90 | 90 | 90 |
| Feed zone (F) | $13\frac{1}{2}$ | 27 | $22\frac{1}{2}$ | 36 | $67\frac{1}{2}$ | 0 |
| Compression zone | $76\frac{1}{2}$ | 18 | 45 | 18 | $4\frac{1}{2}$ | 90 |
| Metering zone (M) | 0 | 45 | $22\frac{1}{2}$ | 36 | 18 | 0 |
| Depth (M) | 0.200 | 0.140 | 0.125 | 0.155 | 0.125 | 0.125 |
| Depth (F) | 0.600 | 0.600 | 0.600 | 0.650 | 0.650 | 0.600 |

the plastic basically in the transition (or compression) section; it is the ratio of the volume at the start of the feed section divided by the volume in the metering section (determined by dividing the feed depth by the metering depth). The CR should be high enough to compress the low bulk unmelted plastic into a solid melt without air pockets. A low ratio will tend to entrap air pockets. High percentages of regrinds, powders, and other low bulk materials are usually processed by a high CR. A high CR can over pump the metering section. Table 4.5 provides a general guide for CR.

A common misconception is that engineering and heat sensitive plastics should use a low CR. This is true only if it is decreased by deepening the metering section, and not having a more shallow feed section. The problem of overheating is more related to channel depths and shear rates than to CR. As an example, a high CR in polyolefins can cause melt blocks in the transition section, leading to rapid wear of the screw and/or barrel. As reviewed in the section on extruding of TSs, the CR is usually 1 so that accidental overheating does not occur and cause the plastic to solidify in the barrel. The barrels used are usually heated using a liquid medium (TSs melt temperatures are usually much lower than those of TPs, so liquid heat control is best) in place of electric heaters so that very accurate control of the melt occurs with no overriding the maximum melt heat. With overheating, a TS melt solidifies. If it solidifies, the CR of 1 also permits ease of removal by just 'unscrewing' it from the screw. CR ratio of 1 are also used for TPs when the rheology so requires.

## Output

The output of a screw is predictable, provided that the melt is under control and reasonably repeatable. With a square pitch screw (conventional screw where the distance from flight to flight is equal to the diameter), a simplified formula for output is [392]:

**Table 4.5** Examples of C/R for different plastics

| | |
|---|---|
| Low-compression screw (1.2 – 1.8 compression ratio) | Acrylics<br>Acrylic multipolymer<br>ABS and SAN<br>Polyvinyl chloride, rigid |
| Medium-compression screw (2.0 – 2.8 compression ratio) | Acetal (Delrin 100)<br>Cellulosics (acetate, propionate)<br>Nylon (low melt index)<br>Phenylene oxide-based resin (Noryl)<br>Polycarbonates<br>Polyethylene (medium to low melt index)<br>Polypropylene (medium to low melt index)<br>Polystyrene (crystal and impact)<br>Polyvinyl chloride (flexible) |
| High-compression screw (3 – 4.5 compression ratio) | Acetal (Delrin 500 and 900; Celcon)<br>Fluoroplastics (Teflon 110)<br>Nylon (high melt index)<br>Polyethylene (high density)<br>Polyethylene (high melt index)<br>Polypropylene (medium to high melt index) |

*Note:* Depending on the melt index and heat (shear) sensitivity of material, compression ratios may differ from those indicated.

$$R = 2.3 D^2 hgN$$

where $R$ = rate or output in kg/h (lb/h), $D$ = screw diameter in mm (in), $h$ = depth in the metering section in mm (in) (for two-stage screw use the depth of the first metering section), $g$ = gravity of the melt, and $N$ = screw rpm. This formula does not take into account back flow and leakage flow over the flights. These flows are not usually a significant factor unless the plastic has a very low viscosity during processing, or the screw is worn out. The formula assumes pumping against low pressure, giving no consideration to melt quality and leakage flow of several worn OD screws.

With all these and other limitations, the formula given above can still provide a general guide to output. If the output is significantly greater that is caused by a high CR, the extruder overpumps in the metering section. This action is sometimes desirable but can lead to surging and rapid screw wear if it is excessive. If output is a lot less than estimated, this usually indicates a feed problem or worn screw and/or barrel. A feeding problem sometimes can be corrected by changes in the barrel heat. More often, the problem is caused by such factors as screw design, shape and bulk density of the feed stock, surface conditions of the screw root, screw heat, and so forth. The problem of the screw/barrel wear can be assessed by measuring the screw/barrel.

An accurate method used to determine output loss due to screw wear is to compare the worn screw's current output with the initial production benchmark that was originally determined by shooting into a 'bucket' to check the weight for a definite time period. Another approach is to measure the worn screw's clearance to the barrel wall (*W*), which is used along with the original measured screw clearance (*O*), and the metering depth (*M*) from the screw root to the barrel wall [392]. Here the approximate percentage output loss (OL), with rpm being constant, is calculated from the formula:

$$\mathrm{OL} = (W - O)/M + 100$$

## Screw torque

The work of melting is done by rotating the screw in a stationary barrel. The rotational force, called torque, is the product of the tangential force and the distance from the center of the rotating member. For example, if a 1 lb (0.454 kg or N) weight was placed at the end of a 1 ft (0.35 m) bar attached to the center of the screw, the torque would be 1 ft × 1 lb or 1 ft-lb (1.36 Nm). Torque is related to horsepower (hp) which equals [torque (ft-lb) × rpm]/5252 or kW = [torque (Nm) × rpm]/7124. The torque output of an electric motor of a given hp depends on its speed. A 30 hp (22 kW) motor has the following torque at various speeds: 87.5 ft-lb (119 Nm) at 1800 rpm, 133 (181) at 1200, and 175 (238) at 900.

The speed of a motor with a given hp is built into the motor. Changes in speed and torque can also be accomplished by changing the output speed of the motor by using a gear train (Chapter 2). The change in torque varies inversely with the speed. As an example, if an AC motor is used, it will develop a starting torque of almost twice the running torque. The screw has to be protected against overload to prevent screw breakage. This is not a problem with hydraulic drives. The drive must supply enough torque to plasticize at the lowest possible screw speed, but not enough to mechanically shear the metal screw. Different torque requirements are used to meet the requirements of the different plastics. As an example, much higher torque is required to plasticize PC than PS. The strength used limits the input hp.

Using little torque to turn the screw indicates the heater bands are providing too much of the energy required to melt the plastic usually as a result of poor or no temperature control. Plastication efficiency suffers in these conditions, and mixing problems and/or long, inconsistent recovery times are symptomatic of this condition.

**Screw cooling**

Various processors, particularly those using PVC, use screw cooling. Some improvement in plastic extrusion is possible by circulating cooling water or oil through the cored center section(s) of the screw, at least the feed section. The amount of cooling required in this 'pipe' is dependent on screw design and operating parameters. Cooling is more critical for larger diameter screws, because the larger volume of melt flow requires more cooling (heat control). Superior extrusion may be achieved by optimizing cooling, but reduced output rates and/or surging may result unless proper processing temperatures are maintained. A definite area for cooling is at the feed entrance from the hopper.

The main objective of screw cooling is to enhance the ability of the screw to advance the solid plastic feed into the extruder at the steadiest possible rate. This is accomplished by providing a more constant and lower coefficient of friction between the screw shank and the plastic. In so doing, the screw is able to rotate inside the mass of unmelted plastic solids while the forwarding of plastic melt takes place inside the barrel surface through the scraping action of the rotating screw flights.

In essence, the extrusion process is a complex heat transfer system in which the temperature in the various locations inside the barrel controls the friction coefficients between the various solids in those locations at any particular moment. It is this friction coefficient that enables the screw to forward the solid plastic to the plastic melt while it is creating a large amount of heat through shear energy coming from the extruder drive motor. The latter is evidenced by the fact that an operating extruder is in the 'cool' mode during most of its operating time.

It is obvious that screw cooling is but a small factor in the extrusion process from the standpoint of heat transfer, but it is a major factor in the performance of the screw insofar as the efficiency of the feed section is concerned. The feed section initiates melting so control of the onset of melting is accomplished by 'tuning' the barrel/screw temperature to attain the optimum feed rate and uniformity. Usually, the best location for the cooling pipe is within about the first five screw flights. It is in this general area that the optimum penetration for most plastics is found. With an adjustable assembly, the cooling pipe can be moved in either directions to find the best location. A good cooling installation should include a closed loop circulating system with a 2°C (1°F) at a flow rate of 15–25 g/m. With proper installation and operation benefits include, more uniform rates, increased output, improved mixing, lower melt temperature/shorter residence time in the extruder, reduced surging, fewer feed problems with variable feed particles/fines, and improved product performance [187, M. Billings].

**Performance**

Single screw extruders have changed greatly over the years (Figs. 4.6 and 4.7). Today's functional modular concepts have been developed mainly for reasons of effectiveness and favorable cost comparisons. Their output rates have significantly surpassed those of the older designs. The performance of machines and production lines (film, profile, coating, etc.) will depend on the many factors that are controlled and synchronized going from upstream through the extruder and the downstream equipment. The type of screw used has always been a major influence in the complete line.

An example is the blown film extruder that typified the new generation of extruders. The most effective screw design, in most cases, has been in L/D of 25. Longer screws with an L/D of 30–33 are chosen for venting or special requirements. High outputs are obtained with LDPE blown film, PP cast film extrusion, etc. The 20 L/D machines now are almost always used only for heat-sensitive plastics. The 25 L/D versions offer the right compromise for obtaining a high output and preventing overheating and damage of thermally sensitive plastics.

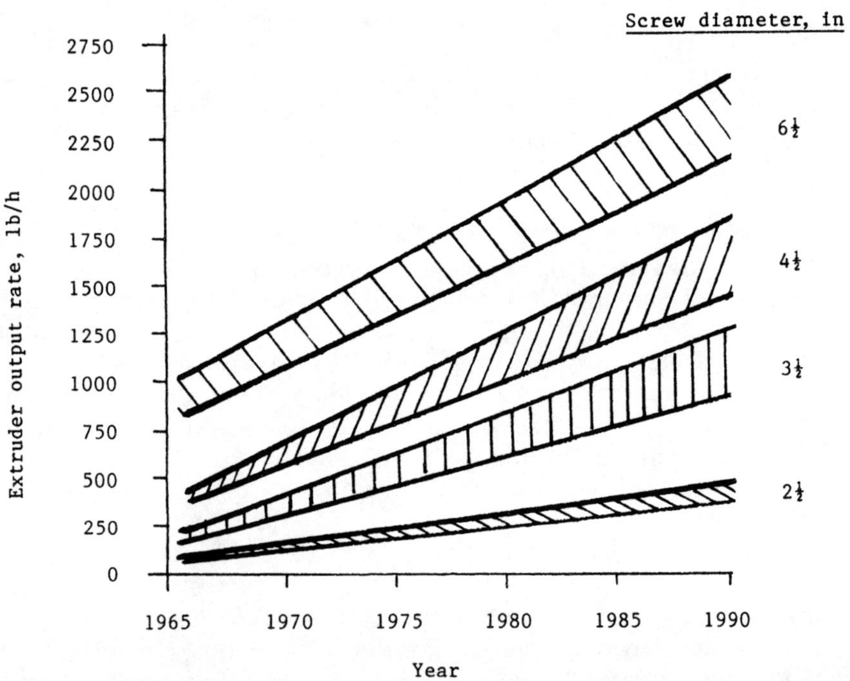


**Figure 4.6** Examples of growth output for various size extruders processing the more commonly used TPs.

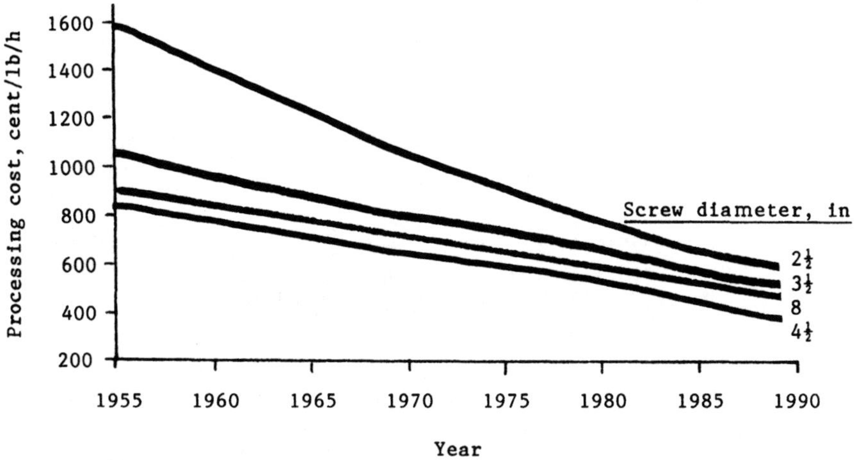


**Figure 4.7** Examples of trends in extruders on the steadily dropping processing costs based on lb/h output of different screws.

Even in today's 'high-tech' world, the art of screw design is still dominated by experienced trial and error approaches, to provide the exact capabilities of the screws for a particular plastic operating under specific conditions. However, computer models (based on proper data input and, very important, experience of the person with a set up similar to the one being studied) play a very important role [152]. When new materials are developed or improvements in old materials are required, one must go to the laboratory to obtain rheological and thermal properties before computer modeling can be performed effectively. New screws improve one or more of the basic screw functions of melt quality, mixing efficiency, melting performance, along the screw, melt heat level, output rate, output stability, and power usage or energy efficiency.

Heating can be controlled by using different machine settings, which involve various tradeoffs. For example, in choosing the optimum screw rotation speed, a slow speed places the melt in contact with the barrel and screw for a longer time via heat conduction. The slower speed produces less shear in the plastic, so that dissipative heating is reduced, and properties of the plastic (particularly of a film) are enhanced. Sometimes an internal heat control is used with a screw. This type of screw is characterized by deeper channels, steeper helical angles, and an internal heating or cooling system. Its internal heating lowers the amount of viscous heating needed to process the plastic. As a result, the melt heat can usually be reduced by 10°C (6°F).

The temperature of the plastic during processing is very important in order to properly melt it. The output zone of the screw and barrel surfaces are 'moistened' (wetted) by the properly melted plastic. External friction between plastic and the flow channel in the plasticator does not influence the melt behavior except when processing TPs of extremely high viscosities, such as rigid PVC and highly filled plastics. The flow of melt in the output section is affected by the coefficient of internal friction (viscosity) especially when the die offers a high resistance to the flow of the melt.

The effective output or pumping of the single-screw is controlled by two opposing factors. With a properly designed screw feed zone, a pressure gradient builds up with plastic filling its feed channel. In addition to generating this pressure, there is also at least the same degree in the transition (compression) and metering zones. The pressure gradient, and particularly the location of the point of maximum pressure, depend on the operating conditions of temperature and rotational speed, type plastic being processed, L/D dimensions, and melt flow resistance of the die. As an example, with screws that have a deeply cut feed zone and a die of very little resistance, the maximum pressure usually occurs in the last third of the screw length. The pressure is zero at both the screw entrance and die exit.

To define screw performance usually starts by comparing with other screws, assuming one exists for the comparison. The parameters that should be considered include: (1) output rate; (2) extrudate melt temperature; (3) extrudate melt quality; (4) extrusion stability/pumping consistency; and (5) energy usage. Different processes will require different levels for each of the parameters listed and these levels should be understood to allow accurate screw design selection and extruder sizing [187, E. Steward].

## MIXING DEVICES

In typical extrusion operations, barrier type mixing devices are used in the screws. Many dynamic mixers, such as those included in Fig. 4.8, are used to improve screw performance. Static mixers are sometimes also inserted at the end of the plasticator (Chapter 2). Proof of their success is shown by their extensive use worldwide. Each type of mixer offers its own advantages and limitations.

These mixing elements are usually installed as near as possible to the end of the metering zone. Where practical, they should not be located in a region where the melt viscosity is not too low. With some of these installations, because extruders may have to operate at a lower speed to avoid problems such as surging, independently driven mixers can be used so machines can operate at optimum speed. Other benefits of independently driven mixers involving feeding. For example, metering pumps can inject with precision liquid additives directly into the mixer.

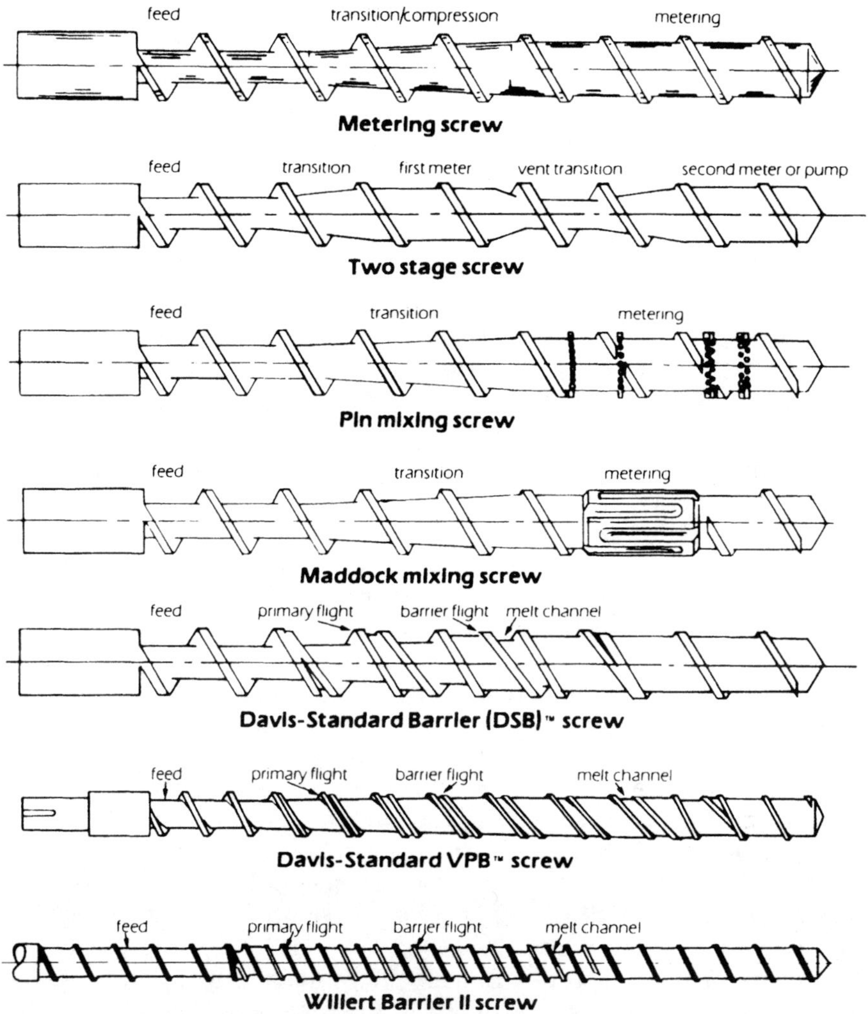


**Figure 4.8** Screw designs with different mixing sections.

Developments continue to be made that can provide good mixing and product uniformity at high production rates without excessive stock temperatures. The answer has been found in a variety of mixing devices that meet the rheology of the different plastics. Some of these mixing devices are described [2].

The Dulmage screw has a Dulmage section incorporated as an integral part of the screw. This design was one of the first mixing screws and was developed by Fred Dulmage of Dow Chemical Co. It has a series of semicircular grooves cut on a long helix in the same direction as the screw flights. There are usually three or more sections interrupted by short cylinder sections. This design interrupts the material flow where the melt follows the screw channel. It divides and recombines the melt many times. In this way, it works something like a static mixer. It is still used on screws that are processing certain materials such as foamed plastics.

Around 1960, several companies started to place radial mixing pins in the screw root. These pins tend to interrupt the laminar melt flow and do a better job of mixing than the regular screw. Because these pins improved mixing, it is also possible to design the screw a little deeper to obtain some more output with the same degree of mixing. Many patterns and shapes (streamlined, etc.) of pins have been used. In general, they are usually placed in rows around the screw. They are located in the metering section after most of the melting has taken place. A typical arrangement would have three rows, with one row at the beginning of the metering section, another row one flight back from the end, and the third row halfway between the other two rows. Pins are usually staggered from row to row. The pins should be hardened and have an interference fit to prevent dislodgment. Pins, unlike other mixing devices, are easy to install as an afterthought. This is usually done after the screw has been running and found to need more mixing capability.

The Union Carbide mixer, also referred to as the Maddock mixer, was given to the public without patent royalty charges. It consists of a series of opposed, semicircular grooves along the screw axis. Alternate grooves are open to the upstream entry. The other grooves are open to the downstream discharge. The ribs or flutes that divide the alternating entry and discharge grooves also alternate. These flutes are called mixing flutes and wiping or cleaning flutes. The plastic is forced over the mixing flute that has an undercut from the screw outer diameter. The cleaning flute is narrower and has a full diameter. This mixer does an effective job of mixing and screening unmelted plastic. The plastic is pumped into the inlet groove, and as the screw rotates, the undercut mixing flute passes under it. The melted plastic ends up in the outlet or discharge groove. As it goes over the undercut mixing flute, it is subjected to high shear but for a very short time interval. The plastic is then pumped out of the discharge groove as new plastic enters over the full-diameter cleaning flute [2].

In the Pulsar mixer (from Spirex Corp.), the metering section is divided into constantly changing sections. These sections are either deeper or shallower than the average metering depth. This requires all the plastic to alternate many times from shallower depth and somewhat higher shear to deeper channels with lower shear. During this changing action, it experiences a gentle tumbling and massaging action. This interrupts the undesirable laminar flow and causes excellent mixing, distribution, and melt uniformity without high shear [2].

BARRIER SCREW

Screw designs to accelerate melting have developed along different and sometimes radically opposed concepts. Usually, the melting rate is controlled by providing a barrier between the solid bed and the melt pool to assure that the solid bed does not break up prematurely and become encapsulated in the melt. Both single and twin screw extruders take advantage of these mixing barriers. Generally, the twin screw designs are significantly superior to the single screw designs.

An example of an opposing concept, introduced by George Kruder of HPM, is called a double wave screw (Fig. 4.9) where:

1. The feed section establishes the solids conveying in the same manner as a conventional screw (Fig. 4.3).
2. At the start of the transition (compression) zone, a second flight is started. This flight is called the barrier or intermediate flight, and it is undercut below the primary flight's OD. Purpose of this barrier flight is to separate the solids channel from the melt channel.
3. As melt moves through the transition zone, melting continues as the solids are pressed against the barrel forming a melt film. The barrier flight moves under the melt film and the melt is collected in the melt channel. With this separation, different functions are performed on the unmelted and melted plastics.
4. The melt channel is deep, giving low shear and reducing the possibility of overheating the already melted plastic. The solids channel becomes narrower and/or shallower forcing the unmelted plastics against the barrel for efficient frictional melting. Break up of the solids bed does nor occur to stop this frictional melting.
5. The solids bed continues to get smaller and finally disappears into the back side of the primary flight.
6. All the plastic has melted and gone over the barrier flight. Melt refinement can continue in the metering section. In some cases mixing sections are also included down-stream of the barrier section. However, generally the melted plastic is already uniform upon exiting the barrier section.

In this double wave design concept, the conventional feed and melting zones are employed until the point at which about 50% of melting is

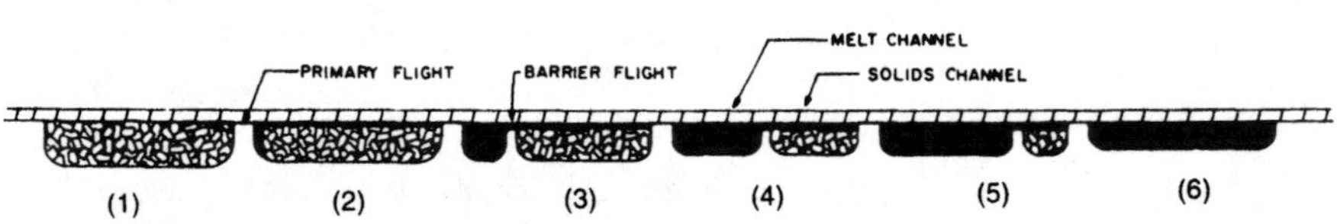


**Figure 4.9** Melt model for double wave barrier screw.

completed. At that section, the melt and solids are mixed together. This is accomplished by varying the metering channel depth in a sinusoidal pattern. The mixing action alternates between very shallow, high shear zones and rather deep, low shear zones. The effect of this action is to promote the distributive mixing of the solid bed melt that has been thoroughly broken up with the melt pool. Thus, this action uses the residual melt pool heat to complete the melting of the remaining solids and in turn results in a very low average extrudate temperature.

Other important developments have occurred that include the Uniroyal mixing screw, Hartig MC-3 screw, Davis-Standard VPB screw, New Castle Industries Efficient screw, Barr II screw, and Willert II screw [2].

## MELTING ACTION

As mentioned throughout this book, the temperature of the plastic melt is important as well as its mechanical condition. What limits many products is the effect that a relatively small change in temperature has on properties such as stiffness or size. Lack of proper melting or mixing of the plastic in the screw is particularly troublesome. Elements in the melt stream that have different mixing histories will appear as defects in the final extruded product. If very small amounts of trapped moisture or encapsulated volatiles of any sort in the melt exists, the final product may not be acceptable by the customer. It could have voids, delaminations, surface blemishes, etc. Procedures to be used in eliminating these type problems will be reviewed latter in this chapter; they have also been include in various troubleshooting tables throughout this book.

In the conventionally designed single screw extruders, melting proceeds generally in the following manner. As the plastic in pellets, powders, or other forms are conveyed down the screw, they are compacted and pressed against the heated barrel wall where a melt forms. Proceeding downstream, some particles of solid material may remain, however they are completely encapsulated by the melt. To develop a completely homogeneous melt, usually some sort of mixing (barrier) device in the screw.

With multiple screws, such as the twin screw extruder, the melting mechanism varies considerably from the single screw. The keying effect of the lower apex and the interaction of the staggered screw flights tend to promote a considerable amount of exchange between the channels of the two screws. The result is that the plastic melt and solid granules become mixed together instead of having a clear separation of the melt pool and the solid bed.

There has been developed an almost universally accepted model of melting in a single screw. This model is the basis for most computer simulations. It has been demonstrated to be correct by many 'freeze tests' (where after the extrudate is providing an acceptable melt, the machine is stopped and screw removed whereby the melted and unmelted plastics are examined). Sketches of these models shows the melting action as shown in Figs. 4.3 and 4.9. This information and their reviews indicate the following relationships between the desired results and metering zone depth: (1) high output by deep screws; (2) low melt temperatures by deep screws; and (3) melt quality by shallow screws.

Models can show how the plastic goes from the solid state to a melt as it moves through the action of the speed controlled screw and the screws melting action on the plastic with the back pressure. The results describe such relationships as high output via deep screws, low melt temperatures via deep screws, and melt quality via shallow screws. For these different situations, solutions have been developed to provide good mixing and product uniformity at high production rates without excessive heat.

## VENTING

During extrusion, melts must be freed of gaseous components that include moisture from the atmosphere and gases from monomers, plasticizers, and other additives as well as entrapped air. The major approach to this plastic degrading situation is by using extruders that have vents in their barrels (Fig. 4.10). It is very difficult to remove all the gases, from particularly contaminated powdered plastics, unless the melt is exposed to vacuum venting (typical of most vented screws, a vacuum is connected to the vent's exhaust port in the barrel). The standard machines operate on the principle of melt degassing. The degassing is assisted by a rise in the vapor pressure of volatile constituents, which results from the high melt heat. Only the free surface layer is degassed; the rest of the plastic can

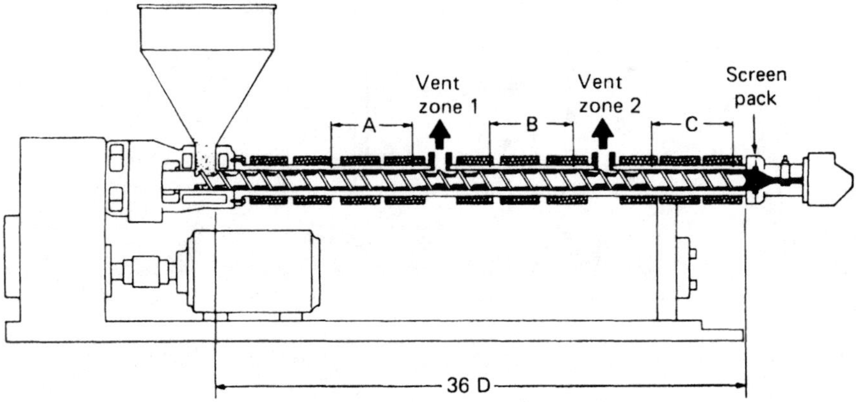


**Figure 4.10** A double-vented extruder using a 3-stage screw.

release its volatile content only through diffusion. Diffusion in the nonvented screw is always time-dependent, and long residence times are not possible for melt moving through an extruder. Thus, a vented extruder is used [1–3, 316].

Gaseous components such as moisture retention in and on plastics has always been a problem for all processors. Many problems develop on products (splay, poor mechanical properties, dimensions, etc.). This situation is very important when processing hygroscopic plastics. As described in Chapter 3, different type dryers are used but in many cases venting is a requirement to ensure proper elimination of the contaminants [1].

Most single-screw vented extruders have two-stages (Fig. 4.11) or three-stages (Figs. 4.10 and 4.12). Those with one vent use a two-stage screw (Fig. 4.11) which basically looks like two single screws attached in series. Where the two meet, there is a very shallow channel section so that when the melt reaches that section, no melt pressure exist. In turn gaseous materials are released. With those having two vents, a three-stage screw is used that provides another stage to eliminate contaminants.

The first stages of the transition and metering zones are often shorter than the sections of a single-stage conventional screw. The melt discharges at zero pressure into the second stage, under vacuum instead of pressure. The first-stage extrudate must not be hot enough to become overheated in the second stage. And the first-stage must not deliver more output per screw rotation at discharge pressure than the second-stage can pump through the die under the maximum normal operating pressure, such as might occur just prior to a screen pack change. This usually means

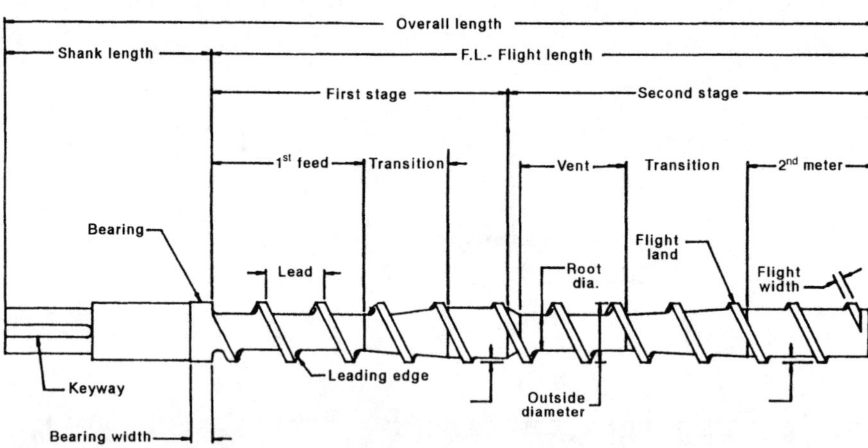


**Figure 4.11** Nomenclature for a 2-stage screw.

**Figure 4.12** Welex sheet line with double-vented extruder used primarily for processing ABS.

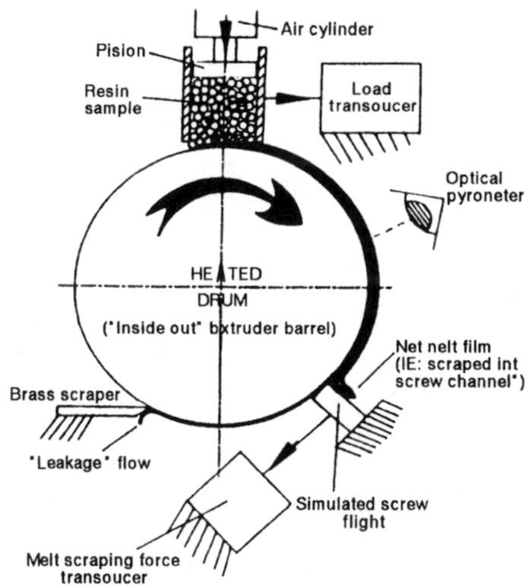


**Figure 4.13** Screw melting simulator.

that the second-stage metering section must be at least 50% deeper than the first-stage.

In practice the best metering-section depth ratio (pump ratio) is about 1.8:1. The ratio to be used depends on factors such as screw design, downstream equipment, feed stock performance, and operating conditions. There is likely to be melt flow through the vent (avoid this situation) if the compression ratio is high or the metering depth ratio is slightly too low. If the metering depth ratio is moderately high, there is a gradual degradation of the output. With the screw channel in the vent area not filling properly, the self-cleaning action is diminished, and the risk of plate-out increases. In any case, sticking or smearing of the melt must be avoided or degradation will accelerate.

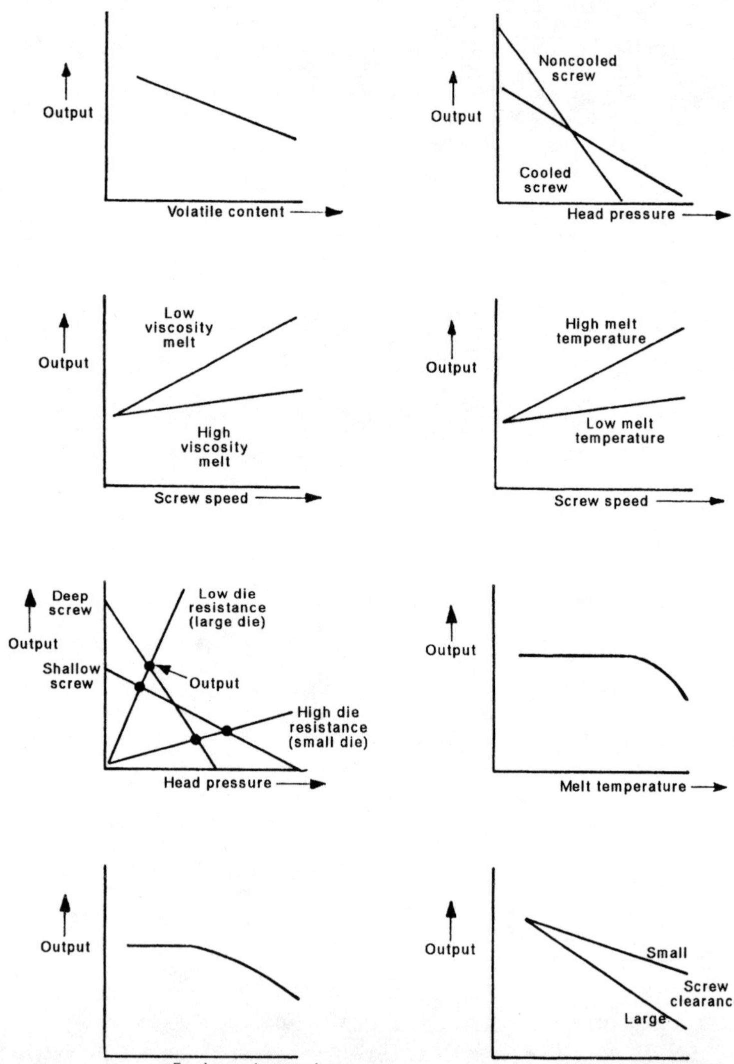


**Figure 4.14** Single-screw extruder effects of machine and plastic variables.

## PROCESS PERFORMANCE

It is rather difficult to see inside an extruder; in the past, some work was done in using transparent glass and/or plastic, as well as 'freeze tests' to study melting action. Recognizing the situation to determine what is occurring during melting, Dr. Chan I. Chung of Rensselaer Polytechnic Institute, NY, developed the Melt Simulator. Figure 4.13 shows a sche-

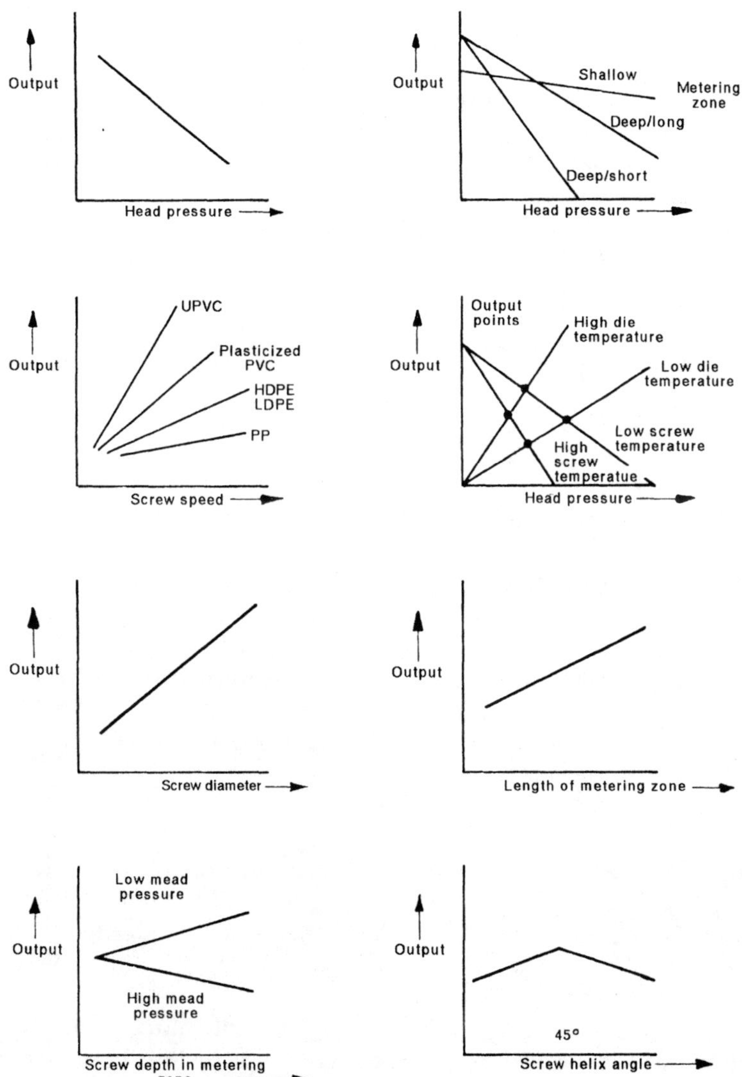


**Figure 4.14** *Continued*

matic of this device. As can be seen, the heated drum surface is an inside/outside extruder barrel. By running the drum's surface speed equal to the peripheral velocity of any size screw and heating it to a temperature equal to an extruder's barrel, a duplication of the melting condition in an extruder is approached.

Returning to the trial and error approach of developing data, review Fig. 4.14. Most of the data presented relates to output from an extruder. The effect of the extruder and plastic material is provided.

## WEAR

Wear in screw plasticators generally causes an increase in the clearance between screw flight and barrel. It often occurs toward the end of the compression section. This type of wear is more likely to occur when the screw has a high compression ratio. Regardless of where this erosion of metal occurs, the plasticator's melting capacity is reduced. If the wear is serious enough, it will effect the process so that products are exiting at a slower rate or more likely a lower quality product at the end of the line.

To circumvent this situation, consider inspecting and measuring the screw upon arrival (or thereafter), and at regular time schedule inspect and recheck dimensions. During any clean-up of the screw is a good time to measure and inspect. By extrapolating to the maximum allowable wear, one can determine when the screw and/or barrel should be replaced or rebuilt. Contact your supplier of screws for details on 'warning' signs.

The mechanism that causes wear include adhesive wear (metal to metal contact under high stress), abrasive wear (galling), laminar wear (thin outer layers of metal interface wear), surface-fatigue wear (micro- or macroscopic separation from the surfaces), and corrosion wear (chemical reaction and mechanical attack of the sliding surfaces [18, P. Colby, J. Frankland, R. Gregory, A. Griff, T. Harrington, R. Knittel, 392].

## SCREW INSPECTION

Screws do not have the same outside continuous diameter. Upon receiving a machine or just a screw, it is a good idea to check its specified dimensions (diameters versus locations, channel depths, concentricity and straightness, hardness, spline/attachment, etc.) and make a proper visual inspection.

This information should be recorded so that comparisons can be made following a later inspection. Some special equipment should be used other than the usual methods (micrometer, etc.) to ensure that the inspections reproduced accurately. Such equipment is readily available and actually simplifies testing, etc., and it also takes less time particularly for roller and hardness testing. Details on conducting an inspection and conducting important processing behaviors has been reviewed in Chapter 2. This chapter describes the Techware Designs (a subsidiary of Spirex Corp.) computer software package called the 'Extruder's Technician' (ET).

# 5

# Die design and performance

## OVERVIEW

A die is a device, usually of made of steel, having a specific shape or design geometry which it imparts to a plastic melt pumped from an extruder (Fig. 5.1). The function of the die is to control the shape of the extrudate. The important word is control. In order to do this, the extruder must deliver melted plastic to the die targeted to be an ideal mix at a constant rate, temperature, and pressure. Measurement of these variables is desired and usually carefully performed [269, 328, 418, 424, 434].

The terms die, tool, and mold are virtually synonymous in the sense that they have female or negative cavity through which a molten plastic moves usually under heat and pressure. However, the term die principally refers to an extruder die in the plastic industry. The design of a die is required to: (1) minimize head and tooling interior volumes to limit stagnation areas and residence time; (2) streamline flow through the die, with low approach angles in tapered transition sections; and (3) polish and plate interior surfaces for minimum drag and optimum surface finish on the extrudate.

Basically, the die provides the means to 'spread' the plastic being processed/plasticated under pressure to the desired width and thickness in a controllable, uniform manner. In turn, this extrudate is delivered from the die (targeted with uniform velocity and uniform density lengthwise and crosswise) to take-off equipment in order to produce a shaped product (film, sheet, pipe, profile, coating, filament, etc.).

The pumping pressure required on the melts entering the different designed die heads differs to meet their melt flow patterns within the die cavities. The pressure usually varies as follows: (1) blown and lay-flat films at 13.8–41.3 MPa (2000–6000 psi); (2) cast film, sheet, and pipe at 3.5–27.6 MPa (500–4000 psi); (3) wire coating at 10.3–55.1 MPa (1500–8000 psi); and (4) monofilament at 6.9–20.7 MPa (1000–3000 ps).

**Figure 5.1** Welex's 'R' style restrictor bar sheet die.

Dies can be categorized by their product performance. Table 5.1 describes sheet dies from Extrusion Dies, Inc. (EDI). There are straight-through, cross-head, and offset dies. To be more specific, they can be classified as: (1) axial or straight through extrusion heads with symmetrical flow channels, particularly tube and pipe heads, circular rod and monofilament dies; (2) angled dies particularly crossheads and angular heads for wire and cable covering, crossheads and offset heads for tube and pipe, and film blowing heads; (3) profile dies that include slot dies for flat film and sheet, and multi-orifice heads for monofilaments; and (4) dies for special products such as netting.

The following general classification may be helpful as a guide to film and sheet thickness selection for a die: (1) film dies are generally applicable for thicknesses of 0.003 mm (0.010 in) or less; (2) thin gauge sheet dies are normally designed for thicknesses up to 0.015 mm (0.060 in); (3) intermediate sheet dies may cover a thickness range of 0.01–0.06 mm (0.040–0.250 in); and (4) heavy gauge sheet dies extrude thicknesses of 0.02–0.13 mm (0.080–0.500 in). Different groups within the industries may have their own thickness definitions, and use names that the buyer or customer is familiar with.

The coupling between barrel and die can be carried out in various ways using bolts or locking devices. They include: (1) flange fitting with a clamp ring on the barrel and a fixed flange on the die; (2) flanges on the barrel and die with tapered links and two bolted half-clamps, or a ring clamp hinged at one side and bolted to the other side; and (3) swing-bolt flange connection between the barrel flange and a die flange. In many extruders,

**Table 5.1** Examples of EDI's different types of dies

| *EDI Model No.* | *Flex Range* | *Restrictor Bar* | *Film 10 ml & Below (254 μm & Below)* | *Coating & Laminating* | *Thin Sheet 10 ml–60 ml (254 μm–1524 μm)* | *Midrange Sheet 10 ml–90 ml (254 μm–2286 μm)* | *Heavy Sheet 60 ml & Above (1524 μm & Above)* | *Lab. Applications* | *Options* |
|---|---|---|---|---|---|---|---|---|---|
| Ultraflex L 40 | 0.040 in (1.0 mm) | | ■ | ■ | | | | ■ | **Lip adjustments** |
| Ultraflex L 75 | 0.075 in (1.9 mm) | | | | ■ | | | ■ | Micro push |
| Ultraflex 40 | 0.040 in (1.0 mm) | | ■ | ■ | | | | | Micro push/pull |
| Ultraflex H 40 | 0.040 in (1.0 mm) | | ■ | ■ | | | | | **Material of construction** |
| Ultraflex H 75 | 0.075 in (1.9 mm) | | | | ■ | | | | Stainless steel |
| Ultraflex H 100 | 0.100 in (2.54 mm) | | | | ■ | ■ | | | Other upon request |
| Ultraflex HM 40 | 0.075 in (1.9 mm) | | ■ | ■ | | | | | **Platings** |
| Ultraflex HM 75 | 0.075 in (1.9 mm) | | | | ■ | | | | Electroless nickel |
| Ultraflex HM 100 | 0.100 in (2.54 mm) | | | | ■ | ■ | | | Polymer impregnated chrome |
| Ultraflex H 40 EPC | 0.040 in (1.0 mm) | | ■ | ■ | | | | | Polymer impregnated nickel |
| Ultraflex H 40 EPC C/L | 0.040 in (1.0 mm) | | ■ | ■ | | | | | **Deckling** |

| | | | | | | | | |
|---|---|---|---|---|---|---|---|---|
| Ultraflex LR 40 | 0.040 in (1.0 mm) | 45° | ■ | | | | ■ | **Removable lips** |
| Ultraflex LR 75 | 0.075 in (1.9 mm) | 45° | | ■ | | ■ | ■ | **Extended lips for** |
| Ultraflex R 75 | 0.075 in (1.9 mm) | 45°* | | ■ | | ■ | | **close approach** |
| Ultraflex HR 75 | 0.075 in (1.9 mm) | 45°* | | ■ | | ■ | | **Roll guard** |
| Ultraflex RC 75 | 0.075 in (1.9 mm) | 45°* | | ■ | | ■ | | **Wrench guard** |
| Ultraflex HRC 75 | 0.075 in (1.9 mm) | 45°* | | ■ | | ■ | | **Insulation jacket** |
| Ultraflex HRMC 75 | 0.075 in (1.9 mm) | 45°* | | ■ | | ■ | | **Lip heaters** |
| Ultraflex R 100 | 0.100 in (2.54 mm) | 45°* | | ■ | ■ | ■ | | **Heat tubes** |
| Ultraflex HR 100 | 0.100 in (2.54 mm) | 45°* | | ■ | | ■ | | |
| Ultraflex RC 100 | 0.100 in (2.54 mm) | 45°* | | ■ | ■ | ■ | | |
| Ultraflex HRC 100 | 0.100 in (2.54 mm) | 45°* | | ■ | ■ | ■ | | |
| Ultraflex HRMC 100 | 0.100 in (2.54 mm) | 45°* | | ■ | ■ | | | |
| Heavy Duty Sheet | 0.550 in (14 mm) | 90° | | | | ■ | | |
| Magnum Sheet | 0.775 in (19.7 mm) | 90° | | | | ■ | | |

* Also available with 90° restrictor bar.

a breaker plate with screens is used between the barrel and die assembly. The purpose of the screens is primarily twofold: (1) to change the melt's spiraling motion, caused by the screw rotation; and (2) to filter contaminants out of the melt (Chapter 2).

If the exit opening of the extruder barrel does not match up with the entry opening of the die, an adapter is used between the barrel and die (Fig. 1.9). The usual arrangement is in the horizontal position. As shown in Fig. 5.2, Battenfeld Gloucester's vertical blown film die uses a 90°

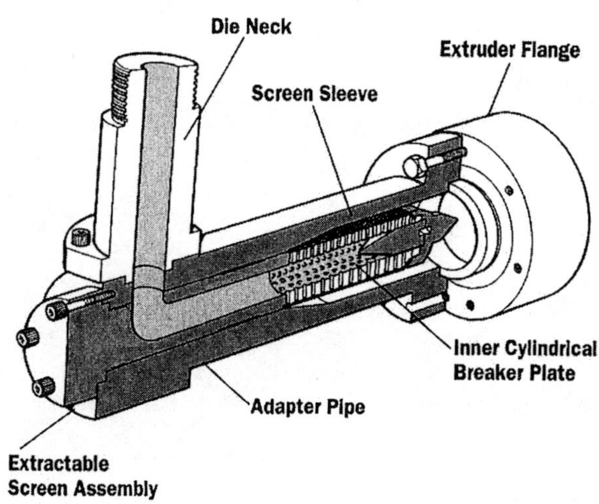


**Figure 5.2** Adapter/screen assembly.

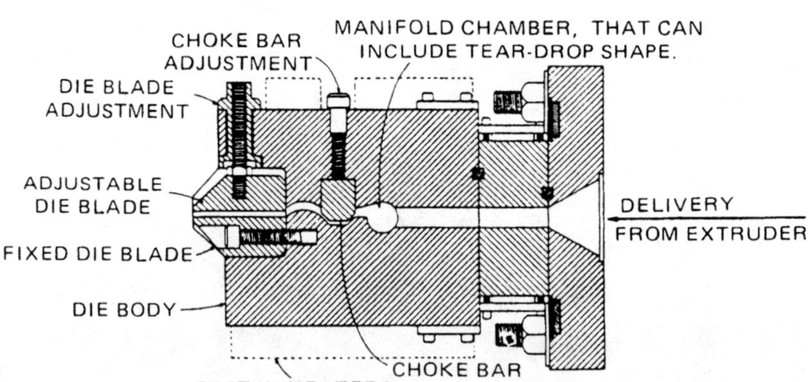


**Figure 5.3** Sheet extrusion die.

adapter. Any adapter used can help reduce complex melt flow patterns entering the die. Dies specifically designed for a certain extruder will usually not require an adapter. The adapter is also called offset adapter, crosshead adapter, or die feedblock. A fish tail die adapter describes the transition from a round block of melt as it exists from the barrel and is then flattened while being spread prior to entering the die.

Basic schematic examples of dies to fabricate different products are shown in Figs. 5.3–5.9. Well built dies can include adjustments to meet the performance requirements for the extrudate. They can include restricter/choker bars, temperature zoning, pressure sensors, adjustment bolt/device, and/or other devices.

Examples of dies will be presented in which the dies are usually designed to process plastics that have specific melting/rheological characteristics. For example, a conventional LDPE blown-film die with 0.8mm (0.030in) die gap will not process LLDPE or mPE (metallocene PE) satisfactorily at a high output rate.

The higher viscosity LLDPE increases back pressure significantly, thereby decreasing the throughput. With this change, there may be melt fracture (shark skinning) which produces a rough surface finish. Processors of LLDPE can overcome these problems with wide die gap openings of about 0.3mm (0.090in). By increasing in the die gap, the head pressure decreases, allowing significant increase in output (Fig. 5.10). Another approach to improve the surface characteristics of film without impacting the cooling is to increase the lip temperature [285].

The mPEs generate a lot of shear heat, so overheating will put them in an unfavorable situation where changes in their viscosities occur. To date these plastics have required an extensive learning period requiring the proper extruder settings. This action is typical of what happens when certain new plastics become available. Eventually equipment, dies, etc., are developed specifically for the new plastics. What makes mPEs very attractive is their higher performance physical properties than what is available. So in time, the mPEs will process with 'ease.'

Figure 5.11 shows a sheet die where the end sections of the melt temperatures are higher than the center area. The basic combination of varying the die gap opening from side to side and controlling the heat profile across the die equalizes the flow rate of the melt as it passes through the die cavity.

Different versions of the popular blown film (Fig. 5.4) spiral feed oscillating die provide an even distribution of the extrudate melt flow. To improve melt distribution for the different plastics, different techniques are used, such as lengthening the spirals and/or increasing the number of distribution points.

LLDPE can be drawn or stretched in the melt with a low induced orientation (Chapter 2), so a wide die gap does not add undesirable film

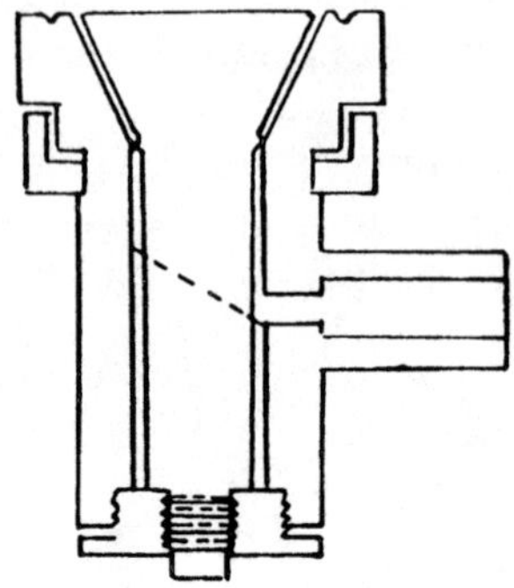

(1) *SIDE FEED DIE*
Advantages:
1. Low initial cost
2. Adjustable die opening
3. Will handle low flow materials

Disadvantages:
1. Mandrel deflects with extrusion rate, necessitating die adjustment
2. Die opening changes with pressure
3. Non-uniform melt flow
4. Cannot be rotated
5. One weld line in film

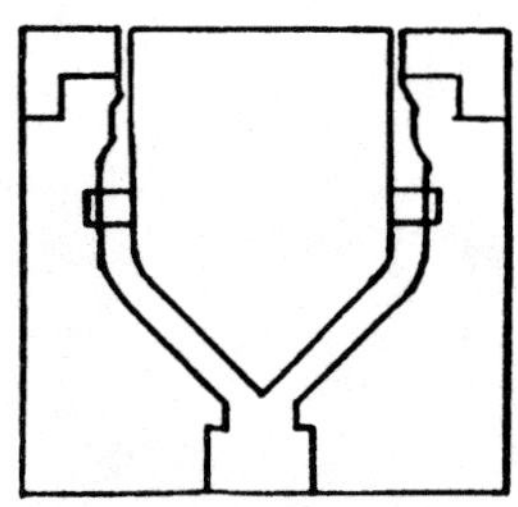

(2) *BOTTOM FEED DIE*
Advantages:
1. Positive die opening
2. Can be rotated
3. Will handle low flow resins

Disadvantages:
1. High initial cost
2. Very hard to clean
3. Two or more weld lines in film

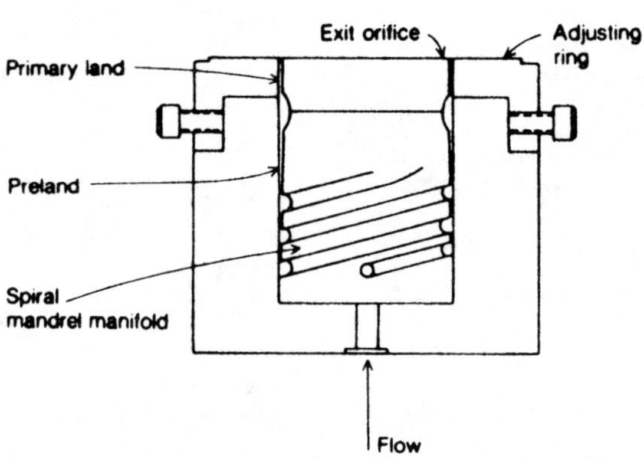


(3) *SPIRAL FEED ROLE*
Advantages:
1. No weld line in film
2. Positive die opening
3. Easy to clean
4. Can be rotated
5. Improved Film Optics

Disadvantages:
1. High head pressure
2. Will not handle low flow resins without modification

**Figure 5.4** Blown-film dies.

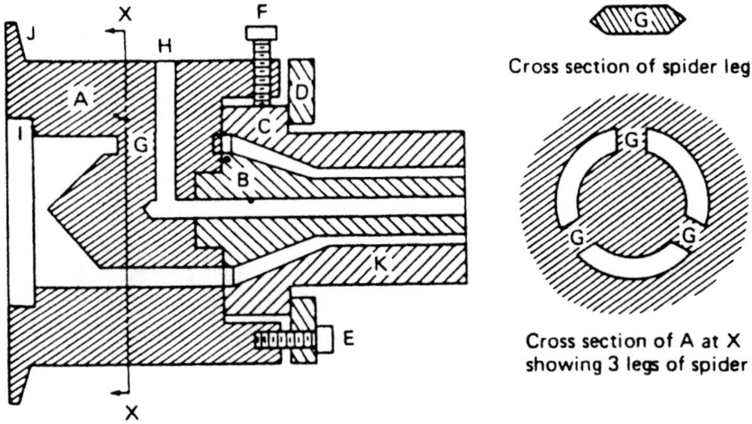


A. Die body
B. Mandrel, pin, male die part
C. Die, die bushing, female die part
D. Die retaining ring
E. Die retaining bolt
F. Die centering bolt
G. Spider leg
H. Air hole
I. Seat for breaker plate
J. Ring for attachment to extruder
K. Die land

(1)

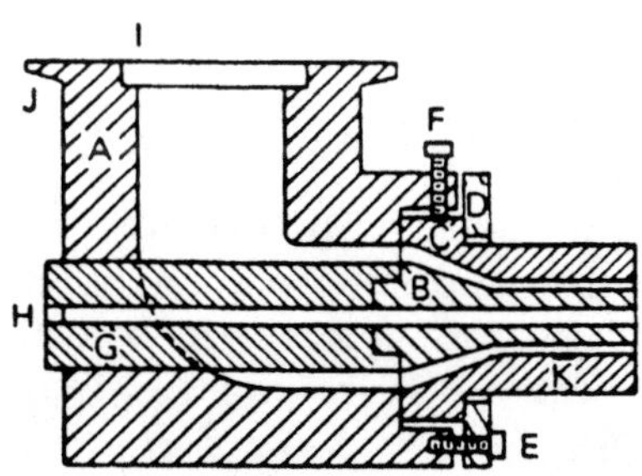


A. Die body, crosshead
B. Mandrel, pin, male die part
C. Die, die bushing, female die part
D. Die retaining ring
E. Die retaining bolt
F. Die centering bolt
G. Mandrel holder
H. Air hole
I. Seat for breaker plate
J. Ring for attachment extruder
K. Die land

(2)

**Figure 5.5** Pipe and tubing die for (1) in-line extrusion and (2) crosshead extrusion.

stresses as with LDPE. The optimum die gap for each application will vary according to the plastic grade, melt heat, and output rate. Only the die mandrel requires modification for conversion to LLDPE. If only LLDPE is to be run, the existing mandrel can be machined. If both plastics are to be run, two separate mandrels are required. As cast film is processed at considerably higher heats than blown film, shear stresses are minimized and no die modifications are required.

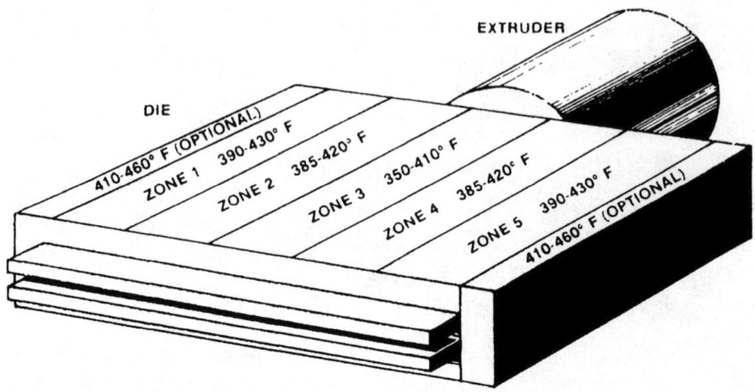


**Figure 5.6** Sheet die using temperature control pattern.

A good PS sheet die can usually run some other plastics. The gauge control capability of the die for PS and the other plastics is largely determined by the flow adjustments. Heavy gauge dies might have a lip land length 75–100 mm (3–4 in) and a relatively coarse method of adjusting the massive lips. In contrast, a film die would have a lip land length of 20 mm (0.75 in) and a lip adjustment capable of extremely fine adjustments. Many film dies and thin gauge sheet dies utilize a flexible lip for extremely close gauge control.

Any analysis of die efficiency must include a careful examination of the compatibility of the die with the extruded products. If a die is designed for sheet thicknesses of 3.8–9.5 mm (0.150–0.375 in), it is extremely difficult to extrude 5 mil (0.001 mm or 0.005 in) film. As there is no die design that could be called a universal die, it is very inefficient to expect an operator to run a die beyond its capabilities. The result would be poor gauge control, and so forth. If the geometry of the flow channel is optimized for a plastic under a particular set of conditions (heat, flow rate, etc.), a simple change in flow rate or in heat can make the geometry very inefficient. Except for circular dies, it is essentially impossible to obtain a channel geometry that can be used for a relatively wide range of plastics and a wide range of operating conditions, such as reviewed for LDPE and LLDPE.

Adjustment capabilities are therefore provided in the die to permit heat and pressure changes. Some dies require the heat profile to be across one direction or in different directions, using individual heating pads, heat pipes, and so on with proper controls.

To simplify the processing operation, the die design should accommodate certain factors. The goals are to have an extruded product with

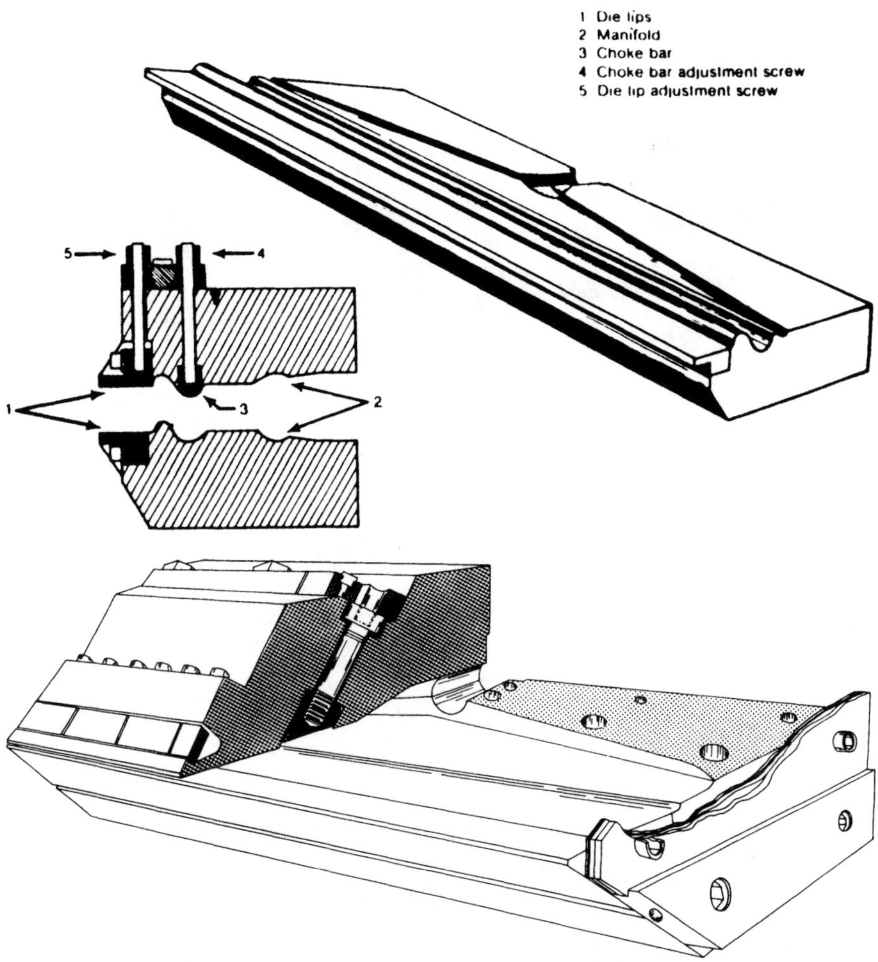


**Figure 5.7** Coathanger die: lower half and cross sections.

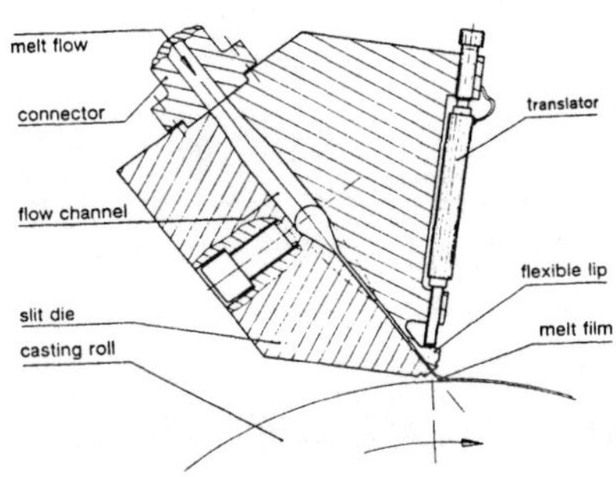


**Figure 5.8** Slit film die with electrical controlled sensors used in a closed-loop control system.

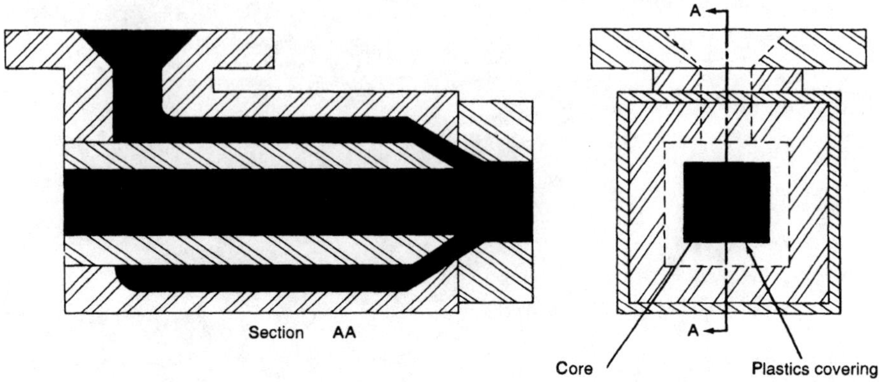


**Figure 5.9** Crosshead coating die.

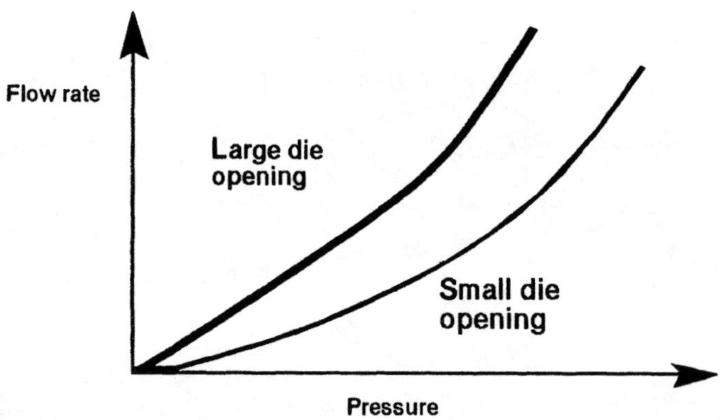


**Figure 5.10** Melt flow versus melt pressure rate based on die opening.

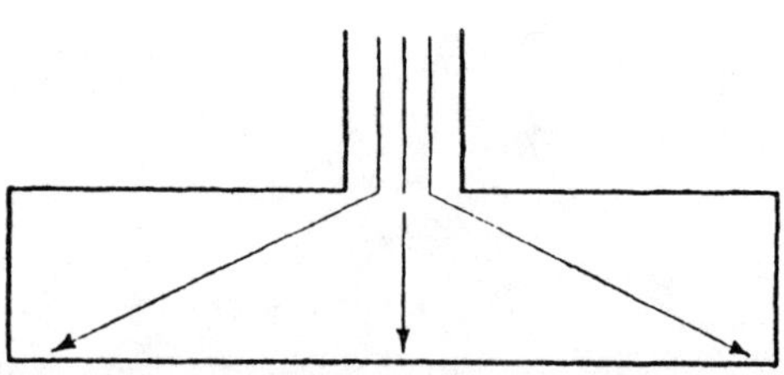

**Figure 5.11** Varying melt flow paths in simple slot die.

uniform wall thickness otherwise the heat transfer problem magnifies as shown in Figs. 5.12 and 5.13. Where practical the design should minimize the use of hollow sections, minimize narrow or small channels, and use generous radii on all corners such as a minimum of 0.005 mm (0.02 in). A so-called impossible or difficult die can be designed; it just requires experience (both practical and empirical) with the usual trial-and-error runs.

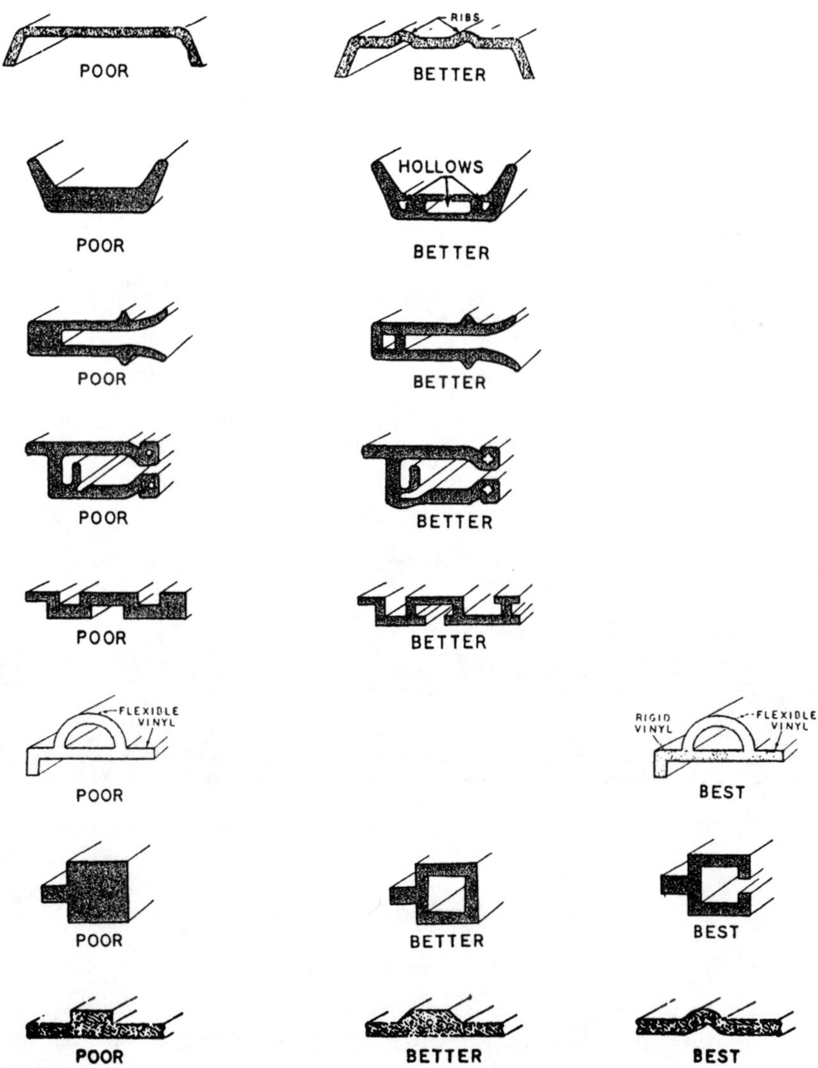


**Figure 5.12** Influence of part design on reducing extrusion profile variables.

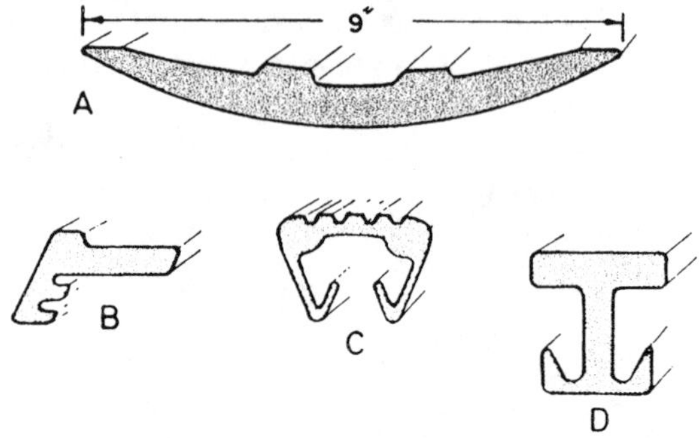


**Figure 5.13** Examples of successful profile shapes with unbalanced walls; (a) rigid PVC shoe, (b) ABS house-trailer trim, (c) flexible PVC bus arm rest, (d) rigid PVC electrical bus bar insulator.

**Figure 5.14** Two dies at the end of one extruder.

Using two or more dies at the end of an extruder, as shown in Fig. 5.14, may not be the most desirable, but it has certain benefits. Economically it may pay to use one extruder delivering to two dies rather than have a separate extruder for each die. Normally setting up one extruder with the usual one die can set up enough problems if not operating at maximum efficiency-to-operating cost. However, a processor could have a 'large' machine that is idle and a 'gigantic' order with an 'impossible' delivery schedule comes into the plant requiring the use of 'small' dies; no 'small' machines are free, etc. So it can make sense to use two or more dies.

## MELT BEHAVIOR

An important approach to melt flow behavior is to recognize that the extruder and die operate as a combined unit. The interaction between screw and die is usually represented by showing the dependence of the output on the melt pressure between the screw and die head. The screw requires that the viscosity of the TP does not change either in the metering zone or in the die. This means that temperature and/or pressure changes and other influences on viscosity have to be avoided as much as possible. The pressure drop through a die varies directly with the land length and inversely with the cube of the gap opening.

The nonNewtonian behavior of a plastic (Chapter 3) makes its flow through a die somewhat complicated. When a melt is extruded from the die, there is usually some degree of swelling (Fig. 5.15, Table 5.2). After exiting the die, it is usually stretched or drawn to a size equal to or smaller than the die opening. The dimensions are reduced proportionally so that, in an ideal plastic, the drawdown section is the same as the original section but smaller proportionally in each dimension. The effects of melt elasticity mean that the plastic does not drawdown in a simple proportional manner; thus the drawdown process is a source of errors in the profile. The errors are significantly reduced in a circular extrudate, such as pipe and wire coating (Fig. 12.10). These errors are corrected by modifying the die and down-stream equipment.

There are substantial influences on the plastic due to the flow orientation of the molecules, such as having different properties parallel and perpendicular to the flow direction. These differences have a significant effect on the performance of the product.

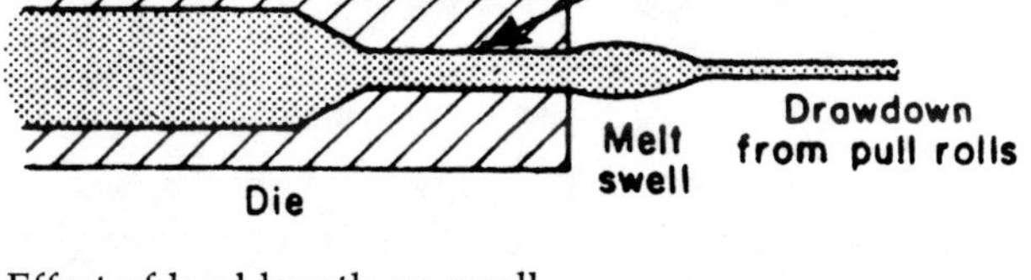


**Figure 5.15** Effect of land length on swell.

**Table 5.2** General effect of shear rate on the die swell of various TPs

| | *Die swell ratio at 200°C (392°F) for the following shear rates* | | | |
|---|---|---|---|---|
| *Plastic* | *10 $s^{-1}$* | *100 $s^{-1}$* | *400 $s^{-1}$* | *700 $s^{-1}$* |
| PMMA-HI | 1.17 | 1.27 | 1.35 | – |
| LDPE | 1.45 | 1.58 | 1.71 | 1.90 |
| HDPE | 1.49 | 1.92 | 2.15 | – |
| PP, copolymer | 1.52 | 1.84 | 2.1 | – |
| PP, homopolymer | 1.61 | 1.9 | 2.05 | – |
| HIPS | 1.22 | 1.4 | – | – |
| HIPVC | 1.35 | 1.5 | 1.52 | 1.53 |

Another important characteristic is that melts are effected by the orifice shape (Fig. 5.16). The effect of the orifice is related to the melt condition and the die design (land length, etc.) with a slow cooling rate having a significant influence, especially in thick products. Cooling is more rapid at the corners; in fact, a hot center section could cause a product to 'blow' outward and/or include visible or invisible vacuum bubbles. The popular coat-hanger die (Fig. 5.7), used for flat sheet and other products, illustrates an important principle in die design. The melt at the edges of the sheet must travel further through the die than the melt that goes through the center of the sheet (Fig. 5.11). A diagonal melt channel with a triangular dam in the center is a way of restricting the direct flow to some degree. The principle of built-in restrictions is used to adjust the melt flow in dies (Fig. 5.7). With blow molding (Chapter 15) and profile dies, the openings require special attention to provide the proper product shape (Figs. 5.17 and 5.18).

Other factors are considered, such as the angle or taper of entry and the parallel length of the die land (Figs. 5.19 and 5.20). For most TPs, the entry angle must be as small as possible to ensure good product quality, particularly at high output rates. The abrupt changes in the direction of melt flow tend to cause rough or wavy surfaces and often internal flaws. This condition is called melt fracture (Chapter 4).

## Die land

A very important dimension is the length of the parallel die land. In general, it should be made as large as possible. However, the total resistance of the die should not be increased to the point where excessive power

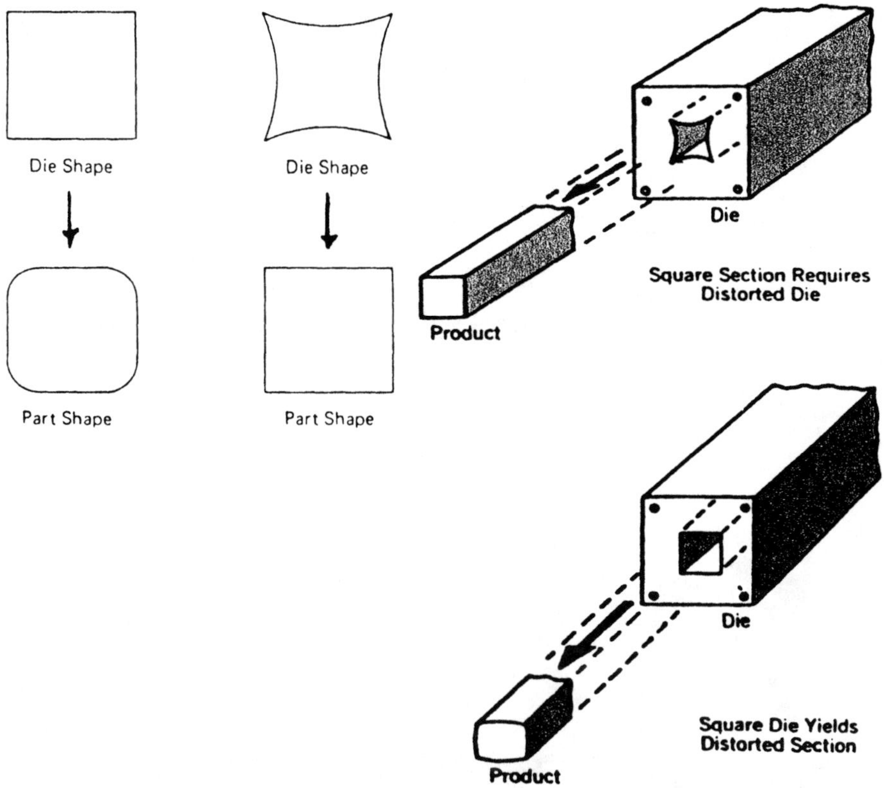


**Figure 5.16** Effect of die orifice shape on 'square' extrudate.

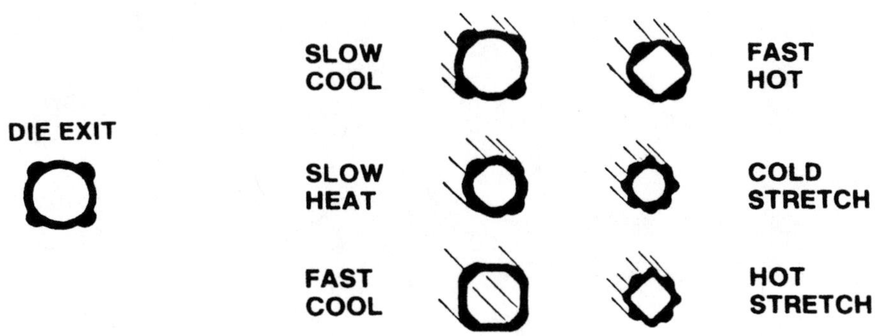


**Figure 5.17** Example of temperature, pressure, and take-off speed (time) variations that can potentially influence the extruded shape.

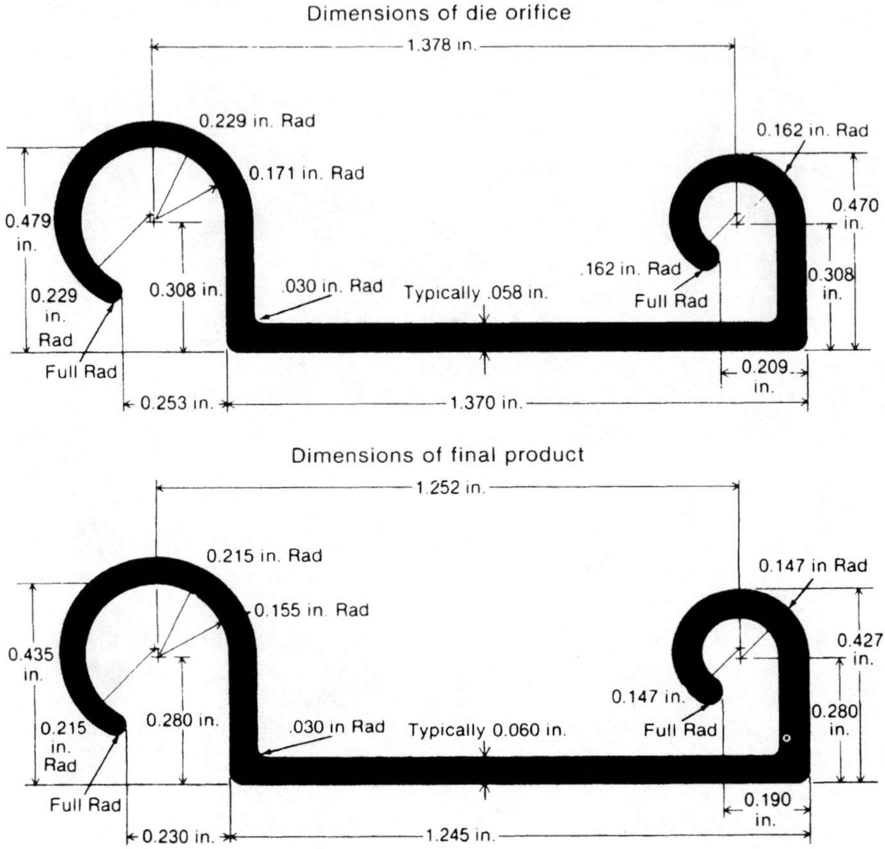


**Figure 5.18** Example of dimensional changes for a PVC profile shape from the die orifice to the product.

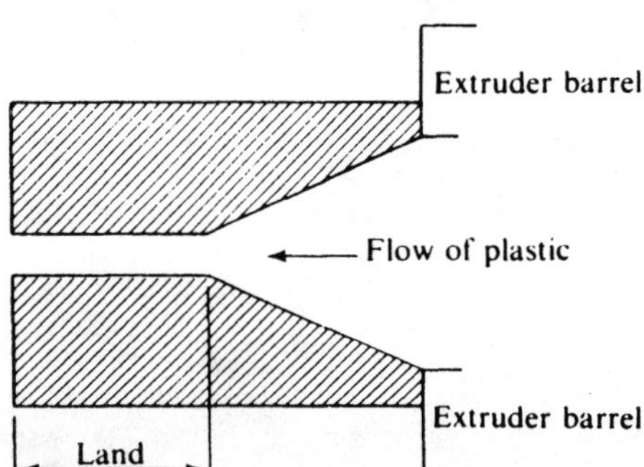


**Figure 5.19** Extrusion die land.

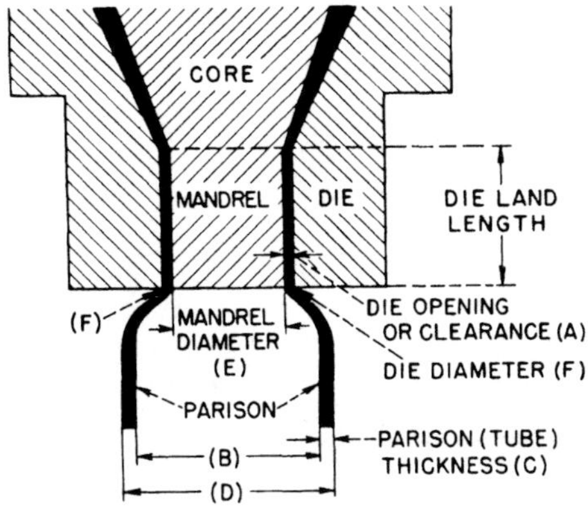


Tube wall swell, % $= \frac{C - A}{A} \times 100$

Die swell, % $= \frac{D - F}{F} \times 100$

Mandrel swell, % $= \frac{B - E}{E} \times 100$

**Figure 5.20** Die land in an extrusion blow molding die: die land length to the die opening is usually from 10–20 to 1.

consumption and melt overheating occur. The required land length depends not only the type and temperature of the TP melt but also on the flow rate. The deformation of the melt in the entry section of the die invariably causes strain which only gradually decreases with time (relaxation). Usually the target is to allow the melt to relax before leaving the die. Otherwise the product dimensions and the mechanical properties may vary, particularly with rapid cooling [205, 285, 362, 440].

There have been various basic rules for the length. One rule states that the length must be sufficiently large for the residence time of the melt in the die at a mean velocity to be at least equal to the relaxation time [370]. Another rule relates to long lands as a means to develop sufficient back pressure, by using a minimum land of ten times the die opening, or (in the case of rod or tubular dies) one diameter of the die opening. A maximum approach angle for tapered sections is 30° and many designs incorporate multiple, decreasing-angle sections or radius sections to minimize drag and plastic hold-up time.

The following flow equations (one of many) provide a guide that should help predict land length of a die. The land length can be estimated by finding first the shear rate and then the pressure drop at the die lips for a particular profile. Using raw material data that can be provided by the material supplier, the following rheological equations have been developed (B.F.G.) for the computation that would be desirable in dies for profile slits, rods, and tear shapes that are use in sheet dies (Fig. 5.21).

**Slit**

Shear rate, $sec^{-1} = 4.61 \times 10^{-2}$ (Flow, lb/h)/(Density, $g/cm^3$)(Die width, in)(Slit thickness, $in^2$)

Pressure, psi = $2.9 \times 10^{-5}$ (Shear rate, $sec^{-1}$)(Viscosity, poise)(Land length, in.)/(Slit thickness, in).

**Rod**

Shear rate, $sec^{-1} = 9.8 \times 10^{-3}$ (Flow, lb/h)/(Density, $g/cm^3$)(Radius, in)

Pressure, psi = $2.9 \times 10^{-5}$ (Shear rate, $sec^{-1}$)(Viscosity, poise)(land length, in.)/(Radius, in)

**Tear drop**

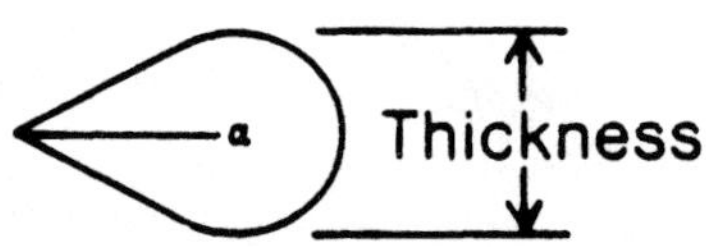


**Figure 5.21** Dimensions of the tear drop.

Shear rate, $sec^{-1} = 0.123 \times$ (Flow, lb/h)/(Density, $g/cm^3$)(Thickness, in.)($\pi/2 \times 1/\tan\alpha$)

Pressure, psi = $1.45 \times 10^{-5}$ (Shear rate, $s^{-1}$)(Viscosity, poise)(Land length, in)($\pi/2 \times 1/\sin\alpha$)/(Thickness, in).

Shear rate relates to the dimensions of the die and the profile's opening with the shear increasing as the volumetric-flow rate increases and decreases while the profile widens; doubling the width of the profile will halve the shear rate. Pressure drop, though not important by itself, basically determines the land length and the thickness of the profile, as seen in these equations. By adjusting the measurements of the die lip and the land length, the designer can adjust the amount of the pressure drop.

A large pressure drop is inevitable in profile extrusion since pressure outside the die is zero, but if the pressure in the die is to high, namely over 28 MPa (4000 psi), the high pressure may cause the die to leak or force the material to back up over the screw. As an example, pressures below 28 MPa (4000 psi) are considered suitable for PVC profile extrusion.

**Temperature**

Provisions should be made to accurately control temperatures in all parts of the head and die. To meet these requirements, it is important to measure melt temperature and pressure in the head via stock thermocouples and pressure transducers. Plus or minus 0.5°C (1°F) is typical using today's control systems. If there is a cold area in the die, the melt flow in that area will be slow and the result will be thin gauge. Hot areas results in more flow, and the potential to burn (degradation) the plastic existing. The die body is usually cored and heated with electric cartridge heaters.

External heaters, liquids, and vapors are also used to heat the die body. Figure 5.22 from Battenfeld Gloucester Co. shows an example of external heaters The wattage used to heat a die is determined by the operating temperature and the mass of the die itself. Insulation of the die should be used whenever practical to protect against burns, to eliminate cold spots, and to conserve energy. Die body cooling, when required, can be done with air or liquid. Heat tubes are also used occasionally to equalize the temperature in the die [418].

Heaters embedded in the die lip can be used to control lip temperature and friction characteristics on sheet lines to aid in controlling transverse thickness. To control thickness, mechanical control of the lip opening is very popular. Figure 5.23 shows a conventional die lip 'A' that is compared to a flexible die lip 'B' where it highlights: (1) streamlined manifold; (2) restrictor bar 'A' and secondary manifold to equalize pressure without

**Figure 5.22** Heating elements around a blown film single layer die.

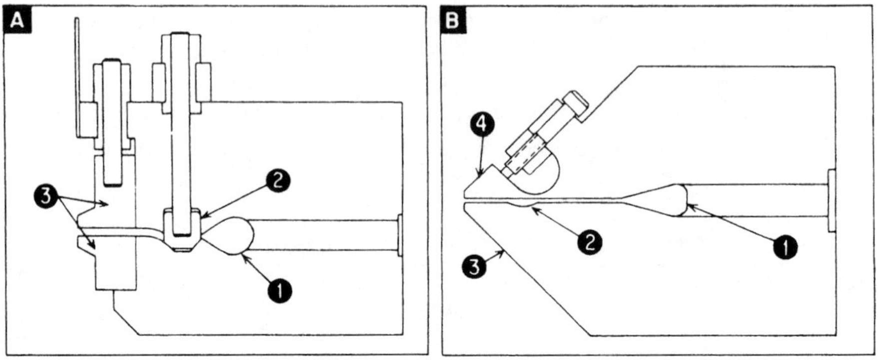


**Figure 5.23** (A) Conventional lip die compared to (B) flexible lip die.

**Figure 5.24** EDI Autoflex die with individual bolt adjustments for the lip opening fitted with heater/air cooling controls.

using a restrictor bar 'B;' and (3) rigid lips 'A' and wedge-shaped rigid front die 'A,' and (4) flexible upper lip 'B.'

A popular worldwide Autoflex patented die system (Fig. 5.24) from EDI has each lip adjusting position fitted with a block containing a

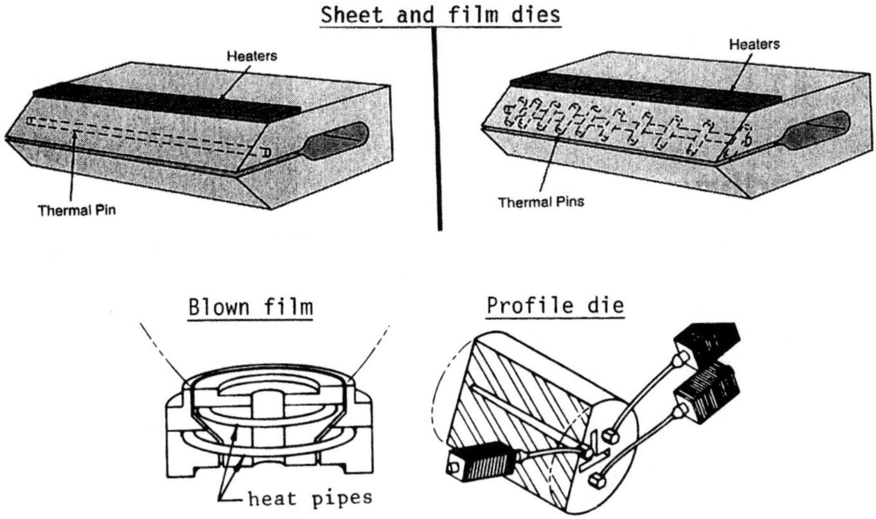


**Figure 5.25** Examples as to where heat pipes are used in dies.

heater and air cool path. Inherent in the system is completely manual override, so that the approximate gauge uniformity is established in the conventional way before the thermal system is engaged. The thermal adjustment system utilizes an electric source for each adjusting point to increase or decrease the temperature of the adjusting block, thereby changing the lip opening. It provides: a very fast, more responsive control; elimination of heat transfer between the die body, lip, and translator block; eliminates individual lock-up bolt on the antibacklash system; and permits use with low temperature or thermally sensitive plastics.

*Heat pipe*

Heat pipes, also called thermal pins, heat transfer devices, or heat conductors, are a means of heat transfer; either to remove or to add heat (Fig. 5.25). Compared to metals, they have an extremely high heat-transfer rate. They are capable of transmitting thermal energy at near sonic isothermal conditions and at near sonic velocity. Sizes range from very small to large diameters with limited lengths; however, pins can be put in series, etc.

They are tubular structures closed at both ends and contain a working fluid. For heat to be transferred from one end of the structure to the other, the working liquid is vaporized. The condensed vapors travel to the opposite end where the condensate becomes a working liquid again and returns to the original end of the pipe.

**Manifold and die**

Different manifold designs for dies are used to improve and control the behavior of melts. As an example, the area of flat die manifold or flow channel design has seen significant improvements in the past decade. As Extrusion Dies, Inc., reports to the industry, aided by computational fluid design programs, state of the art hybrid manifolds tuned specifically to plastic and processing conditions are designed by solving three-dimensional flow equations [418].

The basic manifold for a die is a constant cross section or T-slot design (Fig. 5.26). In all cases, the internal geometry should be streamlined, and care taken to avoid dead spots that could lead to melt degradation. The T-slot manifold relies on a large manifold area and a lip long enough to create a large enough pressure drop to force the melt to the ends of the die. This simple manifold design is used when processing low viscosity plastics that are not thermally sensitive.

The coat hanger manifold design can be tuned for a processing 'window' of different flow rate plastics (Figs. 5.7 and 5.26). This manifold is designed by establishing the same resistance to flow across the manifold and land, at various points across the die. Its shape is typically a tear drop that reduces in size from the center to the ends of the die. This reduction in volume reduces residence time in the die and is of particularly importance when processing thermally degradable plastics. After the land, the melt flows into a secondary manifold, an area of the flow channel that allows the plastic to move laterally again if required. It also is used to control the total pressure drop of the die in conjunction with the final lip land.

When discussing designs, there are always advantages and disadvantages, such as with the coat hanger manifolds. The back line of the manifold is further from the exit of the die at the center than it is on the ends. This could result in a die that will deflect its opening more at the center, causing uneven melt distribution. The so called M and W patterns are common with these manifolds. Cause is a result of the linear flow channels within this type of die.

An advanced manifold design (EDI) is a hybrid which addresses these drawbacks. The tear drop shape is elongated, with an aspect ratio that changes from the center to the end of the die. This approach results in less interface distortion particularly when coextruding. Its land is a nonlinear shape that reduces the M and W patterns. The manifold also has a straight backline that allows for straight line bolt placement. The result is a die

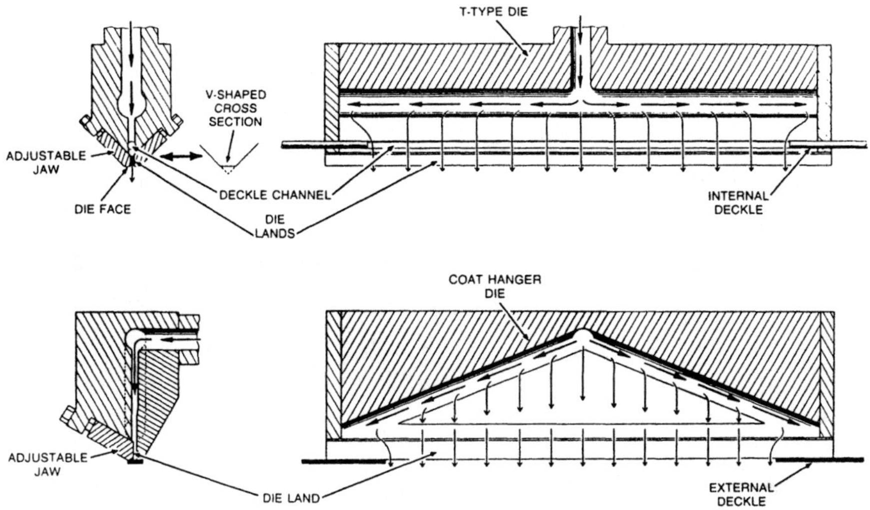


**Figure 5.26** Deckle rods inserted in the T-die and external deckle plates attached to the coathanger die.

that deflects uniformly, further improving the melt distribution in the die.

*Die deckle system*

A deckle rod, plate, or dam is attached to each end of a film or coating die. It can be used to reduce the die slot width and/or used to control the edge bead on the web. The plastic to be extruded and the extrusion process determines whether the die can be deckled. Degradable plastics are not used since deckles dam the melt flow causing stagnate areas where degradation will occur. The most common used is an external type that can be bolted to a die half or permitted to slide (Figs. 5.26 and 5.27). External deckles can be designed to be adjusted manually or driven by power screws or rack and pinion gears.

The disrupted melt flow from behind a conventional external deckle usually causes heavy edges; they prevent good roll contact over the center of the sheet by the polishing rolls and in turn can result in lower sheet gloss. To reduce this undesirable condition, edge beads or internal deckles can be used. The edge bead rod is typically a shaped rod that is inserted into the die stream of the final land. By varying its depth of insertion, the heavy end melt flow can be controlled. A more sophisticated EDI version is the internal deckle that fills some or all of the melt flow channel (Fig.

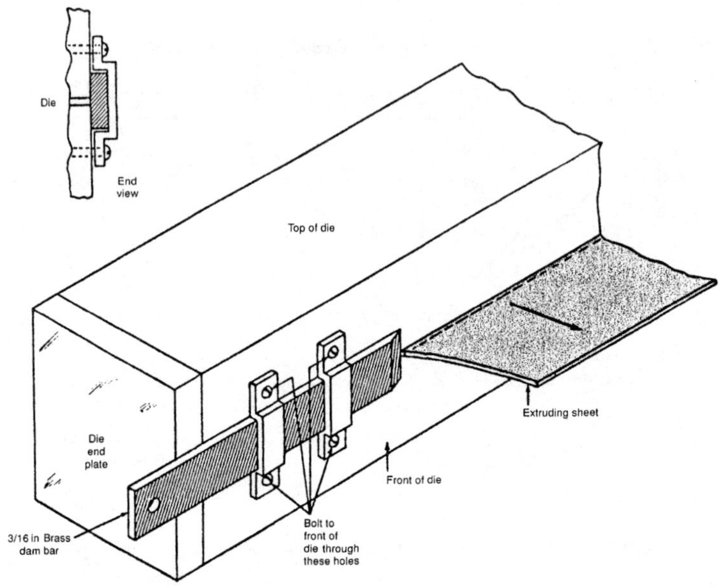


**Figure 5.27** Close-up of die face dam with deckle.

5.28). The deckle flags can be profiled more accurately to produce the desired result. Internal deckles are more difficult to set up, but they reduce melt clutters around the outside of the die at the lip exits, allowing for a closer approach to a nip or casting roll.

*Melt tuning device*

Different approaches are used to adjust melt flow upstream, such as the usual different designs of restricter bars prior to reaching the die lip. An interesting approach is EDI's patented Proteus feedblock developed for cast film, sheet, and coating lines with specific products, such as biaxially oriented PP, coextruding PVC siding, and PETG glazing (Fig. 5.29).

In this design what is being called an early flow tuning (EDI) device basically permits melt flow to be adjusted upstream of the die lips. The approach permits melt flow without completely splitting the melt stream. This control is accomplished through the use of a flexible stainless steel membrane positioned in the die's land. Bolt adjustments are made to position the membranes. With this action, the melt relaxes and orients well before it exits the die [135].

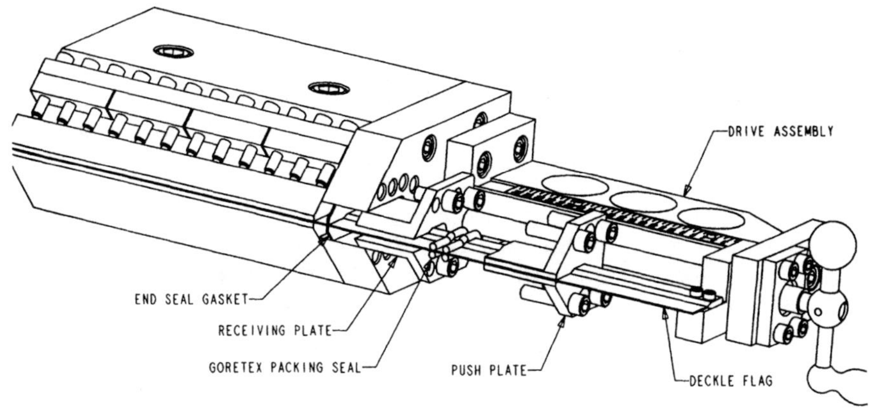


**Figure 5.28** Internal deckle for coating die.

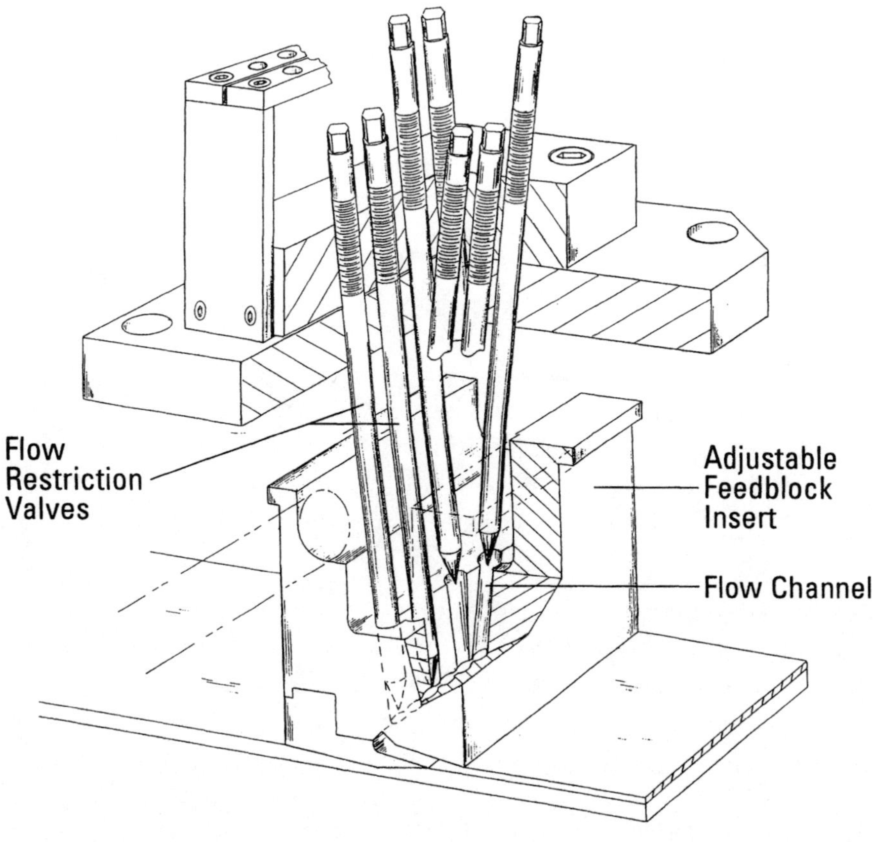


**Figure 5.29** EDI's Proteus feedblock.

*Die shape*

The approach used for shaping cavities in the dies is important requiring three-dimensional evaluations that includes streamlining. Examples for profile dies is shown in Fig. 5.30 and the following description of each die is provided:

1. A method of balancing flow to produce this shape requires having a short land where the thin leg is extruded. This design provides the same rate of flow for the thin section as for the heavy section.

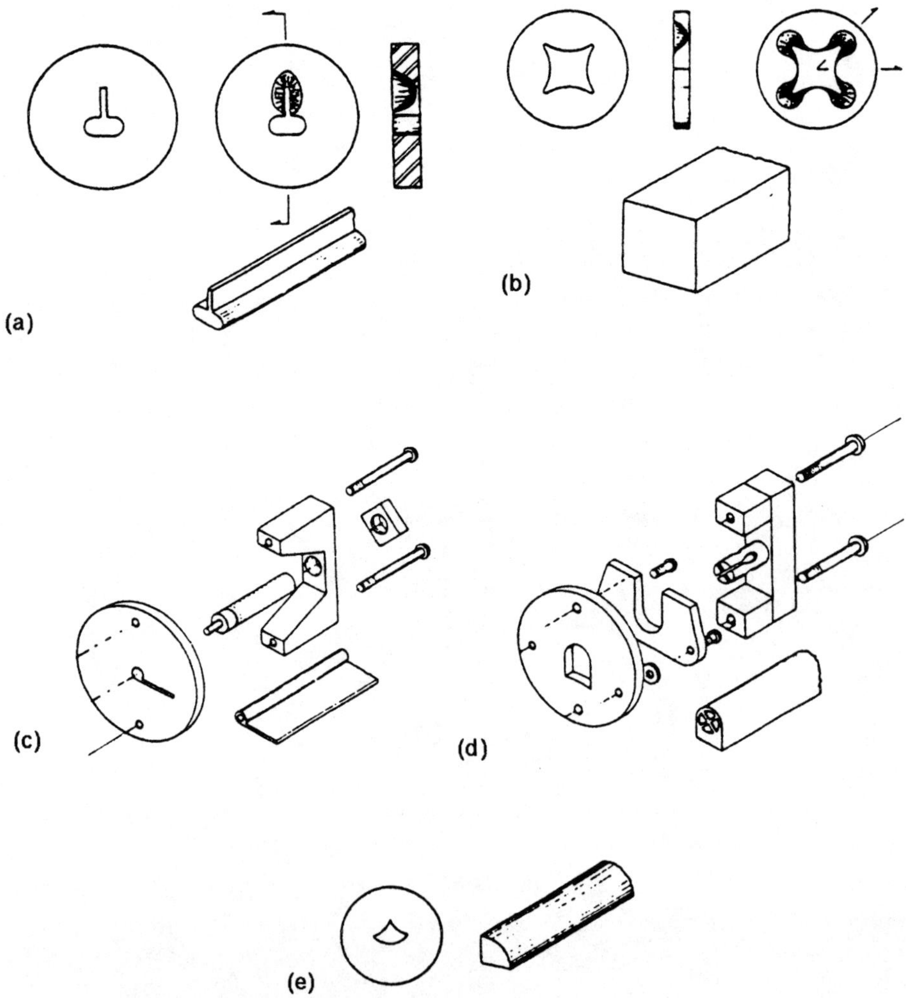


**Figure 5.30** Die designs to produce different profiles.

2. This die for making square extrusions uses complex sides on the die opening so that straight sides are formed upon the melt exiting. The corners have a slight radius to help obtain smooth corners. The rear and sectional views show how part of the die has been machined away to provide short lands at the corners to balance the melt flow. After the extrudate exits the die, cooling is more rapid at the corners.
3. In this die for a P shape, the hole in the P is formed by a pin mounted on the die bridge. The rate of flow in thick and thin sections is balanced by the shoulder dam behind the small diameter section of the pin. The pin can be positioned along its axis to adjust the rate of flow to meet the melt characteristics.
4. In this die to extrude a rather complicated, nonuniform shape, a dam or baffle plate restricts the flow at the heavy section of the extrudate to obtain uniform flow for all sections. The melt flows between the die plate and the dam to fill the heavy section. The clearance between the dam and the die plate can be adjusted as required for different plastics that have different melt behaviors.
5. In this die for extruding a quarter-round profile the die opening has convex sides to give straight sides on the right-angled portion, and the corners have a slight radius to aid in obtaining smooth corners on the extruded product.

Where possible, all dies should be groomed to promote streamlined melt flow and avoid the obvious pitfalls associated with the areas that could cause stagnation. Right angle bends, sharp corners, and sections where flow velocities are diminished are not conducive to streamlined flow. The target is to avoid these design faults. In certain cases, profiles for example, complex shapes do not lend themselves to absolute streamlining. In these instances, the stability of the plastic melt must be watched much more closely than one with a clean flowing die (Figs. 5.31 and 5.32).

There are different approaches to developing the streamlined shapes. They range from totally trial-and-error to finite-element analysis (FEA). The 'trial' method usually involves gradually cutting metal. Between cuts, an examination is made of the extrudate and the metal cavity surface to check on melt hang-ups, melt burning, streaks, and other stagnating problems. With FEA one can easily determine streamline flow patterns for some simple profiles using appropriate rheological plastic data. FEA, even though it is costly, has its place but it requires processing knowledge. Both the 'trial' and the FEA approaches actually require experience. Once the experience is gained, streamlining can follow.

Streamlining can provide a variety of advantages, such as: (1) dies can operate at higher outputs; (2) pressure drops are lower and more consistent over a range of melt temperatures and pressures; (3) generally the melt uniformity across the extrudate is more uniform and shape control is

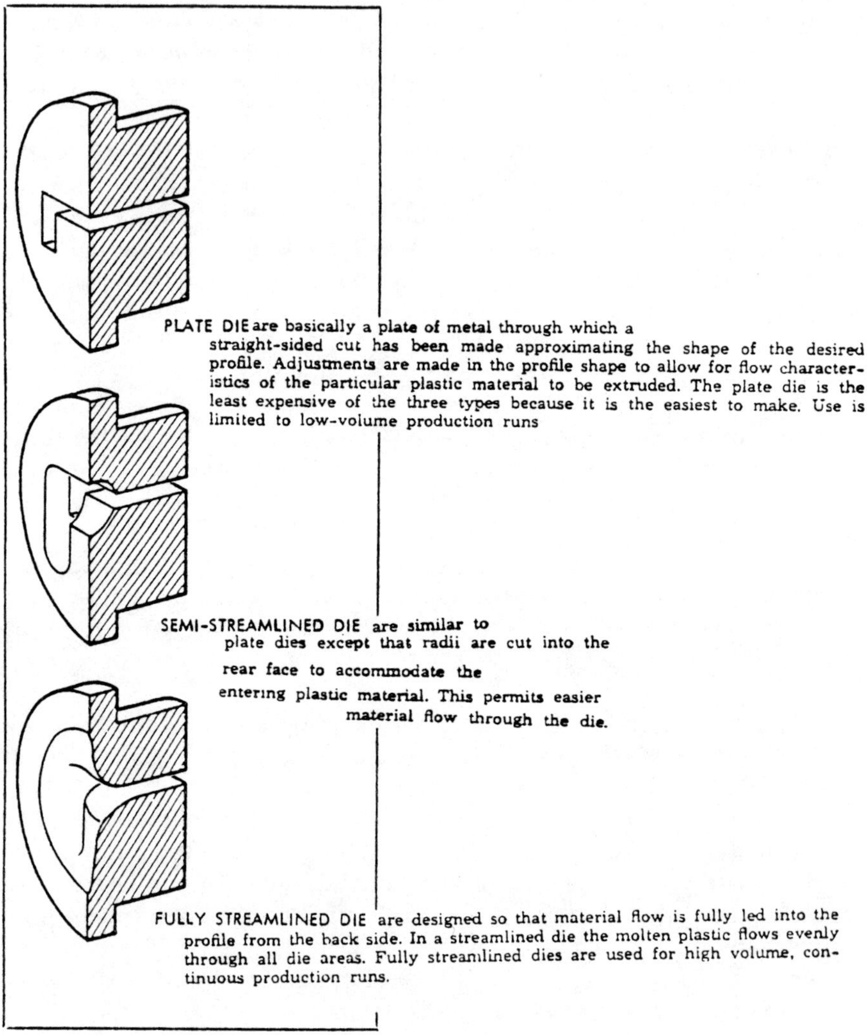


**Figure 5.31** Example of nonstreamlined to streamlined die.

enhanced; and (4) sometimes crucial for high production output rates where plastics have limited stability and causes hang-ups/degradation going through nonstreamlined dies.

## Coextrusion dies

Coextrusion is the simultaneous extrusion of two to at least seven plastic melt streams meeting in a die to produce laminated plastic products.

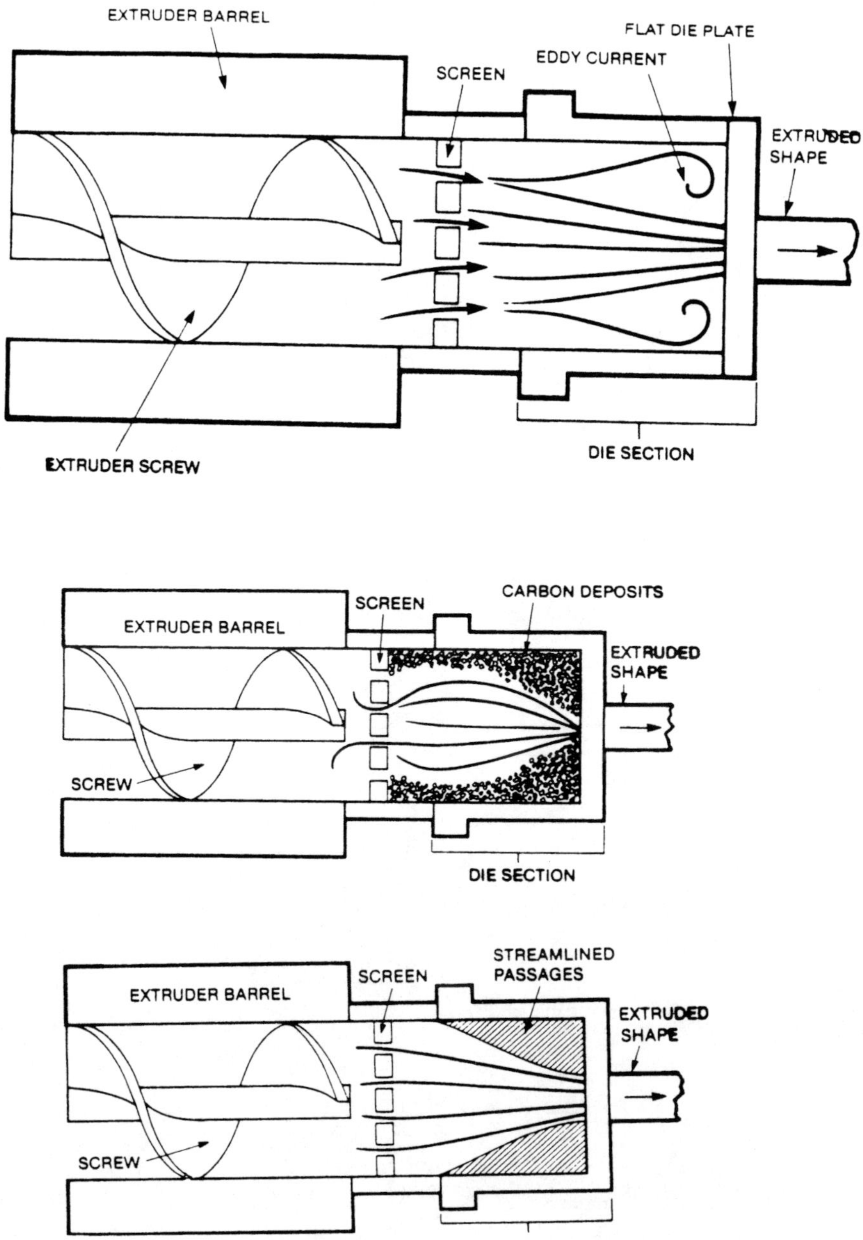


**Figure 5.32** Examples of stagnation areas from non-streamlined to streamlined melt flow in dies.

Figures 5.33–5.37 show different dies. Product performances gained are a combination of what each individual plastic provides. The layup can include recycled material and/or a low cost material(s) that will 'reinforce' more expensive material(s) strengthwise, tear resistance, and so on.

Certain melt processing factors have to be considered in order to eliminate any problems. Some of these factors can be compensated by the available extruder and die adjustments. This interfacial instability is caused by an unsteady balance of shear forces. Examples of these factors include: (1) different melt temperatures of adjacent layers; (2) plastic viscosity differentials that should not be greater than 2–4/1; and (3) minimum thickness of a cap (top) layer because it is subjected to a high shear stress is usually limited to 5–10% of the total thickness. There is a tendency for the less viscous plastic to migrate to the region of high shear stress in the flow channel causing an interface deformation. With a great difference

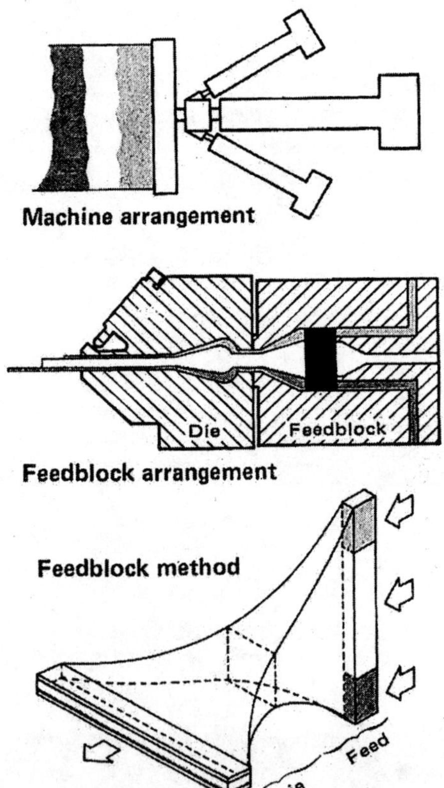


**Figure 5.33** Feedblock being used in a 3-layer coextrusion die.

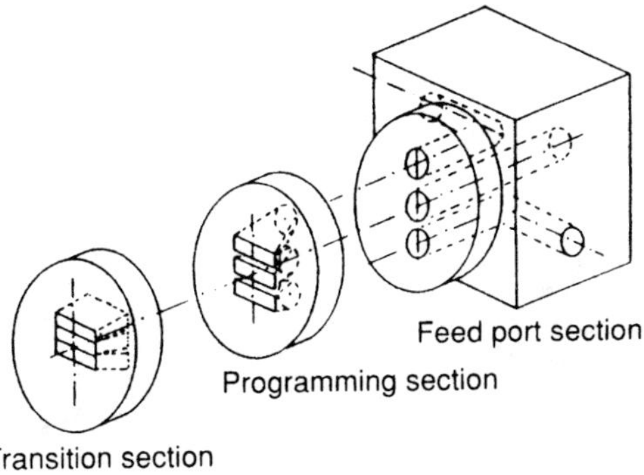


**Figure 5.34** Schematic of a feedblock sheet die.

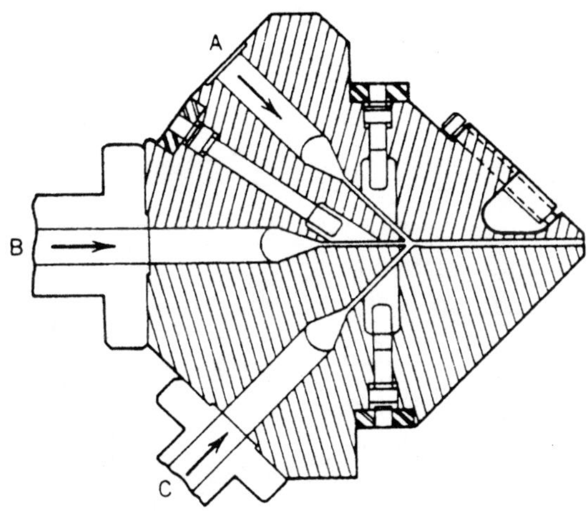


**Figure 5.35** Multi-manifold 3-layer flat-film die.

in viscosities existing between adjoining layers, the less viscous tends to surround or encapsulate the other plastic. Result could be fuzzy interfaces, orange peel, etc.

Many techniques are available for coextrusion, some of them patented and available under license. There are basically the feedblock (single manifold) or the multi-manifold dies (Table 5.3) with a third system that combines the two basic systems. This third system provides processing alternatives as the complexities of coextrusion increases.

With the single manifold die, the plastics meet (combine surface to surface) and spread to a given web width. In a multi-manifold die, the plastics are spread to a given web width and then meet slightly upstream from the die lip exit. Because the plastic layers are kept separate until they

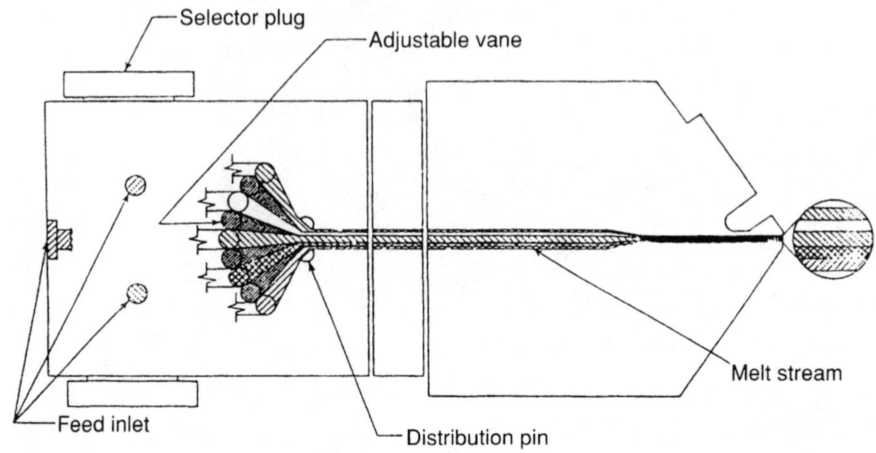


**Figure 5.36** Cloeren 5-layer feedblock with adjustable vanes.

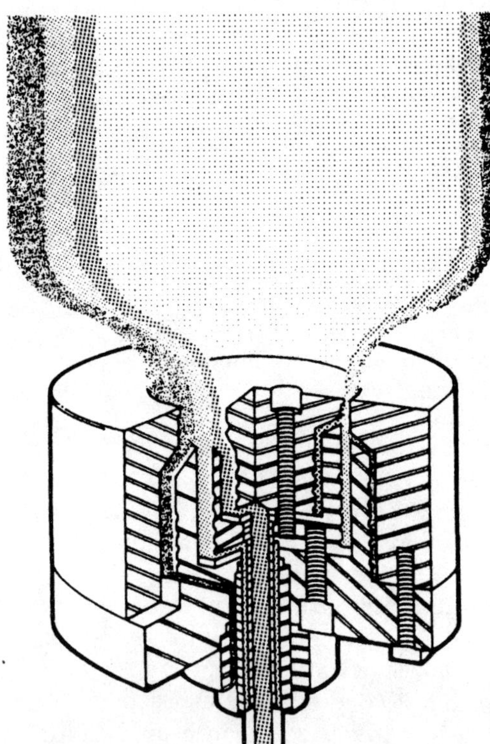

**Figure 5.37** A 3-layer blown film die.

**Table 5.3** Comparison of feedblock and multimanifold coextrusion dies

| *Characteristic* | *Feedblock* | *Multimanifold* |
|---|---|---|
| Basic difference | Melt streams brought together outside die body (between extruder and die) and flow through the die as a composite | Each melt stream has a separate manifold; each polymer spreads independent of others; they meet at die preland to die exit |
| Cost | Lower | Higher |
| Operation | Simplest | – |
| Number of layers | Not restricted; seven- and eight-layer systems are commercial | Generally restricted to three or four layers |
| Complexity | Simpler construction; no adjustments | More complex |
| Control flow | Contains adjustable matching inserts, no restrictor bar | Has restrictor bar or flow dividers in each polymer channel; but with blown-film dies, control is by individual extruder speed or gearboxes |
| Layer uniformity | Individual layer thickness correction of ±10% | Restrictors and manifold can meet ±5% |
| Thin skins | Better on dies >1 m (40 in) | Better on dies <1 m (40 in) |
| Viscosity range | Usually limited to 2/1 or 3/1 viscosity range of materials | Range usually much greater than 3/1 |
| Degradable core material | Usually better | – |
| Heat sensitivity | More | Less |
| Bonding | Potentially better; layers are in contact longer in die | – |

are fairly close to the die exit, interface deformation and interfacial instability are generally not a problem. This result is due to the fact that there is a limited time the plastics are allowed to flow together so that any temperature differential has a short duration. With proper die designs, flow rate can be adjusted for each melt via the usual fixed or adjustable restrictor bars.

Assuming both basic types have good manifold designs, the multimanifold can process a broader range of melt flow plastics. No matter which method is used, it is important to maintain the melt heat of each layer above the 'freezing' temperature of all layers. If temperatures are not properly set, adhesion between the layers will be poor.

The multi-manifold die is usually limited the number of layers. This restriction is due to the physical equipment size required. To compensate for this restriction, one or more feedblocks are used resulting in more layers of materials. Compatible plastics can flow through a single manifold reducing any potential problem down-stream in the multi-manifold. Combinations can be made to provide different laminated designs. However, the final exiting layer thickness distributions can be affected by the amount of die body defection if the die is not properly designed to take the required loads. Any deflection causes distortion influencing the melt flow channel(s).

The RV die (Figs. 5.38 and 5.39) designed by Egan/Davis-Standard is an example of incorporating the versatility of adjustable feedblocks with the repeatability of the modular Dow Chemical style feedblocks. This action is achieved through unique devices know as 'layer control plates' (LCPs). The RV consists of an inlet housing, LCP's transition plate, and division of flow plates. By using this patented design, one can go from 2-layer to 9-layer using the single feedblock.

For certain lines such as film, sheet, and coatings, a dual chamber vacuum box is located just upstream of the die lips (Fig. 5.40). Its purpose is to ensure that the exiting extrudate hot web from the die to its next position, such as a roll, is stable. With proper stability, the final product has quality control at this point in the line. The vacuum box is an evacuated chamber so that it will remove any air carried by the cooling roll and any airborne contaminants that may be near the roll surface. The second chamber controls the vacuum immediately adjacent to the melt curtain; by adjusting this vacuum, the angle of the melt curtain can be independently controlled. With this action, the melt is not dragged over the upper die lip eliminating the potential die lines caused by the die lip [196].

In configuring high barrier coextrusion dies, one is faced with a different set of problems than when designing for commodity plastics. Many barrier plastics, such as nylon and EVOH, as well as some adhesives (tie layers) with their narrow processing window are inherently less thermal stable than plastics such as LDPE and LLDPE. Further aggravating the

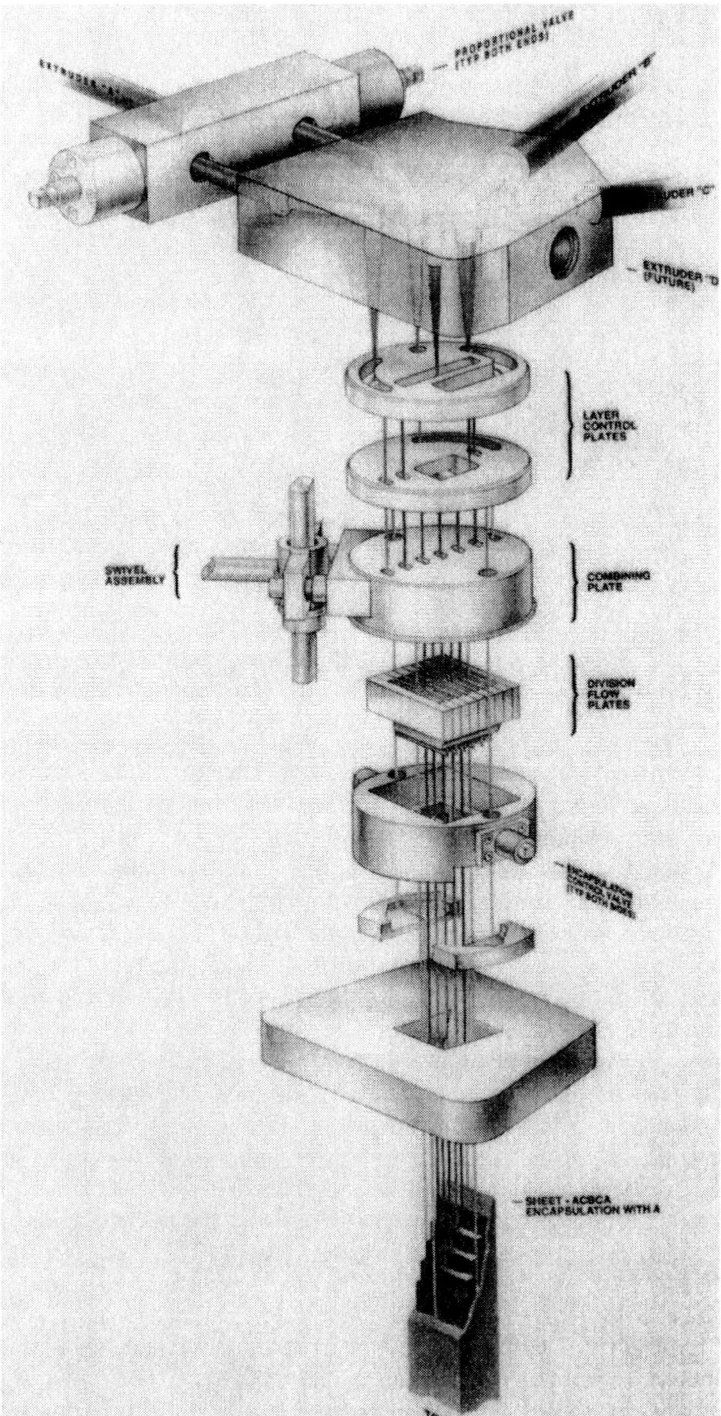


**Figure 5.38** Egan/Davis-Standard RV feedblock schematic showing melt paths through the feedblock.

**Figure 5.39** Egan/Davis-Standard RV feedblock assembled; alternate layer control plates and skin layer flow inserts are in the foreground.

process, particularly over 4-layers, are very low flow rates in the barrier and tie layers. To help meet these different problems, the die is designed to reduce melt residence time.

This action requires tightening all bores, clearances, and passageways. However, this can cause shear rates that increase and can cause other problems, such as melt fracture, frictional heat generation, and preferential flow. With the combination of computer modeling, extensive streamlining and advanced spiral configurations, a die is targeted to develop low residence time and maintain acceptable shear levels. A critical shear rate is maintained that will eliminate or minimize gel formation [120–122].

The extruders used will directly influence the operation of the coextruded die. As there are two or more extruders feeding the separate die channels, the first step is to calibrate them in terms of output rate versus screw speed; gear pumps are usually included. The next step is to establish the required width-to-thickness and the takeoff speed, and to convert them into an output rate for each plastic. These output rates and calibration curves are then used to establish the final screw speeds. When a feedblock is not properly designed, problems such as excessive pressure drops, temperature differentials, shear rates, and residence time, can affect the quality of the extrudate.

All extrusion processes require some form of melt transfer from the extruder to the feedblock or die. The transfer device can be short, such as

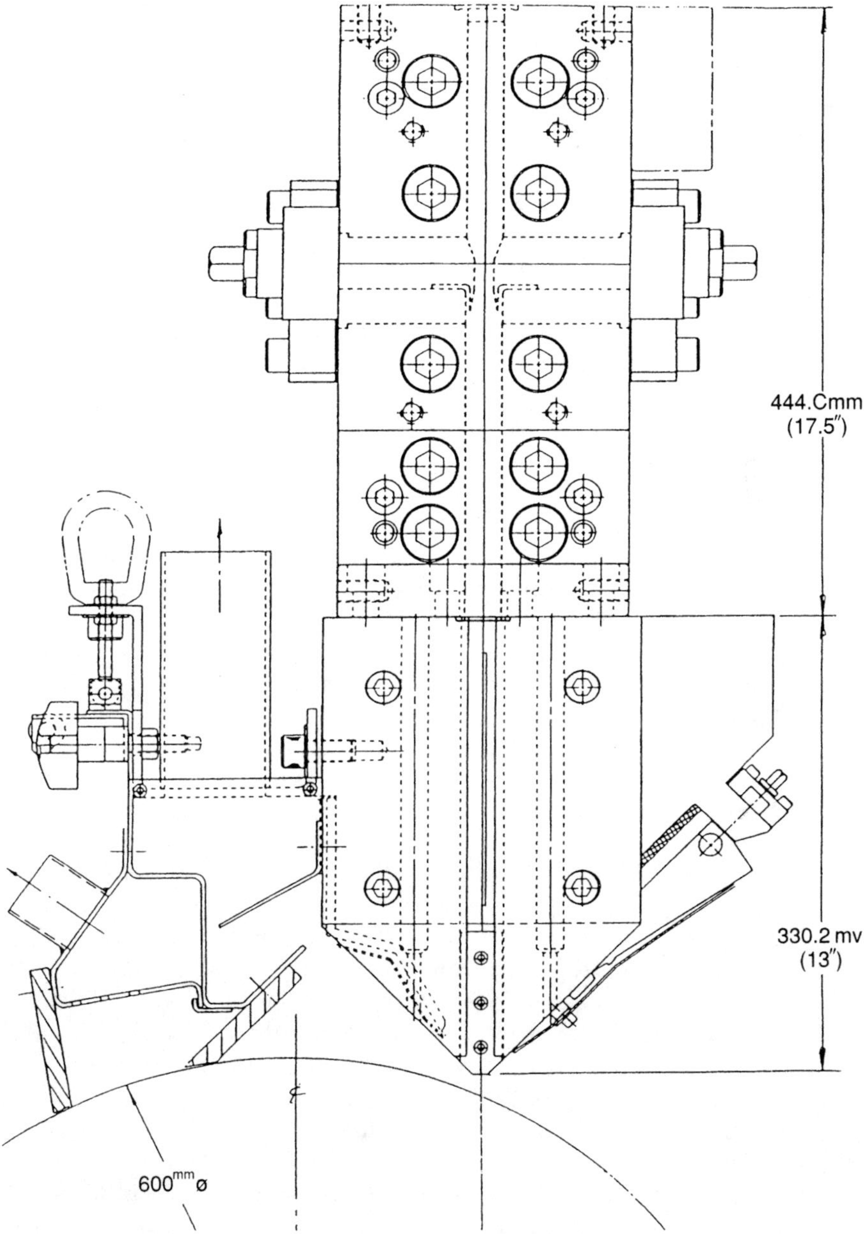


**Figure 5.40** Dual chamber vacuum box, die, and feedblock assembly.

an extruder adapter, or longer, as required in coextrusion 'piping.' The pipe design is integral to coextrusion success. In general, the inside diameter should be large enough to avoid excessive pressure drops, but not so large as to cause extended residence time. Its heavy wall provides maximum thermal stability and heat distribution efficiency. The heat should be uniform to avoid excessive hot and cold spots. Low voltage heaters are desirable; heater tapes can be damaging because of the possibility of nonuniform heat distribution.

Safety considerations should be the determining factor in the design of piping. Pipes have been known to cold pack, depending on shut-down procedures. The location of the control temperature indicator is important. Overheating of a zone, particularly on startup, can cause degradation of the plastic into gas, creating extremely high pressures.

## Special dies

A few special dies are shown in Figs. 5.41 and 5.42. They produce interesting flow patterns that result in tubular to flat netting fabricated products. For a circular output, a counter-rotating mandrel and orifice have semicircular slits through which the melt emerges. If one part is held stationary, it forms a rhombic or elongated pattern. When both parts rotate, it forms a true rhombic mesh. When the slits overlap, a crossing point is formed where the emerging 'threads' are 'welded' [300, 426].

Plastic netting is produced through the counter rotating die lips as a feed stock with strands at a normal 45° to machine direction then oriented into a lightweight tubular netting. A variety of mesh sizes, diameters, and colors are available. They are usually made from PE or PP. Most of the netting produced for flexible packaging started during the 1950s (patent issued) and is now marked throughout the world in forms such as: (1) tubular netting on transverse-wound rolls (commonly called rope by their industry); (2) header-label bags converted from rope, heat-set into open tubes (usually gussets) and sealed by the manufacturer with a folded or sewn printed label; (3) G-bags which are lengths of the net cut with a knife to produce a gathered heat sealed on one end; and (4) sleeving or cartridges where netting is shirred onto a collapsed corrugated board. The board can be opened into a square or rectangle for transfer onto a tube or funnel.

Since the UK patent of 1957, there are extrusion lines producing continuous square mesh using a rotating die with slotted lips machined around its circumference, each rotating relative to each other. The tubular netting can be slit so that a wide flat 'sheet' is produced. In turn, if desired, put through a uni- or biaxially orientation tenter frame (Chapter 2).

The largest market for plastic netting is consumer-sized packaging in the produce industry. Nearly complete ventilation reduces spoilage of

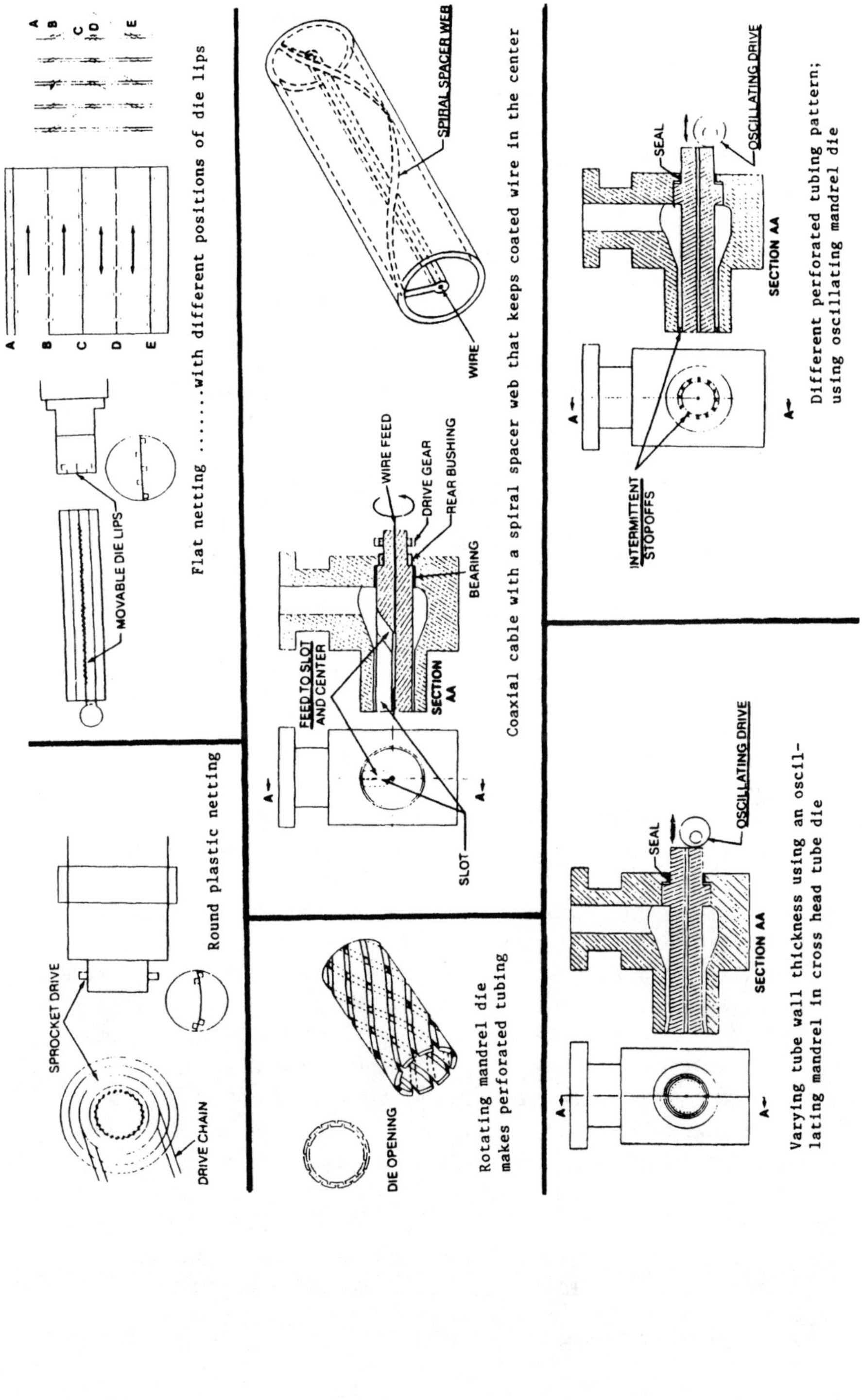


**Figure 5.41** Special-action dies that produce round and flat products.

**Figure 5.42** Netting exits the die and undergoes post-extrusion stretching.

many fruits and vegetables, such as citrus and onions. Automatic packaging equipment using rope shirred onto tubes offer labor and materials costs competitive with other consumer-size packages. Premade header-label bags are also widely used by produce packers and by supermarkets that package in-store. Other markets include pallet stretch-wrap, tools, toys, and protection nets (fencing crop, building protection, etc).

A novel coextruded blown film (or flat film) die produces startling iridescent effects simply by taking advantage of some basic optical principles. Alternating layers of two plastics, such as PE and PP, with at least 115 (and many more) produce an extruded film 0.013 mm (0.5 mil) thick. This Dow Chemical patented process generates hundreds of layers, each one thinner than the wavelength of light. The tubing die generates a large number of layers by rotation of annular die boundaries; the principle is related to earlier work by Dow Chemical on color mixing of PS.

In operation, individual components are pumped through a feedport system into a die in alternating layers extending radially across the annular gap [Fig. 5.43 (a)]. Simultaneously rotation of the die members, i.e., the inner mandrel and outer ring, deform the layers into long thin spirals around the annulus. In effect, the number of layers is multiplied by the increased interfacial surface area. The number of layers and layer thick-

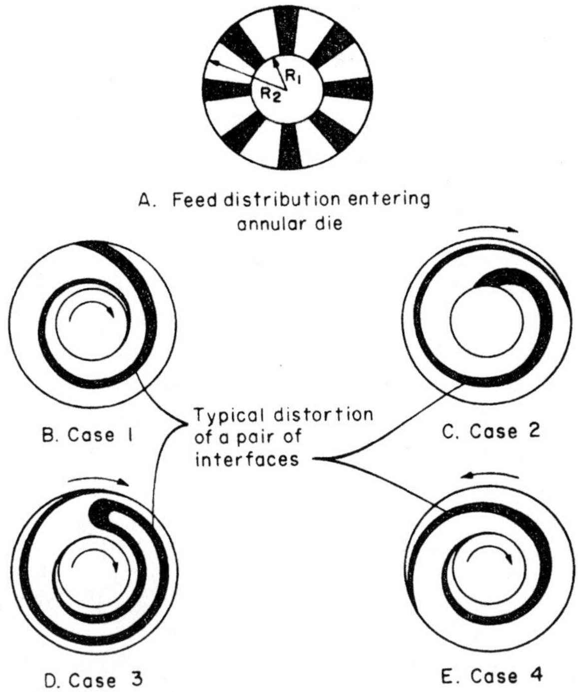


**Figure 5.43** Layer patterns developed by four modes of die rotation.

ness is determined by the dimensions of the annulus, the number of feedports for each phase, the extrusion rate, and the rotational speed of the die mandrel and ring relative to the feedports.

There are four basic layer patterns which may be generated by four modes of the die rotation. They are designated Cases 1, 2, 3, and 4 (Fig. 5.43). With Case 1, the inner die mandrel rotates while the outer ring is stationary where layers are thicker near the outer ring. Case 2 has the inner die mandrel stationary while the outer ring rotates with layers thinner near the outer ring. Case 3 has both inner and outer die members rotating at the same speed and direction with layers of curved open-end loops and thicker layers in the center. Case 4 has inner and outer die members counter-rotating at equal speed generating the maximum number of symmetrical layers with the thickest in the center.

In all cases, layer thickness and number vary with radial location in the annulus at the rotating boundaries. Although the layers are concentric, the deformation is usually so large that the spiral characteristic is indistinguishable when examining the extrudate in the cross section.

## DIE DESIGN

Analyzing melt flow in dies is very difficult. Using available CAD programs can be extremely helpful but what really helps is experience in the design and use of different dies with the different plastics. Industry has specialists in-house or specialty die manufacturers that produce efficient operating dies used to extrude all types of products. What makes it difficult is the nature of the plastic melts that are not 'perfect.' Each melt basically has its own plus and minus capabilities for operating in the die melt channels following its nonNewtonian behaviors (Chapter 3). The extruders (and other equipment) have their limitations, such as heat transfers through metal parts and metal parts which are subjected to wear. Therefore, what exists is an empirical science that continues to work efficiently. The limitations have always existed. But with time as material and equipment developments occurred, designing dies as well as operating equipment continues to improve [187, L. Sansone and others, 265, 269, 297, 346, 397, 457].

As described throughout this book, including this chapter, the melt flow rate is influenced by many variables that starts with the plastic's composition (Chapter 3), extruder capability (Chapter 2), die geometry (Table 5.4) [397], die performance, and down-stream equipment. In studying a complex die system, one must examine each component where the geometry is constant, apply the appropriate formula (approach) to compute factors, such as pressure drop at the desired melt temperature and flow rate, and then sum all the pressure drops to determine the pressure that must be provided by the extruder.

The following review uses a simplified equation obtained through the available (of that time period) high-speed computer study during the early 1960s by G. P. Lahti [265]. He did this work at DuPont and latter went to NASA. It provides an excellent foundation using an empirical approach that pertains to extrusion die channels of several shapes. Flow equations for dies of simple shapes, such as circular or rectangular channels, were known in the previous century; they were first developed by M. J. Boussinesq in 1868 [125]. Formulas for pressure drop through more complex channels had not yet developed. The reason being that the researchers required extremely complicated mathematics. As shown in Fig. 5.44, the following equations can be used:

$$Q = (1/\mu)(\delta P/L)(BH^3/12)(F)$$

or

$$\Delta P = (12\mu QL/BH^3)(1/F)$$

where $H$ = minimum dimension of cross section, mm (in), $B$ = maximum dimension of cross section, $B \geq H$, mm (in), $\delta P$ = pressure

**Table 5.4** Effect of operational and geometrical variables on flow conditions in the die

| *Parameter* | *Effect* |
|---|---|
| Nonuniform exit velocity across the width | Extrudate curling, twisting, rippling<br>Non-uniform drawdown, resulting in residual stresses in the extrudate |
| Nonuniform exit flow thickness across the width | Nonuniform tranversal extrudate thickness before and after swelling<br>Nonuniform exit velocity across the width (see above) |
| Flow thickness varying longitudinally (tapers, steps, choker bars) | Significant extensional effects, with possible melt fracture, and contribution to extrudate swell<br>Flow acceleration along the taper, with velocity instabilities at the entrance and exit of the taper<br>Possibility of stagnation points and corresponding material degradation<br>Relatively large pressure drop |
| Flow thickness of length varying transversely | Pressure drop or flow rate varying transversely (thus extrudate curling, twisting and rippling)<br>Nonuniform residence time and relaxation (affecting degradation and extrudate swell)<br>Differences in shear heating, promoting transverse flow |
| Set (and melt) temperature varying transversely | Transverse components of flow<br>Nonuniform extrudate swell and surface finish<br>Nonuniform shear heating effects<br>Possibility of material degradation<br>Pressure drop or flow rate varying transversely (thus, extrudate curling, twisting and rippling) |
| Transverse flow | Nonaxial exit velocities<br>Changes in the relative exit velocities and swelling<br>Distortion of the extrudate section<br>Possibility of stagnation points and corresponding material degradation |

drop, $Q$ = volumetric flow rate, $L$ = length of channel, $F$ = flow coefficient.

To account for the entrance effect when a melt is forced from a large reservoir, the channel length ($L$) must be corrected or the apparent viscosity must be used, once it has been obtained from shear rate-shear stress curves for the $L/H$ value of existing channel. The entrance effect becomes negligible for $L/H > 16$. This single equation can be used for a variety of flow channels.

Based on knowing the product geometry, plastic viscosity, and pressure drop, the volumetric flow rate ($Q$) can be calculated as shown in Fig. 5.45.

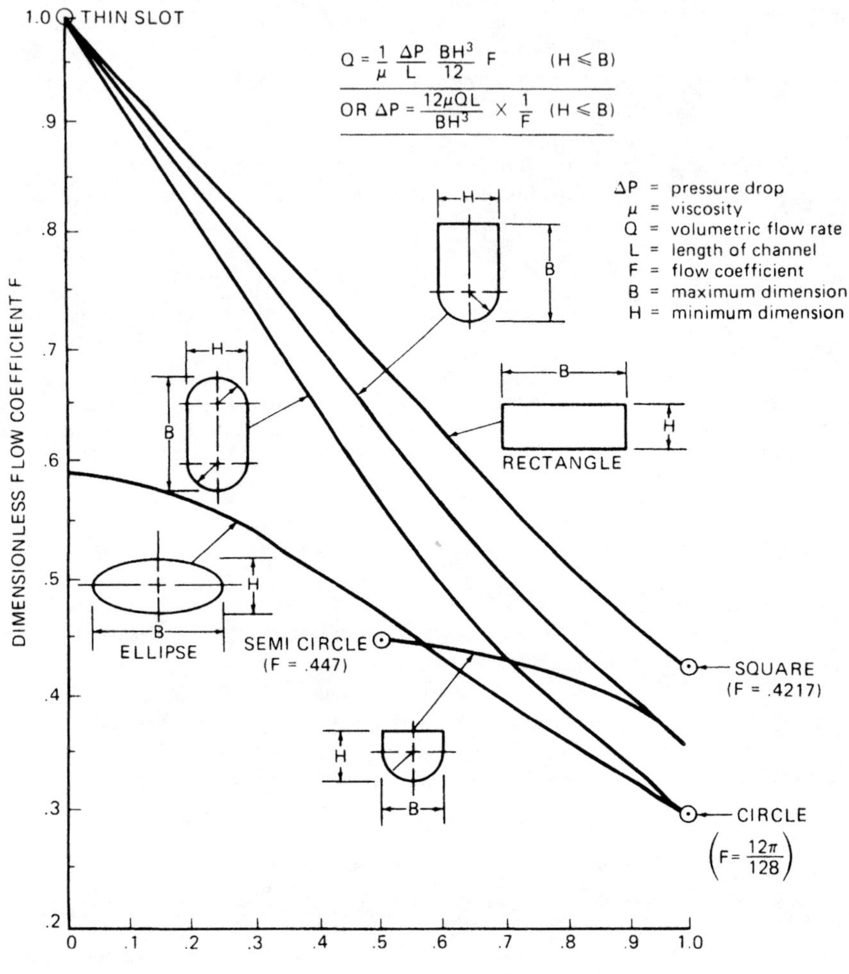


**Figure 5.44** Flow coefficients calculated at different aspect ratios for various shapes using the same equation.

Calculations are also presented for: (1) flow in two or three directions which exists in a tapered; (2) detailed discussion including limitations and assumptions for regular and irregular shapes are made for slow viscous melts; (3) and so forth [125].

To date, various organizations have expanded the capability of CAD software programs for die designs. An example is from EDI where one of their tools to evaluate the flow channel designs uses a computational fluid dynamics computer code. Using this code, EDI prepares flow models

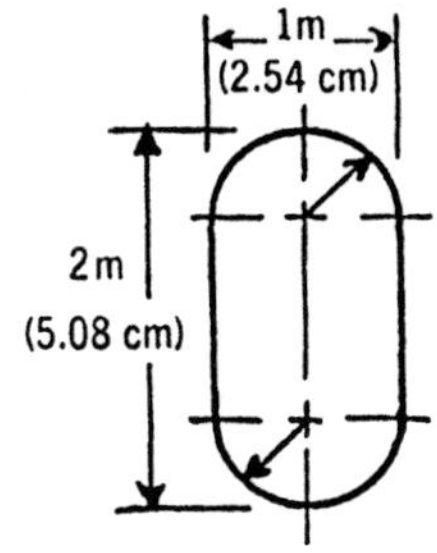


Fór this shape,

where $\frac{H}{B} = 0.5,\ F = 0.565$

$$Q = 10\,in^3/s\ (16.39\ cm^3/s)$$
$$\mu = 690\ poise = 0.01\ lb\ s/in^2$$
$$L = 10\ in\ (25.4\ cm)$$
$$B = 2\ in\ (5.08\ cm)$$
$$H = 1\ in\ (2.54\ cm)$$
$$\Delta P = \frac{12\,\mu\,QL}{BH^3F} = \frac{12(0.01)(10)(10)}{2(1)^3(0.565)}$$
$$= 10.6\ psi\ \ (0.71\ kg/cm^2)$$

**Figure 5.45** Specific shape for flow evaluation.

which utilizes its powerful capabilities. It is basically a finite-element analysis program for calculating the flow of fluids.

## CONSTRUCTION MATERIALS

Dies are 'delicate' instruments and require careful treatments in order to have them continually operate efficiently. These steel precision 'works of art' have at least mirror finishes on the melt flow channel orifice surfaces. The slightest scratches can produce flaws in the extruded products. Great care must be used during their installations, operations, removal, cleaning, and particularly storage. When designing them target is use as few parts as possible, easily lifted for installation, easily disassembled, easily cleaned, and easily re-assembled.

Special metals are used such as for flat film and sheet dies that are usually constructed of medium-carbon alloy steels. The flow surfaces of the die usually have protective coatings (Table 5.5) such as chrome plated to provide corrosion resistance. With proper chrome-plated surfaces, microcracks that may exist are properly covered. The exterior of the die is usually flash chrome plated to prevent rusting. Where chemical attack can be a severe problem (with PVC, etc.), various grades of stainless steels are

**Table 5.5** Protective coatings for dies/molds

| *Material* | *Method of application* |
|---|---|
| Chromium | Plating |
| Nickel | Plating |
| Electroless nickel | Solution treatment |
| Nedox electroless nickel | Solution treatment followed by TFE impregnation; used on copper and ferrous alloys |
| Tufram TFE aluminum | Deep anodizing process followed by TFE impregnation; used on aluminum alloys |
| TFE ceramic | Spray and bake application; used for all die materials that can withstand 250 °C bake |
| Tungsten silicide | Solution treatment; used on steel and ferrous alloys |
| Tungsten carbide | Explosion impact or flame spray with plasma arc; used for all high-melting metals to improve abrasion resistance |
| Aluminum oxide | Plasma flame spray; used for extreme abrasion resistance; used on steel dies but usually limited to small dies because of expansion problems; works best on 18-8 stainless |
| PTFE | Spray and bake application; used for low-friction and low-adhesion application; poor abrasion resistance |
| Polyimide, aramid | Straight organic coatings with high softening points (450–550 °C), which are applied by spray and baked; low friction characteristics against some resins (for example, PVC); moderate abrasion resistance |
| Filled polyimide, aramid | Aramid and polyimide systems containing TFE and other fluorocarbon resins to improve the friction properties |

used with special coatings. Coatings will eventually wear, so it is important that the dies be properly recoated by a reliable platter, usually the original die manufacturer.

Profile, pipe, blown film, and wire coating dies are generally constructed of hot-rolled steel for low pressure melt applications. High-pressure dies can be made of 4140 steel and chrome plating is generally applied to the flow surfaces, particularly when processing certain plastics such as EVA. Stainless steel is used for any die subject to corrosion [450].

The steel used in the manufacture of a die varies, depending on the requirements of the plastic and application. The available spectrum of modern tool steel offers properties in numerous combinations and to a widely differing degrees. Fortunately, the needs of the vast majority of steel users can be satisfied with a relatively small number of these steels, the most widely used of which have been given the identifying numbers of the American Iron and Steel Institute (AISI) [2].

The properties of the die material usually are as follows: (1) wear resistance to provide a long life; (2) toughness to withstand processing and particularly factory handling; (3) high modulus of elasticity so that the die channels do not deform under melt operating pressure and the dies weight; (4) high uniform thermal conductivity; and (5) machinability so that good surface finish can be applied particularly near the die exit.

## MAINTENANCE

The die is an expensive and delicate part of any extrusion line. Great care should be taken in the disassembly and cleaning of components. Disassembly should be done only when the die has had sufficient time to heat-soak or at the end of a run so that it is at operating temperature. Experience has shown a temperature of 232°C (450°F) to be adequate for most nondegradable plastics. For degradables, cleanup should begin immediately after shut-down to prevent corrosion action on the flow surfaces. While the heat is left on, all the bolts are loosened. The heat should then be turned off, and all electrical and sensors removed carefully. While still hot, it is disassembled and thoroughly cleaned with 'soft' brass, copper, or aluminum tools.

If the plastic tends to cling to the flow surfaces, it is usually best to purge the die prior to cleanup (Chapter 3). If any residue is left in the channel, it can cause problems with streaks and lines during production. If a residue remains on other parts, re-assembling could be a major problem. The die bolts should be cleaned, inspected for thread damage and signs of stretching, and recoated with an antiseize compound. During assembly, the die bolts should initially be slightly tight until the die heat is in the normal operating range. Once this heat is reached and a sufficient heat soak has been allowed (could take at least 15–30 min), all bolts should be tightened to the manufacturer's recommended sequence and torque levels.

When a flat film/sheet die is being adjusted, consider that the low areas should be opened, and the high areas closed. The tendency to only close down the high areas results in a die opening that is too tight. The line pressure will then become high, and the die adjustment will begin to feel too sensitive. Each adjustment screw will move a standard flex lip over a wide area, usually 3 to 4 adjustment positions to either side will be affected. This action occurs because the flex lip cross section tends to be stiff. The adjustment should be spread out with small movements made over large areas. If the opposite is done, some of the adjustment bolts will begin to float, taking them out of control of the lip.

After cleaning, it is common practice to treat them with a corrosion or rust inhibitor. Some dies are the property of the customer who buys the extruded product, and the manufacturer usually is responsible for its use and care. If the die is stored disassembled, care should be taken in its handling to prevent damage to individual components and to the flow surfaces. However, dies are usually stored assembled. Storage should be in a dry area. Some plants store them in vacuum-sealed containers to provide maximum protection from the environment.

The movement of dies/tooling through the plant, repair shop, or into position on the processing line can be difficult because of their size and weight so carts and lifts are generally required. There are carts on wheels that can safely move dies. The carts can provide: (1) disassembly and re-assemble; (2) clean with ease (devices for opening, etc. dies); (3) die parts can be rotated to facilitate cleaning; (4) prevents damage while handling large dies; (5) adjustments can be made for different size dies; (6) stores spare parts; and (7) so on. The most commonly used device to assist in moving dies is an overhead crane. However, in many cases the crane is a cumbersome device which contains more capacity than is actually needed. For all-around use a mobile crane offers definite advantages. They can handle at least weights up to 3600 kg (8000 lb).

Eyebolts are used to lift and move heavy dies. They should be used in dies where balanced lifting is accomplished. Holes should be tapped on surfaces perpendicular to the slots. The load capacities of the bolts have to be considered. Forged steel eyebolts usually have the following load-carrying capacities based on their diameters: 13 mm (0.5 in) of 1180 kg (2600 lb); 19 mm (0.75 in) of 2725 kg (6000 lb); and so on.

## TROUBLESHOOTING

As reviewed in other sections in this book, different problems do occur. An example is not operating the extruder properly and/or the down-stream line speed is not properly set. True these type problems are usually corrected. If not efficiently done, someone could assume that the die was at fault. Thus, careful analysis of any potential die problem has to include the complete operation of the up-stream to down-stream equipment. There are obvious problems due to the die such as leakage. Leakage in particularly flat dies can be a major problem (Table 5.6). To avoid leaks proper design and particularly maintenance are required.

After cleaning and installing the die on the extruder, it is important to allow enough time for the heat and adjustments to be made. Let the effect of the adjustments be measured by a proper gauge. Adjustments of the orifice openings do not normally change the output rate, particularly on a flat die; the melt flow is being diverted because of the change of resistance to flow. If the output rate of the extruder is changed, the flow in the die

**Table 5.6** Troubleshooting flat extrusion dies

| *Problem* | *Cause* | *Solution* |
|---|---|---|
| Die seal leaking | Foreign object(s) on seal surfaces | Clean all seal surfaces, soft tools (brass, aluminum) sharpened as a scraper. Do not use power tools. |
| | Seals are not flat | Stone nicks, burrs, or any raised area. If not successful, grind die seals on surface grinder. |
| | Push bolts not retracted | Retract, or better, remove push bolts. |
| | Dowel pins oversized or not aligned | Install proper dowel pins and check alignment. |
| | Seal area too large | Cut seal area down to typical 3/8 in or 1/2 in. |
| | Body bolts bottom out | Re-tap and increase tapped length in the body or use shorter body bolts. Be sure bolt has sufficient thread engagement. |
| | Stretched body bolts | Check if correct body bolts are used. |
| | | Check if high temperature lubricant is used on threads and underneath shoulder. |
| Endplate leaking | Top and bottom half not aligned | Remill sides on top and bottom half and relocate dowel pins if needed. |
| | End plate gasket deformed | Install new end plate gasket. |
| | End plate gasket incorrectly installed | See manual for proper installation of gasket. |
| | End plate gasket incorrect for application | Use different type of gasket, e.g., high temperature, stainless steel, etc. |
| | Insufficient pushing force | Redesign end plate assembly to insure good pushing force. |
| | | Cut seal area in the push plate to increase sealing force. |
| Entrance leaking | Adapter pilot o.d. too large | Cut down adapter pilot OD to insure die seals are touching. |
| | No clearance between die body and adapter flange | Increase adapter pilot length or decrease die pilot depth. |
| | Adapter bolts bottom out | Use correct length of bolts, high temperature lubricant, and proper torque values. |

**Table 5.6** *Continued*

| *Problem* | *Cause* | *Solution* |
|---|---|---|
| Restrictor bar leaking (if applicable) | Bar not aligned | Center bar to be flush with die body at both ends. |
| | Bar too short/too long | Regrind bar and/or die body accordingly. |
| | | Some bars are oversized or undersized depending on thermal expansion. Bar should always be flush with die body at operating temperature. |
| | Bar and/or bar slot damaged or incorrect clearance | Regrind bar and/or slot; built up bar with plating as needed. |
| Removable lower lip leaking (if applicable) | Incorrect seal for application | Use different types of seals, e.g. stainless, high temperature steel, etc. |
| | Lip not aligned | Lip to be flush with die body at both ends. |
| | Lip too short/too long | Insure that lip is up to temperature and at the same expansion as the body. Do not torque lip cold. Regrind lip(s) and/or die body for matching. |
| | Lip seal damaged or not flat | Seals to be ground on surface grinder. |
| | | Remove plating build-up. |
| General leaking | Leaking past lip due to body deflection | Install lip seal(s) in lip or die body. Reduce land length to lower pressure. Do not overadjust flex lip. Install extra bolts. Check torque requirements. |
| | Hot spot | Check for runaway temperature zone. |
| | Cold start-up | Die to be at operating temperature for at least one hour before starting. |

**Table 5.7** Troubleshooting different types of dies

| *Problem* | *Cause* | *Solution* |
|---|---|---|
| ***Blown film extrusion*** | | |
| Thickness variations across the extrudate circumference | Erratic melt quality | Check screw design and/or wear<br>Check set temperatures<br>Check heaters and thermocouples |
| | Surging or feeding inconsistency | Check screw wear<br>Check material in hopper<br>Check regrind percentage |
| | Dirty die | Clean die (check for obstructions at the die lips) |
| | Inadequate die-bolts adjustment | Readjust |
| | Misaligned die/air ring | Centre die to nip rolls<br>Centre air ring to die |
| | Leaky die | Check seals |
| | Temperature fluctuations at the die lips | Check heaters and thermocouples<br>Check set temperatures |
| | Poor air flow distribution in the air ring | Check and clean air ring |
| Lines, streaks and foreign specks | Dirty die | Clean die (check for obstructions at the die lips) |
| | Scratched die lips | Repair, or replace die lips |
| | Inadequate die-bolts adjustment | Readjust |
| | Contaminated melt flow | Check compound<br>Change filters |
| | Melt flow too hot | Reduce set temperatures |
| | Welding lines | Increase melt temperatures<br>Use spiral mandrel die |
| Sharkskin, melt fracture | Melt temperature too low | Increase the die lips' temperature |
| | Friction at the die lips | Repair die lips' coating<br>Modify formulation |
| | Die gap too narrow | Increase |
| Bubble instability/ irregular frost line | Erratic melt quality | Check screw design and/or wear<br>Check set temperatures<br>Check heaters and thermocouples |
| | Dirty die | Clean die (check for obstruction at the die lips) |
| | Excessive air ring velocity | Reduce |
| | Insufficient blow-up ratio | Increase |
| | Excessive melt temperature | Reduce set temperatures |

**Table 5.7** *Continued*

| *Problem* | *Cause* | *Solution* |
|---|---|---|
| Wrinkles | Misaligned die/nip rolls<br>Nonuniform cooling and winding | Align die to nip rolls<br>Check winder |
| ***Cast film extrusion*** | | |
| Thickness variations across the width | Erratic melt quality | Check screw design and/or wear<br>Check set temperatures<br>Check heaters and thermocouples |
| | Surging or feed inconsistency | Check screw wear<br>Check material in hopper<br>Check regrind percentage |
| | Inadequate die-bolts adjustment | Readjust |
| | Dirty die | Clean die lips |
| Insufficient extrudate width | Insufficient die slot<br>Inadequate polymer rheology | Adjust die deckles<br>Check melt temperatures at the die |
| Lines, streaks and foreign specks | Dirty die<br>Scratched die lips<br>Contaminated melt flow | Clean die<br>Repair or replace die lips<br>Check compound<br>Change filters |
| | Melt flow too hot | Reduce set temperatures |
| Sharkskin, melt fracture | Melt temperature too low<br>Die gap too narrow | Increase the die lips' temperature<br>Increase |
| Wrinkles | Flow fluctuations at die exit | Adjust set temperatures<br>Adjust die gap |
| | Excessive winding tension | Adjust line speeds (increase tension gradually to winder) |
| | Non-uniform cooling | Adjust casting unit operation |
| | Misaligned casting unit | Verify rolls alignment |
| Web tear-off | Flow fluctuations at die exit | Check material in hopper<br>Adjust set temperature<br>Adjust die gap |
| | Dirty die | Clean die lips |
| | Melt temperature too low | Increase set temperatures |
| ***Pipe and profile extrusion*** | | |
| Wavy inside surface | Melt temperature too low<br>Low lubricant level<br>Pulsating drawdown | Increase set temperatures<br>Correct formulation<br>Check caterpillar |

**Table 5.7** *Continued*

| *Problem* | *Cause* | *Solution* |
|---|---|---|
| Excessive extrudate thickness | Inadequate die gap adjustment | Adjust or replace die |
| | Inadequate operating conditions | Adjust set temperatures, and/or drawdown ratio |
| | Profile slippage at the caterpillar | Check pulling force<br>Dry profile before reaching the caterpillar |
| Line, streaks and foreign specks | Dirty die | Clean die (check mandrel) |
| | Scratched die lips | Repair die |
| | Drag plug (pressure calibration) too close to die | Correct drag-plug location |
| | Excessive melt temperature | Reduce set temperatures |
| | Contaminated melt flow | Check compound<br>Change filters |
| Sharkskin, melt fracture | Melt temperature too low | Increase the die lips' temperature |
| | Friction at the die lips | Repair die lips' coating<br>Modify formulation |
| | Die gap too narrow | Increase |
| Extrudate velocity with a transversal component | Unbalanced exit flow | Modify die land geometry |
| | Temperature fluctuations in die | Check heaters and thermocouples<br>Check set temperatures |
| | Dirty die | Clean die |

will change. If the plastic is not that uniform (wide and/or uneven molecular weight distribution), there will be an effect on flow through the die. Stiffer plastics will tend to have heavy center flow, whereas the soft plastics will flow more easily to the ends of the die.

With over-adjusting sheet and flat film dies, problems can develop. A flat die has one practical function that is to control transverse thickness uniformity. It cannot do the following: (1) control average thickness (what is required is the proper relationship between the extruder pumping rate and roll speed); (2) correct surges (problem is in the extruder where it is to be corrected); (3) correct poor mix (laminar flow of melt in the die will not provide any mixing action; it has to be corrected prior to the melt entering the die); (4) significantly effect melt temperature (due to the melts very short time of only a few seconds in the die. It is possible to alter skin frictional effects and thereby change temperature to some degree, however it usually indicates a homogenization problem in the extruder); and (5) significantly influencing on extrusion rate (lip opening, restrictor bar clearance, die temperature, and overall die design have small to very moderate effects on back pressure; a combination of these factors usually is an invitation to other problems [187, D. R. Hopkins].

Die lip melt build-up is a common problem that can be minimized by: (1) setting the die lips at a lower temperature; (2) running the melt temperature at a lower temperature; (3) adding stabilizer concentrate to the plastic; (4) each time the line is down, clean the die lips thoroughly; and (5) check the die lips for thin or missing plating. Table 5.7 provides some troubleshooting guides for the different extrusion lines.

# 6

# Process control and computer operation

## OVERVIEW

This is an important area that has to be thoroughly analyzed and studied to obtain the desired performance of the complete line and/or its parts, such as the extruder. The first task is to determine what is required and how to approach any problem(s). Examples of a few problems that could develop are listed in Table 6.1. One example of potentially many product problems is shown in Fig. 6.1. Proper equipment to meet requirements and capabilities in-line, such as the down-stream cutter (Table 6.2), has to be used. (For those not to familiar with problems that can develop, this book includes many of them.) With proper controls, different advantages occur, such as better economics, improved quality of product, maximized output rate, plastic reduction by minimizing tolerances, reduced setup times by means of automated startup aids, savings (electrical and water) by the optimization of the plasticator, and reduced scrap. Table 6.3 provides a guide for comparing important requirements for processes and products [397].

Adequate process control and its associated instrumentation are essential for product control (Fig. 6.2). Sometimes the goal is precise adherence to a control point, other times it is sufficient to maintain a control within a comparatively narrow range. For effortless controller tuning and lowest initial cost, the processor should select the simplest controller (temperature, time, pressure, flow rate, etc.) that will produce the desired results. For the extruder and the complete line, they can range from unsophisticated to extremely sophisticated devices [189, 197, 206, 248, 249, 261, 350–353, 359, 386].

Controls are not a toy or a panacea; they demand a high level of expertise from the processor. There are those that: (1) provide closed-loop

**Table 6.1** Examples of process variables and instrument sensors

| *Process variables* | *Instrumentation* |
|---|---|
| Temperature | Thermoelements<br>Resistance elements,<br>Radiation recorder |
| Pressure | Strain gauges<br>Resistance elements |
| Rotational speed | A-c/d-c-generators<br>Incremental counters |
| Weight | Strain gauges<br>Piezo elements |
| Velocity | A-c/d-c-generators<br>Incremental counters<br>Correlators |
| Torque | Strain gauges<br>Scales |
| Time | Impulse counters |
| Logic state (on/off) | Switches |

control of temperature, pressure, thickness (Figs. 6.3–6.5), etc.; (2) maintain preset parameters; (3) monitor and/or correct equipment operations; (4) constantly fine tune equipment; (5) provide consistency and repeatability in the operations; and (6) self-optimization of the process. Most processes operate more efficiently when functions must occur in a desired time sequence or at prescribed intervals of time. In the past, mechanical timers and logic relays were used. Now electronic logic and timing devices are used based on the programmable logic controllers. They lend themselves to easy set-up and reprogramming.

Microelectronic controls have eliminated many potential human errors in setting the line, such as data entry errors and mistakes due to insufficient extrusion knowledge. They represent a major opportunity to improve the reliability of setting the process while eliminating manual machine adjustments which can penalize the output rate and quality of products. Another important advantage is an increase in the number of controlled variables.

Dedicated input/output devices, such as special keyboards and video displays, provide a significant improvement for the user interface. Mechanical systems have been redesigned to take advantage of the higher precision and faster actuation made possible by the electronic controls. All this action results in a more reliable operating line [115].

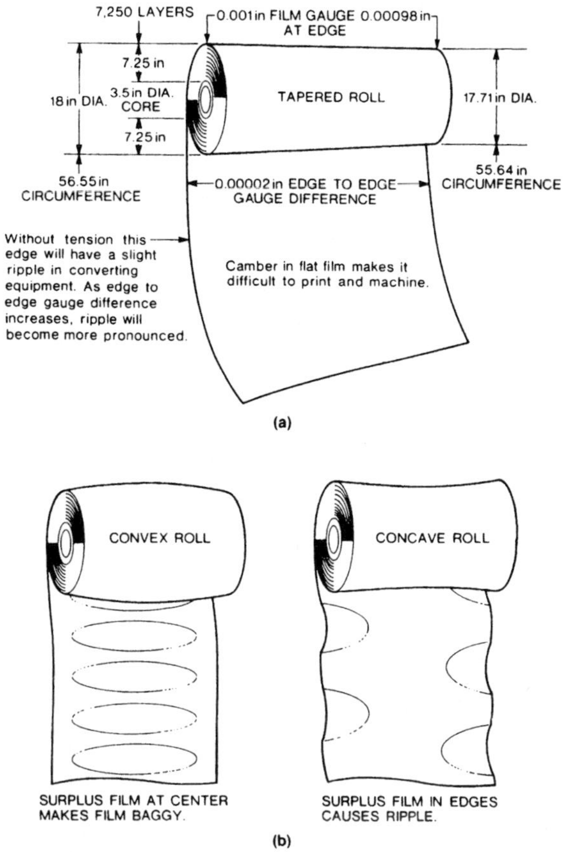


**Figure 6.1** Influences on film performance during windup: (a) effect of tension; (b) effect of uneven thickness.

**Figure 6.2** Welex 3-layer coextrusion sheet system with in-line accumulator controlled by their Ultima 'R' microprocessor.

**Table 6.2** Examples of cutting equipment capabilities

| *Cutter* | *Line speed (m/min)* | *Cuts ($min^{-1}$) maximum* | *Accuracy (±mm)* | *Advantages* | *Disadvantages* |
|---|---|---|---|---|---|
| Saws | 150 | 30 | 0.015 | Easy set-up and large capacity | Requires cleanup via air systems, etc. and uses clamping/travel table units |
| Guillotine | 90 | 50 | 0.015 | Large capacity and angle cuts | Slow blade speed; rigids need profiled bushing and/or blade; high air consumption and few cuts per minute |
| Flywheel | 4500 | 12 000 | 0.004 | High cut rates and high accuracy | Not good under 300 cuts per minute; must adjust blade rpm for speed changes; profile bushings may be needed for rigids; limited to small-angle cuts |
| Die-set stationary traveling | – | 90 | 0.00 | Inline finishing and high accuracy | Price, slow line speed, long set-up time, and long runs only |

**Table 6.3** Comparison of process requirements

| Process | Uniform/constant: Output, $Q$ | Uniform/constant: Pressure, $P$ | Uniform/constant: Melt temperature, $T_m$ | Uniform/constant: Mixing | Uniform/constant: Residence time, $t_{res}$ | Uniform/constant: Specific power, $E/Q$ | Uniform/constant: Elastic memory | Maximum (×) or minimum (✓): Output, $Q$ | Maximum (×) or minimum (✓): Pressure, $P$ | Maximum (×) or minimum (✓): Melt temperature, $T_m$ | Maximum (×) or minimum (✓): Mixing | Maximum (×) or minimum (✓): Residence time, $t_{res}$ | Maximum (×) or minimum (✓): Specific power, $E/Q$ | Maximum (×) or minimum (✓): Shear history | Maximum (×) or minimum (✓): Elastic memory |
|---|---|---|---|---|---|---|---|---|---|---|---|---|---|---|---|
| Rod, section, profile | | × | | | | | × | | | | | | | | ✓ |
| Tube, hollows | | × | | | | | | ×[a] | ✓ | | | | | | ✓ |
| Wire covering, sheathing | | | | × | | | × | | × | × | | | | | |
| Sheet, laminating | | × | × | | | | | × | | | | | | | ✓ |
| Flat film[b] | × | | × | × | | | × | | × | | × | | | | |
| Paper coating | × | | × | | | | | | × | × | × | | | | |
| Tape | × | | × | × | | | | | | | | | | | |
| Fibres, filament, netting | × | | × | | | | | | | × | × | | | | |
| Tubular film | × | | × | × | | | × | | × | | × | | | | |
| Foam | | × | × | × | × | | | | | ✓ | × | | | | |
| Compounding[c] | | | × | × | | ×[d] | | × | | ✓ | × | ×[e] | | × | |
| Blending, degrading | | | × | × | × | × | | × | | ×[f] | × | × | × | × | |
| Degassing, drying | | | | | | | | × | | ✓ | × | × | ✓ | | |
| Filtering | | × | | | | | | × | | ✓ | | | ✓ | | |
| Scrap recovery | | | | × | | | | × | | ✓ | × | | ✓ | ✓ | |
| Dewatering | | | | | | | | × | | ✓ | | | ✓ | ✓ | |
| Polymerizing | × | | × | × | × | | | | | | × | × | ✓ | ✓ | |
| Blow moulding | × | | × | | | | × | × | | ✓ | | | | | ✓ |
| Injection moulding | | | × | × | | | | × | | | × | | | | |

[a] Large pipes.
[b] Incl. multilayer and laminating.
[c] Incl. pumping of additives.
[d] And shear.
[e] Usually.
[f] Degrading (×), Blending (✓).

## TRADE-OFF

Most of these units provide independent control loops and usually only control one major variable. At present, a few units interact different variables so that the operation of a complete line requires the skill of an operator(s). However, the available controls properly installed and used have been extremely useful to operator(s) by simplifying setting-up,

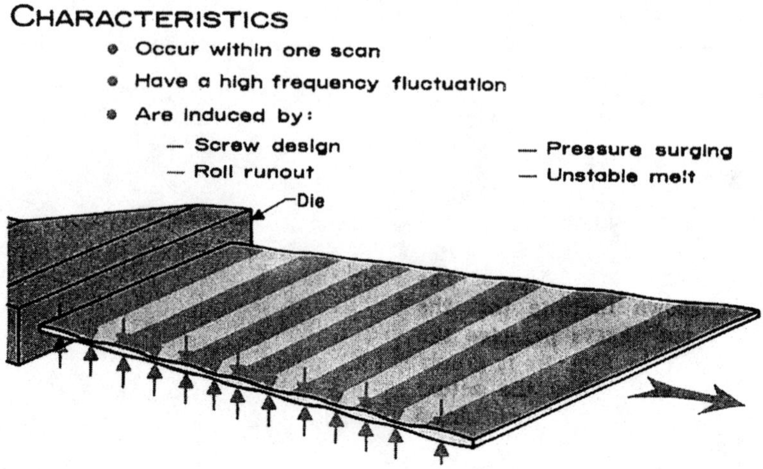


**Figure 6.3** Short-term machine-direction variables.

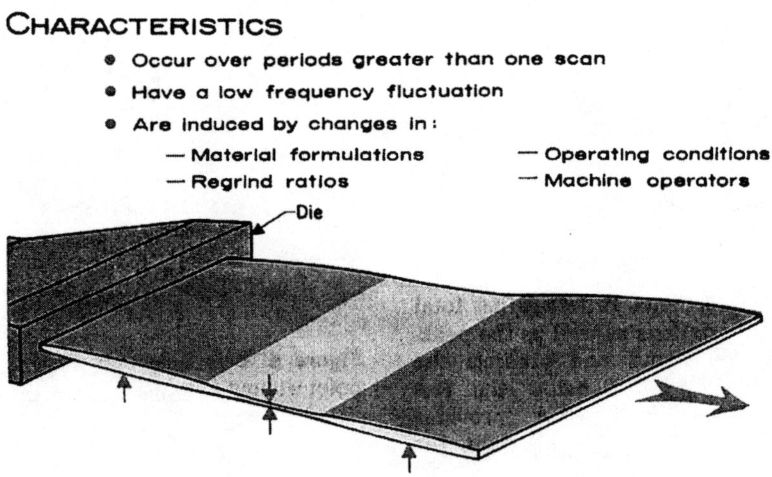


**Figure 6.4** Long-term machine-direction variables.

operating, and shut-down of the line. Trade-offs are inevitable in these complex operations where controls provide a major input on the action to be taken.

Lines have been operating with different degrees of automation via computer-integrated controls providing improvements in operating procedures and quality assurance, with the result that costs are usually reduced. These closed-loop systems maintain long-term repeatability of factors such as melt velocity and pressure. This occurs without compo-

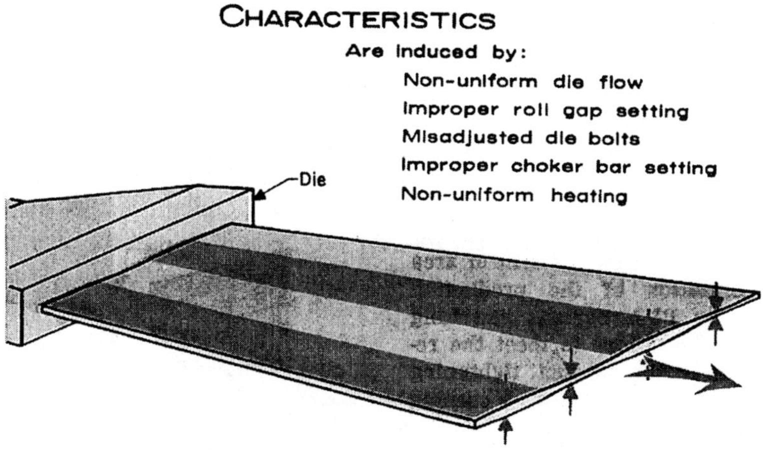


**Figure 6.5** Examples of profile or cross-machine variables.

nent wear, unbalance of equipment in the line, and/or plastic material variations (Chapter 3).

Usually, it is very beneficial to extend noncomputerized equipment capabilities by using computerized operations. Advantages become obvious, such as extending process monitoring and to automatically correct certain processing deficiencies in response to sophisticated sensing devices and control programs. However, some control features are quite expensive and may prove to be inadequate unless the problems are corrected that can extend far beyond the optimization of electronic circuitry.

Usually elaborate control systems cannot correct for problems such as those caused by a: (1) worn screw and barrel; (2) inadequate drive torque; and/or (3) poor screw design. As an example, such systems will not yield good temperature control unless all features essential to good control are well maintained. Obviously, burnt-out heating elements cannot be tolerated. Another common deficiency for liquid-cooled extruders is fouling/restrictions in the plumbing system or inoperative valves. Other factors of these types also exist.

With the best of conventional controls, a person can casually glance at a panel and gauges around the machine and with a fair degree of certainty determine that the extruder is operating normally. With current computerized process monitoring and control features, that person will undoubtedly do much more (and do it more easily) in operating the machine. Automatic alarms can be incorporated to indicate that something is changing even if displays and printouts are available. This approach, if required, provides a means to take action where otherwise something important may go unnoticed.

During data entry on a modern data control system, the operator can 'lose sight' of the vital extruder parameters of melt temperature, melt pressure, and motor load. The reason for this is that the computer documentation may be on 'different pages'. By providing separate display instruments for these values, the extrusion process can be checked quickly without scrolling through several pages of the data control system.

Ideally, extrusion should be a steady-state process, but feedback control loops can be very helpful to compensate for process variations. Automatic adjustments to maintain product thickness control are especially worthwhile in many applications. Procedures should be setup to ensure proper feedback control. For example, it is useful if the process monitoring identified a burned barrel heater instead of adding more heat in adjacent heaters trying to maintain the zone's barrel thermocouple set point [187, G. A. Kruder].

## SENSORS

In the extrusion process, instrumentation is (critically) important to determine what is occurring inside the extruder and to control important process parameters. The minimum instrumentation that should be used are melt pressure before/after the breaker plate, screw speed, melt temperature, temperature along the screw, cooling water temperature of feed throat, and power consumption (heaters, cooling, and screw drive). Table 6.4 reviews different sensors.

The most important process parameter is melt pressure. Monitoring pressure over time allows an accurate assessment of the stability of the extrusion process (output rate, etc.). In order for pressure transducers' information to be accurate, they must be calibrated. Most panel indicators and controllers have a zero and calibration function that should be incorporated into the operator's check lists. The calibration is an electronic function that can compensate for differences in zero point and resistance from gauge to gauge and corrects temperature offset.

True calibration requires the transducer to be tested on a dead weight or similar pressurization device that calibrates to a known pressure. This is to occur off-line on an appropriate apparatus. Alternately, this could be performed by an outside testing service. It should be done at least annually. More frequent checks are usually required in high-precision applications or as required by ISO-9000 systems [187]. Regarding temperature control, the literature has extensive information regarding placing sensors as close to the melt as possible and properly enclosing them [1–4, 187]. Both the machines and plastic materials can set up variabilities in temperature sensing based on where and how they are located [397].

The temperatures along the extruder should be measured close to the inside barrel surface; it is the plastic temperature that is being controlled.

**Table 6.4** Example of performance guide for different sensors

| *Type of sensor* | *Accuracy* | *Dimensional range* | *Transverse measurement* | *Reflex measurement* | *Freedom from interference* | *Material dependency* | *Sensor-size limitations* | *Complexity of equipment* | *Ease of calibration* | *Sensitivity* |
|---|---|---|---|---|---|---|---|---|---|---|
| Rolling-contact | Good | Wide | Yes | No | Good | No | Low | Low | Easy | Med. |
| Air | Good | Wide | Yes | No | Good | No | Some | Med. | Easy | High |
| Magnetic-reluctance | Fair | To $^1/_4$ in | Yes | Poss. | Fair | Some | Some | High | Easy | Med. |
| Sonic | Good | To 1 in | Yes | Yes | Fair | Yes | Some | High | Fair | Med. |
| Optical | Fair | Wide | Yes | Yes | Good | Some | High | Med. | Fair | Med. |
| Laser-intercept | Good | Wide | Yes | Yes | Good | Some | High | Med. | Fair | High |
| Laser-interferometry | Exc. | Ltd. | No | Yes | Good | Yes | Some | High | Easy | High |
| Capacitance | Good | Med. | Yes | No | Fair | Yes | Low | High | Easy | Med. |
| Proximity | Good | Wide | Yes | No | Fair | Some | Low | High | Easy | High |
| Beta-ray | Good | Ltd. | No | Yes | Good | Some | Low | High | Easy | High |

Temperatures can be measured with thermocouple (TC) or resistance temperature detector (RTD) [2]. TCs tend to have shorter response time, while RTDs have less drift and are easier to calibrate.

Traditionally, extruders have used PID control for heating and on–off control for cooling [2]. From a temperature control point of view, the more recent use is the fuzzy logic controller (FLC). One of FLCs major advantages is the lack of overshoot on startup, resulting in achieving the setpoint more rapidly. Another advantage is extruder control is in its multi-variable control where more than one measured input variable can effect the desired output result. This is an important and unique feature. With PID, one measured variable affects a single output variable. Two or more PIDs may be used in a cascade fashion but with more variables they are not practical to use.

With FLC, 5–10 input variables can be utilized to effect one or more output variables. Other than a skilled operator, FLC is the only method of reaching on-line control to compensate for existing variables. Much of the work with FLC has been proprietary.

With pressure, temperature, and other controls, an important parameter to consider is the instrument response loop (which is the measurement of a variable and the corresponding control action) to the precision realized in production cannot be underestimated. As an example, due to the response lag in a pressure sensor, by the time an increase in pressure is transmitted to a control device the actual system pressure continues to change.

The controller that receives this information then must process it and transmit an appropriate control response. This usually takes 'some' time. The active control device, the extruder drive control in this example, then responds to this new output information and again this response has a time factor. The pressure has continued to change at some rate during these time periods. It is obvious that it is impossible to completely eliminate pressure fluctuations without basically using a gear pump (Chapter 2). Reducing these time factors will greatly increase the system's ability to maintain a rather constant or controlled pressure. Different devices are used to measure pressure and each with their own capabilities (Table 6.5).

There are sensors designed to respond to a physical stimulus (temperature, pressure, motion, product gauging, product weight, etc.) and transmit a resulting signal for interpretation, measurement, and/or operating a control. There is a very broad selection of sensors with extremely different sensitivities, capabilities, and repeatabilities (Tables 6.6 and 6.7). To select the correct sensor you should know something about how the different sensors work, and which is used for what application. This is important since not all sensors measure the same way. As an example, there are sensors with different thickness ranges generally for use with

**Table 6.5** Comparison of various pressure transducers

| *Type* | *Robust* | *Temp effect* | *Resp. time* | *Percent error* |
|---|---|---|---|---|
| Bourdon | Poor | Poor | Poor | 3–5 |
| Pneumatic | Good | Poor | Poor | 1.5 |
| Capillary | Fair | Fair | Fair | 0.5–3 |
| Pushrod | Fair | Poor | Fair | 3 |
| Piezoelectric | Good | Poor | Good | 0.5–1.5 |
| Piezoresistive | Good | Good | Good | 0.2–0.5 |
| Optical | Good | Good | Good | 1.0 |

specific plastics. Their degrees of accuracy or resolution are usually different. Other factors to consider are a sensor's measurement range, whether transmission or reflective, whether contact or noncontact on the product, etc.

The three most common sensors used down-stream are nuclear, infrared, and caliper. There are also specialized types such as microwave, laser, X-ray, and ultrasonic sensors. They are available to sense different conditions for operating equipment (temperature, time, pressure, dimensions, output rate, etc.). They can also sense color, smoothness, haze, gloss, moisture, dimensions, and many more variables.

When a manufacturer quotes static accuracy, they are essentially referring to how the sensor will perform under the best conditions. Although it is composed of a number of factors, accuracy should be regarded as just one measure of a sensor's performance; it is not indicative of its overall capabilities. To get a truly representative static accuracy rating, manufacturers can also provide a number of different components or sources of error, such as measurement calibration accuracy, long time repeatability, and noise effect.

The dynamic accuracy is the indication of how well the sensor will operate in the production environment. It can be defined as a comparison of sensor readings, as an example, on a sheet (Fig. 6.6). Dynamic accuracy is a function of a number of components, of which static accuracy is one. Others include flutter sensitivity, air gap temperature, mechanical sensor alignment, and sensor response time. Real time inspections can be made at speeds of at least $400\,m^2/min$ ($4300\,ft^2/min$). Imperfections down to at least $125\,\mu m$ (5 mil) on coated extruded film webs can be found at speeds of at least 250 m/min for 1.5 m width (820 ft/min for 5 ft width).

Throughout this book, different processing variables have been presented. Causes and corrective actions have been presented to eliminate or reduce the variabilities during processing. They concern factors such as output rates, pressures, temperatures, mixing, flatness, and so on. A guide

**Table 6.6** Comparison of various pressure transducers

| *Equipment* | *Web control capability* | *Product application* | *Cost ratio* |
|---|---|---|---|
| Open-loop microprocessor software<br>Low-resolution sensor<br>Slow-response nuclear source | 5–10% thickness tolerance improvement; 10–15% startup time reduction. | Shrink and stretch wrapping film and sheet, industrial liners, disposable thermoform sheet, construction film, heavy-duty sacks. | 15–25 |
| Closed-loop microprocessor, software based on extrusion program<br>Nominal target control<br>High-resolution sensor<br>Predictive control<br>Fast response nuclear source<br>Self-diagnostics | 7–12% thickness tolerance improvement; 15–20% startup time reduction. Streak detection; 2–3% throughput increase. | Appliance liners, diaper backing, clear film, glazing sheet. HDPE deli paper. | 25–35 |
| Target-management software program<br>Predictive control<br>High-resolution sensor<br>Fast-response nuclear source<br>Self-diagnostics<br>Automatic throughput control | 10–25% thickness tolerance improvement; 20–25% startup time reduction. Streak detection; 5–7% throughput increase. | Downgaged films, blister pack sheet, two-color coextruded sheet, medical thermoform sheet, coextruded snack and cereal liner. | 30–45 |
| Automatic profiling software program<br>Automatic profile die<br>Target-management software<br>Predictive control sensor<br>High-resolution sensor<br>Fast-response nuclear source<br>Automatic throughput control<br>Triple CRT resolution<br>Self-diagnostics<br>100% spare part inventory | 40–60% thickness tolerance improvement; 50–75% startup time reduction. Streak detection; 10–15% throughput increase. Detects short term process variations. | Engineering film and sheet, optical grade polycarbonate sheet, barrier coextrusions, polyester capacitance films, die-cut sheet, photographic films, extrusion coating, biaxially-oriented pp film. | 40–100 |

**Table 6.7** Web gauging methods

| *Gauging method* | *Material limitations* | *Applications limitations* | *Type of extrusion* |
|---|---|---|---|
| Roller-contact | None | Requires rigid material | All |
| Air | None | Requires reasonably flat surface | Sheet, profile |
| Magnetic-reluctance | Nonmagnetic materials with thin walls | Position control important | Sheet, tube |
| Sonic | Material with good sound transmission | Material with good sound transmission | Pipe, sheet |
| Optical | None | For dimensional control | All |
| Laser-intercept | None | For dimensional control | All |
| Laser-interferometry | May be transparent in infrared | For dimensional control | Thin tube, sheet |
| Capacitance | Materials with higher dielectric constant | For thickness control | Sheet, tube |
| Proximity | Materials with higher dielectric constant | Requires nonconductive materials | All |
| Beta ray | None | Requires relatively smooth surface | Thin sheet |

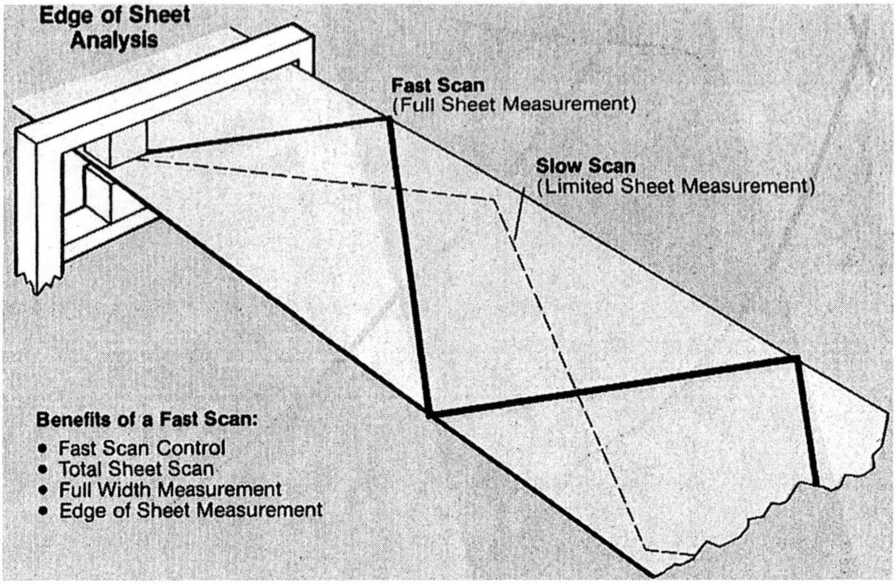


**Figure 6.6** Dynamic sensing results in more accurate information for tighter control and higher productivity.

to processing conditions and degree of control for the different extruded products is summarized in Table 6.8 [397].

## ACCURACY

Accuracy can refer to freedom from making errors or conformity to a standard. An extrusion system can be very 'precise' and have poor accuracy. A manufacturing target of consistent, repeatable fabrication requires more than tight mechanical equipment standards, tight plastic material standards, and precise instruments with fast integrated control response. However, these conditions can go a long way to meeting the target.

What has to be included is calibrating instruments to fixed standards. To achieve accuracy the pressure, temperature, speed, and other control parameters must be calibrated to traceable standards. Physical settings associated with the system operation including system alignment (Chapter 2), distance from the die to the cooling stage, and equipment environment (temperature, humidity, cleanliness) are also important.

By measuring against known standards, the accuracy of the measurements can be determined. There can be parameters that can not be quantified so these will contribute to the variability and limit the accuracy that can be obtained. Variations in molecular weight, pellet size, virgin/

**Table 6.8** Examples of relating product to extruder to control

| Process | Screw diameter (relative) | Screw length L/D ratio | | Screw type | Metering depth | Screw speed (relative) | Melt temperature | Degree of control | | | Comments |
|---|---|---|---|---|---|---|---|---|---|---|---|
| | | | | | | | | Speed | Temperature | Mechanical power | |
| Rod/section and profiles | High | 20/24<br>15/20<br>20/24 | PE, PP<br>PVC<br>Nylon | T<br>T<br>S | Shallow | Low | Medium | ×× | ×× | ××× | Output relative to haul-off speed, for dimensions. Minimum elastic memory. Medium temperature for rigidity while cooling. Add screens to promote mixing. |
| Tube | High | 20/24<br>15/20<br>20/24 | PE, PP<br>PVC<br>Nylon | T<br>T<br>S | Medium | Medium | Medium except die | ×× | ×× | ×× | Minimum elastic memory. High die temperature for gloss. Medium temperature for drawdown of small tubes. |
| Wire covering/ sheathing | Low | 20/24<br>15/20<br>20/24 | PE, PP<br>PVC<br>Nylon | T<br>T<br>S | Shallow | High | High | ×× | × | × | High temperature for good drawing and avoiding melt fracture. High pressure for thin insulation. Output relative to haul-off speed for thickness |

**Table 6.8** *Continued*

| Process | Screw diameter (relative) | Screw length L/D ratio | | Screw type | Metering depth | Screw speed (relative) | Melt temperature | Degree of control | | | Comments |
|---|---|---|---|---|---|---|---|---|---|---|---|
| | | | | | | | | Speed | Temperature | Mechanical power | |
| Sheet/ laminating (including continuous vacuum forming) | High | 20/24<br>15/20<br>20/24 | PE, PP<br>PVC<br>Nylon | T<br>T<br>S | Medium | Medium | Medium | ×× | ×× | ×× | Medium speed is compromise for high output and uniformity of $Q$ and $T$. Steady output to avoid ripples. High output for thick sheet. Uniform temperature for die distribution. Medium temperature for drawdown. |
| Flat film | Medium | 24/28<br>20 | PE, PP<br>PVC | T<br>T | Shallow | Medium | High | ××× | ××× | ××× | High temperature for drawdown. High pressure for die. Medium speed for uniform elastic memory. Uniform output for thickness. |
| Paper coating | Medium | 24/28 | PE, PP | T | Shallow | Medium | High | ××× | ××× | × | Output relative to drum/ haul-off speed for thickness. Medium output for rapid cooling. Very high temperature for flow and surface finish. |

| | | | | | | | | | | | |
|---|---|---|---|---|---|---|---|---|---|---|---|
| Tapes | Medium | 20/24 | PE, PP | T | Medium | Medium | Medium | ××× | ××× | ×× | Medium temperature for crystallization and drawdown. Uniform output for dimensions and crystallization. Uniform temperature for dimensions. |
| Filament/fibers | Medium | 20/24 | PE, PP | T | Shallow | Medium | High | ××× | ××× | × | High temperature for drawing. |
| Tubular film | Medium | 20/24 | PE, PP | T | Shallow | Medium | Medium | ××× | ××× | ××× | High pressure for die. Medium temperature for drawing and blowing. Medium speed for uniform elastic memory. |
| Foam | Medium | 20/24 | | T | Medium | Low | Low | × | ×× | ×× | Temperature depends on blowing agent. Low shear heating for decomposition. Medium pressure and mixing for uniform distribution and pore size. |

**Table 6.8** *Continued*

| *Process* | *Screw diameter (relative)* | *Screw length L/D ratio* | | *Screw type* | *Metering depth* | *Screw speed (relative)* | *Melt temperature* | *Degree of control* *Speed* | *Temperature* | *Mechanical power* | *Comments* |
|---|---|---|---|---|---|---|---|---|---|---|---|
| Compounding/ blending | Medium/ high | 24/28 | PE, PP | T | Deep | Medium | Low | × | ×× | ×× | Deep screw at high pressure for mixing. Large diameter and medium speed for high output. Low temperature for mixing and heat history. |
| | | 20 | PVC | T | | | | | | | |
| | | 20/24 | POM | T | | | | | | | |
| | | | PMMA | S | | | | | | | |
| | | | Nylon | S | | | | | | | |
| Degrading | Medium | 24/28 | PP | T | Shallow | High | High | × | ××× | ×× | |
| Blow moulding | High | 24/28 | PE, PP | T | Medium | Medium | Low | ×× | ×× | ××× | No haul-off. Melt temperature and swell ratio compromised with output by control, e.g., of speed and diameter. |
| Injection molding | High | 24/28 | PE, PP | T | Medium/ shallow | Medium | Medium/ high | × | ×× | ×× | No haul-off. Melt temperature control important. Back pressures low in practice. |
| | | 20 | PVC | | | | | | | | |
| | | 20/24 | PS | | | | | | | | |

*Notes*: The screw type and dimensions are largely dependent on the plastic and temperatures. Screw diameter and length are a compromise with capital cost. Screw speed is a compromise between output and quality. The length/diameter ratio and channel depth depend largely on the plastic used. 'Temperature' under 'Control' refers to set temperatures in the melt pumping section. Mechanical power input is controlled mainly by temperatures in the melting zone. Screw cooling is excluded since it is always very sensitive, requiring precise control.

Suggested *running* limits are:

| Speed: | | Temperature: | | Mechanical power: | |
|---|---|---|---|---|---|
| × | ±1% | × | ±2°C | × | ±3% |
| ×× | ±0.3% | ×× | ±0.5°C | ×× | ±1% |
| ××× | ±0.1% | ××× | ±0.1°C | ××× | ±0.4% |

recycled mixes, etc. (Chapter 3) can effect the process. Unfortunately, these variations are not often recognized or easily identified.

## INTEGRATION

The extruder is usually responsible for the majority of the problems/variabilities in processing lines. In a close second place is usually the plastic materials. These problems have been with the industry for over a century. The important fact is that with time these problems/variabilities have been significantly reduced and better understood. This type of development should continue. Regardless, the extrusion industry from past to the future worldwide continually produces products for the world. Example of extrusion lines with multiplicity of different process sensors and automatic control points interfaced with central computers are shown in Figs. 6.7–6.11.

Use has been made of highly accurate process instrumentation which reports precise information at high frequency. A high-speed processor evaluates the extruder information at a real-time rate with responsive actuation systems to compensate for process variations. This action significantly reduces the mechanical variabilities of the equipment. With this improvement, the process becomes increasingly stable and the amplitude of the variation reduced. At certain operating conditions, the variation becomes small enough that on-line adjustment to down-stream equipment can have a significant effect on the accuracy of the finished product. In fact the result of a fluctuation in one machine may cancel out the effect of another.

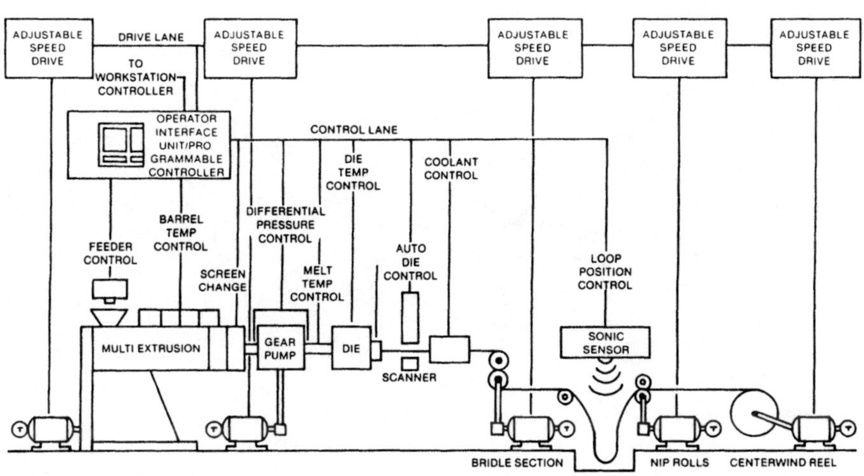


**Figure 6.7** Example of film line with controls.

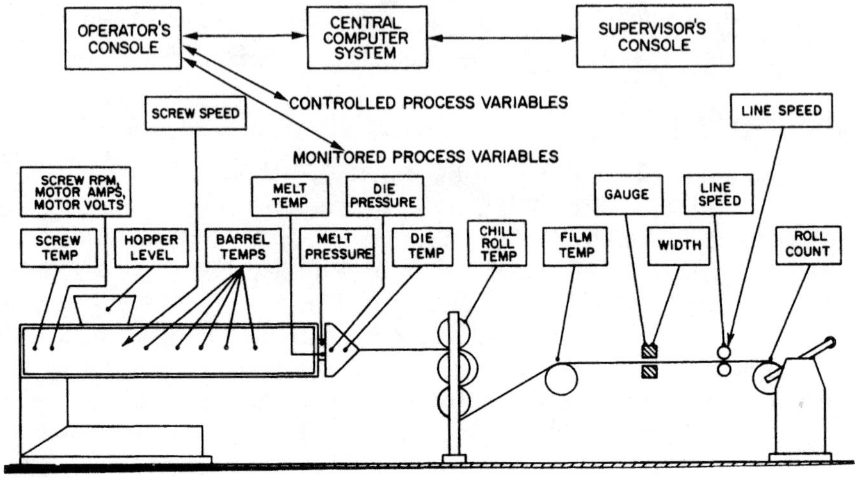


**Figure 6.8** Example of sheet line with controls.

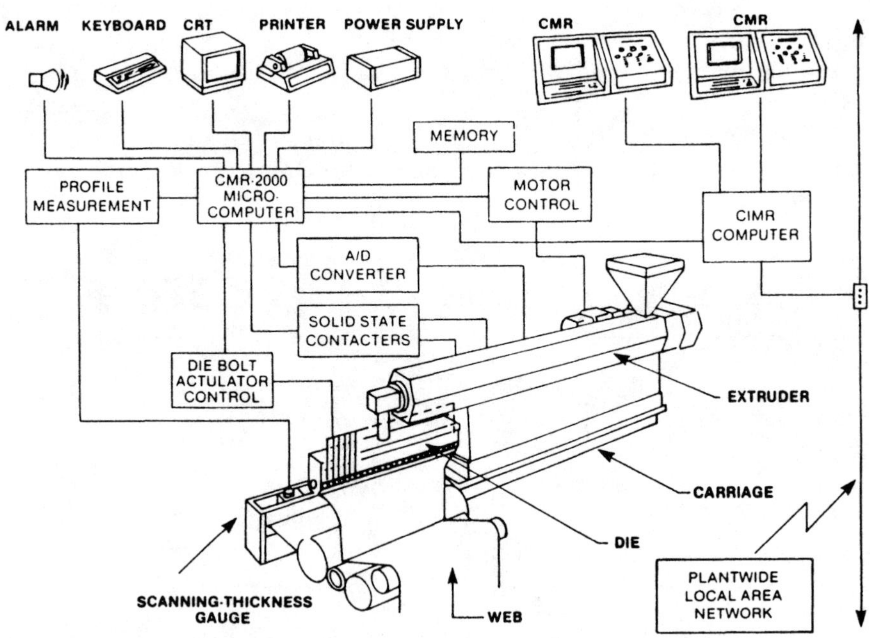


**Figure 6.9** Example of Egan/D-S coating line with controls.

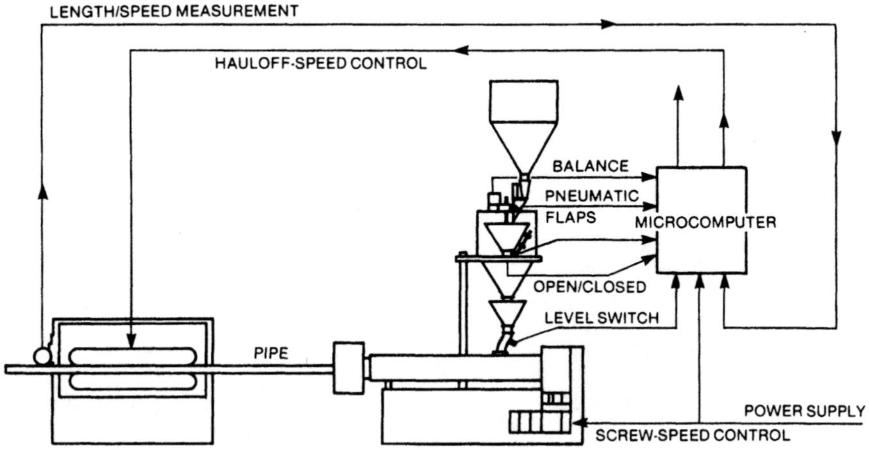


**Figure 6.10** Example of pipe line with controls including gravimeteric feed control.

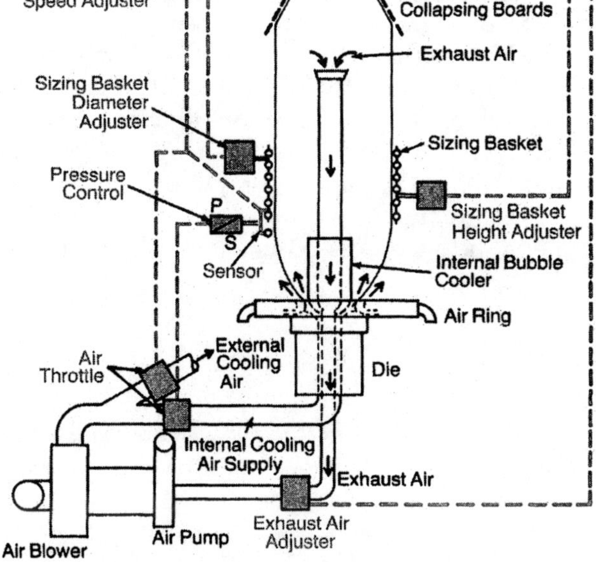


**Figure 6.11** Example of blown film line with controls.

With extrusion lines, particularly where high precision is required or cost reduction is targeted, the up-stream and down-stream equipment should have some integration with the base control system used on the extruder. The extent of this integration is dependent on the equipment's ability to affect the quality of the product. This equipment could be larger than the extruder, like a puller in a pipe line or a turret winder for a film line. If the control system is designed to integrate the dimensions, tensioning, etc., requirements, a control logarithm [procedure for solving a mathematical problem (as of finding the greatest common divisor) in a finite number of steps that frequently involves repetition of an operation] can be created to maintain the requirements.

In most lines, the total variability of the product is from different sources in addition to the extruder and plastic variabilities. It is necessary to identify and quantify each source of deviation in order to create an effective control logarithm to minimize their effect on the product. Integration of all equipment that can affect product quality results in the following capability: (1) to record and analyze variations in the total system; and (2) combine their effects on process stability. As each variable is defined, a control solution can be applied to reduce the effect that the variation has on the product.

## INTELLIGENT PROCESSING

Different approaches are used to cut inefficiency and the costs associated with it; one approach is called intelligent processing (IP). It basically takes the next step (into the future) to what is being reviewed and used in production lines. This technology will utilize new sensors, precision control systems that automatically interrelate 'all' variables, and process models that control all processing conditions to produce products without the need for human control or monitoring. Intelligent procesing involves building in quality rather than attempting to obtain it by inspecting the product after it is made. Being able to change manufacturing processes or types of material being processed is another potential benefit of this technique.

# 7

# Blown film

## OVERVIEW

The technology of TP blown tubular plastic film extrusion originated in 1933. The patent then granted to the Norddeutschen Seekabelwerke AG, Germany, included and was related to extruding PS in tube form, followed with longitudinal and/or transverse stretching at certain temperatures. It used a stretching device or spreader fitted to the core of the die. The product was used by the electrical industry. New applications developed via USA for plastic film when PE became commercial during 1939–45 [370]. The markets that opened and continued to expand included packaging, agriculture, horticulture, building and construction, medical, geomembrane, and so on; practically all markets worldwide [192, 198, 264, 388].

Thermoplastic films are formed by extruders using circular dies for blown tubular single-layer (Figs. 5.4 and 5.22) and multiple or coextrude (Fig. 5.37), flat dies for flat films, calendered films, and other processes, such as solvent casting, chemical conversion, and skiving from solid rolls. Films are distinguished from sheets in the plastics industry by their thickness. A web (film) under 0.254 mm (10 mil) thick is usually called a film. However, under 0.10 mm (4 mil) is also used for film by certain parts of the industry. The 0.10 mm (4 mil) thickness tends to be more used in the packaging industry. Materials over these dimensions are called sheets.

More plastics have been going through blown film lines than any other extrusion lines (Figs. 1.4, 7.1, and 7.2). In this process, the die is usually side fed from an extruder. The melt exiting from a circular or ring orifice is air inflated to the required diameter as it moves vertical. The inflated film is then usually cooled through air cooling, with size controlled by the die and cooling ring sizes, by internal air pressure, and take-off speed. The blown film is directed usually vertically through several guide rolls (different devices are used, for example Fig. 7.3) to keep it aligned with the

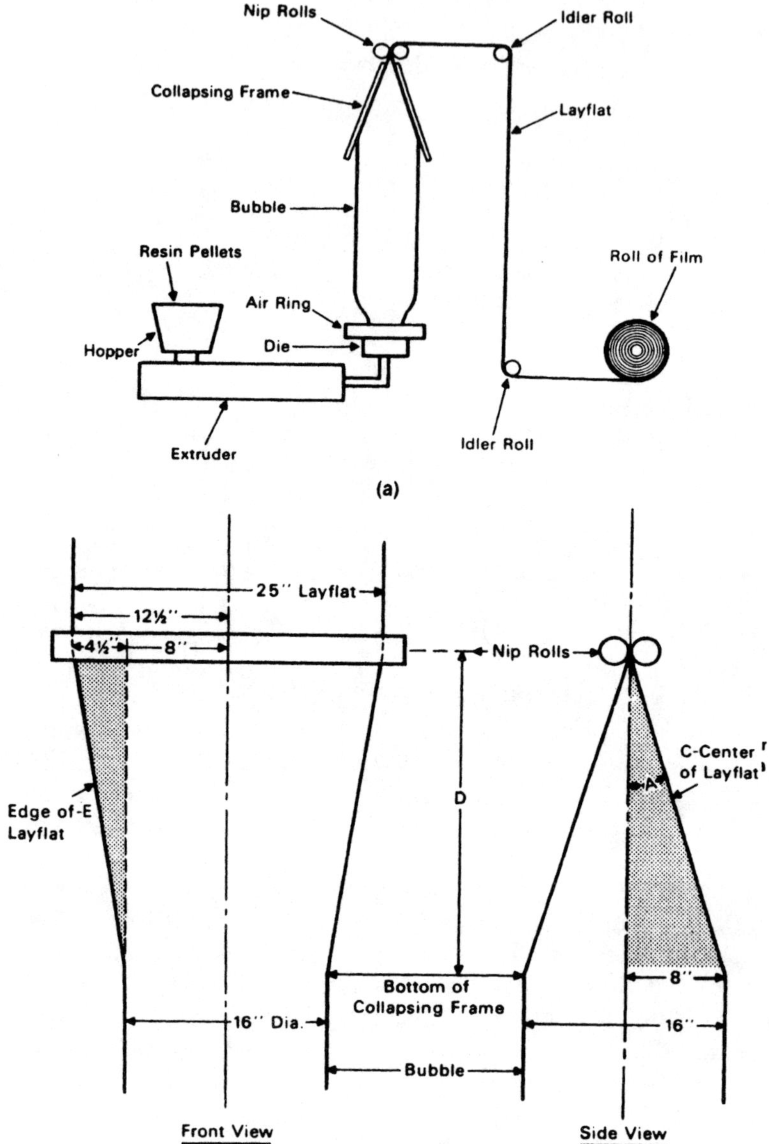


| | | | |
|---|---|---|---|
| Angle A° | 22 | 11 | 5½ |
| Distance D, in. | 20 | 40 | 80 |
| Edge E, in. | 20½ | 40¼ | 80¼ |
| Center C, in. | 21½ | 40¾ | 80½ |
| C>E, % | 5 | 1¼ | 5/8 |

(b)

**Figure 7.1** Basic vertical-up blown film line; geometry of collapsing bubble.

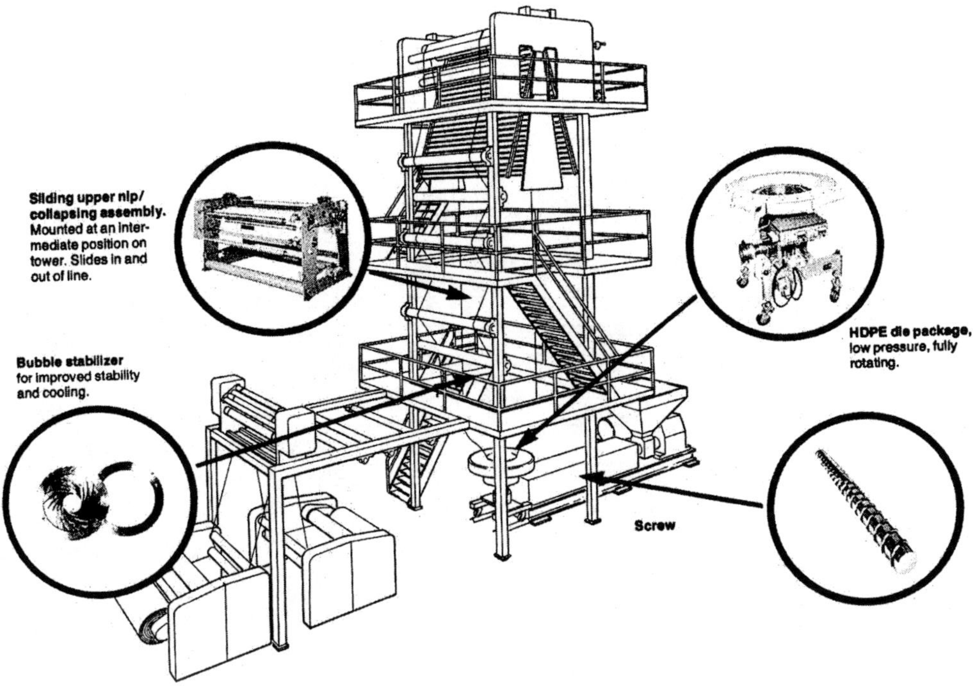


**Figure 7.2** Schematic of line with flat slat collapsing frame.

machine. After a few meters (yards) of free suspension, the film is flattened via some type of collapsing device that directs the flattened film through pressure controlled nip rolls. The rotating speed of the nip rolls is a major tool for controlling the rate with which the bubble is drawn.

While collapsing frames appear simple in concept, refinements in geometry and materials have brought improvements in melt quality. Traditional aluminum rollers tend to transfer heat too readily, which is a major cause of bagginess. Wood does an excellent job in certain applications, such as HDPE. Depending on the film being produced, rollers can be covered with materials ranging from woven glass fibers to different plastic material constructions in order to reduce friction, heat transfer, and surface abrasion. The rolls sometimes include liquid and/or air cooling systems.

At the end of the line, winder technology allows the selection of surface winding, center winding, and a combination of surface/center winding to suit the film being run, as illustrated in Fig. 7.4 where a Battenfeld Gloucester (B-G) complete line is shown. They may be wound directly as a layflat tube (Fig. 7.4), slit at both sides and wound into two flat reels, very wide film slit on one side (so that it can be opened with a visible line

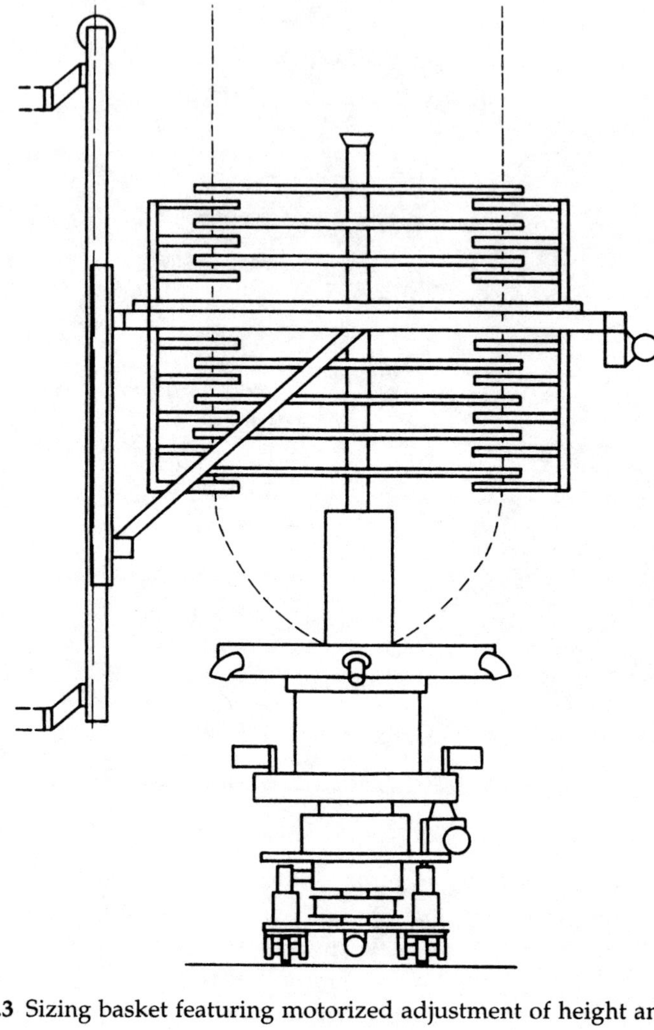

**Figure 7.3** Sizing basket featuring motorized adjustment of height and width.

due to the fold ), or other constructions, such as an in-line grocery bag line (Fig. 7.5).

Small to large high-performance winders are required to ensure the quality of the film rolls and provide trouble free down-stream conversions. Operating are at least 20 cm (8 in) screw extruders with 800 hp drive, 2.25 m (7.5 ft) die (actually referring to the orifice or gap diameter), 15 m (50 ft) flat (opened) PE film, with an output of at least 2300 kg/h (5000 lb/h). Output rates are usually 3.2–9 kg/h (7–20 lb/h) per 2.54 cm (1 in) of die orifice circumference. Lines producing wide film have line speeds of at least 600 m/min (2000 ft/min).

**Figure 7.4** Blown film line rolling with a lay-flat tube on winder.

Trapped air that forms the continuous tube is directed through a mandrel via the die. Once the bubble has been formed, the controlled air pressure required to keep the bubble stable is kept constant. Usual pressure is $1.1\,m^3/min$ ($40\,ft^3/min$).

As the hot tube/melt leaves the die, a cooling system is used to uniformly cool the melt. This cooling action has a major influence on the bubble thickness and uniformity. The usual dual- or three-chamber/lip cooling air ring, located outside the bubbles as it exits the die, has air streams gently cooling the bubble (Fig. 7.6). To speed up lines and improve output performances, internal bubble cooling (IBC) systems are extensively used (Fig. 7.7). They direct cool air at low velocity to enter and exit the inside of the bubble.

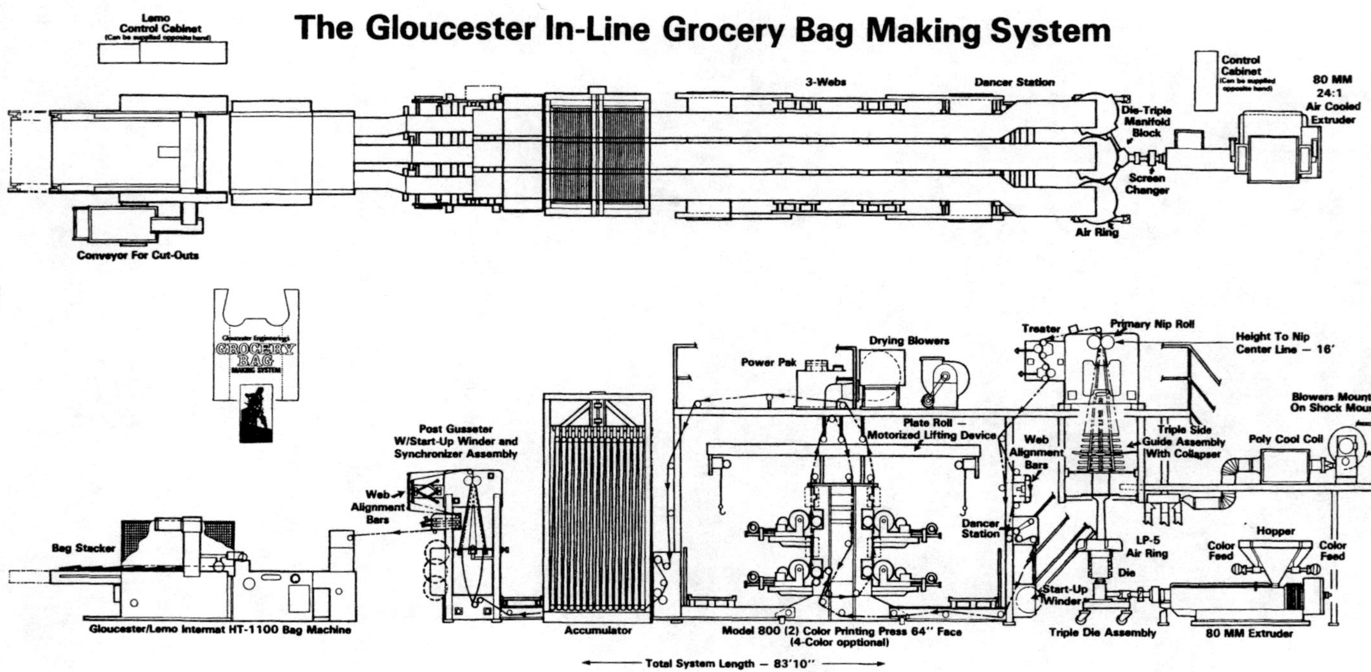


**Figure 7.5** B-G in-line grocery bag making system.

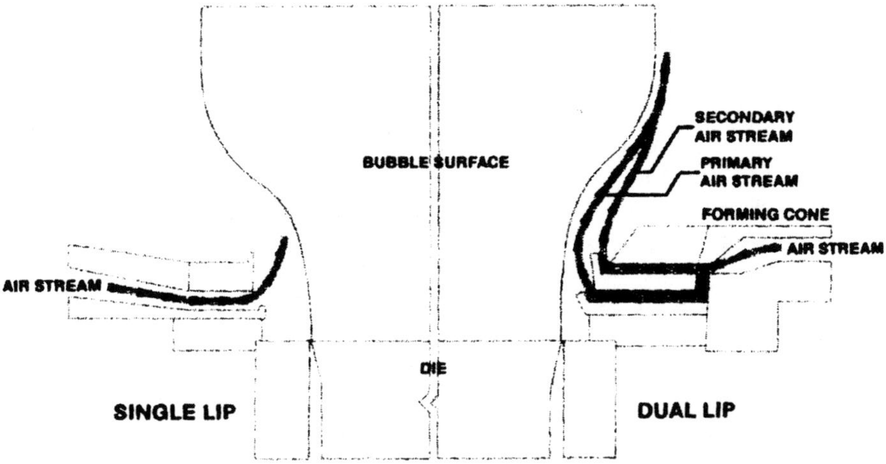


**Figure 7.6** Comparison of single-chamber and dual-chamber air rings for cooling blown film.

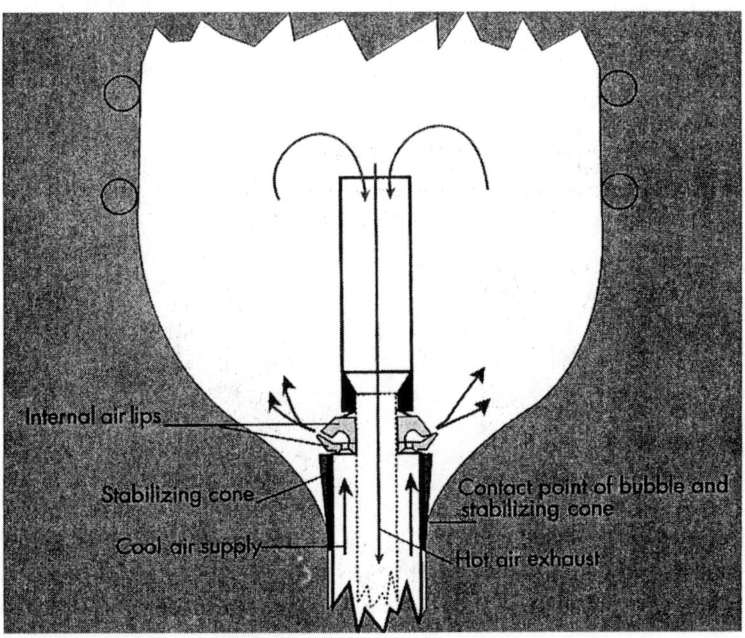


**Figure 7.7** Example of internal bubble cooler.

The bubble diameter is normally always greater than the die diameter. This bubble diameter divided by the die orifice diameter is called the blow-up ratio (BUR) (Fig. 2.20). The bubble diameter must not be confused with the width of the flattened double layer of film between the nip rolls. The width of this double layer is 1.57 times the bubble diameter and called the blown-film width. The blow-up ratio can be determined by taking 0.637 times the lay flat width divided by the die diameter.

Extrusion direction can be (rarely used) horizontal. The usual direction is vertically upwards or vertically downwards. The choice of direction usually depends on the type or melt behavior of the plastic and the desired layflat width. Thermoplastic with low melt viscosity is usually blown vertically upward. The more popular systems are verticals. As a possible guide (since plastic melt strength performance dictates direction), tube diameters up to about 100 cm (40 in) go downward; tubes of greater diameter go upward.

In the horizontal direction, the less viscous plastic expansion zone cannot easily be stabilized with the usual cooling rings. This problem is caused by the abrupt change from the melt to the solid condition that results in stress differences between the underside and upper side of the tube because of the unbalanced thermal conditions. It is also not easy or usually practical to go horizontal for tubing over 15–20 cm (6–8 in). However, this system can provide lower initial cost, high output rate, and rather simple operation. Horizontal operation entails no overhead installation and a low building height, but requires a larger floor space with probable adverse effects of gravity and uneven cooling. Vertical down operation has the advantage of start-up without flooding of the annular die gap by exiting hot melt. The vertical-up operation is the usual method, provided sufficient melt strength exists for any upward startup. Special die heads are designed, such as with a multiple threaded discharging into an expansion space. The tubular melt assumes its final shape in a smoothing-out zone, which is a cylindrical land in a parallel position between the mandrel and the orifice. Its length is about 10–15 times the annular gap width (the lower value applies to the thin film). The gap width is generally 0.5–2.0 mm (0.02–0.08 in).

Blown film dies have been developed with the goals of low pressure consumption, easy self-cleaning, material changes, and ease of maintenance. The automation of blown film plants to reduce film thickness tolerances involves the increased use of the newer plastics with more sophisticated process control elements in the complete line from upstream to down-stream equipment (Chapters 3 and 5).

## PLASTIC MATERIALS

Plastic materials can account for 70–90% of film production cost. Therefore any approach which will reduce the use of excess plastics and produce quality products has always been of importance to the processors. Meeting minimum tolerances will provide major savings. Operating the extruder most efficiently (Chapter 2) is of major importance. Any new technology that reduces scrap, for example during shut-down, is of vital interest to processors [1–4, 324].

As an example, a manually operated modern film line can produce a film with a high degree of homogeneity and finish. The thickness, however, might fluctuate up to 5% due to changes in material bulk density. With on-line weight measurement systems, that continuously measures material consumption through the hopper to the plasticator, thickness changes can be significantly reduced. These gravimetric measurement systems are available that can provide measurement accuracy to ±0.25%.

Better extrusion technology leads to quality film and offers other advantages, such as easier and faster operating lines, better cost-performance ratios, lower scrap rates, and reduced environmental impact. Scrap rates influence every aspect of from cost to profit for the processor and the converter. The better the quality of the film, the more reliably it can be converted and the lower the scrap rate. Scrap rates can typically run between 3–6% for each stage of a typical extruding/printing/converting processing line. For the total line, the scrap rate could be between 9–18% overall. Regardless as to how one analyzes this situation, there is room for improvement; the lines are continually improving their capabilities with the help of new plastic materials (Chapter 3).

In the extrusion process, the major sources of scrap are: (1) edge trim which is usually granulated and recycled with virgin plastic at the machine; (2) thickness variation not meeting the tolerance requirements that is usually a minimum; and (3) production startups, shut-downs, and changeovers. If a convertor has to use gauge variation materials, the customer usually has a major inefficiency (and is 'upset'). If it can be used so that the converter can meet their delivery schedules, during the conversion the film has to run much slower in the presses, laminators, bag machines, etc. With changeovers film extrusion lines can produce scrap at the rate of 227–454 kg/h (500–1000 lb/h). By the time operator(s) set up proper controls for the production line, it could take up to two hours. In the meantime, scrap is accumulating. By recycling, costs go up for the recycling action and the line has to be carefully controlled if more than the usual amount of scrap is blended with virgin plastics (Chapter 3) [433].

Many TP materials, especially PEs, develop a thermal history (Chapter 3) where their properties degrade as they are repeatedly remelted and reëxtruded. A potential more unfavorable situation occurs when coextruding or by most post-extrusion processes because they might be nonrecyclable. However, there are applications where these mixed-up plastics can be used, such as within the coextruded film, sheet, profiles, fabrication of extruded synthetic lumber, utility poles, and so on.

There are TPs that are relatively easy to process, such as LLDPE, LDPE, PP, and PETP. Tighter controls are used when plastics such as HDPE, HMWHDPE, and mPE are processed; there are differences such as the bubble shape. HDPE forms a wine glass shape (Fig. 7.8) with a frost line height (FLH) of 6–10 die diameters using Battenfeld Gloucester equipment. The LLDPE has a FHL of 1–2 diameters. The 'high stalk' shape for HDPE is necessary to obtain desired mechanical properties (toughness, flexibility, etc.). An adjustable inflatable mandrel in the tubular film (Fig. 7.9) can be used to increase bubble stability at high speeds.

Making films from metallocene-based plastics (mPE, etc.) requires changes/adjustments to processing equipment but those changes can be

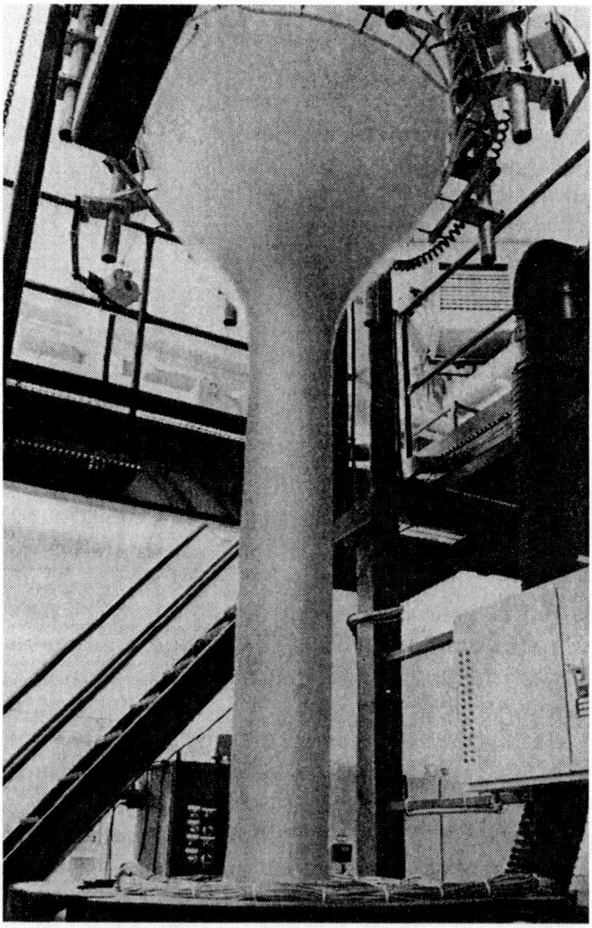

**Figure 7.8** HDPE wine glass shape.

**Figure 7.9** Tube includes inflatable mandrel proving structural support for high density film.

made easily and without great expense [421]. With metallocene LLDPE, nip pressure is slightly increased. Down-stream equipment is adjusted to minimize defects and accommodate the softer films (oscillator on turn bars, haul off, cooling water, etc.). Blends of conventional and metallocene plastics produce improved puncture resistance, tear resistance, load holding force, stretch and resistance to restretching, and processability.

## BLOWN TUBE CHARACTERISTICS

As reviewed in this chapter (and elsewhere in this book), thickness control is of prime importance because it influences properties, amount of scrap, cost, and so on. Table 7.1 provides a guide to the amount of film is produced based on thickness.

There are different thickness sensors used to provide different input controls (Chapter 6). An example is an ultrasonic bubble thickness control sensor. This noncontact sensor in a stabilizing cage measures the location of the film wall and sends feedback to the fans that regulate the flow of air into the bubble. By equalizing the air supply and exhaust of air in the bubble, problems such as surging is reduced. Layflat width can be controlled to within 3.2 mm (0.125 in), eliminating the need to trim extra edge width from the finished roll.

All the improvements that can be made via thickness (or equivalent) sensors depend on the capability of the process computer (PC) control and on having available on the blown film line adjustment capabilities. Modular PCs permit changes to specifications or machine settings. The adjustments can be made on individual control parameters and, in some cases, interrelated parameters (Fig. 6.11). Simplified PC readouts with graphic presentations of most, if not all, functions on line, allow the operators to make complex adjustments with limited formal training. All required functions can be stored in the PC memory so an operator merely opens

**Table 7.1** Guide to plastic yield of films

| | *Yield* | | | |
|---|---|---|---|---|
| *Material* | *($yd^2/lb$) 0.001 in thickness* | *($m^2/kg$) 0.025 mm thickness* | *($lb/1000\,yd^2$) 0.001 in thickness* | *($g/m^2$) 0.025 mm thickness* |
| Cellulose acetate | 17 | 31 | 59 | 32 |
| Cellulose tri-acetate | 16 | 30 | 62 | 33 |
| Nylon | 18.5 | 34 | 54 | 29 |
| Polyethylene | | | | |
| low density | 23 | 43 | 43 | 23 |
| high density | 22 | 41 | 45 | 24 |
| Polyethylene | | | | |
| terephthalate | 15.5 | 28 | 65 | 36 |
| Polypropylene | 24 | 44 | 42 | 23 |
| Polystyrene | 20 | 37 | 50 | 27 |
| Polyvinylidene chloride | 17 | 31 | 59 | 32 |
| PTCFE | 10 | 18 | 99 | 55 |
| PTFE | 10 | 18 | 99 | 55 |
| PVC | | | | |
| flexible | 14.5–17 | 27–31 | 59–68 | 32–37 |
| rigid | 15.5 | 28 | 65 | 36 |

a 'file' aboard the computer to establish proper settings, etc. Feedback can be stored, retrieved, and/or printed for quality insurance, etc. (Chapter 6).

Even though thickness is an important criteria, others exist that are based on product requirements that usually include width, surface and optical properties, mechanical properties, and orientation. Film thickness is controlled initially by the die geometry with its melt temperature and pressure conditions in the orifice. However, once an extrusion melt is running satisfactorily, a very stable melt system exists. Thereafter, thickness is controlled by making adjustments to the blow ratio and haul-off speed. If the die design is good and the die gap is uniform, then uniform cooling should produce a uniform film. However, in practice, there are always variations in the setting of the die gap as well as the cooling device. The variations can cause film that has local thick or thin spots. Rotating die, the usual oscillating die or haul-off devices, and other methods are used to compensate for these variations.

Most tubular film dies in use basically have two comparatively massive pieces of steel. A central core within an annular die lip incorporates adjustments. By reducing the space (film thickness) on one side, the opposite side will increase and so on. A continuously deformable circular die lip is usually used. Provided the circumference can be uniformly heated, a more uniform tube can be obtained. Both film thickness and layflat width are affected by variations in the haul-off speed and blow-up ratio. With the proper speed control device, the haul-off speed is no problem.

Bubble width is temperature dependent and an attempt to minimize the effect can be made either by various methods, such as surrounding the bubble with a thermostatically controlled environment (by air rings, etc.), by ensuring that the air inside the bubble is keep at a constant temperature and pressure, or a combination of these two systems. In the actual drawing zone between the die and the blowing ring, properties are controlled, assuming the plastic has been properly prepared and in turn melted properly in the extruder. Orientation will be reviewed latter.

Film performance can usually be related to the blow-up ratio which is usually 2–4 with as high a freeze line as is consistent with bubble stability. The frost or freeze line is the place where a semicrystalline plastic, usually a polyolefin, starts to crystallize as it cools. An air ring around the blown tube cools and shapes the bubble making a frost line. Usually room temperature air is sufficient for cooling the bubble. The term frost line could be misleading because it tends to imply that solidification occurs in a straight line across the bubble; solidification occurs as it travels toward the nip rolls.

The frost line can be higher on one side than the other side or even be very wavy. This irregularity can be caused by various factors. The temperature of the melt entering the die can be inconsistent or imperfectly mixed. Radiant heat leaving the extruder can effect one side of the bubble. An open window or a stray beam of sunlight can affect ambient conditions around the bubble. Thermal problems are generally the main cause, however many of the variabilities discussed in this chapter as well as others have an influence (i.e., poor material handling, nip rolls not in unison with windup speed, etc.).

There are different systems to take corrective actions. As an example, by measuring via a capacitance gauge the thickness above the frost line, a control system can quickly record these thermal influences. The system corrects the temperature irregularities regardless of the causes by thermal adjusting the corresponding sector of the die. Some nonthermal causes of the gauge variation, such as machine vibration affecting the bubble line below the frost line, can be influenced by thermal means at the die and corrected [433].

The effect of keeping the frost line high occurs at the higher blow ratios and is more marked at the higher melt temperatures, although it may be necessary to make a compromise with haul-off speed. Impact strength is increased by higher haul-off speeds but, at these higher outputs, the bubble shape changes so that orientation in each direction instead of occurring simultaneously is now separated just prior to the frost line. Also an increase in die gap usually has a significant increase in impact strength.

For high tear strength, generally a lower blow-up ratio is preferred. Thus, it tends to become difficult to obtain high impact and high tear strengths since the physical conditions of the bubble formation are different. However, with certain plastic compositions gains can be made in the same directions.

Because of the distance necessary for blowing between the die and air ring (usually the frost line), cooling is relatively slow and the crystallinity for a plastic, such as LDPE film, will be greater than for chill-roll or water bath cast films. This generally results in a stronger and stiffer film, but of poorer clarity than the chill-roll film. Also, slower cooling allows more time for surface defects, such as die lines that originate at the die. Optimum optical properties require a compromise in frost line height. For a fixed frost line height, increased output rate, increased melt temperature, and decreased blow ratio all tend to give slower cooling and increased crystallinity. With HDPE, the rate of crystallization is faster, and with rigid (unplasticized) PVC, the rate is slower than with LDPE. As a result, the crystallinity and film properties are much less effected by these variables [397].

Different methods of bubble cooling exist, each with advantages and disadvantages. For example, because of their different extensional rheologies (flow), LLDPE bubbles are less stable than those of LDPE. Proper cooling is very important in obtaining gauge uniformity. Slow, very cool air has a better cooling effect than high velocity cool air; the slow air helps to minimize bubble instability. Although single lip air rings have proved adequate for some applications, dual orifice and tandem designs provide enhanced cooling which effectively stabilizes the bubble and speeds up the lines (Fig. 7.7). Internal bubble cooling (IBC) with a dual lip air ring is also effective and required in certain lines by increasing stability at high production rates. However, improperly arranged IBC configurations can cause melt fracture due to chilling of the die lip (Fig. 7.8). Tied in with cooling systems are different types of electrical and optical sensors and techniques used to measure the complete bubble circumference and relate it to processing performance. An example is the capacitance sensor.

Heat between the die and the pinch rolls influences the haul-off rate. LDPE, for example, leaves a die at 150–170°C (300–340°F). On its arrival at the pinch rolls, the temperature should have fallen to 40°C (70°F). The film should be wound up at as low a heat as possible in order to prevent excessive shrinkage on the roll, which causes blocking. Thin walled film can be taken off at speeds of at least 20–50 m/min (65–165 ft/min). With film of 150–300 μm (6–12 mil) thickness, rates of at least 10–20 m/min (33–65 ft/min) are achieved.

HMWHDPE thin gauge, blown film high-speed lines take advantage of triple-chamber external cooling ring and internal bubble cooling (IBC) capabilities. Figure 7.10 by B-G shows an 220 cm (86 in) 3-layer die with an external triple-chamber air ring and an IBC unit.

Reifenhauser systems producing 6–8 μm (0.2–0.3 mil) thick, 610 mm (24 in) lay-flat film using a 120 mm (4.7 in) annular die travels at least 245–300 m/min (800–1000 ft/min). The production rate is about 100 kg/h (225 lb/h). This high rate is possible with a rather complete process control line that includes the use of an efficient IBC device. Winders have been designed to roll-transfer and cut over film at the rate of at least 300 m/min (1000 ft/min) with targets that have been reached at 300–600 m/min (1000–2000 ft/min).

Summarizing the blown tube characteristics, the blow-up ratio [Fig. 2.20(a)] is usually 1.5–4.0, depending on the plastic being processed and the thickness required. With crystalline types as reviewed, the melt leaving the die changes from a hazy to a transparent (amorphous) condition. The level at which this transition occurs is the frost line [Fig. 2.20(b)]. The frost line's visual appearance can be a straight/level line or can show a varying line and height that can be related to the processing conditions. Depending on the plastic, it may be acceptable to have an uneven frost line within certain limits. Before or after the layflat operation, some lines

**Figure 7.10** Triple air-chamber air ring with internal bubble cooler is part of this 3-layer coextruded die.

may have a corona discharge pretreatment when using certain plastics such as LDPE. This treatment has the effect of oxidizing and activating the surface so that can be printed, welded, and so on.

### Start-up

To obtain a so-called 'perfect' bubble, the various parts of a blown film extrusion line must be accurately set up. The die must be leveled in all directions after the adapter has been tightly bolted or clamped to the extruder barrel head. Then the nip rolls must be similarly leveled. Any auxiliary equipment, such as an IBC, bubble guides, and sensor devices, must be properly positioned. With a bar or equivalent, plumb from the center of the nip into the hollow mandrel for vertical centering. Set all equipment in the line in their operative positions.

Uneven cooling of the bubble may result in a nonuniform diameter. In turn, this will result in a nonuniform gauge and width, uneven wind-up, and wrinkled rolls. The uneven cooling may be caused by a draft in the plant. If the source of the draft cannot be eliminated, the bubble should be protected by a canopy, such as framework-supported PE film curtains around the entire bubble forming operation. When not in use, the canopy can be cranked up or down.

On actual start-up, have a string-up as the feeder, such as a lead rope or the remains of the last run in position, going from the windup roll to just above the die. Leave the nip rolls opened so the feeder line can travel through it without obstructions. Turn the extruder on at the slowest rate required to obtain a satisfactory melt and have the melt exiting the die. In the mean time, turn on the bubble's internal air system slowly to cause a bubble to be formed just above the die. Remove undesirable melt until it appears that the extrudate is uniform.

During this time, a person wearing a safety face shield and protective gear, including heat resistant gloves, will be grasping the melt to develop and produce an inflated bubble by squeezing the melt. The next step requires the feeder line to be attached to the melt by squeezing them together. In the mean time, start moving the feeder line at a slow speed.

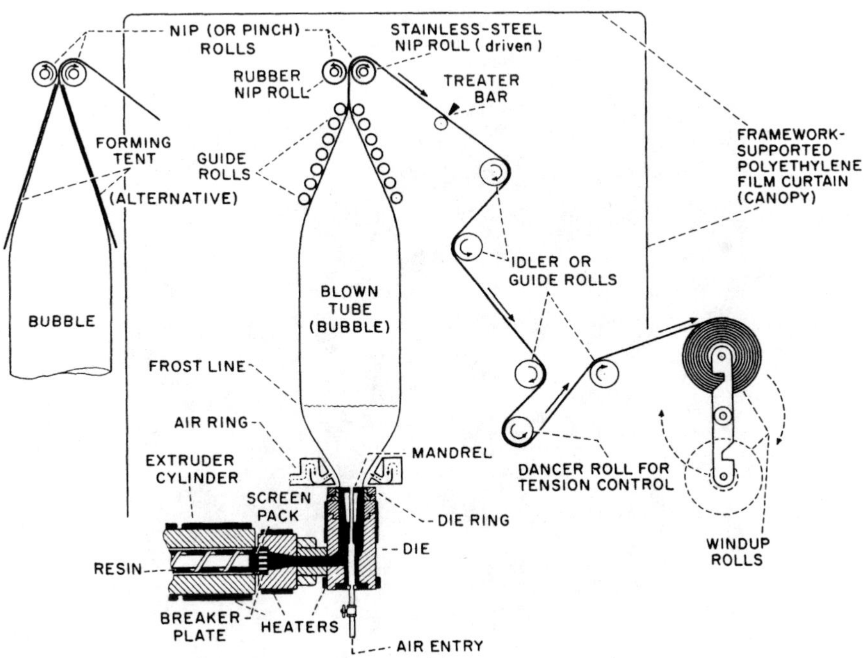


**Figure 7.11** Schematic from start to end of a blown film line.

Figure 7.11 provides a schematic showing the plastic blown film going from the die to the windup roll.

Once the feeder line passes the nip rolls, start turning and closing the nip rolls at a matching speed to get the bubbled melt moving through it. With a successful 'pull' to the nip rolls, which usually takes a few trial runs, start increasing the extruder speed, air supply, and speed of the nip rolls and adjust all systems in the line.

System adjustments include proper tension on any idler, guide, dancer, and/or windup rolls as shown in the Reifenhauser complete line (Fig. 7.12). Typical of these lines, they also set up controls for material handling with recycling trim, drying plastics, cutters for slitting and trimming, and so on. Handling the rolls at the end of some lines can be a gigantic job as the rolls can weigh tonnes and operate at speeds up to 610 m/min

**Figure 7.12** End of line shows a wrapped roll transferred from winder.

(2000 ft/min). In fact, their costs could represent about one-third of the complete extrusion line. So their importance becomes obvious and the fact that they have to be handled very carefully.

There are winders to meet many different requirements that are tied in with the master process control computer. As an example, there can be a winder that produces very large diameter roll, having very low slip and good tension control. This surface winding runs the film over a surface drum and onto the core as the surface drum drives the core. Center winding is used for tackier films, like those that include EVA. With this type film, the surface drum is not used and the shaft holding the core is powered.

There is also gap winding, where a combination is made of the surface and center winding. Both the shaft holding the core and the surface drum are powered. The surface drum is backed slightly away from the roll of film. The result is that the film runs over the surface drum as it makes its way onto the core. The surface drum does not contact the roll. This approach is a way to prevent pockmarks and other blemishes in certain grade materials that have some tackiness [359].

Of course, it is very easy to explain what is required, but rather difficult to start-up. The start-up person(s) requires experience and skill. With today's process control system (Chapter 6) for the complete line, start-up has become easier and quicker.

## ORIENTATION

The blown tubular film process, by its nature, gives orientation to the plastic. Owing to their method of production, all blown tubular films are oriented to a greater or lesser extent depending on the processing conditions. Systems have been designed to increase the degree of orientation in order to obtain films of improved clarity, strength, heat resistance, etc. Except for special applications, where greater strength in one direction may be needed, films are normally made with balanced properties [3, 287].

As reviewed in Chapter 2, orientation is used to improve different performances of plastics that usually provide significant cost benefits. During blown film processing, the blow-up ratio determines the degree of circumferential orientation, and the pull rate of the bubble by the nip rolls determines longitudinal orientation (Figs. 2.20 and 7.13). As an example, the optimum stretching for amorphous plastics (PVC, etc.) is just above the glass transition temperature; for crystalline plastics (PE, PET, etc.) it is just below the melting point (Chapters 2 and 3). During the stretching process, the structure changes because of crystallization, usually necessitating an increase in heat if further deformation is planned. Afterward,

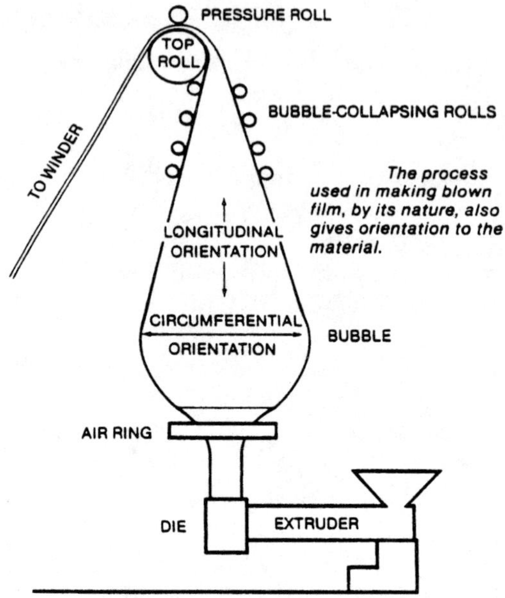


**Figure 7.13** Extruder blown film orientation.

the orientation is 'frozen in' by lowering the heat or, with crystalline types, set by increasing the crystalline portion.

With orientation, film thickness is reduced and surface area enlarged. If a film is only longitudinally stretched in the elastic state, the film thickness and width are reduced in the same ratio. If lateral contraction is prevented, stretching reduces the thickness only.

The direct injection of liquid additives, such as polyisobutylene (PIB), to produce stretched film prevents difficulties in extruding and offers a processor a wider range of materials from which to select. It also provides cost reductions due the use of more economical formulations. This method is suitable for the injection of cross-linking agents, liquid colors, and the like, via the extruder or gear pump.

## PROCESS OPTIMIZATION

There are many different approaches and control devices used to improve the processing of blown film (Fig. 6.11). They range from a simple single machine settings to detailed process controls (Chapter 6). Some of these methods are reviewed in this section. Figure 7.14 curves provide a guide to the influence of some of the on-line machine settings [206].

**Line control**

A basic line could be set-up without a gravimetric feeding system, without the closed loop controls and meet short-run custom shop orders. The operator would provide the control and capability to produce what is required to meet film performance. For the long-run operation, rather complete process control of the extrusion line is normally used (Chapter 6). The system removes the possibility of operator error or chance because there are many elements of control built into the control system. From the plastic feed to the die to the high speed film winder, all machine components are automatically tracked and controlled through a central computer. If color monitoring is involved, the control can provide a layout image of the complete line.

Operators, by means of specific keys on the control panel, can call up the desired line operation details directly on a screen. If some part of the line is not operating to maximum efficiency, based on previous settings during start-up, the control system can provide a warning light and/or buzzer to alert the operator of the slight deviation. In turn, the computer could have means to re-adjust the line or the operator takes the proper action to get back on target. The control can also be set, if desired, to systematically shut-down the line if out of specification film is being produced. Of course, the ideal control system would provide a warning when even though the line is within specification there is a slight deviation, and take action to return to the best settings automatically or manually. Thus, no expensive shut-down occurs.

As summarized in Fig. 1.1 and reviewed in Chapter 2, start-up of the line requires going through different settings to maximize the lines operational efficiency at the lowest cost. During this period, the line will be subjected to variations that will still keep the line producing film within the specification requirements or out of specification. At that time, corrective actions are taken to go back to the best settings to operate the line most efficiently.

The deviations that occurred may be put into the computer system so that during operation the line corrects itself. When this deviation information cannot be incorporated in the computer software, instructions should be prepared for the operator to take the corrective action. The computer can record that corrective action to be taken by the operator and provide the instructions.

Film lines usually can provide point-to-point thickness variations of 10%; and there are lines that can meet at least 2% (or better). So, if a line has to provide a minimum film thickness, consideration should be given to using a line that provides minimum variation whilst meeting the customer's minimum requirement and saving plastic material and processing costs. Thus, the customer can order less plastic without compromising

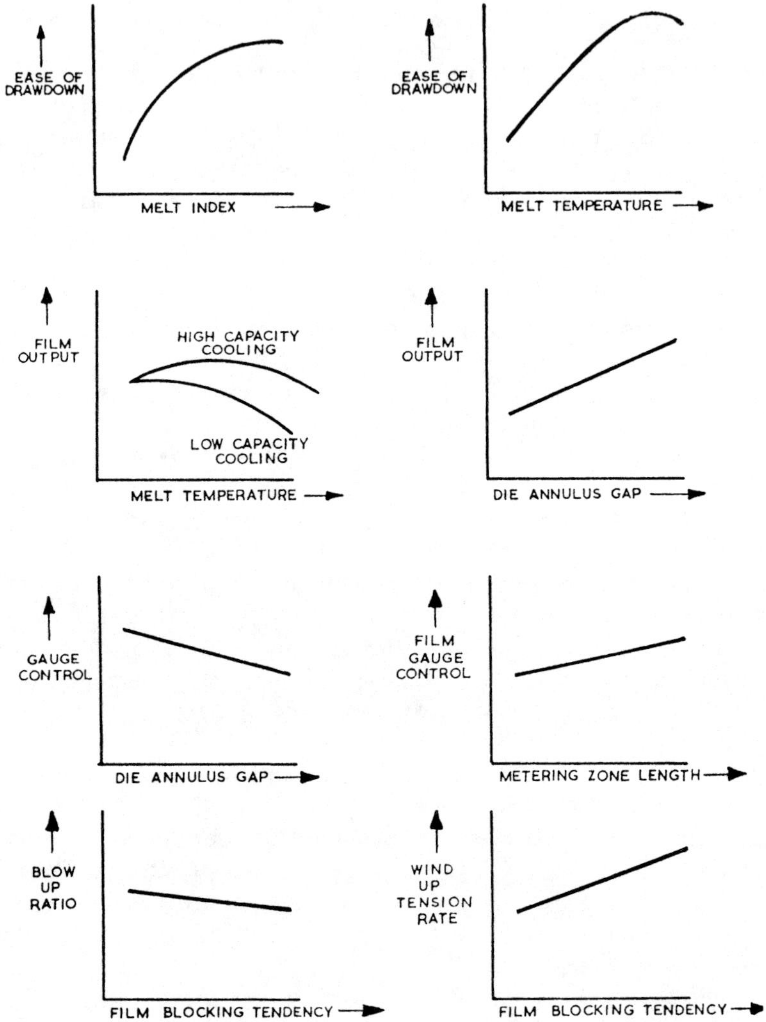


**Figure 7.14** Effect of blown film machine settings on properties of the film.

quality. With the tighter tolerance capability, problems on-line can be reduced since less heat transfer problems will occur, etc.

## Output rate

When analyzing or determining the output rate of a blown tubular film line, various factors need to be considered. To start with, the rate could be

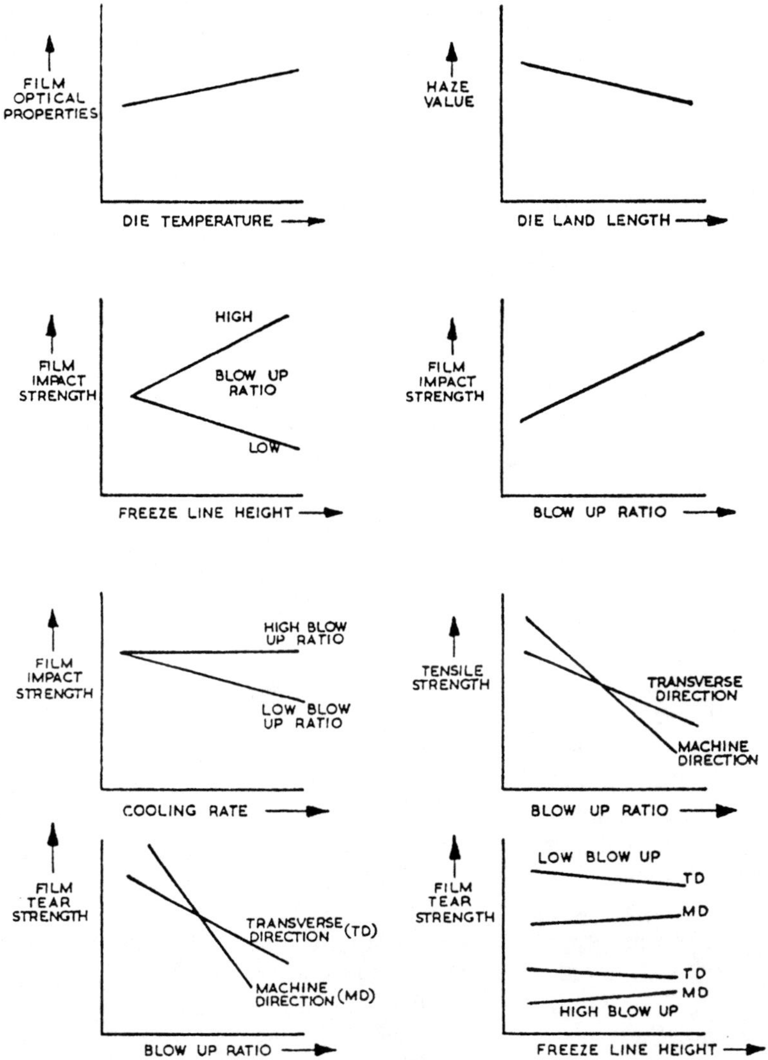


**Figure 7.14** *Continued*

limited by factors such as the material handling limitations, extruder gear speed capability, melt fracture, film blocking, air blower capacity, collapsing frame geometry, air blower capacity, recycled material limitations, and down-stream equipment line speed limitations. The following information provides guidelines that influence the output rate of blown film lines. The values presented are based on empirical observations [386–388].

The plastic melt strength has a major impact on the maximum production rate. A stiffer film allows more air to be blown against the bubble without causing bubble instability. As an example, LDPE can usually be run at 30–50% higher rates than LLDPE because of the higher melt strength. Blends of LDPE/LLDPE rates fall in between those of the two plastics.

Certain materials, such as recycled plastics or low viscosity tackifiers, usually do not have the same melt strength as virgin plastics. Processwise, the line speed will have to be reduced by possibly 10% due primarily to bubble stability limitations.

Most processors use a color masterbatch with a melt index (MI) base plastic to improve the melting and mixing of the masterbatch in the extruder (Chapter 17). The high MI base plastic reduces the melt strength and the rate will be reduced due to bubble stability limitations. With highly pigmented films, the reduction could be about 10%.

An important effect on rate is related to the melt temperature which influences the melt strength in the bubble. A general rule states that for every 10°C (15°F) reduction in the melt temperature will increase the rate by 0.02 kg/h mm (1 lb/h in) processing LLDPE. With the additional cooling of an IBC in the blown film, the rate can increase 25–50%.

For LDPE/LLDPE blends, the highest rates occur when the blow-up ratio (BUR) is about 2.2–2.8. With lower BURs, there is not as much surface area for the area to cool so the rate is reduced. With higher BURs, an increased surface area could result with increased rates. However, because of the potential aerodynamic effects from the increased bubble curvature, there is an increase in bubble stability/shaking that can limit the amount of cooling air and in turn limit the rate. The highest rates are usually at a thickness of about 25–50 μm (1–2 mil). On thinner films, the bubble is softer and cannot handle the higher air flows. With thicker films, the weight of the film in the bubble causes sagging and usually requires that the rate be reduced with the frost line lower.

The plant's temperature influences the bubble cooling rate. With air conditioning, the extra cooling allows for a higher rate while maintaining the same frost line height and cooling air flow. The result could be a 10% increase in rate. If the plant's ambient temperature is over 38°C (100°F), the opposite effect occurs because the rate usually has to be reduced by about 10% in order to obtain bubble stability.

Grooved feed extruders operate at much lower melt temperatures than do smooth bore extruders, especially at the higher head pressures (Chapter 2). The lower melt temperatures usually result in a 10–15% increase in rate over the smooth bore extruder.

There are a wide variety of screw designs (Chapter 4) in use. It is reported that it is not unusual to develop a rate increase of 20% after replacing an older worn screw with a state of the art barrier screw

(Chapter 4). It is possible to achieve even greater improvements if the end product allows the plastic mixing to be reduced to get an even lower melt temperature. If the film requires a higher intensity of mixing than normal, then the rate will be reduced because of the higher melt temperature.

As the screw speed is increased, the melt temperature rises, decreasing the maximum rate. Normally at screw speeds under 25 rpm, the rate can be up to 10% higher than average, while at screw speeds above 70 rpm, the rate can be 10% lower than average. Grooved feed machines can give a fairly constant melt temperature over the range of screw speeds and do not produce any rate variations.

If the extruder torque is not sufficient for the proper operation, the operator will usually increase the barrel temperature profile settings to lower the drive amperage (torque). When the barrel temperature settings are increased, the melt temperature increases, reducing the rate. The loss in rate depends on how much the barrel settings are increased.

If the extruder barrel cooling is operating inefficiently or just not operating, the barrel settings may override their set points, reducing the maximum rate.

When reviewing the size of die circular orifice, film rate/cm (in) decreases with the larger dies. If the same plastic, thickness, BUR, and so on, is run on different size dies, the stress in the film will be similar because the orientation, strain rate, and melt temperature will be similar. Since the stress in the film is the same, the internal air pressure in the larger bubble will be lower than that in the smaller bubble otherwise the bubble would collapse/fail; this action is related to the behavior of thin wall cylinders [3]. The lower pressure along with the larger bubble diameter results in a bubble that is less rigid and cannot be subjected to as much cooling air as a smaller bubble.

With a high-pressure operating die, the rate is reduced in order to compensate for additional cooling time, etc. The higher pressure generates a higher melt temperature in the extruder. Also additional shear heating in the die occurs with the higher pressure drop when the extrudate exits. However, there are some disadvantages to low pressure dies such as: (1) there can be a loss of film properties with certain plastics, notably machine direction (MD) tear strength with LDPE; (2) low pressure dies tend to have a higher residence time and lower shear rate, both of which contribute to gel formation and other losses particularly on coextruded products when a heat sensitive plastic is used; and (3) high pressure usually provides better gauge control.

An interesting development is that when all operating factors are the same, three layer coextrusion film lines will average about 5% more rate than monolayers.

Converting from a single to a dual lip air ring can increase the rate by 30–50%. It is also possible to develop smaller increases by going from an older to the newer dual lip ring design. If chilled air is not available for the air ring, the rate will be reduced by about 15% for an ambient temperature of 32°C (90°F). The air temperature at the ring will be over 43°C (110°F) because of the frictional heating in the blower.

Bubble sizing cages such as those shown in Figs. 7.3, 7.9, and 7.15, provide the required support and stabilization of the blown bubble. This type of support structure allows additional air to be used. Result is at least a rate increase of 10% over a system without bubble support.

At higher elevations the air density is lower, reducing the cooling efficiency. Even if larger blowers were installed, the increase in cfm (cubic ft/min) would cause destabilizing aerodynamic effects that increase by the cfm squared. Result is reducing the rate. Generally rate reduction of 15% occurs at a 1.6 km (1 mile) elevation above sea level.

The end use of the film has a major influence on the maximum rate of

**Figure 7.15** Stabilizing device for blown bubble.

salable film that can be manufactured. At the high rate end of the production spectrum are such items as trash can liner films, where wrinkles in the film do not create a major problem. With these type products, the rate can be increased 15% over normal rate. The low rate end is quality film, where gauge control is critical and wrinkles do not exist. These products may require the line rate to be reduced by 20–30% below what would be considered a normal rate for trash bags.

Wrinkles are problems which have always plagued processors. They can occur intermittently and are annoying as well as costly. Badly wrinkled film rolls are usually scrapped (Fig. 6.1). Wrinkling on the windup roll may be caused by conditions such as the frost line being too high and/or the die ring is out of adjustment. Bias (systematic error, in contrast to a random error) is a condition where the two halves of the blown tube circumference are unequal. This causes excessive friction at the guide rolls, or forming tent, or unbalanced pull at the nip rolls. The result are ruffle-like wrinkles across the center of the lay flat width on the wound roll.

Film may be too cold when it reaches the nip rolls and its stiffness may cause crimping at the nip rolls. Use of higher density plastic will increase the stiffness and its susceptibility to wrinkles. The guide rolls may not be properly aligned with the nip rolls. The use of spreader or expander rolls is often helpful in removing wrinkles caused by uneven or too high a web tension. Surging of the extruder and air currents in the plant are detrimental.

The importance of well trained, experienced operators is extremely important to the successful operation of a line during start-up. It is believed that an inexperience operator, in addition to causing damage to the line equipment, will cause a 20–30% reduction rate. Many plants have on-site process engineering support to ensure lines are running at maximum quality and efficiency. The result is 5–105 increased rate over average. Without this type of personnel, typical loss is at least 10% below average.

There are different important practices to follow in order to ensure meeting production quality and output. These practices should always include maintenance and particularly preventative maintenance procedures with qualified personnel. Plants with poor maintenance get 10–30% lower output than average, whereas those with proper maintenance procedures will keep the lines running with very little down-time, and can gain at least 5% increase in rate over average.

It is reported that where plant management properly monitor and push, production rates can be up to 15% higher output than average. This action by management includes: (1) ensuring that personnel are qualified for their jobs, content, and know what is expected; (2) providing updating for personnel via some type of training particularly technical at the support level; (3) ensuring that all the equipment in-line is up to date; (4) ensuring that equipment is properly maintained; and (5) hiring good people and keeping them.

In determining the output of film and sheet use the following equation [206]:

$$Q = CpWtV$$

where $Q$ = output (kg/h or lb/h), $C$ = constant (6 for metric units or 26 for English units), $p$ = relative density of plastic at 20°C (68°F), $W$ = width of

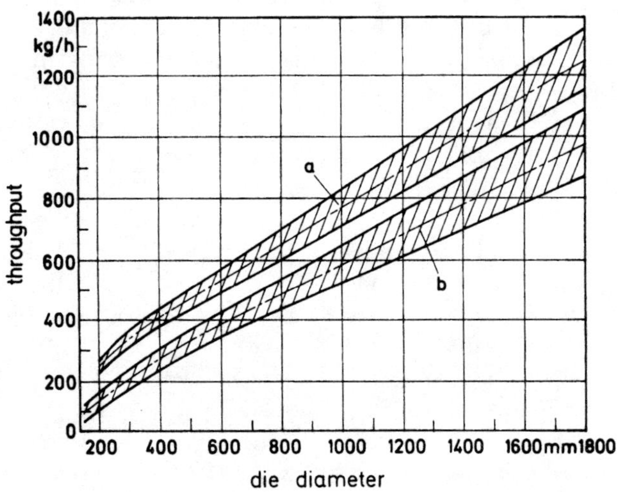


**Figure 7.16** Blown film throughput as a function of die diameter.

**Table 7.2** Example of LDPE blown film thicknesses based on 0.922 density

| *Melt index of plastic* | *Extrusion temperature (°C)* | *Minimum film thickness (in)* | *Minimum film thickness (mm)* |
|---|---|---|---|
| 0.2 | 150 | 0.0020 | 0.051 |
| | 170 | 0.0015 | 0.038 |
| 1.0 | 150 | 0.0012 | 0.030 |
| | 170 | 0.0009 | 0.023 |
| 2.0 | 150 | 0.0010 | 0.025 |
| | 170 | 0.0007 | 0.018 |
| 4.0 | 150 | 0.0007 | 0.018 |
| | 170 | 0.0004 | 0.010 |

**Table 7.3** Effect of die design on haze of blown film

| | | Haze value (%) | | |
|---|---|---|---|---|
| | | Die land length | | |
| | *(in)* | *0* | *0.375* | *0.75* |
| *Die entry angle (°)* | *(mm)* | *0* | *9.5* | *19* |
| 6 | | 11.3 | 9.5 | 10.9 |
| 13 | | 10.6 | 9.0 | 8.9 |
| 33 | | 10.8 | 8.4 | 7.8 |

**Table 7.4** Guide to film thickness die gap setting range

| *Die gap* | | *Film thickness* | |
|---|---|---|---|
| *(mm)* | *(in)* | *(μm)* | *(mil)* |
| 0.5 | 0.02 | 25 | 1 |
| 1.0 | 0.04 | 250 | 10 |

**Table 7.5** Die sizes by product applications

| | | *Diameter Range* | | |
|---|---|---|---|---|
| *Applications* | *Layers* | *(m)* | *(mm)* | *Materials* |
| Form/fill/seal | 3, 5–8 | 6–16 | 150–405 | EVA, LD, LLD, LLD-M, PA, EVOH, ADH |
| Stretch | 1, 3 | 22–40 | 560–1000 | LLD w/PB |
| Lamination | 1, 3, 5 | 12–28 | 305–710 | EVA, LD, LLD, LLD-M, PA, EVOH, ADH |
| Construction and agriculture | 1, 3 | 25–90 | 635–2300 | LD, LLD |
| In-line bags | 1, 3 | 6–30 | 150–750 | LD, LLD, LLD-M, HD |
| High barrier | 5–8 | 6–28 | 150–710 | EVA, LD, LLD, LLD-M, PA, EVOH, ADH |
| Geomembrane | 3 | 70–90 | 1780–2300 | HD, VLD, LLD |

EVA, ethylene vinyl acetate copolymer; LLD, Linear low density polyethylene; PB, polybutylene tackifier; EVOH, ethylene vinyl alcohol; LLD-M, metallocene catalyst LLDPE; VLD, very low density polyethylene; HD, high density polyethylene; PA, polyamide (nylon); ADH, adhesive tie layer; LD, low density polyethylene.

film (tube circumference) or sheet (cm or in), $t$ = thickness of film or sheet (cm or in), $V$ = linear output rate (m/min or ft/min).

As a guide for blown film output rates it generally is about 3.2–9 kg/h (7–20 lb/h) per inch of die circumference. The mandrel's bushing die gap may range from 0.51–1.27 mm (0.020–0.050 in), depending on final clearance. Figure 7.16 provides a guide to output rates based on the die diameter where (a) refers to a wide gap and (b) refers to a narrow gap. Tables 7.2–7.5 provide guides on film thicknesses, die gap settings, and die entry angle.

## DIE

Blown film die technology developments have been very significant during the past decade and continue because they are now understood to perform an important part in the control of film gauge thickness variation. They can adjust for virtually all thickness variations, even those whose problem is down-stream from the die (Chapter 5) [348, 370, 397].

There are systems such as capacitance type thickness gauges that measure film thickness accurate to 0.1 μm (0.004 mil). The gauge revolves once every 2 min around the bubble just below the collapsing frame, taking five readings for every degree of rotation. A controller uses feedback from the gauge to the die. The signals are direct to the die's specific heating elements to turn on–off in response to thickness variations. Cartridge heater elements in the die are narrow and tightly spaced, permitting highly localized variations to be corrected [433].

There are also systems where die bolts adjust the flexible die lip and in turn the gap space. The bolts can be electrically or mechanically driven by stepper motors (Fig. 5.24).

As discussed earlier, about 95% of gauge variations are caused by temperature differentials. The die temperature controller permits adjusting the plastic melt flow behavior as it leaves the die. As the bubble is stretched longitudinally and transversely, the hotter melt sections flow more readily; they stretch to thinner gauge. The cooler melt sections solidify faster and retain more of their thickness.

Since blown tubular films are usually extruded vertically upwards, the extruders are coupled to the tools/dies with central feeding (Fig. 5.2). Conventional mandrel dies can have two basic limitations. One concerns the welding of the individual flow fronts bonding after surrounding the die mandrel which in turn produces a longitudinal weakness in the film. The second relates to development of flow streamlines with different lengths in the down channel direction due to the change of the melt from horizontal to vertical. This action can create pressure drop differences across the die circumference which can cause a gradient in the film thickness [397].

It is possible to equalize the flow path length by diverting or rerouting portions of the plastic. This solution is seldom used because it develops difficulties to manufacture, maintain, and cleaning. There is also the approach of using smooth spherical 'bumps' in place of the spider arms normally used to support the mandrel that cause the weld lines. These bumps serve both to support the mandrel and promote turbulent melt flow with thorough mixing through the die [306].

The usual approach is to allow for some means of spreading the melt throughout the film perimeter, avoiding weak/melt lines or areas. Usually the die, extruder with head assembly, or haul-off device rotate or oscillate. The die rotation requires a drive mechanism and efficient design of the die to provide sealing so that melt only goes through the die lips. If the die does not rotate, an important advantage occurs since the die is shorter, thus providing less residence time.

With all this action a couple decades ago, the spiral mandrel die was designed and successfully put to use (Fig. 5.4). Significant improvements with the basic design continues. In this design, the melt is usually fed either through a central channel connected to a system of radial runners/ports, or through a ring. The latter provides for easier access to the central zone of the die; this condition can be important to provide ease of internal bubble cooling (IBC).

With the central system, the number of radial channels is usually identical to the number of helical channels. There are also designs with one radial channel feeding two spirals using a triangular transition, and also one radial channel feeding half of two spiral channels. It is common practice to use 1–2 grooves per inch of the die diameter [348–353]. Factors to be considered for this type of design are the flow characteristics of the plastic melt (Chapter 3). The target is to have short residence time and low pressure drop. The major part of the pressure drop occurs in the die land. Plastics with high or narrow molecular weight distribution, such as LLDPE, require wider gaps. To aid the melt flow, a relaxation zone between the spirals and the land is generally used.

## MULTI-LAYER OR COEXTRUSION

Dies have become inherently more flexible in their design for multi-layer or coextruded products. With a multi-step extrusion process, each layer is extruded separately, partially, or totally cooled before the next layer is applied. Coextrusion is where the various melt layers emerge simultaneously from the die lips (Figs. 5.33–5.37).

Coextrusion dies can be classified according to their melt flow modes [297]. One method that is popular has each melt separated in its own spiral channel and meeting in an adapter or the parallel zone prior to exiting (feedblock process). The geometry of each channel conforms to the requirements of each melt (rheology) characteristics of velocity, shear stress, and thickness requirements. Individual channels are kept close to each other in order to optimize the temperature control. The die centering system permits ease of rotation. This type die is more popular in flat film dies.

Different designs are used to take advantage of the spiral channels, usually based on the melt flow capabilities of the plastic. As an example, there are dies with longer spiral wraps and more ports are overlapped. This approach makes the plastic, such as HMWHDPE or nylon, less sensitive to changes in their melt flow characteristics. Other dies provide for quick purging where dead spots are eliminated via streamlining. Ports are drilled and reamed so that they intersect at the cylindrical collection chamber. This design permits a more direct feed to the spirals.

With the manifold block process, melts unite in the order that the incoming melts are fed from the extruders. Melt from the extruders to the die are through transfer pipes that replace the manifold block. This design permits the distribution of the melt layers to be changed whenever required. Basically, the melts should have similar viscosities ensuring their laminar flow. When there is a difference, the interlayer cross section will have a convex or concave shape depending on which melt has the higher viscosity. The less viscous melt will 'encapsulate' the more viscous melt. This characteristic is more obvious with flat film or sheet dies.

So called stacked dies are available. These dies provide a processor with a modular design versatility capability to remove or add layers based on what they have to process. By loosening bolts, these plate or disk layers can be quickly and easily removed, exchanged, or assembled in a different sequence. Each layer can be turned and put at any level in the stack. Designs permit space between packs for thermal changes and individual heating using internal as well as the external heating elements. Sealing forces during assembly between individual plates are set with precision, avoiding problems in spiral arrangements of high axial forces generated by large ring surfaces between the plates.

Conventional dies also have versatility. Most dies have interchangeable lips which essentially allow the conversion of die size, such as going from a 15 cm to a 20 cm (6 in to an 8 in) size. There are also those that have the capability to go from coextrusion to monolayer. Entry ports not being used can be insert with adapter blocks.

There are die designs that combine conventional with stacked systems. This combination permits using many layers of plastic and simplifying their flows, particularly inner layer flow instabilities. The conventional section, with concentric annular passages of the innermost three layers, ends in a common annular passage. The upper section of the die acts like a stacked die, providing a lower wetted surface area compared to conventional mandrels. The side fed distribution system reduces melt flow variations.

An example for start-up of a Battenfeld Gloucester radial fed die (Fig. 7.17) is as follows: (1) heat die to operating temperature, allow ample time; (2) bring film into correct gauge, adjust H bolts around the die assembly where needed; (3) if adjustment with H bolts is difficult, then very slightly loosen all D bolts (those which protrude) around the die until adjustment of the film gauge can be made with the H bolts [D bolts are torqued at the factory and should not require initial adjustment. If D bolts are loosened, plastic flow must be stopped due to the possibility of leakage. Do not loosen D bolts any more than necessary for ease of H bolts adjustments; if D bolts are loosened too much die will leak. When adjustment is made, retighten all D bolts to a torque of 34 kg (75 lb)]; (4) use a good brand of high temperature anti-seize compound, such as FEL-PRO (Hi-Temp) C-5A, part number 51007 on all bolt threads when needed. Use medium consistency Dow Corning 44 silicone grease on all Teflon seals when

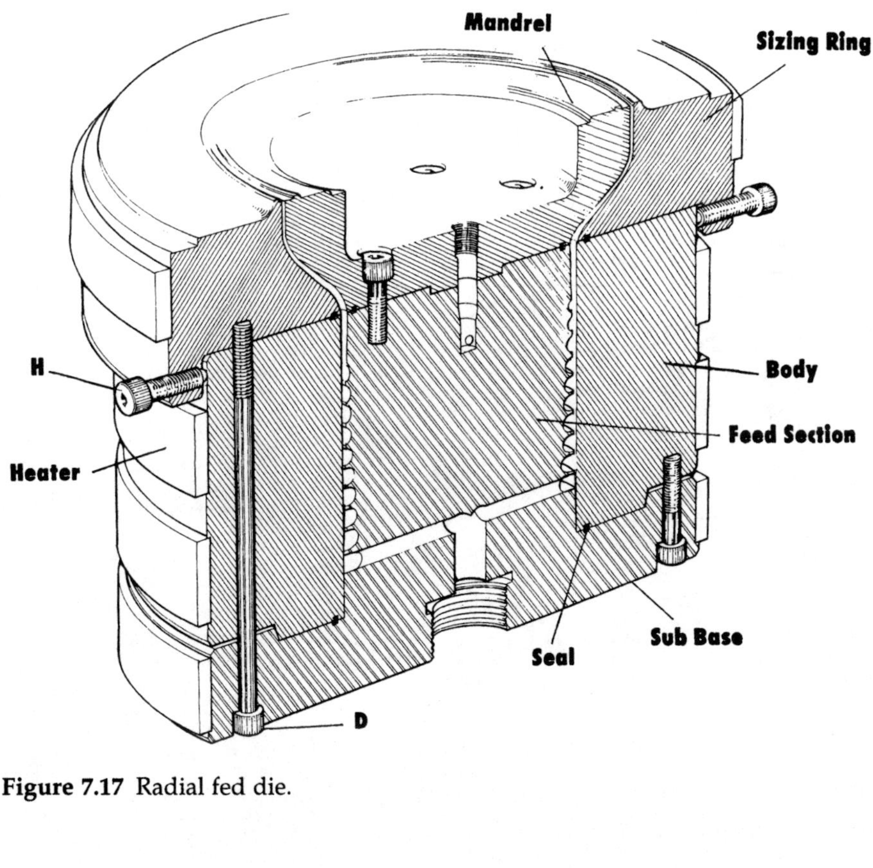


**Figure 7.17** Radial fed die.

needed. These products are not intended for set-up of the die assembly, they are used during reassembly; and (5) keep die clean. Take all precautions not to damage die lips. Use brass or copper tools for cleaning.

## FILM RANDOMIZATION

With blown tubular film, there is always an unavoidable degree of undesirable thickness variation. There are different approaches to this problem, which usually are related to performance requirements and cost. The oscillating haul-off systems provide the best way to randomize film gauge variation for many applications.

Thickness randomization is readily understood and accepted in the industry (Fig. 7.18). Imperfections in circular dies and air rings with process variables cause variations to develop. A major problem with the variations occurs when large rolls are wound in mill roll production. They cause ridges and other roll defects. When the roll is unwound, these

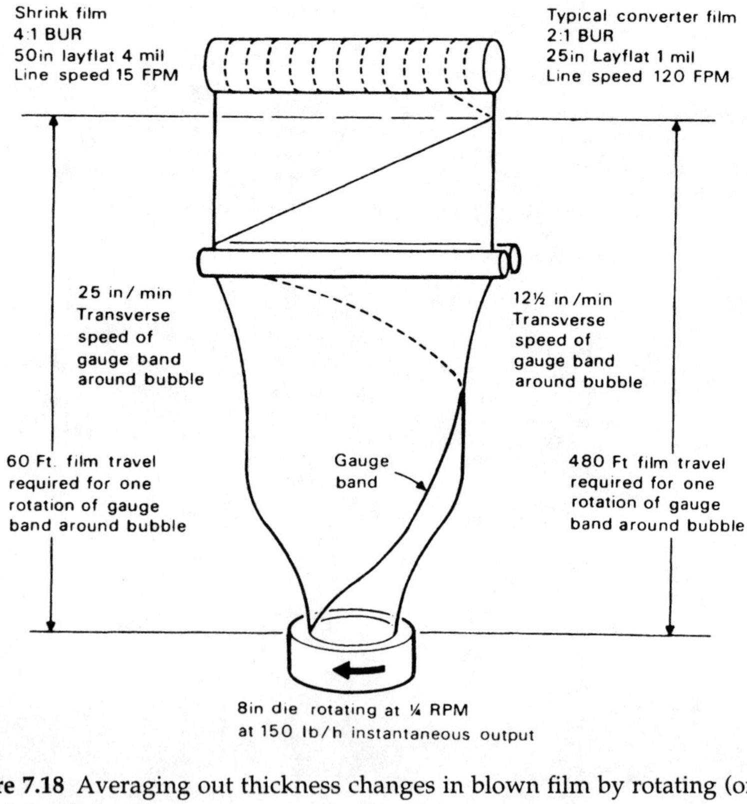


**Figure 7.18** Averaging out thickness changes in blown film by rotating (or oscillating) the diehead.

problems can cause the film to be unsuitable for use in high-speed converting processes.

Film gauge is affected by several factors that include the extruder performance, die performance (output rate uniformity, gap spacing, melt temperature, residence time, drop pressure, etc.), take-off speed, blow-up ratio, and rate of bubble cooling. Among the surface defects, fish eyes are due to imperfect mixing in the extruder or to contamination. Both of these conditions are controlled by the screen pack that creates a back pressure which can improve the melt homogenization (Chapter 2).

An advantage of blown film extrusion over flat film extrusion is the ability to produce film with a more uniform strength in both the machine direction (MD) and transverse direction (TD). In flat film extrusion, particularly at high take-off rates and not using tenter orientation equipment, there is a relatively high orientation of the film in the MD and a very low orientation in the TD. In blown film by balancing blow-up ratios against takeoff rate, it is possible to achieve physical and other properties which are very nearly equal in both directions such as giving a film maximum toughness.

Another advantage of blown film (with a tight tolerance on thickness) is in bag production. It only requires, with the proper size blown tube, a seal across the bottom of the bag, whereas with flat film either one or two longitudinal seals are also necessary. Blown diameters can be produced giving flat film widths that are much wider than anything produced by flat slot-die extrusion; however, tight thickness tolerance is desired to ensure proper performance and minimize the amount of plastic consumed, in order to reduce cost. In addition to packaging, such large width PE film has found extensive use in other markets, such as the building, agriculture, and horticulture industries

The rotating or oscillating die system is by far the most common method of gauging randomization because of its rather low cost and mechanical simplicity. In most cases, the air ring is mounted on the die so that the inner part of the ring rotates with the die. Meanwhile, the inlet chamber remains stationary. In the past, the majority of dies have had a full 360° rotation. Most of the dies now are oscillating, eliminating collector rings and therefore simplifying the maintenance of the equipment. Another gain is improving temperature control in the die.

The distribution across the width of a roll is usually good for many applications. However, the point to point variations in thickness may cause difficulties in certain converting operations, such as bag production. When producing bags, sticking to seal bars is likely to occur when the equipment is adjusted for normal thickness and then relatively thin layers come together under the sealing bars.

Unfortunately, the oscillating die only randomizes the mechanical and melt variables caused from the die to the air ring. It does not address the gauge bands caused by external effects, such as air ring irregularities (lip irregularities, variable air flow and temperature, etc.), ambient drafts, bubble alignment forces (cages, guides, collapsing frame, etc.), tower, drafts, melt channeling, and/or other effects above the air ring. With the proper bubble shielding and careful alignment, these effects can be minimized. However, when wrap-ups are on the larger rolls, even the slightest stationary effects will cause roll defects.

The oscillating die approach is primarily suitable for single layer and certain coextruded structures. It is not the optimal for the more sophisticated coextrusion constructions. Coextrusion dies, being more complex, require additional concessions for the collector ring/oscillator assemblies, rotating IBC air plenums, bearing packages, distribution blocks, die increases in height, increased resident time, increasing pressure drop, and troubleshooting/maintenance. By the substitution of an oscillating haul-off, the die has more freedom in its design. In turn, the relatively simplified die design results in significant benefits processwise and quality of the film [120–122].

In the rotating extruder gauge system, an extruder and die are mounted on a platform which oscillates in a complete circle. This method is better than the oscillating die system in that practically all the gauge irregularities are randomized, except drafts and sags in the collapsing frame. A disadvantage with rotating an extruder is that only works with a small machine. To date, to rotate a large machine would be both impractical and very expensive. Also, because the hopper rotates 360° with the platform on which the extruder is mounted, the plastic can be fed in only one position. Therefore in emergencies, the plastic must be hand fed. Another problem is that the equipment must be tightly packed into a relatively small platform, so maintenance tends to be awkward.

The rotating tower-winder system places a rotating winder on top of the tower. It could be best for randomizing any gauge irregularities, except the sag in the collapsing frame. This method is suitable when converting occurs on the top of the floor of a multi-story building and extrusion is in the basement. It eliminates transportation of heavy rolls from the tower to the converting level, which is relatively expensive if it is done by an elevator. It also becomes dangerous if a hoist is used. Since blown film extrusion is rarely done in multi-story buildings, this method is not widely used.

With a floor level rotating or oscillating winder system, the winder is placed on the ground floor and the extruder on top of the tower. This method employed by the Europeans during the 1970s and 1980s is not widely used today, except for special applications, such as biaxial oriented films.

Placing winders on the ground floor solves the problem of roll handing. However, placing the extruder and die usually 9 m (30 ft) above the ground can complicate accurate process monitoring by the machine operator. Also, due to melt elongation, it is difficult to extrude lower melt strength plastics downward. The stress applied on the melt by the weight of the bubble reduces dart drop and tear strengths of higher molecular weight plastics due to the lesser degree of molecular relaxation. This is the reason why downward extrusion replaces upward extrusion systems when processing plastics such as HDPE.

In the early stages of blown tubular film processing, there was the oscillating cage system. It is still used for special applications, such as biaxial oriented film in combination with a rotating die. In order for this method to work, the bubble must be held firmly above the frost line with a set of rollers that are driven to oscillate 180°. Relatively minor thickness irregularities occur above the die and air ring equipment. These flaws are caused by air drafts usually so minute that they are difficult to measure. They do not rotate around the bubble and create thick or thin sections on the roll surface as they continuously overlap. As roll diameters enlarge, the severe effect of even these minor thickness irregularities increases, latter showing up as hard bands in the roll and limiting the films application.

The oscillating haul-off or nip system was introduced during the 1970s via patents from Windmoeller & Hoelscher Corp., Germany [50]. It eliminates practically all the disadvantages that plagued other methods of film randomization for almost all applications. Aside from extruding downward into a rotating winder, which is structurally impractical, only oscillating haul-off can randomize post-extrusion problems on gauge variation.

Both the extruder and the winder are on the main floor. Extrusion

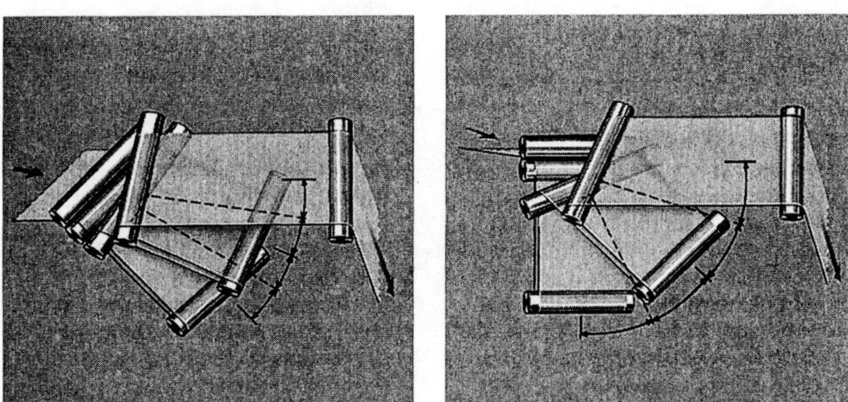

**Figure 7.19** Schematic layout and operational principle of the W&H 360° oscillating haul-off with horizontal arranged turning bars.

occurs upward, and the oscillating nips randomize all gauge irregularities, except those that may occur when collapsing takes place at the nip rolls; none of the systems reviewed achieved this either. Figures 7.19 and 7.20 are examples of this system where the different equipment manufacturers provide different capabilities.

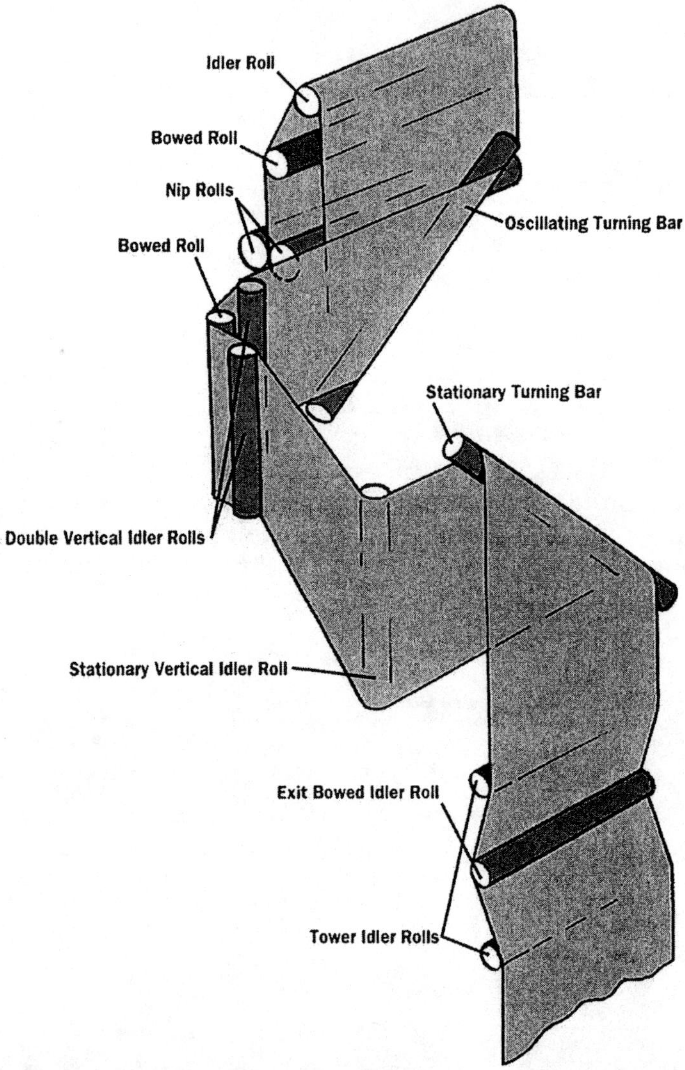


**Figure 7.20** The B-G Traversanip oscillating haul-off eliminates the need for rotating dies.

A short melt flow path resulting in a minimum melt residence time is achievable with a nonrotating die and oscillating haul-off. After leaving the sizing cage, the film bubble is collapsed by the conventional wood slat collapsing frame as it is being pulled by the nip rolls and passes into the oscillating haul-off. The film wraps around air turning bars, etc., so that the haul-off geometry is correct resulting in no stress in the film.

Figure 7.19 shows a Windmoeller & Hoelscher Corp. 360° oscillating haul-off with horizontally arranged turning bars. It has driven roller flattening arrangements. Two of these rollers are air turning bars with anti-adhesive coating, as well as motorized adjustment. The long length of the bubble collapsing section along with the optimized infeed angle ensures good flatness of the film tube. Simultaneously, it minimizes the risk of edge wrinkles. Its modular construction offers the opportunity of adding further modules at any latter time. If production changes to shrink films instead of the more conventional film being processed, quick changes can be made.

The 720° oscillating haul-off system from Battenfeld Gloucester Engineering Co. (Fig. 7.20) provides precision to the collapsing, flattening, and haul-off for smooth film winding. It uses turning bars with fixed angles to eliminate the problem of varying residual stresses. Automatic lay-flat electronic measurement control provides true diameter of the blown tube. If deviation occurs, the control quickly compensates for even the slightest change of setpoint. Variations in collapsing are avoided and scrap is reduced. The oscillating device can be raised or lowered. It can be up when running film that needs extra cooling, such as LLDPE, coextrusions, or when operating close to extruder capacity. It can be lowered when processing wrinkle-prone films that need to be collapsed warm, such as HMWHDPE.

Its slat collapsing frame with interlacing side guides provides the proper geometry to collapse nonextensible plastic films, such as HMWHDPE, without wrinkles or creases. A low-friction plastic slat cover can be used to further reduce drag and bagginess. Segmented roller and air board collapsers can be used for special applications.

The industrial oscillating haul-of or nip systems allow for easy multiple layer extrusion and, with the exception of the rotating die, are the most popular method for gauge randomization. Disadvantages when compared to the rotating die include the cost and greater head-room requirements. When used for coextruded or multi-layer applications, the films produced from the oscillating haul-off system provide better sealing.

There are basically two types of oscillating haul-off systems and, in turn, each has many variations. One has horizontal mounted turning bars that use less head room but the threading operation is more complex. Any little misalignment of the bars creates wrinkles in nonstretchable plastics.

**Table 7.6** Troubleshooting common film defects

| *Problem* | *Cause* |
|---|---|
| Poor strength | Extrusion temperature too low or too high<br>Thin spots in film<br>Low blow-up ratio (in blown-film making)<br>Unequal molecular orientation<br>Excessive pressure or temperature, or both, at the nip rolls |
| Poor clarity | Extrusion temperature too low<br>Inadequate cooling<br>Blow-up ratio too low (in blown-film making)<br>Unsuitable plastic |
| Wide gauge variations | Non-uniform temperature at the die opening<br>Non-uniform flow at the die opening (probably caused by 'surging',)<br>Non-uniform cooling across the film |
| Film structure defects such as 'applesauce' | Extrusion temperature too low or too high<br>Poor mixing<br>Poor screw design. |
| Other film defects such as gels and 'fisheyes' | Poor mixing<br>Flaking caused by a dirty screw or barrel, or both<br>Insufficient purging after changing resins<br>Contaminated resin due to lack of cleanliness in the shop, mixing the resin with too much scrap or reground polymer, faulty start-up or shut-down<br>Plastic hang-up |
| Streaks in film | Inadequate mixing<br>Plastic or foreign matter held up in the die<br>Impurities in the die lands<br>Scratches from the windup |
| Wrinkles on the windup roll | Gauge variations caused by die or cooling defects<br>Insufficient or unequal cooling<br>Non-uniform bubble (in blown-film making)<br>Sticking to the guide rolls or forming tent (in blown-film making)<br>Air currents in the shop, causing film bubble vibration<br>Take-off tension too high or too low<br>Poor alignment of take-off equipment with the die |
| Poor winding (aside from wrinkling) | Non-uniform gauge<br>Full windup tension control – film roll should be *reasonably* tight<br>Excess of slip additive in the resin, resulting in 'telescoping' (generally beyond the operator's control)<br>Air turbulence or drafts around the bubble (in blown-film making) |

**Table 7.6** *Continued*

| *Problem* | *Cause* |
|---|---|
| | Inadequate pressure between the nip rolls (in blown-film making), resulting in air loss from the bubble into the wound-up film<br>Inadequate windup equipment<br>Leak in the valve at the air supply to the bubble (in blown-film making), resulting in increasing film width on the windup roll |
| Excessive blocking (film layers on the windup roll sticking to each other) | Inadequate film cooling (In blown-film making, supplying water to the driven nip roll may help.)<br>Distance between die and nip rolls too small to permit additional film cooling (in blown-film making)<br>Nip roll pressure too high (in blown-film making; it should not exceed 1.4 $kg/cm^2$ (20 $lb/m^2$))<br>Tension at windup too high<br>Build-up of static electricity (especially when making very-thin-gauge film); remedy: install a static eliminator<br>Shop room temperature too low, resulting in warm film shrinking on the windup roll<br>Not enough antiblock additive in resin (beyond the operator's control. – This will probably be the cause for blocking if all previously mentioned causes can be ruled out) |

**Table 7.7** Troubleshooting blown film dies

| *Problem* | *Cause* | *Solution* |
|---|---|---|
| ***Blown film extrusion*** | | |
| Thickness variations across the extrudate circumference | Erratic melt quality | Check screw design and/or wear<br>Check set temperatures<br>Check heaters and thermocouples |
| | Surging or feeding inconsistency | Check screw wear<br>Check material in hopper<br>Check regrind percentage |
| | Dirty die | Clean die (check for obstructions at the die lips) |
| | Inadequate die-bolts adjustment | Readjust |
| | Misaligned die/air ring | Center die to nip rolls<br>Center air ring to die |

**Table 7.7** *Continued*

| *Problem* | *Cause* | *Solution* |
|---|---|---|
| | Leaky die | Check seals |
| | Temperature fluctuations at the die lips | Check heaters and thermocouples<br>Check set temperatures |
| | Poor air flow distribution in the air ring | Check and clean air ring |
| Lines, streaks and foreign specks | Dirty die | Clean die (check for obstructions at the die lips) |
| | Scratched die lips | Repair, or replace die lips |
| | Inadequate die-bolts adjustment | Readjust |
| | Contaminated melt flow | Check compound<br>Change filters |
| | Melt flow too hot | Reduce set temperatures |
| | Welding lines | Increase melt temperatures<br>Use spiral mandrel die |
| Sharkskin, melt fracture | Melt temperature too low | Increase the die lips' temperature |
| | Friction at the die lips | Repair die lips' coating<br>Modify formulation |
| | Die gap too narrow | Increase |
| Bubble instability/irregular frost line | Erratic melt quality | Check screw design and/or wear<br>Check set temperatures<br>Check heaters and thermocouples |
| | Dirty die | Clean die (check for obstruction at the die lips) |
| | Excessive air ring velocity | Reduce |
| | Insufficient blow-up ratio | Increase |
| | Excessive melt temperature | Reduce set temperatures |
| Wrinkles | Misaligned die/nip rolls | Align die to nip rolls |
| | Non-uniform cooling and winding | Check winder |

**Table 7.8** Troubleshooting film turret winders

| *Problem* | *Cause* | *Solution* |
|---|---|---|
| Baggy edges | Bowed roll angle out of adjustment | Adjust |
| Blocking | Excessive tension | Adjust |
| Bumpy roll | Excessive tension | Adjust |
| | Flawed winding shaft | Replace |
| Core collapse | Film wound too tight | Reduce tension |
| | Excessive layon roll pressure | Adjust |
| Floppy web | Insufficient tension | Increase |
| | Rolls out of alignment | Align |
| Fluctuating/ uncontrollable tension | Faulty drive | Replace |
| | Bearing sticking | Check roll-turning resistance |
| | Unbalanced roll | Check linkage and cylinder for triction |
| | Dancer potentiometer malfunction | Check pot and wiring |
| | Moisture in dancer's pneumatic components | Check and replace as needed<br>Filter plant air |
| | Force transducer problem | Check transducer and wiring |
| Fuzzy roll end | Dull blades | Replace |
| | Blades not parallel or perpendicular to web | Check blade adjustment |
| Hard roll | Excessive winder tension | Decrease and/or increase taper |
| Scratched film | Damaged, scuffed rolls | Replace |
| | Rolls not turning | Check roll drag and bearings<br>Check roll speed<br>Check roll balance |
| Soft roll | Insufficient tension | Increase and/or decrease taper |
| | Insufficient layon pressure | Increase |
| Starring | Tension too high | Decrease |

**Table 7.8** *Continued*

| *Problem* | *Cause* | *Solution* |
|---|---|---|
| Telescoping | Tension too low | Increase |
| | Layon roll force too low | Increase |
| | Rolls misaligned | Check all rolls in turret and winding areas |
| | Incorrect taper | Adjust |
| | Incorrect nip drive tension | Adjust |
| Tension differentiation across web | Bowed roll over adjustment | Adjust |
| Uneven film width | Tension varies or is too high | Check tension and taper |
| Wrinkles | Rolls out of alignment | Align |
| | Film too hot (atter treating) | Install chill roll in winder |
| | Overadjusted bowed roll | Adjust |
| | Web tension too low | Increase |

The other type is the vertically mounted turning bars where they are simple to thread and easily maintain alignment. Where space is critical, a turning bar can be mounted downward to save head room space.

## TROUBLESHOOTING

As reviewed earlier, blown film problems have many sources. High on the list are temperature deviations or variabilities, poor tension, and contamination somewhere in the line. Tables 7.6–7.8 lists film, die, and turret winder problems, their causes, and some recommended solutions.

# 8

# Flat film

## OVERVIEW

Flat film is also called chill roll film, roll cast film, cast film, slot cast film, water chill film, etc. A process similar to flat film processing is the casting process to produce film (and sheet) from highly fluid solutions. The usually higher viscosity of TP melts results in differences in machine design as well as in operational techniques. The most popular process used to produce the flat film is with the chill rolls. Water chill or quench films is also a popular process.

The flat film chill roll process usually has low melt viscosity melt extruded through a coat hanger or T-shape slot die (Figs. 5.7 and 5.26) that may be at least 3 m (10 ft) wide. The extrudate is chilled below the melt temperature ($T_m$) or the glass transition temperature ($T_g$) by passing it principally over two or more chrome plated chill rollers (that have been cored for water cooling) or a water bath (Chapter 3). As shown in Fig. 8.1, the extrudate web from the die is made dimensionally stable by contacting several chill rolls prior to being pulled by the nip or pull rolls and wrapped around the windup roll.

Extrudates can leave the die downward as a highly viscous melt that must be pulled away from the die (Fig. 5.8). They may also be led directly onto the upper surface of the chill roll. Cooling in contact with the polished surface of the roll imparts a good surface to the film, while the other surface relaxes as it cools in air. The roll forms a means of haul-off by surface friction and its speed controls drawdown or neck-in and final film thickness [122, 196, 260, 310, 359].

When the hot film is drawn down onto the cooled first chill roll, it will shrink at the edges. The amount of neck-in and beading varies for the different plastics. As shown in Fig. 8.2, this neck-in is the difference between the hot melt width at the die lips and the film width on the chill roll. This diagram also indicates the distance between contact line and

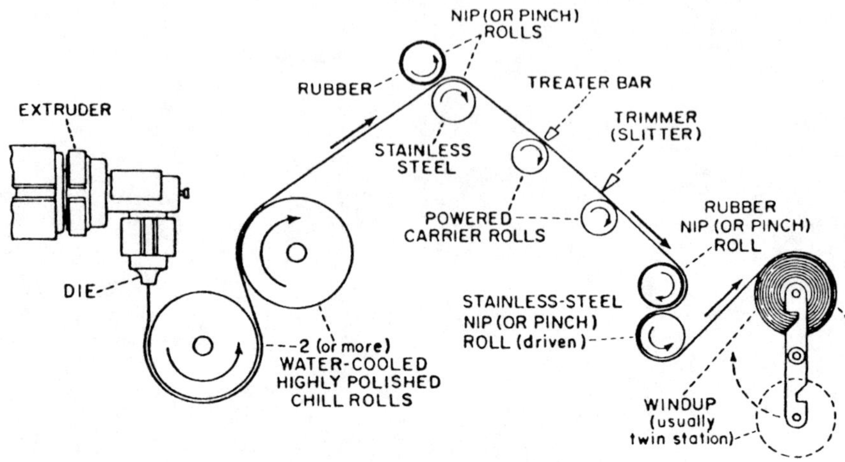


**Figure 8.1** Chill roll system for flat film extrusion line.

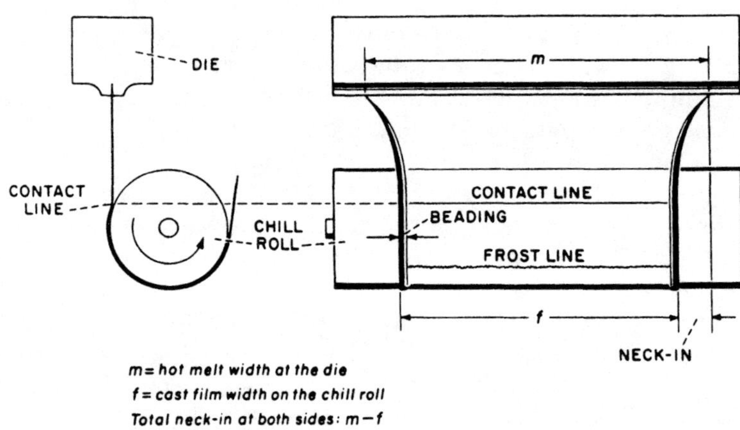


**Figure 8.2** Neck-in and beading on the chill roll.

frost line on the roll. With this neck-in, beading takes place. It is the thickening at both edges of the film. Later in the line, this bead is trimmed. The relatively high degree of drawdown over a short distance requires high elongation with generally high melt temperatures. As many films are manufactured from crystalline (or semicrystalline) plastics, the rapid cooling tends to suppress crystallization. This molecular action is beneficial in providing high elongation of the cooled film for subsequent drawing/orientation processes.

Because stretching occurs, the opening in the extrusion die is from 5–30% oversized to accommodate the drawdown. The speed of the web through the cooling system is controlled by the pull rolls. The space between the die and puller contains various cooling devices, shapers, rollers (Fig. 8.3), slitters (Fig. 8.4), perforators, finishing tools, corona treatments, and/or other devices that impart various qualities to the finished product.

The die is adjustable so that it is positioned as close as possible to the large diameter casting or chill roll. Most of the heat in the film is removed by the chill casting onto the roll. It is the same distance from the first chill roll across its whole width. Adjustment is made so the angle between the die and the roll is constant. The molten web should pull straight out from the die to the surface of the roll. The length of this distance depends on the plastic being processed. As an example, for LDPE the gap could be about 50–75 mm (2–3 in); for HDPE, the gap could be as short as 13 mm (0.5 in). However, as reviewed later, more space is required when using devices such as air knives.

The cooling and annealing of the extrudate requires special attention in order that the web receives the proper cooling without any distortion. Developing final properties depends on the cooling behavior and often determines the performance quality of the product. Suppliers of extrusion lines, as well as water-cooled rolls, provide information on providing the required chilling action necessary for the plastic type and thickness required.

The first cooling or chilling roll is an important part of the process. The roll is made from steel and has a mirror-like chrome plated surface. The rolls can also provide matte or embossed finishes desired for certain runs.

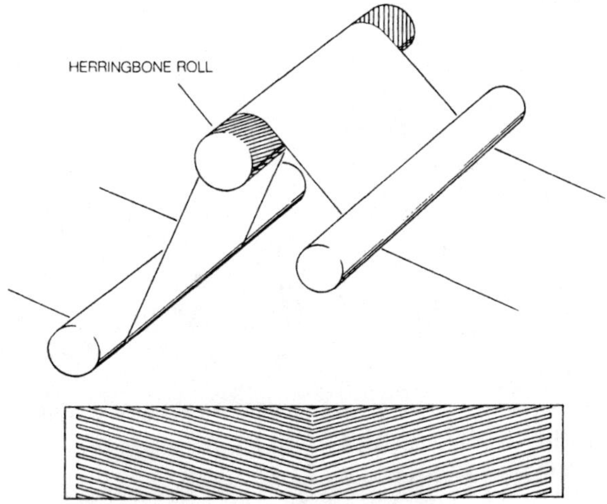


**Figure 8.3** Grooved metal roller using herringbone pattern; this idler roll will assist in reducing the apparent wrinkles occurring from an irregular web.

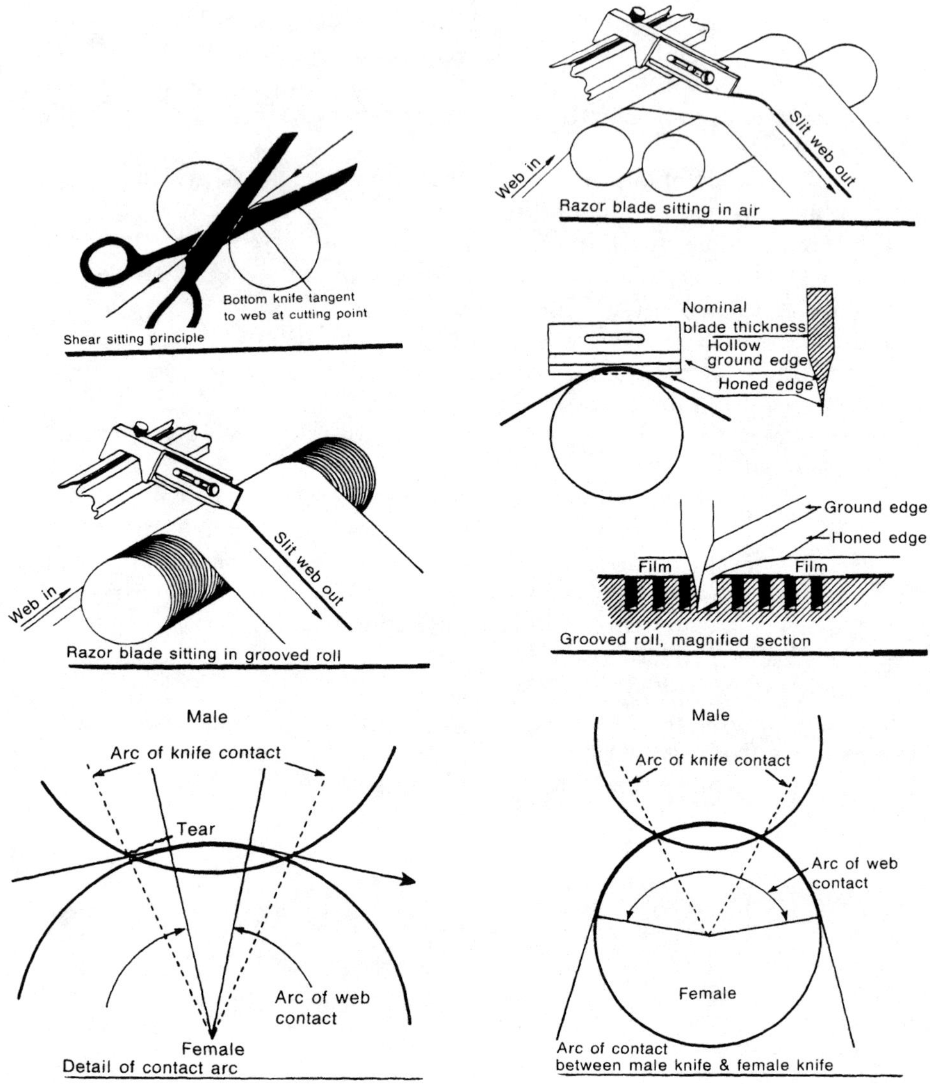


**Figure 8.4** Example of slitting extruded film.

Figure 8.5 is an example where colored PVC is going over a woodgrain engraving cylinder. The product is used in such applications as wall coverings, electronic appliances, and furniture.

During manufacture, this expensive roll is basically ground smooth, polished, and chrome plated. This roll as well as the other rolls have to be treated with care during operation on line as well as when idle so that it

**Figure 8.5** Woodgrain embossing/engraving roll.

is not damaged. Proper cleaning and maintenance is a must. The film reproduces the surface on which it is cast, and any imperfections on the roll are reproduced on the film.

The temperature at which neck-in occurs is critical. The usual practice is to establish a skin on the surface of the web. The result is establishing dimensional stability and maintaining the skin.

Pull roles are generally located as far down the line as possible so that the extrudate will not be marked or distorted by the pressure of the cooling rolls. Certain secondary operations that may be included in the line, such as surface treatments, occur after the pull rolls because the puller might distort the treatment.

Besides the first main chill roll, other down-stream equipment can include additional cooling rolls, carrier rolls, deflecting and transverse stretching rolls, post-cooling rolls, idler rolls, dancer rolls, surface treatment units, edge trimmers, thickness gauges and other measuring devices, static relief devices, and windup rolls. High-peed plants require centrally or combination driven winders. These must be programmable so that they can be matched to the product being wound up. Although it is best to wind the film at the expected storage temperature, this is often impractical. Therefore, it is necessary to take into account the expected shrinkage as well as any degree of post-crystallization.

The windup roll for a cast film five layer coextrusion line for stretch film recently set a world record (Fig. 8.6). This line at the Chapparrel Films Inc. of ITW, Mauriceville, TX., USA is operating a Battenfeld Gloucester turret winder at line speeds of 585 m/min (1920 ft/min). It runs 1.5 mm (0.060 in) film up to 356 cm (140 in) wide which is 20% faster than their other stretch film line [137, 310]. The multi-layer metallocene technology PE film is used mainly for pallet wrap applications.

New winding technology was used to mitigate the two winding effects of deflections and vibrations that influence and limit increasing line speed shaft. The technology includes an overlapping turret design with cantilevered shafts that swing out opposite each other. This dual split design allows the shaft length to be reduced to 183 cm (72 in) on each side. This increases the so-called critical frequency of the shaft, essentially permitting the line to operate with no winding limitations.

This low-noise turret winder provides speed control web tensioning with load cell monitoring, programmable logic control, automatic adhesiveless roll transfer on the fly, etc. It is part of a Battenfeld Gloucester cast film extrusion line that starts with five air-cooled extruders: two 4.5 in, 30/1 units and three 2.5 in, 30/1 units. The casting unit has a 100 cm (40 in) diameter by 3.8 m (148 in) wide primary chill roll and 261 cm (4 in) diameter secondary roll. What is unique, and permits the high speed, is their in-line slitting winder that allows one-second indexing with repeatable

**Figure 8.6** Very large dual turret winder.

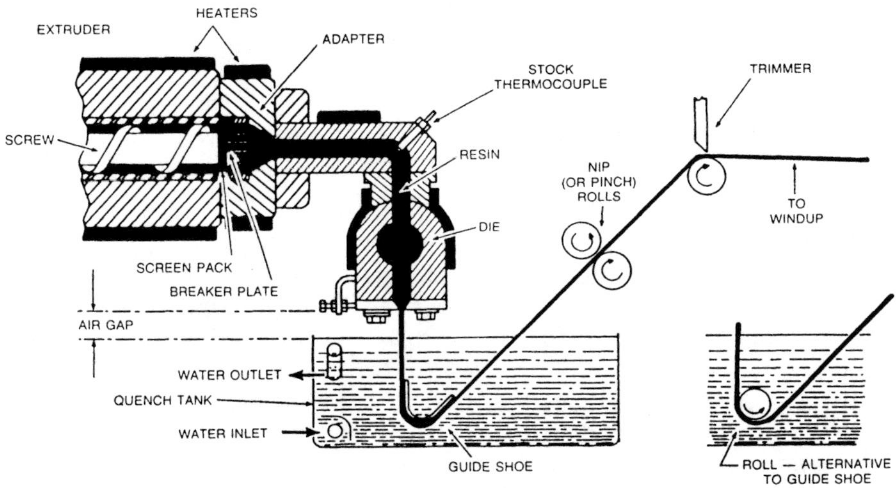


**Figure 8.7** Schematic of water quench tank system for flat film extrusion line.

positioning. The line runs typically six-up 50 cm (20 in). Another feature of this line is that it generally runs at half the neck-in than their other line.

In the past, before the development of adequate chill roll systems, clear flat film was principally made by extruding vertically downward into a water bath. This is a way to produce products such as tapes and packaging straps. Figure 8.7 is a schematic of the water cooling flat film extruded process. From under a polished guide 'shoe', roller, or rod (acrylic, glass, or plastic coated metal) in the bath, the cooled film is pulled out by a pair of nip or pinch rolls.

The rapid water quenching produces good optical properties, and equipment cost is low. However, it can be difficult to get precise control over the water temperature. Vibrations and currents can cause little marks on the film. A critical point is the necessity of maintaining a smooth surface in the water quench tank where the melt first enters. Different devices are used to control the flow of water, such as baffles with openings. It has serious limitations when high production speeds are attempted; the water must be kept from carrying over into any on-line pretreatments and the finished roll. However, these problems can be controlled. This liquid bath system has been used for blown tubular film, where it is inside the blown film to improve cooling characteristics.

Flat film extrusion equipment is used to fabricate thin gauge film usually about 0.025 mm (0.001 in) or less and up to about 0.10 mm (0.004 in); some industries go to 0.25 mm (0.010 in) before it is called sheet. The plastic melt is extruded through a wide die as a thin web followed with a cooling system. If preference is to be given to a flat film extrusion line rather than to the usual less expensive blown tubular film line, the equipment must be very efficient or be capable of making a product whose qualities cannot be achieved by blown film. Flat film is used for making crystalline clear film from plastics that include HDPE, PP, and nylon.

## Other type film

Film is made to meet different requirements. Figure 8.8 shows film tapes fabricated by different processes. There is a film casting process that produces unsupported film or sheet by casting a fluid plastic compound or dispersion on a temporary carrier, such as an endless belt or rotating drum. The plastic is stabilized by heat fusing, evaporation of a solvent, and/or allowing the melt to cool. The final step is removal from its substrate.

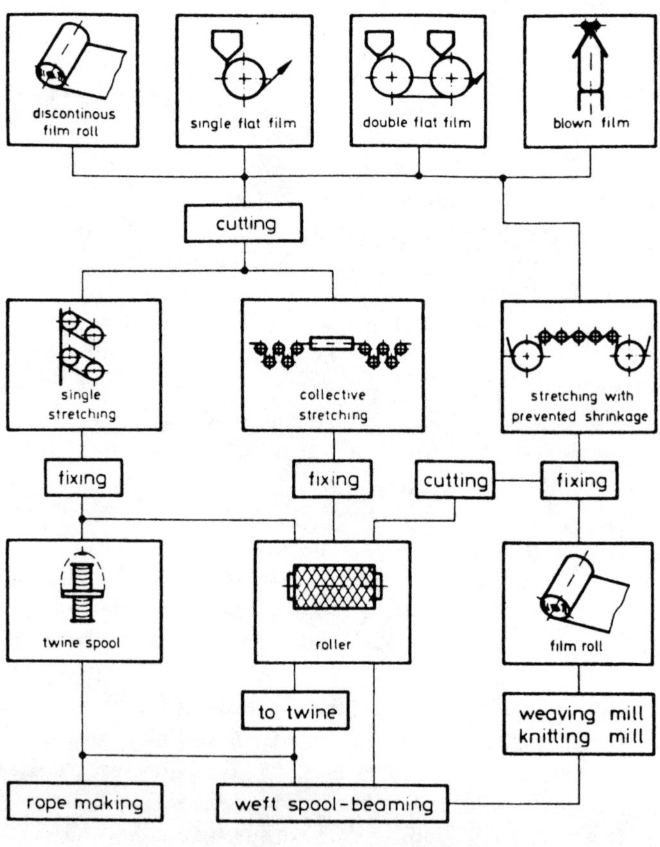


**Figure 8.8** Examples of film tape fabricating methods.

A specialized process for producing film is skiving. It consists of shaving off a thin film or sheet layer from a large block of solid plastic, usually a round billet. Continuous film is obtained by skiving in a lathe type cutting operation which is similar to producing plywood from a tree trunk log. This process is particularly useful with plastics that cannot be processed by the usual plastic film processes, such as extrusion, calendering, or casting. PTFE is an example as it is a plastic that is not basically granulated and reprocessed.

PTFE powder is put into a mold to make billets. Powder is compressed uniformly (carefully) at pressures of 14–34 MPa (2000–5000 psi). This preform is removed from the mold and sintered by heating unconfined in an oven at temperatures 360–380°C (680–715°F) for times ranging from a few hours to several days depending on the size and shape of the billet. Billet sizes go from 1 to 726 kg (2 to 1600 lb), among the largest TP moldings made of any plastics. Time with temperature variation during cure is closely controlled; the target is a final cure without voids and other defects that periodically usually occur internally. Defective billets are not conducive to good film skiving and are not recyclable.

### Flat or blown film

In certain respects, chill roll flat film production is similar to that of blown sheet, though thickness variations due to mechanical distortion of the die assume greater importance. A die deflection represents a greater percentage on a thinner film, and the increased die resistance leads to higher melt pressures. Higher velocities are also required to give similar outputs. This problem is usually resolved by using high drawdown ratios of 10/1 or more on thickness using wider die gaps of about 0.5 mm (0.002 in). The result is lower die velocity and pressure, resulting in improved percentage thickness tolerance. The thinner films are also more flexible, so the unsupported gap between die lips and haul-off/cooling roll must also be reduced.

Whether one uses flat or blown film depends on the applications, and thus performance requirements (transparency, etc.), length of production run, and cost. The main advantage of flat film is the intensive cooling accomplished by means of chill rolls or water. There are few difficulties in the melt and blocking rarely occurs. Another advantage is that there is no flattening operation; the problem of maintaining a constant bubble volume does not exist. Control of thickness across the width of the film is simpler. The winding and printing of the flat film is usually simpler. Basically flat film provides better gauge uniformity, better embossing quality, higher rates of production per total web width, lower density, and softer film.

However, as reviewed in Chapter 7, blown film capabilities can make many of these advantages for flat film less significant. Blown film can process those plastics that go through flat film lines and in addition those plastics with other melt characteristics. The blown film does not require edge trimming or requires less regrind, has better balance of mechanical properties, denser film, stiffer, and improved barrier properties

## PLASTIC MATERIALS

Virtually every TP material can be used in film form either as a monolayer or coextruded/multi-layer. In selection of a film for a certain application, the properties of the TP is considered. The different TPs can have different thermal properties, molecular characteristics, degree of crystallinity of the plastic, etc., that affect processing and film properties. Additives influence extrusion and orientation improve film properties

The melt temperature used for the flat film (or sheet) line processing, such as LDPE, can be 50–70°C (120–160°F) higher than that used in the blown film line processes. The higher temperatures and correspondingly reduced viscosities are permissible because the flat film can be cooled relatively simply and rapidly, as well as being easier to manipulate because no flattening operation is required. The high fluidity of the melt not only facilitates an even flow distribution over the die slot, but also minimizes orientation effects during its elongation.

This higher temperature with rapid chilling also contributes to providing an amorphous plastic with superior optical characteristics when compared to blown film made from the same plastics. The melt in such a flat film line has low elastic forces and the elastic memory is short. This melt temperature behavior helps for making a film with good optical properties.

Certain LDPE flat films when abruptly chilled, have lower rigidity and higher elongation than blown film made from the same plastic. This action can be attributed partly to the more amorphous structure and partly to the absence of transverse stretching. The strength and rigidity can be increased by increasing the cooling temperature.

The density of PE film compounds, the major plastic used, is important since it is directly related to stiffness. This property is important in the major packaging market for PE film that requires stiffness in order to handle the film in packaging machinery. Different applications will require different densities. This casting process produces amorphous material that has lower density then blown film. Another property to analyze is the melt index (MI). With a low MI, such as 1.5, the melt is tougher and less likely to tear; high MI, such as 6.0, the melt will be easier to draw down to very thin film (Chapter 19).

Much of the extruded vinyl film is plasticized PVC. Its plasticizer(s)

must be carefully chosen to meet performance requirements. Composition is important since PVC is always susceptible to degradation. The stabilizer(s) used is a critical additive since the extrusion process adds considerable heat to the compound. Rigid unplasticized PVC compounds have the most severe stabilizer requirements, as they are most viscous and develop heat during extrusion. Particularly for these rigid compounds, lubricants are added to make extrusion easier and better controlled. Also, a small amount of plasticizer, such as 1–5 wt%, may be added without impairing stiffness. Care is required because small amounts of certain plasticizers can have an unexpected embrittling effect on the PVC.

An interesting situation exists with nylon, such as nylon 66. When water quenching, the nylon film leaving the nip rolls has a moisture content of 3.5–6%. Drying is necessary before windup to bring the moisture content down to a level which is typically at 50% relative humidity; for nylon 66, at about 2.5% moisture. An advantage with the water quench process is that the finished rolls of film are essentially in equilibrium with ambient air. With the cast chill roll and blown tubular film processes, film is essentially dry on being wound.

The properties of nylon film produced by the water quench process are similar to those obtained in a cast film process except that its crystallinity produces higher yield strength and stiffness.

To optimize film properties, it is very common to use coextruded composite plastic films. Figure 8.9 shows a B-G 3-layer coextruded cast film line. Each of the layers fulfill separate functions (Chapter 2). This approach permits using thinner and lighter weight film constructions than single material film. An example is a three-layer packaging film. Its outer layer is LDPE that provides good appearance and printability. The center layer is MDPE which ensures adequate stiffness despite the lower overall thickness of the total film. The outer layer is made from a polyolefin produced by metallocene catalysts (mPE) that permits extremely low heat-sealing temperatures and thus fast cycling on automatic packaging machines.

The coextruded films are made to meet all kinds of requirements. Turnkey systems are used to produce from two to at least seven layers of plastics. They can include orientation to provide additional advantages. Chapter 2 provides more information on orientation.

There is a multi-layer film 0.013 mm (0.5 mil) thick/thin of alternating PC and PMMA layers that can range up to a total of 650–1300 layers. It contains no metal yet offers light reflective properties and different optical effects that mimic chrome plating or aluminum vacuum metalization. This Dow Chemical patent includes a special rotating type die (Fig. 5.43).

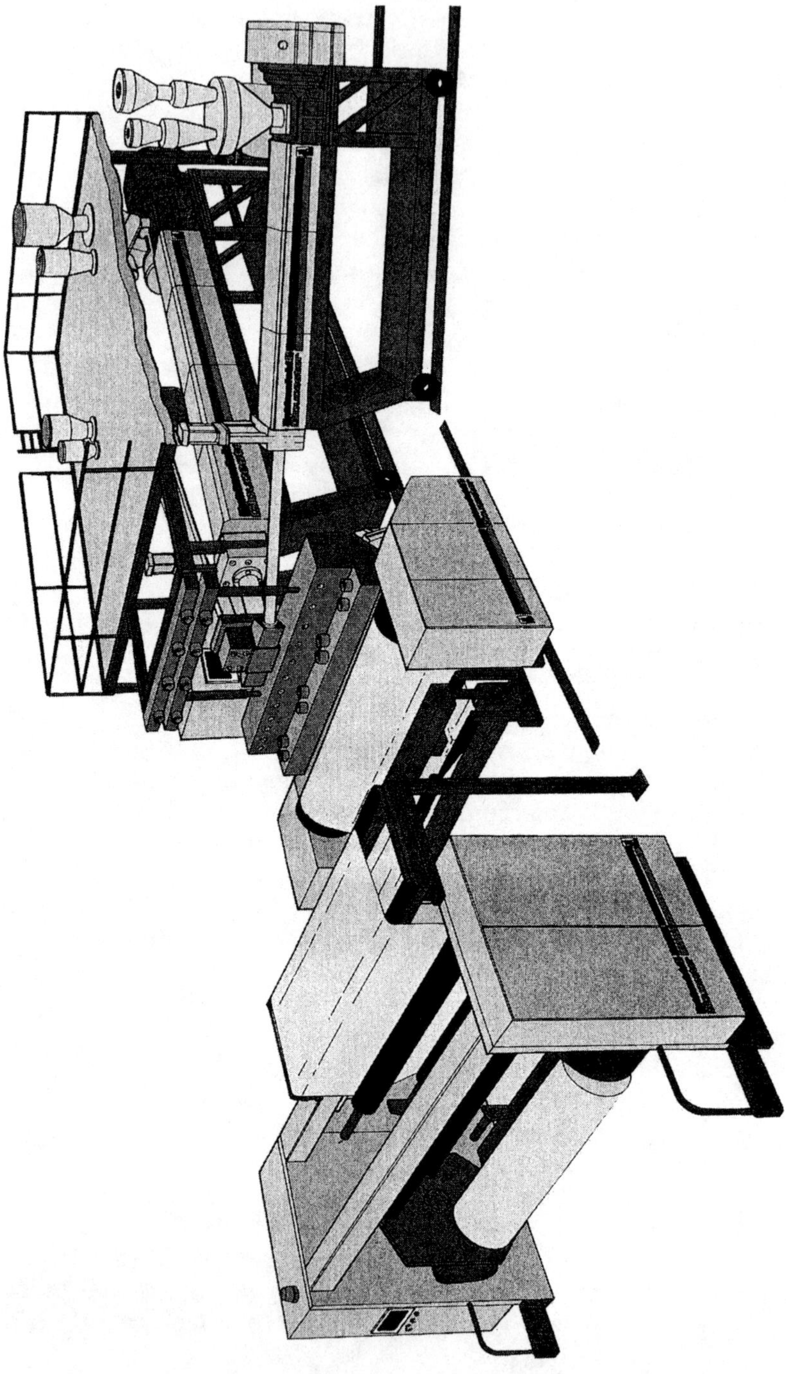

**Figure 8.9** Coextruded cast film line.

FLAT FILM CHARACTERISTICS

There are different chapters in this book that show the effects or relationship of processing and plastic material behavior. Figure 8.10 compares machine and material variables for flat cast film; the data are also applicable to blown film.

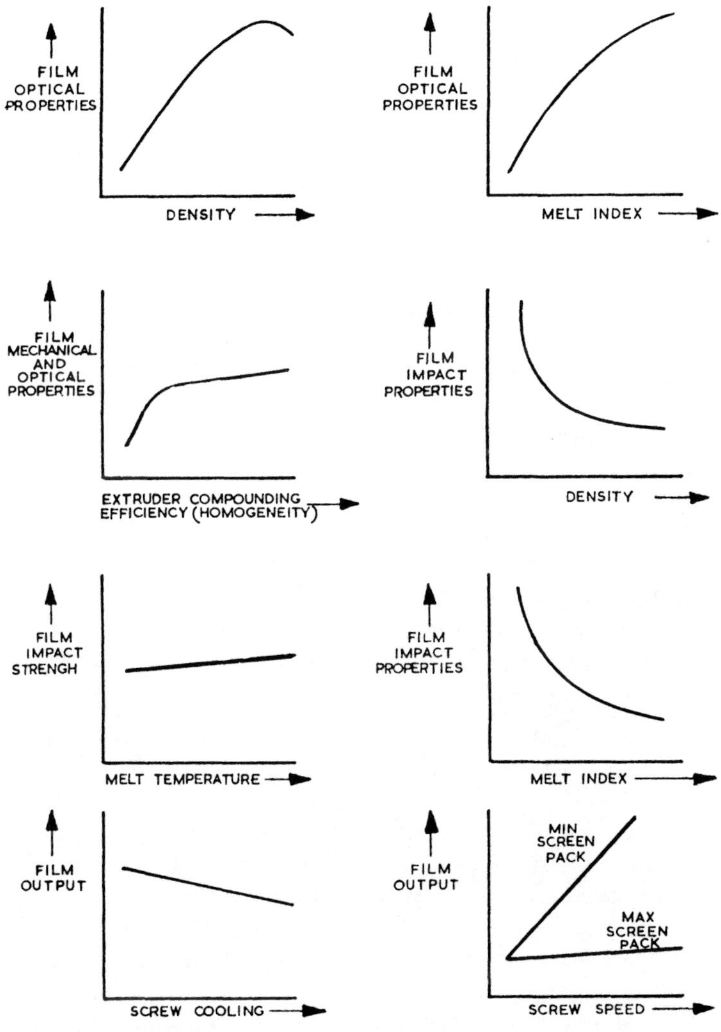


**Figure 8.10** Effect of flat cast film machine settings and material variables.

## Start-up

To get the extrusion film line ready to operate, the extruder barrel-screw-die system is first brought up to temperature. Either the system must be clean already, or the residual from the previous run must be purged. Purging can occur in an extruder (to die) only when the purging material is more viscous than the material being purged (Chapter 3). Once the cleaning is completed and all components (die, etc) are in place, the system is heated to the required temperature and started at below normal speed. During this start-up, if material remains in the system from the previous shut-down, the die is not left hot any longer than absolutely necessary to eliminate plastic degradation.

An approach is to starve feed plastic into the hopper. The first extrudate coming out of the die is generally not fully melted so it usually requires the temperature to be raised even above the normal operating range. When the melt is ready, the hopper is filled for flood filling. Either the melt temperature is measured by a portable pyrometer having a needle probe, or the extruder is equipped with a melt thermometer that gives a constant reading.

As the melt approaches the correct consistency based on instructions or – more important – experience, the extrudate starts to be strung up through the line. It travels through the cooling devices until it enters the pull rolls. It can be carried by suitably protected hands, tongs, rope feeder lead line, or left over web from the previous run. The latter approach, when available, can be used to shorten the start-up time.

Necessary adjustments are made that include web alignment, cooling, sizing devices, and line speed interacted with the extruder output rate until a satisfactory product is produced. This start-up could take relatively a few minutes for a simple narrow web process. For the wider and heavier webs with melt strengths that have small operating windows (temperature and line speed), it usually takes two or more hours. For extremely difficult extrusions, many attempts may be required to get it running satisfactorily.

## Shut-down

In all processes, it is vital to avoid decomposition in the extruder head and die. Such decomposition can be caused by thermal degradation, oxidation of hot plastic in contact with air, or both. For vinyls, special procedures have to be set-up and followed since thermal degradation is always a danger and can easily spread throughout the die ruining the product on start-up.

For all materials, such degradation and decomposition may produce hardened bits of plastic which can break off and lodge in the die lips.

These hardened bits may even be formed in the lips if points of stagnation exist. Such bits cause die lines which are not only unsightly but are also lines of weakness. The bits may also break loose and contaminate the film.

When a film line is shut-down, material is kept moving as the die cools to a temperature depending on material being processed. With LDPE, the temperature is about 130°C (265°F). At that time, the extruder is stopped and the die and head are cooled with air as fast as possible not to inhibit decomposition. PE is frequently left in the extruder barrel as well as in the die to prevent air from entering and oxidizing any plastic bits remaining in the system. Frequently the line is shut-down with the plastic still in position threaded through the system. This makes start-up easier.

## ORIENTATION

In flat film extrusion, particularly at high take-off rates, there is a relatively high orientation of the film in the machine direction (MD) of the extrudate flow and a very low orientation in the transverse direction (TD). A very popular approach is to orient the film either in one direction, usually the longitudinal which is the MD, or bidirectional (BD) in both the MD and TD (Chapter 2).

As an example, biaxial stretch film tapes improve properties as shown in Table 8.1. With this action, thinner films can be used when compared to unoriented film; the result is a major saving in plastic material consumption. Even with the usual high production rates for oriented film, cost of equipment in production usually represents about 5% of cost compared to plastic cost of 70–90%.

The effects of biaxial orientation are important in film manufacture, such as PP, PE, PET, and PVDC. Result is stronger films of high clarity. The orientation induced crystallization of the plastics produce molecular structures that do not interfere with light waves.

When biaxially stretching PS (4/1 ratio in both directions), the higher tensile strength of 70–80 MPa (10–12 ksi) is achieved compared to 40–55 MPa (6–8 ksi) for unstretched PS film.

Tapes (narrow films) for sound and video recording, packaging strapping, etc. (Table 8.1) usually made from PP and PETP plastics are highly oriented in the MD. The initial tape, which may be up to 40 by 1.5 mm (1.5 by 0.06 in), is made by vertical extrusion into a water bath with a limited drawdown (Fig. 8.11). Water temperature is kept low and the distance of the surface from the die is as small as 5 mm (0.2 in) to give rapid cooling and low crystallinity. The tape is then reheated to a temperature well below the crystalline melting point and drawn lengthwise by 10/1 or greater producing a high degree of uniaxial orientation.

It is then heat set by heating to a somewhat higher temperature while

**Table 8.1** Examples of film tapes

| *Ranges of application* | *Demands made* | *Rate of stretching* | *Thermoplastic* |
|---|---|---|---|
| Carpet basic weave | Low shrinkage | 1:7 | PP |
| | High strength | 1:5 | PETP |
| | Temperature stability | | |
| | Specific splicing tendency | | |
| | Matt surface | | |
| Tarpaulins | High strength | 1:7 | PP |
| | | | PE |
| Sacks | High strength | 1:7 | PP |
| | High friction value | | PE |
| | Specific elongation | | |
| | Weather resistant | | |
| Ropes | High tensile strength | 1:9 to 1:11 | PP |
| | Specific elongation | (15) | |
| | Good tendency to splicing | | |
| Twine | High tensile strength | 1:9 to 1:11 | PP |
| | High knotting strength | | PP/PE |
| Separating weave | High strength | 1:7 | PP |
| Filter weave | Low shrinkage | 1:7 | PP |
| | Abrasion resistant | 1:5 | PETP |
| Reinforcing weave | Low shrinkage | 1:7 | PP |
| | Specific elongation | 1:5 | PETP |
| | Temperature resistance | | |
| Tapestry and home textiles | UV resistance | 1:7 | PE |
| | Low static charge | | |
| | Uniform coloration | | |
| | Textile-type handle | | |
| Outdoor carpets | Low shrinkage | 1:7 | PP |
| | Wear resistance | 1:5 | PETP |
| | Weather resistance | | |
| | Elastic recovery | | |
| | Uniform coloration | | |
| | Defined splicing | | |
| Decorative tapes | Effective surface | 1:6 | PP with blowing agent |
| | Low specific gravity | | |
| Knitted tapes, sacks, and other packagings, seed and harvest protective nets | High knotting strength | 1:6.5 | PP |
| | Low splicing tendency | | PE |
| | Suppleness | | |
| | UV resistance | | |
| Packaging tapes | High strength | 1:9 | PP |
| | Low splicing tendency | 1:7 | PETP |
| Fleeces | Fiber properties | 1:7 | PP and blends |

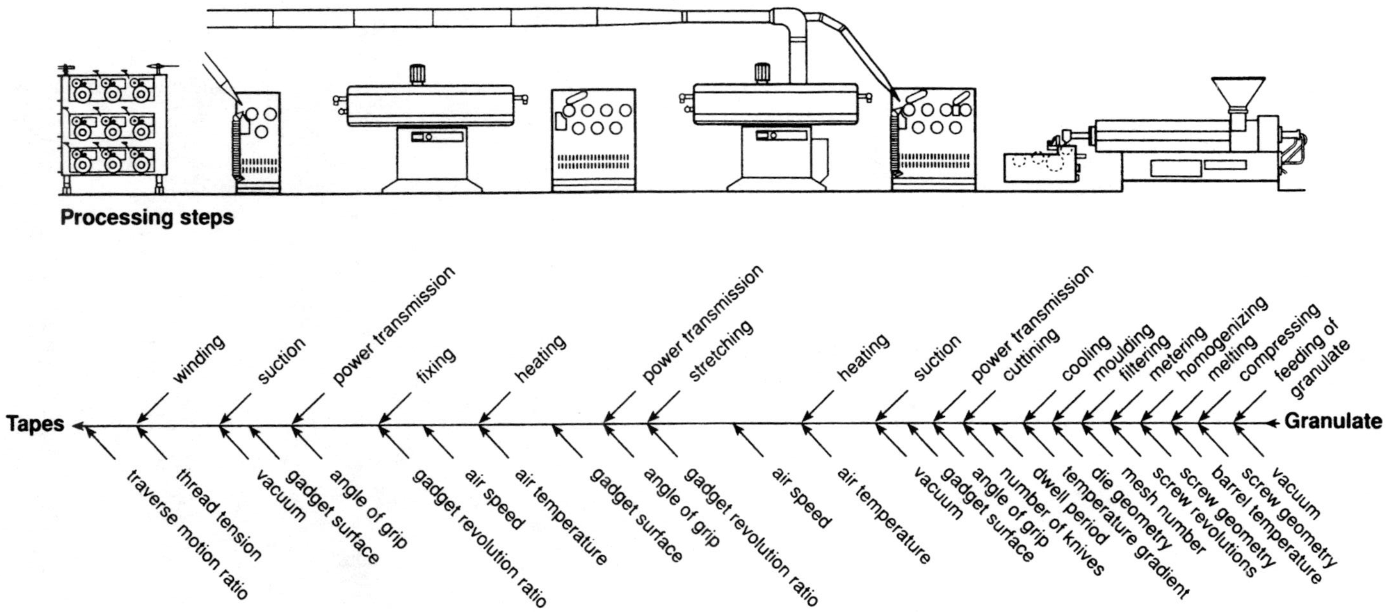


**Figure 8.11** (a) Chill roll process used in oriented film tape line; (b) performance of oriented heat and stretch ratio.

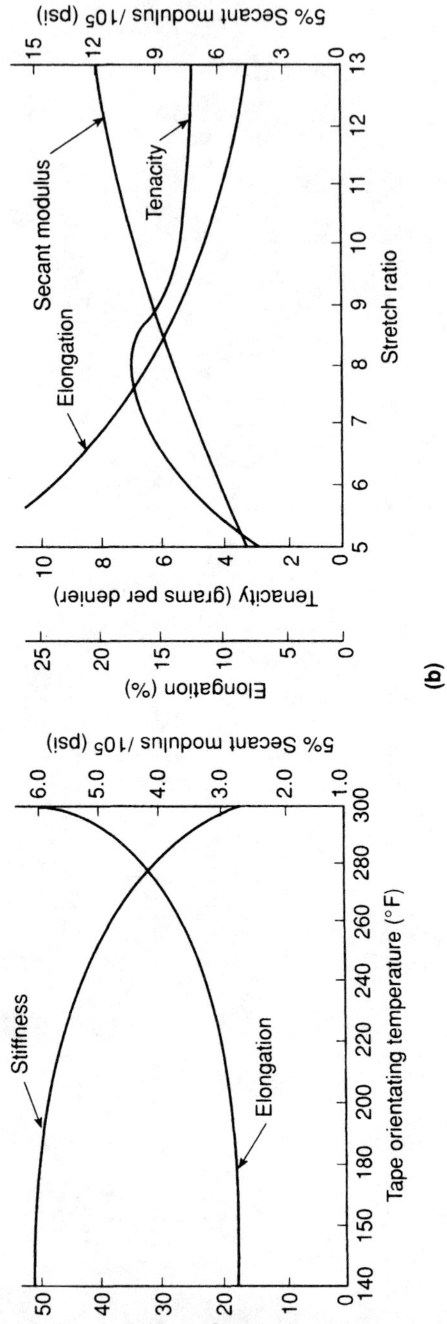


**Figure 8.11** *Continued*

held under tension to crystallize in the oriented state and then cooled. The low initial crystallinity provides high elongation to break which permits sufficient drawing to orient, a low modulus which reduces the tension required to draw at high strain rate, and less tendency to disorientation prior to heat setting.

Eventually the melt temperature at extrusion must be high enough to eliminate crystallinity present in the feed stock, but low enough to permit rapid cooling throughout the relatively thick initial tape web. For consistent properties, melt and water bath temperature must be closely controlled. In turn, to obtain consistent dimensions, extrusion rate and haul-off speed must be kept in proportion [3, 397].

When processing a particular PS, it is first extruded through a slit die at about 190°C (375°F) and cooled to about 120°C (250°F) when the web passes through the chill rolls. As it enters a temperature controlled box/oven, it is grasped on both sides by the tenter hooks. The web is reheated to 130°C (265°F) which is the orientation temperature condition for that specific PS. In the tenter frame, the PS is subjected to a drawing longitudinal (machine) tension and simultaneously it is subjected to a widening tension that produces the transverse or lateral stretching. Stretch ratios of 3/1 to 4/1 in both directions are commonly used.

## PROCESS CONTROL

The on-line equipment are designed to meet tight operating requirements in order to produce quality film. As an example, the cast or chill roll that the extrudate first contacts is designed for efficient and controlled cooling by circulation of water or antifreeze ethylene glycol solutions through double shell constructions. Water flows over the inside diameter of the roll and is pumped at high rates to obtain a uniform temperature across the outside surface of the roll. There may be internal deflector rings and/or special directional flow channels to aid in providing an even roller cooling surface at all times.

Water is piped in and out of the rolls through rotary union joints. For a certain melt, water having a temperature range of 49–66°C (120–150°F) could be used. The 'chill' roll compares the higher temperature melt to its temperature, thereby properly identifying the purpose of the chill roll to cool the melt.

An edge bead is formed during the neck-in of the melt. This neck-in occurs in the air gap between the die and the chill roll. It is the shrinkage in the extruded web. It may be 25 mm (1 in) or less on each side of the web. The exact amount depends on upon the type plastic being processed and the operating conditions. The target is to come as close to no edge bead. The thinner these side strips, the better. This scrap is usually recycled and blended with virgin plastics.

The film can have a rather thick edge bead compared to the film thickness. It is removed by some type of trimmer which also slits the film to the required width. This trimmer, on-line, often is a simple device that holds a razor blade or equivalent. If the edge bead is not removed, it would interfere with providing a uniform roll during the wrap-up at the end of the line.

A most important aspect of this process is to ensure even contact of the hot melt on the first chill roll. At speeds greater than about 14 m/min (45 ft/min), the melt tends to lift away from the roll and ride on a pocket of air. The web also has the tendency to trap air pockets between the plastic and roll at any speed. Air, being an insulator, prevents uniform cooling. To eliminate this major problem, an air knife is used. This device gives a controlled long thin even jet of air which pushes the web against the roll preventing air entrapment. Different air knives are available that provide different capabilities in ensuring the degree of eliminating air entrapment.

For temporary accumulation of film (also wire, tape, calendering, etc.) while in production and on-line, festoon can be used. As an example, it can be used where fast windup rolls cannot be quickly changed from a full roll to an unused roll. Down-stream near the windup, the film is looped between two or more sets of rolls to accumulate an inventory of film. During normal running, the two sets of rolls are close together and permit film just to travel as if they were not in the system. When the film roll is being changed, the distance between the rolls increase usually vertically extending the length of the run and acts as a hang-up station until the new windup roll is in place. Once in place, the rolls return to their nonactive/close setting. When film, plastic coated fabrics, etc., require extra drying, fusing, or cure time, the festoon can be placed in an appropriate oven.

Success in producing flat film (or any other product) depends on analyzing the product requirements. In turn, careful selection of the optimum plastic(s) for the job is required, unless a specific plastic from a specific plastic manufacturer has been specified by the customer. And an important analysis is determining that you purchase the line that is required to meet the customer's requirements and meet applicable standards and specifications. Of course, reliable equipment manufacturers meet all present and many cases future industry standards and specifications.

If you already have a line, the question is whether it can do the job. This is where problems could easily develop. Analyze its capability and the probability that you will be operating at maximum performance to cost efficiency. If proper maintenance for the complete line has taken place, you are going in the right direction. But a certain piece of equipment or roll could become a problem that could shut-down the line.

**Line control**

Different techniques are used to control the operation of the line as reviewed in Chapter 6. Rather complete automation is available from the material handing upstream, through the extruder, and downstream to the windup roll once the proper startup occurs. Settings on the equipment in-line can operate from one control point (Figs. 5.8, 5.40, and 6.3–6.7).

One can relate performance of the line using certain basic observations and in turn set appropriate sensor controls. As an example, a minimum and/or maximum neck-in can be related to line speed. In fact, there may be an opportunity to identify the thickness and/or density of the bead and relate to speed. The effect of the extruder and plastic material variables influence film performances.

**Output rate**

As the schematics show (Fig. 8.1), only a few rolls may be used. However, when the lines are traveling at speeds of over 300 m/min (1000 ft/min) many more gradually cooling rolls are needed to control the movement of the film through the line.

In operating these lines, as in other lines, safety for the operator(s) and equipment is of prime importance. Personnel have to understand the importance of safety instructions as well as following them. It is important to learn the proper procedures for handling the equipment and to know how to use safety shut-off devices. Every precaution should be taken to keep hands, arms, jewelry, and clothing from becoming caught and pulled into the rolls.

By upgrading from epoxy to ceramic corona treating rollers in cast film lines, there are both improved corona-treating up-time, reduced maintenance expenditures, and improved treatment levels. They are more durable and provide higher operating temperatures as well as voltage outputs. The average epoxy roll life is one year and that of ceramic is 4.5 years. When all the factors involved in a line output cost is analyzed, the ceramic rolls are beneficial. Factors include initial costs, recovering, downtime, product waste, etc. Line speeds are increased without fear of roller failures or downtime due to roll changes [245].

Extruding flat film around a chill roll or water bath at continuously increasing rates of output can begin to pulsate rhymically at a critical draw ratio (DR). The DR is the melt's final velocity to its velocity in the die. As the DR increases, the pulsations become more pronounced. This phenomenon is known as the draw resonance. It can occur in any melt-drawing process, such as coating, spinning, and melt embossing, as long as the extrudate passes through the two fixed points of a die and a fixed velocity take-off point.

For certain plastics, the DR occurs under 20 for film. In coatings and embossing, the ratio is about 30. The method used to eliminate DR is by a draw resonance eliminator (DRE). This is directing a fluid medium (air) against the melt between the die and chill roll. Thus, the air knife just reviewed is used.

## DIE

As usual, the extruder and die melt flow must be balanced. Back pressure is required to maintain full capacity in the screw's metering section, sufficient mixing, and steady pumping. Although a separate back pressure valve, choke, or other back pressure device could be used to balance the die and screw, these can create other problems in the extrudate flow (Chapter 5).

Flow characteristics, as well as shape, are considered in the melt flow path die design. They require a long manifold in the die with a large cross sectional area to allow for an easy flow of melt across the die. Streamlining design in the flow channel is desired. Flow considerations are reflected in the cost of the die as well as the quality of the finished product. The die jaw opening has adjustment bolts to set the opening. The end surfaces of the jaws are called the die lips. Together they are called the die face.

There is a T-type die which provides for a simplified flow pattern. It has a distribution channel perpendicular to the main flow direction of the melt entering the die. In turn, this T-shaped cavity distributes the melt across the width of the die head. Although the cross section of the distribution channel is much larger than the thickness of the parallel flow, there are restrictions along the width. The result is that with this relatively low cost die, there is no guarantee of an even flow exiting the die. It works best with low viscosity melts, the types used in coating lines. To help in the flow, adjustments are made via the choker bars and flexible lips.

Designing these dies rheologically by following nonNewtonian flow behavior (Chapter 3), it is theoretically possible to provide an extrudate with an even melt flow distribution (pressure, temperature, and density). The results have been the so-called fishbone and coathanger dies (Figs. 5.7, 5.24, and 5.26). They illustrate an important principle in die design. The melt at the edges of the film (or sheet) must travel further through the die than the melt that goes to the center of the die (Fig. 5.11). Thus, basically a diagonal melt channel with a triangular dam (like a tear drop) in the center is used to restrict the direct flow to the exit. This principle with gradual built-in restrictions is used to adjust the flow in many dies for the different extruded products.

Basically, the die delivers a given plastic melt under specific conditions through a constant die opening at a constant rate and temperature. The melt near the center spends little time in the 'distributor.' It will flow within a narrow gap for most of the length of the die, whereas the melt closer to the periphery will flow along a substantial part of the cavity/distributor. This section has a bigger cross section and in turn less of a pressure drop. However, across the die, rather high melt pressures exist, requiring a structurally rigid die to eliminate or reduce deflection of the die. Pressures of 28–35 MPa (4000–5000 psi) are not unusual. Thus, the optimum die design is based on the melt characteristics (rheological characteristics) for a specific plastic that requires a specific viscosity and output.

Recognize that plastic material performance varies. Temperature, pressure, and melt flow rate fluctuate. In fact, the metal construction has fluctuations in thermal transfer and possibly rigidity (unless built like a 'battleship'). So the target is to approach a die design that could meet the best performance, recognizing that perfection does not exist. Thus, adjustments prior to the melt entering as well as flowing through the die are required to approach perfection. As an example, the fishbone die can have a reduction in its land restriction that makes it basically difficult with many melts for producing a uniform melt distribution. The more expensive and commonly used coathanger die, in comparison to the fishbone die, provides a relatively uniform melt distribution.

The die has relatively thick wall sections on the final lands to minimize deflection of the lips from internal melt pressure. As an example, the die opening for PE may be 0.38–0.76 mm (0.015–0.030 in) even for films that produce less than 0.076 mm (0.003 in) thick film. The reason is that the speed of various driven rolls used for taking up the film is high enough to draw down the film with concurrent thinning.

The die jaw opening is set with adjusting bolts before operating the extruder. After the line is running, these bolts that run all along the opening are readjusted to meet requirements. Depending on the type die, these adjustments can be made manually or automatically. The ends of the die opening are closed off and sealed with fixed or adjustable end plates. As shown in Fig. 5.6, separate temperature control zones are used to provide the temperature control of the melt across the die cavity.

Extensive use is made of producing coextruded/multi-layer cast films. A Cloeren die is shown in Fig. 5.36. The five layer film produced in the high-speed line uses a 370 cm (145 in) wide Cloeren five-layer feedblock that was redesigned to provide very fast adjustment response time. Heaters and push rods are integrated to increase thermal efficiency. It used a special unplated alloy steel that has higher lubricity than chrome plating which contributed to improved line speeds.

Operating with the die is Cloeren's dual chamber vacuum box positioned just up-stream of the air knife controlling the static pressure of the film against the chill roll. Both the vacuum box and air knife are wider than the die instead of being the same width as on Chaparral other line.

The redesigned vacuum box resulted in improved line stability and increased cooling efficiency and film clarity.

## TROUBLESHOOTING

An annoying problem with this process is plate-out where the plastic melt adheres to the chill roll. It is a soft waxy or sticky, whitish solid. Its chemical nature differs from one plastic to another. It may be a natural impurity from the plastic or certain additive. The target is to use plastics that have little or no tendency for plate-out. However, some of these additives are usually included to increase slip and prevent blocking (the mechanical cleavage of one film to another).

It results from the boiling out of some of the plastic while the hot extrudate is in the air gap. The material then condenses on the chill roll. The high melt temperature of this process allows this action to take place. It may cover the roll and hide the smooth surface. Some may come off the roll and stick to the film. Since this problem can easily develop when processing certain plastics, there has to be an operator and/or sensor to catch the problem developing. Periodic cleaning of the rolls is required.

Plate-out is of minimal concern as long as the chill roll is operated at an elevated temperature. The temperature at which sticking occurs will depend on the plastic being processed. There is a temperature range above or below where minimal sticking occurs. Some slight sticking is desirable for uniformity of the cooling/quench.

Another cast film defect which involves cooling is puckering, a slight bulging across the film recurring at regular intervals. Puckering may be reduced by running the first chill roll hot. If the melt flows well, there is little danger of severe puckering.

A general guide on film performances follows.

1. Effect of increasing melt temperature:
   (a) transparency and gloss increases;
   (b) haze decreases; and
   (c) impact strength increases.
2. Effect of increasing chilling temperature:
   (a) yield strength increases;
   (b) haze increases;
   (c) transparency and gloss decreases; and
   (d) impact strength decreases.
3. Effect of increasing air gap:
   (a) haze increases; and
   (b) transparency and gloss decreases.

**Table 8.2** Troubleshooting cast films

| *Problem* | *Cause* | *Solution* |
|---|---|---|
| Thickness variations | Erratic melt quality | Check screw for wear. Check screw design. Check and verify barrel, adapter and die temperatures. Add discrete mixers to promote melt homogeneity. Check heaters, thermocouples |
| | Surging | Check tachometer on motor. Check screw for wear. Check motor amps on auger/crammer if feeding edge-bead trim |
| | | Check screw cooling. Check hopper |
| | Erratic cooling, quenching | Check vacuum seals for leaks/check air knife slots for plugs |
| | Improper die-bolt adjustment | Readjust |
| Die lines, streaks | Dirty die | Clean lips with soft tool |
| | Nicked die lip. Inadequate purging | Change lip/die. Increase purge time between runs |
| Gels | Erratic melting and mixing | Raise barrel settings (if unmelt is detected). Check screw for wear. Add more mixing. Check barrel for wear. Check resin for impurities |
| Stripes | Excessive air-discharge from air knife | Decrease air-knife pressure |
| | Air pockets between film and chill roll | Increase air-knife pressure. Reposition knife |
| Cloudy bars | Melt too hot | Decrease barrel temperature |
| | Erratic cooling | Check air-knife blower speed |
| Wrinkles | Flow variation from die | Adjust barrel/die zones, and die gap |
| | Winding tension too high | Readjust line so that tension increases gradually to winder |
| | Nonuniform cooling (stiff edges) | Adjust web-pinning device further down from die to promote uniform cooling. Or, trim edges sooner |
| Floppy web | Roll tension insufficient | Increase |

**Table 8.2** *Continued*

| *Problem* | *Cause* | *Solution* |
|---|---|---|
| Sharkskinning | Melt temperature too low | Increase. Heat die lips |
| | Die gap too narrow | Increase |
| Film sticking on chill roll | Erratic cooling | Check temperature of chill roll, polymer. Check chill roll passages for blockage |
| Film too narrow (from chill roll) | Die slot too narrow | Adjust deckles |
| | High rate of neck-in | Check position of rolls relative to die exit. Adjust air flow from edge pinners. Relocate pinners |
| Hazy roll | Plate out on chill roll | Add plate-out roll. Use vacuum box, which is equipped with wiper |
| Baggy edges | Excessive bow | Adjust bowed roll |
| Scratches | Rolls not turning at web speed | Check roll drag and bearings. Check roll speed and balance |
| Care collapse | Film too warm | Decrease temperature of second chill roll |
| | Film wound too tight | Reduce tension |
| | Lay-on roll pressure too high | Decrease |
| Hard roll | Winder tension too high | Lower tension; increase taper |
| | Lay-on pressure too high | Lower |
| | Web shrinkage | Decrease temperature of second chill roll |
| Soft roll | Not enough winding tension | Increase tension; decrease taper |
| | Not enough lay-on roll pressure | Increase |

4. Effect of increasing output rate:
   (a) haze decreases; and
   (b) transparency and gloss increases.

Of the above variables, air gap and rate have the greater influence on properties. Table 8.2 provides some information on problems and solutions related to flat film production.

# 9

# Sheet

## OVERVIEW

Sheet extrusion transforms TP materials into roll or sheet stock through a combination of heat and pressure. The extruder plasticizes the plastic and pumps it through a die (Chapters 2–6) and it continues downstream in the sheet line. Figure 9.1 is a schematic of a sheet line and Figs. 9.2–9.4, 2.5, and 2.29 show examples of the type of equipment used in sheet lines. The terms sheet or sheeting normally describe a flat plastic product that is 0.25 mm (0.010 in) or greater in thickness. Some industries use 0.10 mm (0.004 in) as the dividing line between film and sheet. Widths can be up to at least 3 m (10 ft).

Sheet extrusion technology produces single (mono) layer products as well as more complex coextruded or multi-layer (Chapters 2 and 11) engineered products. As equipment and processing technology advances, new sheet products are made available, providing performance and cost advantages. Market demand for increasingly sophisticated finished products has required improved quality so that sheet manufacturing has become more complex for the novice.

As an example, sheets are being produced to meet tighter thickness and width tolerances, improved surface quality, provide increased and more uniform mechanical and physical properties, and line speeds are being increased to help reduce costs. Hence coextruded sheets as well as oriented sheets are now in greater demand in markets such as the chemical and storage container industries. The result of all these developments is the expansion of the present product markets and the opening of new markets [1, 3, 110, 185, 267, 304, 374, 382].

There are three primary techniques used to manufacture plastic sheets: (1) by far the most popular, extrusion through a flat die onto polished rolls (Figs. 9.2 and 9.3) which could include the use of an air knife (Fig. 9.5); (2) extrusion through an annular pipe-like cross section die onto a sizing

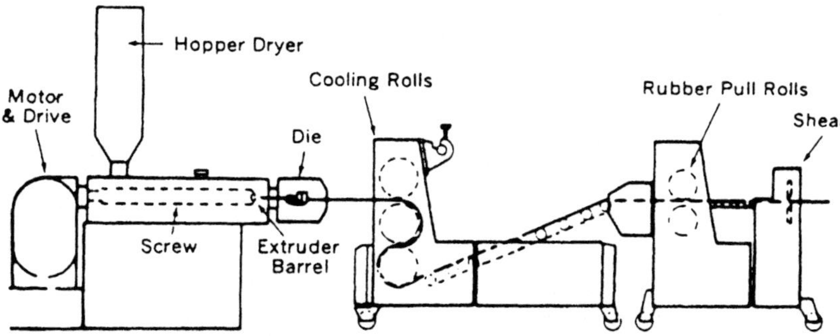


**Figure 9.1** Sheet line using a stand of three cooling rolls.

mandrel, where the extrudate is slit in one or more places and then flattened using rolls into a sheet [very popular in foam sheet production (Chapter 18)]; and (3) calendering (Chapter 10).

There are also other designs, such as water cooled sheet lines. One design has the melt directed vertical downward and being cooled between two water cooled belts [304]. It processes nonnucleated, low haze, high gloss PP sheet that crystallizes between 90–°C (195–265°F). It is reported that cooling occurs in less than 0.1 s compared with 0.25 s in conventional stack roll systems. Another system has a horizontal line where the melt goes around and through a sandwich construction of a large cooling roll and a cooling steel pressure belt followed with a water bath. This high-speed line is being used by various Japanese plants [309].

An extrusion sheet line (train system) consists of a properly controlled extruder (Chapter 2), sheet die, temperature controlled (individually heated or cooled) polishing three roll stack (stand), take away equipment [such as gauge monitors and controls (Chapter 6), edge trimmers, anti static devices, air coolers, cooling tunnels, conveyors, slitters, pull or nip rolls, tension rolls, and/or cross cutting devices], and either a (turret) wind-up or stacking equipment (Figs. 5.6, 5.26, 5.40, 6.2–6.6, and 6.8). Throughput capacities for these lines can be up to at least 400–1100 kg/h (900–2500 lb/ht). Winders can run up to at least 75 m/min (250 ft/min) (Fig. 8.6).

Stack rolls are usually of double shell design, giving internal high-velocity liquid circulation at a controlled temperature. Each roll is equipped with its own individual temperature control system which is built into the take-off unit. The sheet gradually cools as it travels around the rolls becoming sufficiently 'solidified' so that it can continue down the line. After leaving the usual three roll stack (Figs. 9.6 and 9.7), the sheet cools further as it travels down the line being pulled by pull rolls and on

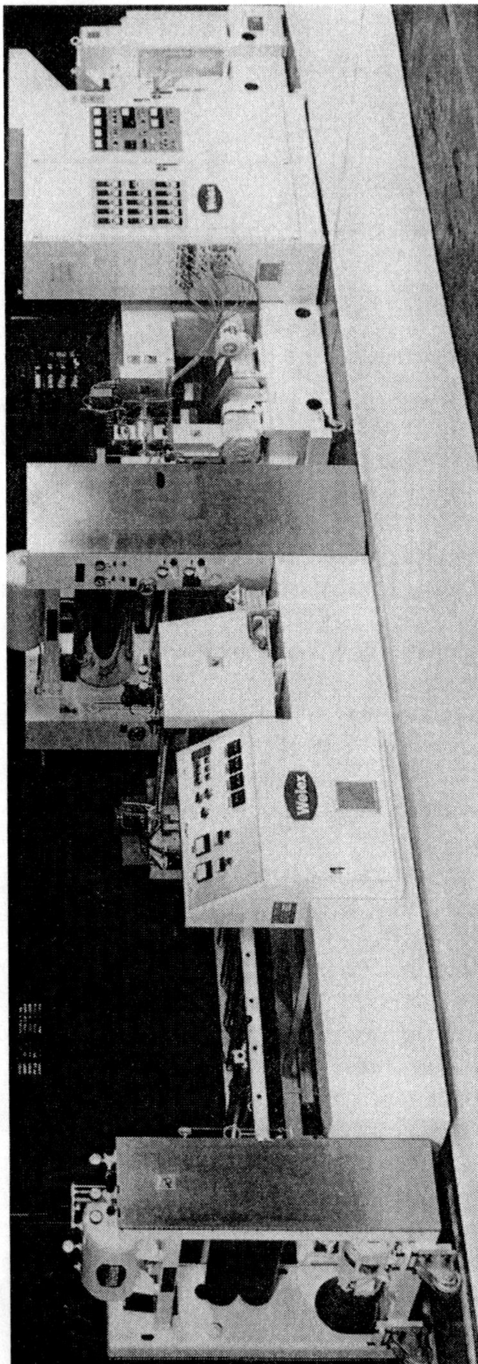

**Figure 9.2** Example of a sheet line from Welex that includes screw venting, screen changer, gear pump, static mixer, sheet die, anti-static bath, and puller.

**Figure 9.3** Example of Welex's sheet take-off cooling system.

**Figure 9.4** Example of Welex's vented extruder to process PET.

**Figure 9.5** Close-up of Welex's air knife system used in extruding of very thin PP sheet on a conventional sheet line.

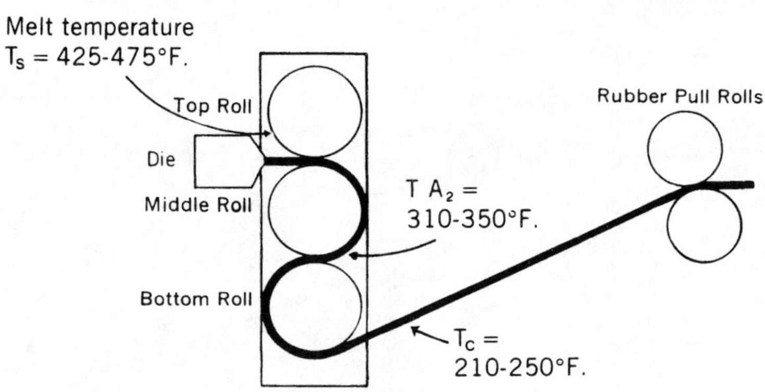


**Figure 9.6** Roll stand down stack.

to the end of the line. This approach permits very fast and efficient temperature control response.

The application for which the sheet is made will to some extent govern the conditions of manufacture. As an example, where the sheet is intended

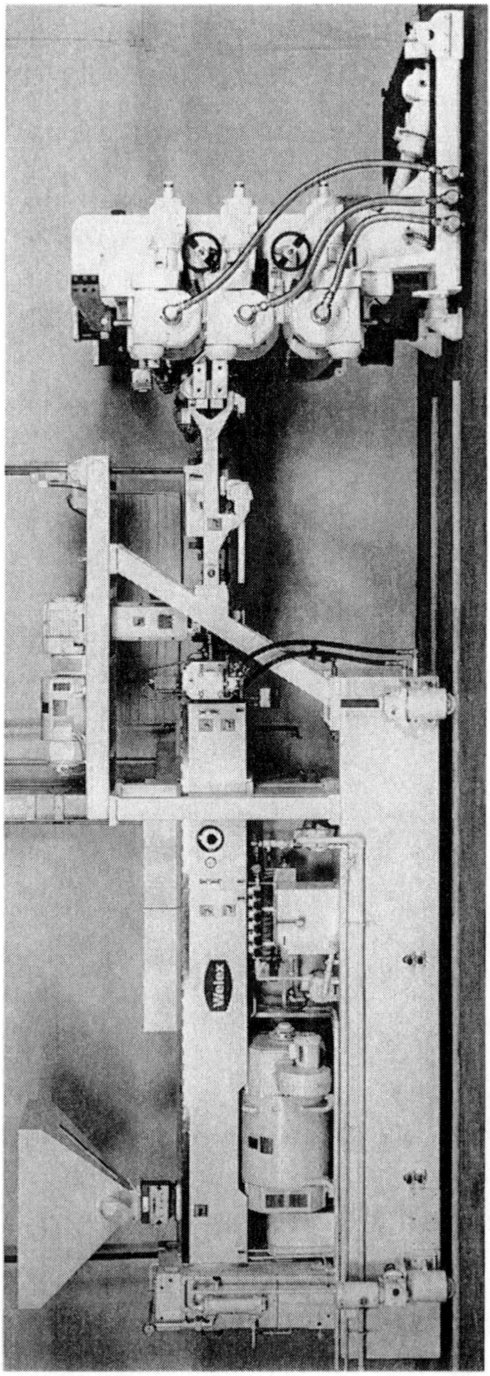

**Figure 9.7** Universal Unistack R sheet take-off from Welex in a down stack position.

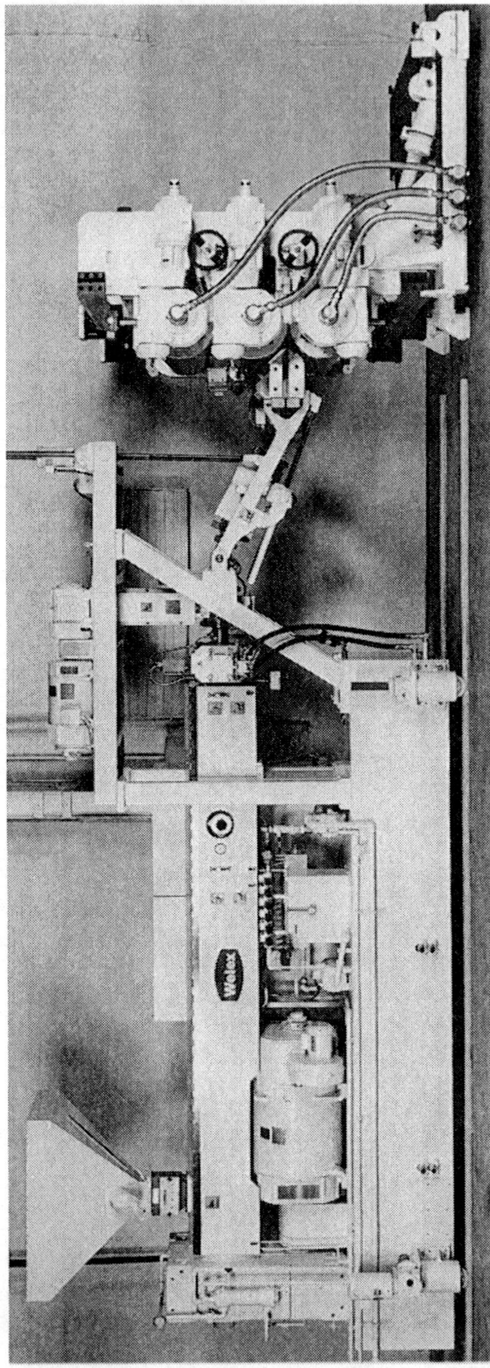

**Figure 9.8** Universal Unistack R sheet take-off from Welex in the up stack position.

for vacuum thermoforming, it is often slightly stretched during manufacture so as to prevent sagging and drooping during the thermoforming process (Chapter 18). Balanced properties are within at least 10% of target, however, tighter tolerance on thickness is most desirable. It is possible that residual memory or frozen stresses will exist in the sheet that should not have occurred during fabrication (Chapter 3).

The residual memory may cause the sheet during forming to bow either upwards (Fig. 9.8) or downwards (Fig. 9.6) in both the machine and transverse directions. This problem is controlled by adjusting the temperatures of the upper and lower polishing rolls, and also by controlling the temperature gradient along the rolls. If need be, the stress relieving can occur when the sheet is properly post cured and/or oriented.

A guide has been prepared that provides general information on the traditional customs of trade for the plastic custom sheet extruder industry. As in every major industry, various commercial and administrative practices have developed over the years which play an important role in the conduct of day to day business. These arrangements expressed in the proposal, the acknowledgments, and the contract forms of individual companies, have been reviewed as constituting 'Business Practices of the Trade.' This manual is designed to identify and explain these practices. The manual was prepared and is periodically updated by the Sheet Producers Division of the Society of the Plastics Industry (SPI).

## PLASTIC MATERIALS

Many different flexible to rigid thermoplastics can be extruded into transparent or colored sheets. They include ABS, AN, PA, PET, PETG, PMMA, PC, PE, PP, PS, PUR, and PVC. Different alloys are processed that include ABS/PA, ABS/PC, ABS/TPU, PPO/PA, and PPO/PS [382]. Monolayered and coextruded sheets are fabricated. Different surface plastics and barrier plastics are used in the coextrusion sheet lines [3].

The following information only highlights the processing of certain plastics that in turn can provide guidelines for the other plastics to be processed. As explained earlier, information on how to process a specific plastic in a sheet line or any other line is available from the material supplier and/or experience.

### Processing ABS

A popular plastic is ABS. This plastic requires predrying to ensure that its moisture content is at a minimum, such as less than 0.03 wt%. This level can be achieved in properly operating equipment, such as with an air temperature of 82–93°C (180–200°F), a dew point of less than −18°C(0°F) or preferably −29°C (−20°F), and a residence time in the hopper of at least

3 h. The equilibrium moisture content of virgin ABS pellets will vary between 0.3 and 0.8% [110] (Chapter 3). Vacuum venting in the extruder provides an additional safety factor to assure the minimum moisture content; just using the vented extruder is usually sufficient to prepare the melt with the minimum required drying (Fig. 4.12). Details on venting are given in Chapter 4.

In processing ABS through an extruder, it usually requires 0.1–0.15 hp/lb h. Table 9.1 summarizes examples of the ranges in drive motors hp and outputs (Dow Chemical). Details of extruder operation are in Chapter 2.

Table 9.2 provides typical readings for the temperature profile of the extruder. These settings with other extruder settings (screw rpm, back pressure, etc.) provides the method to help ensure a uniform melt develops. Details on processing plastics are given in Chapter 3.

The choice of the die to be used is strongly influenced by the quality of the sheet required. The usual type is a streamlined manifold coathanger die with adjustable flexible lips (Fig. 5.7) (Chapter 5). Later in this chapter, details are furnished on the down-stream equipment used in the sheet lines.

Table 9.3 provides typical ABS downstack roll temperatures and Table 9.4 has the temperature readings for an upstack with an embossing top roll. These temperatures are measured with an infrared optical pyrometer. If required, gloss can be improved with moderate to high bottom roll pressure to assure good contact to the bottom roll. Figure 9.9 shows a

**Table 9.1** Example of ABS extruder output rates and hp requirements

| *Machine size (Screw diam. = inches)* | *Output (No./h)* | *Horsepower (hp)* |
|---|---|---|
| 2.5 | 100–250 | 25–100 |
| 3.5 | 300–800 | 50–150 |
| 4.5 | 700–1500 | 100–250 |
| 6 | 1200–2000 | 150–350 |

**Table 9.2** Example of extruder barrel temperatures (°F)

| | *Rear* | *Middle* | *Front* | *Melt temperature* |
|---|---|---|---|---|
| Lustran ABS Resin | 375–425 | 400–450 | 400–475 | 400–475 |
| Cadon Resin | 375–425 | 420–450 | 440–450 | 465–475 |

**Table 9.3** Example of sheet rolls temperatures (°F) on a downstack

| | *Lustran ABS* | *Lustrex polystyrene* |
|---|---|---|
| Top roll | 175–225 | 150–200 |
| Middle roll | 150–175 | 110–180 |
| Bottom roll | 175–225 | 110–210 |

**Table 9.4** Example of embossed sheet rolls temperatures (°F) on an up stack

| | |
|---|---|
| Bottom roll | 150–170 |
| Middle roll (embossing roll) | 200–250 |
| Top roll | 180–220 |

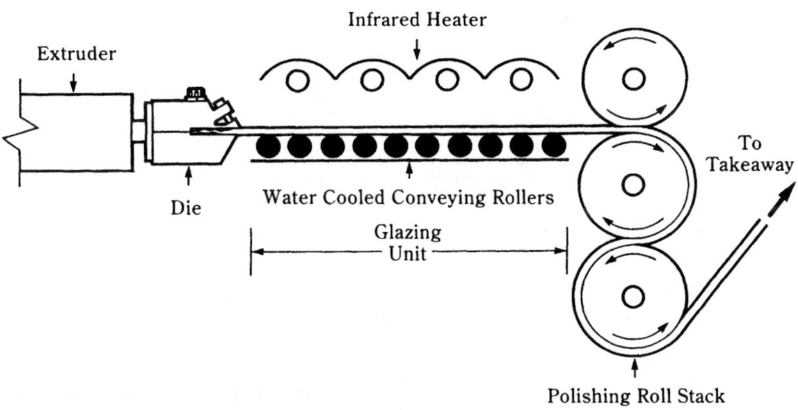


**Figure 9.9** Schematic of a heat glazier to improve sheet gloss.

heat glazing technique for improving sheet gloss (developed by Dow Chemical).

Figure 9.10 translates linear rate for various sheet thicknesses into suggested roll diameter in order to obtain 155–165°C (310–330°F) $A_2$ temperature in Fig. 9.6. As an example, if one estimates a linear rate of 1.5 m/min (5 ft/min) on 5 mm (0.20 in) thick sheet, then the estimated roll diameter would be 460 mm (18 in) and the range about 400–500 mm (16–20 in) [110]. The conversion from lb/h to ft/min is: Rate (ft/min) = [0.0361 × Rate (lb/h)]/($w \times t$), where $w$ = sheet width (die width, in), $t$ = sheet thickness (in).

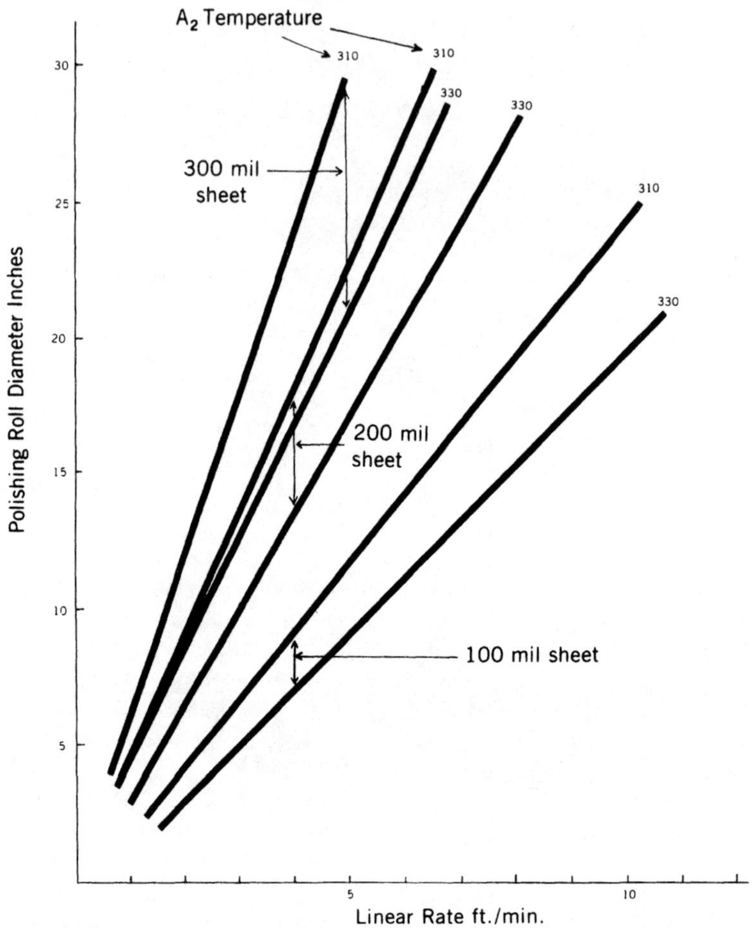


**Figure 9.10** Guide to polishing roll sizes with a 230°C (450°F) melt temperature.

## Processing PET

Running a PET (polyethylene terephthalate) plastic through an extruder sheet line basically does not require a high degree of sophistication. Most PET sheet lines can be regulated by standard discrete control devices. However, when producing thin gauge sheet [500 μm (20 mil) or less] to tight tolerances, a gauge control consisting of an automatic scanning thickness gauge that adjusts a die to correct for deviations is very effective (Fig. 5.26).

When producing optical grade sheet, a three roll stack should be used that has roll alignment to within 0.013 mm (0.0005 in) or less TIR (total indicator reading) should be used. It should also use individual drives which are then speed matched to each other eliminating chatter.

Temperature throughout the line must be kept above the plastic's melting point. A reverse heat profile is used with PET. The highest from the extruder's feed throat to the barrel end could be 316–260°C (600–500°F) that are run on the first two heating zones to obtain rapid melting. Die exit ranges from 300–260°C (570–500°F). If the first two zones are not sufficiently hot, the result will be surging. Too hot a melt temperature results in IV degradation (Intrinsic Viscosity, see Chapter 3).

PET is not difficult to extrude on conventional sheet extrusion equipment provided it was very carefully predried. This is because a small amount of moisture it contains reacts with the molten plastic resulting in the loss of molecular weight and in physical properties. Drying PET can be rather complex (Chapter 3) and costly.

In the last couple of years, vented extruders have operated worldwide that provide the output capacities of conventional single stage screws using predried PET (Fig. 9.4). As an example, an extruder with a 90 mm (3.5 in) screw, extruder has an output of 400 kg/h (900 lb/h), a 120 mm (4.5 in) extruder has an output of 700 kg/h (1600 lb/h), and a 150 mm (6 in) one has an output of 1600 kg/h (3500 lb/h). The IV loss is in the 0.05 range for plastic with an initial IV of 0.70. It is less as the initial IV decreases. The V-PET (vented PET) sheet extrusion line is the same as the conventional predryer line. The difference is that vented extruder with a vacuum pump system [316].

Since recycled 100% PET (post-consumer bottle scrap flake) often absorbs considerable surface moisture, it is best to dry it for one hour at 80°C (175°F) with plain hot air. This action avoids feed problems with the refluxed moisture which can accumulate at the feed end of the extruder. This action is a very minor investment and operating expense (and inconvenience) when compared with conventional crystallization and drying for 5 h at 175°C (340°F) with bone dry air. Feeding of 100% flake scrap (as with conventional dried plastics) also requires the use of a crammer feeder to assure proper feeding which otherwise has a tendency to bridge (Fig. 1.11 and Chapter 2).

## ORIENTATION

Sheet materials are oriented to take advantage of the orientation process that principally provides a combination of improved product performances and lower costs. Details of orientation are reviewed in Chapters 2, 3, 7, and 8 [320].

## PROCESS OPTIMIZATION

There are different processing methods used to enhance and improve the quality of sheet materials. The techniques reviewed elsewhere in this book also apply to sheet extrusion. In this section, the review concerns the down-stream equipment that is vital to successful line operations. It provides guides to limitations that can exist [4, 309].

### Line components

Different and important components of the sheet extrusion lines follow the dies. Just as improper material handling equipment in the beginning (up-stream) of the line can restrict or damage the line's operation, the same is true with the down-stream equipment. Different equipment components are required that are properly aligned and interfaced with each other (Fig. 9.11).

All equipment requires proper matching assuming that the individual pieces of equipment provide the expected performance. All equipment should be studied to ensure that it tends to be maintenance free or can easily be maintained. Equipment should have rigid frame construction, roll bearing assembled units should have the capability to minimize play and deflection, where applicable use self lubrication, and so forth.

### Roll stack

The hot melt from a slit die is directed to usually a three roll stack (Fig. 9.6) that are equipped with quick-acting safety releases where the gap between the rolls open in emergencies. As the extrudate exits the die, it must be supported as it is hot and in a semimolten state. Support is provided almost immediately by the roll stand. Different diameter polishing rolls [such as 20–76 cm (8–30 in)] are used; the larger rolls provide processing advantages with higher output rates.

Typical terminology used to describe the roll stand is roll diameter, roll width, number of rolls, length of conveyor, and hp of drive motors on roll stack and pull rolls. In addition, the cooling capacity of the rolls is important so they are properly sized to handle the capacity of the extruder. Rolls are engineered to ensure quality sheet is produced.

Rolls can be embossed to impart a surface finish on the sheet. They can provide specific patterns or textures, such as prismatics for the lighting industry, grained surfaces for luggage, paneling, and decorative overlays, etc. Embossing can be done continuously during sheet (also profiles, etc.) extrusion with little or no decrease in output. Depending on pattern and plastic used, colder than the usual roll temperatures are used, such as 50°C (125°F) on the embossing roll, to prevent recovery of the surface impressions.

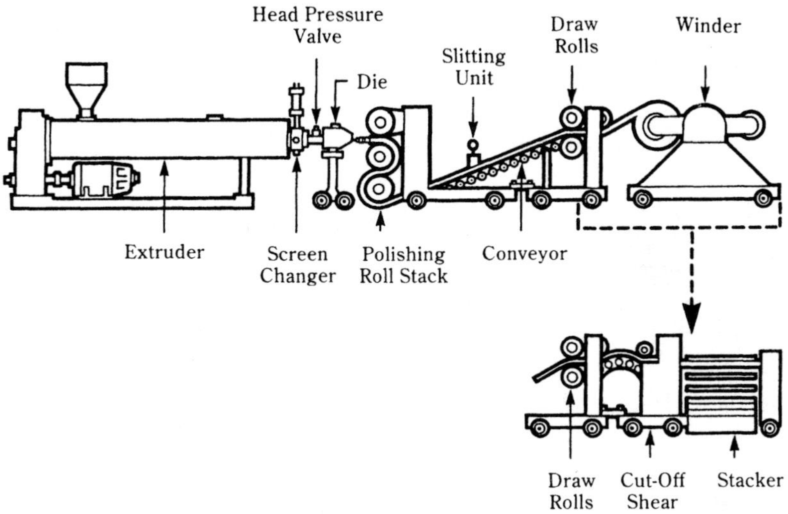


**Figure 9.11** Example of sheet line components.

The behavior of the melt at the first stack roll nip is important to examine. Assume that the die is properly adjusted so that the web will have a uniform thickness when it reaches the nip. During operation, a small roll of melt forms at the nip; it is called a bank or bead. In normal operation, it should be the same size all across the sheet. If it is not even, there is probably a thickness problem requiring roll adjustments to be made.

Proper cooling requires an appropriate roll diameter. If the rolls are too small, the line capacity will be limited. When rolls are too large and improperly cooled, the sheet may not lie flat after cooling or there can be excessive sheet sagging between the die and the roll nip.

It is also important to have the chilling rolls wider than the sheet being processed in order to obtain the proper heat profile on the side edges of the sheet; also the sheet has to be wider than what will exit the line since allowance should be made for shrinkage and trimmings. Note that chilling and cooling refers to a lower temperature than the extrudate temperature.

Polishing rolls cool the sheet, corrects for minor gauge variations, and imparts the desired surface to the sheet. The quality of the sheet is directly related to the surface quality of the stacked rolls. Therefore, chrome plated rolls which have been ground to a surface of 76–150 nm (3–6 μin) are usually desired for good gloss. The rolls must be perfectly round and revolve usually with equal peripheral speed. However, roll deflection could occur when operating under high pressures and low viscosity melts. The mating rolls, particularly down-stream, could belly so center of sheet is thicker than outer sides. To compensate for this action, rolls are made to be bent in the same direction with the resultant sheet having a uniform thickness.

There are different roll drive arrangements for the roll stack. Each provide specific advantages for particular applications. No one drive system is perfect for all applications. One is a silent chain drive (Fig. 9.12) where one motor drives all rolls at the same speed. This system provides: (1) best for all-round use; (2) high-speed line operation; (3) easy operation; (4) easiest maintainence; (5) pneumatic roll closing; (6) limited differential slip drive on the last roll; (7) quick roll change; (8) lowest cost; and (9) other features.

With the helical intermeshing gear drive, one motor drives all rolls, and all rolls are the same size and run at the same speed. This system is best for processing thick industrial sheet, particularly HDPE, PP, and also some optical quality sheets. The system provides: (1) best low to medium operating speeds; (2) more complex operation; (3) high power transmission capability for calendering; (4) hydraulic roll closing; (5) limited differential slip clutch on last roll; (6) quick roll change; (7) minimal chatter; and (8) other features.

The individual roll drive has a separate motor drive for each roll so that they can be run at different speeds. It is recommended for optical quality sheet and thick sheet. This system provides: (1) rolls can differ in diameter; (2) most complex operation; (3) most complex maintainence; (4) high power transmission for calendering; (5) hydraulic roll closing; (6) complete freedom of roll speed differential; (7) quick roll change is standard but more difficult than with the gear drive; (8) lowest possible chatter; (9) potential roll damage if rolls touch each other; (10) best operation for low and medium speeds; (11) most expensive; and (12) other features.

Pneumatic roll actuation is standard on chain driven lines. Most sheet applications are well served by these pressure actuation devices, which provide up to 750 ton/m (400 lb/in) although much less is required. In many cases, the perceived requirement for high pressure is only due to incorrect operating conditions and results in highly stressed sheet (frozen stresses, see Chapter 3).

Regulating pneumatic pressure can be done with much greater delicacy than with hydraulic pressure. Pneumatic regulation allows roll floating for many thin sheet applications. In this operation, the roll opening is not fixed but is allowed to follow the sheet thickness to produce the lowest possible internal stress. In comparison, hydraulic regulation is standard on the above-mentioned gear and individual driven systems. Hydraulic oil closing with higher pressures can be supplied when required.

**Figure 9.12** Side frame of Welex's 3-roll sheet take-off showing standard silent chain driven rolls with cover door guard opened to show details.

A means for setting the roll gaps using positive stops, and precise control of the roll pressures, are important factors. There are micrometer precision manual roll gap adjustments that can be made. Fully automatic roll gap adjustment control uses sensors. A feedback system maintains specific preset roll gaps. Also available are roll gap indicators with analog or digital readout.

The volume of water and oil through the double-shell rolls is a major consideration for cooling and temperature control. The rate is usually at least 40–60 g/m. Treated water in a closed loop is preferred to minimize scale built up on the rolls inner walls. The target is to have the water flow in a turbulent rather than laminar flow pattern. This turbulent pattern permits maximizing, with speed, the proper transfer of heat from the plastic through the roll's steel outer wall and the water [2]. To ensure heat control, temperature indicators should be used on all rolls at the inlet and outlet water circulating lines.

The cooling capacity of the roll stand must be sized to the extrudate exiting rate of the extruder. A built-in closed loop pressurized water system with large-volume, high-velocity circulating pumps can provide accurate and uniform control of roll temperatures. Use is made of a continuous water flow heat exchanger system that eliminates water hardness build-up. A built-in closed loop pressurized water system with the large-volume circulating pumps provides accurate and uniform control of roll temperatures.

Temperature gradients are usually about 0.7°C (1°F) across a roll at full line output. Modulating valves precisely control the temperature for each roll, without the usual damaging pulsation's produced by solenoid valves. A standard cooling system operates up to 135°C (275°F) for conventional plastics. There are plastic melts requiring higher temperatures, such as 200°C (400°F). To provide this extra heat, high-pressure water and thermal fluid systems are used.

Stock temperature, sheet thickness, and linear travel rate are important for optimizing roll diameter. As an example, when processing ABS at 365 kg/h (800 lb/h) using a 11.4 cm (4.5 in) screw extruder sheet line, it would be sufficient to use 30 cm (12 in) diameter rolls. At rates over 910 kg/h (2000 lb/h), rolls could range from 60 to 70 cm (24 to 30 in) in diameter. Usually having rolls too large is not a problem since the roll temperatures can be elevated.

In certain cases, double width dies are used and sheet is split in the center to reduce the linear rate or the sheet is turned 90° in order to reduce the linear rate. There are also occasions where a fourth or fifth cooling roll horizontally immediate follows the three roll stand (Fig. 9.13).

The three roll stands can be in different positions to meet different plastic material requirements and surface finishes desired. Many years

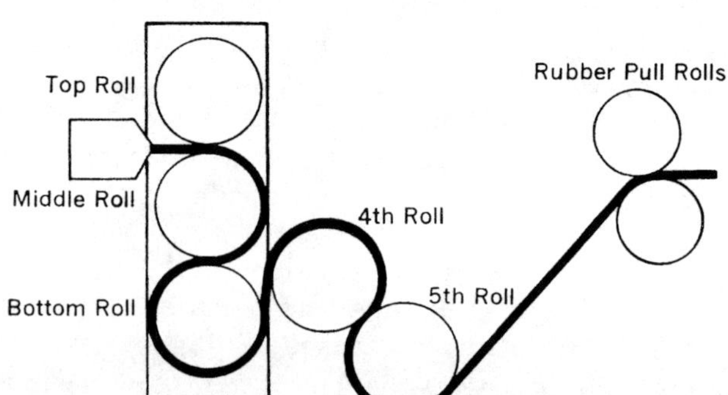


**Figure 9.13** Roll-down stack with five cooling rolls.

ago, virtually all sheet line three roll stands were in the downstack configuration, i.e., the molten web entered the upper-most of the two available nips, and S-wrapped downward. Some years later, the upstack stands were used where the web entered the lower nip and S-wrapped upwards.

There are now roll stands with downstack (Fig. 9.7), upstack (Fig. 9.8), inclined upstack (Fig. 9.14) and horizontal stack (Fig. 9.15) positions. These figures show a system that uses an overhead gantry and universal-jointed die to allow the die to follow the adjustments of a specific roll stack. Sheet take-off with downstack orientation are best suited for most applications. For thick or embossed sheet, upstack orientation is preferred. A somewhat thick sheet, such as for sheet up to 10–13 mm (0.4–0.5 in) thick, processes better since the sheet can 'fall down' to the pull rolls rather than being pulled up. The others are used to meet special requirements. As an example, the inclined upstack can permit enhancing the quality of the sheet since the die can be oriented at a 45° angle to the roll stack, minimizing droop and any resulting air entrapment.

In general, plastic melts will stick to and track the first roll they contact, or the warmest roll if given a choice. Since tracking the center roll of a downstack at the primary nip is the objective, and the central roll is usually the warmest of the three, first contact is usually intentionally made on this roll through the die and nip roll height relationship. With very low viscosity melts of certain engineering plastics and with very thick sheet, there is a tendency for the melt to drape between the die and nip roll. At some point this drape makes it impossible to contact the center roll first on an upstack line. This same situation is true on the inclined upstack lines. On a downstack, the drape naturally goes to the center roll, avoiding potential tracking, premature cooling, and folding problems. Also other advantages exist for downstack.

Advantages in going upstack include that, on larger lines, the extruder barrel/die centerline is lower, minimizing platform height, stairs, etc., unless the extruder is raised or the rest of the line lowered. The primary surface from the center roll faces up on the conveyor, making inspection easier and avoiding idler scratches on this surface. Gravity assists the pull roll drive, as the conveyor is downhill. A rubber roller in the kiss roll (bottom) position can also be helpful with its capability to be surface cooled in a water trough.

In addition to the usual standard mirror finished rolls, a variety of matte, textured, decorative, or engraved embossing rolls are used. The finishes include prismatics for the lighting industry and grained surfaces for luggage, paneling, and decorative overlays. They can be applied on-line during the sheet manufacture with little or no decrease in output. These rolls come in certain sizes, so it may require special provisions if the roll does not match what is in the roll stack. With chain drive, use is made

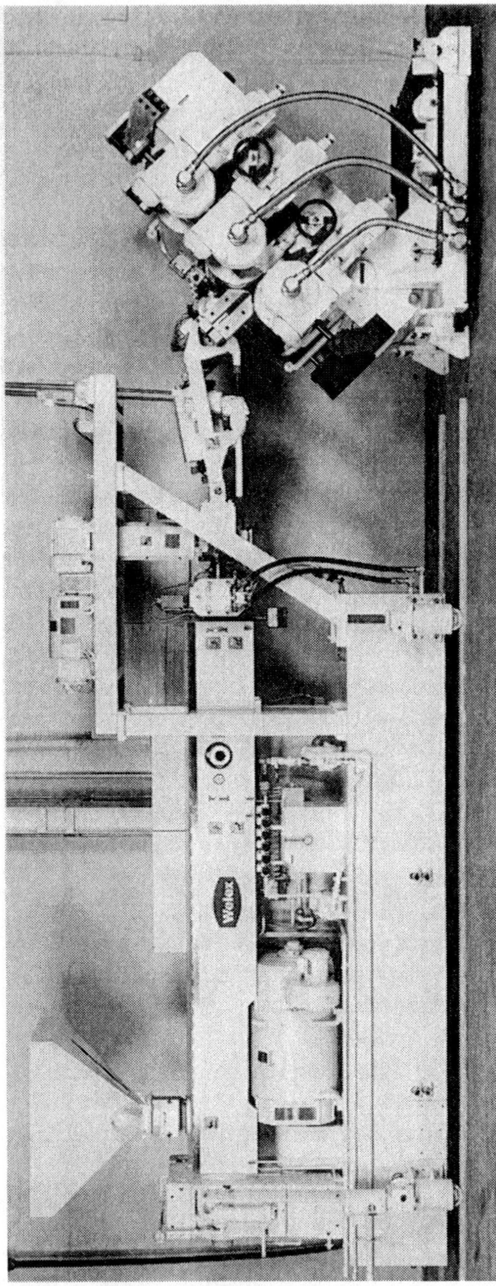


**Figure 9.14** Universal Unistack R sheet take-off from Welex in the inclined up stack position.

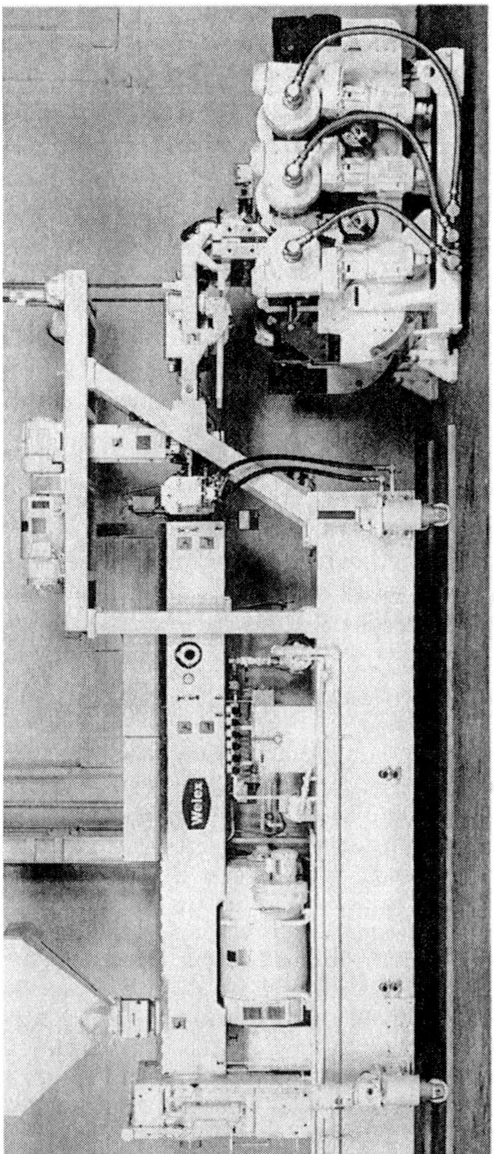

**Figure 9.15** Universal Unistack R sheet take-off from Welex in the horizontal stack position.

of an appropriate sprocket to develop the same roll surface speed. Individual roll drive permits precise speed matching of different roll diameters. Direct drive systems can only use the same diameter roll.

When embossing the usual one side of the sheet, an upstack is used with the embossing roll in the middle. This set-up aids in obtaining the detail of the embossed roll pattern because the melt temperature is highest at the exit of the die. In addition, the embossed surface is up and will not be scratched as the sheet is pulled across the take-off and toward the other equipment down-stream. It is also possible to emboss with two rolls providing impressions on both sides of the sheet.

Depending on the pattern and whether post-extrusion shaping is intended, different roll temperatures are required to permit retaining the surface impressions. The embossing roll is usually maintained at or below 51°C (125°F).

## Coextruded and laminated sheets

As reviewed in Chapter 2, coextruded or laminated material using many different lay-ups can be extruded (Figs. 2.29 and 6.2). In turn, the coextruded melt can be directed into the three roll stack to produce sheets.

**Table 9.5** Examples of coextruded products

| *Composite raw materials* | *Product characteristics and applications* |
|---|---|
| PMMA1/PMMA2 | Ultra-high-transparency sheets with various tints, e.g., for use as glazing. |
| PMMA/ABS | Sheets with scratch-resistant surface for sanitary applications, e.g., sanitary blocks and baths. |
| PMMA/ABS with MB/ABS-scrap | Colored three-layer sheets containing a high proportion of recycled material (70–80%), for sanitary applications. |
| ABS/ABS-scrap/ABS | Three-layer sheets containing a high proportion of recycled material (70–80%), e.g. for the production of cases. Characteristics of the ABS layer are its resistance to Freon and its gloss finish. These sheets are particularly suitable for refrigerators/freezers. |
| PS/HIPS with MB/ HIPS-scrap | Three-layer sheets incorporating regrind from production scrap (65–80%); particularly suitable for refrigerators/freezers. |
| PC(UV)/PC/PC(UV) | Transparent or colored sheets for building applications, suitable for use as safety glazing and billboards. Characteristics of these polycarbonate sheets are their high impact resistance and an integral UV-resistant layer. |

Coextrusion or laminating different plastics produces commercially useful products (Table 9.5, via Reifenhauser). Figure 9.16 shows a close-up of Davis-Standard's high barrier 7-layer sheet line.

Different patterns are used to meet industry requirements. Figure 9.17 provides an example from Welex of a three-color sheet that is thermoformed into a container with an integral hinged lid. Figure 9.18 is their economical approach where the solid side strips can be used to produce thermoformed products from the two-color center; after thermoforming the solid side material can be trimmed and recycled for use again as side strips or as a center layer in a three-layer coextruded composite construction.

Laminated sheet products also can be made using the three roll stacks. In this case, a film or any web material is unwound from a roll (pay-off station) supported above the stack (Figs. 9.19 and 9.20). The film is fed into the nip together with the molten web that exited the extruder die. The film bonds to the sheet and transfers its special surface properties. The pay-off

**Figure 9.16** High barrier Davis-Standard 7-layer coextruded sheet line.

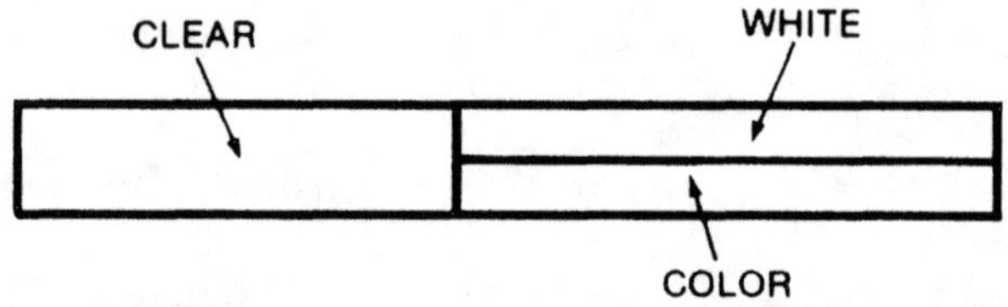


**Figure 9.17** A coextruded sheet that can produce a 3-color thermoformed container with an integrally hinged lid.

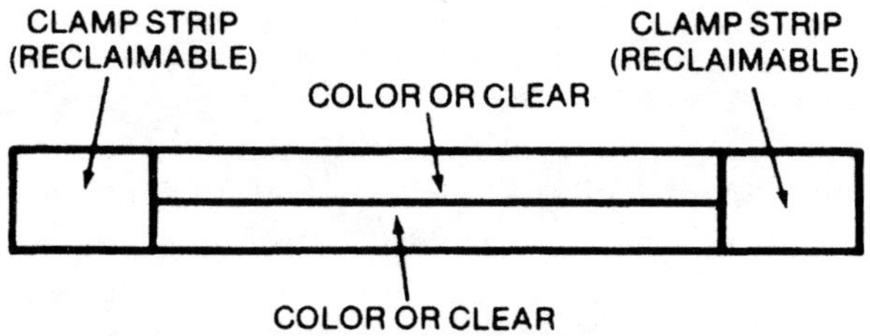


**Figure 9.18** Addition of a single-plastic clamping strip on each side of a coextruded sheet permits scrap reclaim of the thermoformed trim waste.

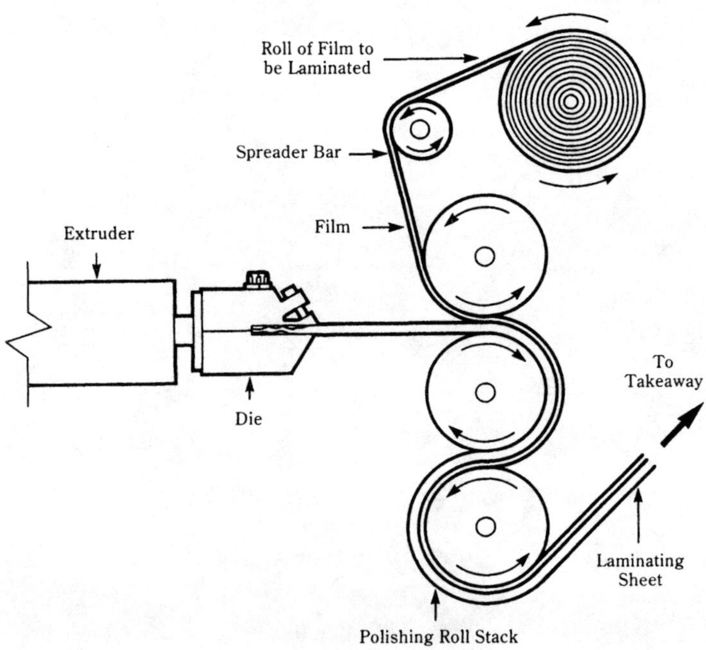


**Figure 9.19** Example of a film-to-sheet lamination.

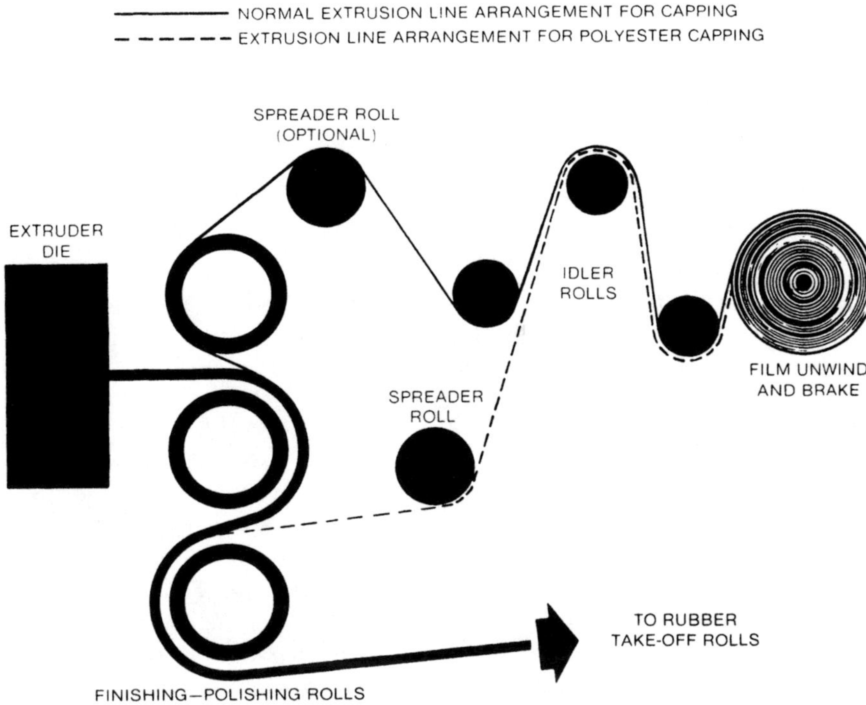


**Figure 9.20** Sheet capping lines are arranged for capping or lamination.

stations maintain the laminating film under adjustable tension, and can provide lateral adjustments. To eliminate problems, it is necessary to maintain uniform sheet thickness, keep orientation low, avoid wrinkles, and prevent air bubbles between laminations.

An example of an application is a protective film that can be applied to optical sheets. Film lamination, originally developed for refrigeration applications, is used to improve gloss and provide excellent oxygen and moisture barrier properties. This more expensive method is used in applications where the added cost is justified.

There is a heat glazing method applied to the surface of the sheet normally being processed for improving gloss that is less expensive (Fig. 9.9). It may not have as good a gloss nor as uniform as the laminated process, it does improve impact strength. Due to the high temperature used, the product could have discoloration and be brittle.

The method of manufacture is to increase the distance between the die and the three roll stack which would normally be using polishing rolls. A support is used to convey the extrudate that is made up of 25 mm (1 in) diameter hollow stainless steel rods that are spaced 50 mm (2 in) apart.

These rods are rotated to balance extrudate take away speed. They are cooled with water to prevent plastic sticking. This approach prevents the melt sagging with output rate and thickness not adversely affected. Glazing occurs with the infrared heaters located over the melt's surface. The number of heaters with their spacing and locations above the extrudate, will determine the degree of gloss.

### Pull roll

A set of pull rolls (usually rubber covered) down the line moves the sheet off the final cooling roll and along the cooling conveyor (Figs. 9.1 and 9.2). The cooling rolls are also driven, but at a slightly lower speed than the pull rolls. These speed adjustments are critical during the set up. This action keeps the sheet flat and taut during the cooling process. The pull rolls keep the sheet moving and feeds it to the end of the line. They provide no cooling or shaping.

The pull rolls' speed is usually automatically proportionally tied into the cooling stack rolls' speed so dual adjustments are not needed. The unit should be equipped with a pull roll differential speed control. The pull rolls can then be adjusted to run slightly faster than the cooling rolls.

For a sheet line, the usual three rolls are actuated by pneumatic or hydraulic cylinders. This precision stand requires the structural support to permit accuracy in providing flat, smooth, roll surfaces, uniform independent temperature control on each roll surface, and very accurate speed (rotational) control.

### Air knife

An air knife provides a means to produce thin PP sheet on a conventional three roll sheet line. (They are also used in other extrusion lines that include cast film, coating, and lamination.) The air knife ensures positive and firm positioning of the sheet against the roll. Automatic die lip adjustment is used. High-efficiency internal design provides precise pressure uniformity and optimum air impingement.

There are designs to quickly position into and retract out of the open upper roll gap of the downstack three roll take-off (Fig. 9.5). Micrometer type adjustments permit optimization of the air position and impingement angle, making the sheet conform to the middle roll for uniform cooling.

### Trim and slitting

Up-stream of the pull rolls, trim knives are used. They provide the means to edge trim both sides of the sheet and also when required slit the sheet into multiple widths using specialty designed razor slitters. These assemblies can be moved to the best locations on the idler conveyor section of the sheet line (Fig. 9.21). Figure 9.22 shows some of the products slit from sheet; also applicable to flat film lines. For very thick sheets, heated rotary slitters can be used (Fig. 9.23). The aim is to process a sheet requiring very little wasted trim even though it is usually recycled.

**Antistatic bath**

An antistatic bath is used when surface application of a slip agent is required to aid sheet product de-nesting, or when required to provide a

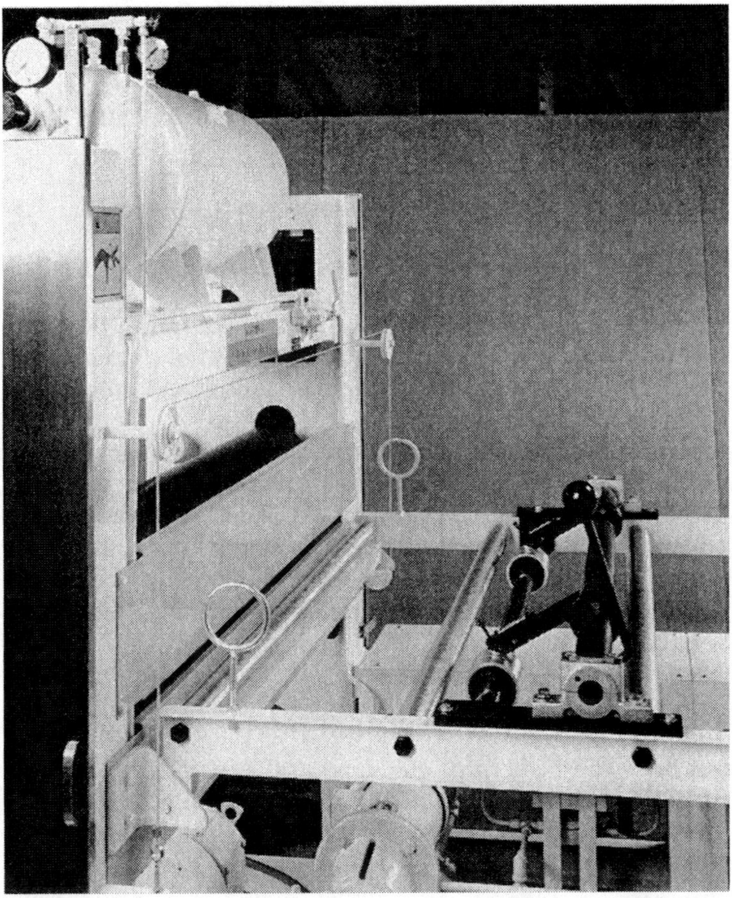

**Figure 9.21** Welex sheet take-off showing detail of standard razor edge trim system.

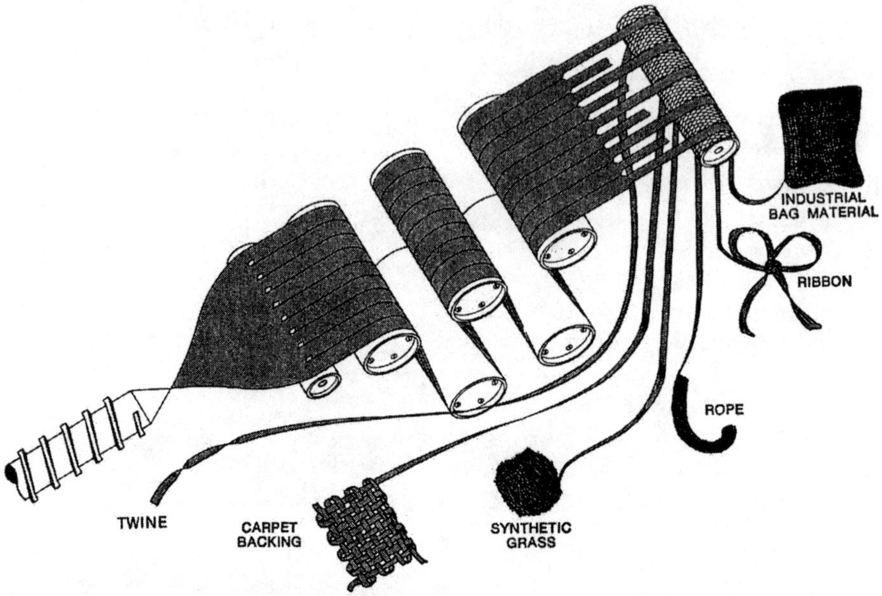


**Figure 9.22** Examples of products from slit tape.

surface treatment on the sheet with antistatic solution to protect against static dust pick-up. Figure 9.24 can be used to treat one side or both sides of the sheet. These baths should be capable of easy and quick installation as well as removal.

This device has a pneumatically operated dip roll for easy threading. Its ultra-soft polyurethane squeegee rolls provides an easy and simplified controlled method to remove excess liquid. When its location is adjacent downstream to the polishing roll stack, it has the capability to provide extra cooling of the sheet. These units are designed with liquid circulating pumps and heat exchangers to maintain optimum bath temperature for high capacity sheet lines.

**Process control**

The sheet lines can be set-up with the latest process control. Simplified line control systems are shown in Figs. 9.25 and 6.8. A single side-to-side motion for a thickness gauge on a Complex sheet line is shown in Fig. 9.26. Information on process controls is in Chapter 6.

The controls can operate from the start to the end of the line. The controls would start by ensuring that the material handling equipment feeding the hopper is properly operating. The extruder can be pumping

**Figure 9.23** Welex sheet take-off with rotary slitting system.

out though the die the desired melt. In turn, all the down-stream equipment to the sheet stackers or winders would be controlled.

A central microprocessor control for the complete line can be used. Software is available that has been preprogrammed providing precise, repeatable, central control of the sheet extrusion line. The control display would provide the information required to set and fine tune the equipment in the line. It provides an easy approach to vary the settings of each operation and interrelate them.

## Cut-off

At the end of line, thick sheets can be cut to length using equipment on casters to fit the take-off tract. There are different types of cutters to handle the variety of different plastic sheets that range from being brittle to nonbrittle (Fig. 8.4). They are designed to cut the sheet uniformly and prevent chatter marks in the sheet. Thickness up to 8 mm (0.30 in) are usually sheared but it depends on the plastic's cutting characteristics.

**Figure 9.24** Sheet take-off showing Welex's anti-static bath (guard cover removed to show design).

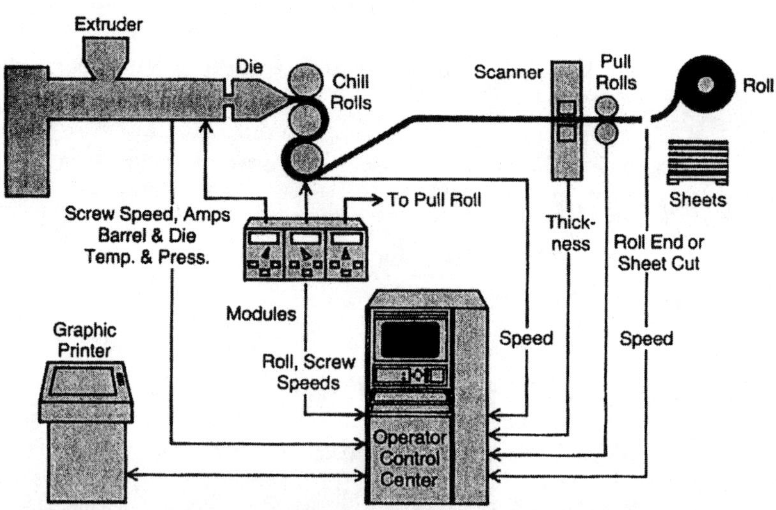


**Figure 9.25** Example of process control in an extrusion line.

**Figure 9.26** Traversing thickness gauge.

Thicker than this amount usually requires sawing, although heavy duty shears are available.

There are cut-off shears that operate at different speeds, thickness capabilities, and accuracy of cut. These shears are automatically precision-controlled by digital-set cut-off length control, such as used in Fig. 9.27 by Complex in one of their lines.

For the thick and very brittle sheet, cut-off saws are used. They usually travel with the sheet during the cutting cycle. The cuts are made with very clean edges. They tend to be dusty and noisy. Hot knife cut-offs are rapidly replacing them. Use of de-ionized air will reduce this dust pick-up and also the dust from static charges generated during processing.

When using heated steel rule dies, they cut through most of the sheet thickness. A thin residual web that remains is broken by a subsequent bending of the sheet. The hot knife cut-offs have a clean cut and are quiet, however they sometimes make a slightly rounded edge.

The cut sheets can be stacked manually or, as done more often, stacked by commercial stacking units. Pallets of stacked sheets are usually moved and stored manually or automatically into storage areas or directly into transportation vehicles. They may be wrapped in 200–250 μm (8–10 mil) PE film to reduce moisture absorption and keep the sheets clean and protected. Moisture pick-up by certain plastics can affect the next step where they are further processed, such as thermoforming.

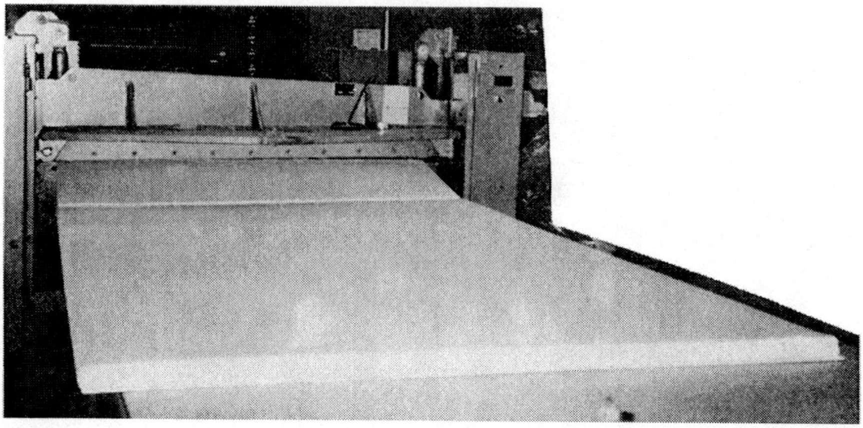

**Figure 9.27** On-line shear cutter with sheets being stacked.

## Winder

If continuous wound rolls are required by the customer, tension controlled winders are used in place of the cut-offs. There are many different types of winders to meet the different wind-up requirements, usually depending on sheet thickness. As a guide, if a sheet is less than 1.8 mm (70 mil) thick and is flexible with high impact and toughness, it can be wound up into rolls of at least up to 900 kg (2000 lb). A minimum core of 200 mm (8 in) is generally used. The plastic has to be sufficiently flexible to avoid skin layer stressing which initiates microcracking. The winders include those with fixed shafts, cantilevers, and turret winders. Example of turret winders are shown in Figs. 9.28 and 8.6.

This type of product can be used for thermoforming (Chapter 18) in a secondary operation or in-line where the sheet goes directly to the former. In-line use permits cost reductions by retaining some of the heat in the sheets and eliminating windup, roll handling, and any potential contamination [288].

To date the best tension control can be obtained with a build-up ratio not exceeding 6/1 from the core diameter to the finished roll size. Thus, a 75 mm (3 in) core diameter is best suited for rolls not exceeding 450 mm (18 in) diameter. Cores with 150 mm (6 in) diameters are suitable for rolls up to 900 mm (36 in), while 200 mm (8 in) cores should preferably be used for rolls up to 1200 mm (48 in) diameter. Square shafts with inflatable quick-release pneumatic core chucks can be used.

The fixed shaft winders are the least expensive and most widely used winders. They can have two or more shafts with each shaft having its own adjustable tension drive. Figure 9.29 is a triple shaft winder. The idea is to

**Figure 9.28** Turret winders operating.

**Figure 9.29** Triple fixed shaft winder from Welex for winding sheet into roll stock.

have one empty shaft for changeover than the number of web widths to be wound. For more that two widths to be wound, the winder can have staggered shafts.

There are fixed shaft winders that can be used to make very large roll stock for in-house use. They are used by processors of sheet so that long uninterrupted runs can be made. An example of a low-cost winder is shown in Fig. 9.30 that provides for winding rolls up to 1800 mm (6 ft) in diameter. The rolls are stored on the mobile winding chart and taken to where it will be used, for example in a thermoformer. Two winding positions, with an overhead feed system, permit easy changeover at high speeds. The winders are powered by plug-in hydraulic motors which remain suspended from the overhead frame. Single or dual width systems can be used.

The cantilever type winders provide from two to, more usual, many shafts. They are supported at only one end of a shaft so when a roll is wound, it can easily be removed. Each shaft is individually driven and has its own adjustable tension control. This type of winder permits winding simultaneously a sheet that has been slit into many sections. These turret winders are extensively used in film extrusion lines.

All types of turret winders are designed and used by the sheet processors that are operating the fastest speed lines. They permit roll changeover

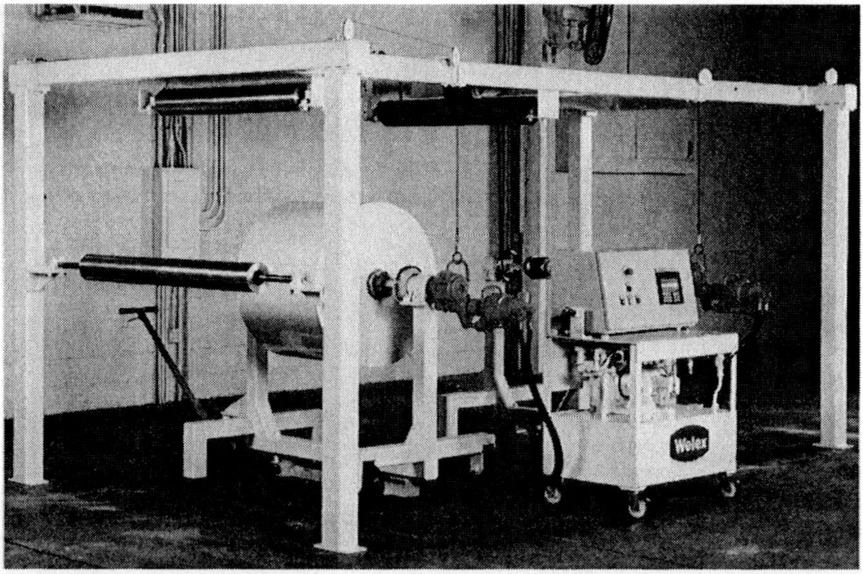


**Figure 9.30** Jumbo sheet roll winder system with hydraulic winding unit from Welex.

by indexing the unit's system prior to change. The result is that there are no loose ends to handle and line speeds do not change.

## DIE

The obvious function of the die is to control the shape of the extrudate. The key word is control. In order to take this action, the extruder must deliver melted material to the die at a constant temperature, pressure, and output rate (Chapters 2–6).

There are many different designs used for sheet dies that range in exit widths of at least 3 m (10 ft). However, the common feature in designing sheet dies is their tapered cylindrical shaped internal die manifold distributor channel. They would have a diameter of about 38 mm (1.5 in), a 10–20 mm (0.4–0.8 in) long die orifice with a gap 0.35–0.60 mm (0.014–0.024 in), an intermediate retarding zone of depth 3.2–6.4 mm (0.125–0.250 in), and enlarged melt compensating zone. All interior surfaces of the die should be chrome plated to minimize flow restrictions and improve the surface characteristics of the sheet.

The preferred feed is central to the axis of the collecting channel, such as the T-type, fishbone, and coathanger dies. End fed or transverse fed dies are rarely used (Chapter 5). Wide sheets are easier to manufacture and, more important, also do a far better job of distributing melt since it has only half as far to travel in the die.

The entrance to the die is an opening of circular or oval cross section. It flares out into the manifold and distributes the melt along the width of the die. Several cross sections of the manifold are used, such as a circular, teardrop, flat sided teardrop, and angular. The circular is the lowest cost to manufacturer, but the teardrop shape has the most desirable performance since it is the most streamlined.

The coathanger's main channel of the manifold angles toward the die lips (blades) to make a triangular preland section. The land length between the main manifold channel and the die lips decreases with distances from the feed point. This results in larger pressure drops on the preland near the feed point and less near the ends of the die. This action compensates for the pressure drops in the main manifold feed channel which increases with distance from the feed point. By proper selection of the angle, it is possible to obtain a fairly well balanced flow at the die lips so that a minimum adjustment occurs to obtain a uniform sheet thickness.

Restricter bar assemblies are an important valving unit since they are a major controller in obtaining uniform sheet thickness. The die lips, while having as large an L/D as possible, are still not an effective enough constriction to act as a flow control for uniformity. The die lips definitely have their place in providing the final shaping element in the process. The basic pressure flow control and, consequently, the overall thickness control are upstream of the die lips. The final adjustment is made by adjusting the die lips. The restricter bar is normally required, although some dies eliminate it and use only the flexible lips for control. For optimum versatility, both features are used in most dies.

An important adjustment concerns the die lip gap. This spacing is generally equivalent to at least the sheet gauge desired. Heavy gauge sheet of over 3.2 mm (0.125 in) may require a setting 5–10% greater than the sheet thickness for optimum gauge control. For the high viscosity thick plastic sheets, such as PC, may require 10–20% greater gap spacing than the sheet. All settings should be uniform across the die width.

There are different designs to aid in controlling the melt flow, such as a flexible preland in the manifold and a flexible lip. They were initially used to process rigid PVC. The length and thickness of the flexible steel must be carefully calculated. It has to permit a sufficient range of motion with the screw adjustment pressure without making the section too thin to hold the melt pressure.

The choice of the die is influenced by the quality of the sheet required. As an example, when processing PS, the melt could be run through dies which have either constant diameter or tapered manifolds. With ABS, the streamlined manifold dies, such as the coathangers with flexible lips and minimum dwell time, are used. Since ABS is more viscous than PS, the ABS die should be constructed to operate at pressures of 10–22 MPa (1500–3200 psi) for 120 mm (4.5 in) extruder. Lower pressures are used with PS. With the higher viscosity plastics, such as PC, the internal pressures require the dies to be stronger so no die deflection occurs during processing. The dies for PC has to withstand pressures of 28–42 MPa (4000–6000 psi). Even higher pressures can be generated when thin sheets are processed. The dies require very narrow die gap settings of possibly less than 0.8 mm (0.030 in).

The die length for ABS or PS should be selected based on the expected sheet thickness. A land length of 51 mm (2 in) should be considered for sheet from 1–4 mm (0.040–0.150 in), while sheet up to 13 mm (0.500 in) may require a land length of 76–100 mm (3–4 in). The longer land increases the internal pressure to allow better control of the melt in the die and aids in the improvement of the sheet's surface quality.

Both sides of the dies should be equipped with heaters that control the die temperatures in several separate zones across the die length and width (Fig. 5.6). This action allows better control of the melt temperature and flow through the die. It also can be helpful in distributing the melt out to both ends of the die. Even though Fig. 5.6 refers to an option on the die's end caps, they should be heated with their separate temperature controls.

The temperature of the die is set approximately at the melt stock temperature entering it. The target is to simply keep the melt constant and not let it go through hot and cool cycles. The end zones of the die are usually set 5–10°C (10–20°F) hotter than the center zones to help distribute the melt across the complete die.

The melt profile exiting the die is balanced by adjusting the restricter bar. The normal settings of the restricter bar would have it bowed slightly in the middle, restricting flow in that section and forcing the melt to the ends. Once properly set, the melt profile for a given plastic will remain constant. Changes should not be necessary, and a need for frequent changes may be a sign of problems upstream of the die.

Narrow sheets can be made on a wide die by blocking (damming) both ends of the die deckle bars (or plates). The deckle bars or dams are sometimes also used to adjust sheet width to minimize scrap and regrinds. There use could cause melt hang up in the die. See Chapter 5 for details. For long runs, a properly sized die should be used to eliminate the stagnant melt flow areas behind the deckles where degradation usually occurs.

In addition to the monolayer sheet die, the coextruded die is also very popular. There are the two usual configurations. One has the separate extruders independently feeding the manifold. The two or more melt streams are combined in layer configuration just before exiting the die. The other configuration makes use of conventional type sheet dies. The separate melt streams are brought together in a feedblock unit before entering the die (Chapters 2 and 5).

## APPLICATIONS

Products for sheet materials are rather extensive. A major user is the converting industry where sheets (as well as films and profiles) are thermoformed. Examples of their products include packaging, boats, camper tops, spas, hot tubs, luggage, aircraft interiors, recreational vehicle parts, credit cards, and so on. Different plastic materials are used to produce monolayer or coextruded sheets for these applications [382].

The different constructions of these sheets provide different performance advantages for the different products. Performances include: (1) high gloss expensive layer over tough nongloss less expensive substrate; (2) thin weatherable, usually expensive, layer over a tough, less expensive, nonweatherable substrate; (3) low gloss layer over a tough, lower cost, substrate; (4) thin expensive decorative (wood grain, marble, etc.) layer over a tough, less expensive substrate; (5) vapor barrier over a tough and strong substrate; (6) chemical resistive layer over a less expensive substrate; (7) soft touch layer over a cost effective tough substrate; (8)

correctly color matched cap over a noncolored less expensive substrate (Fig. 9.20); (9) sheet with a core of recycled plastic with different overlays; and (10) others.

For the record, PVC was the first extruded plastic in the late 1920s used as sheet flooring, insulation, pipes and fittings, electrical fixtures, etc. By the late 1930s, acrylic and PS sheets and films were introduced for signs and glazing. ABS was introduced commercially in the late 1940s, with HDPE in the early 1950s.

## TROUBLESHOOTING

This chapter, as well as most of the others, include a troubleshooting section. The text within these sections contain very useful troubleshooting information applicable to sheets. As an example, if you are going to experiment with thin sheet using a high-melting point plastic (PET, PC, etc.) for the first time, be aware that it is possible for the edge beads to solidify on your roll stack very quickly (Chapter 8). It would put a very high local pressure on the rolls. This pressure can be so high that the roll surface is permanently damaged with a groove on each side.

Even though the following review concerns principally flat sheet extrusion, it provides helpful ideas for the other extruded products. Melt strength effects in a coextruded sheet can be a critical factor in its production. Some combinations of plastics may be critically affected by the melt strengths of skin and core. Slight modifications of skin melt strength can move a product suffering nonuniform draw down from a die to excellent quality. An easy way to determine if melt strength effects are controlling the line is to add a comparable higher melt strength component. In olefin plastics, the addition of LDPE will often raise melt strength. Once determining that this effect is in control, contact your plastic material supplier about possible plastic changes, or if the LDPE does not adversely effect your product's performance, continue to use the blend. If preliminary runs with sample plastics were initial made, the problem would have corrected itself.

Raising melt temperatures can be helpful or could be damaging if not properly limited and controlled. In raising the melt temperature of PC to lower its viscosity in order to match the other plastic, the processor must avoid burning the other plastic which usually has a much lower melt temperature. This is particularly true with EVOH that is commonly used as a gas barrier in various coextruded products.

The primary sources of poor viscosity can arise from several different sources such as: (1) poor melting; (2) nonuniform melt temperature profiles entering the die; and (3) poor plastic selection giving a melt viscosity mismatch between layers. In reality, melt disturbance is due to the poor viscosity match of melt layers. The source of the upset determines what kind of disturbance occurs and, therefore, what is the appropriate troubleshooting approach.

To avoid all kinds of melt disturbances, put the low viscosity plastic on the outside layer, otherwise the layers will rearrange. This is best controlled by plastic viscosity selection or melt temperature levels. The first and second kind of melt disturbance listed above is due to something being broken. In the first kind, unmelted plastic enters the die because the screw is not preparing the proper melt. Process changes may fix the problem, or a new screw may be required. This type is usually the result of an output increase exceeding the capabilities of the screw.

Poor melt temperature profile control is most often due to problems with a melt pipe or perhaps a die heater. Poor temperature zoning or controller tuning, as well as a burned out heater, may be the source of the problem. In order to be effective in resolving these type problems of melt disturbance, one must learn to differentiate between the primary sources of poor layer viscosity control [187, E. Mount].

When running any coextrusion line make provisions to adequately alarm the operators if one of the extruder drives trips out and the other has to be stopped. This action will prevent extrudate from back flowing in the stopped machine and creating a blockage in the feed section. The machines should have adequate controls to have the remaining machine that is operating to automatically stop.

Another situation to recognize is that plastics with broad MWD (molecular weight distribution, see Chapter 3) have a greater tendency to exhibit interfacial instability. This situation is due more to layer ratio than processing conditions or die geometries. In contrast, interfacial instabilities in plastics with more narrow MWDs appear to be related to the stress at the interface, and thus can be affected by whatever affects stress at the interface.

Due to reverse temperature profile (hotter at extruder screw feed end) which is typically used for PP, material can build up in the feed section. This can cause a melt block which will result in a reduction in throughput of the extruder. In heavy sheet extrusion of PP, if this problem should occur, the problem can possibly be resolved by cutting edge trim into ice cube size and dumping them into feed hopper. These large pieces will often times scour the melt block away and eliminate the need to remove the screw to be cleaned.

To keep die lips clean during start-up and initial sheet casting when melt can stick to the face of the die lips, rub the lips with a silicone grease made for high-temperature applications, or a high viscosity silicone oil. Keep the layer very thin otherwise it could damage the sheet.

To maintain sheet gauge control, roll bearings must be in good condition. When bearing problems are suspected such as on the center casting roll, there is a simple procedure for checking it. Put a 0.0025 mm (0.0001 in)

**Table 9.6** Summary review on sheet problems

| *Defect* | *Cause* |
|---|---|
| Continuous lines in direction of extrusion | Die contamination or scored rollers or moisture in the polymer |
| Continuous lines across the sheet | Jerky operation; maybe too low a temperature |
| Curved lines across a colored sheet | Color not mixed well enough |
| Discoloration | Too high a mass temperature or too high a percentage of reground material added |
| Bad gloss | Roll temperature too low or defective rolls |
| Dull strip | Die set too narrow at this point |

**Table 9.7** Troubleshooting guide for sheet and film

| *Problem* | *Cause* | *Solution* |
|---|---|---|
| Applesauce | Resin incompatibility | Check resin compatibility (chemical and rheological)<br>Make sure previous resin is completely purged |
| | Temperature too high (degradation) | Decrease temperature |
| | Temperature too low (increased viscosity) | Increase temperature |
| Bubbles or voids | Moisture in plastic | Dry plastic |
| | Air entrapped | Correct extruder operation |
| Color dispersion | Improper feed selection | Replace material |
| | Poor color mixing | Correct mixer |
| | Clogged screen | Clean screen |
| Draw resonance | Draw ratio too high | Decrease die gap |
| | Temperature too high | Decrease polymer temperature<br>Repair runaway heaters |
| Edge tear | Draw ratio too high | Decrease die gap<br>Increase coating weight |
| | Die end temperature too low | Increase temperature of die ends |
| | Improper polymer temperature | Adjust temperature to optimum |
| Excessive odor | Over oxidation | Decrease melt temperature<br>Decrease air gap |

**Table 9.7** *Continued*

| *Problem* | *Cause* | *Solution* |
|---|---|---|
| Feeding poorly | Feed throat too hot | Lower feed zone heat |
| | Regrind fines or too coarse | Clean up feed |
| | Plastic temperature too high | Decrease melt temperature |
| Gauge bands | Unclean die | Clean die interior periodically<br>Scrape inside of die lips to dislodge foreign material |
| | Improperly adjusted die | Adjust die bolts<br>Measure die gap periodically |
| | Temperature variations | Repair heater malfunctions<br>Keep temperature of adapter, transfer pipe, and die close to temperature of extruder melt temperature |
| | Temperature too high | Lower temperature settings<br>Repair malfunctioning heaters or controllers |
| Gels | Plastic contamination | Keep storage and extruder areas clean |
| | Poor purging | Use proper purging techniques |
| | Poor mixing | Increase extruder back pressure |
| Pin holes | Substrate roughness | Increase polymer thickness |
| Surging | Hopper bridging (improper feeding) | Lower feed zone temperature |
| | Improper screw design | Use different resin or screw |
| | Screw bridging | Reduce temperature on profile or run reverse temperature profile |
| Voids | Moisture | Avoid abrupt temperature changes in product storage<br>Check for leaks in resin handling system<br>Protect hygroscopic resin from humidity |
| | Polymer degradation | Maintain melt temperature below recommended maximum |

**Table 9.8** Guide to troubleshooting PET extruded sheet lines

| *Problem* | *Cause* | *Solution* |
|---|---|---|
| Surging in extruder | Uneven melt conditions in extruder | Boost heat in first two heating zones of extruder |
| | Erratic material feeding | Check feed throat for material bridging |
| Die pressure fluctuating | Clogged screen pack | Switch to new screen and cleanout clogged screen. (continuous screen changer) |
| Orange peel surface | Surface viscosity in the die is too high | Increase die or melt temperature |
| | Uneven surface shrinkage from cooling | Raise chill roll temperatures |
| Haze in sheet | Melt crystallizing due to warm chill rolls; excessive melt bank; too slow processing | Reduce chill roll temperatures; minimize melt bank; boost extrusion rate |
| Gels in sheet | Cold spots in melt | Boost die temperature |
| Pick off of plate-out from roll on sheet | Excessive plate-out on roll due to too cold roll | Increase roll heat to remove plate out |
| Curved lines in transverse direction | Melt bank too high, folding onto web | Maintain small, constant melt bank by controlling extruder output |
| Sheet is brittle | Welt resin | Check and adjust drying conditions |
| Surface roughness | Moisture in resin | Dry the material |
| | Air trapped between extrudate and chill roll | Move die closer to chill roll or alter vertical alignment of chill roll and die |
| | Unpolished low spots | Adjust gauge by decreasing flow to high spots and increasing flow to low spots |

indicator on top of the center roll. Put a brass or wood spacer in the lower nip and close the lower roll. The center roll should not move.

With slitter blades on a sheet line, edge cut can be improved by sharpening blades with a bevel to the outside. This way, the rough cut is allowed on the edge trim, but not on the sheet edge.

Tables 9.6–9.8 provide information on the potential problems that can occur in a sheet line. Also with the problems are included potential causes and potential solutions.

# 10

# Calendering

## OVERVIEW

Calendering is essentially extruding a plastic material between successive pairs of corotating, parallel rolls to form a film or sheet. However, most film and sheet are prepared by extrusion techniques (Chapters 8 and 9). These different techniques can provide different: (1) properties, to meet various product performance requirements; (2) processing performances; (3) delivery schedules; and (4) costs. Thermal properties, molecular characteristics, and degree of crystallinity of the plastics are factors which affect the processability of plastics. Additives can have a major influence on processability (Chapter 3)

As a method of film and sheet manufacture, calendering is widely used. Like extrusion, design and construction of the calendering equipment has become increasingly sophisticated and engineered. The progress made, as with extrusion, has been largely empirical and in the direction of better and faster big machines. Understanding the process and the effect of different processing variables also continues to be studied theoretically.

Calendering was developed over a century ago to produce natural rubber products. With the development of TPs, multi-million dollar, extremely heavy calender lines started using TPs and now process mainly TP materials. Another major product continues to be elastomeric tire-fabric coating.

Calendering is a highly developed art and theory, rather recently elucidated by combining the complexity of melt behavior with the mechanics of screw rotating machines (Chapters 2–17). With this understanding comes the ability: (1) to make calenders more productive by increasing their speed; (2) to produce films and sheets with tighter thickness tolerances and greater uniformity; and (3) to handle thicker sheets more effectively.

The calender consists essentially of a system of three or more large diameter heated rolls whose function is to convert high viscosity plastic melt into film or sheet. The equipment can be arranged in a number of ways with different combinations available to meet different product requirements. Typical line set-ups are shown in Figs. 10.1 and 10.2.

The preparation of the material or compound is important to successful calendering. It is usually done by computer controlled electronic weighing scales that supply precise amounts of each ingredient to a high intensity mixer. It is designed to incorporate any liquids into the plastic particles and to secure uniform distribution of all powdered ingredients (Chapter 17).

Blending is generally done for a specific time and to a specific temperature. The still-dry, free-flowing blend is then charged to a feed hopper where it is screw fed into a continuous mixer, such as an extruder and/or kneader [4]. Under the action of a mixer's reciprocating screw in the confined volume of the mixer chamber, the blend begins to flux or masticate into the required plastic state.

Usually the next step is to force the plastic out of the barrel of the mixing chamber through a die producing strands. The strands can exit as a continuous rope or be chopped into small baseball size buns. This hot plastic material may be passed through a two-roll mill and/or be directly conveyed to the top of the calender.

Preparation of stock material for calendering, conditions on the calender, take-off thickness measurements, windup system, and line

**Figure 10.1** Simplified schematic of calendering.

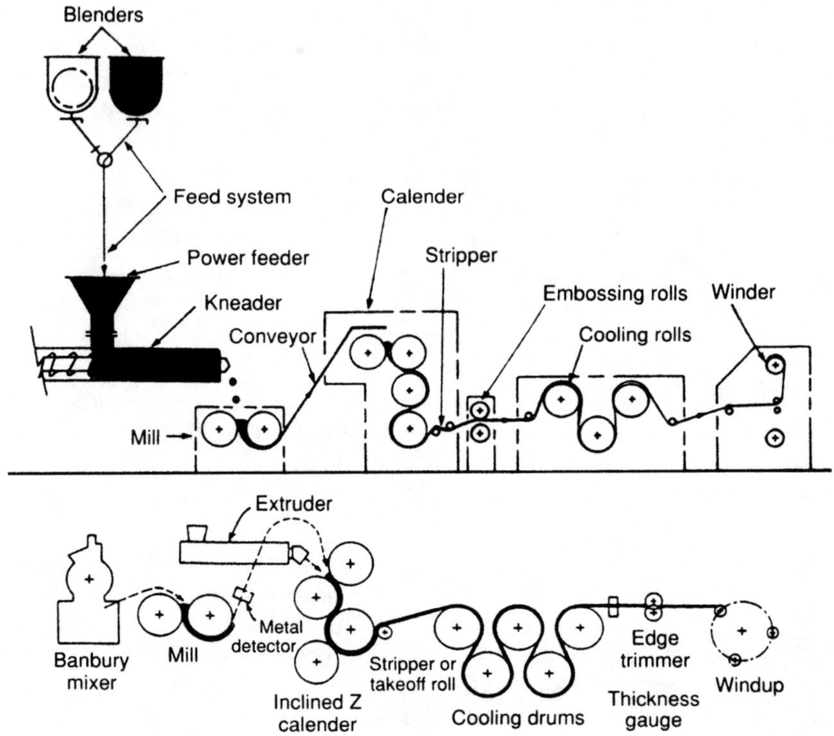


**Figure 10.2** Calendering lines.

control must all be adapted to the plastic compounded system being processed. Other considerations include whether: (1) only the plastic is used; (2) what finish is required, i.e., glossy, semi-matte, or matte; (3) plastic is laminated to a fabric; (4) web is embossed; (5) web is slit in line; (6) very low strain recovery is in the film or sheet; (7) the product needs special properties, such as optical clarity and mono or biaxial orientation; and (8) others. General processing considerations include temperature and pressure controls, calender line speed, dimensions of thickness and width, surface finish, and orientation.

## Calendering or extrusion

Film and sheet can, in principle, be made by calendering or by extrusion. Factors that govern the advantages and disadvantages of each process can interact in a complex way. Factors to be considered include: (1) type of material to be processed; (2) quantity of product to be produced; (3) thickness required on film or sheet; and (4) costs. The capital equipment

**Table 10.1** Guide for producing PVC film and sheet by different fabricating processes

| | *Calender* | *Extruder–calender* | *Blown film* | *Flex-lip extruder* | *Plastisol cast* | *Melt roll* |
|---|---|---|---|---|---|---|
| Lines installed, USA | 155 | 2 | 90 | 40 | 60 | 5 |
| Relative resin cost | lowest | low | higher | higher | highest | low |
| Machine cost ($ million) | 1–6 | 1–2.5 | 0.3–0.6 | 0.3–0.6 | 0.3–0.7 | 0.3–1.3 |
| Rate and range ($lb\,h^{-1}$) | 800–8000 | 500–1500 | 600 (4½ in) | 750 (4½ in) | 750 | 100–1000 |
| Product gauge range (in) | 0.002–0.050 | 0.002–0.005 | 0.001–0.003 | 0.001–0.125 | 0.001–0.012 | 0.0015–0.020 |
| Sheet accuracy (%) | 3 (1–5) | 3 (1–5) | 10 | 10 | 7 | 5 (2–10) |
| Time to heat (h) | 6 | 5 | 3 | 3 | ½ | 3 |
| Time for 'on stream' | 2–5 min | 10 min | 2 h | 5 h | 10 min | 2–5 min |
| Gauge adjust time | seconds | seconds | 5–30 min | 5–30 min | seconds | 1 min |
| Autogauging capability | yes | yes | no | no | no | no |
| Color or product change time | 5–30 min | 10–40 min | 30–60 min | 30–60 min | 15 min | 30–60 min |
| Windup speed ($yd\,min^{-1}$): average (max.) | 80 (150) | 60 (80) | 15 (20) | 15 (30) | 20 (40) | 20 (30) |
| Limitations | High capital cost, heat time | Lower rate, versatility problem | Poor accuracy, long on stream time, low rate, degradation, reduced versatility | | Fumes, inefficiency, high energy cost, resin cost, release paper cost | Reduced rate and range, soft materials only slow manual gauge change |
| Applications and advantages | Versatility, high rate, accuracy, ease and, adjustment ease at reprocess | Accuracy, gauge adjust, reduced cost | Low investment, multiplant capability, thin gauge (0.003 in and under) and heavy gauge (0.050–0.125 in) | | Grain retention (pattern cast in), soft hand and drape | Good on wall covering, thin material, coated fabric, accuracy, reduced investment |

and replacement parts in calendering lines are expensive. The very small unsophisticated lines start about the million-dollar range compared to the much lower cost extrusion lines. Table 10.1 provides a guide comparing different fabricating processes to produce PVC film and sheet.

In general, plastic materials, such as PE, PP, and PS film and sheet, are usually produced through the rather conventional extrusion lines. To produce PVC film and sheet in large quantities, calendering is almost always used since the process is less likely to cause degradation than is extrusion as well as having dimensional and cost advantages. The review in this chapter principally concerns calendering PVC.

A web thickness between 0.05–0.50 mm (0.002–0.020 in) is generally the kind of plasticized film and sheeting produced by calender lines. For extremely light gauges, those under 0.02 mm (0.001 in), calendering could become impractical or damaging to the equipment. The reasons include factors such as, for certain materials, there exists poor strength of the thin webs and also very high forces develop on the matting heavy duty rolls.

For very heavy/thick gauges, such as sheeting over 0.50 mm (0.020 in), calendering may not be the optimum method of production. The reason is that there may not be enough shearing action that can be put into the rolling banks to keep the compound at uniform temperature. In addition, the separating forces on the rolls become so low that gauge variations could become prohibitive.

It can be said that basically the up-stream and down-stream procedures are similar in production lines whether calenders or extruders are used (Chapters 8 and 9). For a given quantity of output, it is usually necessary to have more extruders than calenders. This situation makes the extrusion lines more flexible and more able to handle short production runs. The extrusion flexibility, when compared to calendering, includes ease of changing product thicknesses, widths, and materials.

Calenders are capable of higher production speeds. Thus, there are situations where they provide a favorable situation for long runs. For these long runs, cost advantages exist. Calenders are better at producing sheets and film that have tight thickness tolerances. Calendering also provides product uniformity. Constant in-process monitoring and continuous profile adjustments is usually a significant advantage of calendering over other methods.

## CALENDERING OPERATION

The calendering process prepares plastic materials or compounds. It melts the material and then passes the pastelike melt through the nips of two or more precision heated, counter rotating, speed controlled rolls into webs of specific thickness and width as shown in Figs. 10.1–10.3. The (usual)

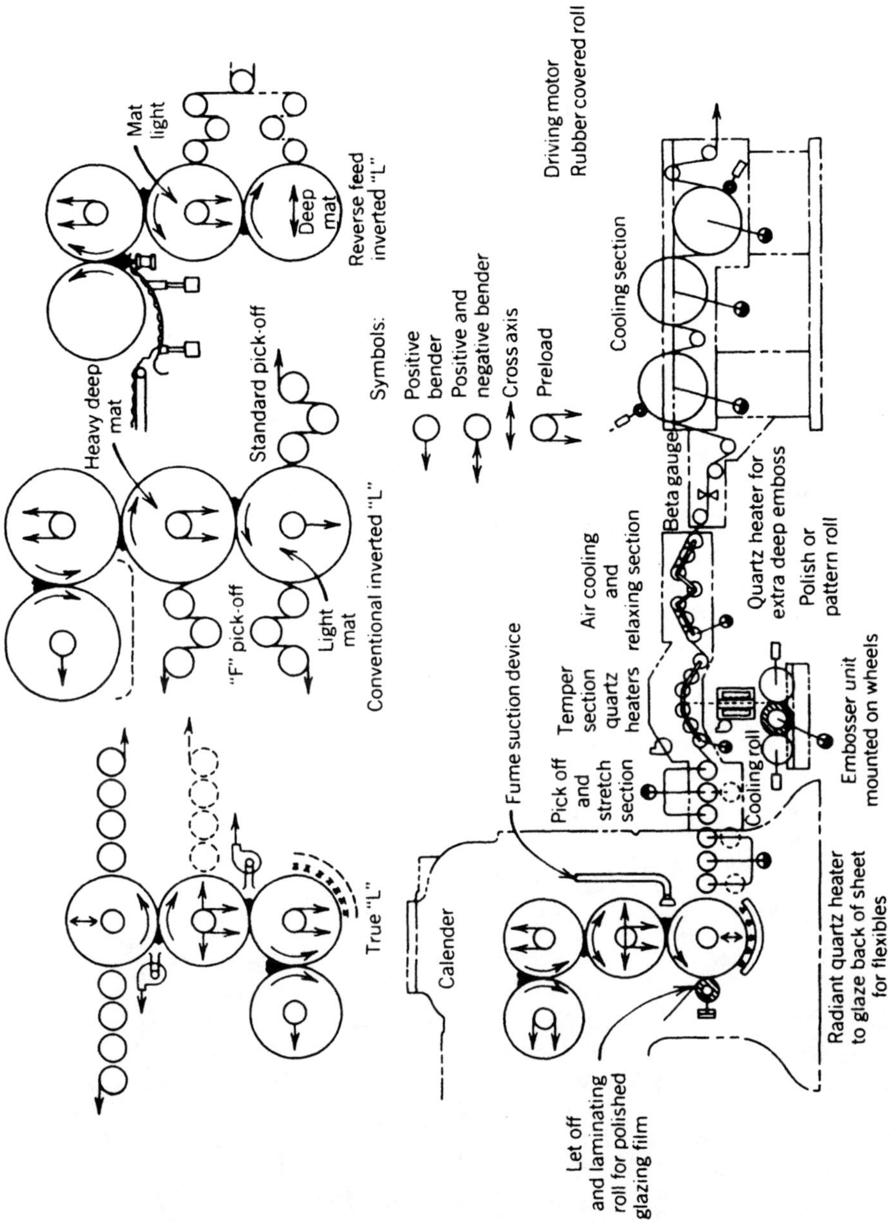


**Figure 10.3** PVC calender operations.

parallel rolls have extremely flat surfaces and rotate at possibly the same speed but usually at slightly different speeds depending on the plastic being processed. Although plastic forming occurs in the calender itself, down-stream precision operating equipment is needed to produce the TP film or sheet.

Up-stream of the calender, a mixer blends the raw materials, usually in powdered form with the desired additives (pasticizers, stabilizers, etc.) and fillers (calcium carbonate, etc.). This pastelike melt passes through a roll mill, such as the multiple roll mill.

As an example, heavy gauge products of 0.6 mm (0.02 in) leave the final calendering roll. In turn, the sheet is drawn down or stretched at take-off speeds consistent with good stripping from the final heated calender roll. Lighter gauge film at 0.18 mm (0.007 in) when leaving the last calender roll is typically drawn down at a ratio of 1.2/1 and wound up at a thickness of 0.15 mm (0.006 in).

Calenders may consist of two to at least seven rolls. They are characterized by the number of rolls and their arrangement. Examples of the layout of the rolls are the true L, conventional inverted L, reverse fed inverted L, I, Z, and so on. Figure 10.4 provides examples of a few of the configurations which are: (1) one-sided coating with vertical 3-roll; (2) one-sided coating with inverted-L using 4-roll; (3) double-sided coating with inverted-L using 4-roll; (4) double-sided coating with two 3-rolls; (5) one-sided coating with Z-configuration using 4-roll; (6) one-sided coating with S-configuration using 4-roll; and (7) one-sided coating with modified-s using 4-roll.

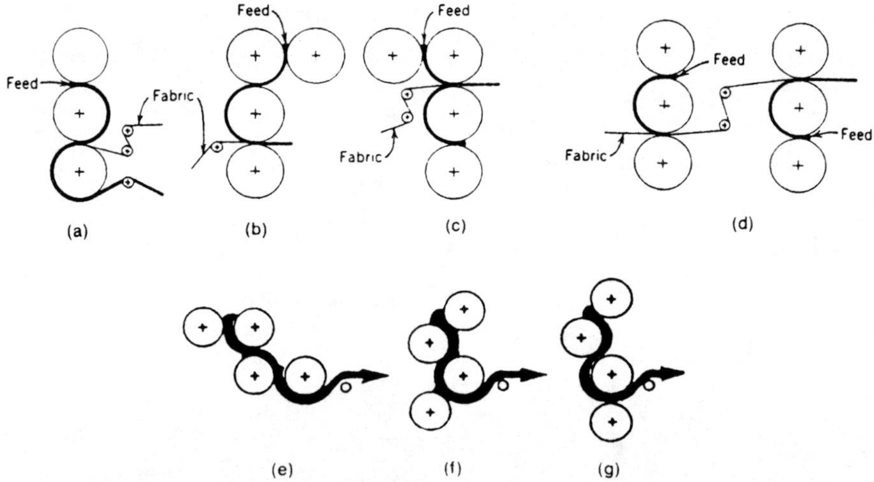


**Figure 10.4** Arrangements of parallel rolls for calender coating.

The most popular are the four-roll inverted L calender and the Z calender. The Z calenders have the advantage of lower heat loss in the film or sheet because of the melts shorter travel and the machines' simpler construction. They are simpler to construct because they need less compensation for roll bending. This compensation occurs because there are no more than two rolls in any vertical direction as opposed to three rolls in a four roll inverted L calender and so on.

The speed of the calendering rolls is rarely the same. They operate at different speeds to provide the best performance of the melt, particularly the required shearing action. A typical relationship of speed settings on an inverted L as shown in Fig. 11.4(b) can be 100% for the top in-line vertical roll, roll on its right side at 0.85%, roll just below at 1.03%, and the bottom roll at 1.06% .

Rigid PVC manufacturers usual prefer the L configuration with four to seven rolls being fed from the floor level. Since there is no disturbing vapors (condensing on rolls and web) from lower calender rolls within the pickoff area, it is preferable to have the pickoff rolls on an elevated level.

Flexible PVC is commonly processed using a four roll inverted L or an F calender. These systems enable the plasticizer-saturated vapors to escape via the usual suction hood located above the calender, where they are filtered before being released to the atmosphere.

A universal five roll L calender is used for rigid or flexible PVC film. It provides heat stability and superior film control with good surface appearance. The major difference between this universal machine and the others is in mounting and placement of the first roll.

There are configurations where the first roll is generally mounted beside the second roll; the more recent design still has the first roll mounted horizontally but now it is next to the third roll, resembling a four roll L calender. The second roll remains in action underneath the third roll, so four gaps are still available. Roll/film wrap length, however, is reduced so dwell time is 40% less, decreasing PVC heat loss and shrinkage. Modern designs for rigid and flexible PVC lines currently show few significant differences. Both actually consist of the same roll groups, but arranged at different elevation levels.

Variations in these multi-million dollar calender lines are dictated by the very high forces exerted on the rolls to squeeze the plastic melt into thin film or sheet web constructions. High forces at least up to 41 MPa (6000 psi) can bend or deflect the rolls, producing gauge variations, such as a web thicker in the middle than at the edges. During calendering, particularly film, roll separating forces in the final nip may be as high as 41 MPa (6000 psi).

This potential problem can be counteracted by different methods that include the following: (1) crowned rolls, which have a greater diameter in the middle than the edged; (2) crossing the rolls slightly (rather than having them truly parallel), thus increasing the nip opening at both ends of the roll; and (3) roll bending, where a bending moment is applied to the end of each roll by having a second bearing on each roll neck, which is then loaded by a hydraulic cylinder. Controls are used to perform any roll bending and crossing of the rolls.

Calenders require high temperatures, with little variations or fluctuations across the rolls during the application of the high forces or pressures. During calendering thin PVC film, the melt stock temperature may be 190°C (375°F), compared to heavier gauge sheeting where 177°C (350°F) stock temperatures are usually required. For the most part, this difference is due to the higher calendering speeds, resulting in generating more frictional heat, required for economical production of thin gauge films.

The high forces with the temperature controls are necessary to squeeze the plastic mass into relatively thin sheets but particularly very thin film. Any unevenness in the forces along the roll, that could include uneven temperature across the melt, is reflected as variations in the product thickness. Mechanically, one cause of pressure fluctuations is too much clearance in the bearings, which can be resolved either by pulling back the rolls against one side of the bearing or by using tapered roller bearings.

Although limited by a number of simplified assumptions, for many years there has been a simplified approach for calculating the calender's rolls separating forces. The equation is:

$$F = 2vVrw\left(1/h_0 - 1/H\right)$$

where $F$ = total separating force, MPa; $v$ = viscosity of stock being calendered, Pa s; $V$ = peripheral speed of rolls, m/s; $r$ = radius of rolls, m; $w$ = width of web being calendered, m; $h_0$ = separation between rolls at the nip, mm; $H$ = height of melt stock at the nip entrance, mm.

Other causes of thickness changes across the calendered web include nonhomogeneous rheology of the stock (Chapter 3), problems with material's lubricity, problems with pressure and temperature sensors, equipment line control malfunctioning, use of damaged calender rolls, and so on.

After forming through these multiple rolls of the calender, the film or sheet is cooled. The sheet or film immediately passes through precision surfaced cooling rolls that are kept at precisely controlled temperatures and/or a cooling tower where the web can be festooned. The temperature of the rolls is gradually reduced as the plastic travels down-stream. There may be up to 19 cooling rolls following an embosser (if used). After leaving the last large diameter calendering heated rolls, the film is 'dropped' vertically into an embosser, usually with three rolls that are the embossing roll itself, a cooling rubber roll, and a contact cooling to the rubber roll.

Diameters of rolls could be 15–25 cm (6–10 in). These rolls are usually driven and temperature controlled in sections between two and five rolls. Temperature accuracy can be controlled within ±1°C (±2°F). In line with these rolls is usually a thickness gauge (such as a beta gauge) controlling the film or sheet thickness.

Thickness gauges are usually located within this cooling section of the production line. Continuous beta scanning equipment can be used that traverses the web constantly. This provides feedback to the upstream larger rolls that makes thickness corrections automatically (Chapter 6). Gauge controls can also be used in the neckdown section, where the hot melt web leaves the last calendering roll and the web is transferred to the first of the cooling rolls. The amount of neckdown, as also reviewed in Chapter 9, provides an important control in the operation of the entire production line.

Proper use of all these controls, and that include speed and stretch in the takeoff train, allows the production of extremely flat sheet with a profile tolerance of ±5% across and down the web. The controls can call for adjustments on different line equipment, such as the nip openings, skew, roll bending, and so on.

After cooling, the plastic is trimmed at the edges and wound. Trim material can account for up to 5% of the width depending on the line's operating efficiency. Of course, the target is to have as little trim as possible. The trim is usually immediately directed back through a granulator and blended with virgin material (Chapter 3). When slitting the webs, typical slit widths are made to the nearest 0.8 mm (0.03 in) on 7.6 or 15.2 cm (3 or 6 in) cores with roll diameters of 36–102 cm (14–40 in).

We have reviewed the typical sequence for a calender line. However, there are many variations to suit the end product and reduce costs. Auxiliary equipment can be included, such as stretching in the machine direction by using the cooling rolls or setup bioriented stretching, water quench tank, annealing, decorating, slitting, heat sealing, festooning, and so on (Chapters 2 and 7–9).

## Surface finishing

Calenders provide styling on films and sheets. PVC surface calendered goods require depths of matte or provide an even roughness, which can be measured with a device such as a profilometer. Different markets use different terms to describe their finishes, such as satin, Sheffield, and light matte. After being processed, a one-sided matte finish may be applied with just an embossing roll. A sharper, deeper, and more precise matte finish can be applied to film or sheet by the calender.

The principle used in embossing PVC film or sheet is to deliver a relatively hot web to a relatively cold embossing roll. This action imparts the desired surface finish to the web and freezes it by rapidly cooling the embossment pattern. Film temperature, roll temperature, and pressure are optimized empirically for each compound formulation and pattern to be embossed. The steel embossing roll and the rubber pickup roll are internally cooled by a liquid medium; both of them require a very uniform surface temperature to develop the desired embossing pattern. But the rubber pickup roll tends to stick to a hot PVC web unless it is cooled externally as well as internally. External cooling in a wet trough is required because most rubbers have poor heat transfer properties, in fact, so poor that a nonexternally cooled backup roll surface temperature would rise almost to the processing temperature of the PVC web.

Quick roll changing techniques are used for the last two calender rolls to provide the capability of changing rolls. Changes from a high gloss to matte, or to different profiles, are usually made in about 10 h or less; in the past, it took at least 80 h. Embosser rolls can be changed in minutes instead of hours.

To make two-sided matte products, such as two-sided credit card stock, the last two calender rolls have matte surface finish. To produce these surfaces, the rolls are usually sandblasted with aluminum oxide grit of controlled particle size such as 120–180 mesh. For the heavier matte, coarser grit and more passes of the sandblasting action on the rolls are required. Proper selection of metal for the calender rolls is also necessary; certain metals are not suitable for producing these decorative, etc., surface finishes.

### *Credit cards*

This is, as we all know, big business. All kinds of data are provided through cards, and in the USA usage could be well over 40 billion cards. In USA, over 300 million are used for health cards alone, and that should at least double very quickly. Reports indicate that Germany has over 80 000 health cards. China, Southwest Asia, Mexico, and parts of Europe show strong growth potential for many types of cards. About 46% of the card-making capacity is in the USA. What makes this interesting to the calendered sheet industry is that practically all of these cards are calendered PVC sheets. The sheets that are produced need to be very carefully processed and handled.

Card manufacturers are not plastic processors, but rather plastic printing and lamination specialists. They apply the features, including the security aspects, to sheets of plastic core stock that is usually rigid PVC.

Most are die cut from sheets 50–58 cm (20–23 in) wide by 66–76 cm (26–30 in) long. Secure card manufacturers generally produce their cards in one of two ways: (1) by applying a 0.045–0.052 mm (1.8–2.0 mil) thick PVC film laminated to both sides of a 0.66–0.67 mm (26.0–26.5 mil) usually homopolymer sheets; and (2) by using the split core method where two 10.34 mm (3.5 mil) thick coploymer sheets are joined through a heat pressure lamination process, followed with a film overlay thinner than that used in the solid core.

Thickness control is important, otherwise problems will develop with card embossing equipment and the magnetic stripe reading machines. Other considerations that influence calendering sheet include ink adhesion, flex capabilities, cleanliness, and embossed-characteristic height retention. Secure cards tend to made from PVC copolymers, while the others with less stringent performance demands often use homopolymers. The life expectancy of cards is about three years. New plastic compounds continue to be evaluated to extend life with cost reductions.

The global card market for 1994 consumption was estimated at above 26 million kg (57 million lb) of plastics with over 90% was PVC. The USA and Canada consumed about 12 million kg (26.5 million lb). While there are over 50 fabricators of rigid calendered PVC sheet worldwide producing about 1.3 billion kg (2.75 billion lb) of products for all markets (cards, refrigerator door liners, etc.), fewer than ten can supply sheets meeting quality levels required in the card market.

The card manufacturing business has relatively low-technical requirements. While the industry is capital intensive, the barriers to enter into the market historically have been low. The result has been intense competition. New minor developments have continued to occur since the use of cards started expanding during the early 1960s, such as unsuccessful replacement of ABS. Most recently, an important new development called 'smart cards' occurred that is targeted to replace conventional magnetic-strip cards. A variety of semiconductor computer chips can be embedded between two preprinted ABS labels by injecting ABS between the labels to form 0.8 mm (0.03 in) thick cards. These cards are already popular in Europe in various applications. Numerous firms compete globally in the production of smart cards.

### *Roll covering*

The application of natural (natural rubbers) and synthetic (plastics) elastomers to the surface of mandrels is called roll covering. Covered rolls are used in a variety of industries, such as extrusion, printing, paper, textile, and many more. Industrial requirements for covering rolls are forever changing. Improving quality through technology has been necessary since the first roller was manufactured. Much of the technology has been directed towards improving compounding materials and processes. Many of the earliest manufacturing methods are still used today. Methods tend to be labor intense and dependent on individual skills.

The diversity of the industry has slowed the evolution toward automation since many roll covering companies may work with over 200 different formulations and limited productions exist. Different methods are used in this versatile industry. The traditional method is based on cutting to size calendered 1.5–3.2 mm (0.06–0.125 in) thick sheets. Layers of the sheet are wrapped around a mandrel with numerous splices. Strips of the calendered sheet are also used by rolling spirally. These methods generally produced good quality but have many negative features, such as poor adaptability to automation, high material costs, etc.

Other methods include using precured tubing (customer purchases for their own installation), multi-pass tape overlap (at different angles including different overlap methods), and crosshead method (similar to extrusion profile covering mandrel). The extrusion crosshead technique has become a popular method for roll covering (Chapter 11).

## PLASTIC MATERIALS

A wide variety of TPs can be used with the majority being PVC, ABS, PE, PP, and PS. However over 80% are flexible and rigid PVCs. When calendering other plastics, there is a tradeoff in economy and quality. The basic limitation of the calendering process is the need to have sufficiently broad melt index to allow as wide a heat range as possible for the process. This behavior of the plastic permits the material to have a relatively high melt viscosity in the banks of the calender rolls; banks indicate where two rolls meet, or the nip of the rolls.

As a result of the viscosity, a shear effect can be developed throughout the process. This shear is of prime importance between the calender rolls. Thus, the calender forms the web as a continuous 'extrusion' between the rolls. Unlike when processing just through a conventional extrusion line, the plastic mass cannot be confined when being calendered. Because of the lack of confinement, the shear effect and a broad melt band are essential aspects of calendering.

The blending or compounding (Chapter 17) of the plastic with different additives and fillers is a critical part of the process. The blending must produce a uniformly colored and stabilized product in an approximate powder form. After blending, the rate of consumption during calendering dictates the temperature of the melt.

The PVC compounds require stabilizer systems. There is an art (based on experience) in preparing these additive systems so they are compatible with the compound. Information on the different ingredients needed to meet different product requirements (that do not interfere with color, etc.) is readily available from the PVC manufacturers and/or the literature. The ideal heat stabilizer system imparts during processing primarily heat stability, as well as adequate lubricating characteristics to reduce or control frictional heat.

These stabilizer systems are also very efficient for plate-out resistance. Plate-out is a condition where the calender rolls and/or embossing rolls become coated with a deposit from the compound being processed. This deposit may start out as a soft, waxy material barely visible on the metallic contact surfaces of the processing equipment. During long runs of over a few 300 m (1000 ft), both the consistency and quality of the plate-out can become prohibitive and interfere with acceptable surface finish of the film or sheet.

When plate-out occurs, the line has to be shut down and the contamination removed. If the contamination is soft, generally all it needs is a wiping. When it is as hard and has an excellent adhesive bond to the steel, some type of mechanical abrasive or wire brush is used (carefully so that will not damage the roll surface).

Because the plastic is processed between the required heat and its critical heat of degradation, the time of heat becomes extremely critical and an important part of the complete process. For example, the processor will minimize the amount of melt in the bank (nip) of the rolls. The residence time of the plastic flux at high heat must be limited. As reviewed throughout this book, PVC is especially sensitivity to heat and time at heat. This situation is not a problem if the machine controls are properly set and operated within set conditions.

The plastic mix fed to the calender may be a simple hot melt, as with PE, PP, etc. However, for PVC it is obtained by premixing the vinyl plastic, stabilizers, plasticizers, calcium carbonate, etc., in ribbon blenders. From the blenders the blend is passed usually through a Banbury mixer where the mass is gelled for a specific time and temperature period depending on the mix. A typical period for gelling is about 5–10 min at 120–160°C (248–320°F).

The gelled 'lumps' are made into a rough web on a two-roll mill and the web is fed to the calender rolls. Another approach is to have the gelled material go through an extruder/compounder to produce a rope when exiting the machine. This rope about 50–100 mm (2–4 in) in diameter is fed to the calender rolls. It is directed over a conveyor belt to the nip of the top rolls. Rather than just being deposited in the center of the nip, a movable 'robot' arm moves along and from side to side feeding the rope from almost one end of the roll to the other end. This method of delivering the rope provide for a more even distribution of the plastic.

Ingredients for PVC film and sheet can include solid and liquid raw materials. Solids are weighed in batches and dropped into a high-speed

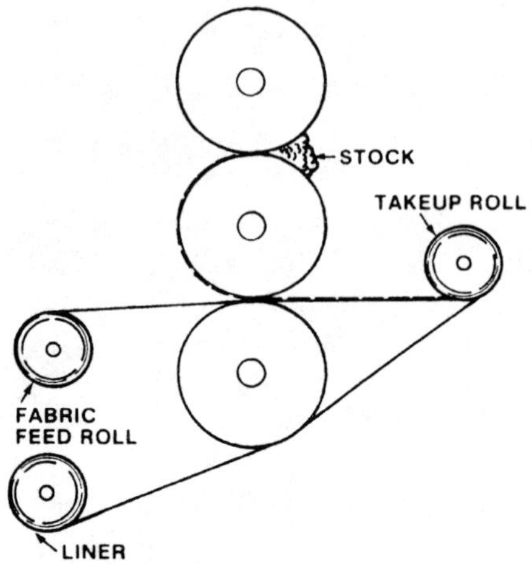


**Figure 10.5** Schematic of calendering with a fabric using a support liner.

mixer. Liquids are metered into the mixer, usually by piston pumps or the volumetric metering devices. A batch mixing system is generally in-line with a continuous fluxing system that starts at a holding bin. Some operations do not use batch mixers for blending. Continuous metering systems, using loss-in-weight feeders and pumps, will meter the mix directly into the continuous fluxing mixer.

If fabric, paper, or other material is feed through the rolls, plastic can be pressed into the surface of the calendered web (Fig. 10.5). When operating the calendering line in this mode, it becomes a coating machine.

## Fluxing and feeding

Fluxing is the preparation of a plastic composition to improve melt flow; fusion is the heating of the vinyl compound to produce a homogeneous mixture. Fluxing units used in calendering lines for preparing TP materials include batch-type Banbury mixers, Farrel continuous mixers (FCMs), Buss Ko-Kneaders (BKKs), and planetary gear extruders (PGEs).

The dry blend is fed into the mixer/extruder. Excellent mixing within a short dwell time and heat transfer control contribute to an improved product. During fluxing, each particle receives the same 'gentle' treatment, generating less heat history and producing more uniform feed rate, color, gauge thickness, web surface, and so on. The feed can discharge onto a two-roll mill. Operating this way, it provides for a second fluxing action, mainly for working in scrap or for convenience as a buffer.

The stock delivered to the first calender nip needs to be well fused, homogeneous in composition, and relatively uniform in temperature. The

**Table 10.2** Example of flexible PVC calendering conditions

| | *Heavy-gauge product* | | | | *Light-gauge product* | | | |
|---|---|---|---|---|---|---|---|---|
| | *Roll temp.* | | *Roll speed* | | *Roll temp.* | | *Roll speed* | |
| *Roll no.* | *°F* | *°C* | *ft min⁻¹* | *m min⁻¹* | *°F* | *°C* | *ft min⁻¹* | *m min⁻¹* |
| 1 | 347 | 175 | 125 | 38 | 338 | 170 | 263 | 80 |
| 2 | 352 | 178 | 128 | 39 | 343 | 173 | 269 | 82 |
| 3 | 353 | 181 | 138 | 42 | 349 | 176 | 279 | 85 |
| 4 | 363 | 184 | 148 | 45 | 354 | 179 | 289 | 88 |

optimum average temperature for good fusion depends on the formulation. A rigid PVC formulation based on medium molecular weight plastic (intrinsic viscosity of 0.90–1.15) (Chapter 3) has a typical optimum temperature of 180–190°C (355–375°F) at the first calender nip. For best calendering, there should be no cold volume elements below 175°C (345°F) and no hot spots above 195°C (385°F). These typical temperature readings are necessary for close control of fusion and mixing conditions; they are based on industry studies of flow effects and their relation with volume and element size. This interaction depends on stock temperature and in turn on the performance of PVC melts.

Flexible PVC is normally calendered at temperatures of 10–20°C (50–68°F) lower than rigid PVC. Typical calendering conditions for unsupported flexible PVC film and sheet are shown in Table 10.2. In flexible PVC production, a short single screw extruder, acting as a strainer, filters out contaminants from stock before reaching the calender. This method is not applicable to rigid PVC because it drastically increases the head pressure and the consequent overheating would cause the stock to decompose.

### Heat sensitivity

The target for the calender is to provide sufficient energy to convert the mass of plastic into film or sheet form without supplying so much heat as to cause degradation. This is a very important consideration particularly when processing rigid PVC. PVC is very temperature sensitive.

As an example, for the manufacture of thin gauge PVC films, very high calendering speeds [over 90 mm/min (300 ft/min)] are encountered. Frictional heat build-up not only affects the compound heat stability in terms of initial color and properties, but it also affects stock transfer from one calender roll to the next. This action can be caused by loss of the lubricant additives at these high temperatures, indicating the need for a lubricant of lower volatility

This situation with PVC has always existed since it became commercially available almost a century ago. However, it is the fourth most consumed plastic material in the world using all the different processes (extrusion, injection, coating, etc.). For those that use it correctly, processing is conducted so that degradation is not a problem. Those that do not process it with care have problems.

The energy required to turn the calender rolls is proportional to their diameter. The frictional heat developed which is transferred to the material is also dependent upon the roll diameter. To control this heating action, the calender operation and control is vital to the successful production of film or sheet.

## Contamination

Very vital to this operation is the 'complete' removal of any metal or hard surface material. This includes microscopic particles. From the start to the end of the calendering process, extreme care has to be taken to ensure there is no contamination of the equipment or plastic being processed. Preventative maintenance of these lines is a continuous operation that includes the operating environments. A relatively clean room has to exist.

This cleanliness is required to ensure that damage to the expensive equipment does not occur, up-stream or down-stream. A micron size piece of metal will destroy the rolls, etc. Replacing these very expensive rolls is expensive and time consuming. Where this type of equipment may not be in the spare parts room, delay in refurbishing or replacing parts is very time consuming.

Special treatments of multiple filtering and screening all raw materials are required. Incoming materials that include all additives and fillers are inspected and/or filtered. Some of the equipment will use duplicate and triplicate screens. The inspection and filtering is made to eliminate both metallic and nonmetallic contaminants.

Magnetic sensors/detectors are positioned practically anywhere the plastic material moves. This includes locating detectors under the conveyor belts that directs the web or rope up to the rolls. This action is taken even though there was a very thorough inspection of raw materials on arrival at the plant and through mixers, blenders, etc., prior to reaching the conveyor belts. Thus, action is taken even though the materials had what one may consider repeated 100% foolproof inspections every step of the way.

## Recycling

Handling PVC scrap and cold trim from the product line can pose a very difficult problem if they are not handled correctly. The scrap and trim could represent 10–40 wt% of a mix with virgin material in certain runs where recycled plastic can be used. The actual amount depends on the width of the calender in relation to the web width and properties obtained.

The flux rate and energy required to remelt the scrap are considerably less than required to flux the virgin plastic. With this differential, there is the potential danger of material decomposition whenever scrap is processed to start up a mix. Optimum uniformity and product standardization requires the proper blend of new and old materials.

Reprocessed material is best added to the blender where a standard can be established, although some plants feed it directly to the fluxing equipment. Careful control of the scrap percentage in the total mix is essential to obtain quality controlled products; control should be somewhat better than that required in other processes. Extreme care is required when handling and/or transporting the plastic to be recycled to ensure that it is not contaminated.

## ORIENTATION

Orientation is applied to the film and sheet. The information already provided in Chapter 2 describes why orientation is used and how to process film or sheet with mono- or by-oriented properties.

## PROCESS OPTIMIZATION

The structure of calender rolls, supports, and other parts have been structurally designed to be extremely sturdy in order for the line to operate under high pressures in addition to the weight of the rolls and other parts. Individual motor drives for the rolls are used and required to provide sophisticated means of controlling line speed.

### Roll

Steel rolls used in the line are very heavy and of intricate design. As reviewed in Chapter 9, these heavy rolls have internal designs to ensure the proper flow of liquid and even surface heating on the roll's surface. Critical temperature control of the massive rolls is required to meet product performance requirements.

The design of the rolls themselves is of great importance, particularly the ratio of length to diameter. The market requirements usually set the product width and roll length. Therefore calender sizes have generally been based on the required separating force needed between rolls. In theory, heavy gauge, high plasticized materials require extensive work in order to maintain frictional heat and uniformity of the bank (nip). To meet this condition, the material should be processed through the larger diameter rolls. However, in practice the separating forces of these type materials are considerably lower and frequently are run through smaller diameter rolls.

With thin gauge, low plasticized materials, the reverse is true in that the frictional heat developed in the bank (nip) is greater on large diameter rolls. Therefore, these materials should be run on calenders with the smallest possible roll diameters. However, there is a 'happy medium' that has to be met. These heavy loads develop high separating forces and considerable gauge changes can occur across the width of the film.

Smaller diameter rolls are used in regions where it is desired to reduce the frictional heat of the melt. These rolls require a different method of roll adjustments. Regardless of the size of rolls, the uniformity of the product will be affected by slight changes in shape in the roll as the equipment starts its heating process.

The concentricity of the rolls when operating the line at the required temperatures during processing is very important. This is not necessarily the same as ensuring that the rolls are concentric at room temperature. In fact, with the larger calender rolls and lines, it is the custom to allow the equipment to run for two to three days. This heating period is actually the beginning of the startup prior to actual production. It is required in order to have all the 'metal' parts reach equilibrium. Once the line is up to equilibrium temperature, this expensive preheat period requires a long production run.

Any deviation in concentricity usually is kept below half the tolerance of the film or sheet to be produced. Some of these products require that deviations be as small as 1/20th to 1/50th of the product tolerance required. As reviewed in the Chapters 8 and 9, roll surface finishes and upkeep is of considerable importance to ensure good quality products.

Roll mountings are another potential source of error. This problem develops due to the possibility of alteration in nip pressure and roll alignment. Also with poor mountings, heavy oscillations can develop. The use of auxiliary rolls before the actual calender rolls has been found to improve the stability of the system.

The methods used in altering the nip arrangement are usually by roll bending or roll crossing. Less power is required to alter the thickness across the web by roll crossing than by actual roll bending. Although each method finds its application in appropriate circumstances, in general it has been found that the nip between the second and third rolls of the calender assembly has the greatest influence on the quality of the product.

Rolls under load suffer a tendency to separate despite their weight mass. The thinner the gauge and the lower the plasticizer content of the

PVC material, the higher the separating force. This type of action also causes a greater amount of roll deflection.

By using special grit-blasting techniques, the third and fourth calenders (last two rolls) may be custom-surfaced to generate a uniform two-sided matte product. Alternately, one or more down-stream embossing stations can be used to produce a custom surface on one or both sides of film. When processing certain plastics continuously around the clock, a satisfactory matte usually will last about five days. It will require down-time of the line to refurbish the roll(s). After separating from the last calendering roll, antistatic, slip agents, and other units may be applied to the surface(s) of the web.

## CONTROL

Many control features in calenders provide versatility, better quality, and higher operating rates. Complete control of friction ratios gives good tracking of stock via individual drives on all calender rolls. Close tolerances on film profiles are obtained from axis crossing, roll straightening, and gap profiling of rolls. Better control over stock temperature is achieved by using internal high rates of circulating liquid flow lines/compartments within the rolls to obtain the roll surface temperature control. Surface temperature can be held steady to within at least 2°C (3–4°F).

In older calenders, severe cases of roll float caused by changes in the balance of forces on a roll were sometimes encountered. Preloading devices reduce roll floating and thus provide a better gap control between pairs of rolls, especially during startup. Web thickness can be measured with improved feedback control.

## APPLICATIONS

Different plastics are formed into support and unsupported webs by calendering. These include PVC, ABS, PUR, SBR, NR (natural rubber), and others. Included are PVC and its copolymers with and without blends of ABS (Chapter 3). Applications are many, including packaging, building and construction (tiles, window shade, etc.), tire fabric coating, swimming pool liners, electrical tape, and so on.

One special and important application is the coating of paper, woven and nonwoven textiles, plastic films and sheets, credit cards, roll coverings, and other substrate. A calender with three rolls is usually sufficient for one-sided coating, but four rolls are used for extremely thin coatings. Double-sided coating can be applied simultaneously on both sides of substrates, such as fabric, using a four roll calender or sequentially by two three roll calenders. Specialized calendering equipment is used for certain products, such as floor tile and window curtains.

## SAFETY

The usual health and safety procedures in plastic operating plants have to be followed. Most of the hazards are associated with the high speeds at which webs normally travel, and the large forces involved in calendering, and/or the high temperatures. Safety devices are located where points of danger exist.

An example is locating a double-action interlocked safety at the nip (bite). As shown in View A–A of Fig. 10.6, a recommended location of the device is shown. The dimensions shown are guides for maximum allowable for safe operation. Distance X will vary in accordance with the roll diameter, but always be great enough to ensure that no part of the operator's hand can reach the bite without tripping the bar.

The full calender view shows the safety bar in position at the feed bite, and also illustrates the recommended location for safety trip cables which are required in addition to the interlocked bite bar. The letters in Fig. 10.6 identify: B, safety trip cable switch; C, vertical safety trip cable; D, adjustable clamp for horizontal safety trip bar; E, horizontal safety trip bar; F, double action electrically interlocked bite bar; J, safety trip cable screw eye tension adjustment; K, safety trip cable guide pulley; and L, calender control switch.

This device can be used but actually the safest way to feed material through the rolls is to use a strip or rope of plastic, or the remains from the last run. The material wound around all the rolls, from start to the end, would be in the open position. Material would logically go through the line with no hands involved. Once the new material starts passing through and/or around the rolls, space between rolls is gradually reduced and adjusted to the operating gap spacings. All this involves no hands near the nip or, actually, the use of a bite safety bar.

## COSTING

The capital cost for a new calendering line will be about $4–5 million, not including building cost and utility services. The sum of fixed costs and the variable costs for operating a new line in USA can amount to $450–550/h. Taking PVC as an example, these lines have a throughput of 400–2700 kg/h (90–6000 lb/h). And the average range of compounding costs for calendering PVC is about $0.49–1.12/kg ($1.08–2.47/lb).

Not including material costs, the expense of operating a large calender line may be $350–560/h, depending on degree of automation, investment, labor, power costs, and overheads. For PVC sheet 2 m (7 ft) wide and

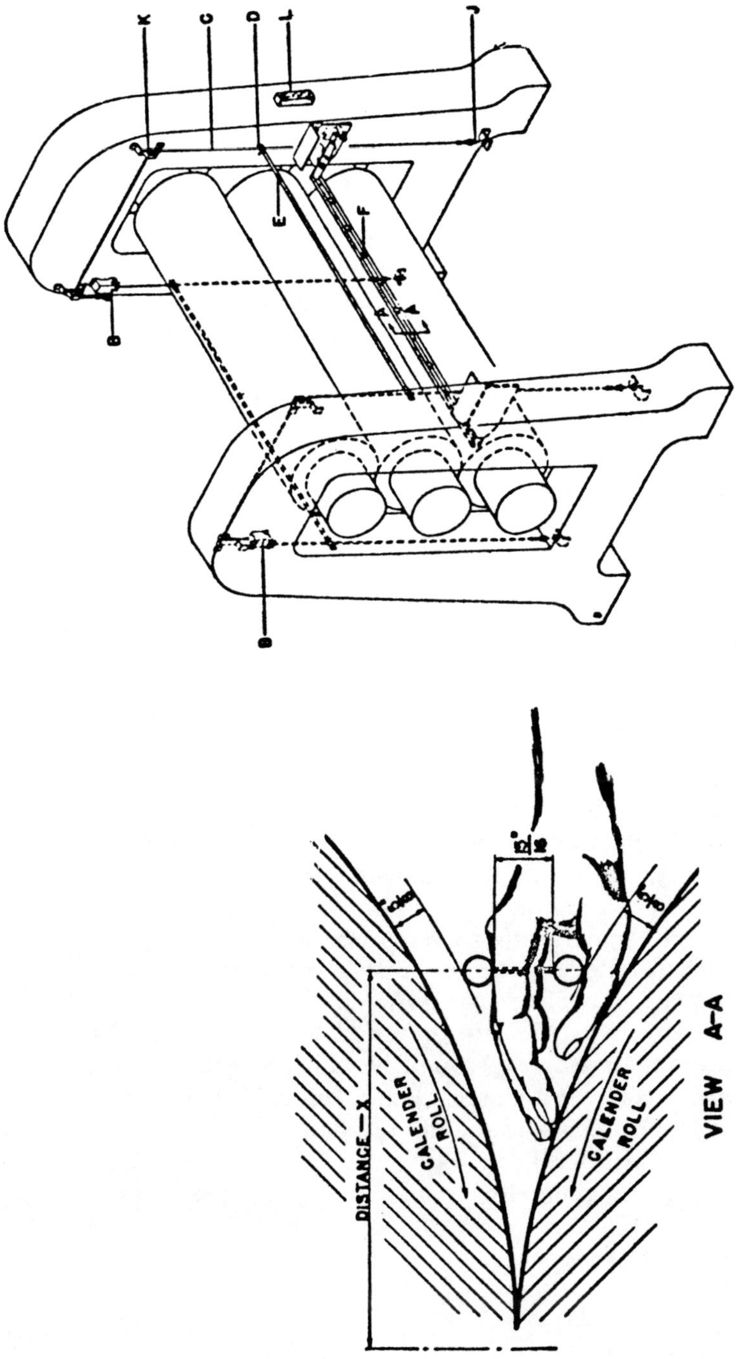


**Figure 10.6** Double action interlock safety bar at bite/nip of rolls.

0.50 mm (0.020 in) thick traveling at 27.4 m/min (91 ft/min) and wound up at a rate of about 489 kg/h (1076 lb/h), the conversion cost is about $0.21/kg ($0.46/lb). These costs are based on PVC plastic at $0.7/kg ($1.70/lb) and plasticizer at $1.10/kg ($2.42/lb). Table 11.2 provides a general manufacturing comparison of calendering with combination extruder/calender, extrusion blown film, extruder with (flexible-lip die) flat cast film, plastisol-cast, and melt roll.

To reduce the cost of products, the logical approach is to increase the line's speed. However, there are numerous factors which limit calender speed. At a given speed, more frictional heat is generated by the more highly filled and rigid stocks. Excessive speed can produce high temperature melt with degradation or sticking of webs to the rolls where stock temperature is limited. Optimal speeds of the final calender roll for double polished, clear melt flexible PVC viscosities at a given temperature degrade at about the same maximum temperatures as the rigid stocks. Another target is to reduce the edge trim if possible.

## TROUBLESHOOTING

This chapter has reviewed all kinds of problems with information on causes and their solutions. Collectively, they provide different types of troubleshooting guides. With most problems comes unexpected expenses. As an example, calendering problems that will affect product costs include: (1) bank marks and shiny patches on the surface; (2) cold marks (crow's feet) caused by the stock being to cold; (3) blistering due to high temperature or a large bank at the roll's nip; (4) pinholes caused by foreign matter or by the presence of unplasticized plastic particles; and (5) watermarking usually due to lubrication contamination.

Bank marks are minimized by optimizing formulations, calendering speeds, and roll temperatures so as to obtain the most orderly behavior of the roll banks of stock at the calender's nip entrances. Proper use of drawdown permits windups to be run substantially faster than the final calender roll on many thin, unsupported film products. Calenders and take-offs are run almost synchronously on heavy gauge products.

Films and sheets with a high gloss taken off a highly polished final calender roll tend to stick to the roll more than their matte counterparts. Very soft webs also tend to stick to the final calender roll. The fastest calender speeds are generally obtained in a median thickness range.

# 11

# Coating and lamination

## OVERVIEW

The basic units of an extrusion coating line, in addition to the extruder, consist of the unwind station, the coating or laminating unit, and the windup unit. Figures 11.1–11.3 provide schematics and equipment views of a coating line and a laminating line. When coating, a plastic material is applied to the surface of a substrate. Laminating involves combining two or more substrates together by 'bonding' them together with a material such as a plastic(s). The basic common principle involved in coating and laminating is that a very hot plastic melt is used. The main purpose of these treatments is to combine the best properties of the plastics and substrates. In this chapter, most of the material concerns coating, however it also applies to laminating [1, 88, 218, 327, 341, 358, 415].

There are variations in the definitions of coatings and laminations in certain industries. An example is in the packaging industry where thick layers of PE are applied to a substrates such as papers/paper boards. Thickness of PE can be applied that is 6–50 mm (0.25–2 mil) with the resulting physical properties significantly increased due to the amount of PE that has penetrated and coated the substrate. This process is a special case and usually considered to be a lamination coating. There is also the process of coextrusion that involves 'laminating' two or more layers of plastics (Chapter 2, etc.). With coextrusion, all the layers can all be 'assembled' in a die; also two or more layers that meet in one die, join just outside the die with one or more layers of plastic to form a complete laminated product. The laminate is at times called a sandwich construction [3].

Many of the different processes used to coat (Table 11.1) or laminate are reviewed in the literature [1, 4, 209, 324]. (There are literally over a hundred different processes used to coat materials.) There are extruding film (Chapter 8) and extruding sheet (Chapter 9) lines that are used to apply coatings or laminating substrates. However, the coating and

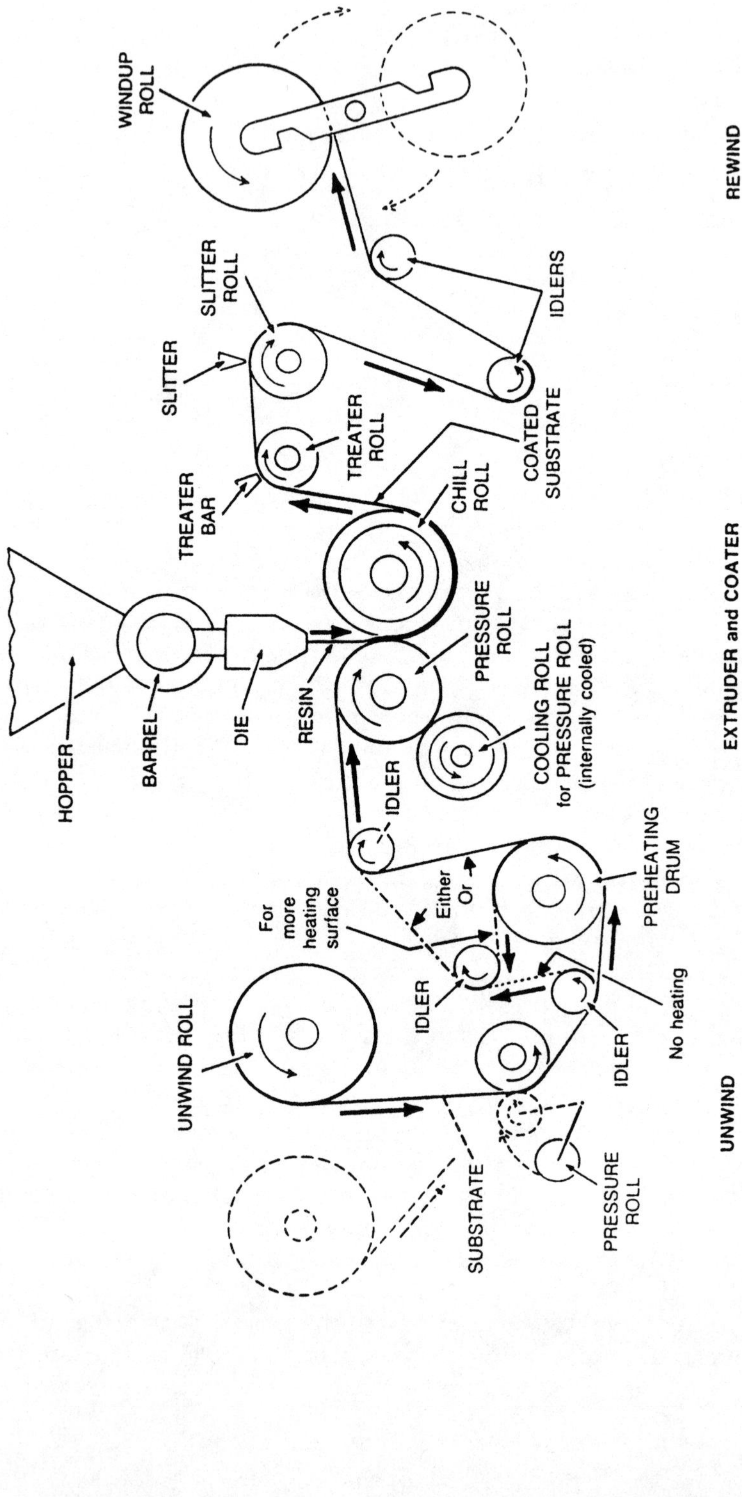


**Figure 11.1** Schematic of an extruder coating line.

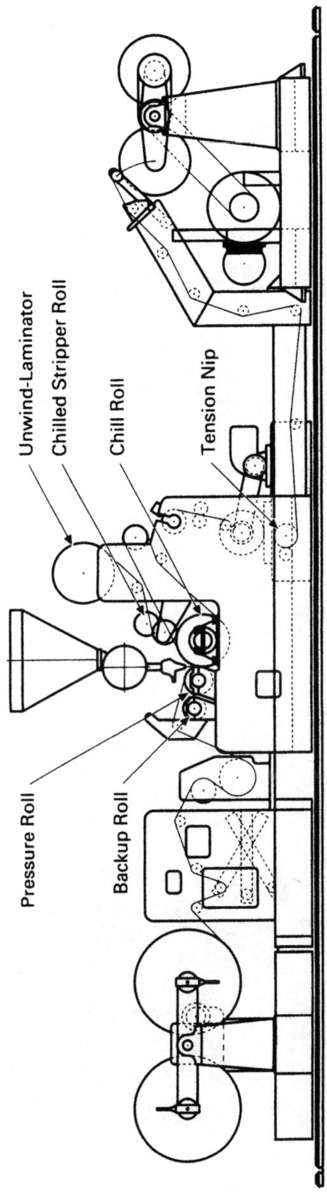


**Figure 11.2** Line with turrets to unwind substrate and wind coated substrate.

**Figure 11.3** Views from opposite ends of a small laboratory coater and laminator that can consume 48 km (30 miles) of paper substrate an hour applying half a ton of PE coating.

laminating lines being reviewed in this chapter use equipment designed specifically for coating and laminating. These dedicated fast lines are extensively used since they provide quality controlled products and are very cost efficient (Fig. 11.3).

**Table 11.1** Examples of coating methods

| *Coating method* | *Base material*[a] | *Coating composition*[b] | *Usual coating speed* ($m\,min^{-1}$) | *Viscosity range,* (*mPa s*) | *Wet-coating thickness range* ($\mu m$) |
|---|---|---|---|---|---|
| Air knife | B, D | P, T, X | 15–600 | 1–500 | 25–60 |
| Brush | B, C, E, F, G | R, S, X, Z | 30–120 | 100–2000 | 50–200 |
| Calender | A, B, D, E | U, V, W | 5–90 | | 100–500 |
| Cast-coating | A, B, D | Q, R, S, T, V, Y | 3–60 | 1000–5000 | 50–500 |
| Curtain | A, B, C, D, E, F | R, S, V, X, Z | 20–400 | 100–20 000 | 25–250 |
| Dip | A, B, D, E, F, G | R, S, V, X, Y, Z | 15–200 | 100–1000 | 25–250 |
| Extrusion | A, B, D, E | T, U, V, W | 20–900 | 30 000–50 000 | 12–50 |
| Blade | A, B | R, S, T, V, X, Y, Z | 300–600 | 5000–10 000 | 12–25 |
| Floating knife | A, B, D | R, S, T, V, X, Y, Z | 3–30 | 500–5000 | 50–250 |
| Gravure | A, B, D, E | R, S, T, U, V, Y, Z | 2–450 | 100–1000 | 12–50 |
| Kiss roll | A, B, C, D, E, F | R, S, V, X, Z | 30–300 | 100–2000 | 25–125 |
| Knife-over-blanket | A, B, D | R, S, T, V, X, Y, Z | 3–30 | 500–5000 | 50–250 |
| Knife-over-roll | A, B, C, D, E | R, S, T, U, V, X, Y, Z | 3–60 | 1000–10 000 | 50–500 |
| Offset gravure | B, D | R, S, T, Z | 30–600 | 50–500 | 12–25 |
| Reverse roll | A, B, C, D, E, F | R, S, T, U, V, X, Y, Z | 30–300 | 50–20 000 | 50–500 |

**Table 11.1** *Continued*

| *Coating method* | *Base material*[a] | *Coating composition*[b] | *Usual coating speed* ($m\,min^{-1}$) | *Viscosity range,* (*mPa s*) | *Wet-coating thickness range* ($\mu m$) |
|---|---|---|---|---|---|
| Reverse-smoothing roll | A, B | R, T, X | 15–300 | 1000–5000 | 25–75 |
| Rod | B, D | R, S, T, V, X, Y, Z | 3–150 | 50–500 | 25–125 |
| Sprays | | | | | |
| Airless spray | A, B, C, D, E, F, G | S, T, V, X, Y, Z | 3–90 | c | 2–250 |
| Air spray | A, B, C, D, E, F, G | S, T, V, X, Y, Z | 3–90 | c | 2–250 |
| Electrostatic | A, B, C, D, E, F, G | S, T, V, X, Y, Z | 3–90 | c | 2–250 |
| Squeeze roll | A, B, C, D, E, F | R, S, T, U, V, X, Y | 30–700 | 100–5000 | 25–125 |
| *In situ* polymerization | A, B, C, D, E, F, G | Y, Z | undetermined | liquid or vapor | 6–2.5 |
| Powdered resin | A, B, C, E, F, G | Q | 3–60 | | 25–250[c] |
| Electrostatic spray | | Q | | | 20–75[c] |
| Fluidized bed | E, G | Q | | | 200–2000[c] |

[a] Key: A = woven and nonwoven textiles; B = paper and paperboard; C = plywood and pressed fiberboards; D = plastic films and cellophane; E = metal sheet, strip, or foil; F = irregular flat items; G = irregularly shaped [9].
[b] Key: Q = powdered resin compositions; R = aqueous latexes, emulsions, dispersions; S = organic lacquer solutions and dispersions; T = plastisol and organosol formulations; U = natural and synthetic rubber compositions; V = hot-melt compositions; W = thermoplastic masses; X = oleoresinous compositions; Y = reacting formulations, e.g., epoxy and polyester; Z = plastic monomers [9].
[c] Dry thickness.

The coating process itself tends to be complex since it contains many controllable variables where each, as well as their interrelation, affects the quality of the coated or laminated product. In general, the processing temperature used in extrusion coating are higher than those in the other extrusion processes. Melt temperatures for the conventional plastics are normally 288–300°C (550–575°F). The higher temperatures are required to affect adhesion to the substrate(s). This higher temperature puts the plastics closer to being degraded.

Large size equipment is used in order to produce products economically. The 160–200 mm (6–8 in) extruders are used in order to deliver through the dies rather large amounts of melt. Dies used are rather wide that go over 3 m (10 ft) wide lips. In turn, large and heavy rolls are used in line to unwind substrates and windup products at the end of the line. Many coatings are used in a thickness range from 0.013–0.025 mm (0.0005–0.001 in) and even some as thin as 0.005 mm (0.0002 in).

The usual objective is to adhere (to coated substrate or between substrates) relatively thin plastic of uniform thickness to a substrate(s). This is accomplished by extruding a thin molten plastic from a slit die very similar to the cast film dies (Chapter 8). This very high-temperature melt film 'curtain' is drawn down to its final thickness in the air gap between the die lips and the point of contact on the substrate(s) in the nip between the rolls. Adhesion between the film and substrate(s) is aided by pressing and quenching between the pressure and chill rolls. This process follows a three or four roll coating line that is similar in principle to the sheeting extrusion line (Chapter 8).

Due to the presence of the relatively stable substrate and the rapid cooling, very high speed can be obtained during processing such as at least 300 m/min (1000 ft/min). In order to obtain these speeds and to avoid fluctuations, the equipment is deigned to be extra sturdy. As in the case of other extruded products, the type of plastic will have a direct effect on product properties, although not strictly a process variable. The important factors to be considered in the process include: (1) adhesion between plies (coating or laminating); (2) gauge variation; (3) neck-in; (4) surface finish; (5) dimensional stability; and (6) heat seal range.

Plastic coatings and laminations are applied in different forms and shapes. This chapter involves only flat products. Chapter 12 reviews wires and cables. Other chapters include coating and laminating different products such film, sheet, profiles, etc. [7–10, 13–15, and 18].

## Substrate

Different substrates are used that are generally flexible. They include different plastics in film or sheet form, paper, paper board, cardboard, woven and nonwoven fabrics, metal foils (usually aluminum), and so on.

Certain combinations of materials only require snug fits. However, they usually require excellent adhesion of the composite materials (coating or laminating). To obtain the proper adhesion usually necessitates some type of pretreatment, such as cleaning, priming, and/or heating substrates.

Overall, there are a number of important controls to set on these lines. They include: different temperatures, line speed, unwind and rewind roll speeds and tensioning, and so on. An important control is the extruder with its up-stream to the die controls (Chapters 2 and 5). Essential operating conditions involve: (1) letting the melt fall first against the pressure roll slightly above the nip; (2) maintaining minimum melt temperature compatible with good adhesion; (3) keeping the chill roll cool; (4) keeping mirror polish on the chill roll unless requiring a matte finish; (5) pressure roll is to have a high finish; (6) preheating the substrate; (7) on line, keeping the substrate cool; (8) priming the substrate; and (9) coating the wettable side of the substrate.

## PLASTIC MATERIALS

Different TP materials are used, such as PE, PP, nylon, ionomer, and EVA. These types of plastics are more popular, based on processing and product performances and cost. Each plastic, these and others, has its own processing requirements and capabilities. As an example, LDPE (a popular material) is more stable than PP, since PP tends to have more fluctuations in extruder output.

However, PP, although it has good heat-sealing strengths, tends to be more sensitive to processing conditions. It bonds well with porous substrates but has problems with nonporous substrates. However, with plastic modifications, these type problems practically disappear.

An example of how properties can change with many plastics, such as certain PE, is in coating flexible substrates with PE. A factor to consider is postpolymerization after leaving the coating line (Chapter 3). This is a phenomenon that occurs usually within 4–24 h. The increase in crystallinity that takes place in the plastics results in dimensional changes, principally shrinkage. This in turn can result in puckering and wrinkling that causes poor roll confirmation. In-line postcuring methods, such as heat, high humidity or both, are effective means of increasing the rate of crystallinity to a level which leads to maintaining good roll configuration.

With recycled plastics, make sure the line can still operate efficiently. Since the plastics have already undergone extreme high temperature when processed the first time, reheating again to the high temperatures may not be satisfactory. The performance of recycled plastics is reviewed in Chapter 3.

## PROCESSING CHARACTERISTICS

The melt from an extruder is forced through a horizontal slit die onto a moving web or substrate (Figs. 11.4–11.6). The rate of application is controlled by the amount of extrudate leaving the die and the pull rate that occurs as soon as the melt is pinched by a set of rolls. The melt stream could be coextruded layers (Chapter 2).

The extruder is located on some type of guide rolls so that it can be retracted from over the coating or laminating line. On start-up of the extruder, the extruder is operated to develop the ideal melt. When the melt is satisfactory, the extruder is moved forward on the rails and over the line to start applying melt to the substrate. When the line is to be shut-down, the extruder is retracted and also shut-down.

As the melt leaves the die lips, the thin melt film is pulled down into the nip between two rolls located directly below the die. The substrate, which is traveling at a speed faster than the extruded film, draws the hot film. The combination of these two different speeds, with the pressure applied between the two rolls, relates to the required thickness to be applied to the substrate. The two rolls are the chill roll and the pressure roll that forces the film onto the substrate(s).

The melt usually contacts the substrate just prior to meeting the chill roll; Figure 11.1 provides a close-up of this delayed contact with the chill

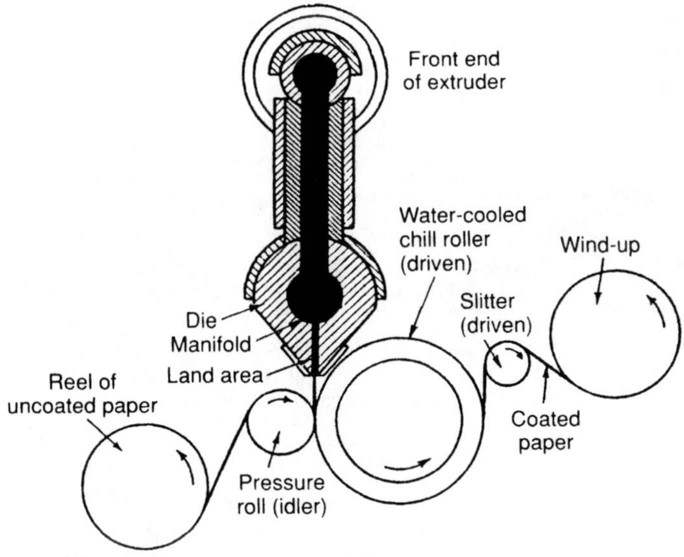


**Figure 11.4** Close-up cross section of coater.

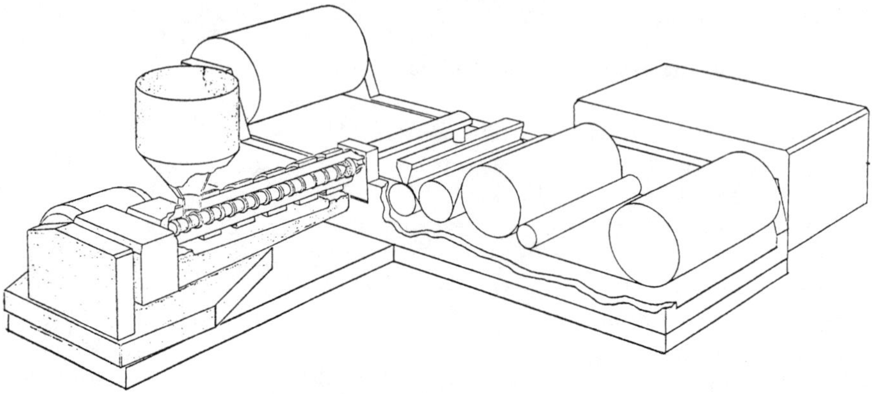

**Figure 11.5** Schematic of coating line including location of an extruder.

**Figure 11.6** View of coating die applying a curtain of hot plastic to a substrate.

roll. For PE, this distance prior to contacting the nip of the rolls is about 6.5–9.5 mm (0.25–0.38 in). The contact time on the chill roll is 0.2 s or less.

The melt web usually extends beyond the edges of the substrate. The chill roll quickly cools (solidifies or quenches) the melt by means of its water-cooled, chrome plated steel roll. The pressure roll is also metal.

To avoid any tendency for the melt to stick on the pressure roll, the roll is covered with an elastomeric sleeve, usually neoprene or silicone rubber; there are also rolls coated with FEP or the equivalent. Sticking, especially of thin film less than 0.025 mm (1 mil), can be probably avoided by cooling the pressure roll. Cooling prevents the roll temperature from rising too high. A water-cooled metal idler roll can be run against the pressure roll. Also, it may revolve in a tray kept full of water. A doctor roll or other device is used to prevent water reaching the composite material. Pressure rolls of the large coater lines may have external and internal cooling capabilities.

Films are generally 0.001–0.005 mm (0.05–0.2 mil) thick. In laminations most films are at least 6.4 mm (0.25 mil) thick, more commonly 0.01–0.05 mm (0.5–2 mil) thick. Line speeds depend on the type and thickness of the coated plastic as well as the substrate being processed. Speeds could range from 15 to 300 m/min (50 to 1000 ft/min). The slower speeds are used in laying on heavier coatings and/or where maximum adhesion is required.

The distance between the die lips and the chill roll should be at a minimum to reduce neck-in. Increasing the extruder screw speed will also reduce neck-in. The balancing of requirements, such as minimum neck-in; high coating speeds, absence of voids, etc., is a matter of experience.

The pressure roll is mounted in such a way that the nip between it and the chill roll can be opened usually up to at least 150 mm (6 in) to make threading of the substrate during startup easier. Quick, wide opening of the nip is also important in an emergency, such as a break in the substrate, and as a safety measure. The position of both rolls is adjustable so that the nip can be moved relative to the die.

Chill rolls must be large to do their job. A usual size is 30–60 cm (12–24 in) in diameter. The temperature of the chill roll is usually maintained between 20–40°C (68–104°F). Depending on the plastic being processed, there is a maximum chill roll surface temperature needed so that performances are not decreased. With PE it could be about 65°C (150°F).

Figure 11.7 is a nomogram that provides guide lines for the cooling water required in chill rolls. In the example illustrated by dashed lines, extruder output is 680 kg/h (1500 lb/h) with PE having an average value for specific heat of 0.61 kg/°C (0.75 lb/°F) between 18–330°C (65–625°F). A straight edge is laid across these two values and a mark made on the reference line. The straight edge is then laid across this mark and the value for the temperature rise, in this example 550°F (620 minus 70°F) or 288°C. Where the line crosses the last scale, read off 2 $m^3$/min (455 gal/min) of water from a refrigerator system of 57-ton capacity.

After leaving the chill roll, the composite material [extruded film and substrate(s)] is drawn over other rolls to be treated by whatever is required for the specific run. Examples include surface treatment, trimming,

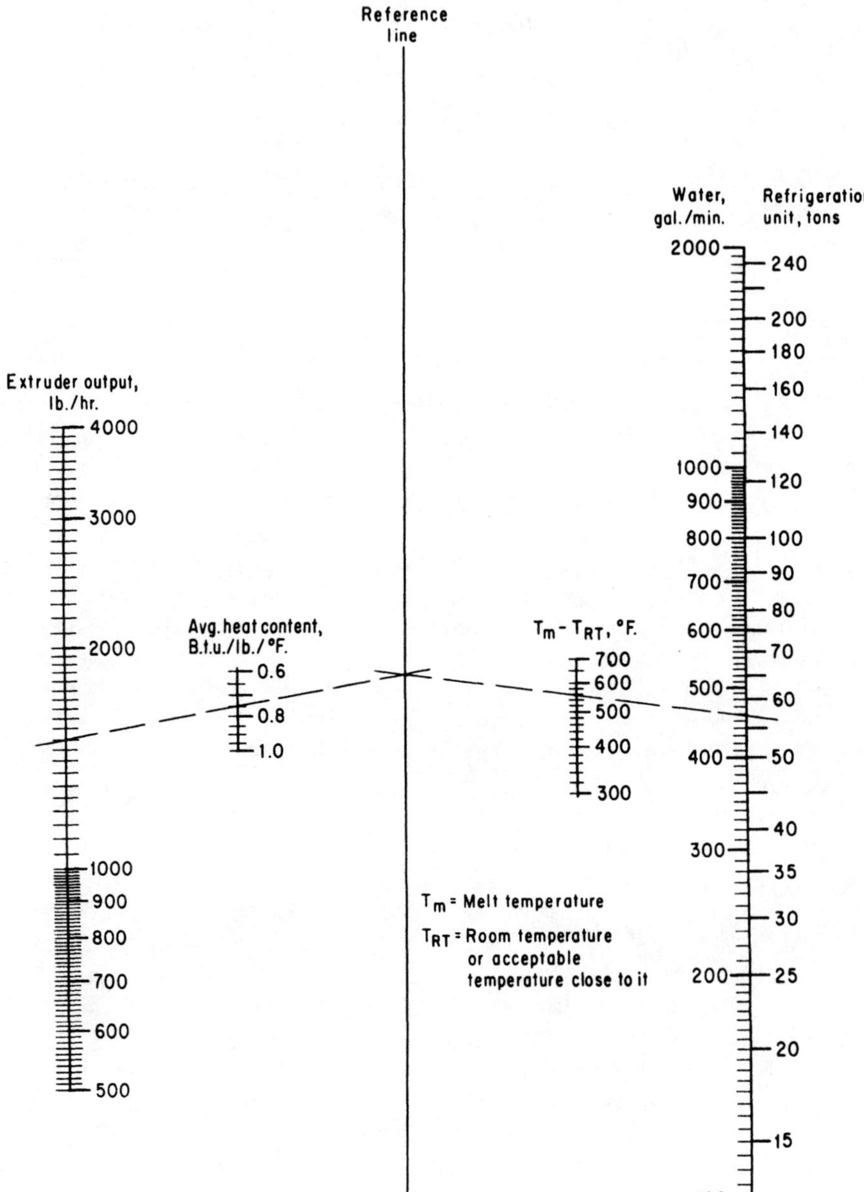


**Figure 11.7** Nomogram for determining quantity of cooling water required.

and/or slitting. The edges are trimmed to remove any uncoated edge and the small bead of plastic which usually builds up at the edges of the composite. At the end of the line, the material is wound up on conventional windup equipment such as those used for film and sheet (Chapters 7 and 9).

In this continuous operation, rolls of substrate material(s) are unwound, automatically spliced on the fly (if required), and surface treated (if required). The surface treatment, such as a chemical priming, is used to make the substrate receptive to the extruded plastic. The substrate normally provides the mechanical and physical strength required for processing and in use.

It is not economical to use narrow substrate widths, since the cost of equipment is high, so die widths are large with a consequent need for large extruders. Higher temperatures are required for extrusion coating and laminating than for unsupported film extrusion. The usual form of heating employed in these extruders and dies is high capacity type electrical heaters.

The weight of the plastic deposited on the substrate is controlled by the speed of the chill roll. For any given extrusion rate, the higher the chill roll speed, the greater the extrudate draw down, thus the lower the coating weight. There are three types of lines: (1) thin film or low tension applications at operating web tension (kilogram or pound force) levels of 4–36 kgf (8–80 lbf); (2) paper and its combinations in the middle range of 9–90 kgf (20–200 lbf); and (3) high tension applications, such as paperboard at 168–680 kgf (50–1500 lbf). Tables 11.2 and 11.3 provides information on surface coverage for LDPE coating based on various thicknesses.

An example of typical surface coverage using PE coating with 100 mm (3.5 in) extruders can coat widths about 600–120 mm (24–48 in); with 120 mm (4.5 in) at 900–2500 mm (35–100 in); with 150 mm (6 in) at 1000–4000 mm (40–160 in); and with 200 mm (8 in) at 3000–5000 mm (120–200 in).

The chill roll has three functions, one of which is to cool the coating to a temperature low enough to enable it to be stripped from the roll. This

**Table 11.2** Example of a surface coverage of PE coating (average density 0.920 g/cm$^2$)

| *Thickness (mm)* | *Mass (g) required by 1 m$^2$ of substrate* | *Length (m) covered by 0.45 kg PE* |
|---|---|---|
| 0.001 | 5.8 | 175 |
| 0.002 | 11.6 | 85 |
| 0.004 | 23.2 | 42.5 |
| 0.008 | 46.4 | 21 |

**Table 11.3** Surface coverage including thinner coatings

| Coating thickness | | | Weight required | | Surface coverage | |
|---|---|---|---|---|---|---|
| *(in)* | *(mil)* | *(mm)* | *for 1000 ft$^2$ (lb)* | *for 1 m$^2$ (g)* | *per lb of resin (ft$^2$)* | *per kg of resin (m$^2$)* |
| 0.00025 | 0.25 | 0.006 | 1.2 | 5.8 | 840 | 10.8 |
| 0.0005 | 0.5 | 0.013 | 2.4 | 11.5 | 420 | 21.5 |
| 0.001 | 1 | 0.025 | 4.8 | 23 | 210 | 43 |
| 0.002 | 2 | 0.051 | 9.6 | 46 | 105 | 86 |

cooling has to be accomplished in about half a revolution of the roll thus necessitating very efficient cooling of the roll. Another is that its speed helps control the coating thickness which in turn controls the economy of the line. The chill roll also determines the surface finish of the film. For high gloss, the surface of the chill roll must be very highly polished; where dull film surfaces are required, the chill roll may be etched or sandblasted.

Adhesion of the film to the substrate is a matter of prime importance and is affected by several factors. High melt temperatures are essential. Other factors include film thickness (thick coatings adhere better than thin ones) and drawdown rate (high draw down rates have an adverse effect on adhesion). Different substrates also have their own individual problems as in the case of paper and paperboard, where a low moisture content is required for good adhesion. Adhesion of PE to aluminum foil may not be an easy matter to achieve and the problems are sometimes overcome by an etching treatment of the foil surface. The coating of regenerated cellulose film also presents problem. All these problems can be solved or eliminated by using the proper (best) startup procedures that are available.

One of the problems encountered in extrusion coating is that of beading at the edges of the hot film where neck-in occurs. This local thickening of the film has to be removed during the trimming operation. Otherwise, when the roll is wound up, the beads at either end only support the weight of the roll, leaving a loosely wound and sagging roll in the middle. The extra trimming required means a reduction in the usable web widths, thus increasing the costs.

In the case of PE, the incidence of neck-in can be reduced by decreasing the melt index or the density, but the risk of voids in the hot film at low coating speeds increases because of imperfect melt flow. So an optimum combination of these two factors has to be found. One should determine the minimum neck-in and bead size based on processing conditions, and use those observations as control parameters. Sensors of neck-in and beads can be used in automatic process control.

Neck-in with nylon is greater than with PE. Typically, the neck-in is about 15 cm (6 in) or 7.5 cm (3 in) per edge. Unlike PE, neck-in with nylon increases with increasing line speed. The amount of neck-in would lead to excessive trimming and waste of nylon. The use of air jets located just below the melt exiting the die will minimize this problem. They are effective in reducing neck-in to less than 5 cm (2 in) per edge and reduce the amount of trim to about 2 cm (0.75 in) per edge. A low flat of air directed down and out has been beneficial. Preheating of the air is effective in reducing the changes in the freeze-off when the web reaches the chill roll.

Air jets also eliminate edge weave; this is the cycling of the edge of the melt web in an 'in-and-out' motion. The edges may move independently or in unison. In severe cases, the magnitude of the cycle can be as much as 10–13 cm (4–5 in). It has also been found that center weaving can be effectively controlled by adjusting the air gap to the correct distance for the particular line speed being run; also required is regulating the die temperatures so that the melt temperature is the same across the width of the entire web. The minimum allowable air gap should be used because any heat loss can adversely affect adhesion. Generally, an air gap of 8–18 cm (3–7 in) is used for running speeds of 30–300 m/min (100–1000 ft/min), respectively.

The uncoated substrate is fed from an unwinding roll or reel into the coater. After being trimmed with or without being slit, the coated stock is rewound on a core roll. There are a variety of types of handling equipment for the unwinding and rewinding stations that are usually turret or center winders. Which type to use depends on the kind of substrate being coated, the volume of production, and other factors. Large production will require fully automatic equipment which must not be stopped to change rolls.

The unwind station, as it unwinds, has its diameter becoming smaller and smaller with tension on the substrate decreasing. The reverse occurs with the rewind station where as the roll becomes larger, the tension increases. Without some type of compensation, this action would result in nonuniform tension resulting in severe coating problems. Thin paper, foil, and textile fabrics require less tension than other substrates at the unwind station, and also at the rewind station, to prevent breakage. To compensate for these tension changes, an automatic tension control is used. Different systems are used that include tension control rolls, friction brakes, etc. See Chapter 6 for information on controls.

Careful control of tension in the windup, especially with flexible substrates such as PE film, is very important, since high tensions may lead to unwanted machine direction orientation and poor roll configuration. Normally, very low tensions of 0.1–0.2 lb/in/mil are used. Higher tensions can be used on slitting and rewinding of the mill rolls.

From the unwind to the rewind rolls, there are a number of rolls over which uncoated and coated substrates travel. They are necessary to guide the web at an accurate position through the line. These rolls are running in ball or other type bearings that are precision ground. They must be kept clean and accurately fitted into the line on sturdy frames. They are kept in perfect alignment in the line.

When the substrate is a very light or a smooth material, it is frequently necessary for several of the rolls to be driven. For paper coatings, however, most auxiliary rolls are idlers.

There are means to consider for increasing output so long as one does not produce rejects that actually reduces production. Some effective steps to be considered (that include limits) for increasing output or to improve quality include the following: (1) increase the extruder barrel temperature to increase output and quality; (2) increase extruder back pressure so quality increases but output decreases; (3) add number of extruder screen packs so quality increases but output decrease; (4) raise extruder screw temperature so output increases and quality decreases; and (5) increase extruder screw speed so both output and quality increase. See Chapter 2 for information on the extruder operations.

### Shut-down

Because the extrusion coating process requires such a high operating temperature, shut-down requires special considerations and care. One can not just flip a switch for an immediate shut-down. This approach would probably result in degraded plastic in the extruder through to the die. Probably the usual approach to purging would not be adequate. See Chapter 2 regarding 'clean-up.'

The proper procedure for shut-down is to keep pumping plastic through the extruder while reducing the temperature from the usual running at 370°C (700°F) to 260°C (500°F). When the temperature has reached 260°C (500°F), it is time to turn off the extruder. At that temperature, the plastic is not usually subjected to rapid, degrading oxidation with its usual formation of hard, discolored plastics which adheres, with exceptional strength, to any adjoining metal.

At the time it is planned to shut-down the complete line, the extruder is pulled away from over the coating line. At that time, the coating line is shut-down by releasing or opening rolls, such as the pressure and chill rolls, with probably coated material in the system so it can be used for startup.

PROCESS OPTIMIZATION

Figure 11.8 illustrates the effect of the extrusion line and plastic material variables on the performance of coatings. The stability of the melt web before it contacts the chill roll depends greatly on having the correct: (1) design of the extrusion die; (2) melt temperature; (3) air gap distance between the die and roll nip; and (4) line speed. A successful coating and laminating line depends on many factors, not the least of which is the skill and experience of the line operator.

Various processing conditions influence the performance of coated and laminated products. These influencing variables can be controlled (Fig. 6.9). However, to be practical and obtain the best overall performances usually requires compromise. A variable could be due to the characteristic of the plastic being processed. It could have damaging effects on these processing lines since it is very critical to operate at the very high temperatures. The plastics instability would be a major problem area. Quality controlling the incoming plastics for these lines could become very critical.

### Gauge variation

Basically, thickness of the extrudate depends on the two independent factors of rate of extruding the extrudate output and surface speed of the chill roll. A uniformity of the extruder output coupled with consistency in the line speed are vital adjustments in order to produce a uniform coating thickness in the machine direction (MD).

Control of die temperatures is particularly important as well as keeping variations in the die lip alignment to a minimum. As reviewed in Chapters 8 and 9, use different temperatures across these wide dies to compensate for pressure drops across the manifold. Raise the temperature at the end zones to develop lower melt viscosities to counteract differences that exist in melt pressures across the die. As with extruding blown film (Chapter 9), local thickness variations may be minimized by oscillating either the die or roll assembly.

### Neck-in

As reviewed in Chapters 8–10, there is a melt film neck-in from the die to the first point of contact on the line. The hot film shrinks on its way down so its width from the die to the chill roll is reduced. At the same time, beading or thickening of the edges occur. To help reduce this situation and make the line easier to run, usually the hot web is wider than the substrate, thus reducing substrate costs since the plastic is usually less expensive. As mentioned earlier, these unwanted edges are trimmed

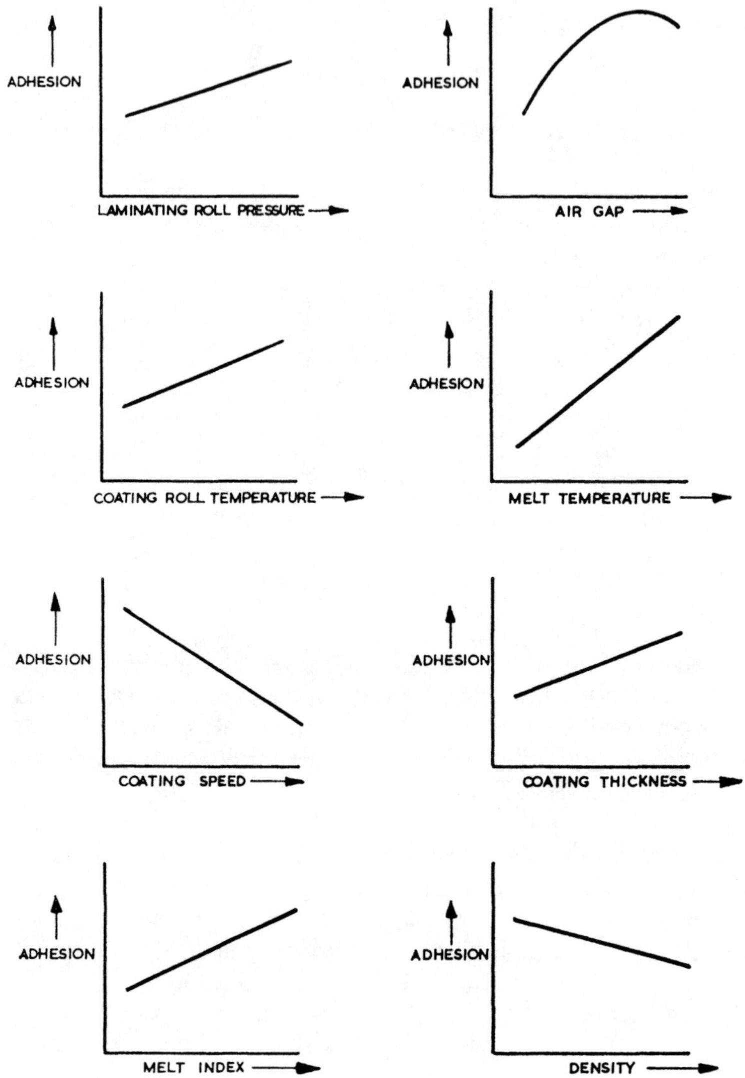


**Figure 11.8** Effect of machine and plastic variables during extrusion coating.

down the line. With the wider web, the edge trims could be all plastic which makes it easier to granulate and recycle.

Neck-in decreases slightly with increasing line speed. A decrease also occurs when lower melt index (MI) plastic, plastic density, or both, are reduced. There is an optimum combination of MI and density below which voids in the hot coating occur at high line speeds owing to improper melt flow.

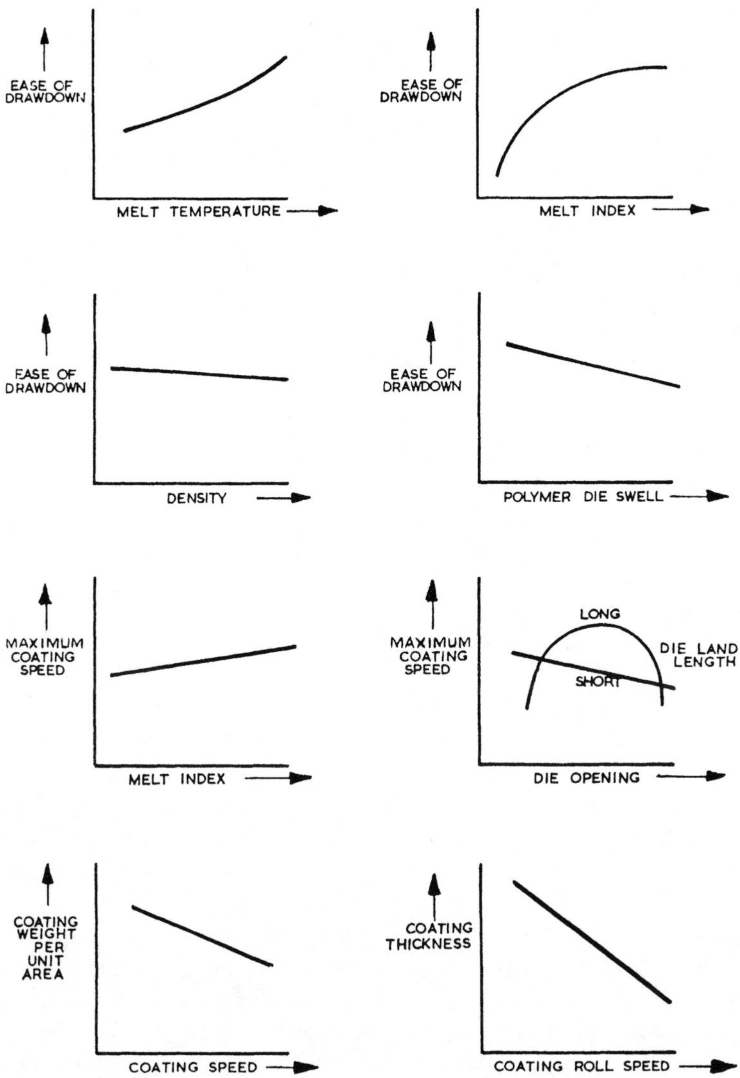


**Figure 11.8** *Continued*

## Substrate

Systems are used to keep the substrates in position and to detect breaks. While deckles (Chapter 5) in the die make sure that the plastic melt does not vary at the edges of the coated substrate, provision is usually made to ensure that the substrate is maintained in the same position relative to the melt flowing into the nip. This edge guide equipment is particularly essential when a flying splice set-up is used at the unwinding station. Mechanical, pneumatic, and/or photo-electric sensing elements are placed at the edges of the substrate. Whenever the web moves sideways, the sensing elements activate a device which readjusts its position.

A similar system can be used to detect breaks in the substrate, sound an alarm, and simultaneously open the nip to prevent damage to the pressure roll by the hot melt. A hot melt covered roll requires costly resurfacing and downtime on the line. Generally, a break in the substrate provides its distinctive 'alarm.' It makes an audible and distinct cracking noise. Without the alarm system, when this cracking noise occurs the operator pushes the necessary 'button' to open the nip.

## Adhesion

As mentioned earlier, good adhesion can be considered the most important consideration in this extrusion process, particularly when the customer complains! The stronger the bond the better will be the heat seal strength, resistance to grease, resistance to crack, or whatever exposure the composite encounters. Good adhesion depends on factors such as temperature, pressure, line speed and weight of plastic, melt flow, but principally on the nature of the substrate.

As mentioned earlier, different plastics, particular different types of a family of materials such as PE, PP, and nylon, each have their own processing characteristics from the extruder to the coating line (Chapter 3). With a nylon, to obtain maximum adhesion, melt temperatures goes from 282–308°C (540–585°F). This melt condition is used when coating 23 kg (50 lb) Kraft paper with one mil 0.025 mm (1 mil) of this nylon. Further increase in melt temperature to as high as 316°C (600°F) results in no substantial improvement in the level of adhesion.

The melting point of plastics being processed can influence adhesion. For example, under identical coating processing conditions, a nylon copolymer having a melting point of 240°C (465°F) can produce a level of adhesion of 80%. With a nylon 66 melting at 258°C (497°F), it produces only 40% adhesion tested under similar conditions. However, these differences become less important with the use of effective adhesion primers or promoters.

Adhesion to porous substrates is very different from that to nonporous substrates. Porous materials, such as paper, paperboard and fabrics, have a natural tendency to retain and hold onto the coating because of the ideal situation where the hot melt 'enters' the substrate and is reinforced. Adhesion in this case is achieved by both physical and mechanical bonds. A problem could develop with PE coated porous substrates. A higher melt index (MI) plastic with their low viscosity adhere better than the PEs with low MI (Chapter 19).

Nonporous substrates that have a smooth surface, such as aluminum foils as well as plastic films or sheets, tend to resist adhesion. There are certain plastic substrates that would naturally permit an excellent bond but they are not the plastics commonly used.

The common plastics, such as PE, PP, and PA as well as the metal foils, require the use of adhesion promoters applied to the surface of the substrates. Usually, the PE surface requires a chemical bond and a trace of oxidation. Oxidation occurs as the result of a high melt temperatures and allowing an adequate distance between the die and substrate. High well-controlled temperatures of the plastic in the die and at the line when it first contacts the substrate are essential for good adhesion.

To improve adhesion between certain materials, priming the substrate surface can be helpful. Primers are adhesion promoters that alter the surface. Priming may be accomplished by three basic methods: chemical, flame, and corona. The method to be used is dependent on the material involved and costs. In any case, priming should be done in-line whenever possible.

There are solvent base chemical primers used on plastic films and metal foils. Water-base chemical primers are used on papers and boards. Both types are usually applied using a gravure system with hot dry air. Flame treatment is usually limited to paper, board, and plastic substrates. Corona priming is used on different substrates including foils.

With high melt temperatures, problems could develop since high temperatures affect sealability adversely and surface finish, and cause odor in the finished laminate. Thus, a compromise is required. With high coating weights and particularly low line speeds, adhesion is increased. This action occurs because more time is available for oxidation to occur.

With thinner coatings, the melt cools faster between the die lips and point of contact, resulting in poorer adhesion. Then there is a sea-saw situation since the lower the coating weight, the higher stock temperature required to obtain good adhesion. Conversely, the higher the plastic weight, the lower stock temperature is required. Also, too low a chill roll temperature will reduce adhesion.

To promote adhesion of plastics to some substrates, particularly porous types, the substrate may be preheated before it reaches the pressure roll. This action makes the surface more receptive to the molten plastic by drying and warming the surface. With nonporous substrates, preheating will help remove moisture as well as foreign substances such as lubricants or sizings, thus promoting adhesion. Preheating is usually done by passing the substrate over a steel drum internally heated to about 175–190°C (350–375°F). Sometimes gas flames or radiant heaters are used, but their control is more difficult.

However, preheating affects only the surface of the substrate; it does not really penetrate into the substrate because of the very short duration of the application of heat. It cannot dry a wet porous web. Adhesion to a wet substrate is always poor. What is required is that the wet material be oven dried, or an equivalent, before being fed into the coating line.

**Printing**

If surfaces, such as PE, are to be printed (or bonded), they may be treated, because certain ink or adhesive will not stick/bond to untreated PE. The treatment modifies an extremely thin surface layer of film. This action makes the surface receptive to ink which then becomes bonded to the coating so that it cannot be removed by rubbing without injuring the coated surface.

Treating is best done in-line, such as between the chill roll and the slitter roll or after the trimming roll. When plastics containing slip or anti-blocking additives are used, there is a potential problem. Treating must be done in-line before the additive in the plastic blooms to the coating's surface.

In-line treatment is by chemical or physical methods; both provide good treatment so that the surface to be coated is a clean, polar surface. Chemical treatments (oxidation, chlorination, etc.) are usually more expensive to use. The physical treatments involve flame and electronic or corona discharge.

Flame treatment has a gas feed flame strike the coating as it passes over a water cooled, chrome-plated drum. Thus, the outer coating surface is subjected to a high temperature while the mass of the substrate is not. Different types of burners are used. For uniform treatment, it is essential that the length of the flame be constant. The burner is directed against a cooling roll/drum at an angle so that hot gases sweep over the coating. The flames may have different intensities, positions in relation to the roll, and distance from the roll. The optimum flame distance for best printability is different for the different PEs and other plastics. Flame treating is usually carried out as a separate unwind–rewind operation from the coating line at speeds that differ. This process is not extensively used since it is usually more expensive and fire hazards exist.

The usual treatment is by electrical discharge equipment. This is relatively inexpensive and operates in a simple, clean manner, and is easily adapted to in line processing. It is less critical than flame treatment and operates at the coating line speeds. The degree of treatment is mainly a function of web speed, web width, and intensity of treatment.

The plastic, such as PE, passes over a ground roll under the electrode which carries a high voltage. The roll is insulated by a dielectric, such as a polyester film covering or rubber covering. The voltage jumps the gap between the electrode and the coated substrate, causing a corona discharge upon the coated side. The air in the gap is ionized causing oxidation of a very thin film of the coating.

The dielectric roll covering prevents the entire discharge from going directly into the metal roll when pinholes or breaks occur. Without the dielectric covering, such defects would result in no treatment, while burning the coating and creating a series safety hazard.

Carefully handle the treaters since they could be hazardous, like driving a car, even though they have safety devices. They use relatively high voltages with open electrodes that must never be touched. Ozone is produced so good ventilation around the line is required, as it is required even without the treaters. Static electricity on the film surface during treatment may give the operator an unexpected shock. Commercially available static eliminators should be used to dissipate the electric charge.

### Block and slip surface quality

The characteristic of coated materials are a compromise between undesirables of blocking and slipping when winding the material at the end of the line. An example of potential problems is with PE coated substrates to be made into bags or pouches that must not block; coated sides must not stick to each other. They must have good slip; coated outsides must glide easily when the bags or pouches are stacked. When paper or board is coated on both sides, later to be sheeted and stacked, good slip and no blocking are essential.

Slipping and blocking are effected by such factors as coating temperature, chill roll surface temperature, over treatment, and the absence or presence of slip and antiblock additives in the plastic. A high coating temperature and a highly polished chill roll, though having beneficial effects on adhesion and gloss of the coating, definitely contributes to blocking. A matte chill roll will improve slip properties considerably, especially when there are two-sided coatings, both of which are to be dull. Increased chill roll temperature will increase slip and somewhat decrease blocking.

### Control

There are different options in setting the variables that exist on-line, such as extruder output rate, pull rate of melt, amount of neck-in, chilling capacity, and so on. Lines include all types of controls with some dedicate only to specific sections.

There are lines with a complete control of the line from up-stream to down-stream. Figure 6.9 is an example of a central control system where a master control operates the complete line. Since the melt is usually at the extreme maximum temperature and pressure, operation of the line requires delicate care. Any slight change along the line could cause a product to be damaged or not to specification – in fact, the line could be shut down. Note that changes in the environment, such as air currents and moisture, can cause immediate down-time; tighter controls may not be necessary if all that is required is not to open a door, particularly a large size garage type door.

### Dimensional stability

Changes in dimensions of the finished products can be caused by: (1) changes in the substrate due to over-stressing during processing due to uneven heating and/or changes in moisture content; and (2) unbalanced forces during the process. The changes usually occur after in storage and/or after a postextrusion treatment which stress relieves the product. Correction can be made by adjusting screw speeds and take-off speeds.

### Heat sealing

PE is extensively used as a coating material because of its ability to develop a strong seal under heat and pressure. This performance is effected by factors such as: (1) additives in the plastic (also effects the temperature spread); (2) amount of adhesion to the substrate; (3) processing conditions (includes size of air gap which influences degree of surface oxidation); (4) coating thickness (with thick coatings, more strength develops; for thin coatings the differences are more pronounced where a small thickness change causes erratic heat seal strengths); and (5) aging effect (it causes raising sealing temperatures, so seal as soon as possible; processing changes can offset this situation).

## DIE

The dies used for extrusion coating are similar in construction to the rugged flat film dies that are described in Chapters 5, 8 and 9. The end-fed dies, most useful in coating narrow substrates, offer better accessibility. The center-fed dies can distribute the melt more uniformly on the usual wider substrates that are used. For extra wide lines, the die may be fed from both ends by two extruders.

As in extrusion of flat cast film, control of the die temperatures is important. A 5°C (10°F) change in temperature can result in as much as a 20% change in thickness. As in cast film, rates of greater than 3.5 kg/h cm (3 lb/h in) of die width are necessary and 4.6–5.8 kg/h cm (4–5 lb/h in) of die width are preferred.

Dies can be over 3 m (10 ft) wide in order to coat a substrate 3 m (10 ft) wide. With a die opening of 0.5 mm (0.02 in), the film thickness after being drawn at the chill roll is 0.013–0.025 mm (0.0005–0.001 in) and even to 0.005 mm (0.0002 in. The latter is a 100/1 draw-down ratio.

With the longer die lands required in coating/laminating lines, they tend to increase neck-in. Deckles can be used to vary the web exiting; internal deckling tends to cause more neck-in than external causes (Chapter 5). Regardless, there is nothing that can be done to eliminate neck-in with or without the deckles. However, since it could be not economically practical to have separate dies for each substrate width, deckles are used.

Design of the die has an important bearing on melt web stability. A high-pressure die with good melt distribution is desirable. It can permit the use of less air gap at a given line speed that can be used with a die having poor melt distribution. Conversely, higher line speeds for a given air gap may be used with a good die design.

Processwise, adjustments can be made to reduce the amount of neck-in but with this gain comes losses as explained earlier in this chapter. So compromise is necessary.

## APPLICATIONS

Coated and laminated products have widespread industrial and commercial applications, and found in many industries. They are basically used for decorating, protecting, and providing different functional treatments that enhance the substrates. Performance enhancements include: heat sealability; impermeability (to moisture, water vapor, and/or other gases); energy barrier (to UV light, heat, etc.); chemical resistance; modify properties such as optics, electrical, mechanical, etc.; and/or alter coefficient of friction to reduce or eliminate blocking of substrates.

Products from extrusion coating and laminating lines have six main market applications. They are liquid packages (milk cartons, aseptics, etc.), flexible packages (foods, medicines, etc.), paper and paperboard packages (detergents, frozen foods, etc.), industrial products (tapes, insulation, etc.), and sacks (tarpaulins, photographic-base papers, etc.).

## SAFETY

As with the other extrusion line, these lines deserve respect and attention. The extruder and melt are very hot; burns will be severe. Its different rolls can be dangerous. The fast-moving substrate(s) can cause deep and painful cuts. Wear the proper clothes around the complete line.

TROUBLESHOOTING

Table 11.4 provides a guide regarding coating problems. If all remedies should fail, you may be using an unsuitable plastic with too low a melt index.

Regarding voids (holes), rears, and breaks, they cannot be tolerated in extrusion coating and laminating lines. They may occur when line speeds are higher than the strength of the hot melt permits, or when hard and

**Table 11.4** Troubleshooting extrusion coating

| *Problem* | *Cause* | *Solution* |
|---|---|---|
| Adhesion | Low melt temperature | Increase. Check heater bands |
| | | Increase back pressure |
| | Hot melt | Take opposite steps |
| | Premature cooling | Decrease air gap. Increase chill-roll temperature |
| | Chill-roll tack | Decrease chill-roll temperature |
| | Substrate surface | Pre-treat or change substrate |
| Applesauce | Resin incompatibility | Conduct rheological and chemical test |
| | | Make sure previous resin is purged |
| | Hot melt | Decrease |
| | Low-melt temperature | Increase |
| Draw resonance | High draw ratio | Decrease die gap |
| | Hot melt | Decrease |
| Edge tear | High draw ratio | Decrease die gap |
| | Die-end temperature too low | Increase |
| Gauge bands | Dirty die | Clean with brass tool |
| | Improperly adjusted die | Adjust die bolts. Measure gap |
| | Melt-temperature variations | Check heaters |
| | | Keep adapter, transfer pipe and die temps close to melt temp |
| | High melt temperature | Decrease. Check for heater malfunctions |
| Gels | Resin contamination | Keep extruder, storage areas clean |
| | Crosslinking | Decrease polymer temperature |
| | | Increase screw speed |
| | | Check for hot spots |

**Table 11.4** *Continued*

| *Problem* | *Cause* | *Solution* |
|---|---|---|
| | Poor purging | Use proper techniques |
| | Poor mixing | Increase back pressure |
| | | Increase number or density of screen packs |
| | | Assure resins are compatible |
| Molten curtain breaks | High draw ratio | Decrease die gap |
| | Incorrect melt temperature | Adjust |
| Odor | Over-oxidation | Decrease melt temperature |
| | | Decrease air gap |
| Oxidation | Degradation | Decrease melt temperature |
| | | Increase screw speed |
| | | Clean non-streamlined areas often |
| | Poor purging | Use proper techniques |
| | | Don't stop extruder when still above 200°C (400°F) |
| | | Purge copolymer with homopolymer before turning off extruder |
| Pinholes | Rough substrate | Flame-treat |
| | | Increase polymer-web thickness |
| Heat sealability | Over-oxidation | Decrease melt temperature |
| | | Decrease air gap |
| | | Check corona treatment |
| | Contamination | Check additives |
| | | Minimize handling of sealant surface |
| | | Store in clean, dry area |
| Surging | Hopper bridging | Reduce temperature of feed zone |
| | Improper screw | Change |
| Voids | Moisture | Avoid abrupt temperature changes in storage |
| | | Check for leaks in resin-handling system |
| | Polymer degradation | Do not exceed recommended maximum temperature |
| | Gels | (see above) |

oxidized plastic particles form in the die, such as gels and fisheyes. Also too high a temperature anywhere in the extruder may break down the plastic such PE into volatile, vaporizing components which cause bubbling resulting in voids.

Rapid changes in temperature or humidity, or both, between the storage area and line may cause moisture to condensate on the plastics. Do not let the plastic become moist or contaminated. With higher density plastics, raise the melt/stock temperature. Proper material handling is required.

The desired condition during processing is to operate with as high a controlled temperature that is feasible when the hot melt leaves the die and travels in air to the nip of the rolls. This heat helps to obtain good bonding of the coating to the substrate. High stock temperature for PE is close to 315°C (600°F) or higher to bond (chemically) to certain substrates. Although high temperatures have their benefits and are critical, there is a possible downside. It may effect sealability and other properties reviewed in this chapter.

Extruded plastic drawn around a chill roll or water bath at continuously increasing rates will begin to pulsate rhymically at a critical draw ratio, namely, the ratio of the melt's final velocity to its velocity in the die [284]. As the drawing rate increases, the pulsations become more pronounced. This phenomenon is known as draw resonance (DR). Eventually the molten plastic tears.

Draw resonance can occur in any extrusion melt-drawing process (extrudate passes between the two fixed points of the die and a fixed velocity take-off), such as coating, filament spinning, and texturing. In extrusion coating, the nip point becomes the fixed velocity take-off point where the extrudate film rapidly cools, assuming its final thickness.

Draw resonance for Newtonian melts becomes critical at draw ratios above 20. For certain nonNewtonian (plastics) melts, resonance occurs at draw ratios under 20. In extrusion coating, draw ratios are higher (usually around 35) since die gaps typically range from 0.5–1.00 mm (20–40 mil (0.5–1.0 mm), film thicknesses are 0.025 mm (1 mil)) or less, and draw speeds equal or exceed 150 m/min (500 ft/min). Reducing the die gap to reduce the draw ratio is not commercially feasible, given the constraints of head pressure and melt fracture.

When LLDPE started replacing LDPE coating lines because of its superior properties and potential for downgauging, DR became a problem. To reduce the problem, different approaches were used, such as blending LLDPE and LDPE, changing line speeds, etc. By far the best solution has been to cool the extrudate gradually rather than expose it to the shock-cooling of a chill roll (or water cooling). Air is directed against the molten LLDPE film at a point between the die and the nip point. The results have included increasing line speeds and thinner gauges. This technology is also applied to other plastics susceptible to DR.

# 12

# Wire and cable

## OVERVIEW

Plastics wire and cable covering or coating in continuous lengths is an important application for extrusion. Large quantities of plastics are consumed that have literally covered billions of miles of these products. This coating process produces many different types and sizes of product. The following three types are examples of coverings: (1) electrical insulation (in conjunction with semiconductive inner and outer layers); (2) internal coverings and fillers (lapped or extruded); and (3) sheaths and external protective coverings [1, 24, 27, 254, 298].

Different plastics are used to meet different performance requirements, both in operation and environment. Because of the potential dangers of the misuse of wire and cables in electrical products, there are many governmental and industrial codes and specifications in use. They include tests for in-line extruding processes.

Extruders have been very successful in the coatings of wire and cable products. These products come in all sizes from very small diameter, single conductors to cables 15 cm (6 in) or more in diameter. They include multi-wire conductors that may be in strap form. A typical single conductor may consist of a copper conductor 0.5 mm (0.02 in) in diameter with an insulating covering of the popularly used LDPE 0.23 mm (0.009 in) thick with an extrusion traveling rate of 2000 m/min (6500 ft/min). For fine wires, high speed is desirable to give a reasonable melt mass flow rate to offset consequent high die pressures, minimized land lengths, and permit greater die swell. Speeds go down as the substrate diameter increases.

Continuity of insulation is essential but, in many applications, adhesion of plastic to wire is also important to reduce breakdown due to corona discharge in high voltage AC applications. As an example, in submarine telephone cables, where a 15 mm (0.6 in) conductor may be covered with a

25 mm (1 in) thickness of LDPE, slow extrusion rates are used. The line downstream will include water cooling in stages. Sometimes cooling will include programmed temperatures to minimize temperature and shrinkage differentials within the insulation and maintain adhesion to the core. With such thick insulation or sheathing, haul-off tension will not be critical. However, haul-off speed will determine thickness and a constant speed is essential to eliminate surface lumps or ripples.

Multi-layer insulation constructions are used. With the high-voltage cable, the center conductor may be covered by three separate layers. The innermost layer will function as an electric screen, the intermediate layer is usually a PE insulation material, and the outer layer serves as an electric screen. The processing technique can be used for primary and secondary coating where the covering is on metallic conductors or coating previously insulated wires.

The melt shear conditions when coating wire or cable differ from processes that use no substrate in that the inner surface is moving with the conductor causing additional drag flow (Chapter 3). Cooling is usually in air, sometimes in a festoon, but water baths have been used for the higher speeds.

Figure 12.1 is a schematic of a wire and cable coating line. The line is usually made up of an unwinding roll for the wire or cable followed with a tension controlled input capstan, possibly a wire straightener, and wire preheater. The wire proceeds to the usual extruder's 90° crosshead die where the plastic coating is applied. It continues through a water and/or air cooling system for thermoplastics (TPs) or a heating system for thermosets (TSs), testers, gauge controllers, tension output capstan, tension controller, and the windup roll. There is also special equipment included for use with certain plastics, such as nylon, to maximize their performance (toughness, stress relaxation, etc.) by using in-line annealing and moisture conditioning equipment.

The high and constant output rates of extruders with engineered upstream and downstream equipment, make possible production of linear speeds very high. The line's output rates can be at least 1300 m/min (4000 ft/min) for certain products. Thin wires, as mentioned earlier, may be coated at up to 2000 m/min (6500 ft/min).

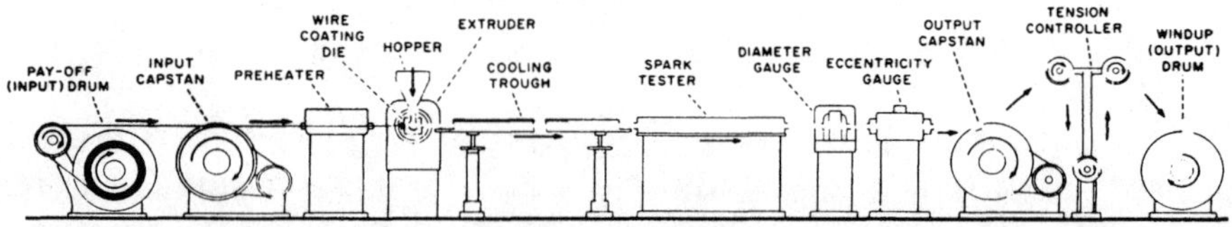


**Figure 12.1** General layout of a wire coating extrusion line using a crosshead die.

The specific equipment used in-line depends upon the kind of wire that is being handled and the product to be made. When plastics covers bare electric wire, the coating is called primary insulation. A secondary layer of plastic extruded over a wire or group of wires is called a sheath or jacket. The sheath may provide only mechanical protection and its electrical insulation quality is of less importance. The cores that refer to either the wires or cables are handled in this extrusion process. The core is the part in the center that is to be coated.

## PLASTIC MATERIALS

There are TP, TS, and TP or PS elastomer plastics used to meet the different requirements. Important TPs based on quantity used include PE, PP, PA, and PVC. The epoxy (EP) and polyurethane (PU) plastics are popular TS plastics. Also extensively used are cross-linked polyolefins such as cross-linked PE (XLPE) and ethylene copolymers that behave similarly to elastomers above the crystalline melting point (Chapter 3).

This chapter provides some highlights on the plastics' capabilities, particularly problem areas, but that should not be taken to indicate that they do not process properly with wire and cable covering as well as the other extrusion processes. Plastics are available to meet your requirements assuming you know what you need. However, they need to be properly processed, starting up-stream to the extruder. Detailed information on processing with specific plastics is readily available from the material suppliers.

Many of these plastic compounds are purchased from the materials supplier. Certain plastics, such as colorants, are compounded in-house. Different reasons exist as to whether precompound or in-house compounding is used. The usual reasons concern costs and/or the use of a special secret formulation or how it is prepared.

Table 12.1 includes plastics in widespread use in cables [263]. Table 12.2 provides extruder output data when coating with LDPE. The bulk of the plastics used are with PEs and PVCs. The PEs offer excellent insulating ability with moisture and chemical resistance, flexibility, and light weight. Compared to the PEs, the vinyls are heavier, very flexible, and are somewhat poorer insulators but still very useful as an insulator, particularly as a jacketing material.

PP is used for its good temperature resistance, other performance, and economics. It must contain a stabilizer/inhibitor additive to prevent any reaction with copper, which would otherwise hasten its degradation. Stabilizers are also needed in PVC to prevent degradation under heat during processing and in service.

Generally nylon (PA) is utilized as the outer cover in wire coating applications. It is extruded over wire already insulated with an insulating

**Table 12.1** Examples of plastics used in wire and cable coatings

| Thermoplastics[a] | | Thermoplastic elastomers | Elastomers | |
|---|---|---|---|---|
| *Polymer* | *Abbreviation* | | *Polymer* | *Abbreviation* |
| Poly(vinyl chloride) | PVC | Blends of | Natural rubber | NR |
| Polyethylene | PE | polyolefins | Butyl rubber | IIR |
| Ethylene–vinyl acetate copolymers[b] | EVA | and rubber | Styrene–butadiene rubber | SBR |
| Ethylene–alkyl acrylate copolymers | | Triple-block | Nitrile–butadiene rubber | NBR |
| ethylene–ethyl acrylate copolymers | EEA | copolymers[c] | Ethylene–propylene rubber | EPM[d] |
| ethylene–butyl acrylate copolymers | EBA | Polyurethanes, | Ethylene–propylene | EPDM[d] |
| Polypropylene | PP | polyesters | terpolymer rubber | |
| Polyamide | PA | | Polychloroprene | CR |
| Ethylene–tetrafluoro-ethylene | ETFE | | Chlorosulfonated polyethylene | CSM |
| copolymers | | | Chlorinated polyethylene | CM |
| Tetrafluoroethylene– | FEP | | Silicone rubber | SR |
| hexafluoropropylene copolymers | | | Epichlorohydrin rubber | ECO |
| | | | Ethylene–vinyl acetate copolymers[e] | EVA |

[a] Plastomers.
[b] Vinyl acetate <30%.
[c] Styrene-diene-styrene.
[d] The collective term for EPM and EPDM is EPR.
[e] Vinyl acetate >30%.

**Table 12.2** Example of extruder output on wire and cable coating with LDPE

| *Extruder size* | | *Output* | |
|---|---|---|---|
| *(in)* | *(mm)* | *(lb/h)* | *(kg/h)* |
| $1^1/_2$ | 38 | 30 | 14 |
| 2 | 51 | 35–40 | 16–18 |
| 3 | 76 | 70–80 | 32–36 |
| $4^1/_2$ | 114 | 150–160 | 68–73 |
| 10 | 254 | 1000–1500 | 454–680 |

plastic, such as PVC, PE, PP, PBT, etc. This nylon sheath or jacket is used when the wire needs protection from mechanical abrasion, toughness, strength, elongation, cutting, chemical attack, etc.

Polysulfone when used requires an exceptional drawdown ratio (DDR is explained later in this chapter). Low-pressure dies are used. Both diameter and wall thickness of the exit annulus are about 20 times the final outside diameter and thickness of the insulation.

This wire coating process usually involves the use of two extruders with two options. The different insulating plastic and jacketing plastics can be coextruded (Chapter 2). The other option is using two separate crosshead dies. The first die applies the insulating plastic followed with the sheathing plastic die. When processing nylon, line speed is usually up to 300 m/min (1000 ft/min).

Plastic materials are used that are precompounded or compounded. An example is a dry-blend of PVC directly up-stream to the extruder hopper. Different dry-blends at the hopper are used that expose the materials to the desirable condition of less heat history. Precompounded or compounded cost of materials plays an important role on which material arrives into the hopper.

When extruding PVC (or other plastic) cable compounds and a problem develops, it may be traced back in the up-stream. There may be blending inefficiency. Perhaps the barrel temperature was not within specification or melt fracture occurred immediately when it was submerged in a cold water trough. So, a good troubleshooting guide is to be used.

A wide variety of plastic materials are used to meet the different performance requirements, since wire and cable products operate in different environments. Examples of requirements the plastic materials must meet include heat resistant wiring cables, very flexible lift cables, mining cables subjected to high mechanical loads, simple installation cables, underground cables, cables for high operating voltages, and cables for maximum transmission capacities. The plastics used have frequently replaced products such as natural rubber and paper insulation. A popular replacement plastic is cross-linked PE (XLPE).

To provide protection or permit these plastics to be used requires having them properly prepared by the plastic manufacture and/or compounders. Very important is the compounding of the plastics with appropriate additives and fillers (Chapters 3, 17). As an example, plastics exposed to high temperatures must be protected by stabilizer additives to give adequate service life. When in use, plastics are subject to chemical changes that adversely affect electrical and mechanical properties. The chemical aging processes accelerate with increasing temperature.

To protect against direct sunlight (in particular, UV radiation), light stabilizer additives are used. As an example, with polyolefins, about 2–3 wt% well dispersed (compounded) carbon black gives effective protection. In the presence of atmospheric oxygen, the chemical aging of many plastics, especially polyolefins, is based on oxidation processes that started or accelerated by heat and light; for protection from 0.1–0.5 wt% of antioxidants are added to the plastics. Other protective measures when using certain plastics are taken when the products are exposed to environmental dangers, such as moisture or animals eating the insulation.

In some specifications for wire and cable designs, a distinction is made between sheaths and protective coverings. The latter afford protection against corrosion in the form of a metal sheath or light mechanical protection for installation cables. Sheaths are provided for higher mechanical stresses or as a protection against the ingress of moisture, solvents, etc. As the properties of plastic coverings and sheaths coincide, apart from their dimensions, the collective term sheath is used.

Different plastics are extruded to produce the sheaths. Plastics used include PVC, PE, PA, PU, and PP. Elastomer plastic sheaths are made from materials such as polychloroprene, chlorosulfonated polyethylene, chlorinated polyethylene, and nitrile-butadiene rubber.

Special plastics have been developed and used with flame retardant characteristics. They have reduced fire propagation even in bundled configurations. Their composition also ensure that no corrosion components are liberated during combustion, low smoke emission, and preservation of the insulation. Extrudable elastomeric plastic polyolefin copolymers are used as basic materials. They include EVA, EPR, or EEA.

## PROCESSING CHARACTERISTICS

Coating of wire and cable products is performed by extruding plastics onto the substrates. Basically the procedure is similar to the other extrusion processes. A critical requirement is that the melt leaving the die is very uniform. Another critical requirement is that the line must be capable of running the wire or cable with uniform tension at a desired but constant speed without variation or drift. The lines are designed for a range of different wires and cables.

There are several different procedures with two being particularly interesting, the first for academic reasons and the second since it typifies what is being used. The first is accomplished, although rarely used, by feeding the wire directly through a hole in the center of the extruder (screw, etc.) so that the coating line is in the direction of the extruder. There is no change of direction in the melt flow from the extruder through the die and onto the substrate occurs. Due to the 'hole' in the extruder, many problems to hardware occurs so that it tends to become impractical to operate.

The other method is by far the more popular that uses 90° crosshead dies similar to those shown in Fig. 12.1. The wire is fed through the die. To improve adhesion and reduce shrinkage of the plastic to the wire, the wire is usually preheated before entering the die. Some lines may also provide special cleaning systems for the substrate. The preheating is accomplished by an electric current, radiant heat, and so on; thick wires or cables can be heated by a gas flame or hot gas.

The process starts with unwinding wire from a reel using what is called the pay-off. There are basically two ways to move the wire from the reels (similar to that used with fishing reels). The whole reel may turn on its center as the wire is pulled off the reel. In the other method, the reel does not rotate or move. The wire is pulled over the side or end of the reel. The equipment to hold the reel for this side method is called a flyer type pay-off. Since the reel does not turn, it can pay out wire at very high rates, unhindered by friction or the weight of heavy wire. These reels are usually good only for single conductors since they put a twist in the conductor for each turn coming off the reel. This twist is harmless to a single solid conductor but it will shorten a stranded conductor.

Dual unwind stations for handling one reel after another are usually used. To have the line operating continuously from one reel to another, the end of the wire on the moving reel is welded to the start of the wire on the stationary reel and so on.

Following the unwinding reel, a capstan provides constant speed, usually followed with a straightening and preheater used prior to having the wire enter the die. The former mechanical device straightens the wire if straightening is needed. Preheating provides a means to improve the performance of the product structure as it moves through the die. It prevents premature shrinkage of the hot plastic that could otherwise cause stresses that makes the plastic susceptible to cracking when warmed. Another benefit is the removal of any moisture or oil on the conductor surface with the flame doing a better job. If preheating is insufficient to clean the wire, a mechanical wipe before the preheater may help.

Heating is accomplished by a gas flame or by the more popular method of passing an electric current through the wire itself. Small wires are best preheated by passing them over metal wheels which make the wire part of an electric circuit. The thin wire then is heated by its own resistance to the current. The result is a clean, closely controlled method of heating. For large wires, this method is not practical because resistance goes down as the conductor mass goes up. Instead, the gas flame system is used. Also used, particularly for the very large conductors, is preheating with high-frequency induction heaters.

Processing temperature for TP melts in the die could range from 160–290°C (320–550°F) and with certain TPs going much higher. The actual temperature to be used is dependent on the plastic being processed and within narrow limits such as 10–20°C (20–40°F).

The line can be operating at a speed of at least up to 1525 m/min (5000 ft/min). However, the rate of movement has to be held extremely uniform (targeting to meet perfection). In order to achieve uniformity, all (expensive) peripheral equipment is carefully controlled and monitored from the pay-off (unwinding roll) to the windup (output) roll (Chapter 6). At high speeds, haul-off speed relative to extrusion speed is especially important for insulation thickness. Thickness must be a minimum for economy and uniform for subsequent lay-up or winding at the end of the line.

Cooling of the TP coated wire usually occurs as soon as it leaves the die by going through air cooling, water spray trough, one or more water cooling troughs which may have cascading heated sections, and/or combinations of air and water. In the high speed lines, the troughs are usually 6–30 m (20–100 ft) long. The TP materials being processed, such as PE, PP, PA, and flexible PVC, are solidified by cooling below their $T_m$ or $T_g$ by going through a cooling cycle (Chapter 3). For certain plastics, such as polyester, the cooling first starts with air cooling (to aid its shrinking) followed with a water trough.

It is very important and usually very critical, in wire and cable coating, to obtain good strength, toughness, and elongation. To achieve an aceptable bond, the preheat temperature of the substrate and the water quench temperature of the coating composite are important criteria to properly set. An example of temperatures is the preheating substrates to at least 93°C (200°F) and the start of hot quenching at 65–93°C (150–200°F).

The design of the water cooling trough allows control of the cooling process (with TPs). Cooling should not be too rapid. From the die to the trough there is an air gap of possibly 10–45 cm (4–18 in) or less. Usually the length of the air gap is adjusted by sliding the trough toward or away from the die; thus, it can be set to provide the required distance so the high shear of the hot melt in the die can be relaxed (stress relieved) prior to entering the trough.

To provide a somewhat uniform and gradual cooling of the plastic, the usual 6–30 m (20–100 ft) trough can be sectionalized. Hot water would be in the first section followed with lower temperature water in the following sections. Temperature could go from 49–54°C (120–130°F) then decreased to room temperature. Following the trough, the wire passes through other equipment, such as thickness control, check center location of the core, electrically check for pin holes, markings, etc.

Wire and cable covering are tested in-line generally more than any other extruded product because it is rather inaccessible for many tests when wound on a reel. Spark testing is very popular. The wire passes through a high-voltage field, and if there are any breaks, pin holes, or thin spots in the covering, a circuit is completed to the conductor and a signal of some type is produced.

Diameter is measured mechanically, electrically, or photometrically. Mechanical devices must be very precisely constructed to operate efficiently. The electrical devices are more accurate, especially for thin wires. They usually measure electrical capacitance of the covering at a given location on the wire. The resulting electrical impulse is fed back to a line speed controller. With the photo device, intersecting beams of light intersect on the wire. Their shadows are picked up by sensors so that when they are not in tolerance, they feed back to a line speed controller.

Measurement of eccentricity is as important as that of diameter. The conductor has to be centered in its insulation. The cathode ray source is a device that projects the images of the conductor and coating on a screen or set-up in a control system that either alerts the operator or is set-up to take corrective action.

Capacitance measurement is also used to determine degree of plastic foamed insulation. The capacitance reading is tied in with the process control to meet the requirements.

The windup at the end of the line, like the unwind at the start of the line, is usually designed with two reels (turret reels) following a puller (capstan). The capstan is a mechanical control device that is used to ensure an even rate of wire or cable movement.

To fill the reel evenly from side to side producing a level wind, the wire has a guide system that goes back and forth (like a fishing reel). Control of the wire all through the line is important, however when a problem develops at the end of the line, there could be a real entangled mess to put back in order. Different windup units are available to provide for different windup systems such as single shaftless reel, single shaft type, dual shaft type, coilers, collapsible coiling reel, continuous dual flyer, etc.

As explained in Chapter 3, TS plastics that include TS elastomers and natural rubber (which is a TS), are not processed as noted above for TP materials; see Figs. 3.1(a), (b) which compare the melting and curing cycles for TS and TP materials. While in the die, the TS temperature is held very carefully between temperature limits, such as 120–132°C (250–270°F); actual temperatures depend on the specific material being processed.

After the coated TS melt leaves the die rather than going through a cooling cycle, it is subjected to higher temperatures than those experienced in the die. When the temperature reaches the TS's curing temperature, which is above the maximum temperature in the die, the plastic solidifies. If the temperature should ever go higher in the die than its set limit, the TS material solidifies in the die causing the line to stop. Cleanup is a real problem, so controlling the die temperature is very critical. To speed up the line, working at the higher end of the temperature range can provide significant advantages but, if not careful, you will have a problem.

Recognize that the term curing, even though it is applied to TS and TP materials, it is a term that refers to a chemical reaction (crosslinking) or change that occurs during its processing cycle. This reaction only occurs with TS plastics or elastomers as well as crosslinked TPs that become TSs. The TP materials go basically through a melting process. However, since the more popular plastics at the beginning of the 1900s (with about 90 wt% of the market and principally the TS phenolic) were TSs and the term 'curing' was correctly used. When TPs came along and took at least 85% of a much larger production, being used in all types of processes (extrusion, injection molding, etc.), the term continued to be used indicating any plastic (TP and TS) that goes from a melt stage to a hardened stage is going through a curing stage.

Different curing system are used usually based on the TS material being processed. They include hot gas systems, vulcanization cures (TS crosslinking action for natural or synthetic elastomers), and others [2, 3, 43, 85, 283].

## Other coating lines

This section covers different methods used to process different plastics to meet different property requirements. Vulcanization is used to cure different type TS elastomers/rubbers, such as SBR, IR, and CR (Neoprene) [4]. Methods of vulcanization or curing include steam, liquid (eutectic mixtures [4]), microwave, and hot air [372]. There are systems used to cure peroxide-based crosslinkable polyethylene (XLPE) compounds. They include steam cure, nitrogen cure, and pressurized-liquid continuous vulcanization. All these methods involve higher heat than the melt heat to cause the peroxide in the plastic to decompose into a reactive radical and initiate the curing cycle (which is similar to the explanation for TS plastics).

Steam continuous vulcanization (CV) has been a common method of curing (Figs. 12.2 and 12.3) [372]. With this method, high-pressure steam contacts the wire and cable as it moves through the CV tube. Typical residence times are from 25 s for thin-walled wire, to several minutes for high-voltage power cables with steam at 190°C (400°F) and 1.7 MPa (250 psi) pressure. Coated products are then water cooled under pressure. Pressure cooling prevents void formation in the insulation due to vulcanization of the peroxide byproducts and/or steam.

There are various tube designs used in steam curing systems. The simplest is the horizontal system, which generally has only one extruder. The line layout is arranged so that the wire in the cooling trough is drawn back in the direction of the payoff reel; therefore, the payoff and take-up reels are located conveniently close together. It is used to fabricate thin wall construction for the automotive, appliance, etc. industries.

Catenary CV lines are used in the manufacturer of large cables, such as 600 V building wire or 15 kV power cable. The lines curve downward from a raised extruder horizontal platform and follow a bowed shape rather than a straight line. One or more extruders can be used to coat cables before curing. The line's curved shape prevents dragging of the cable along the bottom of the tube and aids in minimizing sagging of the insulation. The tension of the capstan is automatically controlled so

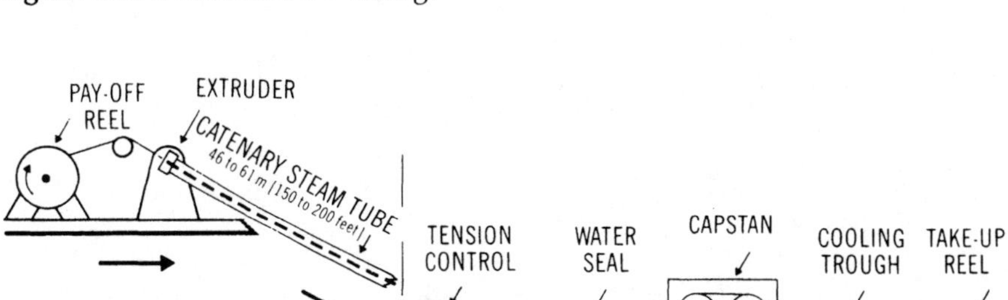


**Figure 12.2** Horizontal CV curing.

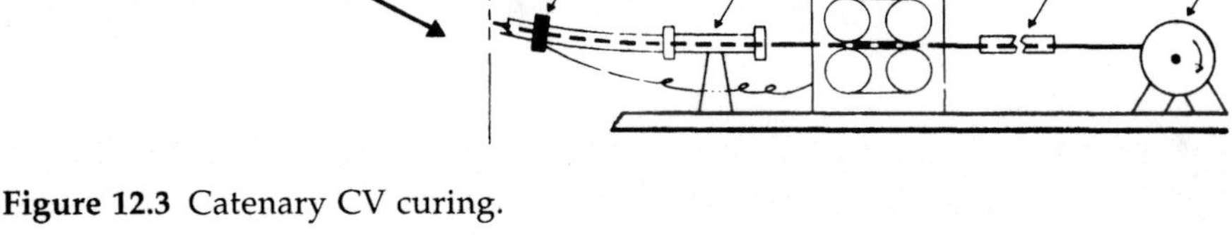


**Figure 12.3** Catenary CV curing.

that the cable is essentially hanging loose as it passes through the steam tube.

For the thickest insulated cables, a vertical CV system is used (Fig. 12.4) which will have a number of extruders supplying the appropriate melts for the overlapping layers. They are on a raised platform directly above the CV tube. The melt is extruded onto the conductor straight downwards to maintain concentricity. Because the cable is vertical, it never touches the sides of the tube. Only a few vertical CV systems have been used.

The first nitrogen lines became operational in the mid-1970s. Dry curing utilizes nitrogen as the heating medium instead of pressurized steam. By using nitrogen, the processor avoids saturating the cable insulation with water. It also reduces the size and number of voids in the cable insulation. Some lines are capable of using either steam or nitrogen as the curing medium.

An example of the nitrogen system is Davis-Standard's DSGC tubular trough line (Fig. 12.5) that is made up of three zones: a hot-gas section where the actual curing takes place; a cooling zone; and a water cooling system. It operates by blowing high-temperature nitrogen over the cable

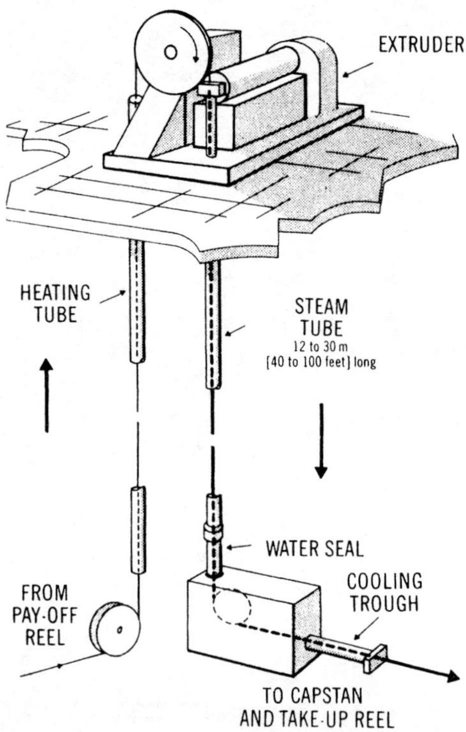


**Figure 12.4** Vertical CV curing.

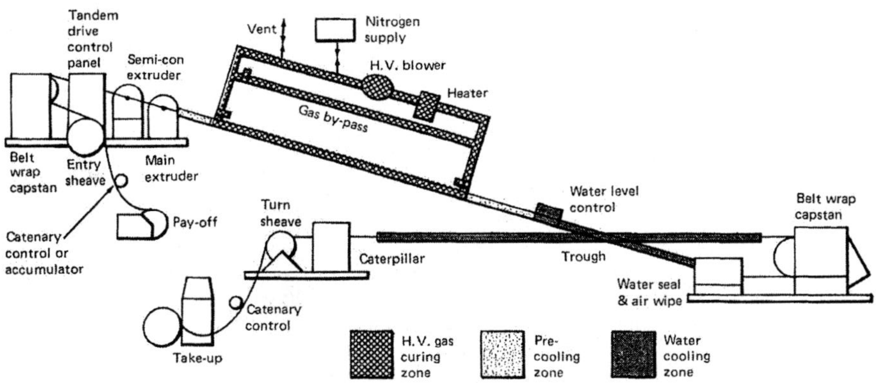


**Figure 12.5** Thermoset gas curing wire and cable line.

surface at high velocities. This action, in effect, breaks up the boundary layer and significantly improves heat transfer into the cable. The result is a fast and high-quality cure.

Other types of dry cure systems have been developed, such as the high-velocity gas cure and radiation cure. The high-velocity gas cure is similar to the steam-cure system. Radiation crosslinking is performed by passing the wire or cable through a beam of electron radiation. Usually, the full dosage is not applied in a single pass because of the rapid temperature increase that accompanies electron beam exposure since it would damage the product. To solve this problem, the wire or cable is festooned on a series of pulleys and passed through the beam several times until the desired dosage is reached to complete the cure.

Radiation curing is commonly used in the production of appliance wires and control cables. These type of applications, competing with other systems, are most suitable because of their relatively thin insulating layers. The materials used for this procedure are similar to those used in peroxide curing, except that the peroxide has been eliminated. In some cases, a crosslinked accelerator is added. The quality and physical properties of radiation crosslinked materials are virtually identical to those produced by peroxide curing.

Figure 12.6 shows a pressurized liquid salt continuous vulcanization line. It has similarities to other lines. The main difference is that a heated salt bath is used to cure the plastics. The salt bath is a molten mixture of sodium, potassium, barium, and calcium chloride or nitrate. Sometimes sodium carbonate and sodium cyanide are added. It provides a heat bath source where the heat is very uniform. In addition to providing heat to the wire or cable line, the heat bath is also used in other lines to anneal the plastic insulation.

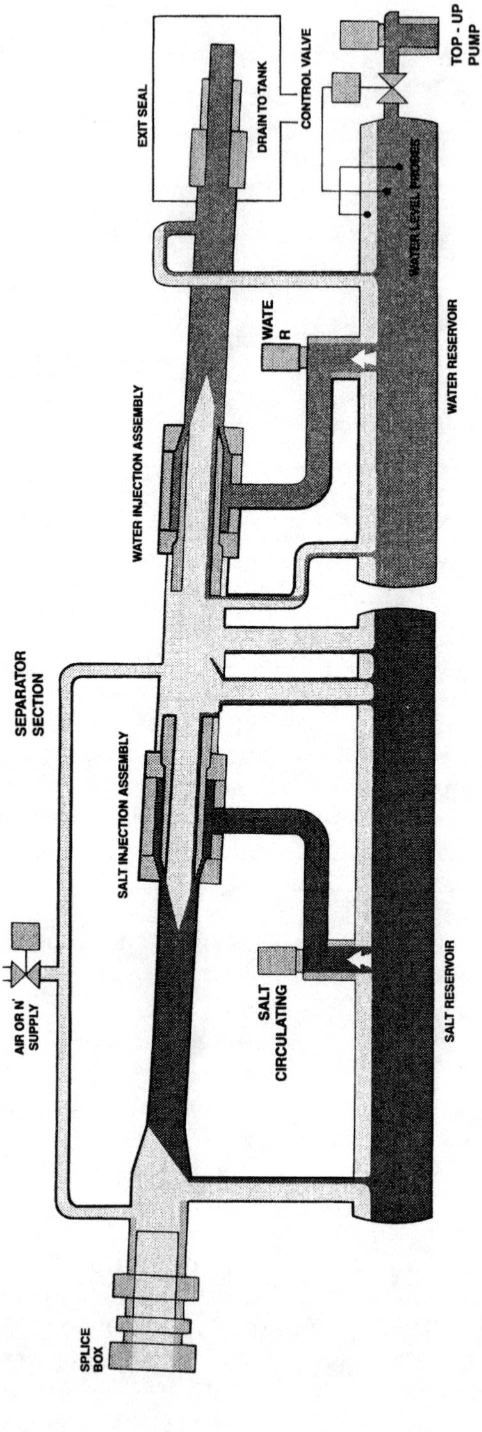


**Figure 12.6** Pressurized liquid salt continuous vulcanization wire and cable line.

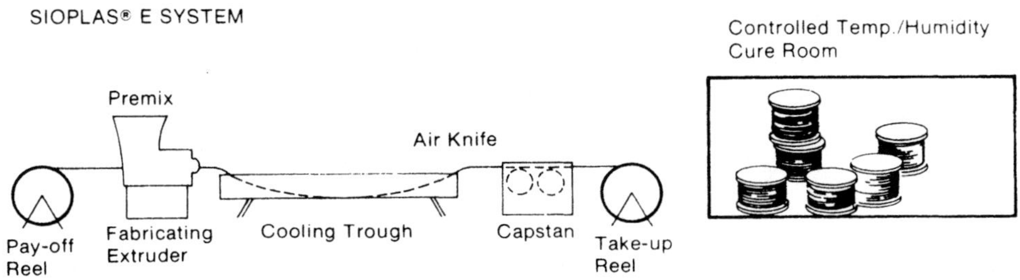


**Figure 12.7** Moisture initiated crosslinking wire and cable line.

As shown in Fig. 12.7, Dow Corning has a design for moisture initiated crosslinking of HDPE, LDPE, and ethylene copolymers (Chapter 3). With all the lines reviewed as well as others not reviewed, the extruder at the start of their lines is an important part of the line. The conventional screw extruder is used (Chapter 2).

In the past, an exception to the use of a screw operating extruder occurred when a very low melt plastic, such as PTFE (polytetrafluoroethylene), was to be used where its specialty properties were needed. The PTFE did not easily melt, particularly in the production of heavy wall PTFE wire coatings, so ram extrusion was used (Chapter 18). Ram extrusion is still being used for special jobs and also for producing rods, tubes, and large round solid 'cakes' that are later skived to make film.

With developments in equipment and the fluoropolymers that have high melt viscosities, the maximum extrusion rate using screws is usually limited by melt fracture (Chapter 3), particularly with the FEPs. However, these plastics have exceptional melt strength. This characteristic makes possible extruding using a die with a large opening and drawdown of the extrudate to the desired insulation thickness.

The coating is applied by using a ram extruder system as explained in Chapter 18 [2]. As an example, the system can coat stranded wire with a jacket 2.3 mm (0.09 in) thick. Also used for heavy coatings is the ram processing of PTFE paste. With the multiple strand ram extrusion die, output can be increased with cost reductions.

## Start-up

As in other extrusion processes, the wire or cable is first threaded through the line. Next step is to start-up the extruder (Chapter 2). Plastic melt will start covering the stationary wire. As the extruder is being properly set (temperatures, back pressure, output rate, etc.), the extruding melt will reach the point where it is producing the desired extrudate. The wire or cable line is started and brought up to speed.

If the melt from the die is not evenly distributed around the substrate, adjustments on the die are made. With selfcentering dies, there is none or relatively an insignificant problem in obtaining the correct melt distribution (Figs. 12.8 and 12.9).

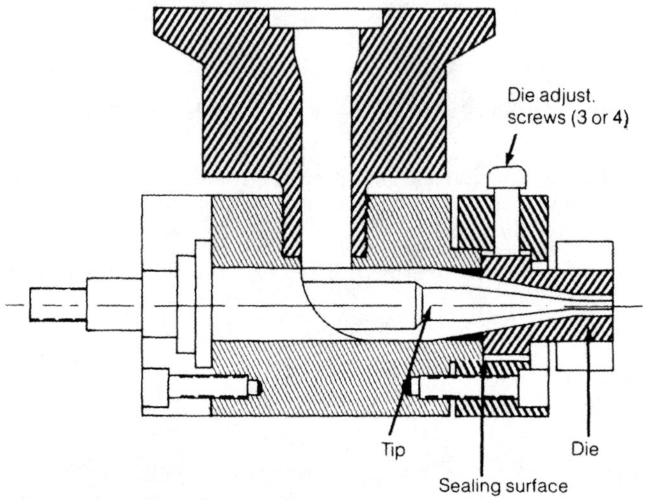


**Figure 12.8** Adjustable for concentricity high-pressure wire crosshead coating die.

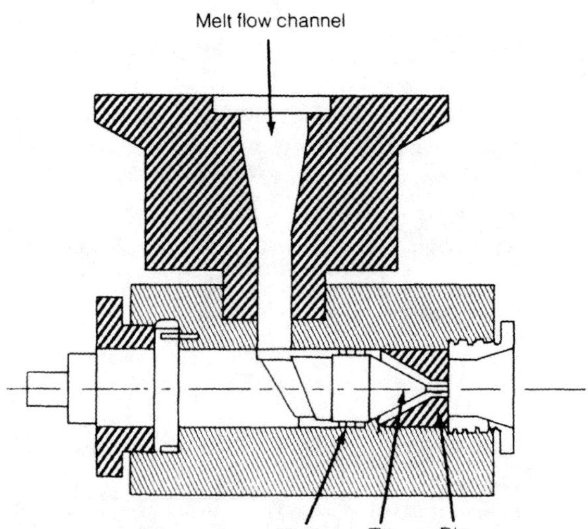


**Figure 12.9** Self-centered fixed in concentricity high-pressure wire crosshead coating die.

PROCESS OPTIMIZATION

Each of the different plastics used usually has different processing requirements to provide the best quality and rate of output as shown in Fig. 12.10 [206]. As an example, with nylon to obtain their best properties, the melt temperature should be 280–300°C (540–580°F). In the extruder, to minimize the total residence time, the rate of travel from the hopper to exiting the die should be 5 min or less.

Nylon 612 is frequently used, principally due to its good moisture resistance, electrical properties, and abrasion resistance. It melts ($T_m$) at 215°C (415°F). With this plastic, the greater the drawdown and the thinner the coating desired on the wire or cable, the higher should be the melt temperature, which ranges from 20–40°C (40–60°F) above the $T_m$ of the plastic.

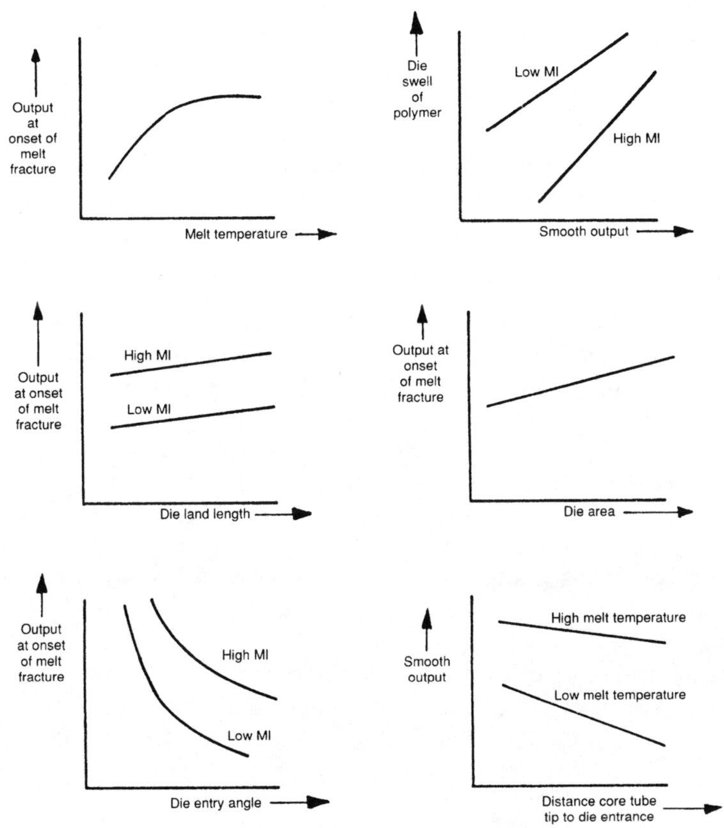


**Figure 12.10** Effect of extruder and material variables during wire and cable coating.

With certain nylons that are preferred for thin jackets, the die land length is equivalent to 2 to 5 times the dies' diameter but not less than 3.2 mm (0.125 in). The drawdown ratio (DDR, reviewed in the DIE section of this chapter) should be 9/1 to 15/1 for line speeds up to 305 m/min (1000 ft/min). With line speeds up to 915 m/min (3000 ft/min), the ratio should be 25/1 to 30/1. Both high- and low-pressure dies are used. To obtain the maximum properties such as adhesion of the nylon a partial vacuum of 130–750 mmHg should be pulled through the die.

Applying the coverings or coatings of any plastic has to be done with precision. Imperfections can occur on the surface and/or in the plastic. They are kept to a minimum by optimizing the quality of the material. The partial electric discharge that occurs at: (1) the high voltage in cavities and gaps in the insulating plastics; (2) the transitions from the conductor to the insulating plastic; and (3) from the insulating plastic to the metal screen, can eventually lead to electrical breakdown.

Water and moisture are prevented from penetrating into conductors and screen areas during the extrusion covering process as well as in storage and transportation. With proper processing conditions, no damage occurs to the outer sheaths. Sheaths, such as PEs, have to be strong. A longitudinal watertight screen area limits water penetration in the event the sheath is damaged. High-voltage cables of at least 36–60 kV include an aluminum composite layer sheath and a longitudinal watertight screen.

## DIE

The usual die used is a cross head with a 90° angle between the wire line and the extruder body axis. With this set-up, the entire length of the extruder projects sideways from the coating line. To help melt flow from developing dead spots in the melt channels, 30° or 45° crossheads can be used. The crossheads provide a more streamlined interior and the extruder location is better adapted to some plant layouts. They are sometimes preferred when processing PVC because of the streamlining and better control of the melts heat profile. Most dies are subjected to very high internal pressures since the uncommon pressure of over 35 MPa (5000 psi) is required.

Precision dies are required. Thus, many dies are required to meet different wire dimensions. Within size limits, there are dies with internal adjustable wire guider lips. And parts of dies may be interchangeable. However, the best performing dies are designed to process a specific wire to plastic configuration.

Unfortunately, a 90° bend of the melt from the extruder to the product covering is not the most desirable situation where melt should flow in a straight, streamline line. This bending of the melt is actually an ideal situation considering the consequences of feeding the wire through an extruder. The basic problem is that when the melt is side-fed with a 90° bend, it can cause different rates of flow around the wire, which in turn could effect the thickness and density uniformities. However, when the die is properly designed, the melt flow can be even all around the die upon making contact with the substrates.

There are two basic types of dies as shown in Figs. 12.11 and 12.12. They are called high- and low-pressure coater dies [179, 236, 372, 348]. With high-pressure dies (also called pressure packing), the melt meets and coats the wire between the die lips prior to exiting the dies. The result is good contact of plastic over the wire, very tight control of the plastic's OD (outside diameter), and the ability to handle plastics that require tight melt control. This latter attribute is particularly important when operating at peak heats and pressures. By pushing plastic against the wire, the potential problem of encapsulating air bubbles is eliminated. Air bubbles would effect the cable's performance.

Pressure dies are usually constructed with the opening equal to the OD of the final product. For certain plastics, such as polyester elastomers, the die should be about 5% larger than the required cover diameter. The land length is usually equal to the final cover diameter; with very thin coatings below 0.5 mm (0.02 in) land length is considerably less. Clearance between the substrate to be covered and the bore of the mandrel, is usually 1–5% of the core diameter depending on the type of plastic used [236].

These pressure dies have more resistance to flow. Tapered flow

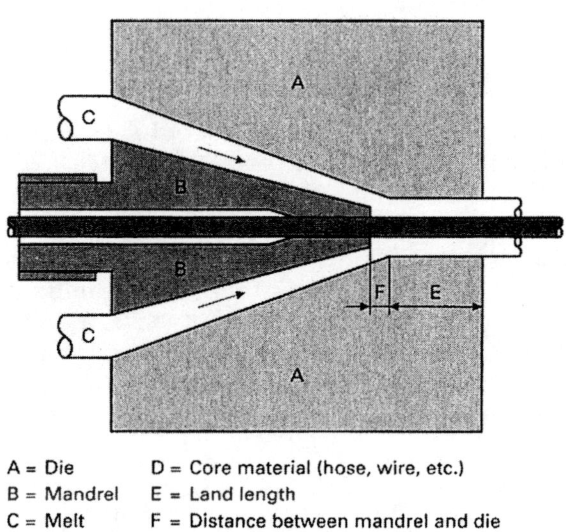


**Figure 12.11** High-pressure coater die.

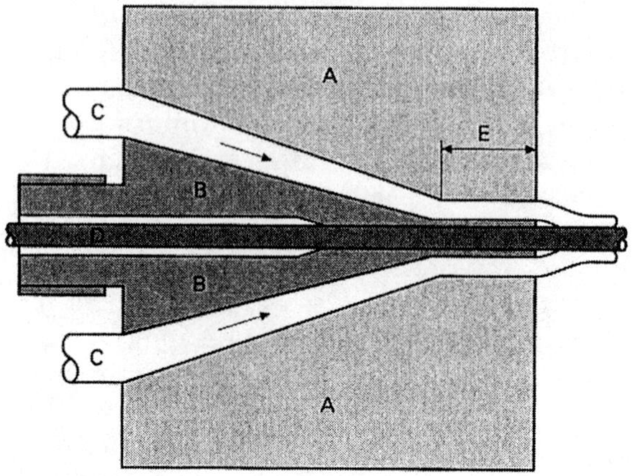


**Figure 12.12** Low-pressure coater die.

channels in the die may be divided into two or more sections so that a decreasing angle of the taper occurs (Fig. 12.13). This action results in reduced resistance to flow while still reaching the required final size.

The plastic swelling as it exits the die will be compensated by draw down and a coating with minimum strains will be processed. The clearance between the wire and its travel through the die's guider tip must be very small such as 0.05 mm (0.002 in). With a rather snug fit, the plastic is prevented from entering and impairing smooth and easy passage of the wire.

In low-pressure dies (also called sleeving or tube dies), the melt makes contact with the wire after they both exit the dies (Fig. 12.11). Literally, a plastic melt tube is pushed and drawn-down usually with a vacuum so that it hugs the wire, with formation of a loose jacket that facilitates removal of the plastic insulation. If spiders are used to support the central mandrel in the die, they are usually thin and streamlined to minimize disruption of the melt's flow and the melt's velocity (Chapter 5). This technique is used in secondary coatings especially where an irregular cross section, such as group wires, is to be covered; also to obtain maximum output with small wires and thin coverings especially with high viscosity plastics.

The optimum drawdown ratio for low-pressure dies is in the range of 5 to 20/1. The advantages of this type die include better control of the cover wall thickness and easier to strip the cover from the core.

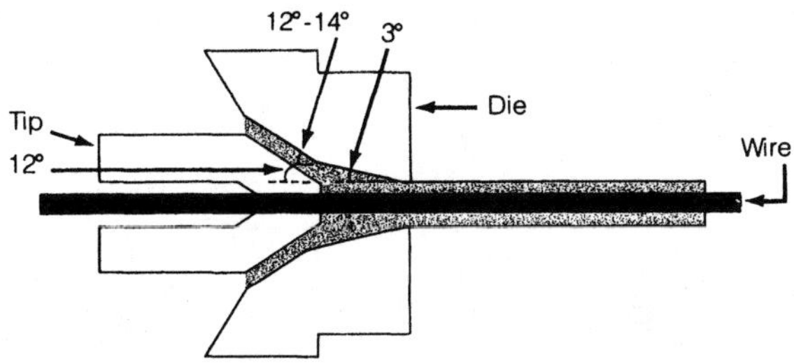


**Figure 12.13** Tapering melt channel in die.

Certain plastics perform better in one type of die although either can be used. Nylon, which flows very freely at extrusion temperatures, is best extruded through low-pressure dies. PVC, used for both primary insulation and jacketing, is run through both types of dies. For very thin wires, regardless of the plastic, low-pressure dies may be used to avoid the increased tension required when traveling through the high-pressure type.

Low-pressure dies compared to high-pressure dies require a larger clearance in the die between the conductor and its guider tip because a vacuum must be drawn through that clearance. An example of a clearance is 0.45 mm (0.010 in) all around. There is no danger of plastic backing up because the plastic is extruded through a separate annulus. The annulus should be as small in inside diameter as possible, although this size is limited by the mechanical strength of the guider tip. The outside diameter is set by the appropriate drawdown ratio.

The drawdown ratio in a circular die, such as a wire die, is the ratio of the cross sectional area of the die orifice/opening to the final extruded shape as shown in Fig. 12.14. To determine the DDR use the following equation.

$$\mathrm{DDR} = \left(D_D^{\,2} - D_T^{\,2}\right) / \left(d_{cw}^{\,2} - d_{bw}^{\,2}\right)$$

where $D_T$ = diameter of guider tip, $D_D$ = diameter of die opening, $d_{bw}$ = diameter of bare wire, and $d_{cw}$ = diameter of coated wire.

If the DDR is too high, a rough surface and/or internal stresses in the coating will exist. As an example, typical DDR values for LDPE is 1.5, HDPE is 1.2, PVC is 1.5, and nylon is 4.0.

Another guide for setting uniformity and best repeatable references is the draw ratio balance (DRB) that aids in determining the minimum and maximum values that can be used for different plastics. To determine the DRB use the following equation.

$$\mathrm{DRB} = \left(D_D / d_{cw}\right) / \left(D_T / d_{bw}\right)$$

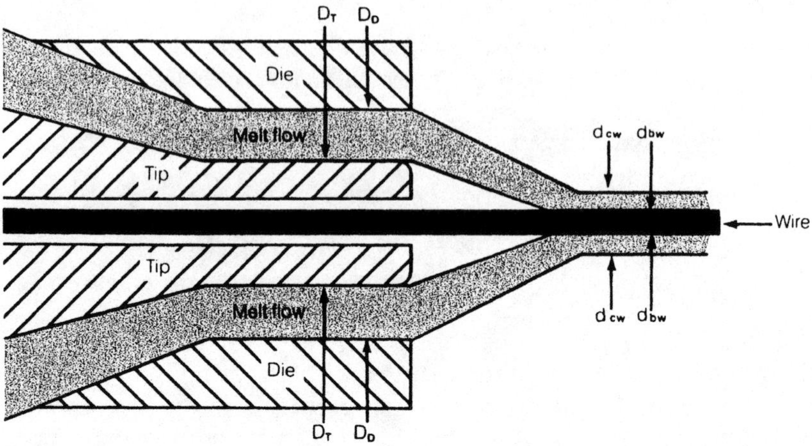


**Figure 12.14** Schematic for determining drawdown ratio (DDR) and draw ratio balance (DRB).

The value of the DRB ranges around 1, and is close to 1. Being outside the set limits can cause at least out of round and melt degradation. Plastics have different DRBs and DDRs which can be used as guides to processability and to help establish their various melt characteristics.

Adjustments of the plastic's wall thickness distribution and concentricity are made with die-centering bolts that are manually and/or automatically controlled (Fig. 12.8). On start-up a preliminary manual adjustment may be used to speed up the adjustment process. During processing, an automatic control can be used that responds to an in-line wall thickness control measurement sensor probe. Figure 12.9 shows a self-centered high-pressure crosshead die. This is a fixed position die that does not require any adjustments.

With high-pressure dies, a vacuum is usually used in the die just prior to the wire going through its snug central support. The result is in obtaining an air free and better bond between the plastic and substrate. This vacuum is adjustable, usually operating at 25 mmHg (1 inHg) or less. A high-capacity pump is required to overcome the leakage around the wire entrance into the die.

A different approach is used with the low-pressure dies to take care of a different situation. A low-air pressure is applied through the center of the die's mandrel tube to prevent collapse of the melt tubing exiting; low pressure also aids in maintaining the plastic OD. This action eliminates melt adherence on the front of the dies, which contamination would require downtime for cleanup.

As reviewed earlier, preheating the wire is very beneficial to melt flow. To help low-pressure designed dies even with preheating, dies can incorporate insulators in the inside the tip to prevent cooling of the tip and melt, large or long tips can be heated to keep melt hot, and using heaters in the core tube to keep the melt at a controlled temperature.

A problem related to the insulation is void formation. As with other extrusion processes as well as other processes such as injection molding (Chapter 18), this action may occur since the outer surface of the plastic solidifies first (very quickly) causing the inner region to shrink freely upon solidification, possibly causing vacuum formed voids. This action causes decreases in adhesion of the plastic to the metal. Also, macroscopic voids can form within the plastic that reduces the dielectric strength of the cable. If this problem develops, the usual solution is to slow up the rate of cooling.

Different dies are used to meet different coating processes. An example is coating multi-wires where two or more wires are used in a single operation. The die is built to take and properly locate the wires. High-pressure dies are preferred, but low-pressure dies are also used. There are also cable-in-duct designs that are often used for burial in the ground. They have a cable inside a continuous extruded duct/tube. This conduit is usually made of PE. Simultaneously, the cable and duct are extruded from one die.

Coextrusion of wire is done to apply two or more coatings simultaneously (Chapter 2). They are used to provide special insulation. An example of a triple head coextruded Davis-Standard die is shown in Fig. 12.15. It can be used for a cable's insulation and jacket using plastics such as PE and XLPE. Different plastics used can all be solid or a lay-up can be used to have one solid and the other a foam structure. Wires may require identifying striping or bands of color. They can be printed or can also be coextruded.

**Figure 12.15** Davis-Standard wire and cable triple crosshead die.

**Table 12.3** Guide to troubleshooting PVC cable compounds

| *Problem* | *Solution* |
|---|---|
| Low output | Repair/replace worn screw and/or barrel<br>Inspect for proper screw design<br>Reduce back pressure<br>Clear clogged screen pack |
| Sparkouts, intermittent lumps | Inspect compound for contamination<br>Inspect for blown screen pack or breaker plate |
| Cold bend, tensile strength, elongation failure | Increase distance between cooling bath and die tip<br>Increase temperature of cooling water<br>Check for blockage causing uneven flow through die<br>Reduce oversized die opening to maintain size without high draw-down |
| Burned material | Clean degraded compound off screw or barrel<br>Reduce zone temperature or shear rates<br>Reduce screw speed<br>Repair or replace worn screw and/or barrel<br>Test thermocouple for accuracy<br>Reduce die temperature |
| Burning on screw tip | Examine screw tip for proper coring<br>Reduce temperature of extrudate<br>Reduce angle of screw tip entering adapter |
| Surging | Check hopper for bridging<br>Check screw feed section for bridging; ensure cool screw surface in feed section |
| Orange peel | Increase screen pack – more or finer screens<br>Decrease line speeds to increase shearing |
| Raised lines on surface | Check die for damage; replace if necessary<br>Thoroughly clean die<br>Polish until all depressions are removed |
| Voids, blisters or bubbles on surface | Inspect substrate for moisture and/or oil contamination<br>Inspect compound for moisture |
| Bumps on extrudate inside diameter | Increase temperature of screw tip<br>Increase rear zone temperatures<br>Increase temperature of screw-cooling media<br>Clean screw tip<br>Increase screw speed |
| Out-of-round extrudate | Increase land length of die<br>Decrease backpressure, temperature<br>Decrease internal die<br>Adjust die system compression ratio<br>Increase air volume for cooling internal die |
| Immediate burn during start-up | Reduce die and/or zone temperature<br>Clean equipment<br>Reduce screw speed and/or back pressure |

## TROUBLESHOOTING

Different problems can develop, such as the separation of the jacket or insulation to the wire. To eliminate this problem, the two basic approaches are to raise the wire-preheat temperature and/or increase melt temperature. With surface roughness or dullness, various solutions can be considered. They include: (1) raise melt temperature and/or die temperature; (2) with plastics containing moisture, such as nylon or hygroscopic plastics, reduce moisture level by further predrying; (3) increase plastic blending/mixing by raising extruder rpm and/or pressure; and (4) melt fracture may be occurring so increase melt temperature or reduce the extruder's rpm and also consider, if practical, reducing the die approach angle.

There can be a variation of the plastic's thickness covering. Solutions include: (1) check die alignment for variation in the die opening; (2) check take-off for consistency of the extruder rpm; and (3) if the extruder is surging increase its pressure, lower melt temperature and screw rpm, use a longer land die, substitute a screw with longer or shallower metering zone for consistent melt pumping (Chapter 4).

A guide to trouble shooting is in Table 12.3. Problems in wire and cable with solid or foam insulation tend not to be resolved by previous action taken; solutions at one time may not resolve the same problem the next time it occurs. This inconsistency results from the many variables in the production line. Experience and keeping good records (trials, errors, and successes) provide the means to simplify troubleshooting [2, 363].

# 13

# Pipe and tube

## OVERVIEW

Plastic pipes (Figs. 13.1, 13.2, and 1.3) and tubings (Fig. 13.3) started commercial production during the late 1930s. There are different definitions that are usually associated with the different industries (plumbing, gas transmission lines, beverages, medical, mining, and so on). A popular definition of a pipe is that it is rigid, hollow, long and larger in diameter than tubes. Tubings are basically the same except they are flexible and smaller in diameter, up to 0.13 mm (0.5 in) or with thin walls up to 15 cm (6 in). In this review, it really does not matter which definition is used because the extrusion line does not make a distinction. In this chapter, 'pipe' will be used to identify pipe or tube unless a specific situation exists for one or the other, such as collapsible tubes (usually not called pipe) [1, 88, 218, 327, 341, 358, 415].

Practically all pipe is extruded using thermoplastics. Single screw extruders (Fig. 13.4) are usually used but with PVC twin screw extruders are also used. Dies in some of the lines use the same basic type dies and plastic melt temperature ranges used in wire coating (Chapter 12).

The basic in-line process (or train) for the production of pipe starts with material handling up-stream from the extruder, the extruder and die, devices for the outside and inside calibration of the pipes cross-sectional area, if required use of air or nitrogen pressure and/or vacuum, wall thickness measuring device, cooling tank, and automatic cutting with pallet equipment for rigid pipe or windup unit for flexible pipe. The line could include a marking device, testing device, etc. An important requirement is to cool the extrudate rather fast near the die while keeping control of dimensions and properties.

Included in the processes are various techniques to control the dimensions/sizes that are either free drawn melts (usually for the small diameter tubes) or sizing fixtures. On line operation, the consumption of plastic

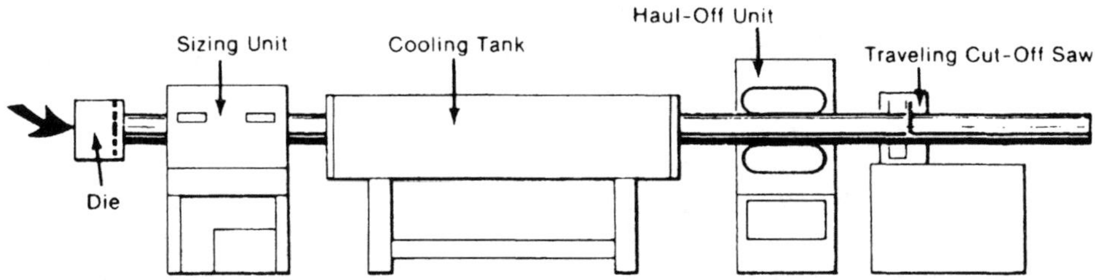


**Figure 13.1** Important down-stream equipment used in pipe and profile extrusion.

**Figure 13.2** Example of a complete extrusion pipe line.

is targeted to be minimized. The cost of producing the pipe could include material costs of up to 80%. The goal is always tighter tolerance control to reduce material consumption. Dimensional and/or thickness calibrating disks of different designs are used.

The operator's expertise in using these calibration/sizing systems is as important as controlling the whole line. A system of feedback and control to the extruder by a microprocessor is used to control wall thickness with gravimetric proportioning of materials at the start of the line. Perhaps it can be said that any equipment from the past is definitely noncompetitive, considering all the new equipment that has been made available from upstream to the end of the line. In the past years, all the equipment has been significantly altered to increase profits, such as using a puller (Figs. 13.5 and 13.6).

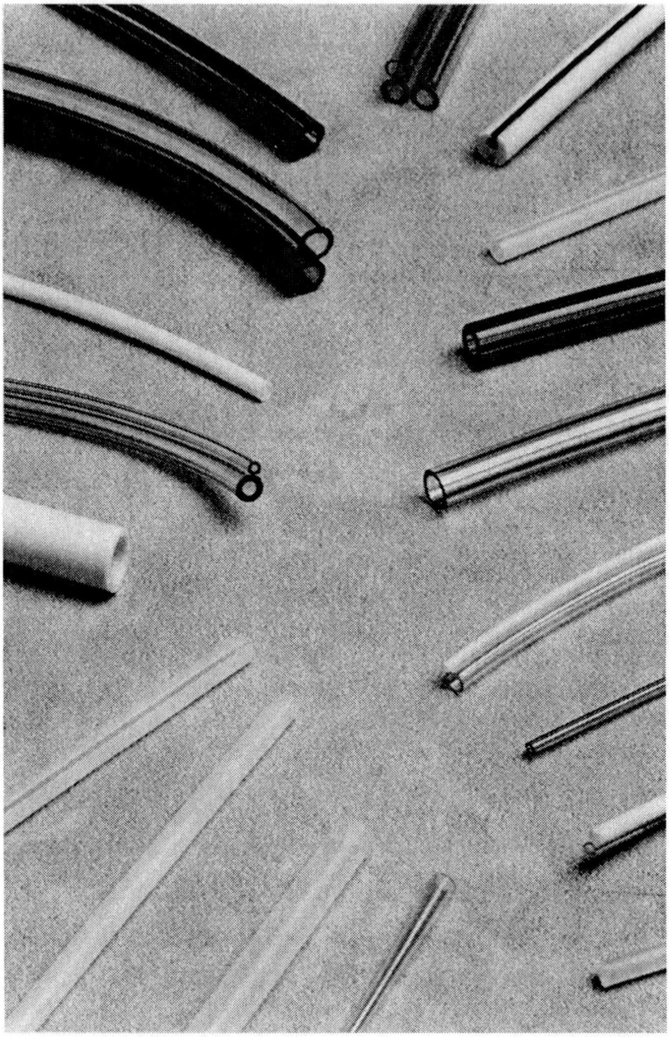

**Figure 13.3** Samples of extruded tubing.

There are small diameter tube lines using draw down control (free extrusion) sizing technology where the extruded tubular melt has no calibrating device after leaving the die. It could have internal air pressure so that the tube does not collapse upon leaving the die. Devices are also used with different designed calibrating/sizing plates or tubes with or without pressure or vacuum assist in and/or outside the pipe (Figs. 13.7 and 13.8).

**Figure 13.4** Welex's tube and profile with 60 mm (2.5 in) diameter screw and 24/1 L/D, with and without the cylinder safety guard removed.

Sizing during cooling is almost universally controlled either on the outside and/or inside diameter (OD and/or ID). This approach provides a means to use the minimum amount of plastic. The dimensions of the die and the relation between its output and haul-off rate determine the final wall thickness and consequently the OD and ID. There are applications where the ID has to be precise, such as disposable hypodermics and beer pumps. Precision OD is required for applications such as the pipe fit into a pressure tank.

Figure 13.9 shows a calibration system for a pipe or tube extrusion line. The top left schematic shows a vacuum tank calibration with cascading, temperature controlled water baths where (a) is the pipe die, (b) vacuum with disks, (c) cascaded heated zone water baths, and (d) caterpillar take-off puller. The top right schematic shows pressure calibration using a plug insert with water spray cooling where (a) is the pipe die, (b) water cooled calibrator (Fig. 13.10), (c) water spray cooling, (d) caterpillar take-off puller, and (e) plug insert to retain internal pressure that helps to control

**Figure 13.4** *Continued*

the pipes OD and ID. The lower schematic shows an example of differential pressure sizing for flexible tubing.

Sizes range from less than 1 mm (0.004 in) in OD for applications such as medical tubing to at least 600 mm (24 in) in OD for applications such as water pipe. When this process involves producing thick walls, it is usually advantageous to cool the extrudate by circulating water through the mandrel as well as running the extrudate through a water bath. With large pipes used for applications such as water, gas, and chemical distribution, smooth internal surface finish with minimum residual stress is required to minimize crack initiation, eliminate or reduce stress cracking, etc.

If there is enough production demand and the extruder has the output capacity, it is common to use multiple dies, each die produces a pipe that is usually of different sizes. There are literally now in commercial production worldwide many hundreds of different series of pipe. Usually each individual size or series has up to 20 sizes and are made of different plastics. Selection of the best product for an application can be difficult. To

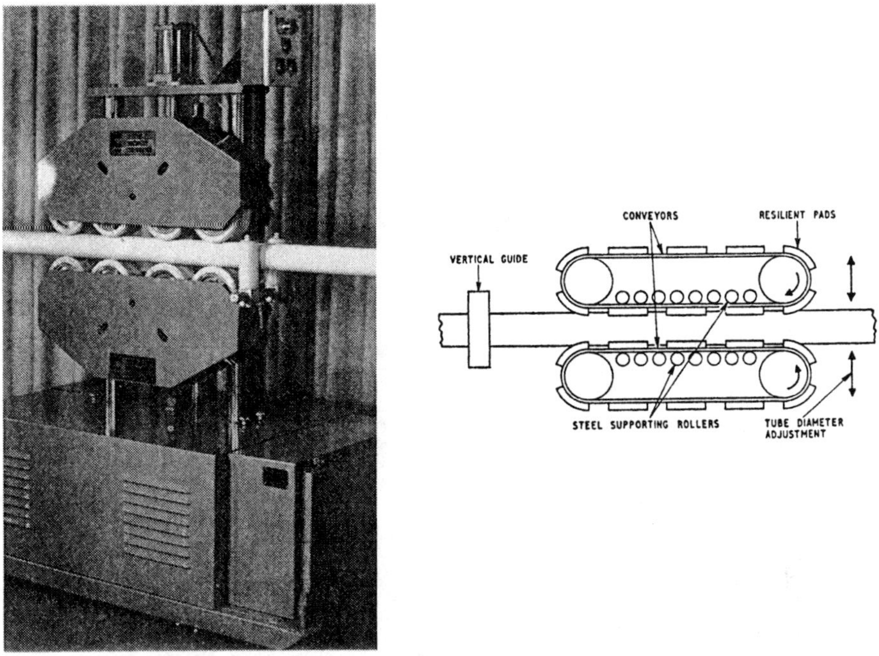


**Figure 13.5** Simple designed caterpillar puller; actual photo and schematic.

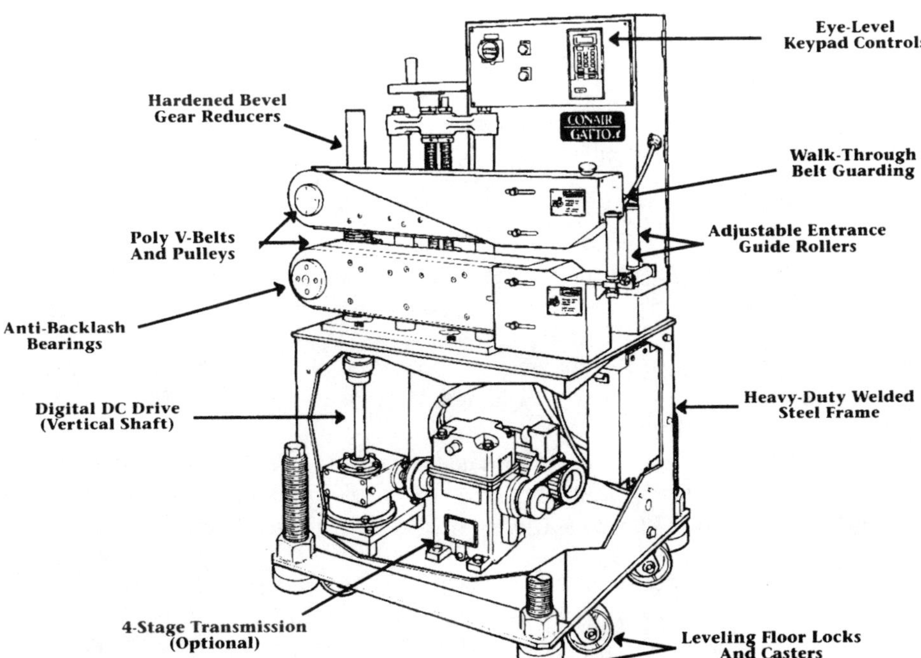


**Figure 13.6** Key features of a Gatto/Conair belt puller.

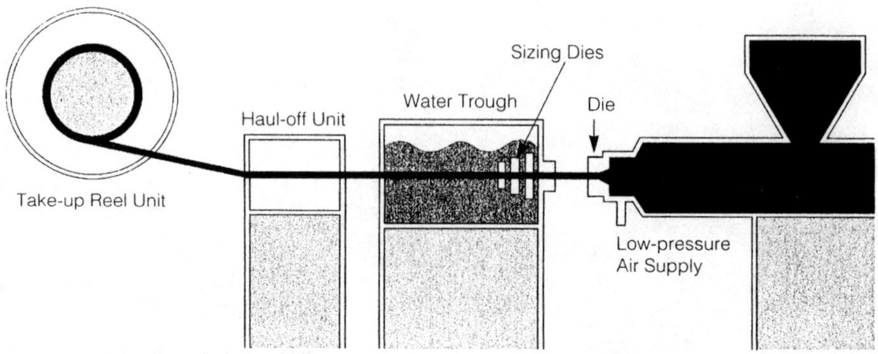


**Figure 13.7** Simplified schematic of tube extrusion.

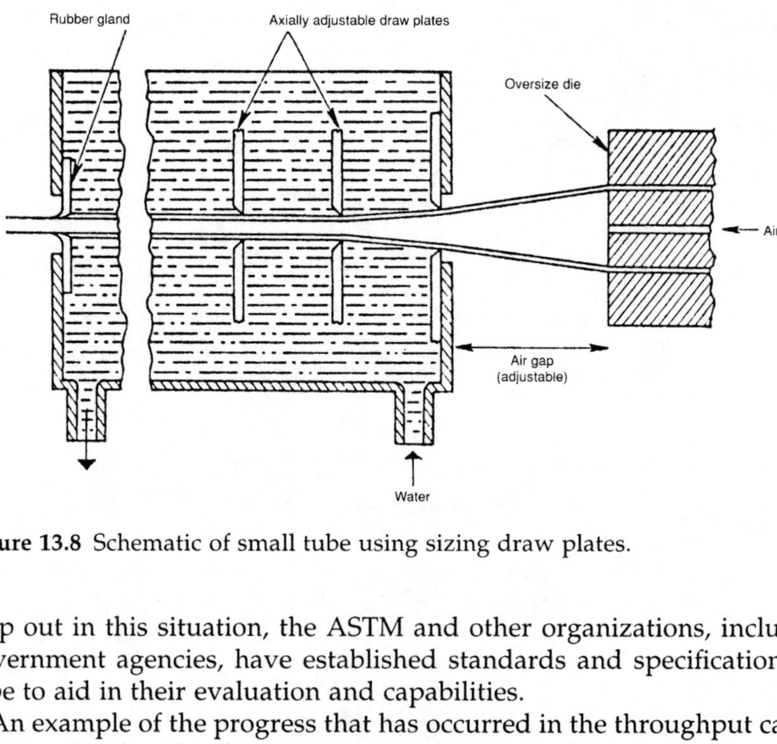


**Figure 13.8** Schematic of small tube using sizing draw plates.

help out in this situation, the ASTM and other organizations, including government agencies, have established standards and specifications on pipe to aid in their evaluation and capabilities.

An example of the progress that has occurred in the throughput capacity in a pipe line has been summarized by Reifenhauser. Taking a pipe extruder of a 90 mm (3.5 in) screw diameter as a typical example, with a corresponding increase in drive power and processing HDPE, the improvements are as follows: 1961 at 90 kg/h (200 lb/h); 1971 at 180 kg/h (400 lb/h); 1981 at 280 kg/h (615 lb/h); 1991 at 450 kg/h (1000 lb/h); 1995 at 600 kg/h (1320 lb/h); and now at least 1000 kg/h (2200 lb/h). During this period, other changes have improved the operation of the extruder,

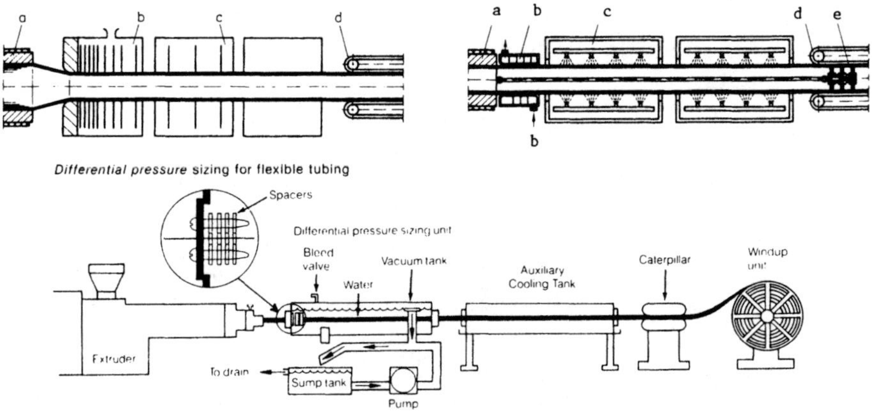


**Figure 13.9** Example of a calibration system for a tube or pipe line.

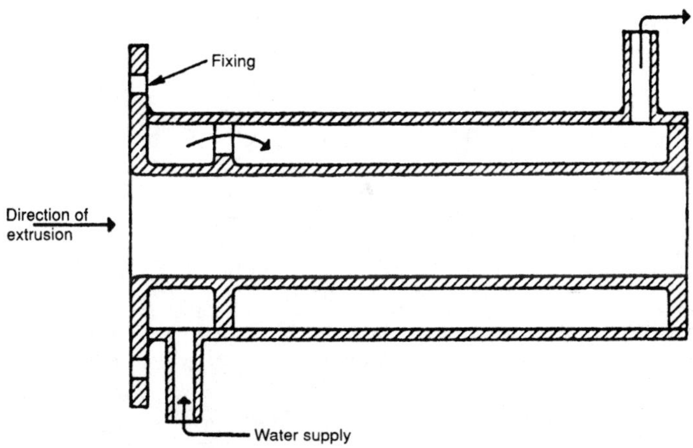


**Figure 13.10** Simplified external water cooled calibrator.

including using grooved barrel feeders, cooled feed sections, etc. (Chapter 2).

## PLASTIC MATERIALS

Virtually all TPs can be made into pipes with some reinforced thermoset plastics. The TPs are extruded but only a few TSs are extruded [3, 7, 12]. This review only concerns TPs other than those TPs crosslinked that become TSs. The largest and major consumption is of PVCs that is almost double to the next which is HDPEs. Others, going from the highest to the lowest consumption, are ABSs, PSs, TPEs, PCs, PPO/PPEs, PAs, ABS alloys, POMs, PBs, and PVDFs. CPVCs are included in the PVC group and XLPE in HDPEs.

Microminiature tubing is extruded using high-performance plastics (PTFE, FEP, ETFE, PFA, PVDF, PEEK, etc.) with IDs as small as 0.025 mm (0.001 in) and walls as thin as 0.038 mm (0.0015 in). There are also heat shrink versions with standard shrink ratios of up to 4/1.

Features of the plastics used include toughness, flexibility, heat fusibility, chemical resistance, etc. The characteristics of the material determines the choice. For example, the solvent resistant polyolefins (PE, PB, and PP) cannot be solvent-cemented. The vinyls (PVC and CPVC) and ABSs tend to be difficult to heat-fuse, however they can be bonded, having properties superior to the pipe material when proper bonding procedure are used.

The choice for a given material is further limited by performance, installation, and maintenance requirements, as well as by the availability of fittings. For applications requiring maximum joint strength and chemical resistance, the following bond methods are used: (1) heat-fusion for PE, PP, PB, and PVDF; (2) solvent cementing for ABS, PVC, and CPVC; and (3) adhesive bonding for reinforced TSs.

In regard to PVC compounds, the single screw extruder predominantly uses granular form but can also process powders and flakes. The twin screw extruders principally handle powders. PVC dry blend mixes allow the processor to take advantage of using their own compounding formulations which should provide cost advantages.

HDPE is used in various applications, such as pressurized gas and water pipe lines that require high mechanical strengths. When processing PE, it requires operating at the lowest possible melt temperatures while maintaining a high level of thermal–mechanical homogeneity in the melt.

General nylon is extruded at as low a melt temperature as possible to provide sufficient melt viscosity and to minimize thermal degradation. Care should be taken to start-up the extruder with its output end at temperatures at −4–10°C (25–50°F) above its melting point.

The toughness, flexibility, and low-temperature properties of TP poyurethanes (TPUR) elastomers make them suitable for reinforced and unreinforced tubing and hose. Uses include lightweight fire hose, unreinforced tubing for low- pressure pneumatic system, etc.

When processing TP polyester plastic, control of the melt temperature is important. If too high, it can it can reduce the melt strength significantly. The result would be irregular wall thickness. Too low a melt temperature can result in a poor tube finish, uneven dimensions, and weld lines. Melt and water bath temperatures are most important because polyester plastic is a crystalline material with a relatively narrow melt–freeze range.

With acetal (POM), the conventional free extrusion process is used going into a water bath with a 25–76 mm (1–3 in) air gap. Draw down ratios from the die has a range of 4–10 providing different wall thicknesses. For maximum thickness tolerance control, quenching in a reduced pressure bath is preferred because of its melt characteristics. Even though the extrusion grades of acetal have a melt viscosity similar to other extrusion grade plastics, they have a low melt elasticity. This means the melt will quickly lose its shape unless it is quickly quenched in water. Viscosity changes very little with changes in temperature. Thus, dies with smaller openings are used.

When traditional plastics are unable to meet performance requirements, candidate plastics are the high molecular weight HDPEs (HMWHDPE). Their significant properties include exceptional high chemical resistance, corrosion resistance, environmental stress crack resistance, and abrasion resistance. Excellent impact resistance goes down in temperature of −50°C (−60°F). Pipe is extruded at least up to 6.3 m (6 ft) in diameter. Lengths range from 6 m (20 ft) to 1520 m (5000 ft) coils. As a general rule, the diameter sizes above 100 (4 in) are not coiled. Sizes below this are coiled on a reel that usually has a diameter of 20 times the outside diameter of the pipe.

The TPs usually include small amounts of additives, such as stabilizers, antioxidants, lubricants, colorants, modifiers, fillers, and extrusion aids (Chapters 3, 17). The material used could be precompounded pellets or a dry blend of the mix. Dry blending inhouse is more economical. Most PVC pipe is made from dry blends.

Fillers and reinforcements provide a means to change and improve the performance of plastics during processing and also in service. Perhaps the most important reason for their use concerns savings in the cost of plastic compound. The most popular fillers used are calcium carbonate, talcum, barium sulfate, and wood dust. The number one in reinforcing fillers is short glass fibers [3, 43]. An example of how polyolefins are modified with fillers and reinforcements is in Chapter 17. Also, adding fillers and reinforcements to the different TP materials provides similar gains [3, 113].

## PROCESSING CHARACTERISTICS

Most pipe is made by straight through extrusion, in-line with the direction of extrusion, die, sizing or calibrating device or tank (Fig. 13.11), water cooling tank (Fig. 13.12), conveyor (Fig. 13.13), cutter, if needed, and take off equipment at the end of the line.

There are different techniques used to control and calibrate the size of the pipe. Basically, there are the free flow usually used for small pipe/tube and the others use sizing devices/dies that are basically calibrating devices as summarized in Figs. 13.14–13.15.

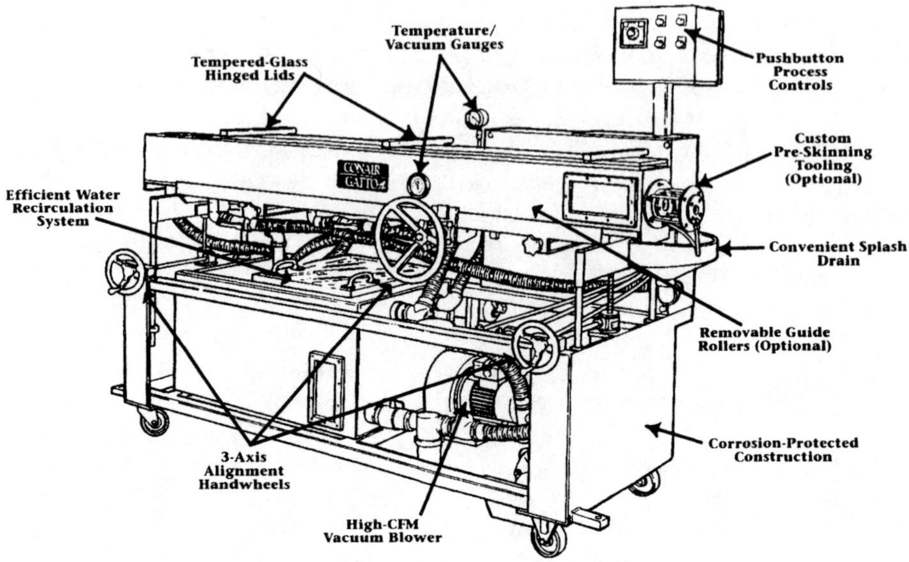


**Figure 13.11** Example of Gatto/Conair's precision vacuum sizing tank.

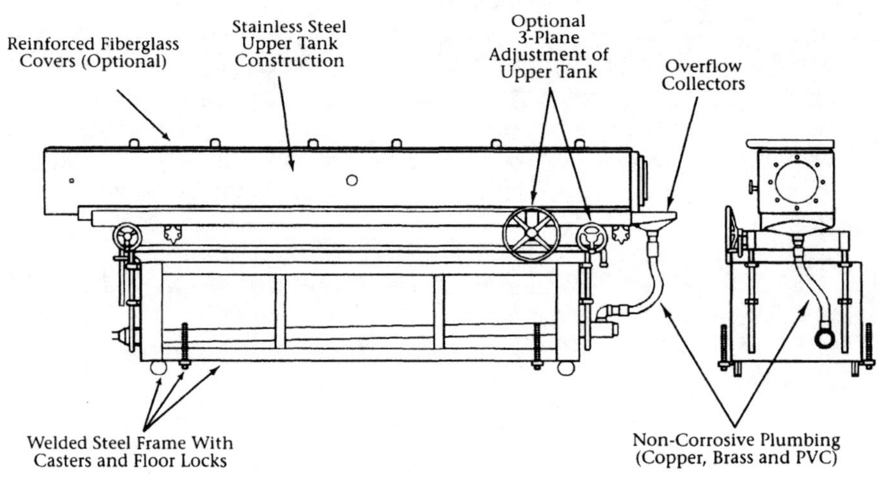


**Figure 13.12** Example of Gatto/Conair's water and spray tanks.

The calibration methods are in most cases used in combination with a circulating water trough/tank cooling system. The main advantage is that the still soft, moldable extruded profile is drawn down or pressed by vacuum and/or air pressure to the wall of the sizing/calibrating device.

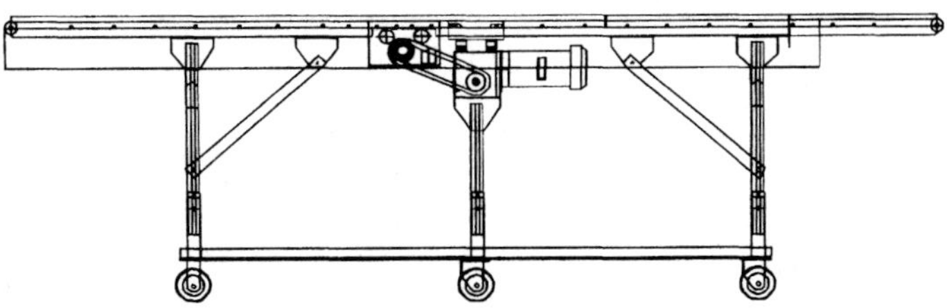


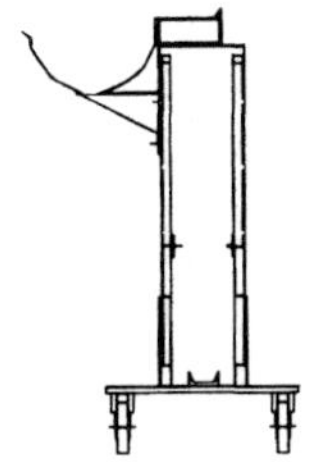


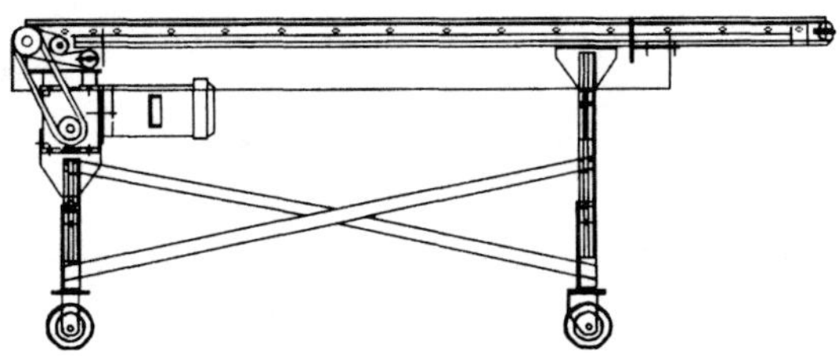


**Figure 13.13** Example of Gatto/Conair's take-away conveyor.

Nitrogen can be used instead of air. Simultaneously, the hot melt is being cooled. Regardless of the approach used, the target is to use as little plastics as possible, only use what is needed.

A typical set-up for free extrusion line is shown on Fig. 13.16 [236]. The first sizing plate is generally 10% larger than the diameter of the finished tube. A succession of decreasing inside diameter plates, usually 4–15 depending on the tubing size, are gradually spaced in a vacuum chamber to lend support and gradually draw the tube down to size.

The front sizing plate, which is mounted on the face of the tank, requires lubrication by the water in the tank or may require spraying or dripping a water stream on the entering tube to minimize possible sticking (Fig. 13.17). This procedure is used with all size pipes where, instead of a sizing plate, a rubber or elastomer ring separates the hole in the metal entrance of the pipe and is lubricated with water.

If required, the ring can include sizing rings to assist in maintaining its shape when dimensions are critical. Numerous tubing sizes can be made from one set of plates and mandrel setup. Usually one simply varies the screw speed, takeoff rate, or internal pressure to change the tubing dimensions. Tubing with dimensions of 0.05 mm (0.002 in) can usually be made up to diameters of 6.4 mm (0.25 in). Larger diameters are produced using sizing devices since the roundness tolerance is difficult to maintain by free extrusion.

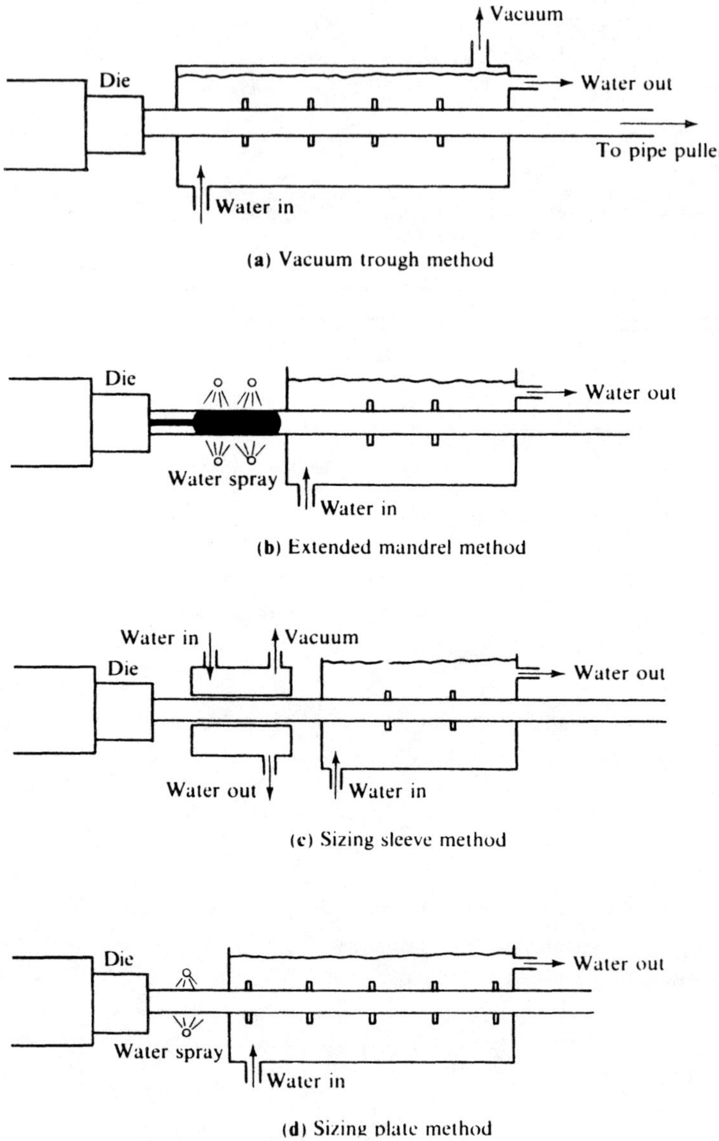


**Figure 13.14** Pipe sizings.

The water level in the tank should be at least 25 mm (1 in) above the top of the tube. Water temperature could have a cascading set-up or be at 15–27°C (60–80°F). At the exit of the tank, a rubber gasket with an ID 1.3 mm (0.05 in) smaller than the 'solidified' finished tube OD. The fit at both ends should permit maintaining the vacuum in the tank.

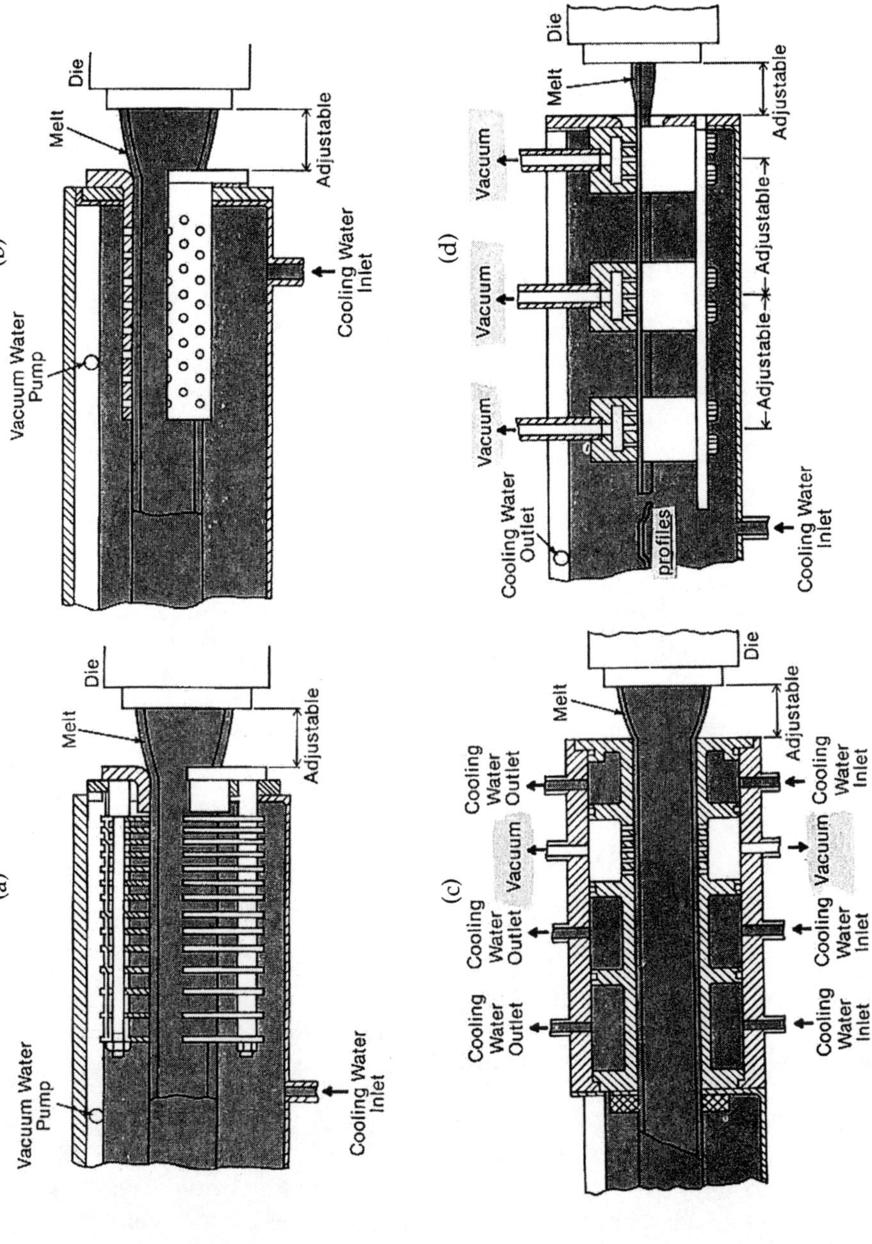


**Figure 13.15** Sophisticated sizing systems.

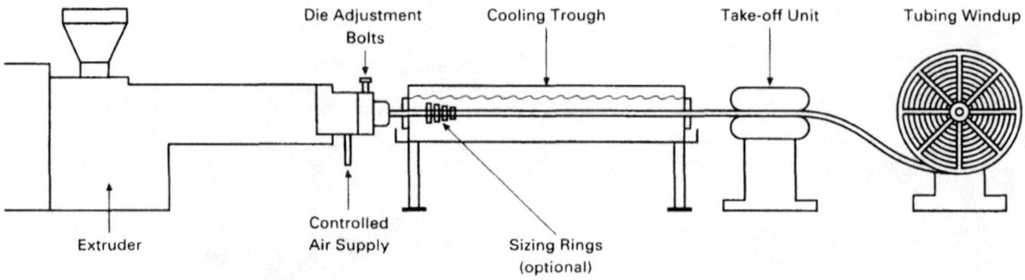


**Figure 13.16** Free extrusion of tubing.

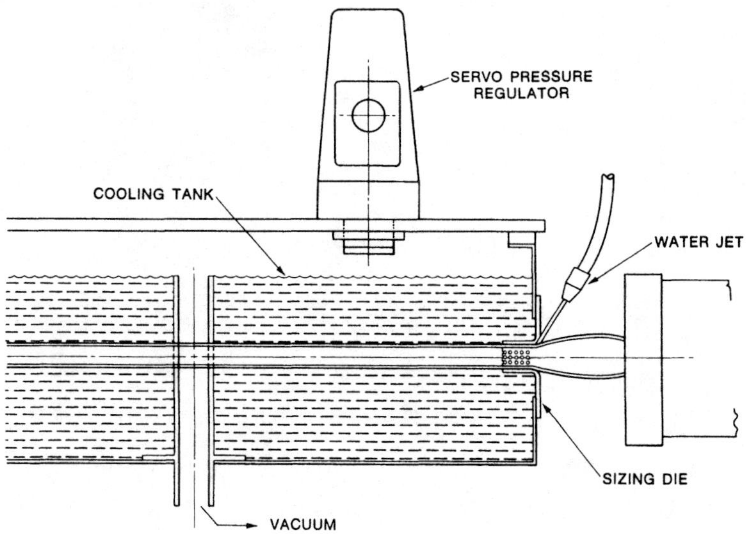


**Figure 13.17** Example of gentle water lubrication when entering a cooling tank.

As shown in Fig. 13.18, after preparing the melt in the extruder, it travels through the in-line die. The hot extrudate goes from the die to the water cooling trough. A take-off unit, such as a caterpillar, pulls the tube and controls the speed it travels [236].

A combination of the melt properties, rate of the line speed, internal air pressure (if used), and rate of cooling influences the size of the tube. The melt drawdown ratio from the die into the water cooling trough is directly relate to the final size of the tube. Depending on the plastics being processed, the ratio ranges from 4/1 to 10/1. Using low ratios helps to minimize extruded-in stresses, which can lead to problems with shrinking and stress cracking (Chapter 3).

An allowance may be required for postshrinkage. Depending on the

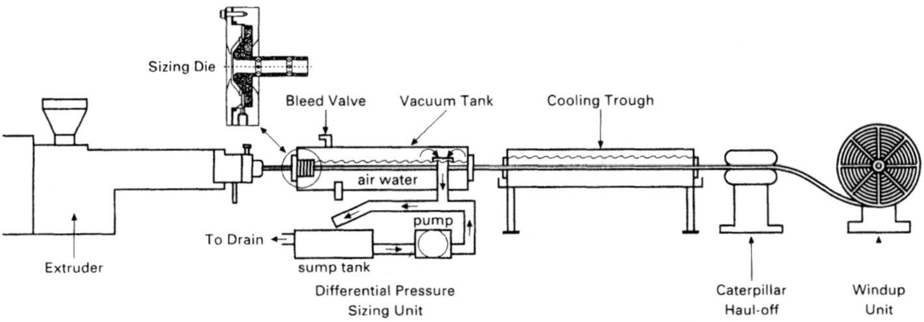


**Figure 13.18** Example of tubing with differential pressure sizing.

characteristics of the plastic being processed, there may be shrinkage, usually within a 24 h period after leaving the processing line. The tube or pipe may also be subjected to annealing and other postconditions that could effect degree of shrinkage.

Selection of the extruder die OD and ID opening sizes for the required tube or pipe dimensions depend on the amount of drawdown to be used. The actual drawdown ratio required is initially dependent on the plastic being processed. As a guide, for tubing up to 12.7 mm (0.5 in) diameter, consider using a ratio of 2/1 to 3/1, i.e., the die ID should be 2–3 times the diameter of the sizing plate. For the larger pipes a 1.5/1 to 2/1 ratio can be considered, with the ratio decreasing with increasing diameter.

If a problem of a collapsing tube or pipe occurs during free extrusion, slight internal air or nitrogen pressure through the die and into the tube or pipe upon leaving the die may be necessary to prevent the pipe from collapsing. This air hole can also be used to direct controlled air pressure and water that will also aid in cooling the pipe. For the air to act as a useful internal support, it must be supplied under pressure control conditions. A sensitive control is used to minimize small pressure variations that can result in variations in the tube or pipe diameter.

Problems can develop when using sizing devices for the manufacture of small pipe (tubing). If the sizing unit is used with a high speed line, cooling problems can occur which result in poor surface finish. If this action is allowed to persist, it can cause sticking of the pipe to the sizing device and the consequent shut-down of the line with time spent to clean up the situation.

The free extrusion system is not generally used for pipe about 6 mm (0.25 in) or larger and wall thickness of 1 mm (0.4 in) or thicker. The reason is that the pipe's roundness is difficult to maintain. For the larger pipe, there are sizing devices with differential vacuum/pressure techniques. As shown in Fig. 13.19, a differential pressure sizing unit is used [236].

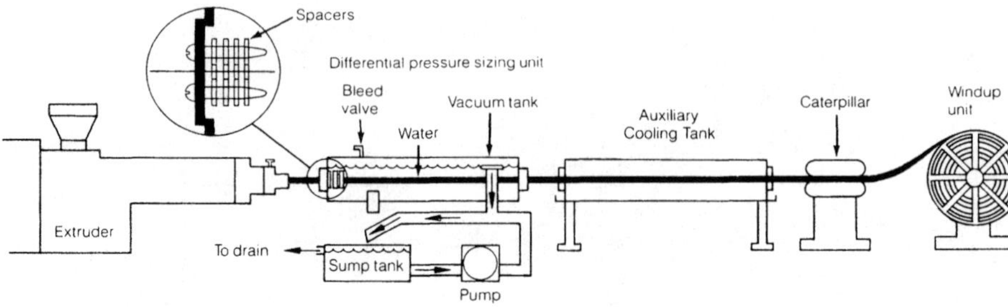


**Figure 13.19** Differential sizing using spacer plates.

Figure 13.15 shows different sizing systems with: (a) vacuum trough sizings/calibrations, using calibrated disks; (b) vacuum trough sizing/calibration, using a perforated sleeve housing; (c) sizing/calibrating system that combines controlled water (temperature and pressure) with controlled vacuum; and (d) sizing /calibrating using just vacuum (profile shape not pipe).

There are different styles of plates, disks, or sleeves used to provide the sizing or calibration of the pipe. The plates and/or disks can be stacked. Each should contain precisely drilled, usually tapered holes to meet the pipe OD configuration that should match the drawdown ratio. These holes will usually be made larger to compensate for the final shrinkage of the extrudate after it cools and probably added shrinkage after in storage. Spacing between them provides a simple means of adjusting the stacked system to meet the drawdown ratio. With the sleeve system, as shown in Fig. 13.15, different designs are used that can include openings to permit the water and/or differential air pressure in making contact with the pipe.

Typical sizing or calibration dies are made larger than the desired OD of the solidified pipe. For line speeds such as 25 m/min (80 ft/min), the usual range used is 3–5% larger. For the smaller diameter pipes (tubes) traveling at higher speeds could require about 10–15% larger openings.

Aluminum or brass materials are used for these devices, with brass usually preferred. The inside surfaces are usually sandblasted and contain straight to spiral grooves to decrease the surface drag of the extruded pipe.

As discussed earlier, lubrication with water is required on the inside of the sizing devices, particularly the sleeve type dies. There may be sufficient water lubrication with the plates or discs. What is required is that a thin layer of water follow the contact area. This can be accomplished by ensuring water flows through the small holes in the tubular sizing die or through an annular water ring at the entrance of the sizing die. Constant pressure lubricating water supply is required, such as being delivered from a header storage tank located about 1.5 m (5 ft) above the water tank.

A general review and summary concerning different sizing methods are presented. There is a method used with an external sizing and/or a sizing sleeve. The pipe is drawn through a water cooled metal sleeve and restricted in the sleeve by air pressure inside the pipe. Alternately, the pipe may be sized internally by an extension of a mandrel in what is called the extended mandrel method. The extended mandrel is an extension to the die mandrel. The shrinkage of the cooling plastic pipe keeps a snug contact with the surface of the metal extended mandrel. To aid in the cooling cycle, the mandrel can be internally cooled. This method is suited for polyolefin plastics or similar plastics that permit ease of sliding over the extended mandrel.

There is the technique of using this extended mandrel for processing thin wall and small diameter tubing, specifically hose. A tapered brass mandrel extension is attached to the mandrel of the die. Depending on the requirements of plastic being processed, the die's mandrel could include electric heaters. The extended mandrel can be either cooled or slightly heated with water via the die to aid in controlling the melt's temperature and prevent sticking. This method offers the advantage of close ID tolerance since the tube is sized on the inside surface.

In the sizing device technique, the pipe takes its shape by passing through a metal sizing sleeve or tube located in the water trough that follows the die. This method can be used in a vacuum trough with circulating water.

One approach has the vacuum drawn over the water in the closed trough/tank to reduce the pressure within the trough. This action permits the pipe being drawn through the sizing tube and/or plates to have a snug fit against the sizing device. The inside of the pipe can remain at atmospheric pressure. This approach permits calibrating and cooling long sections of the pipe.

Another approach involves the use of vacuum ring(s). An example during processing is shown in view (c) of Fig. 13.15. The perforation (for the vacuum) is in a closed sizing device. The vacuum brought about between the ring(s) in the sizing fixture and the still soft pipe pulls the pipe's wall against the sizing tubular fixture. The perforation in the calibrating sizing tube has small holes, such as 0.5–2 mm (0.02–0.08 in) in diameter with holes on the outside smaller to eliminate suction of air from its side direction. The holes are distributed evenly around the ring.

A vacuum sizing tank is generally used for tubing 12.7 mm (0.50 in) or larger. The vacuum in the water cooling trough causes the tube to expand to the sizing die set to the required OD dimensions. In all vacuum sizing operations, there should be provisions for fine adjustments of the vacuum. This approach provides the accurate control and maintaining the OD of the extruded pipe. This is particularly important when extruding small diameter pipe. A vacuum of 2–10 in Hg has been used to maintain dimensional control of the small pipe.

The water tanks in-line operate at controlled water temperatures with or without air controlled systems. The temperatures meet the cooling rate to match the plastics requirement. These tanks, like those being used in wire and cable cooling tanks (Chapter 12), can include sectionalized heat sections and water spray usually located at the end of the cooling cycle. The constant water temperature can be a closed or open system. With the closed system, control can be via a heat exchanger or a temperature controller operating in a closed loop. The open system provides for adequate fresh water makeup and discharge.

Another important aspect in the extrusion line is the precision cutting of the pipe when it is not going around a wrap-up reel. To meet the different requirements, since the plastic properties influence how the pipe is to be cut, different cutters are available that travel with the line speed during its cutting operation. After cutting, they quickly return to their original position to make the next cut.

Cutters are the preferred method of cutting since they produce clean, accurate cuts. The best cuts come from quickly slicing rather than chopping which can distort the plastic. The criteria in selecting a knife are attack angle, force required, and cut time. Attack angle describes the angle between a blade's cutting edge and its axis. A high attack angle means good slicing action; a zero attack angle means a chopped action.

Force required varies with size, density, hardness, and composition of the plastic. Higher forces may require a thicker knife, which in turn increases force requirements to compensate for wedging (force required to work the thick blade down through the product). Cut time describes how long it takes for the knife to pass through the product. This number is usually related to the inverse of cutting time.

Some products, such as medical tubing, require an absolute clean cut without burrs, dust, or so-called 'angel hair.' Lubricating the knife with water, alcohol, or mineral oil can often provide the cut desired. Also knives coated with PTFE or of highly polished chrome are used to reduce friction, resulting in clean cuts. Cutting very small diameter tubing made of slippery, flexible plastics (silicone, latex, PTFE, etc.) is usually best cut with a razor blade knife.

A guillotine cutter uses a wide blade operating in an up–down chopping motion. Its straight downward chop can distort most types of tubing. It is more popular in cutting other extruded products such as film and sheet.

Figure 13.20 is an example of a close tolerance pneumatic-stop clutch-brake assembly. This high-speed rotary knife cutter provides up to

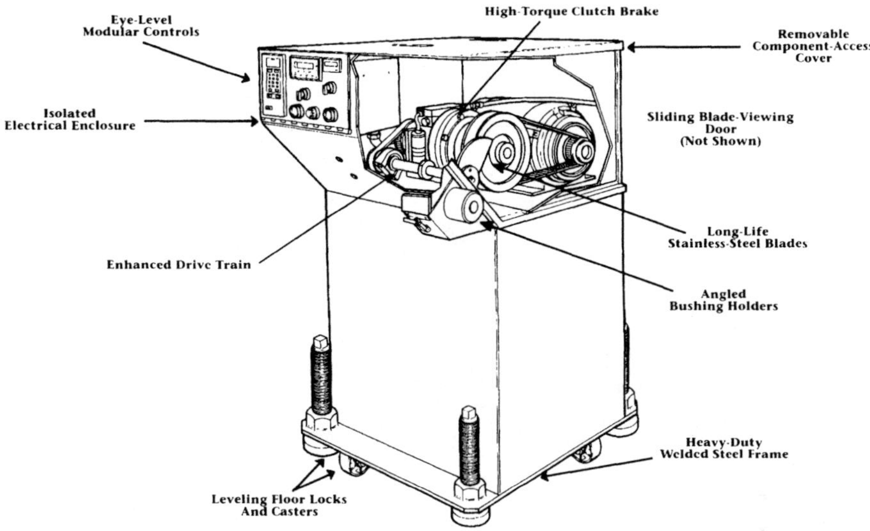


**Figure 13.20** Example of Gatto/Conair's pneumatic-stop rotary knife cutter.

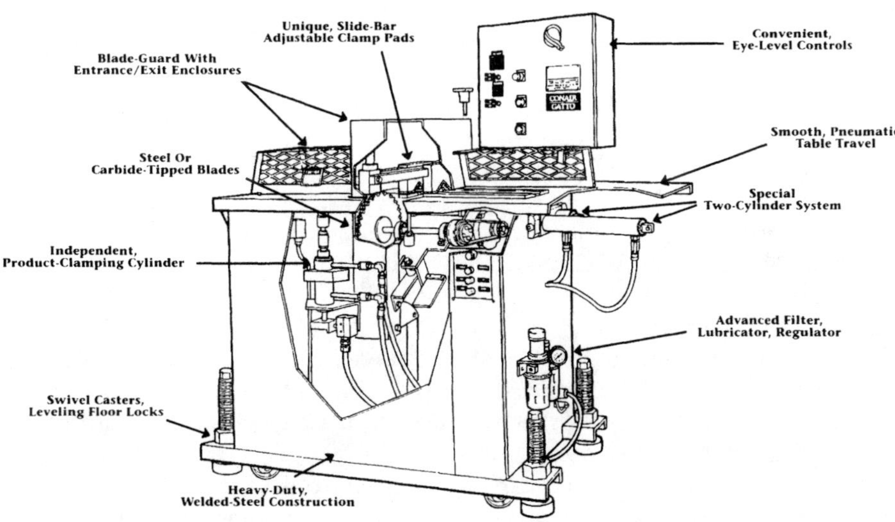


**Figure 13.21** Example of Gatto/Conair's traveling up-cut saw.

350 cuts/min on demand and twice the rate with an optional variable speed DC drive that runs in a continuous mode.

The cost-effective traveling up-cut saw is shown in Fig. 13.21. This versatile cutter has simple adjustable capabilities and can operate

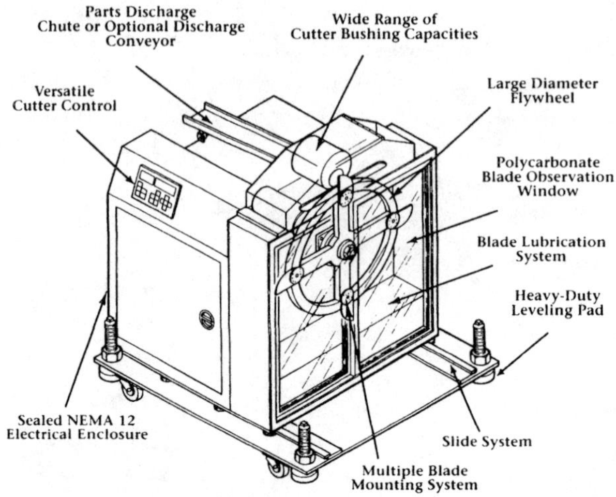


**Figure 13.22** Example of Gatto/Conair's rotary knife cutter.

accurately with electronic control at line speeds where 15–25 cuts/min are made. Figure 13.22 shows a rotary 60 cm (24 in) blade knife cutter that is very versatile for those who have to cut a variety of products (small to large sizes of pipe, profiles, etc.).

## PROCESS OPTIMIZATION

To meet minimal pipe tolerances, it is essential that the die have a high precision adjustment capability. These include the use of controls such as ultrasonic wall thickness measurement, and automatic thermal die centering (Fig. 6.10).

To gain the benefits of multiple plastic layers as reviewed in Chapter 2, coextrusion is used in producing all size pipes. There can also be one or more stripes in the pipe's machine direction that can be include using appropriate coextruded dies.

Tubing and other products can be extruded from a sequence of compatible but different plastics having different durometers or other properties. This technique eliminates the problems and expenses of secondary bonding operations and the potential sacrifice of structural integrity. The multi-plastic or multi-durometer products can be extruded on a single line and without down-time to change from one plastic or durometer to another and be seamless. It can involve coextrusion [218].

Two or more plastics can be combined end to end during continuous extrusion. This technique, in addition to being used in pipe lines, is also used to extrude other products (film, etc.). What makes the system reliable is the tight process controls used for the complete line. These include controlling the complete line, such as screw speed and/or gear pump when required to go from one plastic to another that has different melt flow conditions in the extruder. Its first applications involved small diameter medical tubing so that transitions, for example, went from a rigid plastic to a soft/elastomeric plastic in as little as 50–100 mm (2–4 in). Line speed was at 30.5 m/min (100 ft/min).

With the required extruder output, a single extruder can extrude two or more pipes that are in parallel from a multi-head die (Fig. 5.14). Triple pipe extrusion was used at least during the 1980s [415]. These lines can include coextrusion (Chapter 2). Figure 13.23 shows a single coextruded pipe line and a dual coextruded line with two pipes being fabricated. These schematics show: (a) die and core set; (b) pipe head; (c) coextrusion feedblock; (d) die adapter with optional valve; (e) valves for adjusting inner and outer skin thickness; (f) core layer entry; and (g) skin layer entry manifold.

Multiple extrusion is a balancing act. In a dual system, one pipe is stabilized. With a triple system, the middle line is stabilized first. These operating lines are used as references. Then the rate of the other lines are adjusted by some type of melt flow restricter in the head. After manipulating these adjustments, the cooling tanks are moved back to as close as possible to the die, such as 25 mm (1 in) from the die.

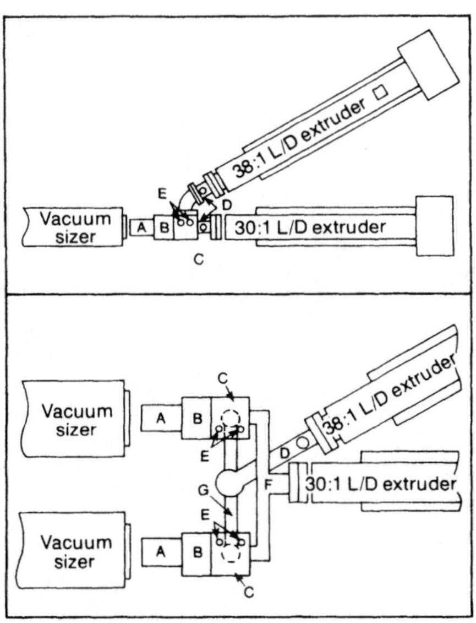


**Figure 13.23** Layout for single and dual pipe coextrusion lines.

While each multi-head die requires its own cooling and take-off system, this method can offer economic benefits. An important factor may be related to production quantity. It may not pay to have different size extruders for different size pipes. If production is limited for different sizes, there could be benefits. It provides a means in reducing labor and down-stream equipment.

It is unfortunate that throughout this book, the statements made are usually general and not specific, so a range of temperatures, pressures, and so on are provided. And, as usually stated, they depends on the plastic used, types of screw, die design, cooling media, and/or so on. The drawdown ratio as it refers to tubing or piping is a good example of this situation where it depends on material, etc.

Calculating drawdown ratio for tube tooling when jacketing a cable requires dimensions to be used before the plastic shrinkage takes place. For semiflexible PVC, this is not a factor to be concerned with. However, with LDPE and LLDPE, material shrinkage from the extrusion temperature may amount to as much as 12%. In turn, the dimensions (Fig. 12.13) for use of drawdown ratio (DDR) or draw balance ratio (DBR) should be those at the completion of the draw, i.e., before the plastic shrinks [115].

As an example, with jacketing grades of LDPE and LLDPE, take the following situations: (1) the 12% plastic shrinkage from 232–20°C (450–68°F); (2) a cable core diameter over wrap of 82 mm (3.23 in); (3) a final jacket thickness of 2.3 mm (0.090 in); (4) an available core tube tip dimensions of 3.50/3.75 (ID/OD); (5) a desired DDR of 3/1; and (6) a DBR of approximately one. The die size calculated, not considering shrinkage, would be 106 mm (4.20 in):

$$\text{Die of } 0.090 = \left[3\left(3.41^2 - 3.23^2\right) + 3.75^2\right]^{1/2}$$

Whereas, if you can make the calculations considering shrinkage to occur after completion of drawdown. The jacket thickness to be used in this case would be 0.090 times 1.12 or 1.01 resulting in a jacket OD for calculation purposes of 3.432. The die size calculated would be 108 mm (4.25 in).

$$\text{Die of } 0.101 = \left[3\left(3.43^2 - 3.23^2\right) + 3.75^2\right]^{1/2}$$

These calculations result in a BBR of 1.06 [187, R. DeBoo].

A popular technique, that relates what has been reviewed, provides air or nitrogen pressure within the pipe during processing from the time it exits the die. The air pressure is generally retained by a plug that is attached to the die with a cord or wire. The pipe slides over the plug that remains in its predetermined distance from the die to help provide the best cooling conditions. Air pressure is directed through the extrusion die. The pressure of the compressed air for thin walled pipe is about 0.02–0.15 MPa (0.2–1.5 atm) and for the thicker walls at about 0.6–1.5 MPa (6–15 atm).

It is essential that the length of this cooling-calibrating system is correctly selected. If the length is too short and the extruded pipe is still soft, the pipe may be easily deformed. However, if the cooling system is longer than necessary, the friction between plastic and metal increases and pulling the pipe through the device will be more difficult or cause the line to stop.

The length of the cooling system's trough or sizing unit is determined arbitrarily in practice or based on experience. If the cooling is not satisfactory, the rate of the water required to flow will have to be increased (at a cost). If too long, water is being wasted

The sizing device technique is popular to hold diameter dimensions. Its important features include: (1) there must be provisions for locating the sizing device in-line and coaxial with the extruder die; (2) the heat transfer from the hot die to the relatively cold sizing device (when they are located next to each other) must be prevented or controlled; (3) the length of the sizing device has to take into account the relation of the line speed to the pipe's diameter and wall thickness; and (4) allowance must be made for the plastic contraction of the pipe on cooling. For this last feature, the amount of contraction will depend to a certain degree on the conditions of the extruded melt, cooling cycle, and line speed. In general, the ID should be at least 2% greater than the required OD.

To simplify start-up and high-speed lines to operate efficiently, dual chamber water cooling troughs/tanks have been used rather than a single trough. They have permitted more accurate vacuum calibration. The vacuum can be adjusted individually in both tanks. In this way the friction between the pipe and calibrating sleeve can be influenced without changing their pressure levels. Also, to accommodate the faster lines, these tanks are 10–12 m (32–40 ft) long replacing smaller units, even those 20 ft (6 m) long.

Water cooling troughs for free flow or sizing lines can handle more than one pipe (or other profile) and retain their close tolerances. Water should be of high quality to minimize rusting and scale build-up. Closed-loop systems are readily available or can be engineered in-house. Closed-loop systems result in savings and improve the quality of the products. Generally, do not use water additives.

As reviewed throughout this book as well as in this chapter, different variables exit that can 'make or break' the extrusion line performance. A few examples are given in Fig. 13.24.

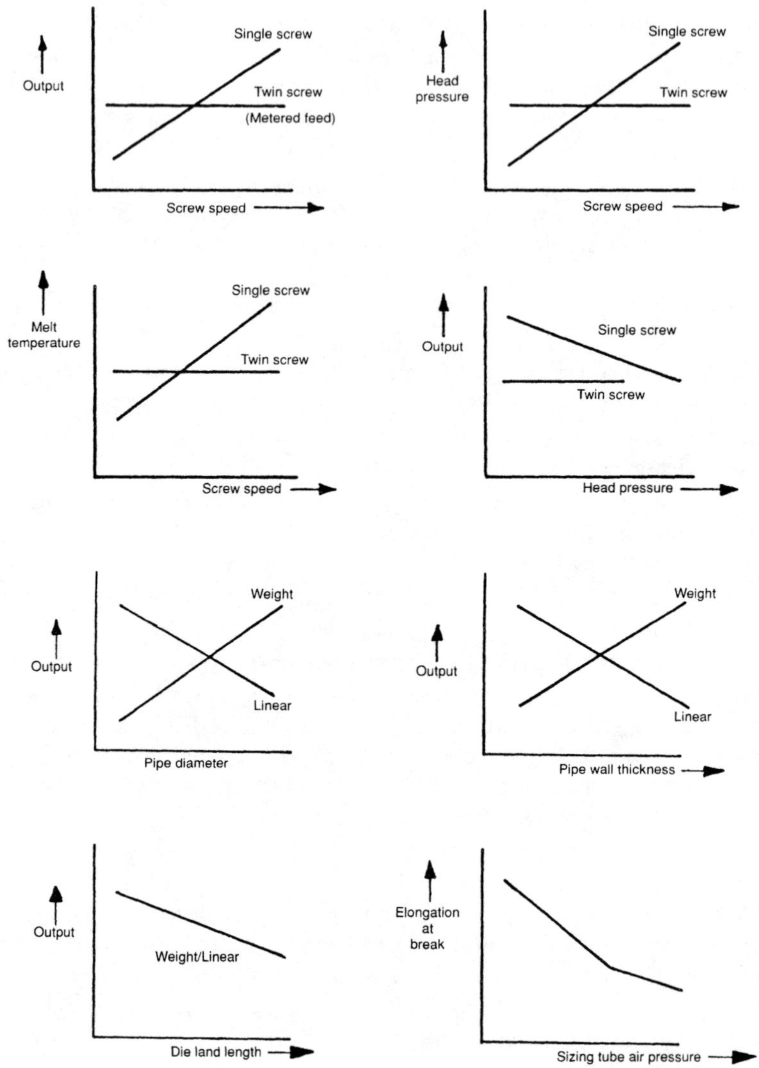


**Figure 13.24** Effect of machine variables in an extrusion pipe line.

## Orientation

To improve properties of pipe rather dramatically during processing, the plastic pipe can be oriented in the uniaxial direction alone or bidirectional [3, 370]. It is a take-off of orienting film or sheet, as described in Chapter 3 which describes the advantages of orientation. With pipe, as with other oriented products, a stronger with lighter weight pipe provides for economical advantages.

Basically, what is required are caterpillars (pullers) on both ends of a reheating tank/chamber located after the pipe has passed from the die through the cooling trough(s). The exiting caterpillar pulls at a faster rate to stretch the pipe while in the heated tank. The amount of differential speed relates to the improved properties.

Bidirectional properties can be related to the action that occurs during plastic orientation via the blown film (Chapter 7) and blow molding (Chapter 15). Within the pipe, a circular tapered plug is held within the pipe. This plug is held within the pipe using a cable or some type of rope attached to the outer face of the die. It is similar to the system used to plug a pipe that will use air or nitrogen internal pressure to calibrate the pipe as shown in Fig. 13.25. This expansion drawing plug may require some heating.

The tapered plug will have the smaller diameter facing the unoriented pipe. The other end will have an enlarged diameter based on the degree of orientation desired or the maximum that can be obtained. The shape of the surface between the smaller entry end to the larger end will be simulating a cone or S curve shape so there is a gradual expansion with no abrupt changes. This action can be called a drawup ratio compared to the comments made elsewhere about drawdown ratio. A smooth transition from the original size of the pipe to the expanded pipe size will occur and provide the expansion require for circumferential expansion.

In the meantime, the exiting caterpillar can operate at a faster speed than the entrance puller to the heated tank providing the pipe's longitudinal expansion. Actually, this plug expansion with the required pulling action may provide the desired finished properties. Temperature to be used during the plug expansion and degree of expansion will depend on

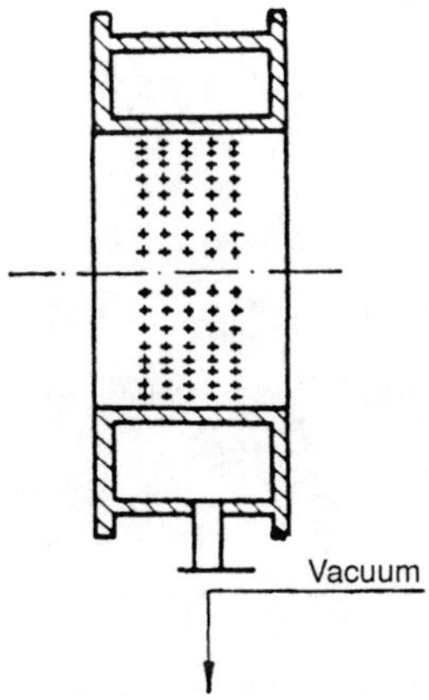


**Figure 13.25** Outline of vacuum sizing calibrating ring.

the type of plastic being processed. Information in Chapter 2 will provide guide lines.

## DIE

Most of the pipe processed is in-line straight-through extruder dies (Fig. 13.26). There is also crosshead extrusion dies that include those without spiders that split the melt opposite the side of the extruder. Figure 5.5 provides cross sections of each type of die.

Reasons that a crosshead die is used include that the processor also extrudes wire coatings that requires a crosshead die with its required floor space (Chapter 12) and to eliminate use of spiders in a die. With one extruder, either pipe or wire coating products can be made. However, in-line systems offer better wall thickness control and uniformity propertywise, dwell time or residence time in the die is less, and areas with dead spots are minimized.

The in-line die system consists essentially of the outer section which forms the outside of the pipe and the mandrel/torpedo which forms the inner surface. The mandrel is mounted on a suitable carrier (such as spiders) within the body of the die.

The important features of the die system are as follows: (1) the extruder adapter and die are designed to target for a smooth streamline melt flow; (2) the mandrel's supporting spiders/webs should be of smooth symmetrical (airfoil/streamline shape) profile beginning and ending with near knife edges; (3) the holdup area of the spiders should be minimized; (4) the cross-sectional area between the supporting spiders of the mandrel should be at least three times greater than the cross-sectional area of the annulus/ring; (5) the approach channel to the final (relatively) parallel lands should taper gradually to maintain melt compression and assist in

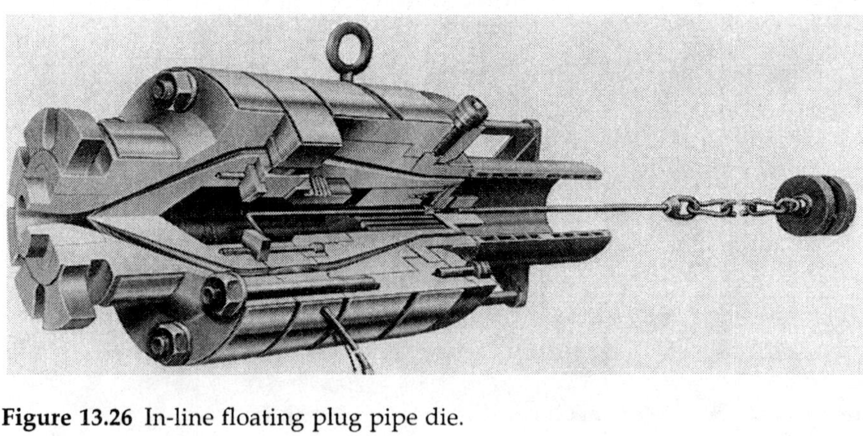

**Figure 13.26** In-line floating plug pipe die.

welding together the melt streams after passing the mandrel spiders; (6) the die lands must be long enough to exert back pressure on the melt and to help eliminate weld lines; (7a) the mandrel tip and the plane of the die face must be in-line; (7b) if the mandrel is projecting or recessed, drag marks may be apparent in the bore of the extruded tube; (7c) drag marks should be avoided since many plastics are highly notch sensitive; (8a) there must be provisions for applying air through the mandrel or alternatively a vacuum chamber for use with sizing devices; (8b) an easy accessible hole of about 6.4 mm (0.25 in) in diameter is usually sufficient; (8c) during processing be careful to keep this hole clean, it is not to be partially or completely plugged; and (9) the die should be snug enough to prevent leaking, yet free enough to be moved for adjustments.

The crosshead die is similar except it does not include spider supports to the mandrel. As shown in Fig. 5.5, a mandrel is used to control the ID of the pipe. The potential weld lines caused by the spiders in the in-line die are eliminated. However, it has a weld line. After the melt splits going around the mandrel, the split makes contact to form the weld line. With both types of dies, the design of dies and careful processing procedure can practically eliminate the problem of the weld line. The weld lines can cause reduction in strength properties, etc. (Chapter 3).

Basically the die cores or mandrels forming the inside shape of an in-line die is supported on some type of spider which divides the flow at the rear of a crosshead (also called side entry) die. There is a thrust due to the drag of the melt on the core that has to be considered in the structural integrity of the die. Fortunately, with these circular dies, drawdown does not distort the extrudate, but evenly reduces the pipe's OD, ID, and thickness. The dimensions of the die and the relation between output and haul-off rate then determines the final dimensions.

Both dies, and particularly the in-line design, takes advantage of using a spider mandrel. It improves bonding of the melt streams, shortens resetting times at raw material and color changes, reduces back pressure in the die, has excellent pipe wall thickness distribution, and optimum quality of the inner pipe surface. These spider type dies can provide outlet rates of 600–1000 kg/h (1320–2200 lb/h). They also permit ease of processing other type plastics such as ABS and particularly recycled plastics that can require tight tolerances.

Since the ideal melt flow is straight, the in-line die has that advantage. With the crosshead die, there is a potential disadvantage with more of a weld line problem. As usual in any processing situation, there is always compromise. The target is to have the melt in the die as short a time so that residence time is reduced (Chapter 3).

An empirical method has been used to design free extrusion tooling. Start by using a die mandrel about 60% larger than the OD of the tubing and follow by calculating the die diameter opening using the derived mandrel size and the drawdown equation. For a 4/1 drawdown ratio, the equation for calculating the mandrel (male part) and the die (female part) dimensions using the above relationships can be reduced to the following form:

$$D_M = 1.6\ OD_T$$

$$D_D = \left(6.56\ OD_T^{\ 2} - 4ID_T^{\ 2}\right)^{1/2}$$

where $D_M$ = OD of mandrel, $D_D$ = ID of die, $OD_T$ = OD of tube, $ID_T$ = ID of tube

With large pipes of over 300 mm (5 in), the linear speed during extrusion tends to be low unless a very large extruder is used. The consequence is the time for degradation in the die is shorter. This situation can be minimized (using UPVC, etc.) by reducing the die length, then rapid changes of cross section occur leading to potential residual stress, excessive die swell, and transverse orientation of the plastic.

Die adjustment can be made to achieve concentricity and uniform wall thickness. Such adjustments should only be used for fine control, since substantial circumferential variations in the die gap and/or land length are likely to lead to nonaxial flow and nonuniform swell. There could also be lateral pressure differences on the die mandrel, which distorts, tending to oppose the adjustment. In such cases, the cause of nonuniform flow or melt viscosity from the extruder should be corrected.

As with choker bars used in flat sheet dies (Chapters 5, 9), a restrictor ring with a narrow annular gap may also be used well before the die lips. This action would remove uneven flow inherent in the design of the die or adapter. Low velocities in the die may be used to give good surface finish and low residual strain and swelling. Low resistance permits use of longer die lips, which also promotes good surface and low swelling. However, pressure may be too low for adequate melting and mixing in the extruder. The consequence is to add screen packs to increase the back pressure [397].

There has been developed by Superex Polymer Inc., Waltham, MA, USA, a counter-rotating die that is used with LCPs (liquid crystal polymers). A biaxial orientation tube/pipe product overcomes structural limitations of conventional extruded LCPs. ACT Medical Inc., Waltham, MA, USA, holds exclusive rights to manufacture tubing for medical and surgical applications [300, 358].

The first commercial tubing had nominal ODs of 2 and 10 mm (0.080 and 0.400 in) and wall thickness of 0.38 and 0.76 mm (0.015 and 0.030 in). Other applications include carrying optical fibers, electric wires to power cauterizing heads, and mechanical wires to manipulate miniature surgical instruments through inch-long incisions.

The LCP used in this development from Hoechst Celanese were two medical grades. Their Vectra A is a monoester used for lower stiffness applications. The Vectra B, a copolyester, is inherently stiffer. They both are compounded with mineral filler and glass or carbon fiber reinforcement for higher strength and stiffness.

LCPs, a TP class of wholly aromatic polyesters, have long rod-shaped molecules that tend to align almost entirely in the flow direction. Unless the alignment is altered during processing such as extrusion, the part will have an essentially uniaxial crystalline orientation with superior mechanical properties in the flow direction and poor transverse properties [3].

Conventionally extruded LCP products are virtually useless because of this unbalanced situation. Pipe can be splintered by squeezing it. Film, sheet, etc., will split lengthwise and so on. The Superex extrusion die from Brampton Engineering, Brampton, Ont., Canada, solved the problem of preventing uniaxial orientation.

As shown in Fig. 13.27, the plastic is forced between two counter-rotating cylinders producing multi-axial stress on the plastic. The resulting biorientation produces an exceptional strong pipe, compared to any other TP extruded pipe, in both the longitudinal and transverse directions.

The degree of difference in their directional properties relates to the cylinder speed and machine direction pumping of the melt. The higher the rpm of the rotating cylinders relative to the tubing haul-off rate, the greater the gain in the circumferential or hoop direction properties. By varying these processing conditions, the pipe can be made to provide specific performance. It can be made to be stronger in bending, have increased torque stiffness, have increased compressive strength in either direction, and so on.

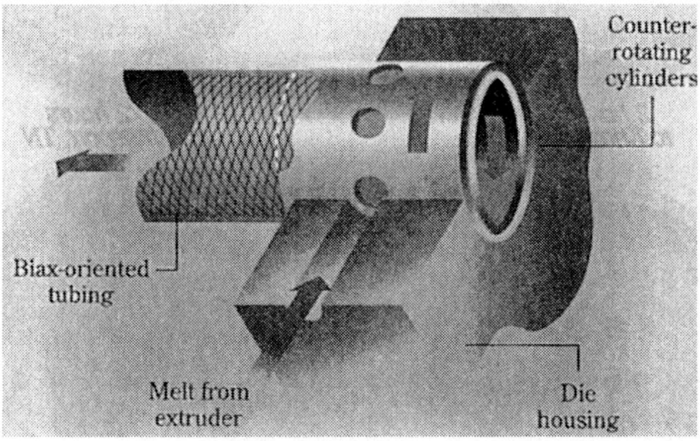


**Figure 13.27** Counter-rotating die creates versatile LCP tubing.

LCPs provides: (1) equal or higher tensile strength of over 140 MPa (20 000 psi) and stiffness with a 0.14 million MPa (2.0 million psi) flexural modulus than any competing plastic [4]; (2) superior dielectric strength of 1100 V/mil; (3) repeatedly sterilizable by autoclave, chemical, ETO, or radiation [4]; and (4) high melting and use in temperatures of 260 and 204°C (500 and 400°F) for the Vectra plastics reviewed, respectively. LCPs even at the present cost of £10–20/lb extrude into tubing/piping can compete with stainless steel and reinforced pultruded types [4].

## APPLICATIONS

There are many different ways pipe and tube products used in moving liquids, gases, solids, and so forth. They can be shaped to provide decorations, safety supports, and so on. In this chapter, different applications have been reviewed.

Corrugated tubing/piping can be produced in-line by shaping operation following or in place of sizing and during cooling; for more information see the section on forming in Chapter 18. Coextruded pipe meets all types of requirements, including barriers by retaining certain gases or liquids within the pipe. The striping of pipe (coextruded) provides market benefits such as identification.

Simulating the coating of wire and cable (Chapter 12), the extrusion line can be used to produce reinforced pipe or hose. The die is a take-off of the wire/cable die. All kinds of applications exist such as internal higher pressure hydraulic lines, fire hoses, and so on [341].

Products, such as coiled pneumatic tubing (coiled electrical cables, etc.), are coiled around a mandrel and then heat set. Heat setting temperature depends on plastic used. It ranges from 107 to 150°C (225 to 300°F) for a time period to be determined. They are cooled to room temperature and should remain on the mandrel until completely cooled. The coil will spring back to some extent when released so the mandrel must be smaller than the desired final diameter of the coil. Trial and error approach and/or experience is used to set meeting the requirements.

An interesting example is the collapsible squeeze-tube. The first patent was issued on September 11, 1841, that was from tin to hold oil paints [4]. Now, this giant market, uses about 30 wt% of metals of which 95% are aluminum, 40% laminated materials that include paper, aluminum with predominantly plastic films, tapes, etc., and 30% of predominantly PE and PP extruded tubes with molded caps. Some of the aluminum tubes have to be internally coated with a plastic barrier material to protect aluminum from certain packaged products and their exteriors may be coated with plastics for decorations, etc. [1].

## COST

Steps to reduce the cost to process plastic pipes include: (1) operate to minimize OD and ID tolerances; (2) improve in pipe quality and properties via studying gains with compounding ingredients and blending procedures used; (3) reduce set-up time by means of start-up aids and automation systems; and (4) develop savings in electricity consumption by the optimization of the extruder (Chapter 2) and the use of efficient dies, cooling tanks, pullers, and take off equipment. The tolerance control of pullers is critical to minimizing OD and ID dimensions, particularly the smaller tube sizes. A very slight variation in the puller speed can drastically influence their dimensions and waste of plastics.

**Table 13.1** Troubleshooting pipe and profile

| *Problem* | *Cause* | *Solution* |
|---|---|---|
| Wavy inside surface | Melt temperature too low | Increase set temperatures |
| | Low lubricant level | Correct formulation |
| | Pulsating drawdown | Check caterpillar |
| Excessive extrudate thickness | Inadequate die gap adjustment | Adjust or replace die |
| | Inadequate operating conditions | Adjust set temperatures, and/or drawdown ratio |
| | Profile slippage at the caterpiller | Check pulling force |
| | | Dry profile before reaching the caterpillar |
| Line, streaks and foreign specks | Dirty die | Clean die (check mandrel) |
| | Scratched die lips | Repair die |
| | Drag plug (pressure calibration) too close to die | Correct drag-plug location |
| | Excessive melt temperature | Reduce set temperatures |
| | Contaminated melt flow | Check compound |
| | | Change filters |
| Sharkskin, melt fracture | Melt temperature too low | Increase the die lips' temperature |
| | Friction at the die lips | Repair die lips' coating |
| | | Modify formulation |
| | Die gap too narrow | Increase |
| Extrudate velocity with a transversal component | Unbalanced exit flow | Modify die land geometry |
| | Temperature fluctuations in die | Check heaters and thermocouples |
| | | Check set temperatures |
| | Dirty die | Clean die |

**Table 13.2** Troubleshooting tubing for medical use

| *Problem* | *Cause* | *Solution* |
|---|---|---|
| Gels, fisheyes | Cold material | Adjust temperature profile |
| | Screen pack too coarse | Change |
| | Poor die design | Adjust let-down from breaker plate to die lips |
| | Poor compression ratio | Change/modify screw |
| Black specs (flex PVC) | Material degradation | Check screw/barrel for wear |
| | Dead spots in die | Switch to more streamlined design |
| | Screen pack too coarse | Change |
| Clarity (flex PVC) | Melt too cold | Adjust temperature profile |
| | Finish on die-flow surfaces | Refinish |
| | Poor die design | Switch to more streamlined design |
| Poor burst strength | Too much draw down | Adjust |
| | Wet material | Check drying recommendations |
| | Melt temperature too cold | Adjust profile |
| | Poor screw design | Replace |
| Diameter fluctuations (long distance) | Puller speed variation | Check drives and belts |
| | Variation in cooling water temperature | Check and adjust |
| | Variation in vacuum level | Check and adjust |
| | Screen pack blockage | Check and replace |
| | Drag caused by improper sizing devices or seal gaskets | Replace |
| Diameter fluctuations (short distance) | Feeding problem | Check for pellet-size variation |
| | Bridging | Adjust feedthroat cooling |
| | Oversized tooling | Replace |
| | Incorrect draw-down ratio | Adjust |
| Spider lines | Poor die design | Switch design (blisters, tear-drop style spiders, high polish) to improve flow |
| | Melt temperature too cold | Adjust profile |
| | Die head pressure too low | Adjust |
| Tube oval | Poorly designed rollers | Change |
| | Lack of vacuum-assist (flex PVC, urethanes) | Change process to include |
| | Excessive belt pressure | Adjust |
| | Too few rollers | Add |

**Table 13.2** *Continued*

| *Problem* | *Cause* | *Solution* |
|---|---|---|
| | Improper calibrator design | Check size, spacing, length, spray-ring cooling, etc. |
| Thin spots, holes in tube | Not enough rollers to prevent flotation | Add |
| | Water level not high enough | Adjust |
| | Material contamination | Check and replace |
| Small air bubbles in wall | Moisture in resin | Check drying recommendations |
| | Overheating | Adjust temperature profile |
| | Poor compression ratio | Change/modify screw |
| Lumen collapse | Internal air too low | Increase pressure |
| | Internal air blockage | Check and clear |
| | Internal die leakage | Check mandrels |
| Lumen blockage | Poor die design | Check and change |
| | Melt temperature too low | Adjust profile |
| | Internal die leakage | Check mandrels |
| Inconsistent lumen size | No individual air regulators | Add |
| | Imprecise air regulators | Check pressure and flow |

## TROUBLESHOOTING

This chapter has reviewed many problem areas with approaches to solutions. A number of variables exist, including not having the ideal situation in the extruder, die, and so on. Each plastic or compound, features typical operating windows that influence the lines operation and maximum capability. Tables 13.1 and 13.2 provide some examples.

# 14

# Profiles

## OVERVIEW

Profiles of a very wide variety of shapes are extruded. They are sometimes called shaped extrusions and, in fact, they could include all the extruded products reviewed in this book. Pipes and tubes could be included but the industry has them classified as a separate category because they represent a major and large market on their own. Profiles are normally shapes that are noncircular or are not symmetrical. However, there are exceptions where a circular or symmetrical extruded product, such as capillary tubing, is usually called a profile. Profiles can be solid, hollow, or a combination of solid and hollow [111, 426].

Popular shapes include many hollow sections, such as in window frame profiles, tapes, edgings, and gaskets, as well as a combination of rods with different cross sections, and structural shapes in the form of a T, U, I, or H, squares, etc. (Figs. 14.1–14.4, 1.3, 5.12, 5.13, 5.17, 5.30, and 13.15). The product shapes and sizes are as limitless as the number of applications.

Another important profile product line is the coating of different profile-shaped materials. These substrate materials include wood, aluminum, steel, and other plastics. These types of profiles can also be called coated products (Chapter 11). There is also the use of extruders to feed reinforced plastic (RP) pultrusion (profile) dies. RP pultrusion, in contrast to extruding profiles, is a process in which a combination of liquid plastic and reinforcement (continuous fibers, ropes, tapes, etc.) is pulled continuously through a heated die having a specific structural shape (T, U, I, H, squares, etc.) [1, 3, 43].

The fabricating process for the very big markets, where only plastic profiles are produced, is a take-off of the method used in processing pipes and tubes reviewed in Chapter 13. Most profiles are extruded horizontally through dies similar to the part's cross section. The extrudates are cooled and sized with air jets, water troughs, cooling sleeves, mechanical aids,

**Figure 14.1** Window profile extrusion line (B-G).

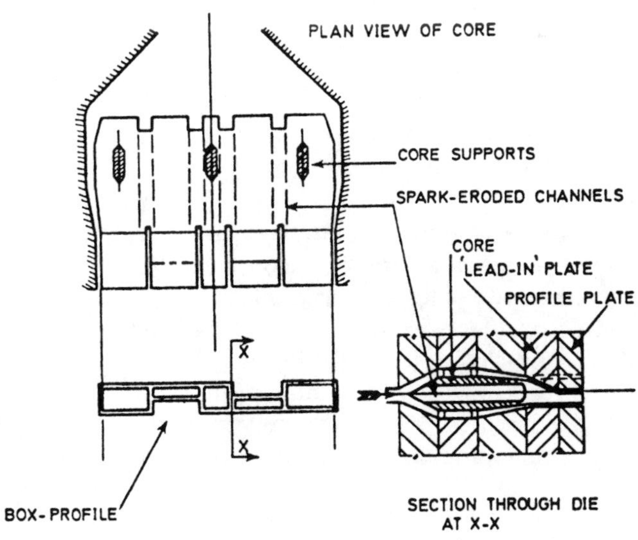


**Figure 14.2** Profile die with core/mandrel supports (spiders).

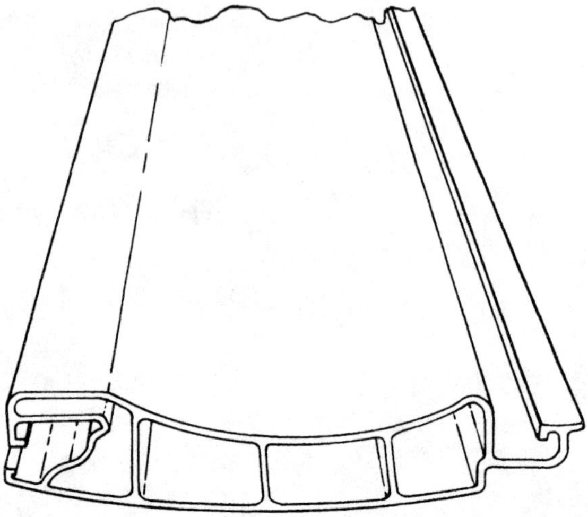

**Figure 14.3** Example of profile used for slats in roll-up blinds.

specially designed frames, dimensional measuring devices, or their combinations. Flexible to rigid shapes are produced which, in turn, at the end of the line are coiled or cut to required lengths. Line speeds tend to be slow rather than fast in order to hold accurate part dimensions. Lines have been set-up using a double die at the end of an extruder for complex window frames that have an output of 390 kg/h (850 lb/h) [156]. It is rare that the torque or power output of extruders limit production rates.

For simple shapes such as solid and V-profiles with no close tolerance requirements, a free extrusion technique can be used. The plastics which can be easily processed are those with high viscosity and softer grades, particularly when extruding at the lowest possible melt temperature. They can be used for short runs and prototyping with inexpensive dies that can, in many cases, be made from flat plate dies. For long runs, these plate dies have the disadvantage of encouraging a build-up of stagnant plastic on the rear of the plate and eventually degrade the plastic producing unacceptable profiles. Usually no melt degrading problem develops before the short run is completed. Or if the problem starts developing or the run is completed, the die is dismantled and cleaned.

For the coated profiles, the processing method is a take-off of the method used in processing wires and cables (Chapter 12). Crosshead dies are used.

Many of the profile manufacturing lines are in small specialized plants. An important and rather large market exists in the short runs. The profiles may be small and only require small extruders. These plants usually design their dies and have their own machine shop to make the dies. If

**Figure 14.4** Plastic siding profiles extensively used worldwide.

required they may also do their own plastic compounding since many runs are limited but exchanges in profile shapes may be extensive. To set-up these lines for the small operations generally requires personnel with experience and skill.

There are also operations where large production runs are made of small to large size profiles. These plants operate like those that have been already reviewed (sheet, film, etc.) where long production runs are made, providing reduced product costs. Large volume lines that usually are continually running include those that produce drinking straws, strapping, window frames, and sheds.

A recent example of a potential large production is the development of the PVC extruded wall profile house concept; profiles are supplied by the Royal Building Systems Ltd, Weston, Ontario, Canada. Since at least the

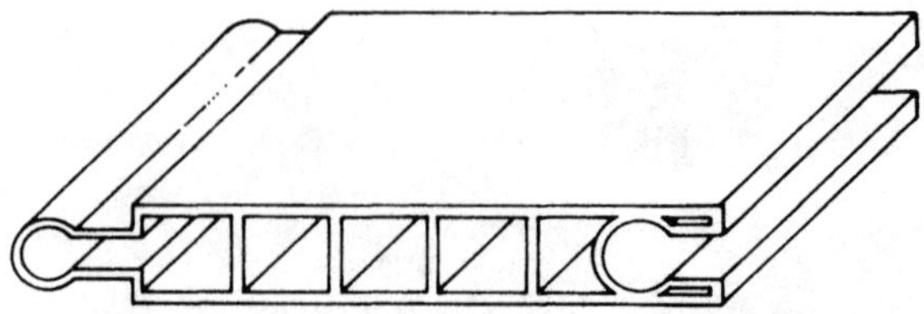

**Figure 14.5** Example of a profile used in building constructions.

1940s, various organizations have used PVC profiles (solid, foamed, and their combinations) in different style buildings worldwide. With the Royal design, PVC becomes the finished exterior and interior surface of the home's walls. They are easily cleaned and provide excellent weather resistance. The homes are built by erecting interlocking profiles, such as shown in Fig. 14.5, which in turn are filled with concrete.

The basic plastic components of these homes are profiles having rectangular cross sections of about 15 by 76 cm (6 by 30 in) and cut to the lengths desired for the walls. Ribs tie the sides of the profiles together, dividing it into five cross-section chambers.

Grooves allow these profiles to be interlocked to form walls of any length. These interlocked, hollow walls are erected on a concrete slab to form the shell of the house. Special profile shapes form the door and window openings. Then concrete is poured to fill the space within the hollow extrusions. Their light weight and interlocking design allow a complete 152 $m^2$ (500 $ft^2$) home to be erected in less than three days by three low-skilled workers, for a final cost to the buyer of about \$16 000 (US). The ribs tying the sides of the profiles together prevent the sides from bulging under hydrostatic pressure of the poured concrete.

Based on testing conducted by independent laboratories, the homes exceed the requirements of all Canadian and US building codes for conditions such as wind lift and seismic specifications. The concrete filling is appropriate for warm climates. However, for cooler climates, an insulating material, such as foamed plastics, can be poured into the hollow profiles (Chapter 18). During 1993 over 6700 of these plastic homes were ordered by builders in different countries after successful initial constructions of 400 homes.

## PLASTIC MATERIALS

Flexible and rigid PVC are the most popular plastics used. Other major plastics processed are HDPE, LDPE, PP, PMMA, and PS. However, different TPs are used to meet certain performances and cost requirements [176].

The TP materials shrink during processing. If you have a choice, use a plastic with the least shrinkage. The amount of shrinkage depends on the materials being processed and the operating parameters set-up during extrusion (back pressure in the die, drawdown ratio, rate of line speed, cooling conditions, etc.). To compensate for the shrinkage, the die opening is usually made oversize; the drawdown ratio and other operating conditions can influence minimizing the shrinkage. However, with improper machine control settings, these lines can cause problems, such as developing unwanted internal stresses in the profile, changing mechanical properties, and so on.

The more practical approach to the shrinkage situation is to oversize the opening/orifice of the die, even though it is very difficult to know what shrinkage factor should be used. The best guideline is having experience with processing the same material, extruding a similar shape and size profile. A general guide for oversizing is as follows: nylon at 20%, HDPE at 20%, LDPE at 15%, flexible PVC at 15%, rigid PVC at 10%, CA at 10%, and PS at 10%. Table 14.1 tolerance provides examples of tolerances to expect when extruding profiles with the more popular type plastics.

Even though transparent amorphous PMMA and PC have low viscosities, they are successfully extruded into transparent profiles. Their excellent light transmission are used in different industrial products. In nonpigmented condition, partially crystalline materials, such as PE and PP, are translucent or opaque and are used as profiles for applications requiring scattered light.

In terms of viscosity and melt rigidity, PMMA and PC differ considerably from standard extrudable plastics. These plastics exhibit practically no melt rigidity, which considerably affects the design and processing. They are processed with extreme care and tight control. As soon as the melt leaves the die, it is almost immediately in a sizing fixture so that it becomes stable. It continues downstream being gradually cooled to the end of the line [176].

## PROCESSING CHARACTERISTICS

Some profiles only require free air-floating motion during cooling into a water trough. However, most profiles pass through a shaping device that can include a vacuum 'box' that is built specifically for the profiles (Fig. 14.6). The profile continues into a water trough, under a cascading water stream, and/or over a conveyor belt with or without fans to cool in air. For certain complex shapes, including shapes such as rods, U-channels, and profiles with tight outside dimension requirements, air and water cooling calibration sleeves or vacuum calibration sizing plates are used. The sizing sleeve/tool can be tapered to match the shrinkage of the plastic being processed (Chapter 13) [156, 221].

**Table 14.1** Some dimensional tolerances for plastic profile extrusions

| *Dimension* | *Rigid vinyl (PVC)* | *Polystyrene* | *ABS* | *Polypropylene* | *Flexible vinyl (PVC)* | *Polyethylene* |
|---|---|---|---|---|---|---|
| Wall thickness | ±8% | ±8% | ±8% | ±8% | ±10% | ±10% |
| Angles | ±2° | ±2° | ±3° | ±3° | ±5° | ±5° |
| Profile dimensions. ± mm (in.) | | | | | | |
| 0–3 (0–1/8) | 0.18 mm (0.007 in) | 0.18 mm (0.007 in) | 0.25 mm (0.010 in) | 0.25 mm (0.010 in) | 0.25 mm (0.010 in) | 0.30 mm (0.012 in) |
| 3–13 (1/8–1/2) | 0.25 mm (0.010 in) | 0.30 mm (0.012 in) | 0.50 mm (0.020 in) | 0.38 mm (0.015 in) | 0.38 mm (0.015 in) | 0.63 mm (0.025 in) |
| 13–25 (1/2–1) | 0.38 mm (0.015 in) | 0.43 mm (0.017 in) | 0.63 mm (0.025 in) | 0.50 mm (0.020 in) | 0.50 mm (0.020 in) | 0.75 mm (0.030 in) |
| 25–38 (1–1 1/2) | 0.50 mm (0.020 in) | 0.63 mm (0.025 in) | 0.68 mm (0.027 in) | 0.68 mm (0.027 in) | 0.75 mm (0.030 in) | 0.90 mm (0.035 in) |
| 38–50 (1 1/2–2) | 0.63 mm (0.025 in) | 0.75 mm (0.030 in) | 0.90 mm (0.035 in) | 0.90 mm (0.035 in) | 0.90 mm (0.035 in) | 1.0 mm (0.040 in) |
| 50–75 (2–3) | 0.75 mm (0.030 in) | 0.90 mm (0.035 in) | 0.94 mm (0.037 in) | 0.94 mm (0.037 in) | 1.0 mm (0.040 in) | 1.1 mm (0.045 in) |
| 75–100 (3–4) | 1.1 mm (0.045 in) | 1.3 mm (0.050 in) | 1.3 mm (0.050 in) | 1.3 mm (0.050 in) | 1.7 mm (0.065 in) | 1.7 mm (0.065 in) |
| 100–125 (4–5) | 1.5 mm (0.060 in) | 1.7 mm (0.065 in) | 1.7 mm (0.065 in) | 1.7 mm (0.065 in) | 2.4 mm (0.093 in) | 2.4 mm (0.093 in) |
| 125–180 (5–7) | 1.9 mm (0.075 in) | 2.4 mm (0.093 in) | 2.4 mm (0.093 in) | 2.4 mm (0.093 in) | 3.0 mm (0.125 in) | 3.0 mm (0.125 in) |
| 180–250 (7–10) | 2.4 mm (0.093 in) | 3.0 mm (0.125 in) | 3.0 mm (0.125 in) | 3.0 mm (0.125 in) | 3.8 mm (0.150 in) | 3.8 mm (0.150 in) |

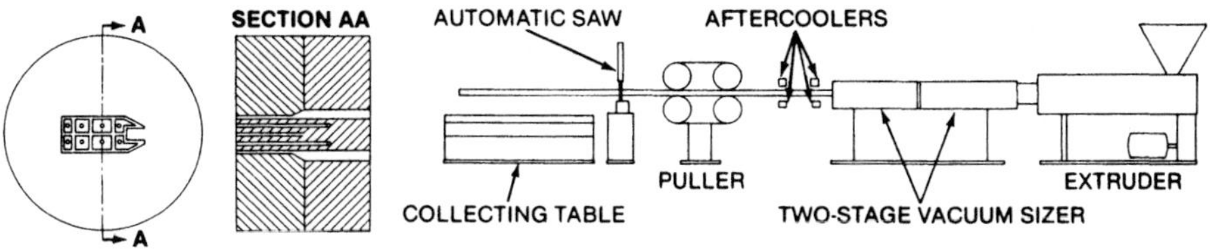


**Figure 14.6** Profile die in the extrusion line.

These devices can be horizontally split for ease in start-up. They can operate like opening a book where the top is left open until after the profile is threaded to a set of pull rolls to be cut or the take-off reel/drum. After threading, the top is put into its proper position to obtain the required shape. As reviewed in Chapter 13, vacuum cooling sleeves can be used. Adjustable brass rods are used to guide the profile into its final shape. During the sizing, air jets and/or water sprays can be used to support the shaping action with the cooling action.

Final sizing may still be taking place during the cooling period so this change has to be taken into account at the die and the line's operational control settings. Different plastics behave differently during cooling. Some only require air cooling particularly when they have a simple and thin wall shape such as a tape or belt. Most others require water cooling with some using extensive time in the water particularly when thick cross sections exist.

Typical of many extrusion lines that require specific lengths, pullers direct the profile to various types of cutters that are used to meet different requirements based on the plastic being processed and profile shape. They include traveling saws, guillotines, high-speed rotating blades, and razor-type cutters (Chapter 13). With profiles that are going to be wrapped, power driven wrap-up reels or drums pull the profile so the line operates at a controlled uniform rate. Also, downstream after cooling, different equipment is included, such as product identification codes, decorative patterns, printer, embosser, surface holes or cuts, coatings, flame polishing (increase surface gloss), etc.

## Coated profile substrate

Wood, as well as other materials (plastic, aluminum, steel, etc.), can be easily coated in profile shape. Procedures used would be similar to wire and cable coating including possibly cleaning and preheating as reviewed in Chapter 12. Crosshead dies would be used (Fig. 5.9). For noncontinuous profiles, equipment is available. Short-length wood or other material is fed automatically from a storage/feeding magazine, through a crosshead die, and finally through a cooling of air and/or water (Fig. 14.7).

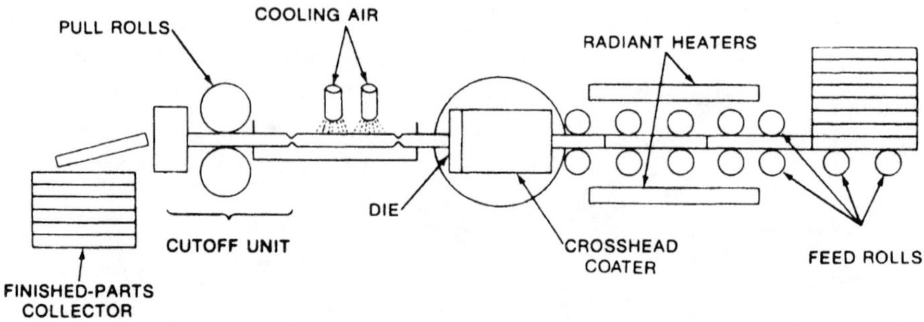


**Figure 14.7** Noncontinuous profiles being automatically coated with plastics.

## PROCESS OPTIMIZATION

At times processors may want to use a different plastic to meet specific requirements such as improved production rate, etc. As an example, some processors have reported that the same profile die can be used for PVC as well as HIPS. However, when extruding a profile for siding applications with 15–20 cm (6–8 in) flat in the center and curved edges for structural strength, HIPS shows waves on the flat sections. With PVC, they stay perfectly straight.

These waves can be three-dimensional, forming bumps and sink holes along the panel, which is an indication of too much flow in the center part. The reason for this action can be explained with the difference in rheology between both plastics. HIPS is more shear thinning than PVC, which means that in areas of high shear rates (such as the flat-sheet part), the viscosity of HIPS drops more than PVC. This will allow more flow since, as a result of lower viscosity, the pressure gradient is less.

### Cooling profile

To aid in obtaining maximum dimensional control of profiles, additional cooling is usually required. Air or water cooling open or closed box around the profile exiting the die may be required. This action provides better support of the extruded profile, faster controlled cooling of the profile, and start a more uniform cooling cycle in-line with rest of the line.

The box will have controlled cooling air, water, or combination, circulating the profile. Other devices include matching or a single cooled roller(s), draw plate that the profile travels through with slight reduction in size, and post holding and support systems for air cooling profiles. With shapes, like a clip or U, a support mechanism is used while the profile travels from the die. This support holds the part shape as it is being cooled by air and/or water.

**Window frame precision fast line**

An example where dry sleeve calibration is very useful is when extruding very complex window frame PVC profiles (Fig. 1.6). Figure 14.8 shows cross sections of single and multi-chamber designs. Rigid mechanical and physical requirements, that include dimensions, have to be meet. The system for a precision profile line is basically as follows: (1) a single or twin screw extruder; (2) an extremely well designed streamline extrusion

**Figure 14.8** Examples of PVC cross sections for window frame fabricating.

line; (3) a series of 2, 3, or 4 vacuum calibrators that are similar to a sizing sleeve in a vacuum tank, which are mounted on a specially designed calibrator stand to supply specifically metered vacuum and cooling water to the calibrators; (4) air cooling and radiant heating stands to keep the profile straight; (5) a puller with pulling force capability from 2–4.5 tonne; (6) a reliable cutoff saw; and (7) a stacking system.

The key to this dry-sleeve calibration system is the design of the extrusion die, its size relationship to the calibrator, and the size of the calibrator which is designed to allow for the shrinkage of the specific compound being extruded (Fig. 14.9). The extrusion die is typically constructed of stainless steel and designed to afford maximum internal streamlining. No dead spots or eddies can be tolerated. The die is designed to distribute the plastic to fill all the walls and internal webs uniformly. Often webs are not just for structural strength, but to accomplish balancing of the flow of plastic melt to the orifice opening.

Flow directors and die lands are designed into a series of plates in the die to effect streamlining of the die and pressure drop which ultimately succeed in a uniform volume of plastic flow to all portions of the die orifice opening. The target is to minimize the die swell and drawdown of a given compound. This is accomplished by allowing most of the die swell to occur inside the extrusion die by having a final melt restriction internal to the die and die lip land orifice larger than the restriction.

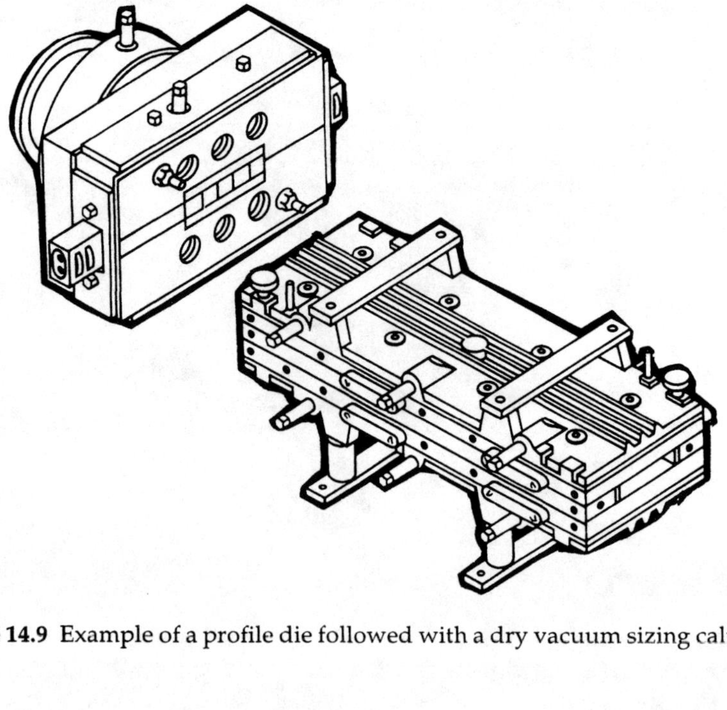

**Figure 14.9** Example of a profile die followed with a dry vacuum sizing calibrator.

The plastic expansion typical of an elastic mass released from compression occurs in the final die land and delivers melt to the calibrator, which is usually positioned 8–20 mm (0.300–0.800 in) from the extrusion die orifice opening. Die swell control and proximity of the calibrator to the extrusion die have a predictable effect on the shrinkage and straightness of the final profile because such treatment minimizes strains in the PVC due to drawdown and any probable nonuniform cooling.

A typical main frame window profile extrusion die would only be 1.5–2.5% oversize of the final dimensions of a profile, allowing for little or no drawdown of the extrudate prior to its entry into the calibration sleeve. The main frame requires the most footage in a window system at the highest obtainable output consistent with keeping tolerances desirable. The higher outputs are obtained through use of multiple calibrators with each one in turn cut smaller on its OD and finally to a dimension which allows the part its final shrinkage into the dimensional specification requirements.

The dry calibrator sleeve design is critical in that it must deliver a means of thermal energy removal from the extrudate uniformly to assure straightness and dimensional uniformity in the final product. The location of holes and slots on the internal path of the calibrator is critical to assure surface contact of the extrudate to cooling surfaces in a pattern consistent with a given profile, its wall thickness, and areas of maximum tolerance.

The location of cooling coring in the calibrator is critical to uniform cooling. The calibrator table usually has a multiple source of both vacuum and water so that each source can be varied in a way to assure optimum performance in obtaining the required extrusion rate and precision dimensional tolerances. Lines with four calibrators with the very complex profiles obtain line speeds at least of 2.5–3 m/min (8–9.75 ft/min) of rigid PVC [187, L. Fow].

## Rod

Extruding thick/large rods can set up problems. They require special care in order to eliminate voids. Some plastic rods, particularly small diameters, have no problems. Different techniques are used. One well-established approach is to ensure that the plastic melt is packed at a sufficiently low output rate so the relationship of pressure, time, and output rate is closely controlled.

A die with a sizing tube is attached to the die face. This sizing tube is also called a cooling die or forming die. The sizing tube is surrounded by a jacket that continually has water flowing through and around the jacket. It could include baffles and restrictions to direct the water's entrances and exits in different sections so that a temperature cascading flow exists.

Plastic melt is pumped into and through this sizing tube. It starts to freeze/solidify as soon as it touches the inner cool wall of the sizing tube. A hardened skin forms and slides along the metal wall surface. While it moves, the skin becomes thicker. In the meantime, the plastic shrinks; however the cooling action has not reached the rod's core. With the proper rate of flow, the extruder pumps more plastic melt to pack the interior with more plastic.

The packing action eliminates the voids. The plastic continues to go downstream while it starts with surface cooling and the internal packing. To ensure cooling action from the sizing tube internal wall to the plastic's wall surface (i.e., avoiding shrinking away from the jacket wall), the jacket wall diameter is reduced down-stream to permit contact of the plastic. This action is usually not required if the packing action causes the plastics to continually press and slide along the straight bored jacket hole. Before the rod leaves the sizing tube, it is sufficiently cooled along the complete cross section. This packing action is similar to what happens when filling a mold during injection molding [2].

Rod extrusion is slow because the plastic is a poor conductor of heat. Output rate may be about 9–23 kg/h (20–50 lb/h). After extrusion, these rods are usually annealed for stress relief and stabilization of dimensions. If required by design, they can be centerless ground to final size.

### Robotic profile

A different extrusion profile has been developed that uses a robot to deliver the hot melt over or around a substrate. This patented robotic extrusion process can uses a flexible, heated, high-pressure hose that is connected to the front of the extruder The hot melt travels through the hose. At the end of the hose is a nozzle; the tip of this nozzle is, in effect, the actual profile die.

The positioning of the profile nozzle die is controlled by a computer-regulated six-axis robot. The nozzle/die is guided by the robot to deposit the profile's hot melt on a substrate which is on a multi-station rotating table supporting other substrates to be covered. An application is coating edges of an auto window pane [138]. After the substrate is covered, the nozzle temporarily is moved off-line; its bled off-line is returned to the extruder. In the meantime, the substrate with the deposited plastic is subject to cooling while it moves so an untreated substrate gets into position to receive a coating, and so on.

### Mechanical strength

An advantage that is used in profile product designs takes advantage of the property gained by using certain shapes. Different designs of shapes

| Case | Shape | Change | Moment Of Inertia | Increase In I | Increase In Weight | Ratio I / Wt. |
|---|---|---|---|---|---|---|
| 1 |  | Base 2"×1/4" | 0.0026 | | | |
| 2 |  | Double Height | 0.0208 | 700% | 100% | 7 |
| 3 |  | Add1/8" wx 1/4"H Rib | 0.0048 | 85% | 6.25% | 14 |
| 4 |  | Add1/4"wx 1/4" H Rib | 0.0064 | 146% | 12.5% | 12 |
| 5 | 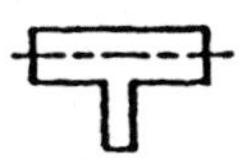 | Add1/8"wx 1/2"H Rib | 0.0118 | 354% | 12.5% | 28 |
| 6 | 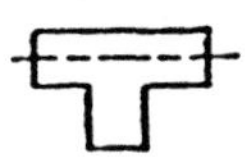 | Add1/4"wx 1/2"H Rib | 0.0194 | 646% | 25% | 26 |

**Figure 14.10** Examples of ways of using ribs to increase rigidity and reduce weight of profiles.

exist in the literature [3]. An example is shown in Fig. 14.10 where the rib shape has different mechanical performances based on its size.

## APPLICATIONS

One might say that plastic extruded profiles are all around us. They are used in many markets, such as building and construction, electrical and electronic devices, transportation vehicles (auto, aircraft, etc.), medical devices, lighting devices, etc. Specific parts include window channels, rain gutters, building sidings, auto door trims, garage overhead doors, gaskets, seals, and so on. To expand the market for the already complex window frame profiles, some processors have extruded more complicated profile designs to eliminate such potential problems as internal humidity buildup due to PVC window frames not fitting too tightly [156].

An interesting and unusual application for extruded strips involves a technique for producing very large pipes using a relatively small extruder. A heated solid or hollow profile strip of PVC, 2.5–5 cm (1–2 in) thick and usually about 25 cm (10 in) wide is extruded tangentially onto a round mandrel that revolves. The angle of the strip to the axis of the mandrel and the rotation of the mandrel can be set for the strips to overlap and move the wrapping forward.

As a result, a type of butt welding takes place that forms the pipe. Welding is a consequence of the heat of the profile and the pressure at which the newly wound profile sections come in contact with the previously wound strip. Altering the rotational speed of the mandrel increases or decreases the amount of pressure. In order for the wrapped pipe to move on the mandrel, the mandrel could be tapered inward. The amount of taper would depend on the shrinkage of the plastic. This type of fabricating is a take-off postforming as reviewed in Chapter 18.

## DIE

There are what the industry identifies as standard and custom dies. The standard dies are well established, producing stock profiles such as rods, T-shapes, belts/tapes, straws, etc. They include simple to complex shapes. The custom dies are generally the types with limited production and usually owned by the customer. Since this business includes the use of short runs of many different shaped profiles, many of the dies (plate dies) are of simple design and manufacture, and not streamlined.

Even though most profile dies are complicated to design, there are some basic rules to follow in order to make them perform more efficiently. Consider the following: (1) keep it simple – is the complicated cross section really required? (2) eliminate design where a melt hang up could occur; (3) target for a symmetrical profile; (4) the plastic to be used and/or profile should be able to retain its shape when exiting the die sideways or downwards; (5) all corners should not be square and have radii at least a quarter of the profile wall thickness to avoid stress concentrations; (6) keep outside the wall thickness uniform; and (7) interior supports/walls should be uniform and about 20–30% thinner than the outside wall. Cooling the interior of these complex profiles is difficult, that is they cool at a slower rate then the outside wall so these suggestions can help to unify the cooling action (Figs. 5.12, 5.13, 5.17, 5.18, and 5.30). An example of a modular complex die that produces window profiles meeting very tight tolerances on the finished profile, made by Technoplast (Kunststofftechnik GmbH, Am Kreuzfeld 13, 4563 Micheldorf, Austria), is shown in Fig. 14.11.

Start from the dimensions of the final product, i.e., the sizing die and work backwards. The only real limiting factor is the plastic's elastic memory properties. For example, PE is usually very unsuitable for those profiles where high dimensional accuracy is necessary since its elastic memory characteristics make it impossible to obtain sharp definition.

The design of the type of die depends on whether the product is hollow or compact/solid. Also whether the cross section is simple or complex. The target is to establish the most appropriate melt flow channel that is to be streamlined and meet minimum and even residence time. It also has to

**Figure 14.11** Modular designed window profile die.

meet the need to compensate for the down-stream postextrusion dimensional changes which occur. With all these conditions, add that the exiting melt is at a uniform speed and temperature.

Fortunately to meet these requirements for a rectangular or circular cross section is relatively easy compared to the more complex shapes. With the usual complex profiles, it becomes difficult; it requires understanding and application of the behavior of the plastic melt to the profile shape.

Some processors confronted with a poorly designed die try to compensate by changing the line's operation. Drawdown ratio and cooling conditions are changed, which may be beneficial. However be aware that more problems can develop, such as residual stresses and distortions of the profile.

Conical-shaped die adapters provide a streamlined melt flow. The die length may vary to ensure uniform flow when the melt exits the die. When the cross section reduces linearly, the flow speed will be initially slow and then increases significantly. Sometimes this can cause overheating of the melt. Trumpet-shaped channels are designed to increase the speed linearly, or to maintain a constant accelerating flow in the axial direction. However, the trumpet-shaped channel is more difficult to machine and also more costly than the conical shape.

Most dies are in-line with the extruder. For very fluid plastics, such as nylon, a crosshead die is used directing the melt downward. There are also crosshead dies that permit covering different shapes made from wood, aluminum, etc. These dies extrude sideways to the extruder like the wire and cable lines reviewed in Chapter 13.

In-line dies with hollow sections that have a single or multiple of hollow sections require cores or mandrels that are supported by spiders. The design is similar to that of the dies reviewed for pipes and tubes in Chapters 5, 12, and 13. However, they are usually much more complex because the profiles with multiple hollow sections are small and temperature controls become more sensitive. For all these dies as well as others (as with other dies reviewed in other chapters), the target is to have the melt channels in the dies streamlined up to their lands.

Profiles with one or more hollow cross sections require (small) mandrels, a take-off to the requirements for pipe and tube designs discussed in Chapter 13. With the mandrels, flow is interrupted. Streamlined spiders are important to use. Flow rates where two or more individual flow channels unite, the average flow velocity becomes higher. To help reduce this situation, when possible, different geometries are used that are take-offs of coathanger dies (Chapters 5 and 8) to balance the flow.

Since the profiles can have complex shapes, undesirable unequal cooling can occur. Cross sections, when possible, should not have sharp corners or edges. The cooling and melt flow is made easier. However, most of the profiles use sharp corners and edges. They are processed but require trial and error and/or experience to obtain the desired shapes.

All sharp corners and edges will cool faster since more surface area exists. With the faster cooling, the first section to shrink is thinner. So obviously that section is made thicker to balance off the uneven shrinkage. As reviewed in Fig. 5.16, to obtain a square profile the die opening has a different shape that is not square.

As reviewed in the other chapters on die design, the melt should flow in a laminar pattern. With the profiles having all kinds of cross-sectional shapes, the melt flow will change. The target is to eliminate hang-up areas so the melt does not degrade and cause unacceptable profiles.

The land lengths, constant cross sections before the orifice openings, the angle of entry into the final land section, and the percentage oversize of the orifice to the dimensions of the profile are all critical. A land is needed to provide sufficient back pressure on the melt so that the extrudate is controllable. With proper pressure control, the weld lines caused by the spiders can be more properly sealed and literally eliminated.

The land length are generally ten times the width of the lip/orifice opening. This ratio is reduced for very viscous melts and lengthened for very fluid/low viscous melts. The resistance to melt flow that produces the back pressure varies as the cube of the changing thickness. Small openings of 3.2 mm (0.125 in) are very effective in developing back pressure. With large openings, the land length may require a more costly long die.

The streamlined internal entrance to the die is important. For small dies, the entrance angle may be as small as possible usually using up to a 60° angle that smoothly tapers to the lands. With larger dies, a similar entrance can exist requiring the die to be longer. Tradeoff is usually required so that residence time in the die is not long enough to cause the melt to start degrading. Compromising is required, particularly where the longer die (for short runs) is more expensive.

As in other dies, melt fracture can occur particularly with small opening dies (Chapter 3). The high shear rates cause the melt fracture which in turn produces a rough surface on the profile. Reducing the entrance angle with less sharp bends of the melt flow can reduce or eliminate the melt fracture.

Swelling occurs when the melt exits the die (Chapter 5) and shrinkage of the extruded plastic occurs as it is cooled. The degree of these two conditions vary according to the plastic being processed. Different plastics can produce very little to a high degree of swell. Long entrance angles and short lands produce more swell. Pressure drop upon the melt exiting the die, its rate of being pulled, and rate of temperature change from within the die to the cooling cycle controls the letdown due to the swell.

To compensate for any swelling when the melt exits the die and the shrinkage that occurs down-stream, the die orifice for a new profile is made undersize. With a new shaped profile (also with experience), one can use the practical approach where the opening is made as small as might be required. In turn if the profile dimensions are oversized (as they should be using practical experience), it can be machined to increase the exiting extrudate.

With oversizing, the drawdown ratio is greater and can induce more stresses in the plastic. Also, with the usual unbalanced profile (not circular), there will be uneven stretching in different parts of the profile's cross section. These conditions can cause additional dimensional changes other than just those due to the size of the die's orifice opening. As reviewed at the start of this chapter, personnel are required that have experience and skill to produce acceptable profiles.

Crosshead dies are used for the low viscosity melts, such as nylon, since their melt strength is not sufficient to permit their proper shape if exited from an in-line die (Fig. 5.9). The melt is usually directed downward into a water trough and on 'down' the line. The crosshead dies are also used to coat substrates where they extrude sideways to the extruder. These dies are designed as streamlined as possible with their 90° angle that bends the melt (Chapters 12 and 13).

Profile dies for long production runs are usually very complex to design and manufacture. More than one piece of durable steel is used requiring precision machining. Other materials are used in die construction for short runs since they usually have limited life, are easily scratched, etc. These materials include aluminum, brass, and beryllium–copper. Designs should permit assembly and cleanup to be as simple as possible. For long runs, some type of coating is applied in the melt channels to provide smooth melt flow and long life. Chrome plating provides a hard surface layer that can be polished to a fine finish. PTFE impregnated nickel or chrome plating adds a slippery surface. However, as reviewed in other chapters, certain coatings can influence heat transfer through the metal.

For short runs and/or prototype runs, low-cost tooling can be used. However, prototyping for long runs can start with the low-cost dies but, prior to manufacturing the final die, consider a step in between where the die simulates what is required. Again, experience really dictates the approach to be used in developing the die for long runs. The low-cost dies have also been used for long runs even if they require time for the die cleanup or using a second die and so on.

Low-cost dies are the flat or plate dies that are simplest to design and manufacture. They can be made from aluminum plates and are bolted over the adapter. The adapter has an exit hole just slightly larger that the plate opening. The required orifice opening is cut out of a plate. It follows simple designed flow channels in the die body. The result is a minimum, if any, streamlining. It is susceptible to dead spots or stagnate zones in the die.

A guide to plate dies is that they can have a thickness of 6–12 mm (0.25–0.5 in), depending on the profile size. They should be reduced locally to 3–5 mm (0.12–0.2 in) where the profile may have a section that is significantly thinner than the main profile. The actual cut-out dimensions of the die orifice should be approximately twice the dimensions of the finished profile. This relationship for the commercial grade plastics will provide enough drawdown of the melt from the die's exit to the water cooling trough and/or sizing fixture.

The drawdown is necessary to produce sufficient tension in the extrudate to prevent sagging and fluctuations in dimensions. Normally some modification to the orifice plate is necessary after initial trials, in order to achieve the required distribution of plastics in the cross section of the profile. It is usually to open out the orifice, or chamfer any internal corners at the back of the die opening by carefully hand filing or grinding, to encourage more flow in those areas.

As a guide where necessary, the design of more permanent streamlined dies can be developed from the shape of the plate die prototype. However, if the plastic material is changed or the line conditions, such as line speed or melt temperature, change then this guide may become useless.

These dies are usually used for short runs that process thermal stable plastics so that before any unacceptable profile problem develops, the line is stopped. The die is cleaned and ready when the next run has to be made. Depending on the plastic being processed, there may never be any real problem. Many of these type dies exist as stock items.

The dies can also be made to be semistreamlined by using a series of plates. Orifices in each plate are cut in a cascading pattern so that the outer plate meets the necessary requirements. These dies can provide for longer production runs. The bodies of these plate dies provide simple streamlined channels whereby they can be used with different surface plates. The minimum to maximum output capacity of the extruder can permit using a variety of profile sizes and shapes.

To heat any die different methods are used depending on the shape, size, and requirement of the die. There are band-type electrical resistance heaters in cylinders and strips, or custom made to fit a specific contour. There are also cartridge and aluminum block electric heaters. The heater must have a tight contact with the die in order to have proper heat transfer and control. A loose fit with its well-insulated air space interferes with providing the maximum and uniform heat to the melt.

Popular in use are coextruded dies to enhance properties of the profiles (Table 14.2) [111]. Decorative stripes of different color can be added. Flexible and rigid plastic coextruded provides for special effects. The advantages and designing dies have been reviewed throughout this book starting with Chapter 2.

Figure 14.12 provides information on design tips for coextruded profiles where: (a) a dual extrusion for a modular cabinet wall panel; (b) if the flexible sealing portion wears out from abrasion, a replacement flexible insert can be slid into the rigid plastic slot; (c) a cross section of a dual extrusion used as a ball-return trough for a billiard table; (d) a bowling-ball return trough made from 15 cm (6 in) diameter extruded tube with one or more layers where the tube is slit while still workable and guided over a forming die; (e) typical dual extrusions of rigid and flexible PVCs;

**Table 14.2** Example of compatibility of different plastics in dual coextrusions

| *Plastic materials* | *Rigid vinyls* | *Flexible vinyl* | *ABS* | *Polyethylene* | *Polypropylene* | *EVA* |
|---|---|---|---|---|---|---|
| Rigid vinyls | Yes | Yes | Yes | No | No | No |
| Flexible vinyls | Yes | Yes | No | No | No | No |
| ABS | Yes | No | Yes | No | No | No |
| Polyethylene | No | No | No | Yes | Yes | Yes |
| Polypropylene | No | No | No | Yes | Yes | No |
| EVA | No | No | No | Yes | No | Yes |

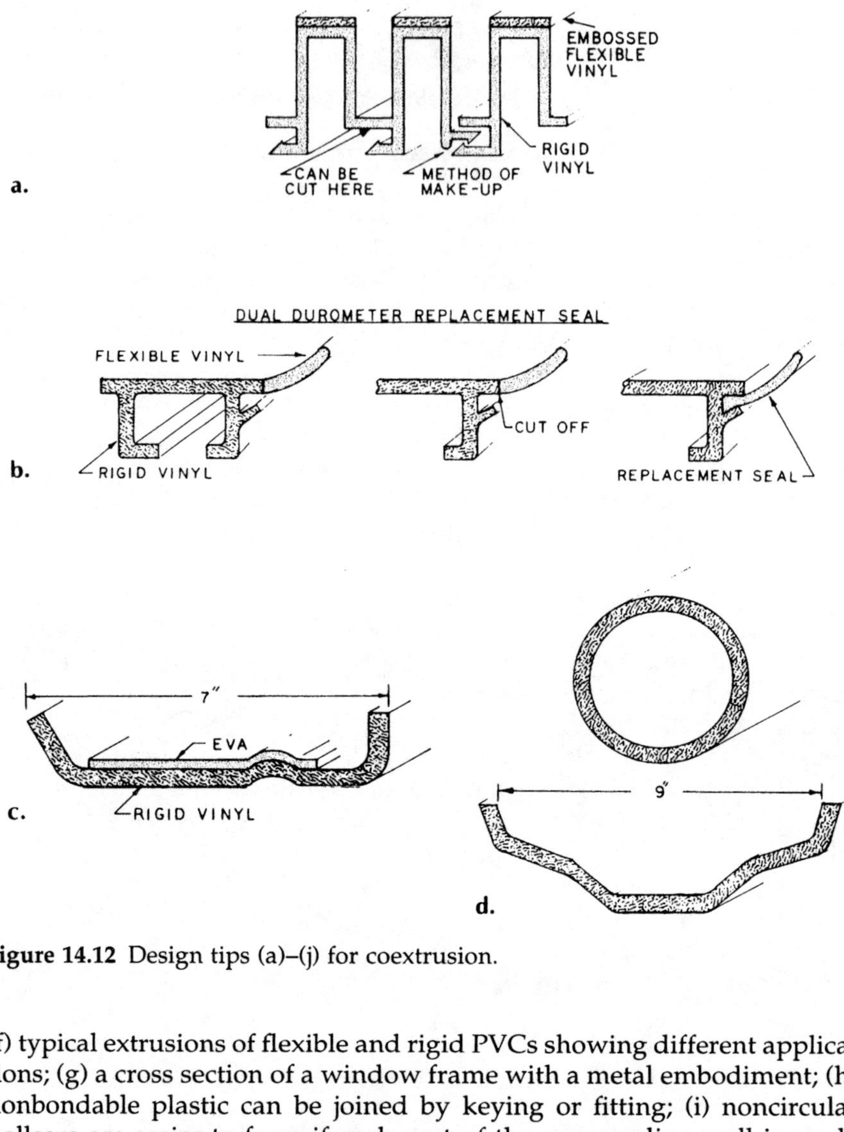


**Figure 14.12** Design tips (a)–(j) for coextrusion.

(f) typical extrusions of flexible and rigid PVCs showing different applications; (g) a cross section of a window frame with a metal embodiment; (h) nonbondable plastic can be joined by keying or fitting; (i) noncircular hollows are easier to form if each part of the surrounding wall is made from the same family of plastic: (A) the rigid PVC base will remain flat and not bulge and (B) the air pressure inside the hollow section will cause the flexible base section to bulge; and (j) different applications for metal-embodiment extrusions.

To aid in the design of dies GE Co., Pittsfield, MA, USA, has developed a computer software program for predicting the profile dimension for engineering plastics. It predicts the die dimensions based on the desired finished profile dimensions. GE's semi-empirical methodology uses both

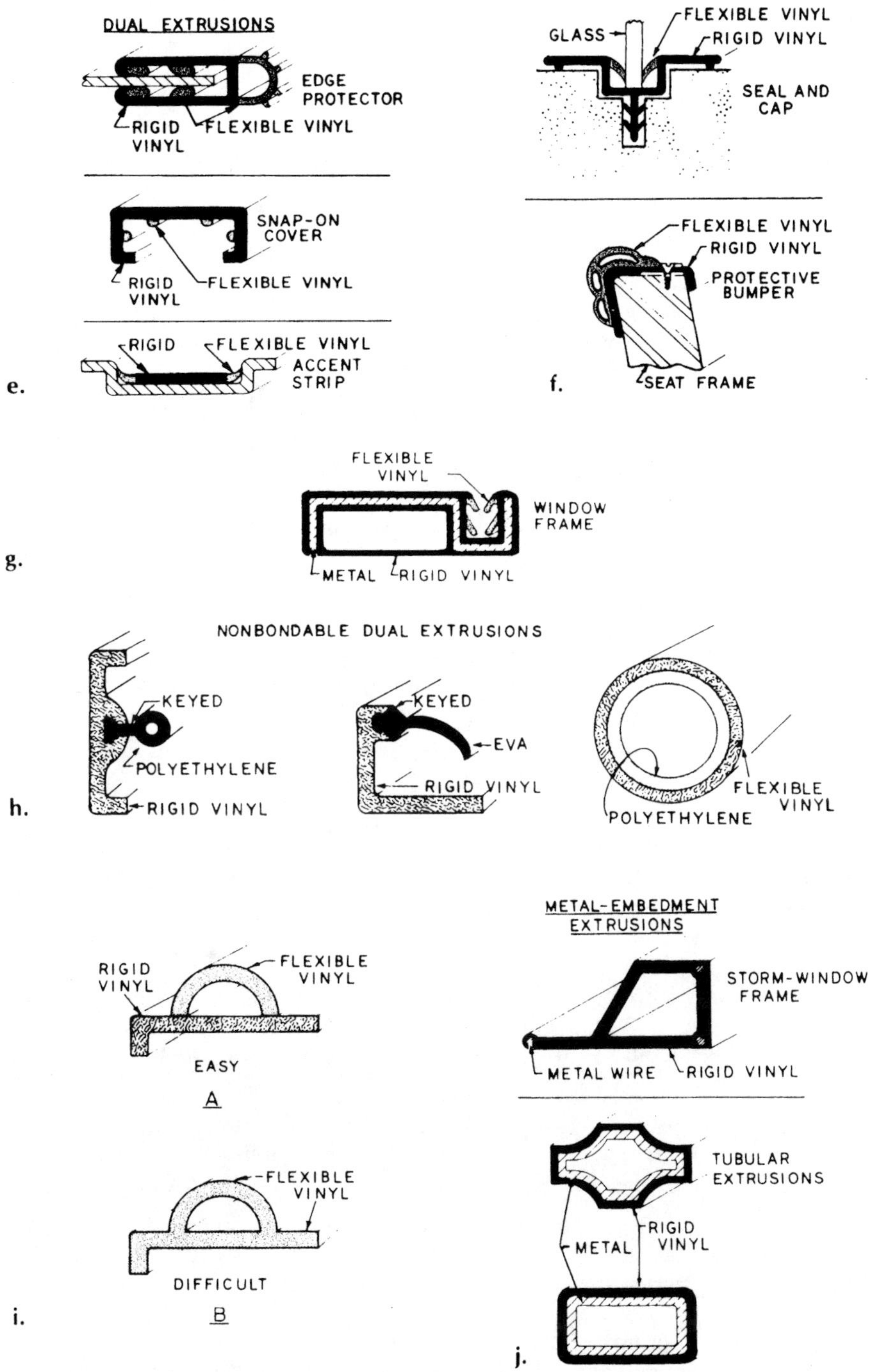


**Figure 14.12** *Continued*

**Table 14.3** Troubleshooting profiles

| *Problem* | *Cause* | *Solution* |
|---|---|---|
| Die lines | Contamination | Clean tooling thoroughly |
| | Tooling | Check for proper assembly |
| | Material quality | QC incoming resin |
| | Processing temperatures | Correct processing window |
| | Regrind | Correct regrind |
| | Screens | Change if dirty |
| | Additives | Compatible with resin |
| Poor impact | Drying | Check set point, beds, residence time |
| | Contamination | Check for dirt, regrind bland |
| | Melt homogeneity | Check processing window, screw rpms |
| | Rapid cooling | Water temperature of cooling tanks |
| | Material quality | Correct resin grade |
| | Processing temperatures | Check recommended temperatures |
| | Regrind | Correct % and grade (processing cycles) |
| | Additives | Check for compatibility |
| Delamination | Contamination | Cross contamination by another grade |
| | Melt homogeneity | Check screw design, correct screw for grade |
| | Processing Temperatures | Check processing log |
| | Regrind | Correct regrind grade |
| | Additives | Compatible with resin |
| Dimples/ craters | Drying | Check set point residence time, beds |
| | Contamination | Check for cross contamination |
| | Water marks | Water circulation in the cooling tanks |
| | Volatiles | Proper operation of vacuum vent |
| | Material quality | QC incoming resin |
| | Additives | Compatible with base resin |
| | Vacuum vent | Open, clean and proper operation vacuum pump |
| Warpage | Tooling design | Improper internal streamlining |
| | Rapid cooling | Water temperature in cooling tanks |
| | Part design | Check intersecting walls for uniformity |
| Pits/grit | Contamination | Check for dirt, cross resin contamination |
| | Material quality | QC incoming resin from supplier |
| | Processing temperatures | Proper processing window for selected resin |
| | Regrind | Correct grade, % usage, etc |
| | Screw efficiency | Check screw geometry and wear |
| | Dirty screens | Change if dirty |

**Table 14.3** *Continued*

| *Problem* | *Cause* | *Solution* |
|---|---|---|
| Gels | Melt homogeneity | Correct screw geometry |
| | Material quality | Check incoming resin |
| | Processing temperatures | Correct processing window |
| | Dirty screens | Check and change regularly |
| | Additives | Compatible with base resin |
| Sticking in sizer | Processing temperatures | Excessive processing temperatures |
| | Vacuum level too high | Lower vacuum level |
| | Excessive draw down | Check calculations, resin supplier and tooling mfg. |
| | Setup | Proper alignment of downstream equipment |
| Dimensional variation | Contamination | Cross contamination by another resin |
| | Non-uniform cooling | Check water temperature and cooling lines in tanks |
| | Processing temperatures | Check processing logs |
| | Vacuum level fluctuation | Check operation of vacuum pump |
| | Surging | Check regrind %, particle size |
| | Puller/haul-off slippage | Change broken belts |

computer analyses and experimental data to address the two critical elements in profile die design, namely die swell and flow balance. The software program is called PEDAP, short for Profile Extrusion Die Design Analysis and Synthesis Program [426].

## TROUBLESHOOTING

This chapter has reviewed many problems with solutions when extruding different types of profiles. Table 14.3 provides a summary review.

# 15

# Blow molding

## OVERVIEW

Blow molding (BM), the third most popular method for processing plastics, consumes about 10 wt% of all plastics worldwide after extrusion and injection molding that are in first and second places, respectively. Blow molding offers the advantage of manufacturing molded parts economically, in unlimited quantities, with virtually no finishing required. It is principally a mass production method. The surfaces of the moldings are smooth and bright, or as grained and engraved as the surfaces of the mold cavity in which they are processed [1, 3, 35, 38, 40, 274].

Blow molding can be divided into three major processing categories: (1) extrusion BM (EBM) with continuous or intermittent melt (called a parison) from an extruder and which principally uses an unsupported parison; (2) injection BM (IBM) with noncontinuous melt (called a preform) from an extruder and which principally uses a preform supported by a metal core pin; and (3) stretched/oriented EBM and IBM to obtain bioriented products providing significantly improved performance-to-cost advantages. Almost 75% of processes are EBM, almost 25% are IBM, and about 1% use other techniques such as dip BM [2]. About 75% of all IBM products are bioriented. These BM processes offer different advantages in producing different types of products based on the plastics to be used, performance requirements, production quantity, and costs [38].

Blow molding requires an understanding of every element of the process, starting with the extruder (Chapter 2). With EBM, compared to IBM, the advantages include lower tooling costs and incorporation of blown handleware, etc. Disadvantages could be controlling parison swell (Fig. 5.20), producing scrap, limited wall thickness control and plastic distribution, etc. Trimming can be accomplished in the mold for certain designed molds, or secondary trimming operations are included in the production lines.

With IBM, the main advantages are that no flash or scrap occurs during processing, it gives the best of all thicknesses and plastic distribution control, critical bottle neck finishes are molded to a high accuracy, and it provides the best surface finish, etc. Disadvantages could include its high tooling costs, only solid handleware, it was somewhat usually limited to small products (however large and complex shaped parts can now be fabricated), etc. (Table 15.1). Similar comparisons exist with biaxial orienting EBM or IBM. With respect to coextrusion (Chapter 2), the two methods also have similar advantages and disadvantages, but generally have advantages over extrusion.

Basically, the BM lines have an extruder with a die or mold to form the parison or preform, respectively. In turn, the hot parison or preform is located in a mold. Air pressure through a device will expand the parison or preform to fit snugly inside their respective mold cavities. Blow molded products are cooled via the water cooling systems within the molds. After cooling, the parts are removed from their respective molds.

Auxiliary equipment used for in-line molded support functions uses equipment applicable to molding processes [2]. Up to the die, BM lines are similar to the lines reviewed in other chapters. Thereafter, the lines include their respective equipment, such as conveyors, trimmers for EBM (if flash is not removed during molding), dimensional and/or weight testers, inspection devices, labeling/decorating equipment, and some

**Table 15.1** Injection versus extrusion blow molding

| *Injection blow molding* | *Extrusion blow molding* |
|---|---|
| Use for smaller parts | Used for larger parts, typically $\geq 237\,cm^3$ (8 fl oz) |
| Best process for GPPS and PP; most resins can be and are used | Best process for polyvinyl chloride; many resins can be used provided adequate melt strength is available |
| Scrap-free: no flash to recycle, no pinchoff scars, no postmold trimming | Much fewer limitations on part proportions, permitting extreme dimensional ratios: long and narrow, flat and wide, double-walled, offset necks, molded-in handles, odd shapes |
| Injection-molded neck provides more accurate neck-finish dimensions and permits special shapes for complicated safety and tamper-evident closures | Low-cost tooling often made of aluminum; ideal for short-run or long-run production |
| Accurate and repeatable part weight and thickness control | Adjustable weight control; ideal for prototyping |
| Excellent surface finish or texture | |

type of collecting equipment at the end of the lines; sometimes bottles are filled and capped on-line.

The nature of these processes requires the supply of clean compressed air to 'blow' the hot melt located within the blow mold. Other gases can be used, such as carbon dioxide, to speed up cooling of the blown melt in the mold. The gas usually requires at least a pressure of 0.21–0.62 MPa (30–90 psi) for EBM and 0.55–1 MPa (80–145 psi) for IBM. Some of the melts may go as high as 2.1 MPa (300 psi). However, stretch EBM or IBM often requires a pressure up to 4 MPa (580 psi). The lower pressures generally create lower internal stresses in the solidified plastics and a more proportional stress distribution; the higher pressures provide faster molding cycles and ensuring conforming to complex shapes. Lower pressures or lower melt stresses goes with improved resistance to all types of strain (tensile, impact, bending, environment, etc.).

## PLASTIC MATERIALS

Originally, nearly all EBM plastics used commodity types and latter the engineering plastics were used (Chapter 3). Typical melt heats used for some of the plastics are given in Table 15.2. The polyolefins (PE and PP) and rigid PVC have proved to be the most suitable materials performance-to-costs. Its heat control and rheology allow PE and PP to be processed relatively easily.

**Table 15.2** Guide to processing temperatures of plastics for blow molding

| *Plastic* | *Temperature (°C)* |
|---|---|
| LDPE | 130–180 |
| MDPE | 150–200 |
| HDPE | 160–220 |
| HMWPE | 180–230 |
| PVC | 190–205 |
| PP | 200–220 |
| PS | 280–300 |
| PA | 240–270 |
| POM | 150–280 |
| SB | 170–210 |
| ASA | 200–230 |
| ABS | 180–230 |
| ABS/PC | 230–250 |
| PPE | 240–250 |
| PBT | 245–260 |
| PBT/PC | 240–260 |
| PUR | 180–190 |

The thermal sensitivity of PVC and the reprocessing of the flash can cause several difficulties if not handled properly; however it is easy to process. In retrospect, PVC in particular which imposed important and high performance requirements on the processing operation, provided momentum to the further development of EBM technology. It also provided the impetus for a thorough engineering analysis of melt flow through the extruder and blow heads.

EBM, although well suited for most plastics, is best with PVC. PVC can degrade rapidly if overheated slightly so controlled care is required when it is being processed. The relatively slow uninterrupted flow of plastic melt in this process reduces the tendency for hot spots to occur, which would damage the plastic.

IBM originally was used to produce specialty small products, such as for the pharmaceutical industry and cosmetic bottles. These type products frequently require small and precise neck finishes; here IBM is more efficient than EBM. The plastics most commonly used are HDPE (a very inert, low cost, forgiving plastics), PP, and PS. The PS receives a degree of orientation which enhances impact strength.

IBM has been used for many decades to fabricate these type of products; it was not used with PVC until the late 1970s. The use of PVC had to await the development of a process where the heat did not cause degradation. Development of machinery was also a factor [3, 38].

New plastics used and improved operational machinery allowed plastics such as PET to grow in importance and expand IBM in new and very large markets. This action occurred originally for the one-liter packaging carbonated beverage bottles with stretched IBM. Although PET usually lacked the required melt strength for EBM, it could be processed when coextruded with other plastics. In the mean time PETG was developed and used with EBM.

Any BM scrap (flash, rejects, etc.) can be recycled. It is vital to granulate the material properly and prevent severe reduction in performance and/or prevent contamination (Chapter 3). The effect of increased use of regrind with virgin plastics, can result in the reduction of melt viscosity, parison swell changes, performance properties of the blown product may be reduced or unacceptable, etc.

HDPE is the dominant plastic used for EBM and PET for IBM. PP and PVC are also major users. LDPE is processed by both techniques, but applications are not as common. UHMWPE is processed by EBM, especially where environmental stress-crack resistance is important. Like PVC, its heat sensitivity suggests continuous rather than noncontinuous EBM.

Nylons are available for EBM and IBM. They are used alone and also as barrier layers in coextrusion. Automotive under-the-hood temperatures for BM products have been used. The under-the-hood environment are gradually reaching temperatures 204°C (400°F) with plastics such as nylon used.

**Table 15.3** Average parison swell for some commonly used plastics

| *Plastics* | *Swell, present* |
|---|---|
| HDPE (Phillips) | 15–40 |
| HDPE (Ziegler) | 25–65 |
| LDPE | 30–65 |
| PVC (rigid) | 30–35 |
| PS | 10–20 |
| PC | 5–10 |

An important factor in EBM is the effective diameter swell of the parison. Ideally, the diameter swell would be directly related to the weight of the parison and would require no further consideration. In practice, the existence of gravity, the finite parison drop time, and the anisotropic aspects (the parison has directional properties) of the BM operation prevent reliable prediction of parison diameter swell directly with the weight.

Parison swell tends to be the most difficult property to control in efforts to produce low cost and lightweight products. One can usually see it actually shrink even after it stretches. If it is shrinking in length, the wall must be thickening, and the parison is heavier per unit length, a behavior known as weight swell (Fig. 5.20). Table 15.3 gives swell action of some common plastics.

### Coextrusion

All the BM methods can process using coextrusion or coinjection melts (Fig. 15.1). As explained in Chapter 2, these multi-layer constructions provide advantages in using the combination of different plastics. As an example, automotive EBM fuel tanks include use of 6-layers to meet new US Clean Air Act setting tighter hydrocarbon emissions standards. The layers include HDPE (MI 5), EVOH, and up to 40% regrind from EBM multi-layer fuel tank scrap.

## PROCESSING CHARACTERISTICS

### Extrusion blow molding

#### *Continuous method*

In EBM, the melted plastic from the extruder through a die head is continuously extruded as a parison (also called a tube) vertically down into

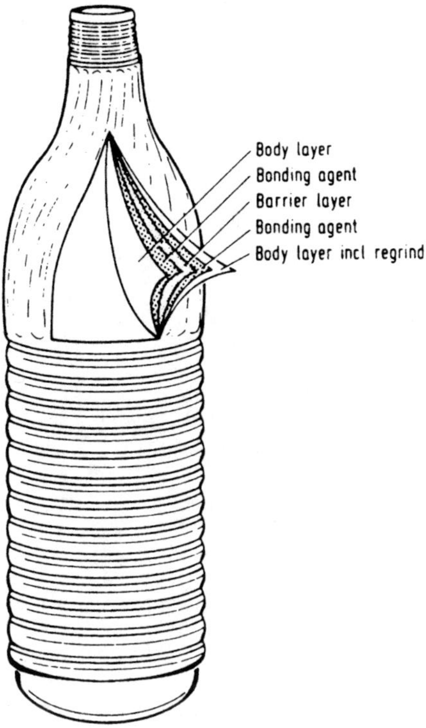


**Figure 15.1** Coextrusion blow molding provides flash-free multiple layers with easy, high speed production; six or more layers can be produced at a time.

air. It is located between the two halves of a mold (Fig. 15.2). The melt flowing through the die can form different cross sections with or without changes in the parison's wall thickness as it exits the die (Fig. 15.3). When the parison has reached its required length, long enough to cover the height of the mold cavity, an open mold closes around the hot parison. A blow pin is inserted through the parison melt, permitting air to enter. Different molds and blow pins (with different locations around the mold cavity) are designed to meet different requirements.

Unlike IBM, when the mold closes flash exists normally only at the top and bottom of the mold cavity. This excess plastic is formed when the parison is pinched by the mold's 'pinchoff' usually at the top and the bottom of the cavity. As an example, with a bottle, the top has its threaded opening with flash around it (Fig. 15.4) simultaneously parison is sealed to contain the blown air. The bottom of the bottle's pinchoff closes the other end of the parison to be blown with flash attached. Molds can be designed where automatically all the flash is removed or the line will have

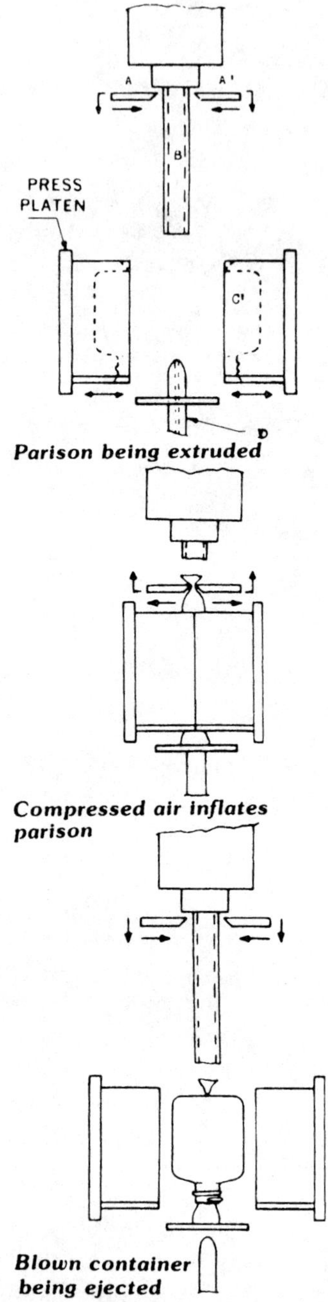


**Figure 15.2** Basic continuous EBM process: A = parison cutter; B = parison; C = blow mold cavity; D = blow pin.

**Figure 15.3** Truck fascia extrusion blow molded PP.

a secondary operation to remove the flash after the cooled part leaves the mold.

In the EBM machine, a die can have one or more parisons exiting (Fig. 15.4). This multi-parison approach uses a mold with the number of cavities equal to the number of parisons. This multiple approach increases production provided the extruder output capacity is adequate [202].

With this continuous EBM process, the closed mold with the parison is moved downward from the continuing dropping parison. This rising method has the parison continuously extruded. When the parison reaches the proper length, the open mold located around the parison quickly closes pinching the parison, and quickly returns to its lower position (there are also machines where it positions itself sideways to its blow station) so that the parison continues to extrude with no interruption. After the part is blown and cooled, the mold opens, the part removed, and the process repeats.

In addition to the rising method, there are other modes of operations to increase production. Two other popular modes of operation are the rotary

**Figure 15.4** Multiple continuous extrusion die head (three parisons) BM three containers simultaneously in a shuttle clamping system.

wheel and shuttle modes. The rotary wheel method uses at least 2–20 clamping stations with molds. These stations are mounted to a vertical or horizontal wheel. One approach is where the die with its parison moves around in the path of the molds. A mold is opened while the parison is moving through it. The mold closes pinching the parison and starts its cycle of blow, cool, and eject by opening the mold. In the meantime, the next mold is opened and the parison is pinched, etc. This system is timed so that when the parison drop returns to the 'first mold', which is an open mold, and the rotary system continues. The other approach is having the molds move with the parison remaining in a fixed location.

The third mode is the shuttle method where usually two or more sets of molds are used. Each set of molds can have two or more molds. Their blowing stations are around the periphery of the extruder die head and parisons. One set of molds in the open position is located under the die. With proper length of the parisons (a parison for each mold), the open molds underneath close. After the molds are closed, parisons are cut usually with an electrically charged hot wire, and quickly shuttle to its blow station where blow pins are inserted into the parison openings. BM parts solidify and are released from the molds when they open. In the meantime, the parisons continue to be extruded as another set of open molds are positioned around these parisons. Thus, the molds alternately shuttle producing molded parts.

Another way to increase production is to use one extra-long parison to cover two cavities located vertically in the mold. In fact, one parison can extend the multi-parison with two or more vertical cavities. All that is required is a machine with the capacity to handle the output from the extruder to the clamping capability.

*Intermittent method*

With an accumulator located above the die, the flow of the parison through the die is cyclic, permitting intermittent or discontinuous EBM (Figs. 15.5–15.7). These systems can fall into three modes. The most common system is with an accumulator head and is used to mold small to

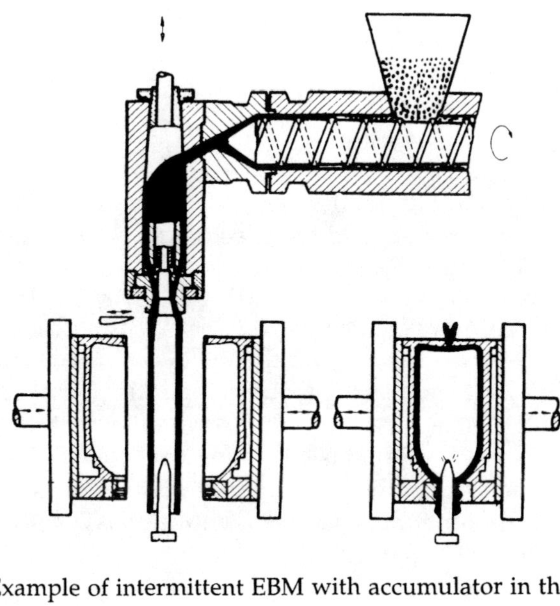

**Figure 15.5** Example of intermittent EBM with accumulator in the die.

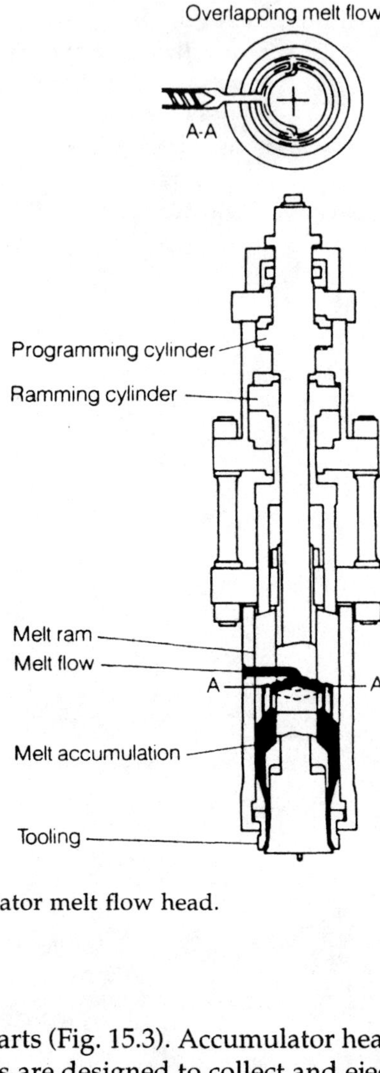


**Figure 15.6** Accumulator melt flow head.

particularly large parts (Fig. 15.3). Accumulator heads attached to the exit end of the extruders are designed to collect and eject a measured amount of plastics (Figs. 15.5 and 15.6).

A reciprocating screw unit can be used. It is a take-off from the single stage injection molding machine (IMM) (Chapter 18). Plastic is conveyed and melted by the screw turning. As the melt accumulates in the front of the screw in the barrel and has the required quantity (shot size), the screw stops turning and pushes forward (ram) forcing the melt through a die to form a parison. Basically all that is needed is an IMM having the required shot size with a die to form the parison rather that the usual IM mold [2].

The ram type machine incorporates a continuous rotating screw that delivers melt into a chamber (Chapter 18). A ram in the chamber then

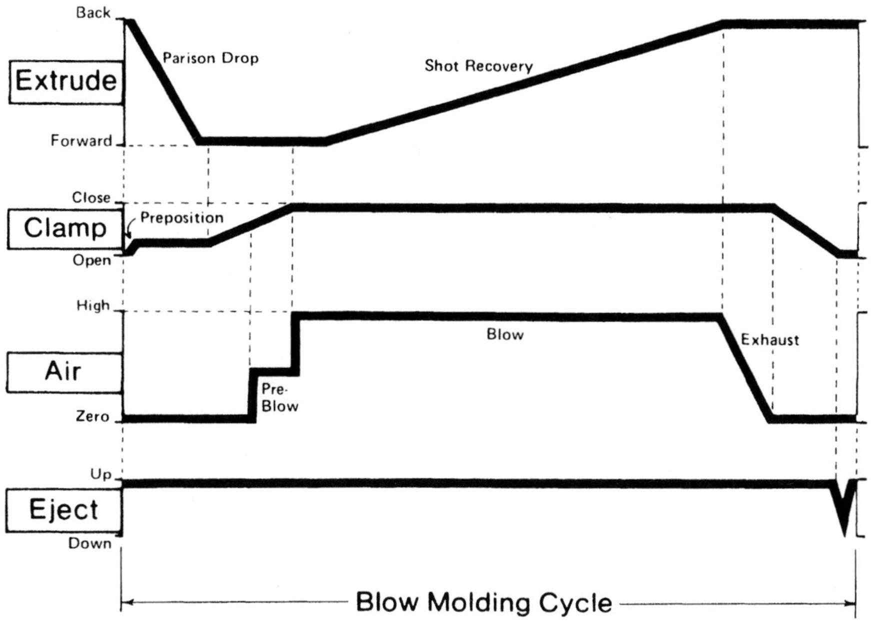


**Figure 15.7** BM using an accumulator head.

forces the hot melt from the chamber through the parison forming die. This system uses a two-stage IMM [2].

*Air pressure*

The air used for blowing serves to expand the parison tube against the walls of the female mold cavity. It is usually required to enter the parison at very low pressure during extrusion of the parison to eliminate its collapse. When the mold closes, full air pressure is applied (Table 15.4), forcing the hot melt to assume the shape of the mold and forcing it into the surface details such as raised letters and surface designs. The air performs the three functions of expanding the parison, force the melt into corners, etc., of the cavity, and aids in cooling the hot melt.

During the expansion blowing phase, it is desirable to use the largest available volume of air, so the parison expands against the walls in a more uniform and/or the shortest possible time. A maximum volumetric flow rate at a low linear velocity can be achieved by making the air inlet orifice as large as possible.

A blow pin is usually located opposite the pinched closing end of the parison. It is not long enough to blow directly on the parison which would result in freeze-off and stresses at that point of contact. However, the pin

**Table 15.4** Guide for air blowing pressure

| *Plastic* | *Pressure (psi)* |
|---|---|
| Acetal | 100–150 |
| PMMA | 50–80 |
| PC | 70–150 |
| LDPE | 20–60 |
| HDPE | 60–100 |
| PP | 75–100 |
| PS | 40–100 |
| PVC (rigid) | 75–100 |
| ABS | 50–150 |

can be located in any position and usually around the mold's parting line. Air can enter through the extrusion die head (as with pipe lines, Chapter 13) and through a blow pin over which the end of the parison has dropped. The blow pin can be located at the bottom of the mold (Fig. 15.2). Air can enter through blow pins or needles that pierce the parison. It is possible to avoid the blow pin mark when using EBM by employing hypodermic needles and pulling them out before the plastic solidifies (this has been done for over a century with Christmas ball decorations, etc.).

Small orifices may create a venturi effect, producing a partial vacuum in the tube and causing it to collapse. For certain plastics, if the inner velocity of the incoming blown air is too high, its force can actually draw the parison away from the extrusion head end of the mold, producing an unblown parison. The air velocity must be carefully regulated by control valves placed as close as possible to the blow tube. Too high a blow pressure will often 'blow out' the parison. Too little pressure will result in at least a lack of adequate surface details. The optimum blowing pressure is generally determined by trial and error on the BM machine and/or experience.

General guidelines for determining the optimum diameter of the air entrance to the orifice during blowing are: (1) up to 1 quart (0.95 $dm^{-3}$) use 0.06 in (1.5 mm); (2) for 1 quart to 1 gallon (0.95–3.8 $dm^{-3}$) use 0.25 in (6.4 mm); and (3) for 1–54 gallons (3.8–205 $dm^{-3}$) use 0.5 in (12.7 mm).

The blowing time differs from the cooling time, being much shorter thasn the time required to cool the thickest section to prevent distortion on ejection. A guide to the blow time of a product may be obtained by using Table 15.5 and the following equation.

**Table 15.5** Discharge of air at 14.7 psi (101 kPa) and 21°C (70°F)

| | *Discharge of air ($ft^3 s^{-1}$) for specified orifice diameter* | | | |
|---|---|---|---|---|
| *Gauge pressure (psi)* | *1/16 in (1.6 mm)* | *1/8 in (3.2 mm)* | *1/4 in (6.4 mm)* | *1/2 in (12.7 mm)* |
| 5 | 0.993 | 3.97 | 15.9 | 73.5 |
| 15 | 1.68 | 6.72 | 26.9 | 107 |
| 30 | 2.53 | 10.1 | 40.4 | 162 |
| 40 | 3.10 | 12.4 | 49.6 | 198 |
| 50 | 3.66 | 14.7 | 58.8 | 235 |
| 80 | 5.36 | 21.4 | 85.6 | 342 |
| 100 | 6.49 | 26.8 | 107.4 | 429 |

$$\text{Blow time, } s = \left(\text{Mold volume, m}^3/\text{m}^3\text{s}^{-1}\right)$$
$$\left(\text{Final mold pressure, kPa} - 101\,\text{kPa}/101\,\text{kPa}\right)$$

This is free air; but there will be a pressure buildup as the parison is inflated, so the blow rate has to adjusted. The value of $m^3 s^{-1}$ is obtained from Table 15.5, according to the line pressure and the orifice diameter. The final mold pressure is assumed to be the line pressure for purposes of calculation. Actually the blow air is heated by the mold, raising its pressure. Calculations ignoring this heat effect will be satisfactory when blow times are under 1 s, the air will have time to pick up heat, causing a more rapid pressure buildup and blow times shorter than calculated.

*Cooling*

As much as 80% of the blow molding cycle is cooling time. Several methods are used to reduce cycle time. A part is normally cooled externally by the moving water liquid within the mold/next to the mold cavity based on thermodynamic studies [2]. This forces heat to travel through the entire wall thickness as is done in injection molding.

There are systems using air chillers that reduce the temperature of the blown air to about −70°C (−95°F) and blow pins that permit heated air in the blown part to exit. This means that a continuous flow of cool fresh air enters the part as it is being cooled. With such a system, the output of molded parts can increase by 10–30%.

Liquefied gas systems, such as liquid carbon dioxide ($CO_2$) or nitrogen ($N_2$), can be used. Immediately after the initial air blowing action, the gas is atomized through a nozzle in the blow pin into the interior of the blown part. The liquid quickly vaporizes. This precise control action, like the chilled air, continually pushes fresh gas in and heated gas out. The cost of this system requires high production but it provides an increase of 25–35% in production. Other systems, such as supercold air, are used.

Methods to speed up cooling used also include postcooling of blow molded parts that can shorten the blow molding cycle. Shuttle machines, which maximize production in continuous EBM, are preferred. They can produce finished containers in the machine. Trimming cannot proceed until the scrap areas where usually the thickest walls of the part have been cooled sufficiently, so the cycle depends on getting parts cool enough to trim.

Plastics vary in cooling requirements. As an example, it is not usually necessary to postcool PVC; it gives up its heat much more readily than the polyolefins (making it more appropriate for a dedicated operation than for a custom blow molder). Also the bigger the part, the more cost-effective its cooling.

### *Clamping*

The mold clamping methods are usually hydraulic and/or toggle, similar to, but less sophisticated than, those used with IMMs [2] since BM molds are not subjected to high internal pressures. Clamping system vary depending on machine operation (Fig. 15.8), part configuration, and the location of the parting line.

Size platens and sufficient daylight (maximum space between platens when opened) are needed to handle the size of the molds with its movements and maximum opening capacity to remove blown parts, accommo-

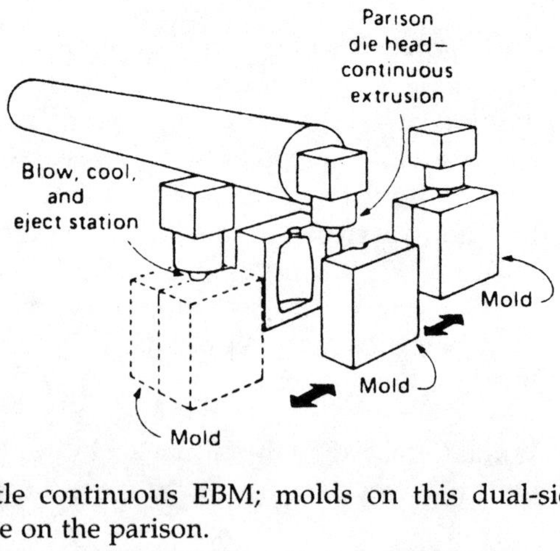


**Figure 15.8** Shuttle continuous EBM; molds on this dual-sided system move alternately to close on the parison.

date the parison systems, ejection systems, possible unscrewing or insertion equipment, and/or other special equipment.

Controls are used to operate the clamps. Examples include: accurate timing and speed in opening and closing; if required using a delay closure action to aide pinchoff weld formation; flash removal for EBM, and so on (Chapter 6).

*Shrinkage*

The shrinkage behavior of different plastics and the part of geometry must be considered. Shrinkage is generally the difference between the dimensions of the mold at room temperature of about 22°C (72°F) and the dimensions of the cold blown part, usually checked 24 h after manufacture. The elapsed time is necessary to allow the part to shrink. Trial and error and/or experience determines how much time is required to ensure complete shrinkage.

Differences exist between the amorphous and crystalline plastics (Chapter 3). The crystalline plastics have greater shrinkage in the longitudinal than the transverse directions, whereas the amorphous plastics can balance themselves. Certain plastics, such as PEs, have higher shrinkage with higher densities and thicker walls.

Shrinkage of the blown part depends on many factors, such as the plastic density, melt heat, mold heat, part thickness, rate of cooling, part geometry, and pressure of blown air. A guide to typical PE shrinkages is as follows: LDPE at a thickness up to 0.075 in (1.9 mm) has a tolerance of 0.010–0.15 in, and at a thickness over 0.075 in (1.9 mm) has a tolerance of 0.015–0.030 in; whereas HDPE at a thickness up to 0.075 in (1.9 mm) has a tolerance of 0.20–0.035 in, and at a thickness over 0.075 in (1.9 mm) has a tolerance of 0.035–0.055 in. Once the operating conditions are established, tolerances of 5% are easy to attain with tighter tolerances achievable. When fillers are used in the plastic compounds, it is a different 'ball game'; they have less shrinkage. Other gains can be lower material costs.

## Injection blow molding

IBM has basically three stages as shown in Figs. 15.9–15.11. The first stage injects hot melt through the nozzle of an injection molding machine [IMM which is a noncontinuous extruder (Chapter 18)] into a mold with one or many more cavities to produce the preform(s). There is usually more than one cavity. An exact amount of plastic enters each cavity. These molds are designed as in regular IMM [2] to meet the required BM melt temperatures and pressures.

After injection of the melt into the mold cavity(ies), the two-part mold opens and the core pin(s) carry (counterclockwise in Fig. 15.10) the hot

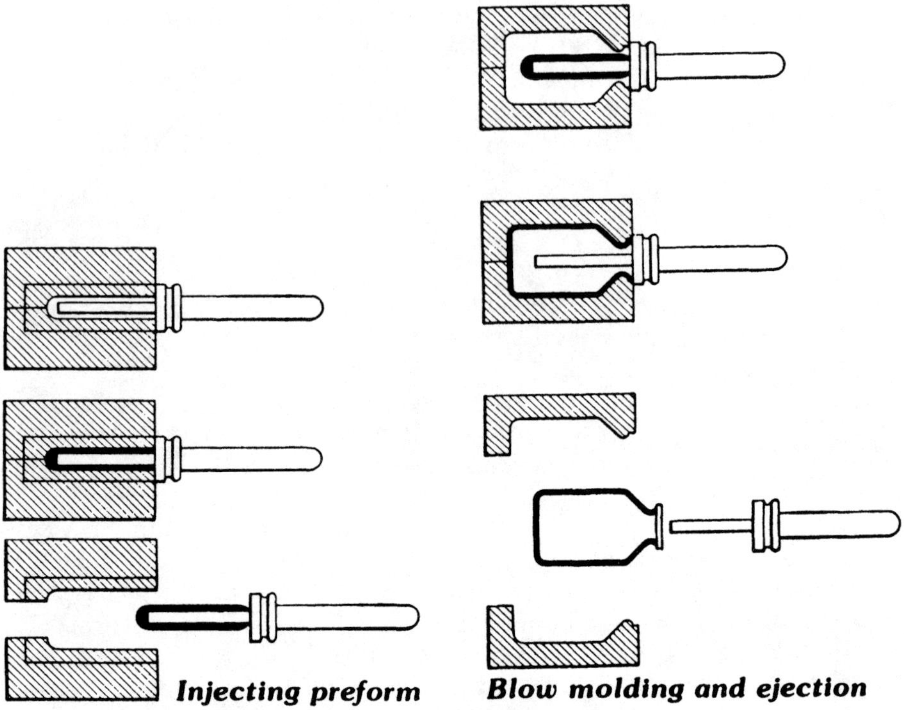


**Figure 15.9** Basic injection blow molding process.

plastic preforms to the second stage for blow molding. Upon the mold closing in this second stage, air is introduced via the core pins. Controlled chill water, usually 4–10°C (40–50°F) circulates through predesigned mold channels around the mold cavities and solidifies the blown parts [38].

This two-part mold that did the blowing opens when the part(s) solidify. In turn, the core pins carry the blown parts to the third stage. In that stage the parts are ejected. Ejection can be done by using stripper plates (Fig. 15.10), air blowing, combination of stripper plate and air, robots, and others.

IBM can have three or more stations (stages). A station can be located between the preform stage and the blowing stage to provide extra heat-conditioning time for the preform(s). Between the blow and ejection, a station can be used to apply decals, decoration, testing dimensions, etc. After ejection, a station can be used to add an insert for decoration, reinforcement, etc.

The process parameters that determine the quality of the blown parts are the screw melting capability, injection pressure, holding (packing)

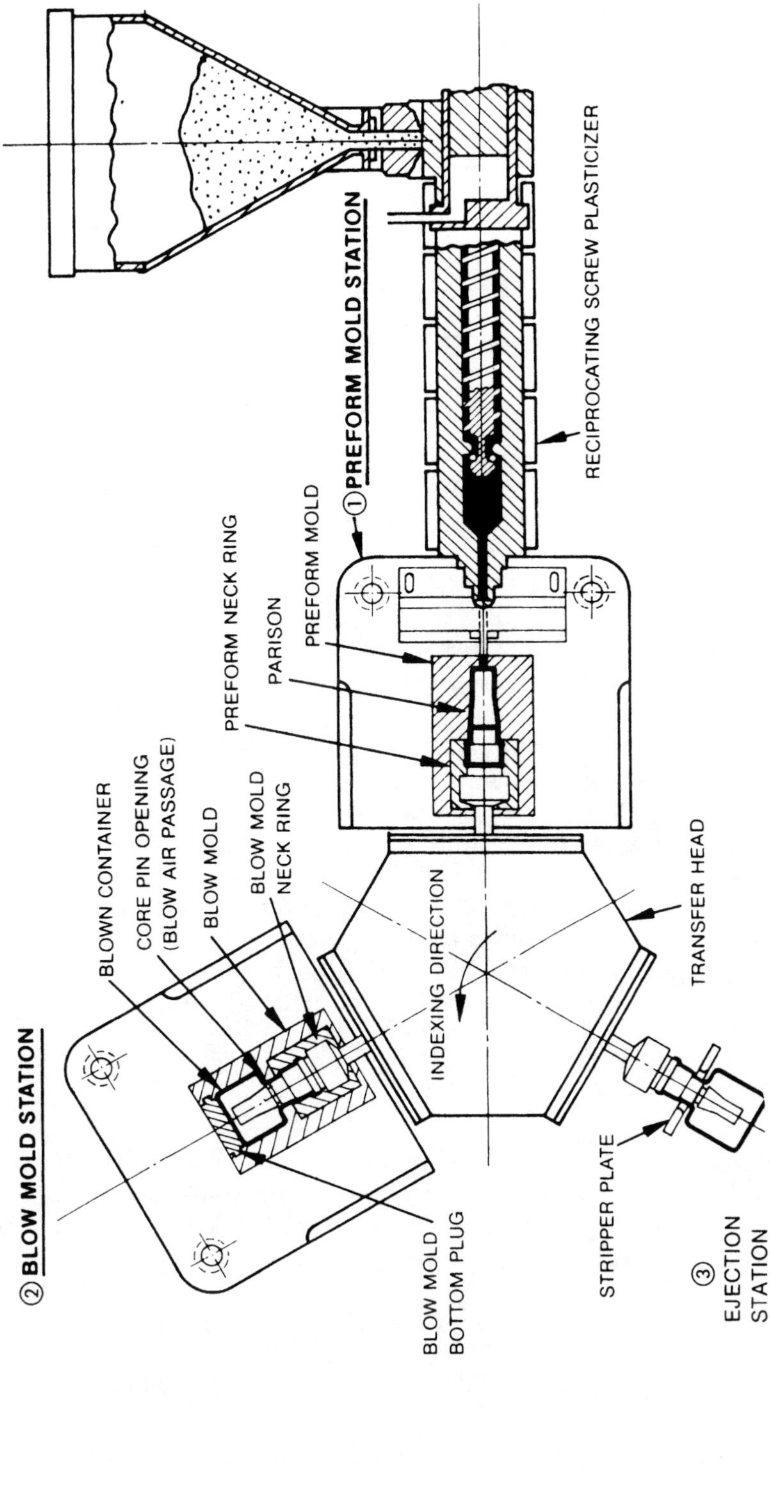


**Figure 15.10** Three-station IBM.

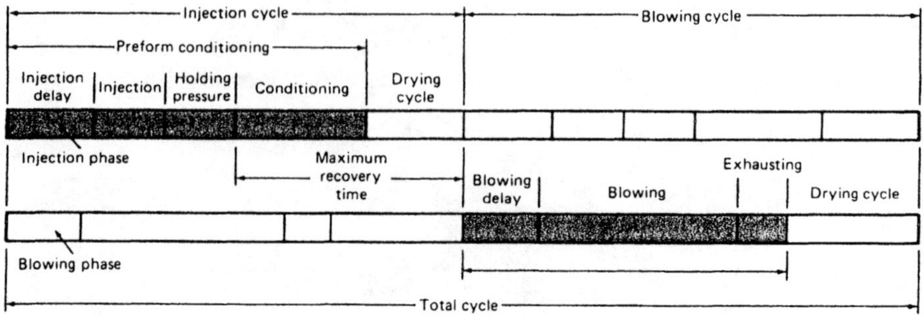


**Figure 15.11** IBM complete cycle begins with injection molding of the preform followed by the blowing cycle.

pressure when melt is in the cavities, heat control of the preforms in all the stages, and cooling rate of preforms.

This process permits the use of plastics that are unsuitable for EBM (unless modified), specifically those with no controllable melt strength, such as the conventional PET, which is predominantly used in large quantities using the stretch IBM method for carbonated beverage bottle (liter and other sizes). The information on blowing parisons, cooling, clamping, and shrinkage that was presented for EBM is also similar for IBM.

Several different methods of IBM are available, each with different means of transporting the core rods from one station to another. These methods include shuttle, multi-parison rotary, etc. These blow molded products have precise dimensions. This action occurs since the initial preforming cavities were designed to have the exact dimensions required after blowing the plastic melt as well as shrinkage that may occur. Another advantage is that no flash or scrap exists. Neck finishes, internally and externally, can be molded with an accuracy of at least 0.10 mm (4 mil). It also offers precise weight control in the finished product accurate to 0.1 g [38].

## Stretch blow molding

High-speed EBM and IBM take the extra step in stretching or orienting. As an example, orientation in a bottle is made almost simultaneously in both the longitudinal and hoop directions. Figure 15.12 shows a schematic for stretched IBM; with EBM the stretching action is basically similar. With EBM, the parison can be mechanically gripped at both ends of the hot tube in the mold, stretched, and blown (it occurs during the 'compressed air inflation').

This process definitely advanced IBM from its past unimportant posi-

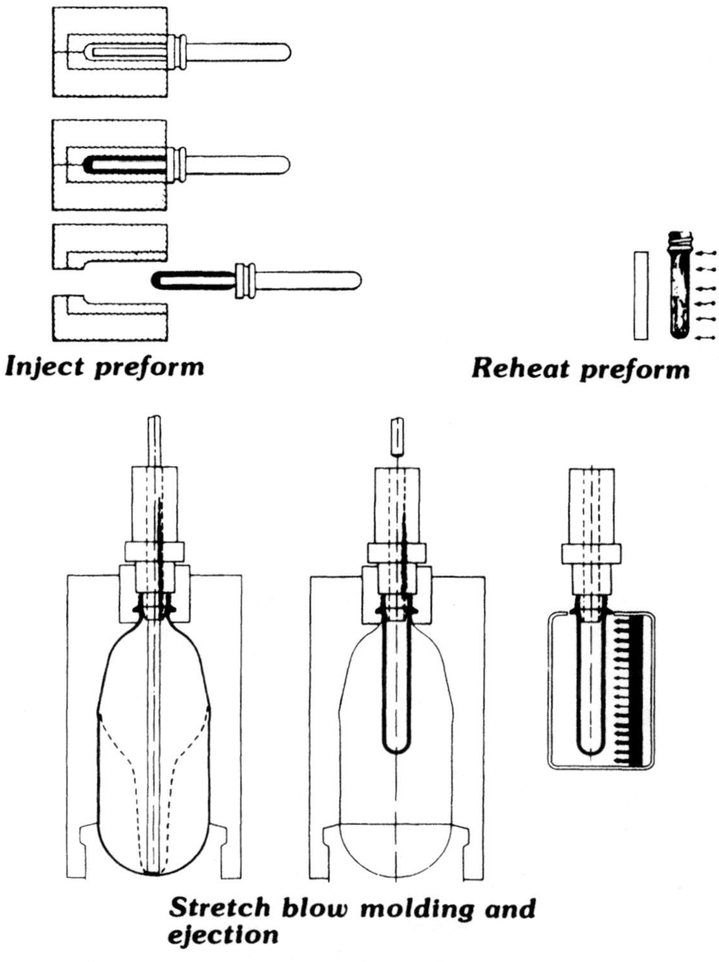


**Figure 15.12** Stretch IBM using an internal (longitudinal) expanding rod.

tion. Immediately, when commercially developed and accepted by the market just a few decades ago, the stretch BM take-off with most of the action with IBM. Prior to that time, the stretching process was about to take off but since AN was used, it unfortunately (when it should not have occurred) became a 'dead' issue [2, 4].

By biaxially stretching the extrudate before it is chilled, significant improvements can occur with savings in heat energy. Chapter 2 provides information on the processing and performances gained with orientation. This technique allows the use of lower grade plastics or thinner walls with no decrease in strength, both approaches reduce plastic material costs.

Stretched BM gives many plastics improved physical and barrier properties (Tables 15.6 and 15.7). The process allows wall thicknesses to be more accurately controlled and also allows weights to be reduced.

Draw ratios used to achieve the best properties in PET bottles (typical 2- to 3-liter carbonated beverage bottles) are about 3.8 in the hoop direction and 2.8 in the axial (longitudinal) direction. These ratios will yield a bottle with a hoop tensile strength of about 200 MPa (29 000 psi) and an axial tensile strength of 104 MPa (15 000 psi).

Stretch blow is extensively used with PET, PVC, ABS, PS, AN, PP, and acetal, although most TPs can be used. The amorphous types, with a wide range of thermoplasticity, are easier to process than the crystalline types such as PP (Chapter 3). If PP crystallizes too rapidly, the product is virtually destroyed during the stretching. Clarified grades of PP have virtually zero crystallinity and overcome this problem.

The stretching process takes advantage of the crystallization behavior

**Table 15.6** Volume shrinkage of stretch BM bottles

| *Type of bottle* | *Percent* |
|---|---|
| Extrusion blow molded PVC | – |
| Impact-modified PVC (high orientation) | 4.2 |
| Impact-modified PVC (medium orientation) | 2.4 |
| Impact-modified PVC (low orientation) | 1.6 |
| Nonimpact-modified PVC (high orientation) | 1.9 |
| Nonimpact-modified PVC (medium orientation) | 1.2 |
| Nonimpact-modified PVC (low orientation) | 0.9 |
| PET | 1.2 |

Seven days at 80 °F (27 °C).

**Table 15.7** Gas barrier transmission comparisons for a 24 fl. oz (689 $cm^2$) container weighing 40 g

| | *Rate ($m^2$ $day^{-1}$)* | |
|---|---|---|
| *Type of bottle* | *Oxygen ($cm^3$)* | *Water vapor (g)* |
| PET (oriented) | 10.2 | 1.10 |
| Extrusion blow molded PVC | 16.4 | 2.01 |
| Stretch blow molded PVC (impact-modified) | 11.9 | 1.8 |
| Stretch blow molded PVC (nonimpact-modified) | 8.8 | 1.3 |

At 38 °C (100 °F).

**Figure 15.13** Easy to operate and control in-line stretch injection blow molding machine by Cincinnati Milacron.

of the plastics and requires the preform or parison to be temperature-conditioned then rapidly stretched and cooled into the product shape.

There are in-line and two-stage processes. In-line processing is done on a single machine (Fig. 15.13), whereas two-stage requires two machines with one injection molding the preform or an extruder producing the tube/parison. The second machine takes the preforms or tubes, reheated and blown.

In the beginning, most lines used the two-stage since the plastic's temperature processing conditions were not that stable for the in-line. Now, more are in-line with easy-to-use plastics, machine improvements, and so on. The in-lines are more economical in the production of stretched blown products.

With either type of process, a specific heat profile is required on the

**Table 15.8** Stretch BM processing characteristics

| Plastic | *Melt temperature* °C | °F | *Stretch orientation temperature* °C | °F | *Maximum stretch ratio* |
|---|---|---|---|---|---|
| PET | 250 | 490 | 88–116 | 190–240 | 16 |
| PVC | 199 | 390 | 99–116 | 210–240 | 7 |
| PAN | 210 | 410 | 104–127 | 220–260 | 9 |
| PP | 168 | 334 | 121–136 | 250–280 | 6 |

preform or parison tube. With the in-line system, the hot, firm plastic passes through conditioning stations that bring it down from the melt heat to the proper orientation temperature (Table 15.8). A rather tight heat profile is maintained in the axial direction which is required for the based wall thickness and amount of stretching. Advantages of this approach are that the heat history is minimized (crucial for heat sensitive plastics), the preform or parison can be programmed for optimum plastic distribution, etc.

With the two-stage process, cooled preforms or parisons are conveyed through an oven (usually using quartz lamps) that reheats them to the proper orientation heat profile. The last step is the stretching action. This two-stage provides a means for molding preforms for storage. When parts are needed, they go into the second-stage machine.

## PROCESS OPTIMIZATION

As shown in Fig. 15.14, there are examples of how machine and plastic variables influence each other that include melt behaviors. Melt properties are of critical importance to BM, particularly EBM. It may be said that this is more so than for conventional extrusion (it depends on who is in the discussion). Melt viscosity determines whether sagging or lengthening of a parison can be minimized and/or controlled, particularly in noncircular parisons (Fig. 15.15) [3, 100, 206, 370].

Because engineering plastics have so far been used mainly with injection molding (IBM), most processors attempt to use easy flowing, low molecular weight IM-grade plastics (Chapter 3). But in BM, particularly EBM, the objective is very different. The melt should be viscous and of high molecular weight (high melt strength). This requirement also generally insures another important feature of better impact strength. The melt viscosity should be nearly independent of the shear rate and the processing heat.

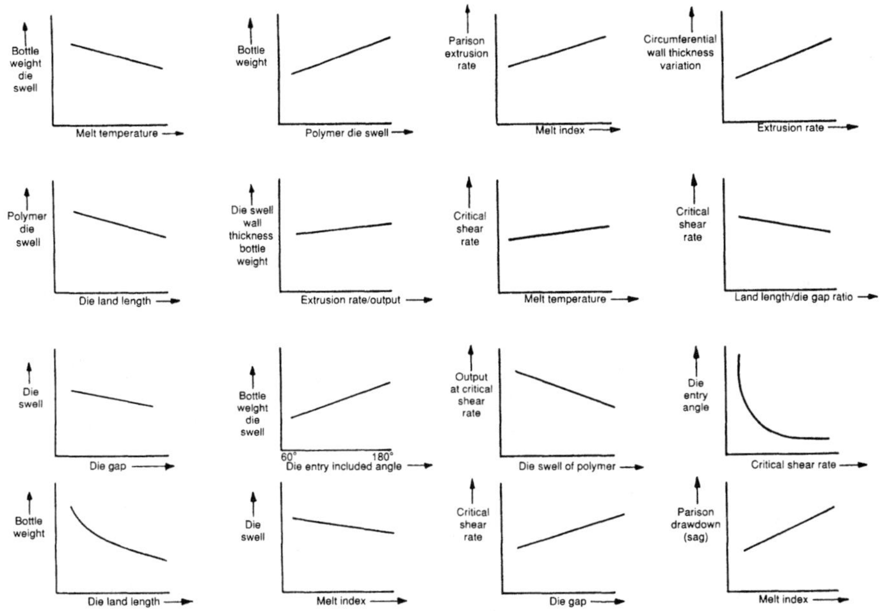


**Figure 15.14** Effect of machine and material variables with blow molding.

For EBM, the parison thickness control is very important to processing and reducing the amount of plastics consumed. The control and monitoring functions range from extremely simple ones to expensive, but very useful, complete microprocessor systems. Some machines use electric relays that permit a certain degree of control. However, to produce good quality parts with the least plastic resulting in lower product costs, the more sophisticated are required.

The most common method is orifice modulation (Fig. 15.16). The die is fitted with a hydraulic positioner that allows positioning of the inside die diameter during the parison drop. The OD and ID relationship of the tapered die orifice opening is varied in a programmed, repeatable manner to increase or decrease the parison wall thickness. The programmer uses a closed-loop servosystem supplying proper signals to control the amount, direction, and velocity of the movement of the hydraulic positioner. Programmers are told the number of program points required; they can be from 5 to 100. Consider a blown shape, such as the Dawn soap bottle, with a wide base and very narrow center. When not controlling parison thickness, in order to provide enough thickness on the edges of the bottom corners, the center section will have over four times the thickness required with lots of useless plastics. With parison thickness control, you obtain the thickness where you want it.

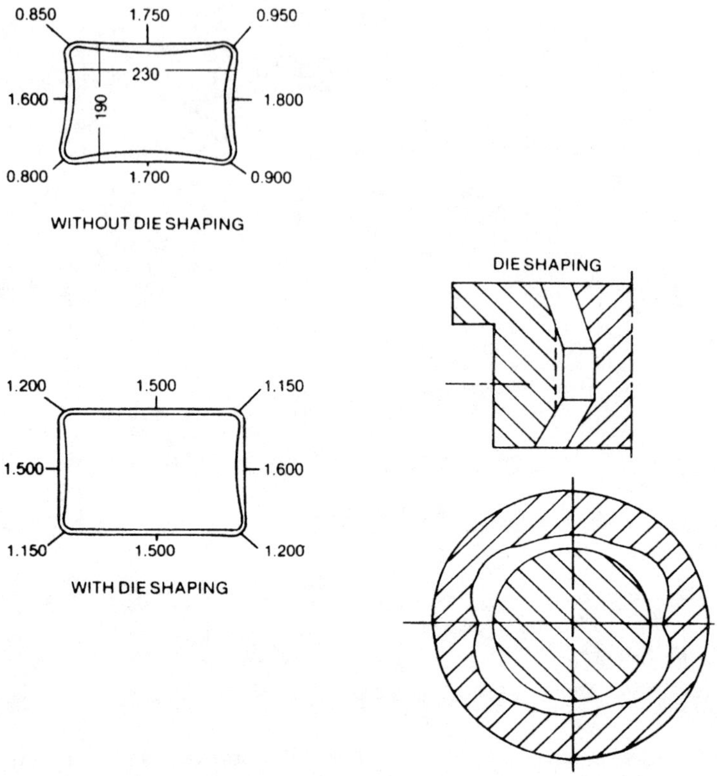


**Figure 15.15** Noncircular BM die with and without wall thickness die shape (dimensions in mm).

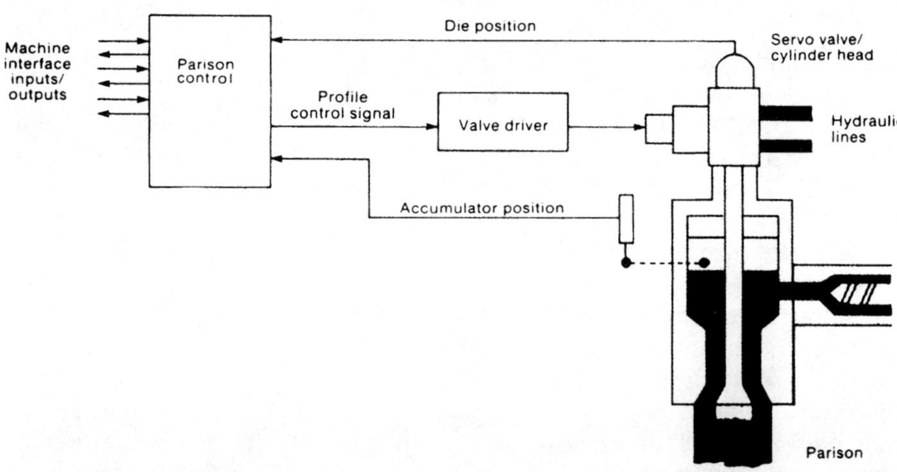


**Figure 15.16** Accumulator head with programmable process control for rate of forming parison and its wall thicknesses.

With a large or long parison, the wall thickness will vary as the weight of the plastic increases and it sags. Parison control can be helpful, such as a method to increase melt pressure in the die, either by regulation of the extruders back pressure or possibly by pressure variations via the ram when an accumulator is used. In addition to this longitudinal control, there are also circumferential distribution controllers.

Different types of microprocessor-based modules control BM machines and melt parameters, ranging from single to multiple functions. The modules interact at high speeds, coordinating process variables, such as heats, timings, parison or preform molding speed, melt wall thickness, and air pressure.

Control technology is used to improve machine production cycle rates, as in employing proportional hydraulics to safely speed up mold movements. In addition, production monitoring systems have become part of some BM plants, helping managers make effective decisions. These improvements in monitoring and controlling have contributed significantly to the manufacture of products with zero defects and to profits.

## DIE/MOLD/TOOL

The terms dies, molds and tools are interchangeable with dies being more descriptive for an extruder. A die, as used with EBM, takes the melt from the horizontal extruder and changes its direction to have the melt exit the die vertically downward. The die can be designed to permit a change in the thickness of the exiting hot melt. As shown in Table 15.9, different die designs are used to meet different processing requirements.

Figures 15.17 and 15.18 represent the continuous EBM dies. As it shows, the hot melt leaves the extruder and through the die with no interruptions. The result is a continuous moving parison, as already reviewed.

Figures 15.5 and 15.6 represent the intermittent EBM dies. The connecting channels between the extruder and accumulator, as well as the accumulator itself, are designed to prevent melt flow restrictions that might impede flow or cause the melt to hang up. Flow paths should have low resistance to melt flow to avoid placing an unnecessary load on the melt.

To ensure that the least heat history or residence time (Chapter 3) is developed during processing, the design of the accumulator ensures the first melt into the accumulator is the first to go out when its 'ram' literally empties the accumulator chamber. The target is to have the accumulator totally emptied on each stroke. Plastics that are not heat sensitive permit some relaxation in their heat history during this action.

Molds with female cavities only, are made for all the types of BM ranging from simple to complex shapes (Figs. 15.19 and 15.20). The terms molds, dies and tools are interchangeable and can be used but molds are more descriptive with the BM part shape.

**Table 15.9** Examples of different performing EBM dies

| *Die type* | *Feature* | *Advantage/disadvantage* |
|---|---|---|
| Simple die | Fixed die gap | Simple; inexpensive; no adjustment facility |
| Die profiling | Premanently profiled; preferred in die land area | Fixed circumferential wall-thickness change; time-consuming; complex |
| Die centering | Can be permanently shifted laterally to correct parison drop path | Compromise between required drop path and equal wall thickness |
| Open-loop axial die-gap control | Can be axially shifted during extrusion | Equal circumferential wall-thickness change possible; no feedback |
| Servohydraulic closed-loop axial die-gap control | As above, with greater speed, accuracy, and flexibility | Equal circumferential wall-thickness change possible, with feedback |
| Stroke-dependent die profiling | Permanently ovalized die gap | Fixed, unequal circumferential wall-thickness change possible affects entire parison length |
| Die/mandrel adjustable profiling | Settable adjustment of die-gap profile | Settable, unequal circumferential wall-thickness change possible; rapid optimization |
| Servohydraulic closed-loop radial die-gap control | Programmable ovalization and shifting of die gap | Programmable circumferential wall-thickness change possible, independent of parison length |

With commodity plastics, a sandblasted cavity surface can be used to aid in air venting (between the parison and cavity wall) and also to provide a smooth surface on the blown part; a characteristic of most melts generally prevents penetration of the 'rough' surface. With engineering plastics, the surface of the cavity is generally reproduced precisely, so sandblasting does not aid venting. When venting is required, vents are located on the mold's parting line. For certain molds, holes or slots are located where needed. They are kept as small as possible so the blown melt does not have an impression of the opening. Their sizes can start with a range of 0.05–0.10 mm (0.002–0.004 in). If necessary, they are made larger. Different plastics behave differently so actual sizes is based on experience and/or trial and error.

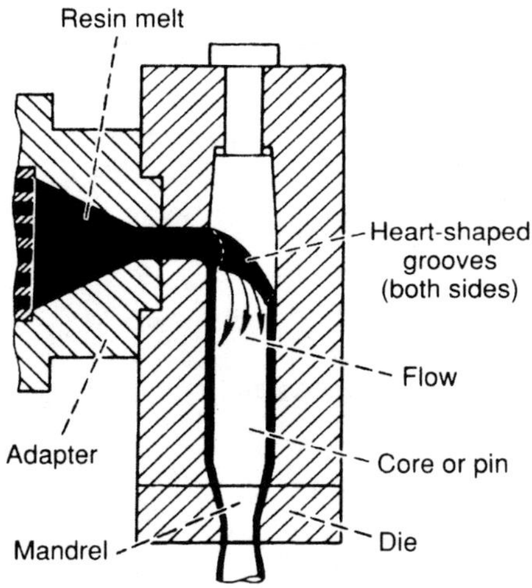


**Figure 15.17** Side fed or radial flow head around the core; die fed with heart-shaped grooves.

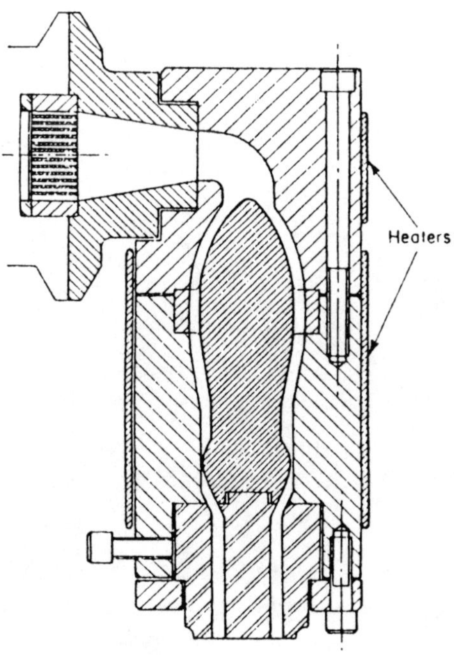


**Figure 15.18** Continuous EBM head having a spider-support core.

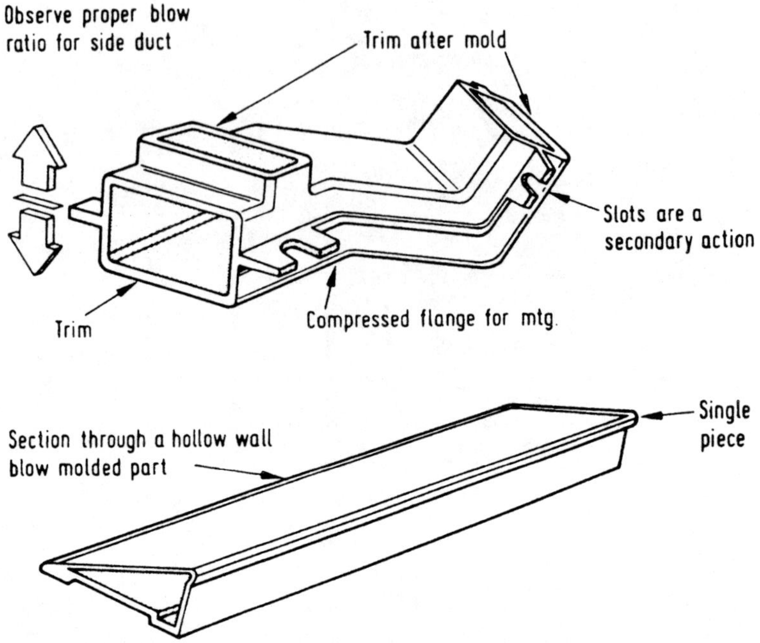


**Figure 15.19** BM air duct for an auto spoiler.

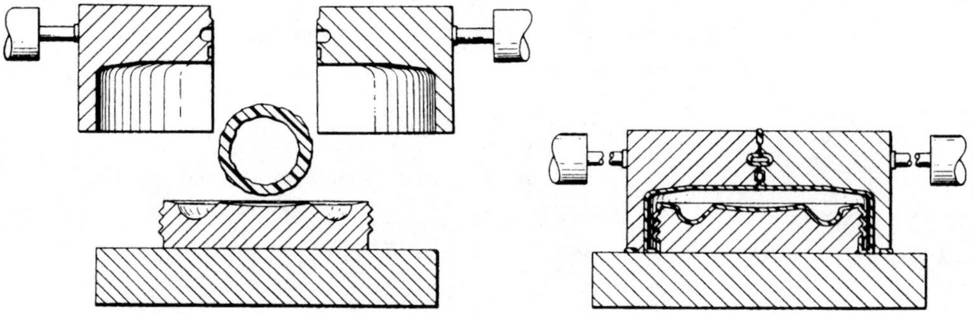

**Figure 15.20** Complex shaped EBM mold includes threaded forming core; views of this 3-part mold shows it in the open and closed positions with blow pin located in the top two sections of the mold.

The terms molds, dies and tools are interchangeable with molds being more descriptive with the BM part shape. Blow molds are principally made from aluminum or steel. Aluminum provides for faster cooling since its heat transfer is faster [2]. Materials of construction for molds are shown in Tables 15.10 and 15.11.

**Table 15.10** Examples of materials used in the construction of flow molds

| *Material* | *Hardness*[b] | *Tensile strength* psi | *Tensile strength* MPa | *Thermal conductivity* (Btu in. $ft^{-2}$ $h^{-1}$ $°F^{-1}$) |
|---|---|---|---|---|
| Aluminum | | | | |
| A356 | BHN-80 | 36975 | 255 | 1047 |
| 6061 | BHN-95 | 39875 | 275 | 1165 |
| 7075 | BHN-150 | 66700 | 460 | 905 |
| Beryllium copper | | | | |
| 23 | RC-30 | 134850 | 930 | 728 |
| 165 | (BHN-285) | | | |
| Steel | | | | |
| 0-1 | RC 52-60 | 290000 | 2000 | |
| A-2 | (BHN-530-650) | | | |
| P-20 | RC-32 (BHN-298) | 145000 | 1000 | 257 |

[a] BHN = Brinell hardness; RC = Rockwell hardness (C scale).
[b] Specific gravities (lb $in^{-3}$) Al = 0.097, Be/Cu = 0.129–0.316, steel = 0.24–0.29.

**Table 15.11** Guide to selecting construction materials for blow mold parts[a]

| *Property* | *Steel* | *Machined* *Aluminum* | *Machined* *Be/Cu* | *Cast* *Aluminum* | *Cast* *Kirksite* | *Cast* *Be/Cu* |
|---|---|---|---|---|---|---|
| Pinch life | 4 | 3 | 2 | 2 | 1 | 3 |
| Cavity life | 4 | 3 | 4 | 2 | 1 | 3 |
| Surface finish | 4 | 3 | 4 | 2 | 1 | 3 |
| Heat control | 2 | 4 | 4 | 2 | 1 | 3 |
| Mold modifications | 2 | 4 | 2 | 1 | 1 | 2 |
| High volume | 4 | 3 | 4 | 2 | 1 | 2 |
| Mold lead time | 2 | 3 | 2 | 4 | 4 | 3 |
| Low cost | 2 | 3 | 1 | 4 | 4 | 3 |
| Prototype cost | 1 | 3 | 2 | 3 | 4 | 3 |
| Complex shapes | 3 | 4 | 3 | 3 | 2 | 2 |
| Moving mold parts | 4 | 3 | 3 | 3 | 1 | 1 |

[a] 4 = best, 1 = poorest.

The pinchoff is a critical part of the EBM mold, where the parison is squeezed and welded together, requiring good thermal conductivity for rapid cooling and good toughness to ensure long production runs. The pinchoff must have structural soundness to withstand the plastic pressure

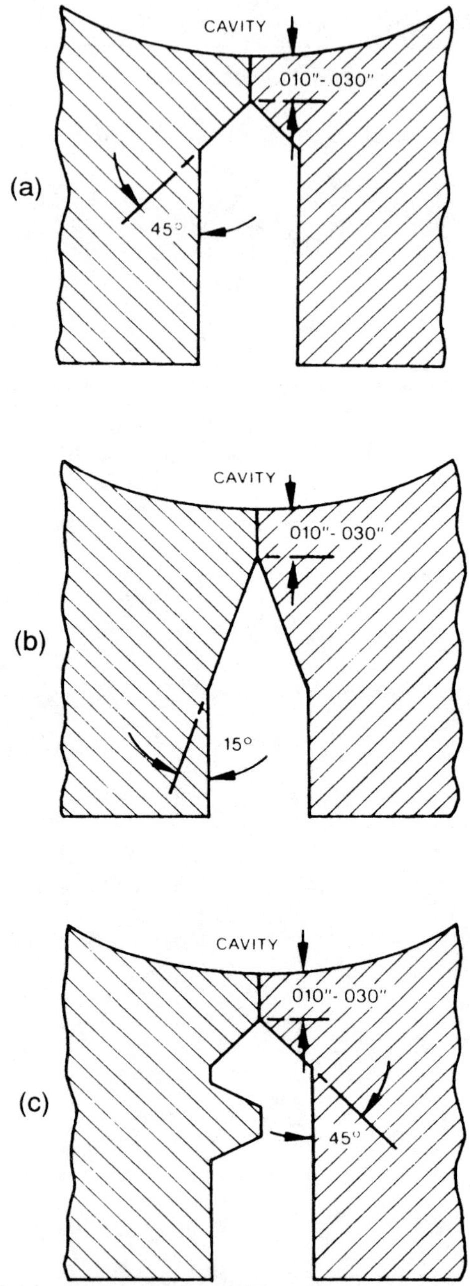


**Figure 15.21** Typical pinchoff double-angle designs.

and repeated closing cycle of the mold. It must usually push a small amount of plastic into the interior of the part to slightly thicken and reinforce the weld. It can also provide a cut through the parison to remove the flash.

Figure 15.21 identifies typical pinchoffs designated (a), (b), and (c). Most molds use a double-angle pinchoff (a) with 45° angles and a 0.25 mm (10 mil) land. When a blown part is large relative to the parison diameter, the plastic will thin down and even leave holes on the weld line requiring pinchoff (b). Using shallow angles of 15°, (c) has a tendency to force the plastic into the inside of the blown part.

A gross miscalculation of pocket depth (which must be learned through experience) can cause severe problems. For example, if the pocket depth is too shallow, the flash will be squeezed with too much pressure, putting undue strain on the mold, mold pinchoff areas, and machine clamp press sections. The molds will be held open, leaving a relatively thick pinchoff, which will be difficult to trim properly. If the pocket is too deep, the flash will not contact the mold surface for proper cooling. In fact, between molding and automatic trimming, heat from the uncooled flash will migrate into the cool pinchoff and cause it to heat up, creating problems like sticking to the trimmer. During trimming it can stretch instead of breaking free and 'clean.'

The knife edge cutter width of the pinchoff depends on the plastic used, the wall thickness, the size of the relief angle, the closing speed, and the time when blowing starts. As a guide for small parts up to 0.025 mm (10 mil), the width is 0.10–0.30 mm (4–12 mil). When processing LDPE, one uses the narrowest edge.

It is necessary to provide a heat control system for the mold to obtain the required part finish (Table 15.12). The mold surface heat depends on the plastic being processed and is usually 40–50°C (70–85°F) below the softening temperature. A higher mold heat means a longer cooling time,

**Table 15.12** Examples of recommended temperatures for cavities in blow molds

| | *Temperature* | |
|---|---|---|
| *Plastic* | *°C* | *°F* |
| PE and PVC | 15–30 | 59–85 |
| PC | 50–70 | 122–160 |
| PP | 30–60 | 85–140 |
| PS | 40–65 | 105–150 |
| PMMA | 40–60 | 105–140 |

although engineering plastics may require the higher heat to provide their highest quality performance. But the effect of this heat control is not great enough to compensate for the extruder's and/or the die head's ineffective operations causing defects.

## APPLICATIONS

BM is versatile. It is no longer just confined to the very popular production of bottles and other containers. It offers and has produced different processing advantages, such as fabricating extremely irregular (reentrant) curves, low-stressed parts, produces variable wall thicknesses, use of plastics with high chemical resistance (etc.), favorable processing costs, and so on. Reentrant curves are the most prominent features, so much so that it is difficult to find examples without them. They combine esthetics with strength and cost benefits. Examples of the many products that have been BM are shown in Figs. 15.22–15.27 and Table 15.13.

## COST

Table 15.14 provides a cost comparison guide of BM techniques for PVC and PET plastics. This information is to be used only as a guide.

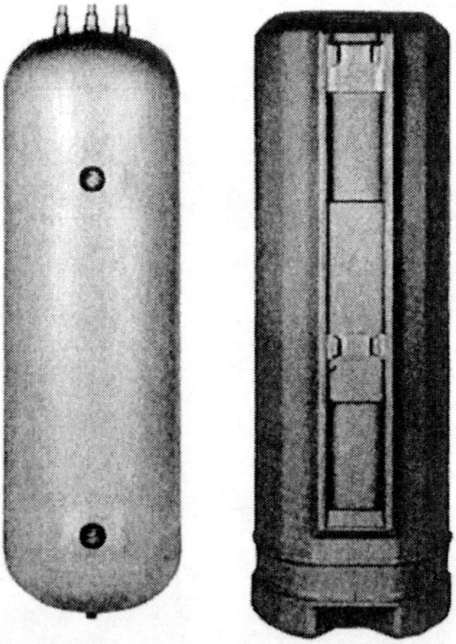

**Figure 15.22** EBM 25 gallon (200 $dm^3$) electric hot-water heater tank.

**Figure 15.23** EBM floating pontoons made from PP.

**Figure 15.24** EBM auto panels have generous radii at their corners and edges.

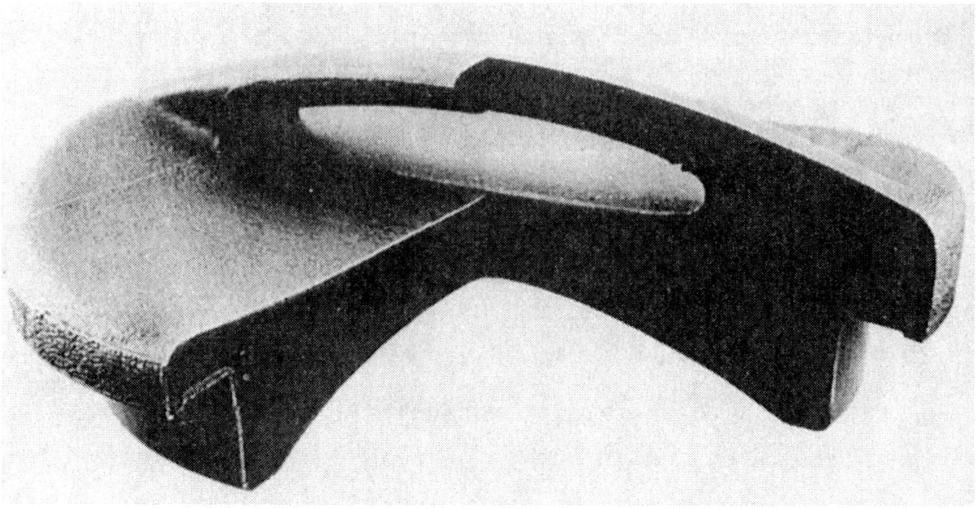

**Figure 15.25** EBM of HDPE integral handle for a container lid.

**Figure 15.26** EBM of PP aquacycle wheels included paddle fins on their sides.

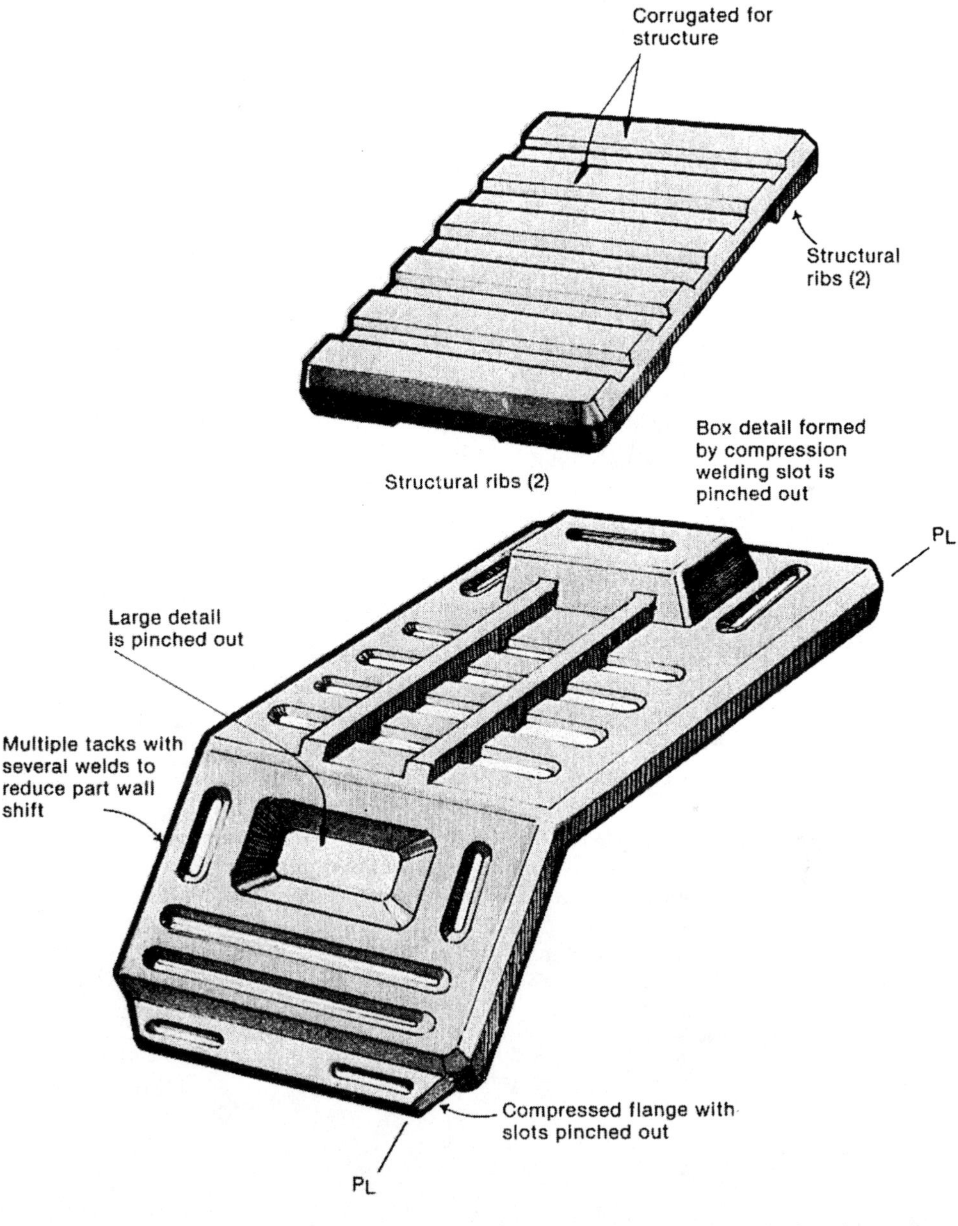


**Figure 15.27** Single multilayer/coextruded EBM part can often replace several different injection molded parts.

**Table 15.13** Hollow and structural BM shapes

| *Industry* | *Application* | *Required properties* |
|---|---|---|
| Automotive | Spoilers | Low temperature, impact, cost |
| | Seat backs | Heat distortion, strength/weight |
| | Bumpers | Low temperature, impact dimensional stability |
| | Underhood tubing | Chemical resistance, heat |
| Furniture | Workstations | Flame retardance, appearance |
| | Hospital furniture | Flame retardance, cleanability |
| | Office furniture | Flame retardance, cost |
| | Outdoor furniture | Weatherability, cost |
| Appliance | Air-handling equipment | Flame retardance, hollow |
| | Air-conditioning housings | Heat distortion, cost |
| Business machine | Housings | Flame retardance, cost |
| | Ductwork | Cost |
| Construction | Exterior panels | Weatherability, cost |
| Leisure | Flotation devices | Low temperature, impact strength cost, weatherability |
| | Marine buoys | Low temperature, impact strength cost, weatherability |
| | Sailboards | Low temperature, impact strength cost, weatherability |
| | Toys | Low temperature, impact strength cost, weatherability |
| | Canoes/kayaks | Low temperature, impact strength cost, weatherability |
| Industrial | Tool boxes, ice chests | Low temperature, impact strength, cost |
| | Trash containers, drums | Low temperature, impact strength, cost |
| | Hot-water tanks | Low temperature, impact strength, cost |

**Table 15.14** Guide for fabricating cost comparison of 16 fl. oz (454 g) BM bottles

| | *Standard Extrusion blow molding: two-parison head, fourfold* | *Stretch blow molding PVC: two single-parison heads, fourfold* | *Stretch blow molding PET* |
|---|---|---|---|
| **1.0 Machine cost ($)** | | | |
| Including head, molds, ancillaries (license fee, stretch PVC and PET) | 270 000 | 450 000 | 850 000 |
| **2.0 Hourly machine costs ($ h$^{-1}$)** | | | |
| Five-year depreciation (30 000 h) | 9.00 | 14.85 | 28.33 |
| Five-year financing, cost at 12.5% | 2.80 | 4.65 | 10.20 |
| Labor (1 worker) | 13.00 | 13.00 | 13.00 |
| Energy at $0.06 per kWh | 2.50 | 5.35 | 11.00 |
| Floor space | 1.50 | 2.00 | 4.00 |
| Maintenance and consumables | 2.25 | 3.75 | 4.50 |
| Total | 31.05 | 43.60 | 71.03 |
| **3.0 Bottle specs (hourly/annual production)** | | | |
| 3.1 16 fl.oz finish weight (454 g) | | | |
| Regular 37 g (1.3 oz) | | | |
| Stretch PVC 20 g (0.7 oz) | | | |
| Stretch PET 20 g (0.7 oz) | | | |
| Cycle time (Bottles per hour) | 8.4 s (1714) | 7.5 s (1920) | (4000) |
| Bottles per year (millions) | 10 286 | 11 520 | 24 000 |
| **4.0 Annual costs ($ y$^{-1}$)** | | | |
| 4.1 16 fl.oz (454 g) | | | |
| *Resin* 37 g | 585 200 | | |
| $0.70 lb$^{-1}$ ($1.54 kg$^{-1}$) | | | |
| 20 g | | 334 950 | |
| $0.66 lb$^{-1}$ ($1.46 kg$^{-1}$) | | | |
| 20 g | | | 634 360 |
| $0.60 lb$^{-1}$ ($1.32 kg$^{-1}$) | | | |
| *Machine costs* | 186 300 | 261 600 | 426 180 |
| Total | 771 500 | 596 550 | 1 060 540 |
| Annual royalty to Du Pont (PET) | | | |
| Cost per thousand | 75.00 | 51.78 | 45.44 |

[a] Figures are not be to considered as absolute costs, but rather **reflect comparisons** between various machine options. All calculations are based upon 100% efficiency. All bottle weights are finish weights (flash being considered as 100% reusable).

**Table 15.15** Guide to common BM problems

| *Problem* | *Cause* | *Solution* |
|---|---|---|
| Rough parison; orange peel | Melt fracture; melt temperature too low | Polish all tooling<br>Raise melt temperature |
| Poor gloss | Mold too cold | Increase die surface temperature |
| Black specks in part | Contamination from degraded material | Purge to clean system<br>Keep materials clean |
| Gels in parison | Excessive fines in regrind<br>Moisture in resin<br>Screw too deep | Screen out regrind fines<br>Dry material before use<br>Use higher-shear screw and lower barrel temperatures |
| Bubbles in wall | Moisture in trapped air | Increase extrusion pressure<br>If moisture, lower screw speed; reduce feed-zone temperature |
| Uneven wall thickness circumferentially | Pin not centered in die ring | Adjust die-pin position |
| Parison hooking | Head temperature not uniform | Stagger heater-band gaps on head |
| Incomplete blow | Extrusion rate too high<br>Blowup air pressure<br>Blowup time too short<br>Parison is cut at pinchoff | Reduce screw speed<br>Increase blow-air pressure<br>Reduce mold-closing speed |
| Holes in parison and/or bottles | Contaminated or degraded resin<br>Trapped air<br>Moisture in resin | Purge and clean tooling and screw<br>Let extruder run for a few minutes<br>Dry the resin |
| Parison stretches | Resin melt index too high<br>Melt temperature too high | Use lower melt index<br>Reduce melt temperature<br>Increase screw speed<br>Boost extrusion rate |
| Parison blowout | Blowup too rapid<br>Melt temperature too high<br>Pinchoff too sharp<br>Blowup ratio too high | Program blowup start with low air pressure and increase<br>Align molds<br>Use larger parison |

**Table 15.15** *Continued*

| *Problem* | *Cause* | *Solution* |
|---|---|---|
| Die, weld, and spider lines in parison | Damaged die ring | Repair or replace die tooling |
| | Mandrel spider legs cause improper knitting | Streamline spider legs<br>Reduce die temperature to increase back pressure |
| | Contamination from material | Clean diehead |
| Webbing in handle | Parison walls touch when mold closes | Align parison closer to handle side of mold |
| | Wrong parison diameter | Increase die diameter<br>Reduce melt temperature |
| Rocker bottoms | Blowing air not vented before mold opens | Increase air exhaust time |
| | Insufficient cooling | Clean cooling channels of mold<br>Increase blow time |
| Tails not pulled | Parison is too short | Lengthen the parison by increasing extruder speed |
| | Plastic or foreign matter holding mold | Clean mold parting surfaces |
| Bottles thin in various areas | Parison curling | Adjust die ring concentricity |
| | Parison too long or short | Increase/decrease extruder speed and adjust parison temperature<br>Reduce head temperature |
| Molds not separating from neck finish | Cutting ring is dull | Sharpen or replace cutting sleeve |
| | Poor contact between cutter ring and striker plate | Increase overstroke and downward pressure of blow pin |
| Weak shoulders on bottles | Parison sag<br>Parisons too long or short | Reduce melt temperature and decrease/increase extrusion rate |
| | Container too light | Program increased weight |
| Slanted neck finish | Blow pin/cutter entry too deep | Raise blow pin until it just cuts |
| | Parison folding over | Replace dull knife blade<br>Adjust knife-cut delay timer |
| Parts sticking in mold | Mold too hot | Improve mold cooling |
| | Cycle too short | Lengthen cycle |

**Table 15.15** *Continued*

| *Problem* | *Cause* | *Solution* |
|---|---|---|
| Mold parting line indented in part | Blowup air introduced prematurely | Delay blowup |
| | Hooking parison | Reduce mold temperature |
| Handle missing | Insufficient die swell | Position parison closer to handle |
| | | Use larger tooling |
| Sink marks | Air trapped in mold | Improve venting |
| | | Lower mold temperature |
| Parison tails | Parison is too long | Reduce extruder speed |
| | Pinchoff improperly designed | Design pinchoff to compression cool tail |
| Poor detail definition | Blow-air pressure too low | Increase blow-air pressure and blow time |
| | Poor mold venting | Improve venting |
| | Cold mold | Increase mold temperature |
| | **Coextrusion blow molding** | |
| | **Most of the above tips also apply to blow molding multilayer containers** | |
| Skips in barrier layer | Temperature of barrier material too high | Reduce barrier material temperature |
| | Pressure fluctuations at extruder | Maintain constant pressure at extruder screw tip |
| | Degraded material in head | Purge head and/or extruder |
| Barrier integrity of handle breached | Too little material in handle<br>Poor pinchoff | Program more material into handle and pinchoff area |
| Layer separation, blistering or bubbles in container | Adhesive layer too cold, did not flow around structure; adhesive too hot to stick to adjacent layer | Adjust temperature of adhesive material up or down |
| | Adhesive layer cooled too fast | Raise mold temperature to prevent fast cooldown |
| | Moisture in materials | Dry materials |

## TROUBLESHOOTING

In addition to the problems and solutions reviewed in this chapter, Table 15.15 lists some of the common BM problems with information on causes and solutions.

# 16

# Fiber and filament

OVERVIEW

Thermoplastic fibers or filaments are produced by screw extruders. They are manufactured using the three common methods of melt spinning, dry spinning, and wet spinning. There are many variations and combinations of these basic processes. Other types of fiber-forming processes include: (1) reaction spinning; (2) dispersion, emulsion, and suspension spinning; (3) fusion-melt spinning; (4) phase-separation spinning; and (5) gel spinning (397, 413).

The processes basically force molten plastic using an extruder and/or gear pump through fine holes in a spinneret/spinaret (die). In turn, they are immediately stretched or drawn (oriented), cooled, and collected at the end of the line. During this process they may be subjected to other operations, such as: (1) thermal setting and thermal relaxation processes to provide dimensional stability; (2) twisting and interlacing to provide cohesion of the filaments; (3) texturing; and (4) crimping and cutting to provide staple products. The speeds of certain lines using the melt and dry spinning processes can go from 2000 to 4000 m/min (6600 to 13 000 ft/min).

Numerous techniques for producing fibers without using the spinneret have been devised. They include centrifugal spinning, electrostatic spinning, tack spinning, and solid-state extrusion (SSE). The SSE process extrudes through a capillary rheometer with a conical die; the processing temperature is close to the melting point of the plastic.

Fiber is a general term used to refer to filamentary materials. Often it is used synonymously with filament, monofilament, whisker, and yarn. It is any material in a form such that it has a minimum length of at least 100 times its diameter. Diameters are usually 0.10–0.13 mm (0.004–0.005 in). Fibers can be continuous or reduced to short lengths (discontinuous), where the industry lists fibers as having a specific length, such as 3.2 mm (0.125 in).

A filament is the smallest unit of a fibrous material and is usually not used alone. Filaments are the basic units formed during manufacture which are gathered into strands of fiber. Their diameters are less than 0.025 mm (0.001 in). The fineness of a fiber is also identified by its denier. Denier is a unit of weight expressing the size or coarseness but particularly the fineness of a continuous a fiber or yarn. The weight in grams of 9000 m (30 000 ft) is one denier. The lower the denier, the finer the fiber, yarn, etc. Sheer women's hosiery usually runs from 10 to 15 denier (Table 16.1).

Another type of product is staple fibers which is obtained by cutting

**Table 16.1** Conversion of yarn length to weight

| *Denier* | *Number of filaments* | *Yield (yd/lb)* | *Yield (m/kg)* |
|---|---|---|---|
| 55 | 25 | 81 175 | 163 636 |
| 195 | 90 | 22 895 | 46 155 |
| 195 | 134 | 22 895 | 46 155 |
| 200 | 134 | 22 320 | 44 997 |
| 380 | 180 | 11 749 | 23 684 |
| 380 | 267 | 11 749 | 23 648 |
| 400 | 267 | 11 160 | 22 500 |
| 720 | 490 | 6 200 | 12 500 |
| 750 | 490 | 5 952 | 12 000 |
| 840 | 534 | 5 314 | 10 714 |
| 1 000 | 666 | 4 464 | 9 000 |
| 1 140 | 768 | 3 916 | 7 895 |
| 1 420 | 1 000 | 3 144 | 6 338 |
| 1 500 | 1 000 | 2 976 | 6 000 |
| 2 160 | 1 000 | 2 097 | 4 228 |
| 2 250 | 1 000 | 1 984 | 4 000 |
| 2 840 | 1 333 | 1 572 | 3 169 |
| 2 840 | 1 000 | 1 572 | 3 169 |
| 3 000 | 1 333 | 1 488 | 3 000 |
| 4 320 | 2 000 | 1 048 | 2 110 |
| 4 560 | 3 072 | 979 | 1 974 |
| 6 000 | | 744 | 1 500 |
| 7 100 | 5 000 | 630 | 1 268 |
| 8 640 | 4 000 | 524 | 1 057 |
| 10 800 | 5 000 | 413 | 833 |
| 11 400 | | 391 | 789 |
| 15 000 | 10 000 | 298 | 600 |

continuous filament into 12.7–50 mm (0.5–2 in) lengths and 1–5 denier. They are processed like natural fibers, such as wool and cotton, with which they can be blended.

## PLASTIC MATERIALS

Different plastics are used to produce fibers. The more important materials are PP, nylon 66, polyester, and PETP; other plastics are also used. Each family of a plastics has different grades to provide different properties during and after processing. Their plastic fiber structures have different levels of molecular organizations with each relating to certain aspects of fiber behaviors and properties (Chapter 3).

As an example, the organochemical structure defines the chemical composition and molecular structure of a fiber. This molecular structure is directly related to the fiber's chemical properties, dye ability, moisture sorption, swelling characteristics, and indirectly related to all physical properties.

The physical properties of fibers are influenced by the processing techniques used on-line, from melt conditions to windup speed. However, they are also strongly influenced by the plastic's morphology (Chapter 3). All plastic fibers that are useful in textile applications are usually semicrystalline, irreversibly in an oriented pattern.

Nylon or polyamide (PA) was the first synthetic fiber of importance. Although a PA was first prepared in 1899, it was the fundamental research by W. H. Carothers at DuPont during the 1930s that led to nylon fibers. Its process of manufacture is by melt spinning.

Polyamide fibers based on aliphatic monomers are called nylons. Those with monomers that are at least 85% aromatic are referred to as aramids. The arimid fibers are characterized by: (1) flame resistance and thermal stability; (2) high strength and stiffness; or (3) combinations of these.

The polyolefin fibers, also called olefin fibers, have at least 85 wt% of ethylene, propylene, or other olefin units. There are different polyolefins that are capable of forming fibers, but polypropylene (PP) and to a much lesser extent polyethylene (PE) are the major materials. Crystalline olefins, having melting points lower than PP, have very limited use in textile applications. There are those with the higher $T_m$, but they also have high costs. The low $T_g$ (glass transition temperature) of polyolefins may result in compressive creep (Chapter 3).

Regardless, the polyolefin fibers continue to be one of the most important in the growing fiber market. In the past, the primary process for PP fibers was melt spinning. Now, there are tailor-made short-spin systems for low-volume staple production. The spin systems (slit-film, blown, etc.) take account of the fact that polyolefins, because of their higher molecular weight and broad distribution, have much higher viscoelasticity than traditional melt-spun plastics, such as nylon and polyester.

With polyester fibers, as with nylons, high-speed melt spinning is used. In the past, a two-step spinning and drawing process was used. Dyeing these fibers is difficult because of the lack of hydrophilic sites and their inherent stiffness. However, techniques have been developed and are used.

The acrylic fibers are manufactured by either dry- or wet-spinning. In dry spinning dimethylformamide is a common solvent.

Elastomeric fibers are principally made from segmented polyurethanes (spandex) and polyisoprene (natural rubber). The elastomeric fibers consist of plastics with a main $T_g$ well below room temperature. This criterion excludes some fibers with elastic properties.

Two principal fiber types identify vinyl fibers. Those fibers that contain at least 85 wt% vinyl chloride are called vinyon fibers. Fibers with at least 50 wt% vinyl alcohol are called vinal fibers. Other fibers in this category are based on vinylidene chloride or contain fluorine.

Spandex fiber is produced primarily by dry spinning and wet spinning with a few producers using melt spinning. For the natural rubber, a latex mixture is continuously forced through a capillary tube into an acid bath, where it is coagulated; the thread-like coagulum is pulled from the bath followed by washing, drying, and curing.

## PROCESSING CHARACTERISTICS

In the manufacture of fibers, a relatively isotropic plastic with properties similar in all directions converts into an orthotropic plastic where most of the plastic's strength is in the direction of the fiber axis [3]. This desirable effect provides a certain degree of fiber strength in the longitudinal direction, but usually not enough. So the fibers are made stronger by stretch-orientation during or after processing (Chapter 2).

These spinning lines can include a variety of operations useful for the fibers different applications. A finish can be applied after cooling in-line. Rather than keeping them straight, texturing techniques are used. Texturing introduces crimp, whereby the straight filaments are given a twisted, coiled, or randomly kinked structure. A yarn made up of these filaments is softer and more open in structure; it is more pleasing to the touch.

Finishes are used to improve the processing and handling of fibers. The finishing mix includes a lubricant. The finish is usually applied during the spinning processes before windup in order to reduce friction against guides, improve interfilament cohesion, and eliminate filament separation arising from electrostatic charge. The lines can include a relaxing stage where hot-air drying is used.

**Melt spinning**

This process was developed in the late 1930s to process nylon 6 and nylon 66. Since that time other plastics have been used that include polyesters and polyolefins.

In this process, molten plastic usually goes through an extruder, mixing device, and/or gear pump (Figs. 16.1 and 16.2). The plastic and ingredients (such as primarily stabilizers, pigments, and rheological modifiers) are fed into a screw extruder. The gear pump (Chapter 2) accurately meters melt through a filter pack of graded sand or porous metal and a spinneret. The multi-hole spinneret represent the die.

Upon leaving the spinneret, the molten filaments pass at very high speed usually vertically downwards into water or a counter current of air where they are cooled and solidified. At the same time, after leaving the spinneret, they are stretched to the desired diameter. Finish can be applied prior to the fiber reaching the end of the line where it is wound on bobbins or other windup rolls. To obtain the required high performance properties, fibers are oriented (Chapter 2) by reheating and drawing. This operation is usually a separate operation since it requires much higher linear speeds than melt spinning.

Pressure drops through the filter and spinneret are high. However, as the filter pack blocks up during a run, its pressure drop will increase, resulting in a decreased extruder output. To maintain proper drawdown,

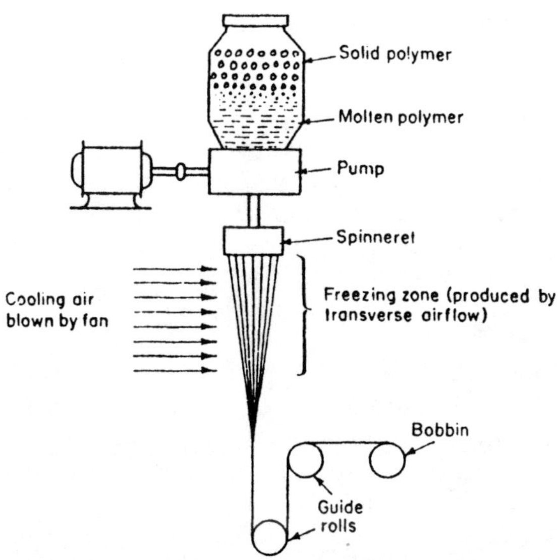


**Figure 16.1** Melt spinning of filaments to fibers using only a gear pump.

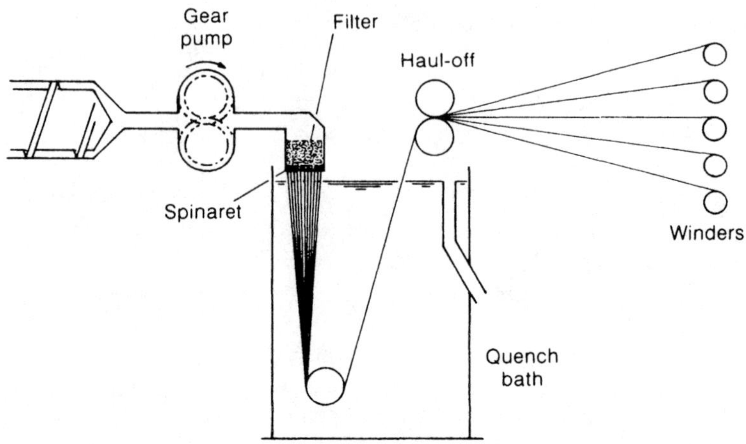


**Figure 16.2** Melt spinning with extruder and gear pump.

properties, and to minimize tension (that would break the filaments), the extruder output is controlled (Chapter 6) to meet the haul-off and winding equipment requirements. To maintain proper control, it is difficult to adjust the extruder screw or the winders speeds. By using the gear pump's positive metering action, the extruder is just a source of delivering melt to the pump. The pump provides a constant and high melt pressure with constant output to minimize viscosity and pressure drops [397].

## Dry spinning

In dry spinning, a plastic solution is extruded (metering pumped) through a spinneret. The filaments exit the spinneret through a gas-heated cabinet where the solvent is rapidly removed from the plastic filaments. The suitable solvent is filtered and recovered for further use in-line. Filaments end up at the driven haul-off roll.

## Wet spinning

In this process, also called reaction spinning, a plastic solution is extruded from a spinneret and immersed into a spin bath tank containing a circulating nonsolvent solution that coagulates (precipitates) the plastic. In some cases, both the solution and the precipitation stages involve chemical reactions. After passing through the spin bath tank, the fibers are washed prior to the windup.

Conventional wet spinning has the slowest line speed compared to the other lines. However, since it permits very short distances between the holes in the spinneret face, a single spinneret may carry a very large number of holes. With this single spinneret, high production rates can be achieved [263].

## Other processes

PP is processed through the melt-spinning system followed with orientation. Systems are also used that are tailor-made for producing stapled PP. A substantial volume of PP fiber is produced by the slit-film or film-to-fiber technology. Fibers are produced by cutting or slitting a film. Here the fiber is oriented by stretching before or after the cutting proces; also used are mechanical or chemomechanical fibrillation. In this procedure, film anisotropy is created by stretching before fibrillation.

Fibrillation is the phenomenon wherein filament or fiber shows evidence of basic fibrous structure or fibrillar crystalling nature. It occurs by a longitudinal opening up of the filament under rapid load with excessive tensile or shearing stresses. Separate fibrils can then often be seen in the main filament trunk. The whitening of the plastic when unduly strained at room temperature is a manifestation of fibrillation. Applications of this product are primarily for carpet backing, rope, and cordage.

Unusual polyolefin fibers are produced by a spurted or melt-blown spinning technique. A variety of directly formed nonwovens with excellent filtration characteristics are produced. Original development was by Exxon Corp. that produced a very fine, submicrometer filaments. Pulp-like olefin fibers are produced by a high pressure spurting process developed by Hercules, Inc. and Solvay, Inc.

A high modulus commercial PE fiber with properties approaching those of aramid and graphite fibers is prepared by gel spinning. Higher tensile strengths are also available from gel spinning or fibrillar crystal growth [263].

There are different extruder systems to produce monofilaments to small 'rods.' Different plastics are used. Fig. 16.3 shows an example.

## Fiber twist

Twist in most fiber (strands, etc.) is an important condition to apply during processing. In fibers, yarn, and other textile strands, it is the spiral turns about its axis per unit of length. Twist may be expressed as turns per inch (TPI) or 0.4 turns/cm. The letters S and Z indicate direction of twist, in reference to whether the twist direction conforms to the middle section slope of the particular letter (Fig. 16.4).

The S twist exists when the fiber is held in a vertical position and the spirals conform in slope to the central position of the letter S; it has a Z twist when the spirals conform to the central portion of the letter Z. Fibers

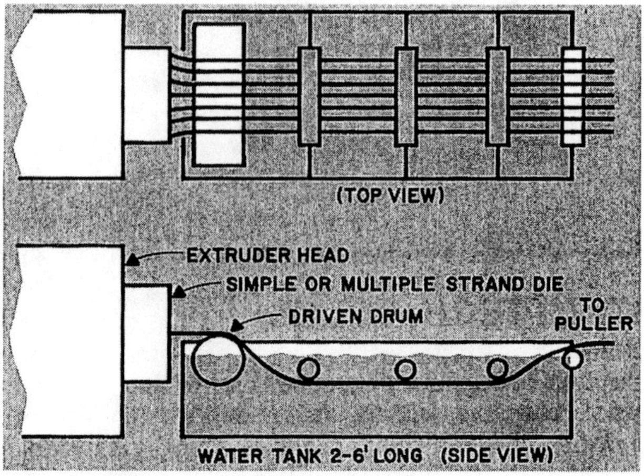


**Figure 16.3** Monofilament and small rod extrusion.

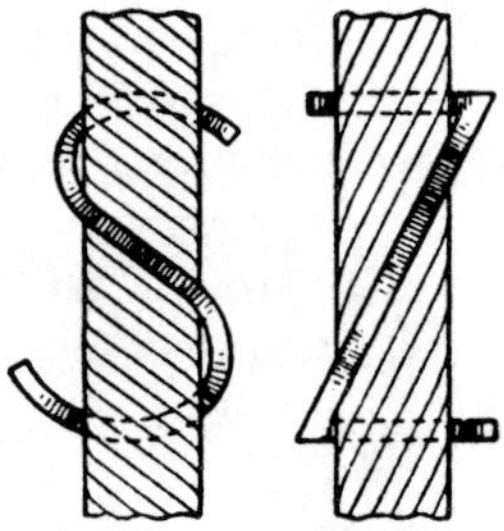


**Figure 16.4** Twist directions (S and Z) in a yarn or other textile strand.

that are simply twisted greater than one turn/in will kink, corkscrew, and unravel because of their twist in one direction only. The plying of fibers operation normally eliminates this problem. For example, single S twist fiber plied with a Z fiber results in 'balancing' the fibers. The amount of twist with plying provides many different combinations useful in different applications.

## Filtration

Many processes require a plastic melt free of contaminates larger than a specific size. The fiber processors usually filter down to 5 μm (0.2 mil) particle size to protect the melt spinning machines from filament breaks. The fiber process typically operates at very high speeds. A filament break at this speed is costly to both the product quality and the process efficiency.

The media used for filtration has included sand packs, wire screens, sintered metal powder sheets, and sintered metal fiber sheets. There are also different sandwich combinations, such as wire screens and sintered metal fiber sheets, which in theory provide the best properties of each component.

In a gear pump, running clearance can be as low as 0.006 mm (0.00025 in) about its periphery and on either side of the metering gear. Any slight burr, nick, or particle of any foreign matter will cause scoring and possible seizure of the pump. Recognize that 0.025 mm (0.001 in) equals 25 μm (microns), so filtration down to just the pump has to be down to 6 μm or less.

Wire screens have dominated the extrusion filtration systems with square woven screen being the most important media. A 500 mesh square woven screen, fine by most standards, will usually have an opening of approximately 0.025 mm (0.001 in) on a diagonal opening of requirement for particle retention. If wire screen is preferred, the appropriate construction may be a twilled Dutch weave.

Fine mesh screens are, of necessity, manufactured from fine wire. They are therefore weak in tension. A satisfactory fine mesh screen must be supported be a series of sequentially more coarse mesh screens in order to reduce the unsupported fine mesh without wire strength failure. In addition, the pressure drop across the total screen pack must be monitored carefully to prevent a pressure differential which will rupture the fine mesh screen (Chapter 2).

Sintered metal powder and fiber filter sheets are considered an in-depth media. Gel capture and holding capacity of screen packs is primarily a function of surface area. Sintered powders of fiber media are better for gel capture since they can have, when properly designed, a significantly higher area in contact with the plastic due to the thickness dimensions.

Porous media usually has a strength-to-pressure drop ratio which is considerably higher than that of most equivalent screen packs. The advantage of small pore size coupled with the greater thickness leads to potentially a higher allowable pressure drop than does a screen pack on an equal area basis [187, R. B. Gregory].

## Nonwoven

These are fibrous sheets made from fiber-spinning material. They include spin-bonded, flash-spun, and melt-blown nonwovens. Reifenhauser's spin-bonded system is shown in Fig. 16.5 [357]. It consists of material handling at the hoppers, extruders, melt spinning systems for filament forming, filament drawing and web forming with the Reifenhauser

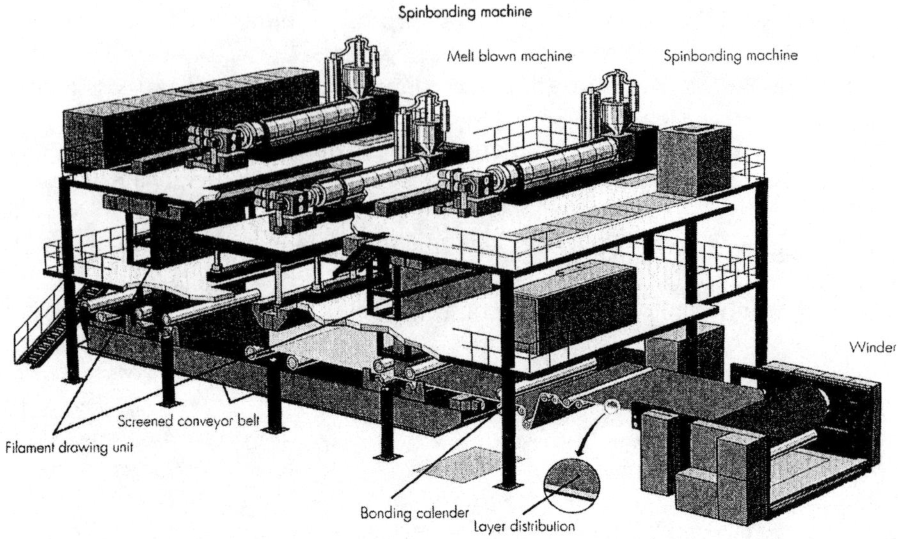


**Figure 16.5** Advancement in spinbonding technology by Reifenhauser.

'Reicofil' process, screened conveyor belts, bonding calender, and winding station.

This turn-key process has filaments emerging from thousands of capillaries being stretched, then deposited at a uniform rate in a random pattern on a screened conveyor belt. The nonwoven web is bonded in a calender, slit in-line, and wound into rolls.

## SPINNERET

The spinneret (or spinaret) is a type of die; it is usually a metal plate with many small holes (or oval, etc.) through which a melt is pulled and/or forced. Conventional spinneret orifices are circular and produce a fiber that is round in cross section. They can contain from 50 to 110 very small holes.

A special characteristic of their design is that the melt in a discharge section of a relatively small area is distributed to a large circle of spinnerets. Because of the smaller distance in the entry region of the distributor, dead spaces are avoided, and the greater distance between the exit orifices make for easier threading.

Precision machining of the orifices is required in order to avoid differences in thickness between the filaments being pulled. Note that the volumetric discharge from a cylindrical die increases with the fourth power of the diameter. An error of 10% in diameter will cause a 47% error in output. Because these differences in spinneret heads cannot be balanced out by adjusting the individual filament hauloff speeds, the diameter of the monofilament is altered by 21% [370].

There have been improvements in machining methods, advances in fiber production technology, and progress in the study of the effect of profiled fibers on fabric characteristics. These developments has increased the importance of modified cross section or noncircular filaments.

## PROCESS OPTIMIZATION

The cross section of the fiber is an important variable in designing the fiber properties for a specific use. If maximum strength is required, a round cross section with the fewest possible surface flaws is the target. Another example is of antisoil carpet fibers, the cross section of which should be free from crevices that could help trap dirt particles, so rounded edges of square or triangle fiber may be preferred. For upholstery fabrics requiring high luster, a triangular cross section may be desired rather than flat surfaces that enhances light reflections. An elongated cross section, such as an ellipse, improves bulk and covers more than round fibers.

Most of the fiber productions go through water/liquid quenching baths/tanks that are temperature controlled. The required cooling medium temperature depends on the type of plastic being processed, the melt temperature, the filament diameters, the line speed, and the length of the bath. With crystalline plastics, rapid cooling results in small crystalline and a higher amorphous content, so that the threads retain the ductility required for the stretching process.

A general guide to cooling bath temperatures follows: PVC copolymers 15–20°C (60–68°F); LDPE 20–30°C (68–86°F); rigid PVC 50–60°C (122–140°F); HDPE 40–70°C (105–160°F); PS 110–120°C (230–250°F); and PP 115–125°C (240–250°F). Some of the plastics may use air cooling alone or in series.

Water ripples in the bath can cause thickness variations, particularly for plastics with a sharp melting point such as nylon. To protect the fibers entering the bath, some type of guard is used such as safety rings. The other end of the tank may require insurance (wetting agent in water, whippers, etc.) that water is removed from the fibers where certain downstream operations, such as stretching or orientation, are to take place. Drops of water on the fibers could cause uneven reheating during orientation causing uneven stretch.

The products obtained from spinning is usually relatively weak and not commercially suitable. This situation occurs because very little or no orientation occurs during the extrusion process. To increase their tensile strength and other properties, the filaments are subjected to orientation during extrusion or usually as a separate operation.

The filaments are usually stretched between two rolls or two sets of rolls. The fibers are exposed to a draw ratio which is the ratio of the stretched length to the original unstretched length. The size of the draw ratio is dependent on the plastics and what properties are desired. The stretching is controlled by adjusting the different surface speeds of the rolls. Heat is applied to the fibers just before and/or during the stretching. With nylon the fibers are heated and stretched (drawn) to 30–600% of their original length, depending on spinning speed. This action results in oriented crystallization and increased tensile strength. The difference in the draw ratio produces fibers with different properties. Where special properties, such as a very low shrinkage, may be required, as in tire yarn, a heat relaxation step will be included. For more details on orientation, see Chapter 2.

Carbon fibers (CFs) are the predominant high-strength, high-modulus fiber used in the manufacturer of advanced reinforced plastic composite products [3, 43]. These fibers are made by the pyrolytic degradation of a fibrous organic precursor. Most CFs are obtained by the pyrolysis of polyacrylonitrile (PAN); this old technology subjects the PAN to temperatures up to 2000°C (1080°F).

Solid-state extrusion is a means for the deformation and evaluation of uniaxial molecular chain orientation and product extension for a wide range of plastics. Developments produced HDPE drawn into fibers with some of highest specific tensile moduli and strengths. A two-step drawing process was used for the preparation of polyoxyethylene, PP, and PE fibers. The plastic is first drawn to their natural draw ratio at a fast rate and subsequently slowly super-drawn at a temperature that depends on the crystalline dispersion temperature.

A highly oriented extrudate can be obtained by extruding through a capillary rheometer with a conical die at temperatures close to the melting point. Initial work led to the development of transparent and fibrous linear PE extrudates. These were obtained by extruding HDPE from the molten state [132–136°C (55–58°F)] above a critical shear rate in a capillary rheometer and through a conical die.

This procedure was subsequently modified by processing HDPE exclusively in the solid state where the plastic is semicrystalline before extruding through the die. This modification produced continuous transparent fibers with moduli in the range of 30–70 MPa (4400–10 200 psi).

## THEORETICAL VERSUS ACTUAL VALUE

Through the laws of physics, chemistry, and mechanics, theoretical property values of fibers have been determined for different materials. These

**Table 16.2** Comparison of theoretically possible and actual values of fibers (1944 data)

| | *Modulus of elasticity* | | | *Tensile strength* | | |
|---|---|---|---|---|---|---|
| *Type of material* | *Theoretical, N mm$^{-2}$ (kpsi)* | *Fiber, N mm$^{-2}$ (kpsi)* | *Nonfiber, N mm$^{-2}$ (kpsi)* | *Theoretical, N mm$^{-2}$ (kpsi)* | *Fiber, N mm$^{-2}$ (kpsi)* | *Nonfiber, N mm$^{-2}$ (kpsi)* |
| Polyethylene | 300 000 | 100 000 | 1 000 | 27 000 | 1 500 | 30 |
| | | (33%)[a] | (0.33%) | | (5.5%) | (0.1%) |
| | (43 500) | (14 500) | (145) | (3 900) | (218) | (4.4) |
| Polypropylene | 50 000 | 20 000 | 1 600 | 16 000 | 1 300 | 38 |
| | | (40%) | (3.2%) | | (8.1%) | (0.24%) |
| | (7 250) | (2 900) | (232) | (2 300) | (189) | (5.5) |
| Polyamide-66 | 160 000 | 5 000 | 2 000 | 27 000 | 1 700 | 50 |
| | | (3%) | (13%) | | (6.3%) | (0.18%) |
| | (23 200) | (725) | (290) | (3 900) | (246) | (7.2) |
| Glass | 80 000 | 80 000 | 70 000 | 11 000 | 4 000 | 55 |
| | | (100%) | (87.5%) | | (36%) | (0.5%) |
| | (11 600) | (11 600) | (10 100) | (1 600) | (580) | (8.0) |
| Steel | 210 000 | 210 000 | 210 000 | 21 000 | 4 000 | 1 400 |
| | | (100%) | (100%) | | (19%) | (6.67%) |
| | (30 400) | (30 400) | (30 400) | (3 050) | (580) | (203) |
| Aluminum | 76 000 | 76 000 | 76 000 | 7 600 | 800 | 600 |
| | | (100%) | (100%) | | (10.5%) | (7.89%) |
| | (11 000) | (11 000) | (11 000) | (1 100) | (116) | (87) |

[a] For an experimental value the percentage of the theoretically calculated value is given in parentheses, e.g., (33%).

are compared to actual values in Table 16.2. With steel, aluminum, and glass, the theoretical and actual values have been practically the same. However for PE, PP, and PA, they are far apart.

The plastic fibers have the important potential to reach values that are far superior to those of other materials. But what makes the values in this table very important and significant is that the table was prepared during 1944 [68]. The 'normal' plastic properties reported in the table are based on plastic material available at that time. With the passing of time, plastic fiber properties have significantly increased but have not reached their potential [3].

## APPLICATIONS

Different fibers or filaments are used in many different markets. The fibers are individual filaments held together by a slight twist or by intermingling and packaged as a tube or cone. The monofilament is a single fiber or filament of definite length that is strong enough to function as a yarn. The popular nylon monofilaments have been in use for a long time in applications such as in brush bristles, fishing lines, sewing thread, fabrics, etc.

Staple fibers are made up of a very large number of discontinuous, randomly oriented, individual fibers normally shipped in a box or bale. They are usually subjected to a series of processes, culminating in textile spinning to yarn.

The precursor of staple fibers is tow, which consists of large numbers of roughly parallel, continuous filaments. They are converted by cutting or breaking into staple fibers or directly into a slivers, intermediate stages between staple fibers and yarns. In the latter case the filaments remain parallel.

Substantial amounts of fibers are sold as monofilament, which is single-filament yarn of substantial greater diameter than those in continuous filament and staple yarns. They are also available in the form of tapes, fibrillated tapes, and slit-film products.

There are certain types of so-called nonwoven fabric that are directly formed from continuous filaments without isolation of a yarn. Products of this type include melded and spun-bonded fabrics.

Recycling fibers has been in existence for at least a half century. An interesting application using recycled plastic fibers is shown in Fig. 3.12 where a man's suit and tie were made from recycled PET during 1979.

## TROUBLESHOOTING

Information concerning troubleshooting and applicable to this chapter are reviewed in different chapters. Producing filaments can be complex (for the novice) because the speed of operations is very fast, so it is important to locate problems as soon as possible.

Examples of problems include: (1) filament breakage where the cause could be that holes in the spinneret are blocked, melt too hot or cold and uneven, drawdown too great, moisture or contamination in melt, oxidation or weakening where the gap between die and quench tank is too long, too many orientation nicks or abrasive areas on roll(s), and poorly controlled or erratic drive; (2) all filaments with varying diameters caused by surging, nonuniform melt flow in die, etc.; (3) some filaments differing from others in size or strength due to temperature gradient in the extruder die, unequal melt flow in the die, etc.; (4) poor surface due to melt fracture, moisture in the melt, poor grade plastic, etc.; (5) low strength due to insufficient orientation, poor grade of plastics, nicks or cuts in filaments, degraded plastics, etc.; and (6) noncircular cross sections due to spinneret die holes not round (due to wear, etc.), too much tension during wrapping, filaments too hot while passing over rolls, etc.

# 17

# Compounding

## OVERVIEW

Of all the 17000 plastics available worldwide that are used in all the processes (extrusion, injection molding, etc.), most undergo compounding in order to meet product requirements [241]. It is estimated that over 65 wt% of the 150 million tons (340 billion lb) consumed worldwide are compounded in one of many different designed extruders and batch type mixers that have a prominent place in this industry. Compounding is a very large and important market.

The thousands of different additives and other ingredients used in the compounds improve the plastics performance during processing and/or meet the many different product requirements, including cost restraints (Chapter 20). The composition of the compound is dictated by the requirements of the converter's application. The method used to compound these plastics can be by dry blending, melt mixing, or both using batch or continuous methods. Figure 17.1 provides a guide to the different compounding methods [373]. Batch systems tend to be more labor-intensive and less complicated to operate and control. Continuous systems are more consistent and easier to instrument for statistical process control [1, 136, 164, 242, 301, 373].

Extrusion equipment, predominantly twin and other multi-screw types, provides the major source for compounding (Figs. 17.2–17.4). They usually operate at rates ranging from 230 to 2300 kg/h (500 to 5000 lb/h). Compounding is usually carried out at high rates because costs are reduced. Large extruders are required where screws could range up to at least 300 mm (12 in) in diameter (Chapter 2).

Compounding is performed by some plastic material suppliers, in-house by product processing (converters) plants, and companies that specialize in compounding. Usually the compounding is for a specific product application. The compounding line may be arranged several

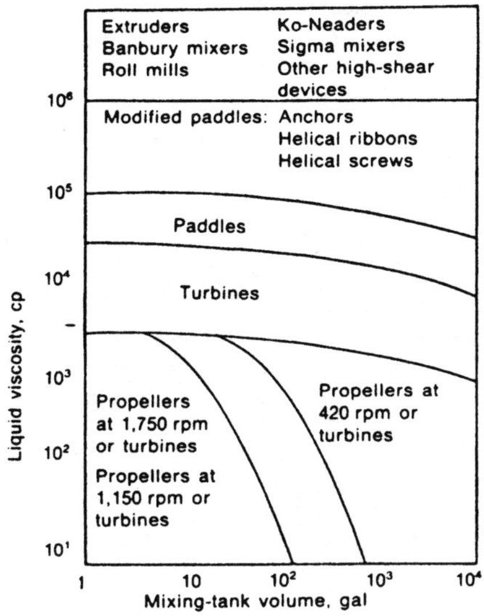


**Figure 17.1** Guide to mixing basics.

ways; it involves many operations. Each operation can be performed on more than one specific piece of equipment. This offers numerous choices, the best of which is frequently the simplest.

## PLASTIC MATERIALS

In compounding, a plastic is modified physically and/or chemically by additives that change its properties. Table 17.1 provides an example of a few of the changes that occur with polyolefins [113]. The ways in which a plastic can be modified are many. Because of an unlimited number of combinations, a plastic is usually available from a compounder in a wide range of grades.

Compounding also permits the mixing of two or more plastics to form plastic blends. In turn, these blends can include the many different available additives. Prior to compounding, a carefully developed mix (recipe) is used that could include the order to be followed when being compounded in the extruders. The better the mixing/compounding, the better the materials performances (Fig. 17.5). The plastic being used is usually in the form of powder just as it comes from the plastic manufacturer.

What makes compounding interesting and rather complicated is the fact that the many different plastics and their additive compounds behave many different ways (Table 17.1). The plastics and additives have different melt flow characteristics and controls. Recycled plastics are included in mixes [2, 3, 367].

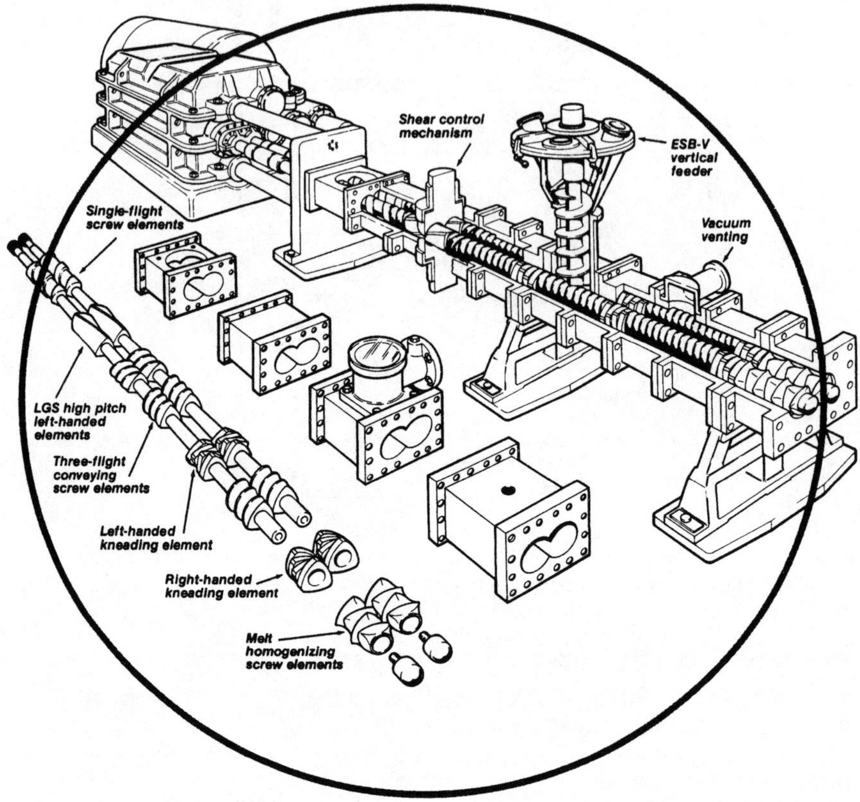


**Figure 17.2** Modular design compounding extruder by Werner & Pfleiderer that can easily be adjusted to meet changing material requirements.

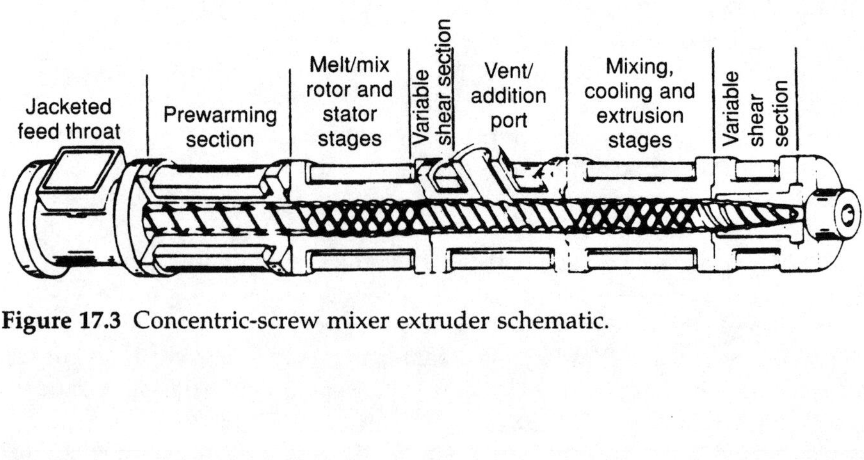


**Figure 17.3** Concentric-screw mixer extruder schematic.

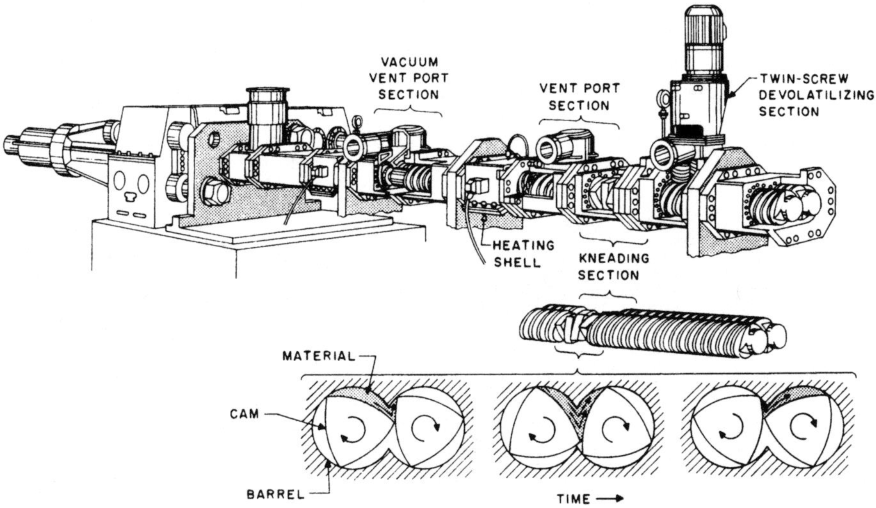


**Figure 17.4** Twin-screw W&P extruder that operates in different modes.

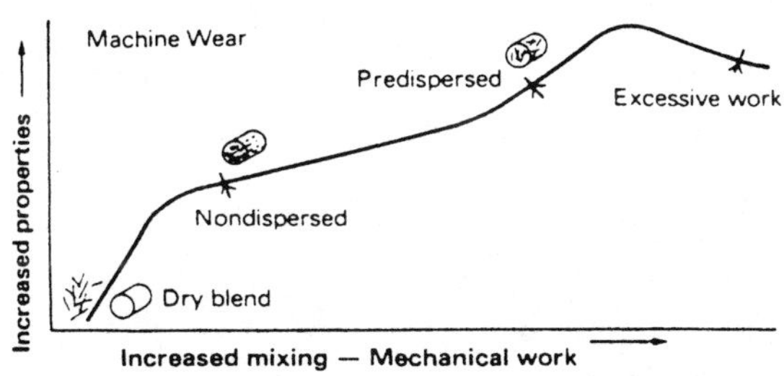


**Figure 17.5** Advantages of properly predispersing plastic compounds.

Color requirements are important to practically all processors. In the past, they used colored pellets. However, the time arrived when the processor could easily add color at the hopper. Concentrate or masterbatch compounds are used. They are acceptable but usually the most accurate method is to use colored compounds. Concentrates other than color are also used, such as lubricants and blowing agents, however it is best for the processor to study the best type of compound that should be used based on the plastic to be processed.

Flexible and rigid PVC compounds uses different additives to meet different requirements (Fig. 17.6). The major ingredient, as an example, for flexible PVC is the plasticizer(s) that provides flexibility. There are almost an unlimited number of plasticizers that provide from low to very high performances (low to very high cost). Selecting and ensuring the best

**Table 17.1** Example of property changes due to fillers and reinforcing agents used in polyolefins

| | *Rigidity* | *Heat resistance* | *Dimensional stability* | *Tensile strength* | *Impact resistance* | *Thermal conductivity* | *Acoustic properties* | *Price reduction by direct extrusion of the individual components* |
|---|---|---|---|---|---|---|---|---|
| Chalk ($CaCO_3$) | + | + | + | − | + | + | + | + |
| Talcum | + | + | + | − | − | + | + | + |
| Barium sulfate ($BaSO_4$) | + | + | + | − | | + | + | + |
| Wood dust | + | | + | − | − | | + | + |
| Glass fibres | + | + | + | + | − | | | |

\+ Property improvement.
− Property loss.

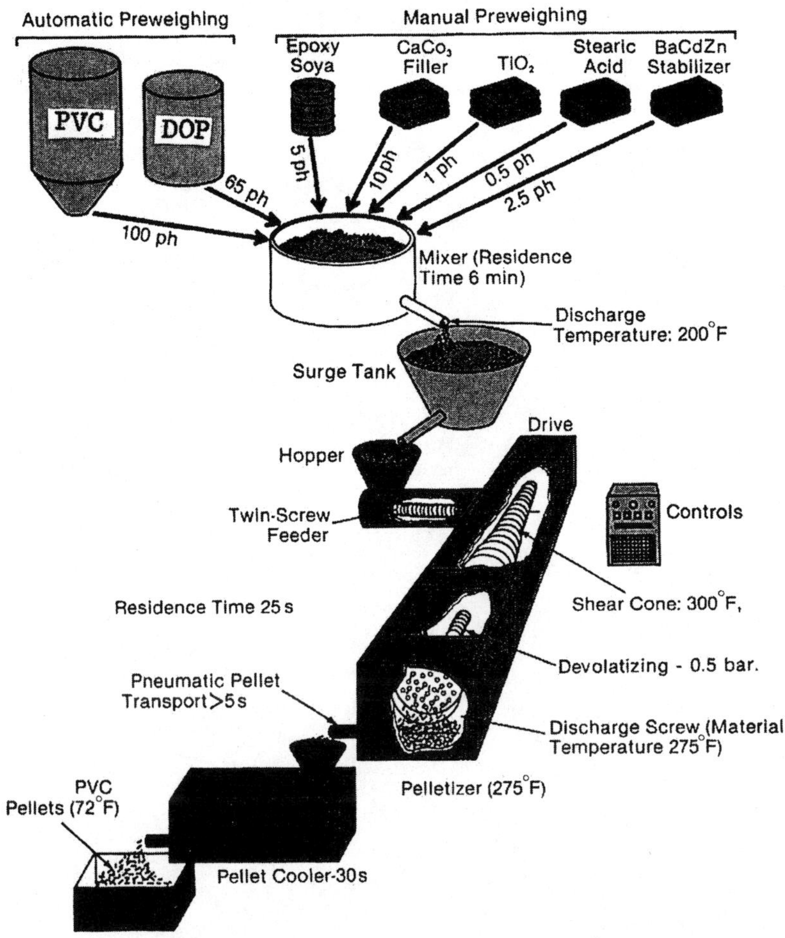


**Figure 17.6** Example of PVC compounding sequence for producing white shoe-sole compound.

grade to use becomes important, particularly since the compounds usually come in both the low to high cost types. Other additives include heat stabilizers and lubricants.

PVCs are regarded as the most versatile and suitable plastic used in many applications. In order to provide these capabilities, it is important that the actual mixing of this very heat-sensitive plastic be carefully handled. The credibility of the compounder (including in-house operations) becomes important, just as much as setting up the correct quality control procedures. This comment also applies to compounds other than PVC.

Figure 17.7 provides a nomograph that shows how to determine the weight of filled compounds. To use the nomograph, the only data needed is the specific gravity of the plastic and filler with the weight fraction of filler in the proposed compound. The example shown is based on a formulation that contains 30 wt% of a filler with a specific gravity of 2.5; the plastics has a 1.04 specific gravity. The nomograph is based on the following equation which can be used if more accuracy is desired:

$$\text{Sp. gr., compound} = 1/\left[\left(X/\text{Sp. gr., filler}\right) + \left(1 - X/\text{Sp. gr., plastic}\right)\right]$$

$$\text{Part weight, filled} = \left[\text{Part weight, unfilled}\right]\left[\left(\text{Sp. gr., compound}\right)/\left(\text{Sp. gr., plastic}\right)\right]$$

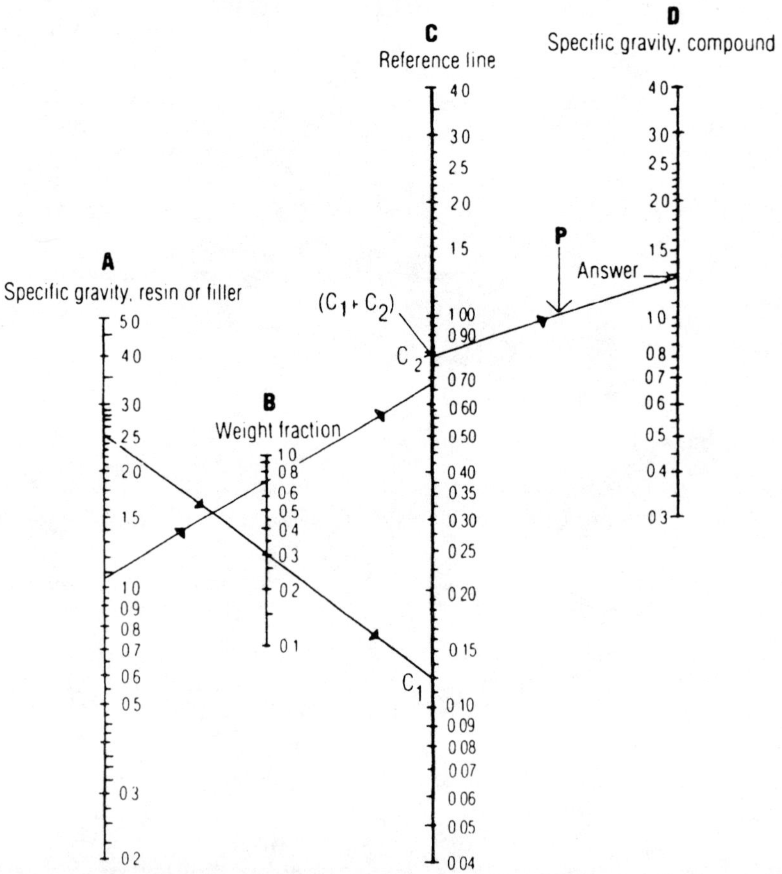


**Figure 17.7** Nomograph for determining the specific gravity of filled compounds by using various fillers and reinforcements.

$$\text{Plastic weight} = (1 - X)(\text{part weight, filled})$$

where $X$ = weight fraction of filler in compound, expressed as a decimal

## Blend

When two or more plastics are mixed, several factors should be considered. One factor is whether the phases of the plastics are ultimately miscible thermodynamically. In most cases where they are immiscible, which is predominantly the case with plastics, the minor component(s) become the dispersed phase(s) and the major one becomes the so-called matrix or continuous phase. Therefore, another relevant factor is the ultimate desired phase morphology (spherical, lamellar, co-continuous, etc.) [136, 151, 163, 164, 175, 222, 253, 289, 301, 308, 353, 373, 393, 411].

Other important factors include: (1) the order of feeding (mixing protocol), the thermal ($T_g$, $T_m$) and rheological (viscosity, melt elasticity) properties of each plastic (Chapter 3); (2) whether the feed streams are solids or melts, or both; (3) the average stress levels; (4) number of flow reorientations experienced by the melts; and (5) anticipated phase stability of the final morphology (resistance to coalescence) with respect to subsequent flow and heat histories downstream of the extruder mixing elements (pins, gears, kneading paddles, pelletization operations, static mixers, etc.) (Chapter 4).

When feed streams consist of separated bulk phases, irrespective of whether they are miscible or not, mixing processes in general, and extrusion compounding of plastics in particular, involve two basic mechanisms: the size reduction (break up into smaller regions) and the spatial rearrangement (uniform distribution of these regions). These mechanisms are commonly called dispersive and distributive mixing, respectively.

With all feeds in particulate solid form, a gross distributive mixing is partially built-in from the outset, which is an advantage. In this situation, attention must be paid to which component melts first and what will be the viscosity of the melt. Generally in melts: (1) dispersive mixing is enhanced by high stress levels (narrow gaps, converging channels, etc.); (2) distributive mixing is enhanced by stream splitting and recombination; and (3) both are enhanced by stretching and folding motions. This is the reason that plastic mixing and compounding machines are usually twin screw continuous or batch devices where extensional flows, stretching, and folding are predominant.

Single screw extruders, by contrast, are characterized by shear, nonreorienting flows, which may result in slow, incomplete, and inhomogeneous distributive and dispersive mixing. Specialized mixing elements incorporated onto single extruder screws correct this situation and improve processability. However, single screw extruders are better pumps than they are mixers.

The most favorable conditions are when the properties (viscosity, melt elasticity) of the dispersed and continuous phases are comparable at processing conditions (temperature, shear rate) or when the dispersed phase is slightly 'weaker.' The reverse is a much more difficult situation.

It is important to remember that the rheological/mechanical properties of all plastics are time rate dependent. In this condition, the stress increases with strain rate. This time rate has to be considered when attempting to predict the potential effect of extruder geometries and operating conditions on dispersed phase break up and homogenization [197, J. A. Biesenberger].

## PROCESSING CHARACTERISTICS

The basic functions of a compounding extruder is to melt the plastic(s) and evenly disperse additives, fillers, etc., to obtain the compounds performance requirements [338, 456].

For extrusion compounding, feeding the extruder often presents special difficulties. Many times powders may tend to cake in a hopper and not flow freely. Powders may also produce uneven pick-up by the screw. To overcome such difficulties, special feeding equipment is used that include screw stuffers, vibrators, and crammers (Chapter 2). They force material down the feed throat and into the screw at a controlled rate, usually weightwise. Devices are also used to reduce rates so that starve feeding can be used when required.

### Batch compounding

These are an important part of every compounding operation, primarily in the premixing and post compounding areas. Some of them can be called noncontinuous extruders. There are three intensities of batch mixing: low, medium, and high. Low-intensity mixers are the most commonly used because they are the least expensive, are relatively easy to clean, and can be employed for a multitude of tasks where time and dispersion are not of prime consideration.

Medium-intensity mixers usually consist of low-intensity mixers fitted with attachments called intensifiers. The intensifiers are normally a high-speed mixing head added to the low-intensity mixer.

The high-intensity mixer is a high-speed unit with specialized blade designs. Mixing times are typically measured in minutes, as compared to half an hour or more for the other type mixers. High-intensity mixers are used most often in the areas of pigment dispersion and premixing of the compound.

### Continuous compounding

These mixers use single and multi-screw extruders. The single screw extruders are simple in design and reasonably priced. They are extensively used and can perform certain desirable functions (Fig. 17.8). They subject melts to relatively laminar flow. Processing is normally limited to the barrel surface and conveying is only by drag flow. This action minimizes the use of single screws for low viscosity mixing. Many of their shortcomings can be overcome by grooved barrel walls, barrier screws, and shortened screw design (Chapter 2).

Important and extensively used to compounding are many different multi-screw designs with twin screws meeting special requirements (Chapter 2). They have differently designed screws to meet many different mixing requirements (Chapter 4). Some screws are built in sections so they may be taken apart and put together to provide different mixing actions (Figs. 17.9 and 17.10).

Twin screws provide much higher torque and volume than the single screw extruders. They are used to give a low melt temperature and very little decomposition of the compound. The usual screws rotate up to 400–600 rpm and are typically starve-fed. There are higher-torque extruders using oversized motors that have screws rotating at 1000 rpm and permitting more plastic input. Result could be doubling outputs. These extruders when operating at the same rpm can be compounding at least 30% more plastics.

**Figure 17.8** Typical Welex single-screw compounding extruder.

**Figure 17.9** W-P twin-screw compounding extruder.

An important action is that they produce good mixing by forcing the melt back and forth from one screw to the other screw. This action breaks up laminar melt flow patterns. Venting is very efficient giving good exposure to the melt in the vent zone (Chapter 4). Often the twin screw has self-wiping screws which can be purged rapidly and efficiently.

The nonintermeshing twin-screw has many advantages over the single screw. These include better distributive mixing and feeding because of the presence of higher volumes and the downward motion when using counter rotating screws. The fully intermeshing twin screw is by far the most versatile and the most popular design currently used. The intermeshing screws allows the mixer to approach plug flow, creating uniform heat and shear mixing. This type is a high-shear machine, permitting both distributive and dispersive mixing [301, 136].

Other mixing systems includes distributive mixing or the homogenization of the ingredients without particle size reduction. As will be reviewed, dispersive mixing is a high shear operation that reduces particle sizes.

### Feeder and blender

As with other processing equipment, suppliers of dryers, blenders, and metering equipment are increasing the feeding accuracy using different

**Figure 17.10** Theysohn Maschinenbau GmbH twin-screw extruder with a clamshell or book-type split barrel design.

devices such as microprocessor controllers. Also materials are being reduced in size with more uniformity to significantly improve uniformity in compounding. Processors can use blenders mounted on extruder hoppers that target for precise and even distribution of materials.

In volumetric blending, variable speed metering augers feed multiple components. Operators can calibrate the actual volume by occasionally diverting (manual or on automatic equipment) the mix to a sample chute and weighing the sample.

Gravimetric blending improves accuracy and requires less operator (if any) involvement in calibration, particularly when running processes where great accuracy is required. Metering by weight eliminates over feeding expensive additives. The principle of gravimetric feeding with throughput or metered weight control is well established. Equipment can at least provide an accuracy of at least ±0.25–0.50 wt% for ingredient and blend ratios of 2Σ (two standard deviations). With gravimetric metering, coextruders have a simple means of constantly maintaining the average thickness of individual films and overall thickness at better than 0.5 wt%. By comparison, volumetric and quasi-gravimetric blenders that use batch operation usually have accuracy variations of 2–10 wt%.

This mixing and metering activity is part of the overall conveying process within the plant. Conveying includes not only mechanical and pneumatic transport throughout the plant or within each individual processing line, but also filtration of the exhaust air and safety engineering to prevent dust explosion. With so much mixing required, color changes become increasingly frequent. Table 17.2 explains the advantages of central coloring versus coloring at the hopper throat. The processor should determine what is required and obtain equipment to meet the 'specification.'

## Reactive compounding

Reactive extrusion is one of the occasions when an extruder is used as a chemical reactor instead of just a melt processor (Fig. 17.11). This technique is also called reactive compounding and reactive processing where it refers to the performance of chemical reactions during extrusion of plastic compounds. The most common reactants are plastic or preplastic melts and gaseous, liquid, or molten low molecular weight compounds. A particular advantage of this extruder as a chemical reactor is the absence of solvent as the reaction medium. No solvent-stripping or recovery process is required, and product contamination by solvent or solvent impurities is avoided [91, 256, 266, 323, 297, 448, 452].

The chemical reaction may take place in the melt phase or, less

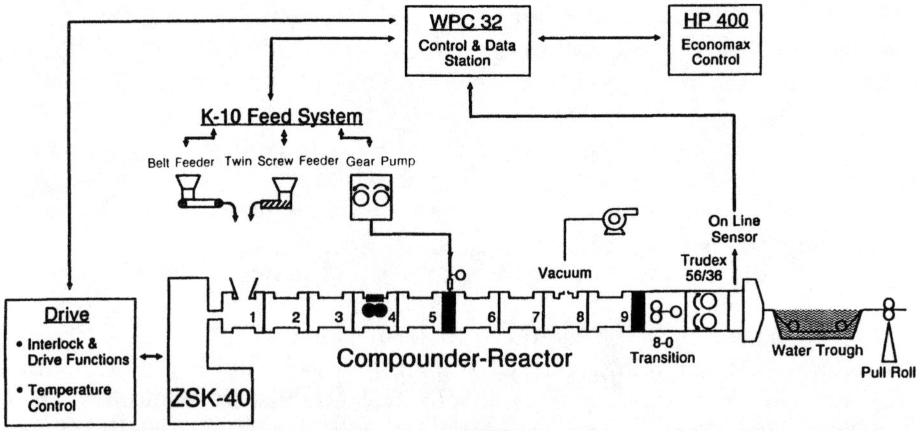


**Figure 17.11** Process flow diagram illustrates the W&P reactive process equipped with co-rotating, closely intermeshing, self-wiping screws.

**Table 17.2** Comparison of central blending vs. coloring in the hopper throat

| *Central blending* | *At-the-throat coloring* |
|---|---|
| 1. If the color blend is not correct (parts too light or more color than necessary in the mix), the mix must be emptied from the hopper and reblended, a costly and time-consuming process. Even then, parts may not have the exact depth of color you desire. | 1. Adjustments to color are made with immediate results. If parts are light in color, you can make an adjustment and see the results in minutes. |
| 2. Although some parts, as a cost-saving measure, could be with less color than others, the mixing of several batches with different let-down ratios is generally considered impractical. | 2. Every part can be custom blended while in production for optimum color usage and cost savings. Settings are recorded and used for exact repeatability on future runs. |
| 3. Conveying of blended material often produces separation of the color pellets from the natural material due to the much heavier bulk density of color pellets. The result is inconsistent coloring of parts. | 3. Material is conveyed without color; separation is not a factor. |
| 4. Normal machine vibration and flow patterns in the hopper cause some separation of the color pellets during residence time in the hopper. Again, the result is inconsistent coloring of parts. | 4. Color is metered only as parts are fabricated. There is no residence time for separation to occur. |
| 5. When a production run is over, any inventory of blended material left in the hopper must be held for use at another time. Storage space, contamination, spillage, mislabeling and tracking of this valuable inventory are all problems. | 5. All inventory of blended materials is **eliminated**. Unused color is returned to the container it came from, uncontaminated. |
| 6. When a production run for one color is over, your hopper must be emptied in preparation for the next color. Time is lost. Sometimes material is lost. If hopper and conveying system are not adequately cleaned, color contamination and rejected parts will result. | 6. Color changes require no emptying of the main material hopper. Production does not stop. Transition time to the new color is several minutes. |
| 7. If your material must be dried before use, the drying of color blends presents many logistical problems: more driers, careful planning for transition to the next color, cleanout of dryers, etc. | 7. You dry only natural material without having to concern yourself about color changes. |

commonly, in the liquid phase, as when bulk polymerization of monomers performed in an extruder, or in the solid phase when the plastic is conveyed through the extruder in a solvent slurry. The types of reactions developed include bulk polymerization, graft reaction, interchain copolymer formation, coupling or branching reaction, controlled molecular weight degradation, and functionalization or functional group modification.

The extrusion device as a reactor combines several chemical process operations into a single piece of equipment with accompanying high space time yields of product. An extruder reactor is ideally suited for continuous production of plastic(s) after equilibrium is established in the extruder barrel for the desired chemical processes.

Because of their versatility, most extruder reactors are twin screw extruders that posses a segmented barrel, each segment of which can be individually cooled or heated externally. In addition to external heating, a molten plastic may be shear heated by the resistance of viscous plastic to the conveying motion of the screw; these processes provide energy for chemical reaction. Extruder screws often have specialized sections or configurations, such as high shearing sections. Twin screw extruder screws may be equipped with interchangeable screw elements that provide different degrees of mixing and surface area exposure by varying the depth between screw flights, the individual flight thickness, and direction and degree of flight pitch.

In a typical reactive extrusion process, the reactants are fed into the extruder feed throat where the plastic is usually heated to initiate reaction or increase the reaction rate. The reactant mixture is conveyed through sequential barrel segments where a degree of mixing and specific energy input brings the reaction to the desired degree of completion within the limits of residence time in the extruder. At this stage, the reaction may be quenched by cooling or addition of a catalyst quencher where applicable and volatile by-products or excess reactants may be removed. Melt is forced from the extruder through the die with one or more openings. The design of the die openings is one factor determining the pressure against which the extruder has to pump by the conveying motion of the internal screw, as is the case with conventional extrusion processing. Melt from the die is usually rapidly cooled by contact with a fluid medium such as water.

## Pellet

In the production of TPs finished products, the primary form of materials to be processed are compounds in pellet form. Pellets offer the following advantages, particularly when they are uniform in shape and size: (1) simpler feeding system with fewer feeders; (2) dust-free handling; (3)

easier cleaning between changes in feedstock; (4) fewer unsatisfactory finished products; (5) greater extrusion (injection molding, etc.) capacity; (6) lower shipping costs as a result of higher feed-bulk density and (7) others. Disadvantages with pellets could be the plastics additional heating cycle (extending residence time) and the additional cost. Overall the advantages provide more gains to the processor.

Compounding extrusion lines usually produce pellets using dry-face dicers or the more popular strand pelletizer that uses underwater, water-ring, and water centrifugal pelletizers.

The strand pelletizers use two basic methods. One method produces strands 3.2 mm (0.125 in) in diameter and is called the strand pelletizer. The hot extruded strands after cooling in a water trough are cut in lengths of about 3 mm (0.125 in) (Figs. 17.12–17.16). The other method is called underwater pelletizing. The die openings are actually underwater. A fast rotating knife almost touching the die face cuts the strands of melt into almost spherical shapes as the strands exit the die openings (Figs. 17.17 and 17.18).

The strand pelletizer is a simple process. Generally, each strand will produce about 23 kg/h (50 lb/h) of pellets. So if one wants a 454 kg/h (1000 lb/h) total output, the die will have 20 orifices. These die openings are usually 3.2–6.4 mm (0.13–0.25 in) in diameter to produce a strand

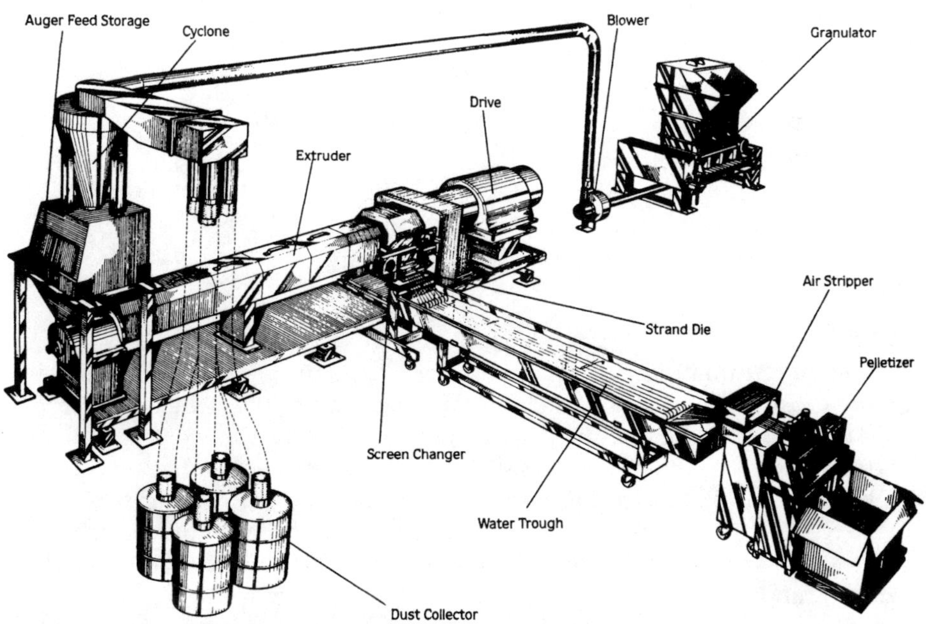


**Figure 17.12** Example of a complete pelletizer system.

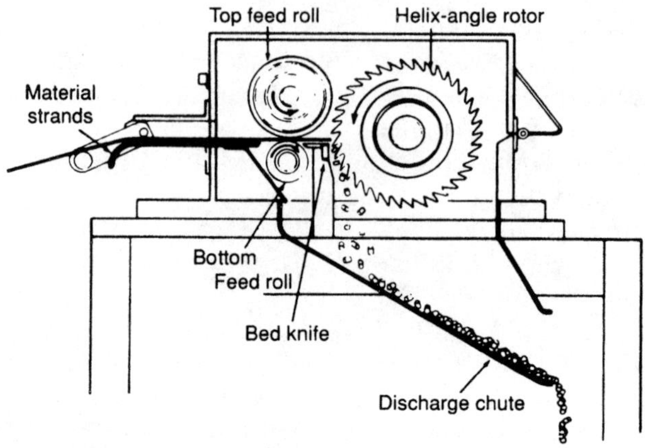


**Figure 17.13** Example of continuous strand pelletizer.

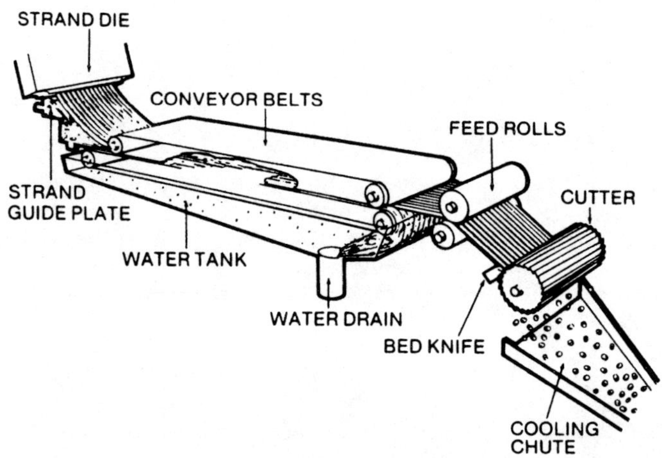


**Figure 17.14** Example of strand wet-cut pelletizing.

diameter of 3 mm (0.125 in). A drawdown ratio is used. This ratio is based on the cross section (area of a circle) of the strand to the pellet. Thus, for a 6.4 mm (0.25 in) die opening, the DDR is 4 (Chapter 12).

As reviewed in other chapters, the die is streamlined so that the same melt flow rate will exit each opening. Depending primarily on the strand's melt strength, the die can face from straight out to straight downward. As short an air gap from the die to the water trough as possible is desired. When strands leave the water bath, some type of system, such as blowing air, is generally used to remove the water from the strands. Strands are pulled by rolls at the entrance of a cutter. They are cut while the strands

**Figure 17.15** Example of a strand pelletizer with water-cascade guide pan.

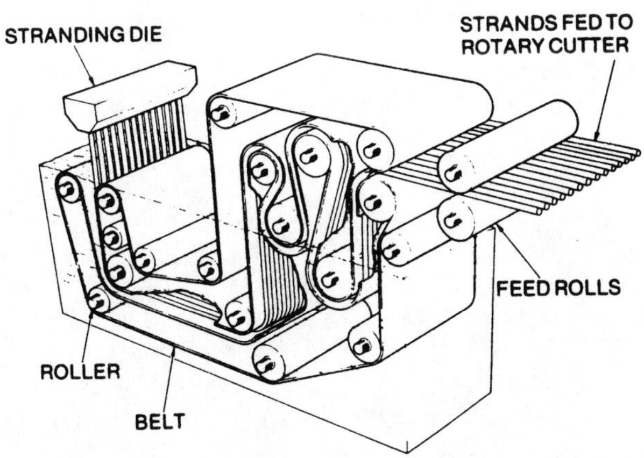


**Figure 17.16** Example of a forced-conveyance strand pelletizer.

are still hot so proper precaution is taken to ensure no sticking occurs and blades are changed before they become dull and not provide the proper cut. Pellets are dried, may be separated from shorts and longs, blended and possibly coated with additives prior to being packaged.

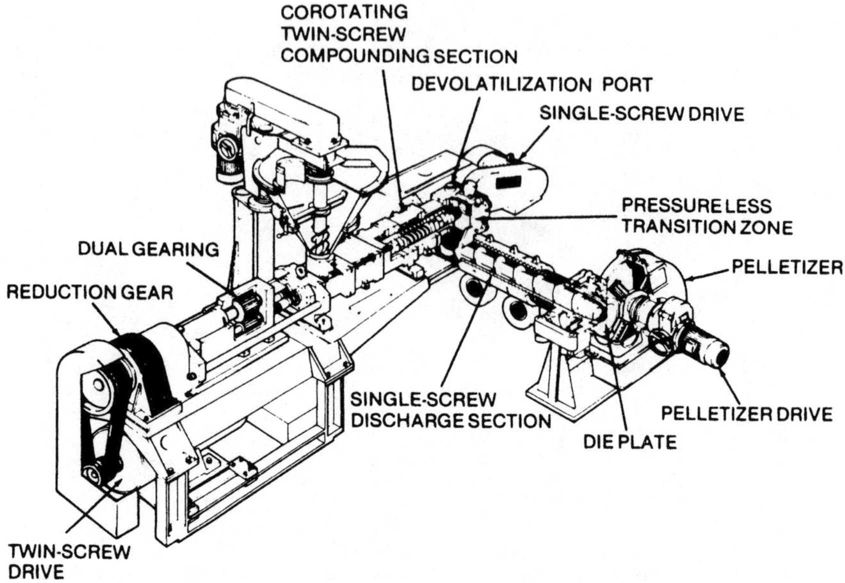


**Figure 17.17** Schematic of a two-stage compounding WP extruder.

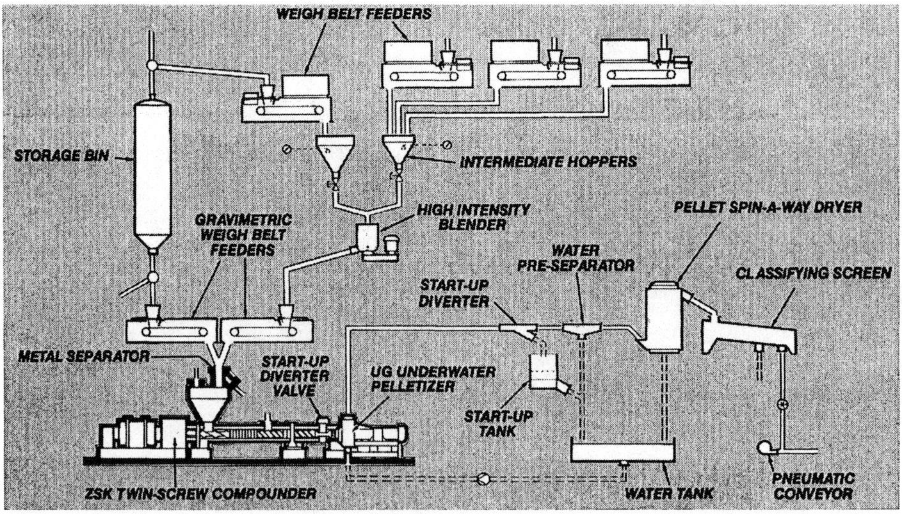


**Figure 17.18** Example of a compounding operation for polyolefins.

During this process some so-called fines and longs are produced. They can cause conveying and feeding problems by the processor. Fines caused by the cutter may result in dusting problems. Longs are pieces of strands up to 100 mm (4 in) long which go through the cutter.

The underwater pelletizer was developed to eliminate handling a large number of strands as well to eliminate fines and longs. When you study this system, recognize that a major accomplishment exists. There is a die with the very hot melt on one side and cool water on the other side of the die with a pressure drop in the die going back to the breaker plate and screen pack [411]. With improper operation, the die is frozen solid by the water resulting in the lines down-time and a clean-up mess. Start-up procedures have to be prepared and strictly followed.

This method provides high rates of production requiring the larger extruders. The dies can have many more openings than the strand pelletizer (without having handling problems of the strands). Pellets are cut at the die face and are carried away by water circulating past the die face at high rates. In turn, they are collected on a conveying belt and are blown free of water by air. The final operations are drying, blending, and packaging.

Most extruded pellets must conform to a few critical dimensions and shapes. Generally, the pellets can vary in size more than extruded film, sheet, etc., and yet be perfectly acceptable. However, closely controlled pellet sizes makes it easier for the user to process. There are processors that require the more expensive, very tightly controlled pellets or else they cannot produce acceptable products.

## Dicer

Instead of producing pellets, diced shaped material can be produced. After a plastic compound has been prepared, it can go through a slit flat die opening extruding a narrow or wide flat strip or sheet (Chapter 10). Roll mills are also used to produce the strip or sheet. Like the strand pelletizer, this strip or sheet is cooled usually through a water trough. The flat material is pulled by rollers that direct it into the cutting/dicing chamber.

Dicers can produce square, rectangular, parallelogram, and other shapes; Figures 17.19 and 17.20 are examples of the dicers. As the plastic enters the cutting chamber, size reduction is achieved through the action between rotating knives on a rotor and a stationary bed knife. The design configuration of the knives, strip-fed rate, rotor speed, and the number of rotating knives determines pellet size, configuration, and their output rate. After the strip is diced, pellets or cubes fall through a discharge chute onto a vibrating screen for separating stringers, etc.

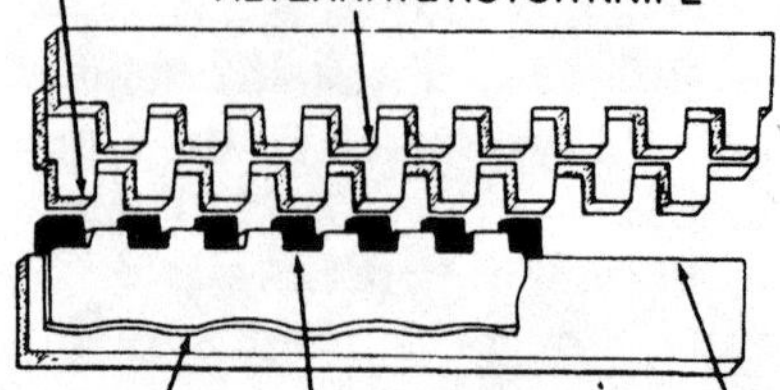


**Figure 17.19** Example of a notched-knife dicer.

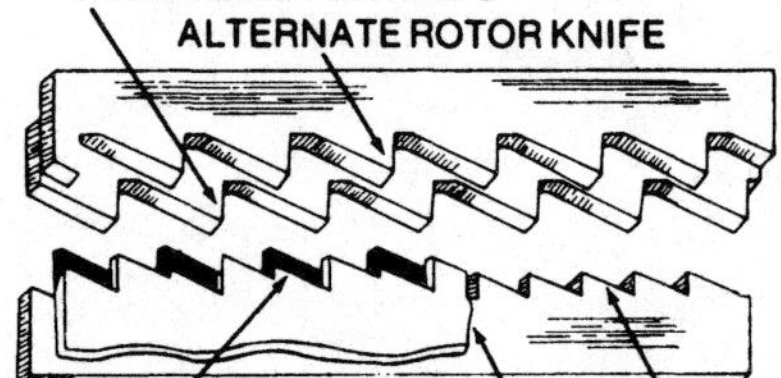


**Figure 17.20** Example of a ratchet-tooth dicer.

### Type of mixer

In addition to the extruder compounding mixers, there are other types. Figure 17.1 shows the type of equipment as a function of viscosity and vessel size [373]. Extruders are included at the top of the graph. Most in the graph represent batch type mixers.

## PROCESS OPTIMIZATION

In a compounding operation with multiple lines and diverse products averaging runs from 454 to 45 000 kg/lot (1000 to 100 000 lb/lot), cross contamination can become a real problem. When changing the compound material in a line, it is essential to take time for thorough clean-up of the equipment in the immediate peripheral area. This includes not only the feeders, through the compounder, to the downstream equipment, but also paying attention to parts of the building structure.

It is not uncommon to have settling dust, fines, and powders accumulate on the beams, ducts, pipes, and conveying lines. If not properly cleaned before start-up, these contaminates can break loose due to conditions, such as vibration, air movements, or wind. In turn, they can land in the gaylord, the equipment mechanical devices and controls, and/or on the compounded product.

Good manufacturing practice requires that attention be given to these details. It is good practice to have containers with processed materials covered at all times [187, S. Jakopin].

**Control**

Significant improvement in compounding has been through process control that typically monitors the equipment and the compound, including the temperature of the motor, lubrication system, torque, oil level in the gear box, etc. The compound is monitored for conditions such as material mixing ratios with feed rates, melt temperature profile, back pressure, die pressure, etc. (Chapters 2 and 6).

With data acquisition, the compounder collects data and trend variables over the entire run to ensure a consistent product is being prepared. Statistical process control can be added to the system to inform the operator when the processing variables are going out or have gone out of specification.

Most compounding systems are starve-fed, so accurate control of the feed streams is required. An accurate feeder is the loss-in-weight designs. In these systems the entire feeder, including the hopper, is placed on a load cell or is sometimes suspended from the load cells. The weight can be measured as often as 10 times per second. This information is then processed through the controller to increase or decrease the feed rates, based on previously set points.

## DIE

As reviewed earlier, dies used for extruder compounding are of a relatively simple deign. Die design is reviewed in other chapters. When compared to the other dies reviewed in the other chapters, a basically simple streamline approach is used for the pellet and dicer operations (Chapter 5).

## TROUBLESHOOTING

For a compound to be useful, any additives should generally enhance a specific property(s) without significantly degrading the other properties of the plastic. Additives should not produce discoloration over the expected lifetime of the finished product. Because additives are usually complex chemicals, they can interact with each other or with the bulk plastic in various ways, producing unexpected, deleterious effects.

A common problem could be improper compounding. Incomplete mixing can produce small, hard particles 0.5 mm (20 mil) in diameter scattered throughout the plastic part; these particles are known as flecks. Another problem can be that insufficient stabilizers (to prevent decomposition) are used, or its nonuniform distribution. This situation may result in small, discolored particles occurring throughout the part if processing temperatures exceed the upper stability limit of the plastic.

Small bubbles can be formed with the additives decomposing that generate water vapor or other gases products. Bubbles can also be caused by additives that oxidize or dehydrate at the processing temperature. Also, bubbles form when moisture accompanies the additives or plastic. Hygroscopic plastics not properly dried will cause bubbles and degradation of the plastics (Chapter 3).

Improper use or using the wrong additives cause all kinds of problems with the plastic compounded finished product. Different chemical and mechanical destructive results can occur. These results may include: (1) bleeding of the additive(s) to the surface; (2) blooming; (3) plate-out where additive is deposited on the converters processing equipment such as rolls and/or on the plastic; and (4) chalking where pigment or filler particles surface forming a dust-like coating.

# 18

# Other processes

## INTRODUCTION

In the world of extrusion, the extruder is used in different manufacturing operations in addition to the usual operations that have already been reviewed in this book. This chapter will present some of these other processes. They basically include continuous and noncontinuous operating machines [1–4, 95].

## INJECTION MOLDING

Injection molding (IM), which is a continuous or noncontinuous extruder, is a large business worldwide consuming approximately 32 wt% of all plastics. It is in second place to the conventional extrusion which consumes approximately 36 wt% of all plastics. In the industry, IM is not identified as an extruder. It has a screw plasticator which is an extruder that prepares the melt as explained in Chapters 1 and 2. This screw extruder plasticator is also called an injection unit [2].

The injection molding machine (IMM) can be identified by the two popular methods of operation: the one-stage (Fig. 18.1) and the two-stage (Fig. 18.2) systems. The one-stage system is more popularly known as the reciprocating screw IMM, which is actually a noncontinuous extruder. The popularly known two-stage system has other names, such as the piggy-back IMM, which is a continuous extruder.

Both these machines utilize through the following stages:

1. Plasticizing: heating and melting of the plastics in the usual way, as reviewed in earlier in this book;
2. Injection: injecting under pressure a controlled volume shot of melt into a closed mold with solidification of the plastics beginning on the mold's cavity wall;

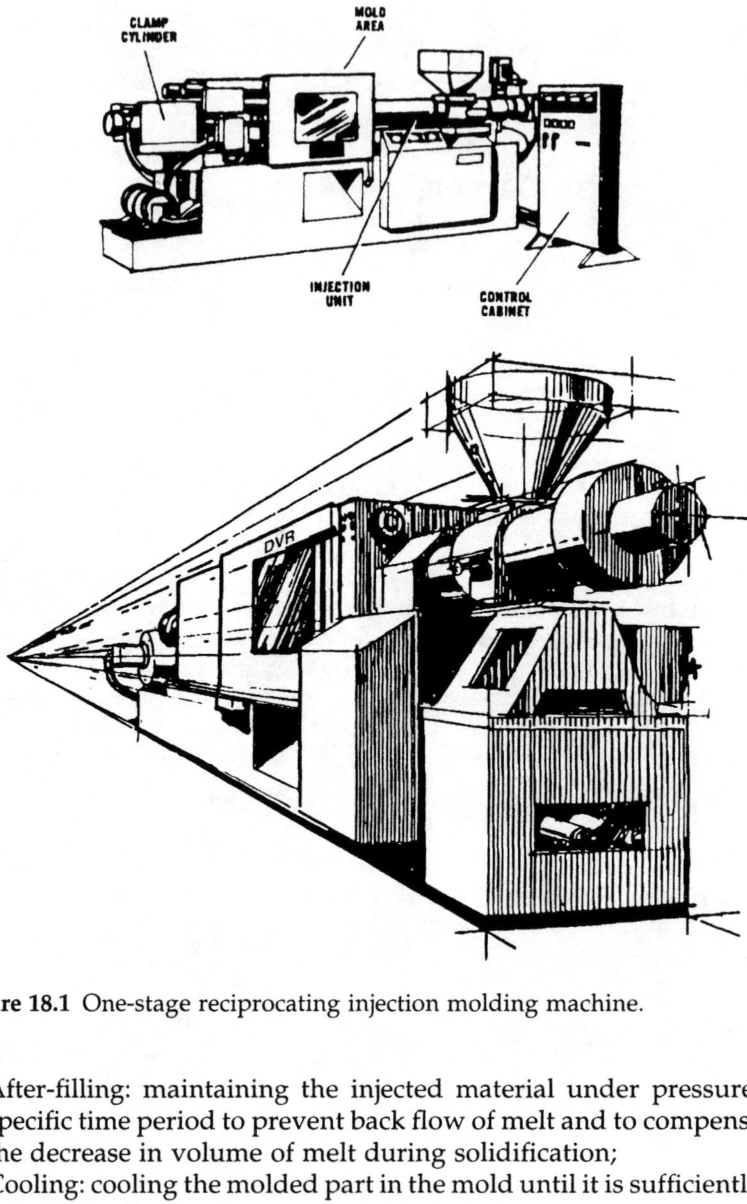


**Figure 18.1** One-stage reciprocating injection molding machine.

3. After-filling: maintaining the injected material under pressure for a specific time period to prevent back flow of melt and to compensate for the decrease in volume of melt during solidification;
4. Cooling: cooling the molded part in the mold until it is sufficiently rigid to be ejected;
5. Molded part released: opening the mold, ejection of the part, and closing the mold so it is ready to start the next cycle with a shot of melt.

The IMM features the three basic components of an injection unit, mold, and clamping system. The injection unit, which is an extruder, prepares the proper melt and transfer the melt into the next component which is the

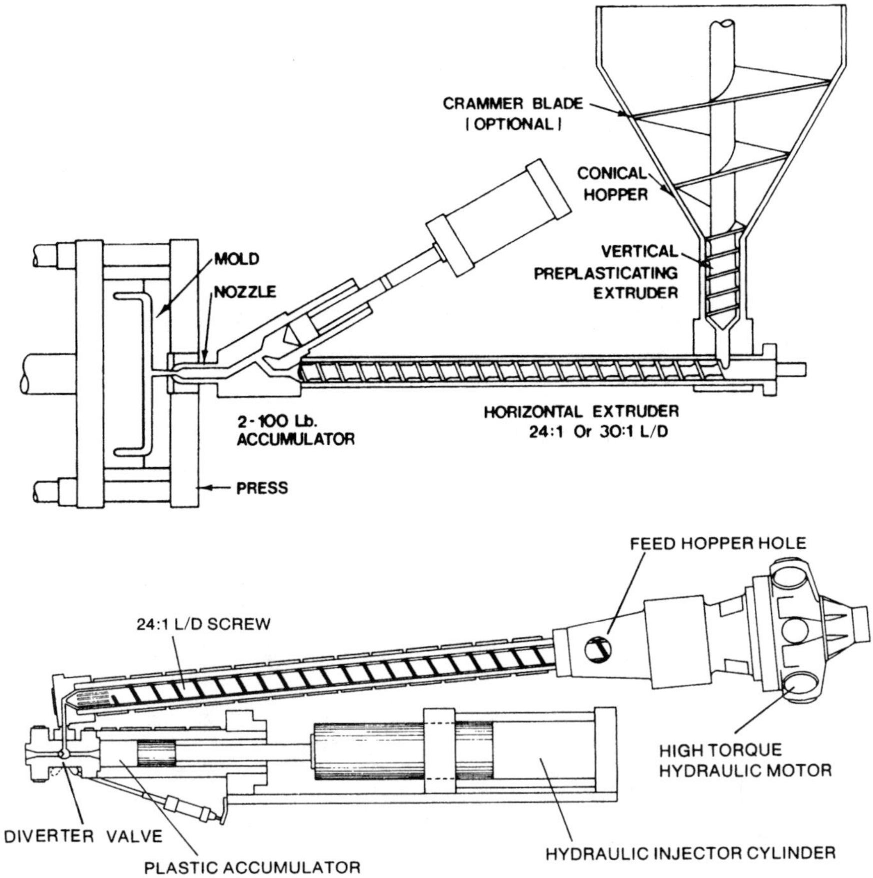


**Figure 18.2** Plasticator with two-stage injection molding machine.

mold. The clamping unit closes and opens the mold. This operation tends to be more complex than other processes since it involves moving and stopping the melt into the mold, rather than having a continuous flow of melt without interruptions. Examples of how the IMM effects product performance are shown in Fig. 18.3 [2].

The IM process is extremely useful since it permits the manufacture of both simple and very complex and intricate three-dimensional shapes. When required these shapes/parts can be molded to extremely tight tolerances. The IM process needs to be thoroughly understood in order to maximize performance and mold parts at the least cost. Examples of the controls required are summarized in Figs. 18.4 and 18.5. They relate to controlling temperature, pressure, and cycle time as summarized in Fig. 18.6 [2].

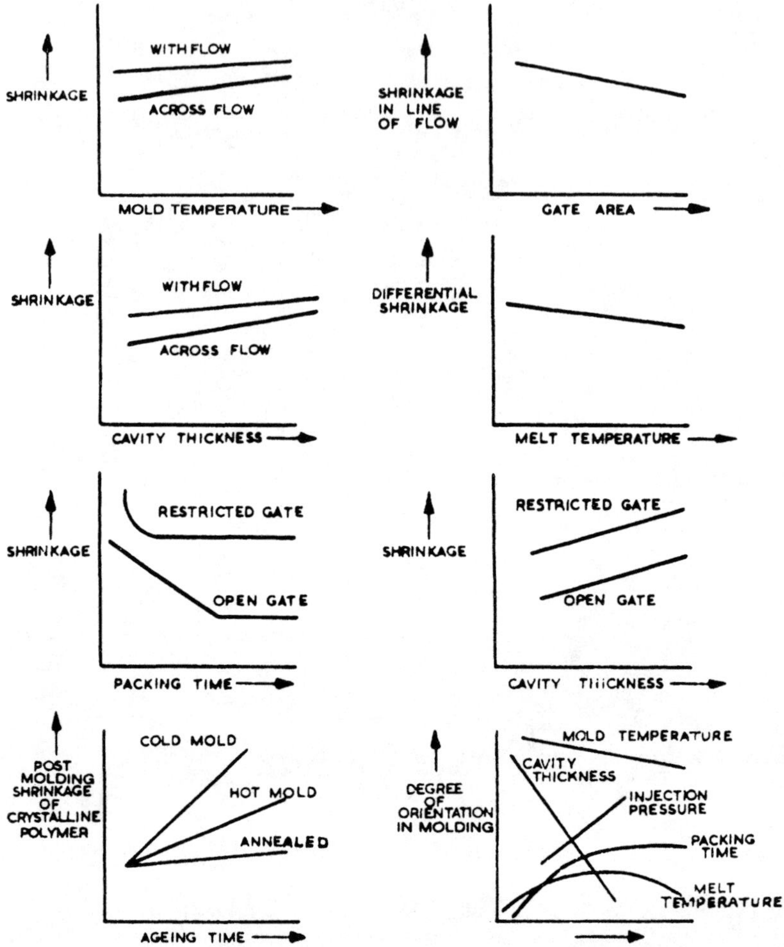


**Figure 18.3** Effect of IMM variables on product performances.

IMMs are characterized by their shot capacity (shot represents the volume of melt required by the mold cavity or cavities), rate of injection, injection pressure, and clamp pressure with mold-locking force. Shot size capacity may be given in terms of the maximum weight that can be injected into a mold cavity(ies), usually quoted (as a standard) in ounces or grams of general purpose PS (GPPS). A better way to express shot size is in terms of the volume of melt which can be injected into a mold at a specific pressure. The rate of injection relates to its speed and also the control capability of cycling the melt to move fast–slow–fast, slow–fast, etc.

Injection pressure in the barrel when the shot is injected can range from

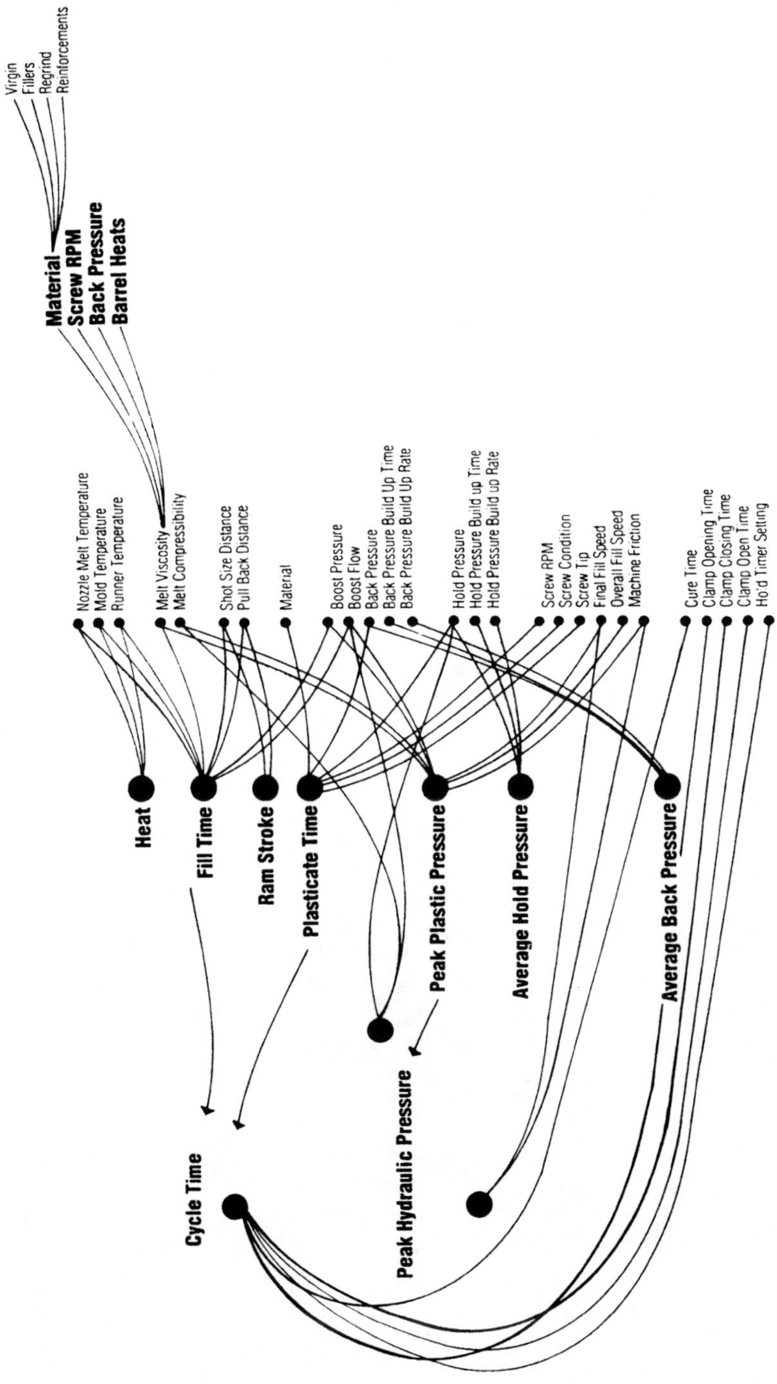


**Figure 18.4** IMM controls.

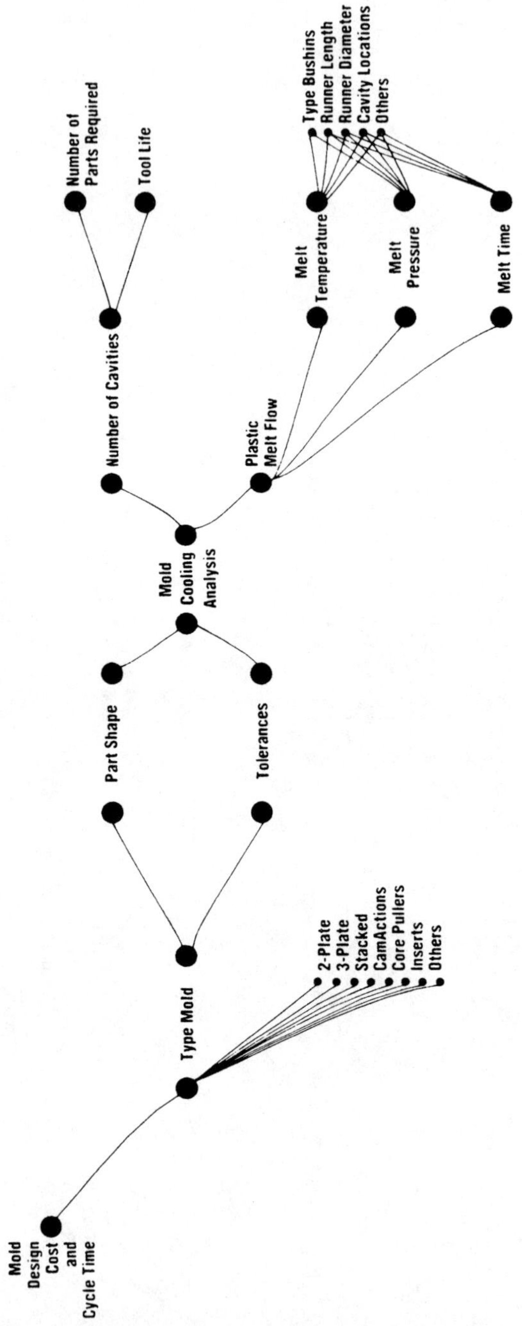


**Figure 18.5** Mold operation: important control factors.

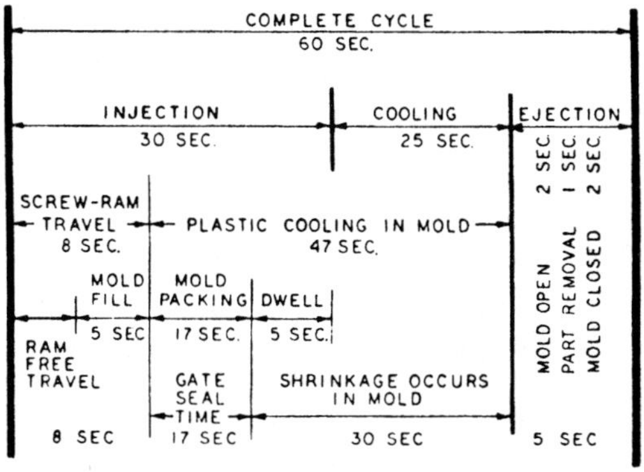


**Figure 18.6** Example of IM cycle for processing TPs.

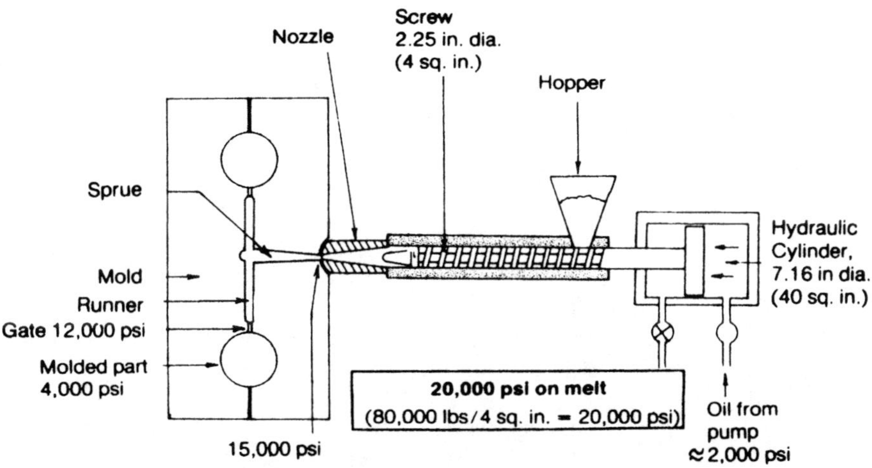


**Figure 18.7** Example of pressure loading on a melt during IM.

14 to at least 205 MPa (2000 to 30 000 psi). The characteristics of the plastic being processed defines the pressure is required in the mold to obtain good parts. Based on what cavity pressure is required, the barrel pressure has to be high enough to meet pressure flow restrictions going from the plasticator into the mold as shown in Fig. 18.7. This figure provides an example where the plastic required a pressure in the mold of 28 MPa (4000 psi). The shot size in this example, which uses a TP cold runner mold, includes what is required in the mold cavities and from the IMM nozzle to the cavities. If a TP hot runner mold is used where the sprue and runners do not solidify, after the first shot is made, the shot size is only what is required in the cavities. Thus, the next shot moves from the

runner's gates into the cavity [2]. Table 18.1 provides an example of the effect of machine operation with product performances [2].

The clamp pressure required in IMM depends on plastic being processed (that requires a specified pressure in the mold cavity) and basically the cross sectional area of any melt located on the parting line of the mold (that includes the cavities and the TP cold runner that is located on the parting line; whereas the TP hot melt is located within the mold and not in the parting line). By multiplying the pressure required on the melt and the cross sectional area, the clamping force required to keep the mold closed during injection is determined. To provide a safety factor, consider a 10–20% increase.

Thus, depending on what is being molded, the IMM clamping force may be from less than 18 tonne (20 ton) to thousands of tonnes. The average machines used provide pressures from 90 to 360 tonne (100 to 400 ton), with large machines that provide thousands of tonnes' clamping pressure. A force is also required to open the mold that is usually much less than the clamping force. So one has to ensure that available opening clamping pressure is available. Usually this requirement is not a problem. Clamping predominantly uses hydraulic systems. Also becoming popular are all electric drive systems and/or with hydraulic systems. The actual mechanical mechanisms range from toggle to straight ram systems. Each of these different systems have their individual advantages.

Even though most of the literature on IM processing specifically identifies or refers to TPs, some TSs are IM. Approximately 90 wt% of all IM plastics are TPs. During IM, the TPs reach maximum heat during plastication before entering the mold. The TSs reach maximum temperatures in the hot molds. If too much heat develops during plastication, the TSs solidify and the IMM stops requiring clean-up, etc. See Chapter 3 regarding temperature files for TPs and TSs.

**Noncontinuous extruder**

The reciprocating screw IMM prepares the melt and injects the melt into the mold cavity all in one-stage. Plastic is melted in the machine's barrel and transferred toward the nozzle end of the machine by the rotating screw. The accumulation of melt at the screw tip forces the screw toward the rear of the machine until enough melt is collected for a shot. The required low back pressure on the screw is used during this plasticating action and when the shot size is produced, the screw stops rotating, thus the process is noncontinuous. As soon as the mold has ejected its molded part(s) and closes, the nonrotating screw acts as a plunger and 'rams' the melt into the cavity(ies) using very high injection pressure [2, 16, 28, 48, 65, 255, 311, 331].

This reciprocating sequence of operations are shown in Fig. 18.8 where:

**Table 18.1** Troubleshooting IM: simplified approach for cause due to plastic material

| | Possible causes | | | | | | | | | | | | |
|---|---|---|---|---|---|---|---|---|---|---|---|---|---|
| *Problems* | *Too high a moisture content* | *Too-little lubricant* | *Too-lmuch monomer* | *Contaminated granules* | *Too high a proportion of material to be ground* | *Too-long preheating* | *Too high a drying temperature* | *Uneven addition of colorant* | *Uneven granule size* | *Too-much fines* | *Uneven granule feed* | *Variations in granule preheating temperature* | *Variations in moisture content* |
| Sink marks | × | × | × | | | | | | | | | × | |
| Flow marks | × | | | × | | | | | | | | | |
| Brittleness | × | | | × | × | | | | | | | | |
| Discoloration | × | | | | | × | × | × | | | | | |
| Surface blemishes | | | | × | | | | | × | × | × | × | |
| Varying shrinkage | × | | | | | | | | | × | | × | |
| Varying dimensional | | | | | | | | | | | | | |
| Stability | × | | | | | | | | × | | × | | |
| Sticking to the mold | × | | × | | | | | | | | | | |
| Varying strength | | | | | | | | | | | | × | × |

(a)

(b)

(c)

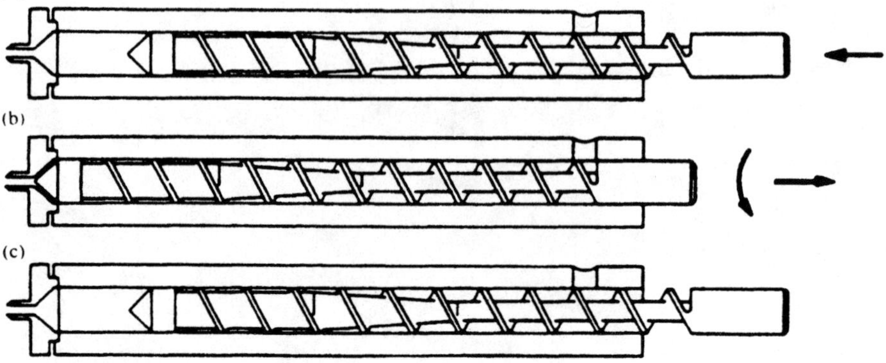

**Figure 18.8** Reciprocating screw sequence of operations.

(a) nonrotating screw moves axially forward, forcing the hot melt into the mold cavity(s); (b) after no more pressure is required and the molded part starts solidifying, the screw starts turning and retracting to prepare the next melt shot; and (c) with the shot size prepared, the screw stops rotating and gets ready for going into the (a) part of the cycle.

The advantages of the reciprocating screw IMM compared to the two-stage IMM include the following: (1) reduced residence time; (2) self-cleaning screw action; and (3) possibly more accurate and responsive injection control. These advantages are the key to processing heat-sensitive plastics.

**Continuous extruder**

A two-stage IMM uses a fixed plasticating screw (first-stage) to feed melted plastic into a chamber (second-stage). After the second-stage receives its shot size, a plunger then forces the melt into a mold. Its first stage is a continuous extruder similar to those described in Chapters 1 and 2. A diverter valve has three positions. In the first position, it is closed and facing the extruder while the melt is being prepared. Next position directs the melt from the extruder into an accumulator. The third position directs a shot of melt from the accumulator into the mold cavity. During this fast action, which occurs within seconds, the extruder continues to operate, producing melt.

When the valve is not directing melt into the accumulator, the melt remains in the barrel and could be building up slight pressure for a short time period. The extruder is designed so that the screw can have some motion in moving back for melt to accumulate in the front of its barrel without any major build-up of pressure. It is also designed so that controls can be set whereby no damage occurs to the melt.

When compared to the reciprocating screw IMM, the advantages of the nonreciprocating IMM include: (1) consistent melt quality; (2) ram action in the accumulator provides high injection pressure very fast; (3) very accurate shot size control; (4) product clarity; and (5) easily molding very thin-walled parts. Disadvantages include higher equipment cost and increased maintenance.

## Mold

Optimizing the IM process to reach higher productivity includes the design and operation of its mold. A mold can be a highly sophisticated/ expensive piece of machinery. It can comprise of many parts requiring high-quality metals and precision machining (Fig. 18.9). To capitalize on the advantages of IM, the mold may incorporate many cavities, adding further to its complexity.

Many molds have been pre-engineered as standardized products that can be used to include cavities, different runner systems, cooling lines, unscrewing mechanisms, etc. As Fig. 18.5 shows, it requires controls to operate efficiently [2].

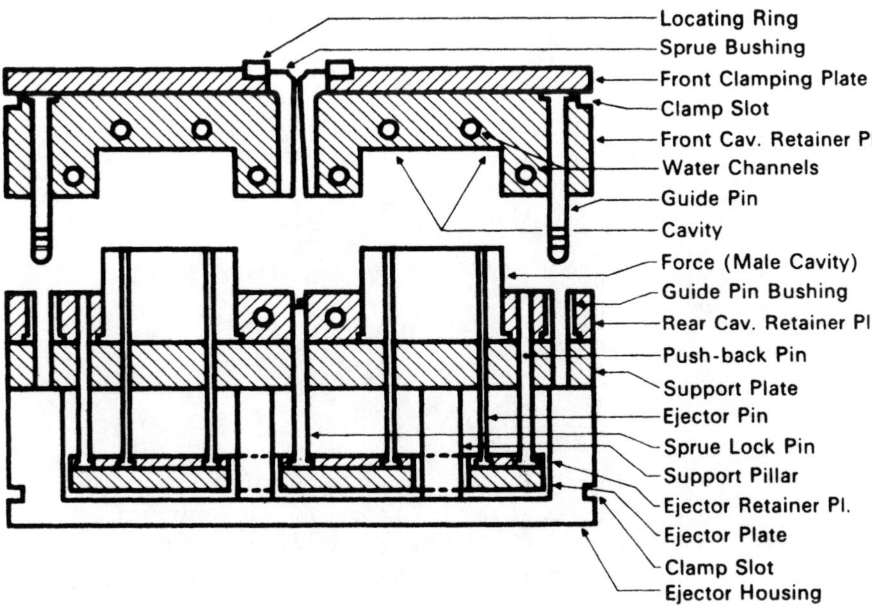


**Figure 18.9** Example of a preengineered standard mold.

## INJECTION–COMPRESSION MOLDING

Coining, also called injection stamping and more often injection–compression molding (ICM), is a variant of injection molding thermoplastics (Fig. 18.10). The essential difference lies in the manner in which the thermal contraction of the molding during cooling (shrinkage) is compensated. With conventional IM, the reduction in material volume in the cavity due to thermal contraction is compensated by forcing in more melt during the pressure-holding phase.

By contrast with ICM, the melt is injected into a cavity that has is a short shot in a compression mold where a male plug fits into a female cavity [1] rather than the usual flat surface parting line mold halves for IM. The melt injected into the cavity is literally stress-free; it is literally poured into the cavity.

Prior to receiving melt, the mold is slightly opened so that a closed

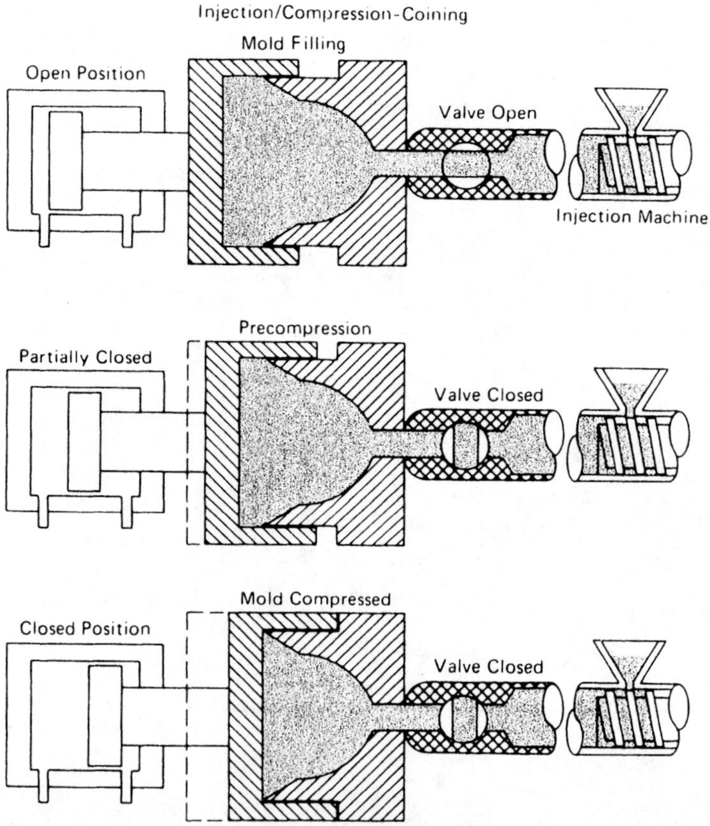


**Figure 18.10** Coining combines injection molding with compression molding.

cavity exists; the male and female parts are engaged so the cavity is closed. After the melt is injected, the mold automatically closes, producing a relatively even melt flow. Upon controlled closing, a very uniform pressure is applied to the melt. Sufficient pressure is applied to provide a molded part without stresses. This type of molding has many advantages in favor of molded part performances [2].

The ICM process has been used for parts of different sizes, particularly thick-walled parts with tight dimensional requirements such as optical lenses and compact disks. The short-shot melt in the cavity literally flows unrestricted resulting in stress-free parts.

## INJECTION/BLOW MOLDING WITH ORIENTATION

This process called molding with rotation (MWR) combines injection molding and injection blow molding (Chapter 15) with melt orientation (Chapter 2). Both the IM and IBM machines are 'extruders.' The process as shown in Fig. 18.11 is patented by Dow Chemical. The equipment used is what is commercially available except the mold is modified. The male or female part of the mold rotates, usually the male core rod is rotated.

This process provides a practical approach by which multi-axial orientation can assure maximum properties in the molded part that can be a bottle or other shape having polar axis of symmetry and reasonably uniform wall thickness. Chapter 2 describes the advantages and processing requirements for obtaining the proper orientation characteristics.

Parts can be molded directly from the IMM or can be used in a two-stage fabrication where the parison is first molded and later blow molded in another machine. Cooled parisons fabricated with MWR can be reheated and blown without problems of eccentricity. Original R&D with MWR included the use of AN plastics producing beer bottles.

The first step is injection molding a preform in an injection molding machine (IMM) the conventional way. With core rotation, the pin self-centers resulting in excellent part wall uniformity. Orientation of multiple layers of melt occurs when the core rotates. The next step involves blow molding by transferring the hot oriented preform by means of its core pin to a blowing station (Chapter 15).

Temperature control and profile of the melt during rotation near its melt temperature are very accurately controlled with respect to temperature and time. Controls are required regarding the usual machine operations with special emphasis on relating temperature to injection rate into the cavity, rotational speed, and temperature profile of the preform in the axial direction just prior to blowing.

The molder using the MWR process can control the molecular orientation. During fabrication, two forces act on the plastic melt that are the injection (longitudinal) and rotational (hoop). The targeted balanced

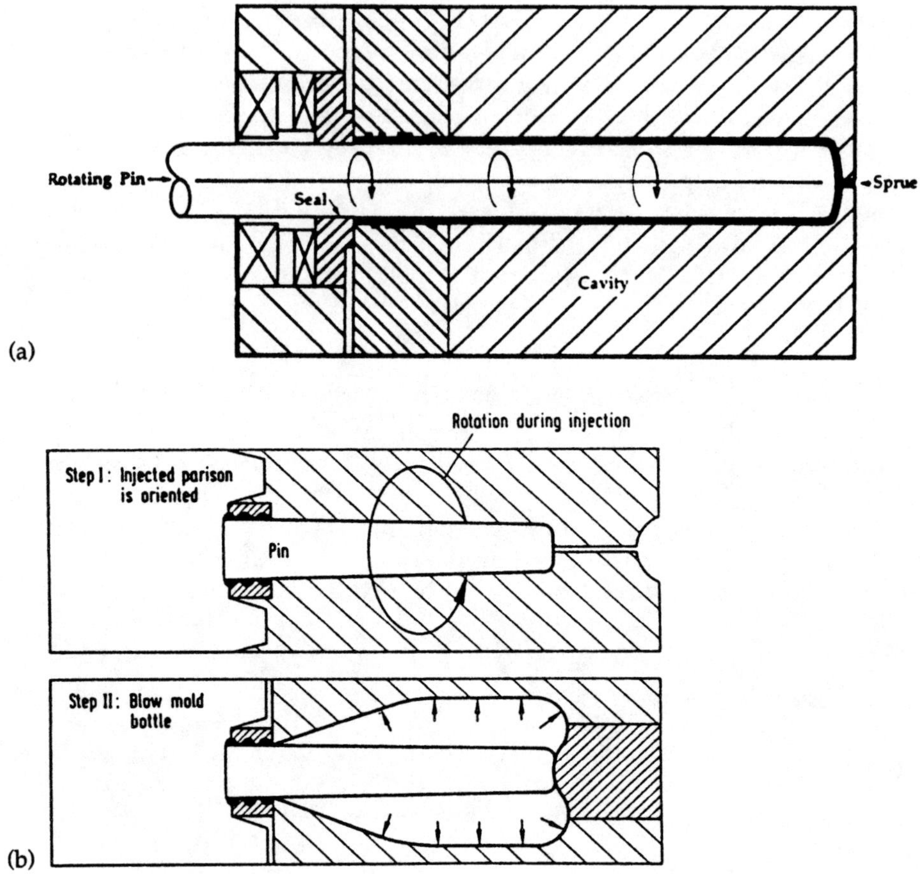


**Figure 18.11** Schematics of injection molding with rotation: (a) example of pin rotating; and (b) example of cavity rotating.

orientation is a result of these two forces. As the part wall cools, additional high-magnitude, cross-laminated orientation is developed that is frozen in and throughout the wall thickness.

Orientation on the individual molecular planes, from the male and female cavity walls, occurs as each layer cools after injection. This orientation can change direction and magnitude as a function of wall thickness. The result is analogous to plywood with strength improvements being as dramatic.

In this process, there are infinite layers, each of which has its own controlled direction of orientation. By appropriate processing conditions, both the magnitude and direction of the orientation can be varied and controlled throughout the wall thickness.

MWR technology produces parts having greatly increased tensile strength compared to the parts produced conventionally by IM. With MWR, the internal structural bonding of the plastic part is greatly improved through the multi-axial laminar orientation of the molecules. This results in a measurable improvement in stress cracking resistance of the molded part. Other gains occur, such as resistance to crazing, excellent dimensional properties, reduced molding cycle time, etc.

## CONTINUOUS MOLDING WITH EXTRUDER

Machines to mold (like injection molding) using a continuously operating extruders have been designed, built, and used in different major production lines. The extruders are conventional types such as those reviewed in Chapter 2. They melt the plastic and have different techniques to delivery the melt into (injection molding) cavities. These continuously rotating machines use many molds. The molds are usually located on rotating circular table that can operate as Ferris wheels and carousals [2].

Feeding a melt onto a rotating mold is not new. The concept has been applied since at least the 1940s. Products made include shoe soles, shoe sandals, boats, and so on.

### Velcro strip Ferris wheel molding

The Velcro nylon fastener consists of two mating strips. One strip is covered with nearly microscopic hooked or barbed plastic spines, the other with tiny plastic loops. When the two strips are pressed together, their projections become entangled to produce the gripping action.

Peeling the strips apart deflects the splines, disengaging them from the loops. Because of the resilience of the plastics (usually nylon), the projections snap back to their original geometry so that the strips can be used repeatedly.

For many years in the past, Velcro had been made by a slow, complex textile process in which the loops were woven through the back of a flexible base strip, for the male strip. For the mating side, loops were first made and in turn cut to make the hooks. Seeking a more economical alternative, Velcro USA (Manchester, NH., USA) had Foster-Miller Assocs. (Waltham, MA., USA) work on a successful injection molding process.

A few details about the fastener will underscore the formidable molding problems that the designer had to solve. The splines are almost too small to see. They project about 1.5 mm (0.06 in) from the 0.25 mm (0.010 in) thick base strip and are 0.50 mm (0.20 in) wide at their base tapering to about 0.30 mm (0.012 in) at the tip (Fig. 18.12). They are very

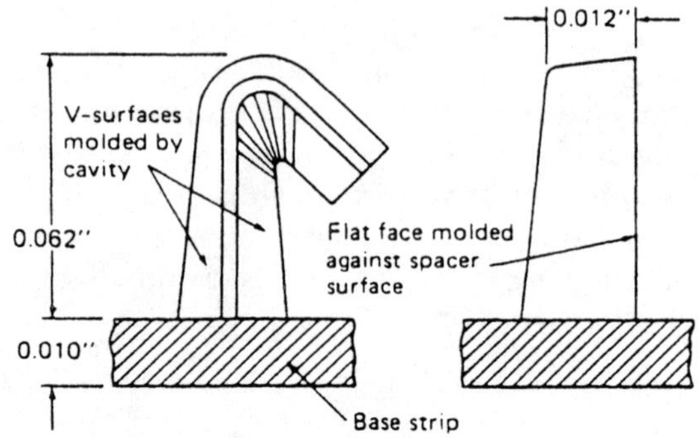


**Figure 18.12** Size and shape of Velcro spines.

closely spaced on approximately 1.3 mm (0.050 in) centers; a single square inch (25.4 mm$^2$) of the strip contains more than 250 of these projections. Moreover, they are not simple, needle-like shapes, but are triangular in cross section. They have a microscopic hook or other type of barb at the tip. Dimensions must be held to 0.38 mm (0.0015 in) and no flash is permissible anywhere on the strip.

*Molding line*

The molding is in a continuous process. The equipment molds the strip from the plastic, trims it, conditions it for flatness, applies an adhesive backing, and winds it on a reel. The key elements are a conventional extruder and a rotating Ferris wheel mold. The extruder runs continuously, feeding the melt into the continuously rotating mold through a special adapter, mounted on the extruder barrel melt outlet. As an example a 61 cm (2 ft) diameter mold turns at about 10 rpm, delivering Velcro at 18–21 m/min (60–70 ft/min).

The mold contains more than 15 000 cavities meeting at least the dimensions already listed. These cavities are in parallel rows around the mold's circumference [Fig. 18.13(a)]. Besides the task of designing a tool to mold Velcro, a method had to be developed to strip it from the mold. Each spine has a hook or other projection that must be disengaged without damage from the undercut at the base of its cavity as the strip is being peeled from the mold [Fig. 18.13(b)].

A wheel-shaped mold was designed consisting of several dozen thin, 0.015 mm (0.060 in), round plates bolted together. The plates are two types, which alternate across the thickness of the 'wheel.' There is the cavity plate that contains a ring of molding cavities for the projections on both sides of the plate at its outer edge. Between each cavity plate is a

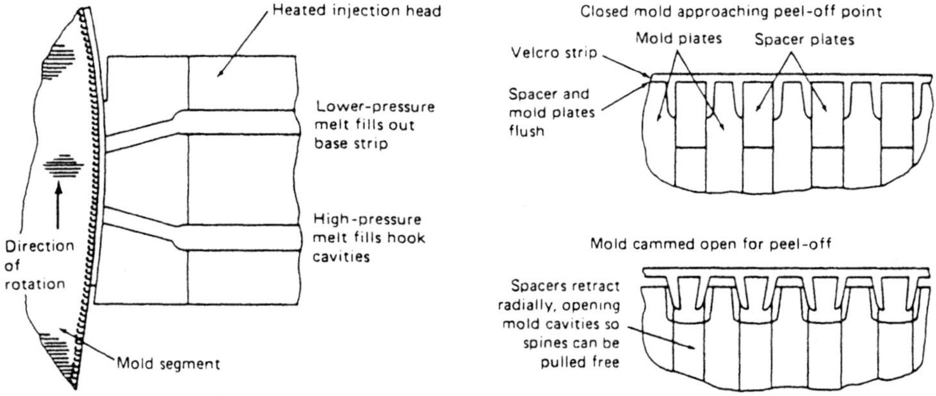


**Figure 18.13** Molding Velcro: (a) two orifices feed melt onto rotating mold; and (b) method used to peel Velcro strip from mold.

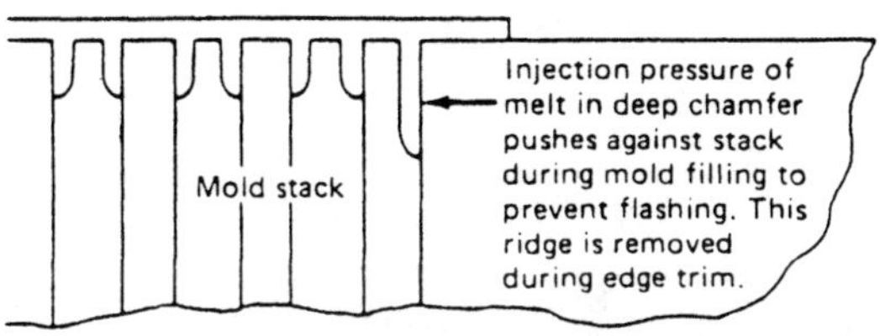


**Figure 18.14** Built-in 'melt piston' compresses mold.

blank spacer plate. Being in intimate contact with the cavity plate, the spacer plate acts to seal off the open side of the cavities. The set of alternate spacer plates are designed to slide in and out (radially) as a group; the cavity plates have no radial motion. During most of the cycle, the spacer plates are extended to the full diameter of the mold so that their edges line up with those of the cavity plates. This alignment creates the flat surface that molds the inner face of the Velcro strip (Fig. 18.14).

*Injection process*

As the mold rotates past the extrusion-injection die head, the melt flows onto the circumference of the mold and is forced by the melt pressure into the cavities. Enough additional melt is supplied to create the base strip at the same time. By the time the mold has completed about one-half a revolution, the plastic has been cooled and solidified. At that point, the spacer plates are pulled down by a circular cam.

This action is like the effect of a piano keyboard with every other key pushed in. The retracted plates open up the sides of the cavities to give the spines room to deflect outward and release from the cavity undercuts as the Velcro strip is pulled off the mold by the Velcro winder. After passing the peel-off point, the spacer plates move back out to their original position (Fig. 18.15).

Multiple separately mounted segments, each individually free to move radially, make up each spacer plate. For manufacturing convenience, the cavity plates likewise are constructed from individual segments. To make electrodes of such small size and high precision, Foster-Miller had to develop special techniques, with which the working electrodes were made by electro-forming.

Because of its complicated construction, the mold is cooled externally. Mold temperature is controlled by an air plenum located about one-quarter of a revolution before the injection point; the molded Velcro itself is cooled by another plenum ahead of the strip-off point. The air flow is adjusted to keep the nylon sufficiently warm and flexible so that the projections can pull free from the undercuts in the cavities without damage during strip-off.

**Figure 18.15** Rotating 'mold' to produce Velcro strips 'continuously' in a special designed IMM.

*Extrusion-injection diehead*

The extrusion-injection diehead contains two orifices, each fed by a separately geared pump (Chapter 2). The lower orifice, which fills the projections, operates at a relatively high pressure that is equivalent to a typical nylon injection pressure in a closed mold (Fig. 18.13). This pressure forces the hot melt into the blind cavities. The second orifice, just above the other orifice, supplies additional hot melt at lower pressures to fill out the base strip.

The clearance between the die head and the mold, which determines the strip's thickness, is set by a fine gear-reduction drive and is measured by four air gauge sensors, one at each of the four corners of the head to ensure parallelism. A 0.25 mm (0.01 in) error in the 0.025 mm (0.10 in) gap can result in as much as a 50% variation in injection pressure. Cartridge heaters in the die head maintain the melt at the property conditions (viscosity, etc.); temperature control is within ±3°C (5°F).

The sides of the gap between the diehead and the mold are not enclosed. Seepage is prevented by a careful balance of melt temperature, mold temperature, injection pressure, mold cooling, and mold velocity. The as-molded edges are uneven, but the edges are squared off by trimming the strip in a downstream operation after conditioning.

One of the most critical requirements in the mold design was to prevent flash. Two approaches were used to avoid the problem. One approach was to control the geometry of the mold plates to extremely close tolerances. Every plate in the mold stack is surface-ground to be flat within 0.050 mm (0.002 in) across a 0.6 m (2 ft) diameter. Also, plate thickness which determines the spacing between adjacent rows of splines as well as the quality of the seal between adjacent plates, is controlled to within 0.0025 mm (0.0001 in).

The second approach against flashing was to prevent the edges of the mold plates from flexing outward as the melt is injected into the cavities. To supplement the tie-rods to hold molds (typically used in IMMs), a different approach was used. The melt itself is used to supply hydraulic squeezing action on the mold stack (Fig. 18.14).

At the outermost mold plate in the stack, and beyond the nominal width of the strip, deep chamfer cuts are made around the edge of the plate. As the mold is being filled, the pressurized melt also flows into this chamfer. The resulting sideways force against the side of the plate tends to compress the mold stack and prevent the plates from spreading.

Another key to reliable production is ensuring trouble-free radial movement of the spacer segments during mold opening, and particularly during closing. Any significant galling or binding between the spacers and the mold plates could prevent the spacers from returning to their original position, flush with the edges of the mold plates. The resulting offset would produce thickness steps across the base of the Velcro strip, and probably flashing as well. This problem is avoided by applying a low-friction coating in the form of an internally lubricated plastic to the sliding faces.

### Carousal molding

Different machines have been designed and built using many molds on a rotating carousal and, like the Velcro molding, use extruders to deliver the hot melt. One example is where the molds are stacked next to each other so it appears that only one continuous mold exists. There are no spaces between adjacent molds.

In one basic design, the melt entrance for each mold is facing outward from its side located at the molds parting lines. A fixed extruder is located so that its diehead matches the radius of the rotating molds. It feeds melt under pressure through each mold opening as the mold opening passes the diehead. In between the melt shots into the cavities, the melt is restricted to remain in the diehead. It can perform like the first-stage of a continuous two-stage injection molding machine (Fig. 18.2). Some of the designs have used a tract that the diehead follows to ensure no melt leakage.

Once the melt is injected into the cavity, a mechanical device, such as a cam-operated shutoff or plate located at the mold opening, closes off the opening. At the same time another cam-operated ram from the top of the mold applies pressure to the melt in the cavity. As the filled mold rotates, it completes its required cooling cycle, mold opens, part ejected, mold closes, and is in position to receive another injection.

The other basic design has the melt entering the mold's opening from the top. All openings are lined in a circular pattern so as the carousal table rotates, all openings are in-line. This method could make it easier to fill cavities and apply a plunger mechanism through the same opening. Some of the tables have different size mold cavities. A programmed system is used to inject the shot size according to what is needed.

## THERMOFORMING

Thermoformed or shaped plastics provide a great variety and quantity of marketable products, in many sizes, ranging from drinking cups and small containers to boat hulls and so on to complex shapes. Thermoforming has many advantages over other manufacturing methods. Extrusion is very important to thermoforming because it uses very large amounts of extruded sheets and films that are produced in-line with thermoforming equipment or use rolls or flat sheets of materials to feed the thermoforming equipment [1, 191, 267, 288, 337, 409].

**Figure 18.16** A high-speed continuous thermoformer.

Many different techniques are used to form/shape plastics [1]. Figures 18.16–18.18 show major types of in-line systems. These systems form products such as food packaging containers (Fig. 18.19). Extruding sheet or film in-line requires dedication and control to ensuring that the extruder is operating efficiently as well as the thermoformer. For those with this type production, advantages exist including cost savings.

The end of the line in Fig. 18.20 is an example of a complete in-line 'push button' start-up. It is producing a waffle pattern (knobby sheet) using HDPE sheet that process sheets 0.40–50 mm (0.015–0.020 in) thick. This rotary thermoforming, at the end of the extrusion line, uses special vacuum drum technology to make 3 m (10 ft) wide thermoformed sheets. The drum's capacity can handle up to 1400 kg/h (3000 lb/h) of plastic. They are used in applications such as load-bearing barrier protecting building foundations and roofs against water and/or soil. The sheet's strength is due to the shaping of the sheet with raised knobs or truncated cones of equal shape and height. It is used as an alternate to tar. To meet different requirements, coextruded sheets also go through this line.

(a)

**Figure 18.17** (a) Welex/Irwin rotary sheet thermoformer in-line with extrusion system; and (b) close-up of rotary thermoforming drum.

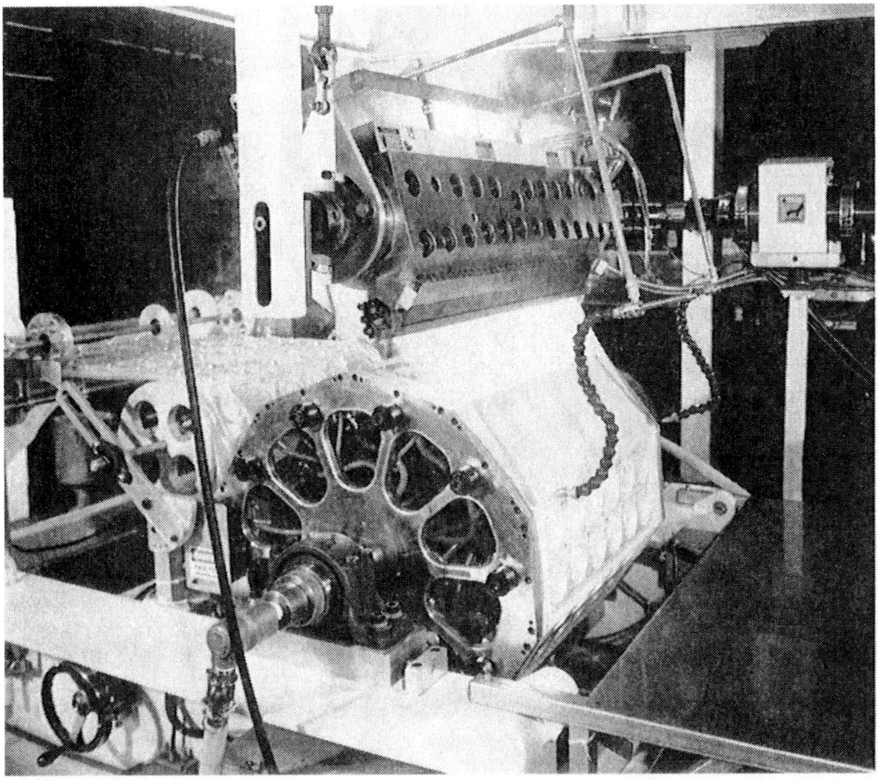

(b)

**Figure 18.17** *Continued*

**Figure 18.18** Compact in-line Welex sheet extrusion system features shortened sheet take-off to reduce overall length (retains suitable heat in sheet) when used in-line with thermoformer.

**Figure 18.19** PET products thermoformed on Welex extruder sheet line.

**Figure 18.20** An example of Welex's rotary thermoformers for in-line high volume production.

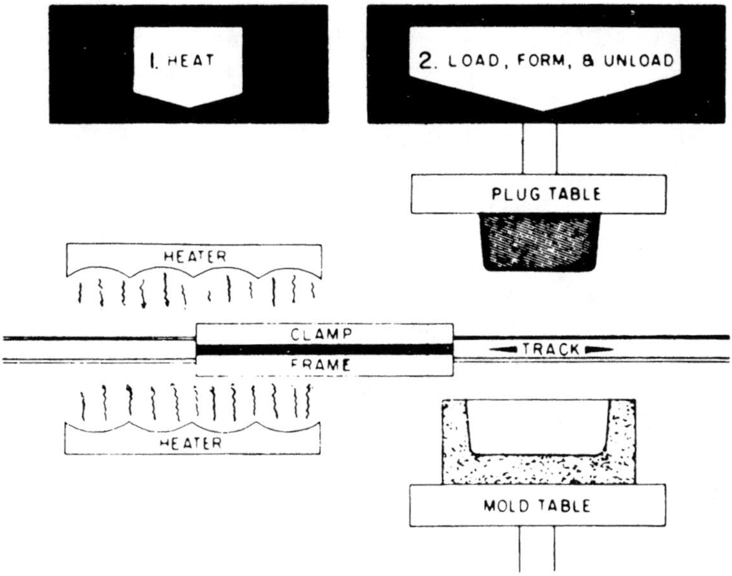


**Figure 18.21** Single-stage, shuttle-type thermoformer.

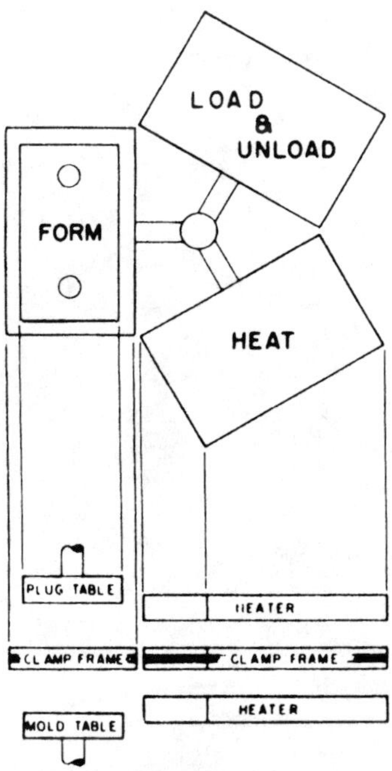


**Figure 18.22** Three-stage rotary carousel-type thermoformer.

Other rather conventional type thermoformers are shown in Figs. 18.21 and 18.22. Figure 18.23 shows sheets being thermoformed into boat hulls using the Fig. 18.21 system. In Fig. 18.24, a tube (mono or coextruded) is extruded and is in turn flattened, followed by going over a multiple mold ferris wheel with book-type closing action thermoforming mold units. The still hot flattened tube moves over the open mold. In turn the mold closes, vacuum applied, and parts such as lids are thermoformed. Output rates are at least 12 000/h (microporous air-permeable, molds can be used).

Figure 18.25 shows two thermoformed parts, made from coextruded sheets. They are bonded together to create a fuel tank. The tank's inside has a nylon surface to resist gasoline deterioration; the outside uses PP to provide the required support strength while keeping the cost low.

Thermoforming usually begins with plastic sheets or film that are monolayer and coextruded. Sheet thickness tend to be 0.25 mm (0.010 in) and greater and film thickness are normally less than 0.25 mm (0.010 in) with most being 0.08 mm (0.003 in) or less. Practically any TP material can

**Figure 18.23** Heating sheets in an IR oven; heated sheets are thermoformed into a boat.

**Figure 18.24** Example of a drum thermoformer; schematic and machine view.

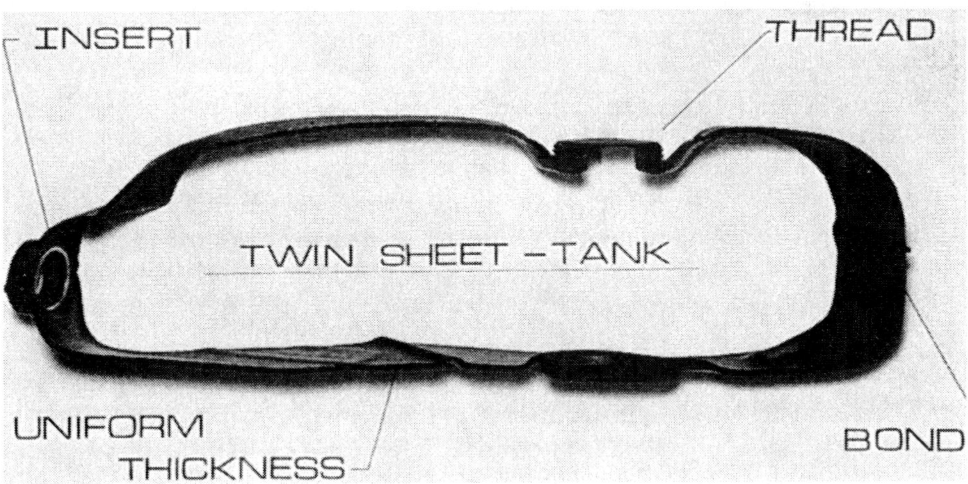


**Figure 18.25** Two thermoformed coextruded sheets create a gasoline fuel tank.

**Table 18.2** Troubleshooting guide for thermoforming

| *Problem* | *Cause* | *Solution* |
|---|---|---|
| Blisters or bubbles | Overheating | Lower the heater temperature<br>Increase distance of heater from sheet<br>Attach masks or baffles |
| | Wrong sheet type or formulation | Obtain correct formulation |
| | Poor storage conditions | Do not remove material from moistureproof wrap until ready to use |
| Blush or change in color intensity | Insufficient heating | Lengthen heating cycle |
| | Mold is too cool | Warm the mold |
| | Assist is too cool | Warm the assist |
| | Sheet cools before it is completely formed | Speed the drape action<br>Add vacuum holes |
| | Too deep a draw | Use heavier-gauge sheet |
| | Poor mold design | Use mold of proper design |
| Sticking to the mold | Rough or improperly designed mold | Make mold smoother<br>Increase the taper of male plugs<br>Use mechanical release assists<br>Use air pressure to blow piece from mold<br>Use mold-release agents |
| Incompletely or improperly formed pieces | Sheet is cold | Lengthen the heating cycle<br>Bring heater closer to sheet |
| | Insufficient vacuum | Check vacuum system |
| | Vacuum holes are plugged up | Clean, relocate, or add vacuum holes |
| Warped or distorted pieces | Poor mold design | Redesign mold using proper tapers and ribs |
| | Sheet removed while too hot | Increase cooling cycle<br>Use water-cooled molds |
| Webbing or bridging | Insufficient vacuum | Check vacuum system<br>Add more vacuum holes |
| | Sheet is overheated | Shorten heating cycle<br>Increase heater distance from the sheet |
| | Long parallel molds with extrusion direction parallel | Move sheet 90° in relation to mold |
| | Poor mold layout or design | Use mechanical drape or plug assists |

**Table 18.2** *Continued*

| *Problem* | *Cause* | *Solution* |
|---|---|---|
| | Sharp corners on deep draw | Increase radius |
| Bad surface markings | Markoff (due to trapped air) | Slow draping action<br>Add more vacuum holes |
| | Markoff (due to accumulation of plasticizer on mold) | Use a temperature-controlled mold<br>Have mold as far away from the sheet as possible during the heating cycle<br>Shorten the heating cycle (if too long)<br>Wipe the mold |
| | Mold is cool | Warm the mold<br>Bring the heater closer to the sheet |
| | Mold is too hot | Provide cooling for mold |
| | Improper mold composition | Avoid phenolic molds with clear transparent sheet |
| | Mold surface too highly polished | Remove high surface gloss from mold |
| | Mold surface too rough | Smooth surface |
| Excessive post shrinkage | Sheet removed from mold while still hot | Increase cooling time |
| Pinholing or rupturing | Vacuum holes too large | Partially plug up holes with wood or solder or completely plug and redrill |
| | Uneven heating | Attach baffles to the top clamping frame |

be used. The major high volume plastics are high impact and high heat PS, PVC, ABS, HDPE, LDPE, PP, PET, PETG, PUR, CPET, PA, and PMMA. Other plastics of lesser usage are transparent styrene–butadiene block copolymers, acrylic multi-polymers, cellulosics, ethylene–propylene, etc.

The thickness tolerance and the plastic distribution are decisive quality criteria for thermoformed products. Thicker parts are formed differently from thin sections. These differences can significantly effect the cost. Table 18.2 provides a guide to troubleshooting thermoforming.

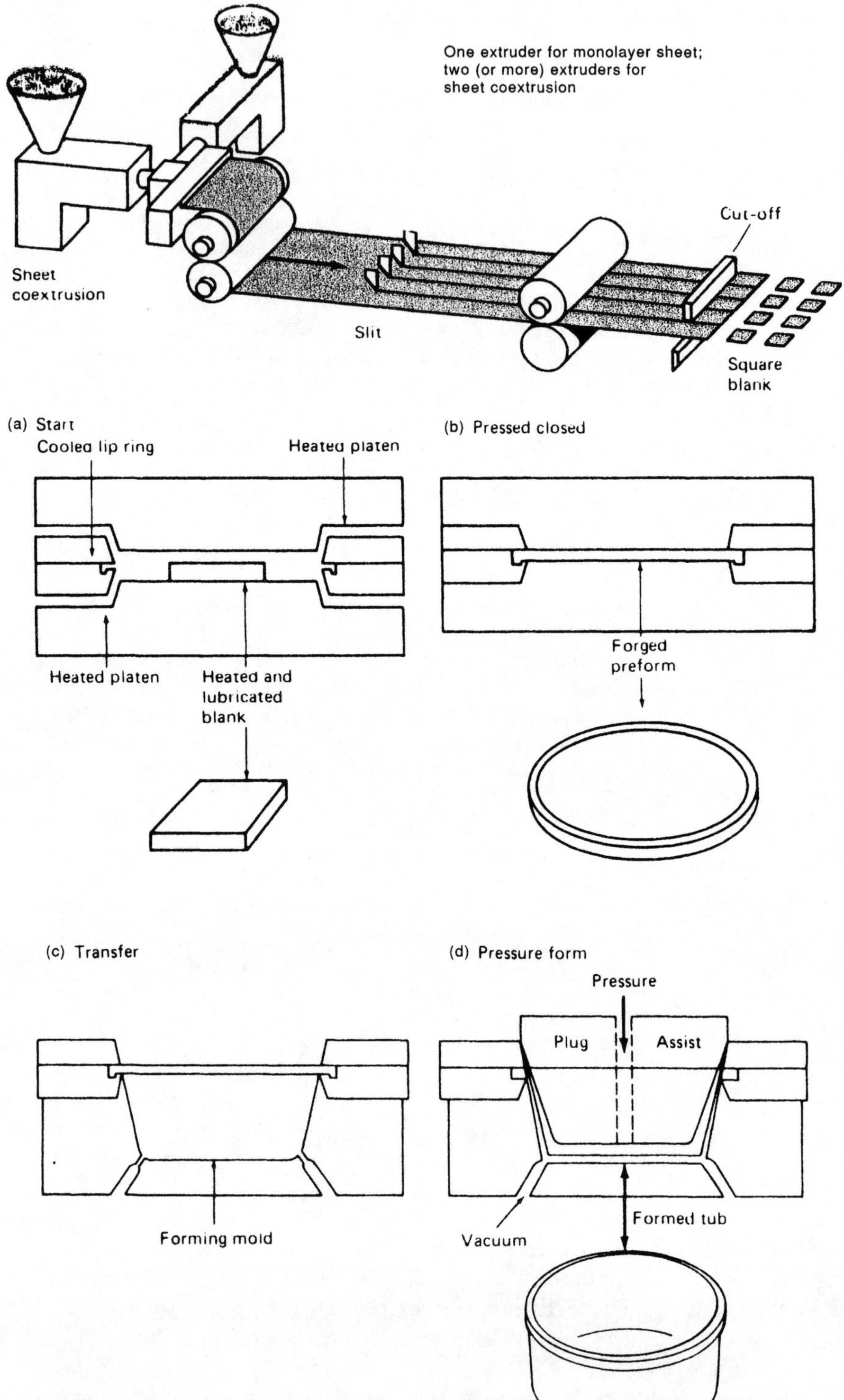


**Figure 18.26** Scrapless, solid-phase forming process.

## SCRAPLESS FORMING

A specialty forming technique patented by Dow Chemical is what is referred to as the scrapless process. As shown in Fig. 18.26, it is a relatively simple process. An extruder or extruders produce a sheet. In turn, the sheet is slit and cut into square blanks. These blanks are heated and pressed into circular disk with a lip. The disk is thermoformed into a cup. This process basically eliminates scrap.

## POSTFORMING

The information presented here has been around and used at least since the 1940s. In-line postforming or postextrusion refers to the special processing that may be done to the extrudate, usually just after it emerges from the die but before the plastic has a chance to cool. It provides performance and cost advantages, principally for long production runs or customer's special orders.

The process is used with different products such as tubes, profiles, sheet, film, etc. Upon leaving the extruder's die and while it is retaining its heat, the plastic is continually postformed (Figs. 18.27–18.31). As an example, Fig. 18.29 is a system that can be used with different profiles, such as small or large [up to at least 1.2 m (4 ft)] corrugated tubing or

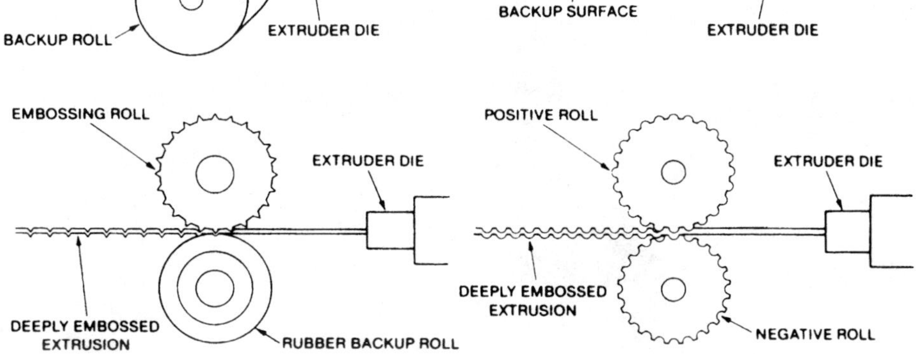


**Figure 18.27** Examples of in-line postforming; shallow and deep embossing patterns.

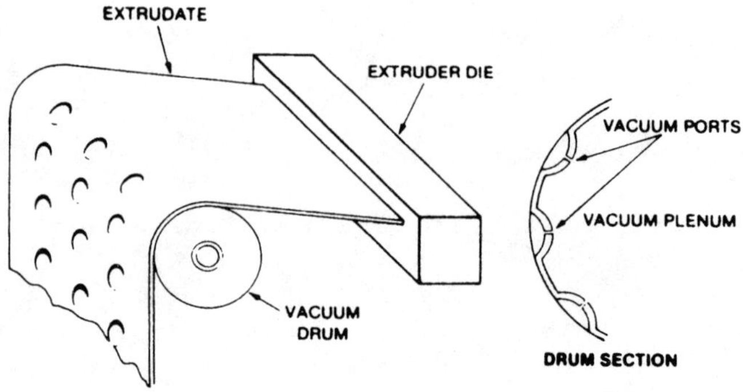


**Figure 18.28** In-line vacuum-forming embossing roll with water cooled temperature control.

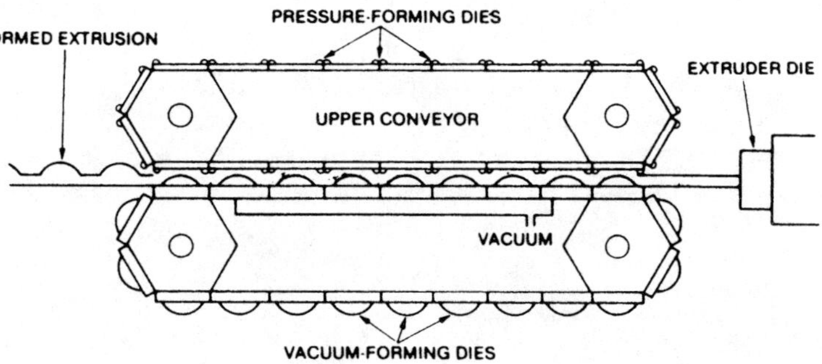


**Figure 18.29** In-line vacuum/pressure former for sheet with matched, water-cooled forming molds on a continuous conveyor.

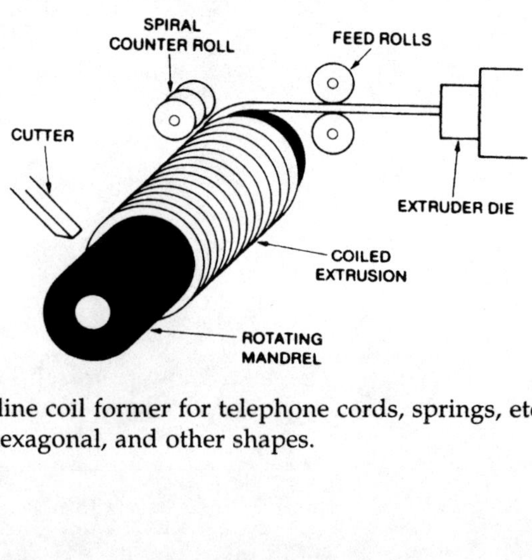


**Figure 18.30** In-line coil former for telephone cords, springs, etc. using extruded round, square, hexagonal, and other shapes.

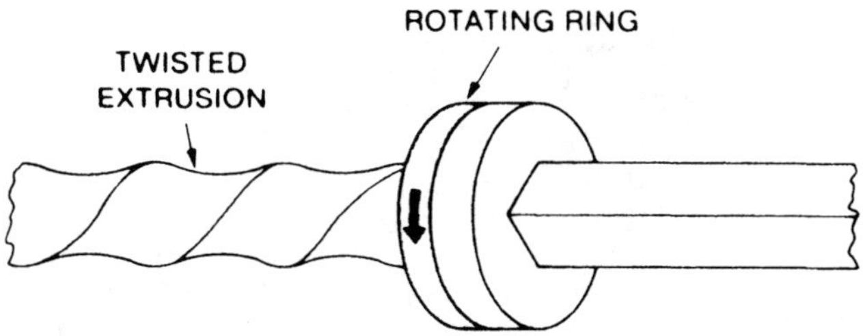


**Figure 18.31** In-line fixed/rotating rings used to twist extrudate.

piping. Moving split molds (Fig. 18.32) operating like a caterpillar has the moving corrugated mold cavities closed while the tube/pipe passes with each mold half providing vacuum pressure and required water cooling through internal tubing. Near the end of the line, the mold positions start separating to start releasing the corrugated product and gradually open the two part molds. Figure 18.33 is another approach to forming corrugations.

With this type of in-line, the hot plastic is reduced only to the desired heat of forming [1]. All it may require is a fixed distance from the die opening. Cooling can be accelerated with blown air, a water spray, a water trough, or their combinations. This equipment requires, like others, precision tooling with perfect registration. When the plastic is prepared as reviewed, it is in-line, as opposed to forming the same shapes after extruding and products are cooled.

## NETTING

Extrusion dies (Chapter 5) can be designed to produce different flow patterns such as tubular to flat netting types (Figs. 5.41, 5.42, and 18.34). For a circular output, a counter-rotating mandrel and orifice can have semicircular shaped slits through which the melt flow emerges. The slits can be of any shape [159].

If one part of the die is held stationary, then a rhomboid or elongated pattern is formed. If both parts of the die rotate, then a true rhombic mesh is formed. During the time when the melt extrudes through the orifice and the slits overlap, a crossing point is formed where the emerging threads appear to be welded but it is really a uniform melt flow through the matching/aligned slits. For flat netting, the slide is in opposite directions.

## FOAM

Almost all plastic fabricating processes and plastics are used to produce foamed products. All the extruded systems discussed in this book can

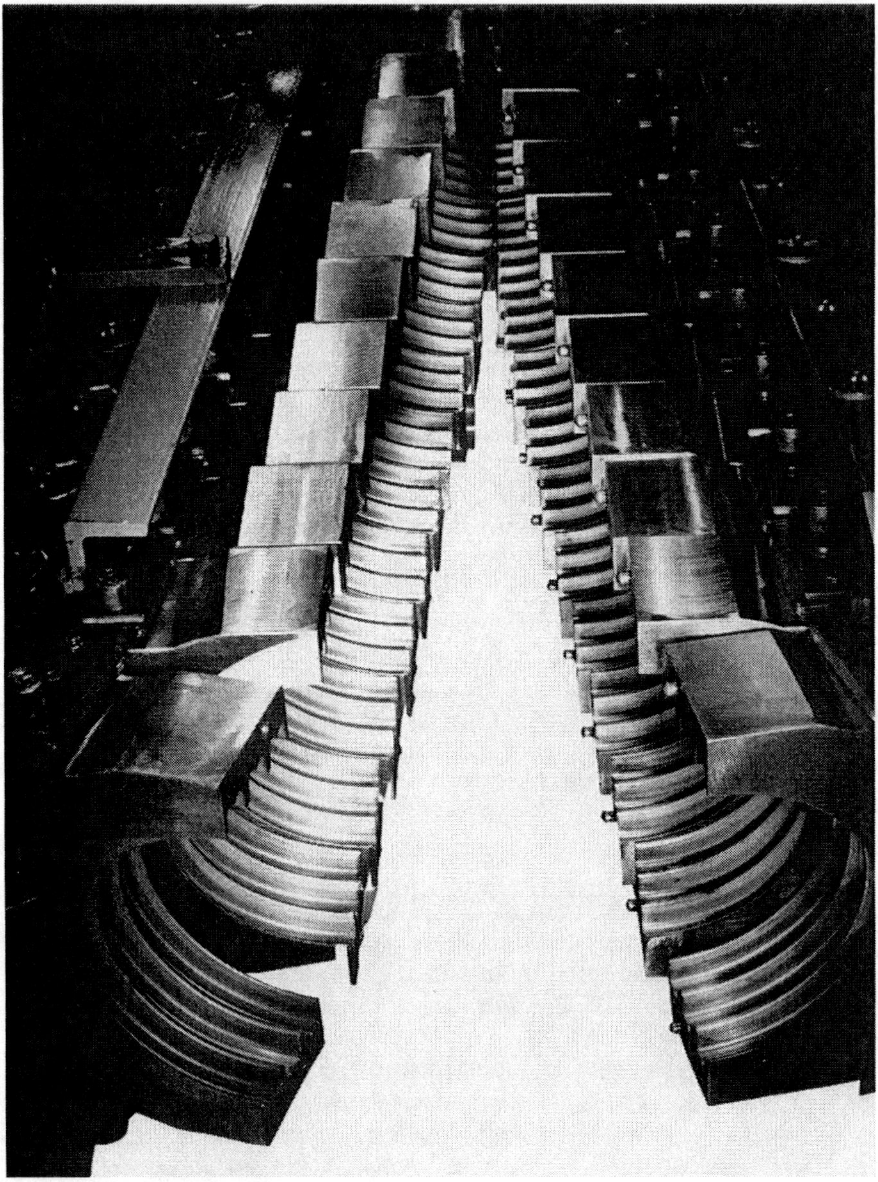

**Figure 18.32** Split molds on a caterpillar used to postform corrugated pipe.

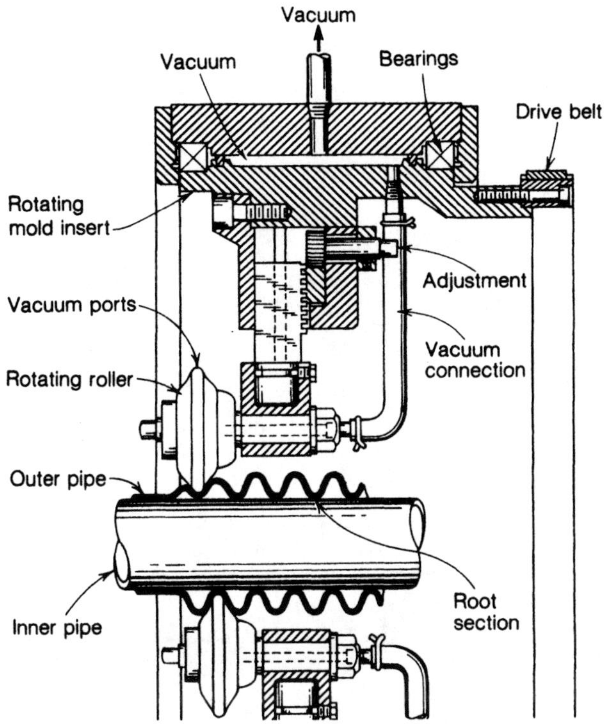


**Figure 18.33** Simplified postformed corrugated pipe.

process foamed products, such as foamed sheet and film. Figure 18.35 shows a foamed sheet tandem line with two extruders. The first extruder prepares the melt that is fed into the second extruder that provides a uniform temperature to the melt prior to forming a foamed tube through its die. This tube is slit into two halves as shown in Fig. 18.36. Two flat foamed sheets go down-stream to windup rolls. Figure 18.36 shows a single extruder that is not as efficient in the melt temperature final control when certain plastic mixes with blowing and additives are used or performance requirements exist. However, this single-stage extruder is widely used because of its economic advantages [348].

Other foamed products include foamed pipes and profiles, foamed blow moldings [38], injection moldings (Fig. 18.37) [2], and so on. There are monolayer extruded and coextruded products (Figs. 18.38 and 18.39).

For about a century foamed products, whether TPs or TSs, have been a large part and special category within the plastics industry. They are known by different names such as cellular plastics, expanded plastic foams, structural plastic foams, and just plastic foams. A plastic foam is a

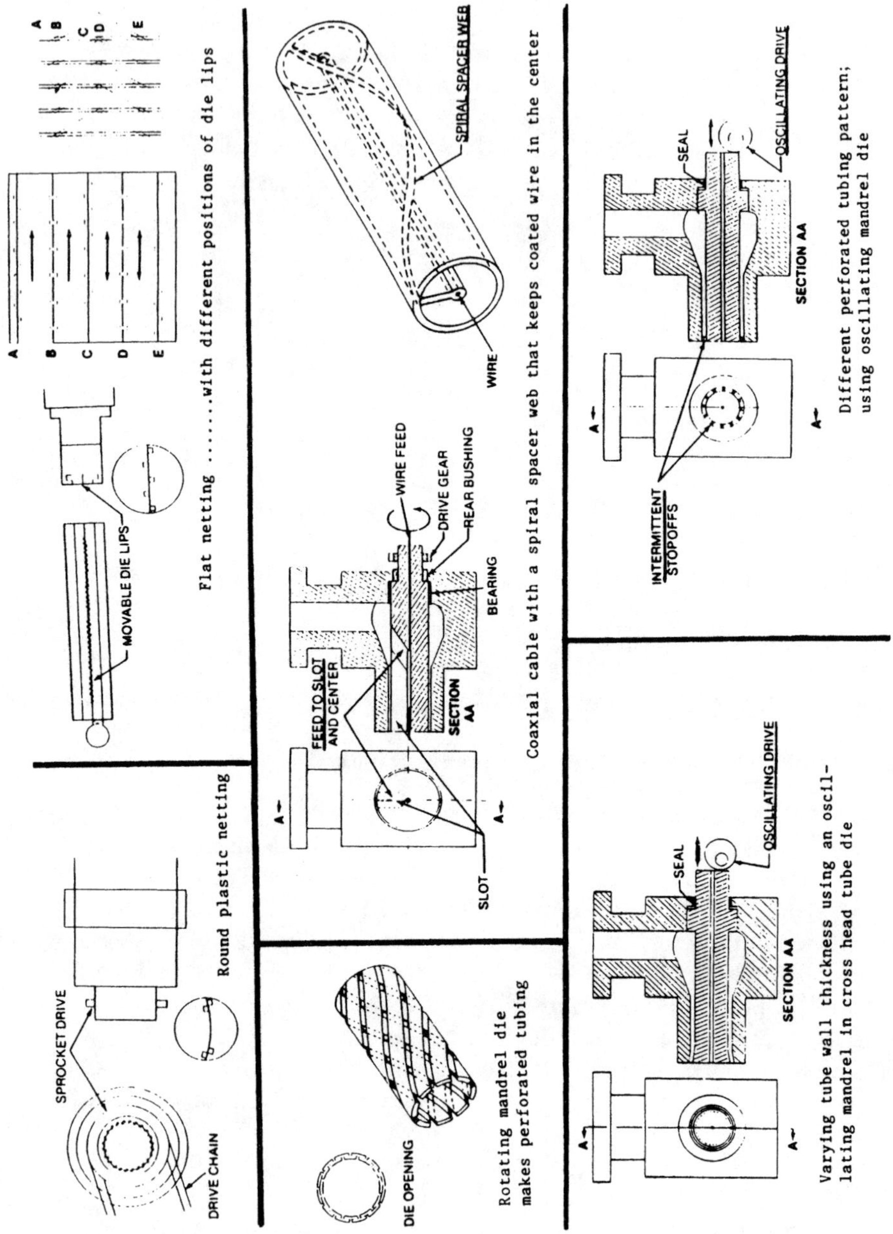


**Figure 18.34** Schematic cut away for extruded tubular netting.

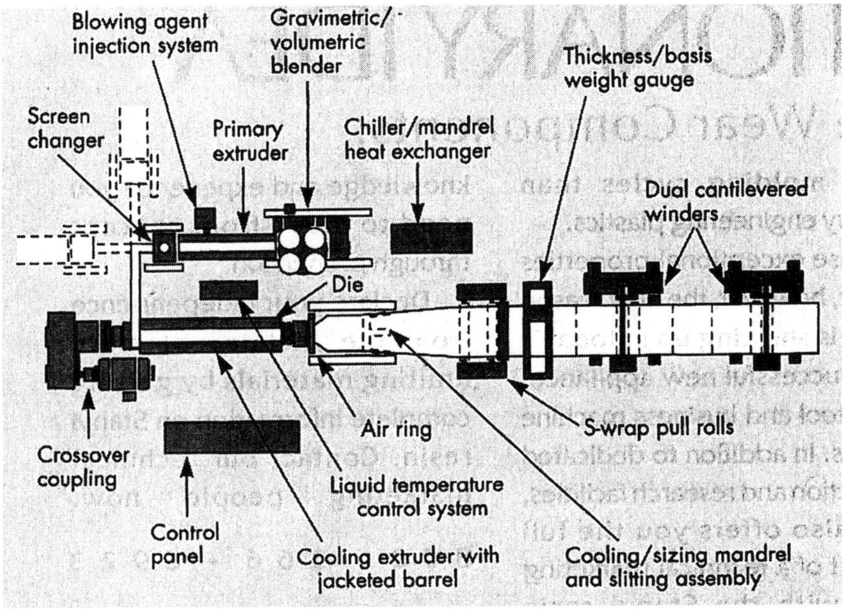


**Figure 18.35** Example of a tandem extrusion sheet foam line.

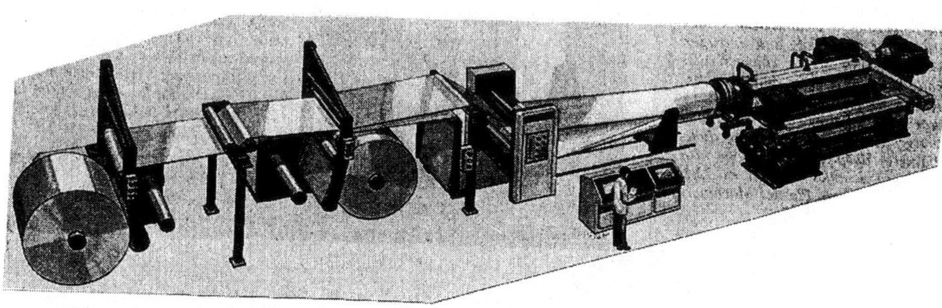

**Figure 18.36** Overview of single extruder sheet foam line.

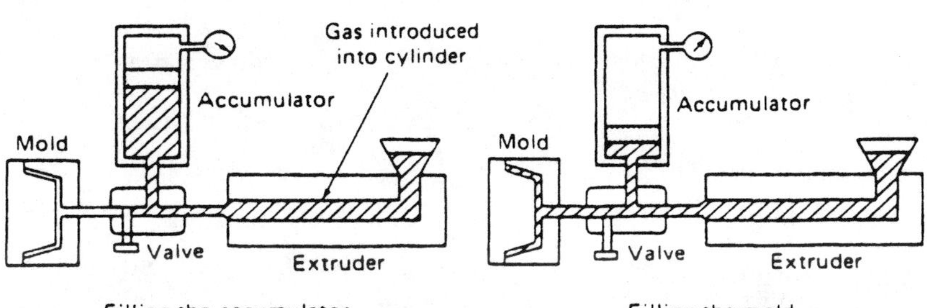


**Figure 18.37** Low-pressure foam IM.

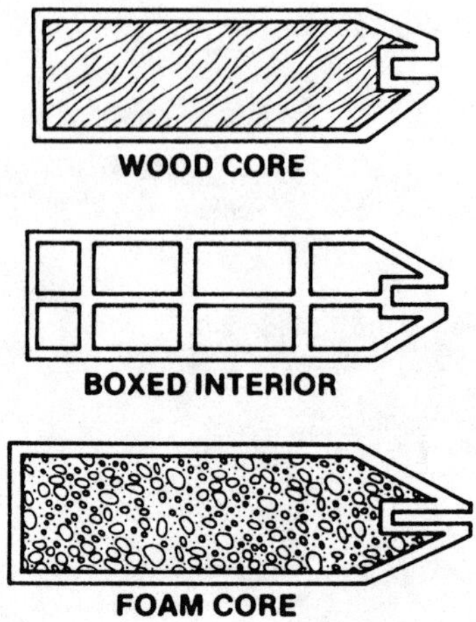


**Figure 18.38** Window frame TP profiles filled with TP foam.

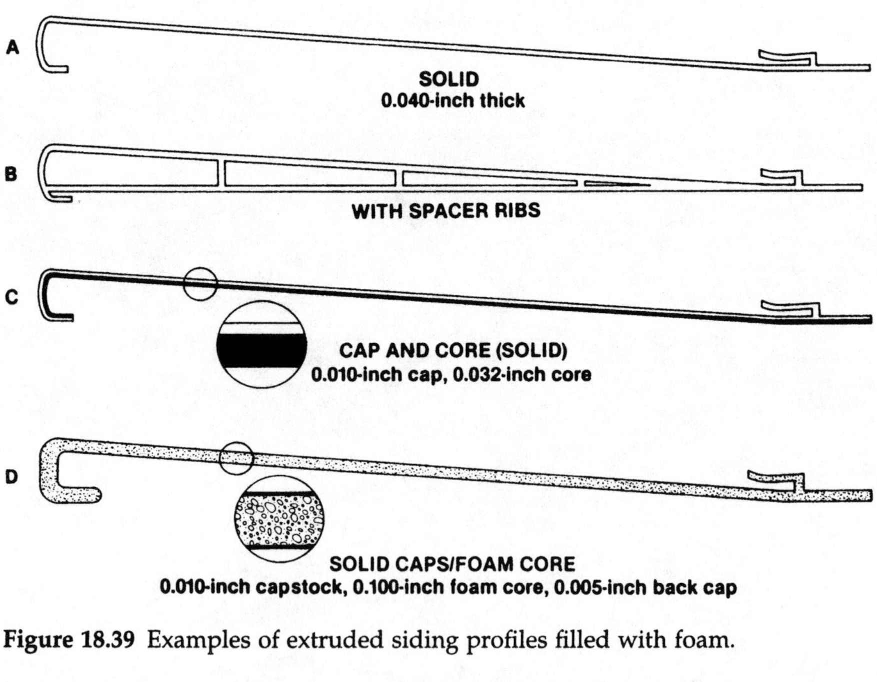


**Figure 18.39** Examples of extruded siding profiles filled with foam.

plastic whose apparent density is decreased by the presence of numerous cells throughout the mass; it is a two-phase gas–solid system in which the solid is continuous and composed of plastic material.

Depending on how they are processed, foams are available with open-celled construction, closed, interconnecting construction, or their combinations. Their densities usually range from 1.6 to over 960 kg/m$^3$ (0.1 to over 60 lb/ft$^3$). There are micro-cell PS sheets with densities of 0.035–0.1 g/cc (2–6 pcf) with uniform closed cells measuring less than 50 microns. Cell sizes of 150–250 microns are typical for conventionally formed PS sheet (MIT). They can be rigid, semirigid, or flexible as well as colored or plain. They offer all kinds of advantages, such as insulation, cushioning, and many more [2, 4, 229, 243, 250, 275, 302].

Foamed plastics, like their solid counterparts, can be used for almost an unlimited range of products. An architectural example was an R&D project in 1966 (Figs. 18.40–18.42). Dome-shaped buildings in Lafayette, IN., USA, were built using PS foamed boards that used the Dow Chemical patented 'spiral generation' technique. Figure 18.40 is a model of the clinic fabricated. Boards were heated on site so they could be formed and bonded to each other in layers, forming a continuous structure (Fig. 18.41). The self-supporting domes required no internal or external support during or after manufacture; they provided their own insulation, and it was easy to cut out sections to create openings for doors, windows, and connecting halls from dome to dome. Their outsides were covered with steel wire mesh and concrete then waterproofed (Fig. 18.42). The inside walls were covered with appropriate plaster and paneling.

The foaming methods vary widely. For the production of foams, depending on the basic plastic and process, different blowing agents, also called foaming agents, are used to produce gas and thus to generate cells

**Figure 18.40** Model of PS foamed dome-shaped buildings.

**Figure 18.41** Dome-shaped buildings using PS foamed boards.

**Figure 18.42** Protective roof coverings.

or gas pockets in the plastics. They are divided into the two broad groups of physical blowing agents (PBAs) and chemical blowing agents (CBAs).

PBAs are represented by compressed gases and volatile liquids. The compressed gases often used are nitrogen or carbon dioxide. These gases

**Table 18.3** Troubleshooting guide for plastic foam film

| *Problem* | *Cause* | *Solution* |
|---|---|---|
| Random, poor cell structure | Low melt pressure | Increase screw speed<br>Reduce die-lip temperature<br>Decrease gauge of screen packs<br>Use resin of lower melt index<br>Reduce die gap |
| | Hangup in die | Reduce land length<br>Clean die |
| | Stagnant low-pressure areas in head | Increase screw speed |
| | Irregular cells in spider area or opposite die-ring feed | Use bottom-fed spiral die<br>Increase head pressure |
| Poor skin formation | Too much blowing agent | Reduce blowing-agent level |
| | Linear skin speed too low | Increase screw speed |
| | Loss of melt pressure in die land | Reduce land length<br>Increase $L/D$ ratio<br>Increase screw speed |
| | Die-block temperature too low | Increase temperature |
| Pinholes in film or bubble burst on surface | Too much blowing agent | Reduce blowing-agent level |
| | Die temperature too high | Reduce die temperature |
| | Resin melt index too high | Decrease resin melt index<br>Reduce processing temperature |
| | Blowing agent decomposing too soon | Reduce processing temperature<br>Increase screw speed<br>Reduce blowing-agent level level |
| | Poor flow within polymer skin | Improve flow in head and die |
| Cells collapsing | Resin melt index too high | Decrease melt index<br>Reduce processing temperature |
| | Cooling too fast | Reduce cooling rate |

are injected into a plastic melt in the screw barrel under pressure (higher than the melt pressure) and form a cellular structure when the melt is released to atmospheric pressure or low pressure. The volatile liquids are usually aliphatic hydrocarbons, which may be halogenated, and include materials such as carbon dioxide, pentane, hexane, methyl chloride, etc. Chlorofluorocarbons were formerly used but they have now been phased out due to the reported environmental problems.

Chemical blowing agents, generally solid materials, are of two types: inorganic and organic. Inorganics include sodium bicarbonate, by far the most popular, and carbonates such as zinc or sodium. These materials have low gas yields and, compared with organics, the cell structure they create is not uniform. Organics are mainly solid materials designed to evolve gas within a defined temperature range, usually called the decomposition temperature range. This is their most important characteristic and allows control over gas developments through both pressure and temperature.

This increased control of the CBAs produces a finer and more uniform cell structure as well as better surface quality on the foamed plastic. There are over dozens of different types available that decompose at temperatures from at least 105–370°C (220–700°F) and possible higher. Many of these CBAs can be made to decompose below their decomposition temperature through the use of activators.

Recognize that only certain CBAs can be used with certain plastics. They have to be compatible chemical and start gassing at the required temperature. If they are not compatible, different problems develop, such as discoloration, property losses, etc. A CBA with a temperature over the melting temperature of the plastic will not gas to form gas, etc.

Information on foaming is available from the literature [1–4, 348]. Processing includes preparation of materials, how to mix materials, processing conditions, handling the foams being extruded, for certain plastics and blow agents soaking period after produced such as foamed sheets, storage method, safety procedures, etc. Table 18.3 provides a guide to troubleshooting plastic foam film.

# 19

# Testing and quality control

## INTRODUCTION

Testing yields basic information about plastic, its properties relative to another material, and its quality with reference to standards. Most of all, it is essential for determining the performance of plastic materials to be processed and of the finished extruded products. The words 'testing' and 'quality' may conjure up terms such as inspection, statistical process control, auditing, and standards. While quality of the raw materials and extruded products includes these principles, they also include customer satisfaction [1–4, 37, 79].

Testing and quality control (QC) are the most discussed but often the least understood facets of business and manufacturing. Many companies spend a high percentage of each sales dollar on QC. Usually it involves the inspection of components and products as they complete different phases of processing. Products that are within specifications proceed, while those that are out of specification are either repaired or scrapped. The workers who made the 'out-of-spec' products are notified that they have produced defective products, and that they should correct 'their' mistakes.

The approach just outlined is after-the-fact QC; all defects caught in this manner are already present in the product being processed. This type of QC will usually catch defects, and it is necessary, but it does little to correct the basic problems in production. One of the problems with add-on QC of this type is that it constitutes one of the least cost-effective ways of obtaining high-quality products. Quality must be built into a product from the beginning as illustrated in Figs. 1.1 and 19.1; it cannot be inspected into the process.

The closest any add-on or after-the-fact quality control can come to improving the quality built into a product is to point out processing defects to the departments or persons responsible for them. Instead, the object should be to control quality before the product becomes defective.

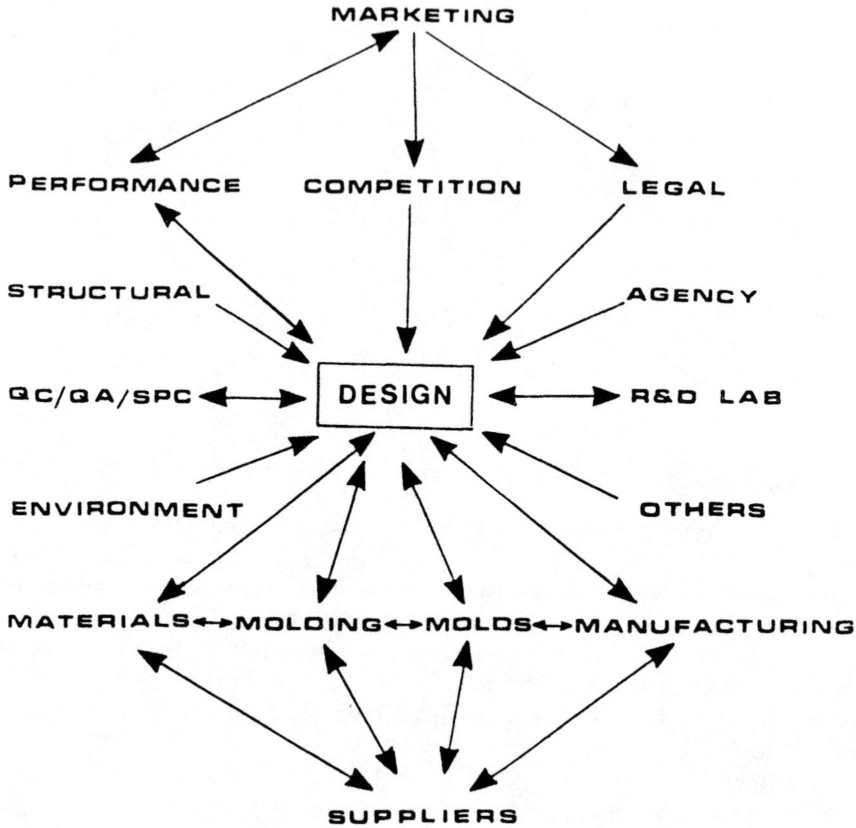


**Figure 19.1** Designing fundamentals.

Properties, such as the physical, mechanical, and chemical properties of plastics, are governed by the molecular weight (Fig. 19.2), molecular weight distribution, structure, composition, and other molecular parameters (Chapter 3). Also involved are the nonplastic additives, such as stabilizers, antioxidants, plasticizers (Fig. 19.3), flame retardants, colors, calcium carbonate, wood fibers, milled glass, and many more. These additives and fillers are used in the plastic compounds (Chapters 3 and 17) to enhance certain processing and/or product performance characteristics.

## Quality assurance

QA is all the planned and systematic actions necessary to provide adequate confidence that a processing facility and/or product will perform

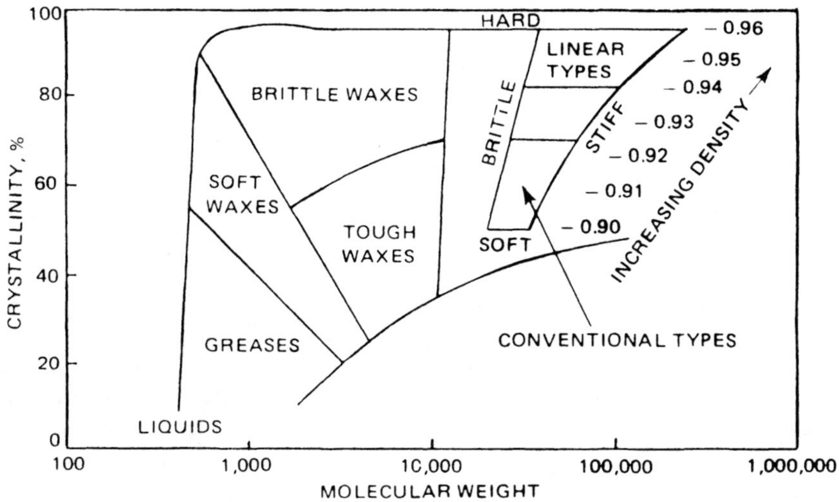


**Figure 19.2** Relationships of PEs between crystallinity, molecular weight, and properties based on different polymerization reactions.

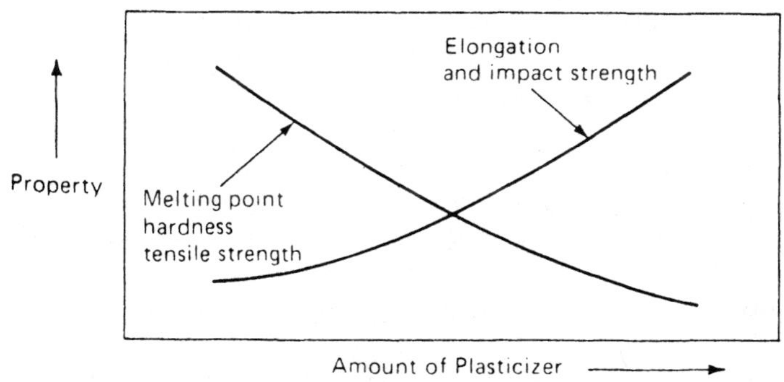


**Figure 19.3** General effect of plasticizers on properties.

satisfactorily in service. It includes quality control, quality evaluation, and design assurance. A good QA program is a coordinated system, not a sequence of separate and distinct steps.

## TESTING

Testing refers to the determination by technical means of properties and performances. This action, when possible, should involve the application of established scientific principles and procedures. Testing requires specifying what requirements are to be met. There are many different tests that can be conducted, relating to practically any requirement [2, 4, 98, 185, 216].

The properties of plastics are directly dependent on temperature, pressure, time, and environmental or service conditions. In turn, these properties can be related to raw material, processing, and product performance. An important testing phase is that done on the finished extruded product. However, testing the plastic materials prior to and during processing is very important to ensure meeting the product performance.

Unfortunately, there is no single set of rules designating which tests are to be conducted in order to manufacture a part repeatedly with zero defects. The tests depend on the required performance, which is usually set by the customer. There are many standard, industry accepted tests that are conducted on raw materials, such as the melt index (MI) tester and other rheological test equipment that can be used in-line as the melting occurs in the extruder [3, 4, 38]. On-line there are practically an unlimited number of ways to run dimensional tests (Chapter 6).

Personnel or operators involved in testing from raw materials to the end of the extruder line develop capability via proper training and experience. Experience and/or developing the proper knowledge is required to determine the tests to be conducted. At times, with new problems developing on-line, different tests are required that may not be available or have to be developed. Unfortunately a great deal of 'reinventing the wheel' can easily occur so someone should have the responsibility to be up to date on what is available. An uncertainty can exist when a viable test has been developed and used within the industry. In time, the test may be changed many times by different companies and organizations (ASTM, etc.) to meet their specific requirements. Someone studying the potential of that particular test may not have the access to the basic test which probably is all that is required. However, just the reverse could be true.

If a part is to operate where any type of failure could be catastrophic to life, then extensive and usually expensive testing is necessary. How deeply one gets involved depends on the performance requirements. If all that is required is to check the weight of the part, that test is all that needs to be performed. However, it is usually not so simple. As reviewed earlier, where large production runs of products are being made, extensive testing is required that involves inspection of the raw plastic materials and in-line equipment operations.

In order to determine the strength and endurance of a plastic under stress, it is necessary to characterize its mechanical behavior. Moduli, strain, strength, and toughness, in addition to the conventional methods, are measured microscopically [3, 43, 163]. These parameters are useful for raw material inspection as well as process and product evaluation. The underlying mechanisms of deformation and fracture have to be understood because of the viscoelastic properties of plastics.

The fracture behavior of plastics, especially microscopically brittle plastics, is governed by the microscopic mechanisms operating in a heterogeneous zone at their crack or stress tip raising flow [190]. In TPs, craze zones can develop that are important microscopic features around a crack tip governing strength behavior. Fracture is preceded by the formation of a craze zone, which is a wedge shaped region spanned by oriented microfils. Methods of craze zone measurements include optical emission spectroscopy, diffraction techniques, scanning electron beam microscopy, and transmission electron microscopy [282]. Properties are affected by their previous heat history since the transformation of the plastics through a process involves heat, pressure, and time. This residence time (Chapter 3) can cause definite undesirable variations in properties of the plastic during processing and/or the finished product. These variations can occur even when the same plastic (from the same source) and same machine are used. Different thermal tests are available and used to meet specific requirements.

## Type of test

A wide variety of test equipment are available. The equipment can characterize plastic properties in the raw material stage, during processing, and the final fabricated product. Simple to complex tests are used depending on what needs to be determined or analyzed. Thermal, optical, chemical, mechanical, physical, electrical, tests, and others, can be conducted. All kinds of properties can be determined, such as density, morphology, molecular structure, hot melt strength, and many more. Examples of tests are given in Table 19.1.

Industry specifications and standards provide very useful information on the many different types of test, with some including purpose, use, and evaluation of results. They are updated regularly and will prove useful to anyone who must choose tests and QC procedures. Table 19.2 lists sources, each with certain areas of interest as their names indicate. Important are ISO and ASTM who cover practical all types of tests.

## Density and specific gravity

The density of any material is a measure of its mass per unit volume, usually expressed in grams per cubic centimeter ($g/cm^3$) or pounds per cubic inch ($lb/in^3$). Specific gravity is the ratio of the mass in air of a given volume compared to the mass of the same volume of water, both being measured at room temperature (23°C or 73.4°F). Since this is a

**Table 19.1** ASTM test methods by subject.

| *ASTM No.* | *Subject* |
|---|---|
| Mechanical Testing | |
| D 638 | Tensile Properties of Plastics |
| D 695 | Compressive Properties of Rigid Plastics |
| D 2344 | Apparent Horizontal Shear Strength of Reinforced Plastics by Short Beam Method |
| D 3039 | Tensile Properties of Oriented Fiber Composites |
| D 3518 | In-Plane Shear Stress–Strain Response of Unidirectional Reinforced Plastics |
| D 732 | In-Plane Shear |
| D 785 | Rockwell Hardness |
| D 790 | Flexural Properties of Plastics and Electrical Insulating Materials |
| D 953 | Bearing Strength |
| D 2344 | Short Beam Shear |
| D 3410 | Test for Compressive Properties of Oriented Fiber Composites |
| Fatigue | |
| D 3479 | Tension-Tension Fatigue of Oriented Fiber Resin Matrix Composites |
| D 671 | Flexural Fatigue of Plastics by Constant Amplitude of Force |
| Impact | |
| D 256 | Impact Resistance of Plastics and Electrical Insulating Materials |
| D 1822 | Tensile-Impact Energy to Break Plastics and Electrical Insulating Materials |
| D 3029 | Impact Resistance of Rigid Plastic Sheeting or Parts by Means of Tup (Falling Weight) |
| Creep | |
| D 2990 | Tensile, Compressive, and Flexural Creep and Creep-Rupture of Plastics |
| D 2991 | Stress Relaxation of Plastics |
| Physical Properties | |
| D 570 | Water Absorption |
| D 792 | Specific Gravity and Density of Plastics by Displacement |
| D 1505 | Density of Plastics by the Density-Gradient Technique |
| D 2734 | Void Content of Reinforced Plastics |
| D 3355 | Fiber Content of Undirectional Fiber/Polymer Composites |
| Thermal Properties | |
| D 648 | Deflection Temperature of Plastics under Flexural Load (HDT) |
| D 746 | Brittleness Temperature |
| D 3417 | Heats of Fusion and Crystallization |
| D 3418 | Transition Temperatures |
| Thermal Expansion | |
| D 696 | Coefficient of Linear Thermal Expansion of Plastics |
| E 228 | Linear Thermal Expansion of Rigid Solids with a Vitreous Solica Dilatometer |
| Thermal Conductivity | |
| C 117 | Steady-State Thermal Transmission Properties by Means of the Guarded Hot Plate |
| Electrical Properties | |
| D 149 | Dielectric Breakdown Voltage and Dielectric Strength of Electrical Insulating Materials at Commercial Power Frequencies |
| D 257 | Electrical Resistance |
| D 495 | Arc Resistance |
| D 150 | AC Loss Characteristics and Permittivity (Dielectric Constant) of Solid Electrical Insulating Materials |
| Wear Resistance | |
| D 673 | Mar Resistance of Plastics |
| D 1242 | Resistance of Plastic Materials to Abrasion |
| Chemical Resistance | |
| C 581 | Chemical Resistance of Thermosetting Resins Used in Glass Fiber Reinforced Structures |
| D 543 | Resistance of Plastics to Chemical Reagents |
| Flammability Tests | |
| D 635 | Rate of Burning |
| D 2843 | Smoke Density |
| D 2863 | Oxygen Index |
| E 662 | Smoke Emission |
| Weatherability Tests | |
| D 1499 | Operating Light- and Water-Exposure Apparatus (Carbon-Arc Type) for Exposure of Plastics |
| D 2565 | Operating Xenon-Arc Type (Water-Cooled) Light- and Water-Exposure Apparatus for Exposure of Plastics |
| D 4141 | Conducting Accelerated Outdoor Exposure Testing of Coatings |
| E 838 | Performing Accelerated Outdoor Weathering Using Concentrated Natural Sunlight |
| G 23 | Operating Light-Exposure Apparatus (Carbon-Arc Type) With and Without Water for Exposure of Nonmetallic Materials |
| G 26 | Operating Light-Exposure Apparatus (Xenon-Arc Type) With and Without Water for Exposure of Nonmetallic Materials |
| G 53 | Operating Light- and Water-Exposure Apparatus (Fluorescent UV-Condensation Type) for Exposure of Nonmetallic Materials |

dimensionless quantity, it is convenient for comparing different materials (Table 19.3).

Like density, specific gravity is used extensively in determining part cost versus average part thickness, initial sheet thickness, product weight, and quality control. It is frequently used as a means of setting plastic specifications and following product consistency. In crystalline plastics (Chapter 3) such as PE, density has a direct effect on properties such as stiffness and permeability to gases and liquids. Changes in density may also affect some mechanical properties.

The ASTM D 792 standard provides the relationship of density to specific gravity at 23°C (other tests include ASTM D 1505 and ISO 1183):

$$\text{Density}\left(\text{g/cm}^3\right) = \text{specific gravity} \times 0.9975$$

also

$$\text{Specific gravity} \times 0.0361 = \text{lb/in}^3$$

**Table 19.2** List includes organizations involved in preparing or coordinating specifications, regulations, and standards

| | |
|---|---|
| ISO | International Organization for Standardization |
| ASTM | American Society for Testing and Materials |
| UL | Underwriters Laboratories |
| DIN | Deutsches Instut, Normung, Germany |
| ACS | American Chemical Society |
| AMS | Aerospace Material Specification of the Society for Automative Engineers (SAE) |
| ANSI | American National Standards Institute |
| ASCE | American Society of Chemical Engineers |
| ASM | American Society fo Metals |
| ASME | American Society of Mechanical Engineers |
| AWS | American Welding Society |
| BMI | Battele Memorial Institute |
| BSI | British Standards Institute |
| CPSC | Consumer Product Safety Commission |
| CSA | Canadian Standards Association |
| DOD | Department of Defense |
| DODISS | Department of Defense Index & Specifications & Standards |
| DOT | Department of Transportation |
| EIA | Electronic Industry Association |
| EPA | Environmental Protection Agency |
| FMRC | Factory Mutual Research Corporation |
| FDA | Food and Drug Administration |
| FMVSS | Federal Motor Vehicle Safety Standards |
| FTC | Federal Trade Commission |
| IAPMO | International Association of Plumbing and Mechanical Officials |

**Table 19.2** *Continued*

| | |
|---|---|
| IEC | International Electrotechnical Commission |
| IEEE | Institute of Electrical and Electronic Engineers |
| IFI | Industrial Fasteners Institute |
| IPC | Institute of Printed Circuits |
| ISA | Instrument Society of America |
| JIS | Japanese Industrial Standards |
| MIL-HDBK | Military Handbook |
| NADC | Naval Air Development |
| NACE | National Association of Corrosion Engineers |
| NAHB | National Association of Home Builders |
| NEMA | National Electrical Manufacturers' Association |
| NFPA | National Fire Protection Association |
| NIST | National Institute of Standards & Technology (previously the National Bureau of Standards) |
| NIOSH | National Institute for Occupational Safety and Health |
| NIST | National Institute of Standard Testing |
| NPFC | National Publications and Forms Center |
| NSF | National Sanitation Foundation |
| OFR | Office of the Federal Registrar |
| OSHA | Occupational Safety & Health Administration |
| PLASTEC | Plastics Technical Evaluation Center of DOD |
| PPI | Plastics Pipe Institute of the Society of the Plastics Industry |
| QPL | Qualified Products List |
| SAE | Society of Automotive Engineers |
| SPE | Society of Plastics Engineers |
| SPI | Society of the Plastics Industry |
| STP | Special Technical Publications of the ASTM |
| TAPPI | Technical Association of the Pulp and Paper Industry |

For greatest usefulness, density needs to be measured to an accuracy of at least 0.001. Sample shape is not an issue, but the sample must be completely wetable in the liquid and contain no voids (per the standards using a gradient or liquid displacement method).

### Melt index

A low cost, easy to operate, widely used test is for melt index (MI) which is basically a ram extrusion plasticator (Fig. 19.4). This so-called rheological device is used for examining and studying plastics in many different plastic fabricating processes including extrusion. It is not a true viscometer, in the sense that a reliable value of the viscosity cannot be calculated from the flow index. However, it does measure isothermal

**Table 19.3** Specific gravity and density comparisons of different materials

| *Materials* | *Specific gravity*[a] | *Density (lb in$^{-3}$)* |
|---|---|---|
| Thermoplastics | | |
| ABS | 1.06 | 0.0383 |
| Acetal | 1.43 | 0.0516 |
| Acrylic | 1.19 | 0.0430 |
| Cellulose acetate | 1.27 | 0.0458 |
| Cellulose acetate butyrate | 1.19 | 0.0430 |
| Cellulose propionate | 1.21 | 0.0437 |
| Ethyl cellulose | 1.10 | 0.0397 |
| Methyl methacrylate | 1.20 | 0.0433 |
| Nylon, glass-filled | 1.40 | 0.0505 |
| Nylon | 1.12 | 0.0404 |
| Polycarbonate | 1.20 | 0.0433 |
| Polyethylene | 0.94 | 0.0339 |
| Polypropylene | 0.90 | 0.0325 |
| Polybutylene | 0.91 | 0.0329 |
| Polystyrene | 1.07 | 0.0386 |
| Polyimides | 1.43 | 0.0516 |
| PVC, rigid | 1.20 | 0.0433 |
| Polyester | 1.31 | 0.0473 |
| Thermosets | | |
| Alkyds, glass-filled | 2.10 | 0.0758 |
| Phenolic, GP | 1.40 | 0.0505 |
| Polyester, glass-filled | 2.00 | 0.0722 |
| Rubber | 1.25 | 0.0451 |
| Metals | | |
| Aluminum SAE-309 (360) | 2.64 | 0.0953 |
| Brass, Yellow (#403) | 8.50 | 0.3070 |
| Steel, CR Alloy (Strip and Bar) | 7.85 | 0.2830 |
| Steel, Stainless 304 | 7.92 | 0.2860 |
| Magnesium AZ-91B | 1.81 | 0.0653 |
| Iron, Pig, Basic | 7.10 | 0.2560 |
| Zinc, SAE-903 | 6.60 | 0.2380 |

[a] The number of grams per cubic centimeter is the same as the specific gravity. For example, if the specific gravity is 1.47, that substance has a density of 1.47 g cm$^{-3}$.

resistance to flow using an apparatus and test method that are standard worldwide. Standards include ISO, ASTM (D 1238; methods A and B), DIN, and JIS.

In this instrument, the unmelted-solid plastic is contained in a 'barrel' (Fig. 19.4) with a temperature indicator. The instrument is surrounded by

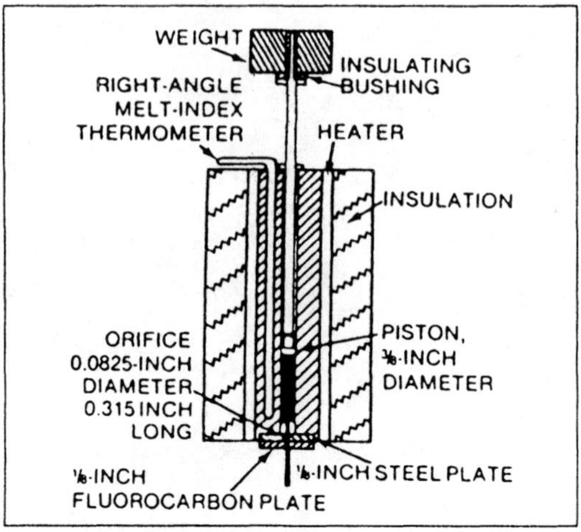


**Figure 19.4** Schematic of melt index tester.

an electrically controlled heater that melts the plastic. A weight drives a plunger that forces the melt through the die opening. Its construction and operating procedure follows the standard's requirements. The usual procedure involves the determination of the amount of plastic extruded in 10 min (after initial flow starts). The flow rate expressed in g/10 min. More than one test is conducted and, in turn, an average is reported. As the flow rate increases, the viscosity decreases.

This MI makes a single point test that provides information on the resistance to flow only at a single shear rate. Examples of MI used by extruders are shown in Figs. 19.5–19.7 and Tables 19.4 and 19.5. Note that the extruder normally requires a low MI when compared to injection molding since it requires melt strength when extruding into 'open' space. Whereas, IM is forced into a closed mold (Chapter 18).

Because variations in branching or molecular weight distribution (MWD) of the plastic can alter the shape of the viscosity curve, the MI may give a false ranking of plastics in terms of their shear rate resistance to melt flow. To overcome this situation, extrusion rates are sometimes measured for different loads and other modifications are made to the instrument, such as changing the size of the orifice.

## Melt rheometer

For a 'real' method to characterize plastics melt behavior, the capillary melt rheometer is used. This instrument is fully automated and covers a very wide range of shear rates. When used on-line with extruders, many operating benefits, such as improving/insuring quality, reduced scrap,

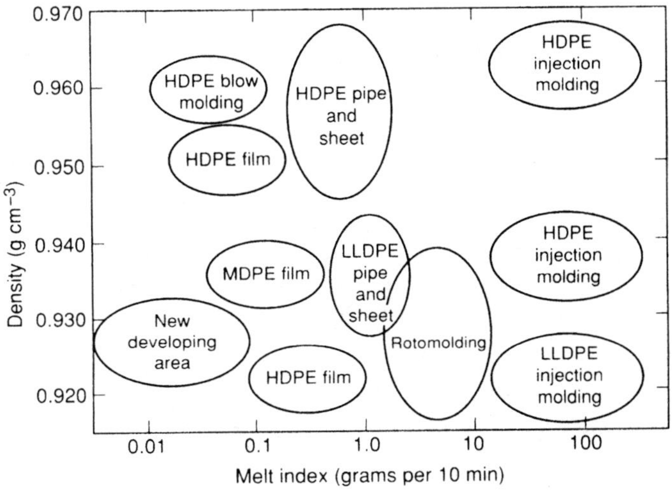


**Figure 19.5** Examples of PE markets based on melt index (MI) versus density.

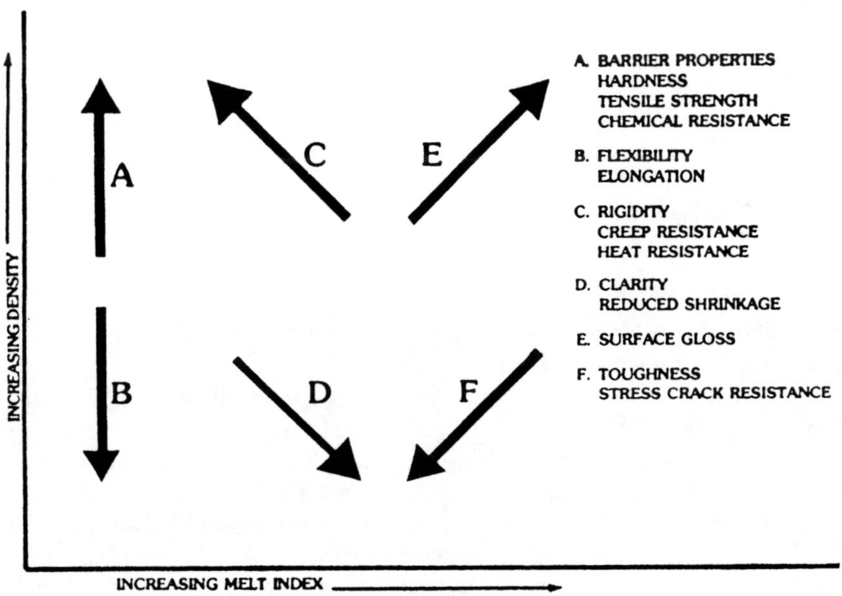


**Figure 19.6** melt index (MI) versus density.

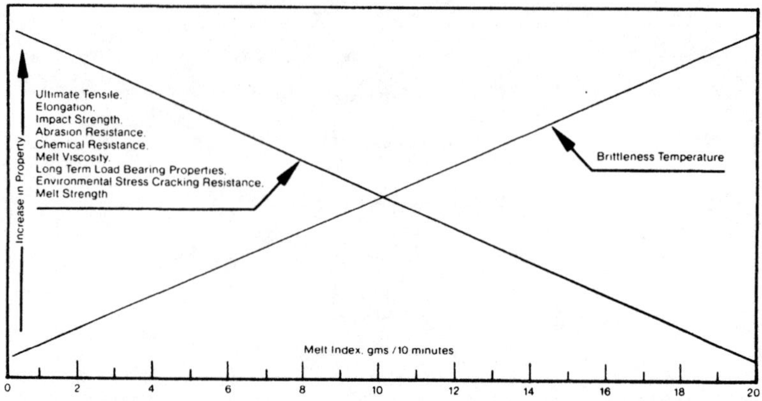


Basic Physical Property Changes with Increasing Molecular Weight

| Property | Increase or decrease with increasing molecular weight |
|---|---|
| Hardness | Increases |
| Toughness | Increases |
| Melt Viscosity | Increases |
| Tensile Strength | Increases |
| Brittleness | Increases |
| Chemical Resistance | Increases |
| Yield Strength | Increases |
| Elongation | Increases |
| Abrasion Resistance | Increases |
| Solubility | Decreases |
| Adhesion | Decreases |
| Melt Index | Decreases |

Typical Specific Property Changes in Materials

| Property | Material | Change |
|---|---|---|
| Toughness | Polyethylene | Increases with an increase in molecular weight. Increases with narrowing of molecular weight distribution. |
| Solution Viscosity | Polyethylene | Decreases with a decrease in molecular weight. |
| | Polystyrene | |
| Fiber Strength | Polyesters | Increases with an increase in molecular weight. |

**Figure 19.7** PE properties versus melt index (MI).

and lowers costs, are gained [104]. This instrument is far superior in understanding melt behavior to others.

An example of an on-line rheometer from Rheometrics, Inc., Piscataway, NJ, USA, is shown in Fig. 19.8. Its real-time process control provides a quick response time where the sample is immediately sent from the extruder's plasticator through a tube to its sampling chamber (Fig. 19.9). In a typical operation, a small stream of melt is diverted from the plasticator. A gear pump directs the melt flow into the sample chamber. The melt fills the space between two concentric cylinders. A torque motor drives the outer cylinder in an oscillating manner. This action imposes a sinusoidal strain on the melt. The corresponding oscillatory stress is measured by monitoring the torque on the inner cylinder (follows Hagen–Poiseuille law) [4].

**Table 19.4** MI versus density

| *Properties* | *With increasing melt index* | *With increasing density* |
|---|---|---|
| Rigidity | – | Increases |
| Heat resistance | Decreases | Increases |
| Stress crack resistance | Decreases | Decreases |
| Permeation resistance | – | Increases |
| Abrasion resistance | – | Increases |
| Clarity | – | Decreases |
| Flex life | Decreases | Decreases |
| Impact strength | Decreases | Decreases |
| Gloss | Increases | Increases |
| Vertical crush resistance | – | Increases |
| Cycle | Decreases | Decreases |
| Flow | Increases | Decreases |
| Shrinkage | Decreases | Increases |
| Parison roughness | Decreases | Increases |
| Parison sag | Increases | Decreases |
| Pinch quality | Increases | – |
| Parting line difference | – | Increases |

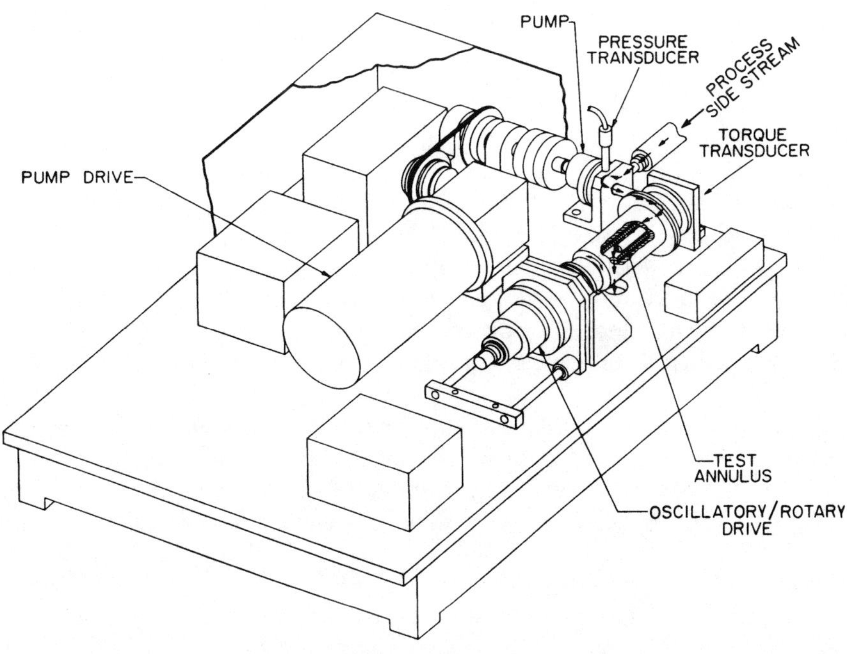


**Figure 19.8** Schematic of on-line rheometer.

**Table 19.5** Examples of MI for different processes

| *Process* | *MI range* |
|---|---|
| Injection molding | 5–100 |
| Rotational molding | 5–20 |
| Film extrusion | 0.5–6 |
| Blow molding | 0.1–1 |
| Profile extrusion | 0.1–1 |

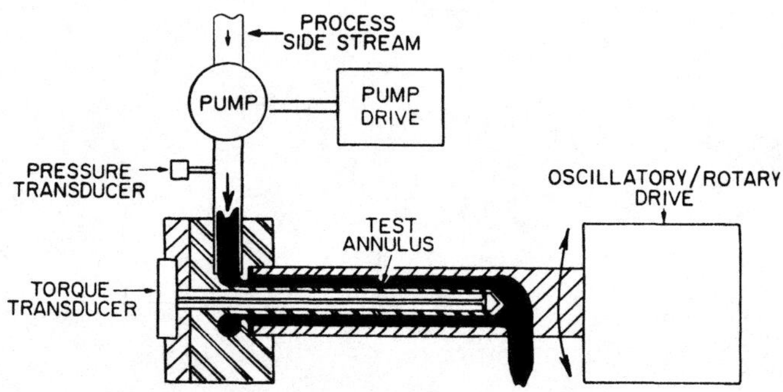


**Figure 19.9** View of on-line sample chamber.

## QUALITY CONTROL

Processors should keep quality under control and demand consistent materials that can be used with minimum of uncertainty. Plant quality control (QC) is as important to the end result as selecting the best processing conditions with the correct grade of plastic, in terms of both properties and appearance. After the correct plastic has been chosen, its blending, reprocessing, and storage stages of operation need to be frequently or continuously updated. The processor should set up specific measurements of quality to prevent substandard products reaching the customer [342, 387].

QC involves those quality assurance actions which provide a means to control, measure, and establish requirements of the characteristics of plastic materials, processes, and products. As an example, although care is taken by materials manufacturers to assure consistency, subtle variations exist in their product that previously was unnoticed and/or considered of no importance. In most general applications, these variations have little effect on processing and/or finished product. But in more stringent cases, these irregularities can present processing problems.

To simplify the task of assuring that the physical properties of a system are in specification, simple techniques can be used in incoming, on-process, and outgoing QC [3, 163]. Use of these procedures by companies concerned with maintaining critical properties can keep a tight rein on product quality and provide documented qualification.

Acceptable quality level (AQL) is a quality of the product, expressed as a percent defective, such that a lot having this percent defective will have a probability of rejection by the customer. An ideal sampling and inspection plan would accept all lots or better quality and reject all lots of lower quality. Any practical plan can approach this ideal. AQL is the process average at which the risk of rejection is called the producer's risk [4].

There are different ways to apply QC on-line. One way is with infrared measurement. The ability to record IR spectra of plastic melts provides for process monitoring and control in the manufacture process. Precise information on quality can be obtained rapidly. Furthermore, it is also possible to make measurements on unstable intermediates of importance. Although spectroscopy on melts is considerably different from that on solid materials, this does not limit the information content.

IR has for many years been an important aid to investigating the chemical and physical properties of molecules. It gives qualitative and quantitative information on chemical constituents, functional groups, impurities, etc. As well as its use in studying low molecular weight compounds, it is used with equal success for characterizing plastics. It is a highly informative method of applying testing.

## STATISTICAL PROCESS CONTROL

Statistical process control (SPC) can be used as a statistical method of process monitoring to meet quality assurance during fabrication. Basically, it is concerned with information about the stability and reproducibility of a process; it concentrates on a specific sequence of key tests on the characteristics of a product being extruded, etc. Such results presented graphically as a QC chart, provide information on the progress of the process and if there is any need for intervention.

The proper use of control charts with plotted data and graphs will help to continuously boost quality by tightening control limits. SPC is targeted to decrease scrap or rework and increase the first-time yield of higher quality cost-effective products that meet designed requirements [428].

A process can be identified as being in statistical control when only random variation (or common cause) is falling within acceptable limits are exhibited. A process is statistically capable when products can be consistently produced within specification. Both condition should be present for optimum process operations. SPC analysis can prove the balance and offer means for problem solving if special cause variations take a process out of control or capability. The first thing to know is what natural variation of the process exist so that it is known when to intervene, and when to simply leave the situation alone.

A process out of control does not necessarily mean that bad products are resulting, but the SPC analysis can indicate a variation that shall be tightened. The application of SPC to short-run production can cause problems when there is not enough data to calculate limits in a timely manner. However, the competitive power of SPC can be realized even by those who produce complicated products, such as extruded profiles in low volume.

SPC seeks to closely control the manufacturing process and permits the manufacture of tighter-tolerance products by indicating when a process is starting to drift away from the ideal set points. There are two basic approaches for real-time SPC. The first approach, done on-line, involves the rapid dimensional measurements of a product or a nondimensional 'bulk' parameter such as weight, and is the more practical method. Other dimensional measurements of the precision needed for SPC are generally done off-line, which is the second approach, and produce a response that is too slow. Obtaining the final dimensional stability needed to measure a product may also take time; amorphous plastics require at least 30–60 min to cool and stabilize (Chapter 3).

There are four common features of SPC: (1) raw plastic material characterization; (2) inplant plastic material handling (drying, blending, mixing, etc.); (3) extruder machine operation; and (4) implementation of SPC (supported by management, etc.) [351].

**Monitoring process variables**

There are three phases in the evolution of most QC systems: (1) defect detection where an 'army' of inspectors tries to identify defects; (2) defect prevention where the process is monitored, and statistical methods are used to control process variation, enabling adjustments to the process to be made before defects are produced; and (3) total quality control where it is finally recognized that quality must extend throughout all functions and it is management's responsibility to integrate and lead the various functions towards the goals of commitment to quality and customer-first orientation.

When using the defect-detection approach to quality control certain problems develop. Inspection does nothing to improve the process and is not very good at sorting good-from-bad. Also, sampling plans developed to support an acceptable quality level (AQL) of 5%, for example, a company may be content to deliver 5% defects.

**Assessing statistical practice**

A fabricator's assessment of its use of SPC tools should consist of a formal, documented examination of current statistical practices and procedures, and an evaluation of future plans for improvements of the company's QC. To be useful, this assessment should go beyond the compliance-oriented approach that is commonly seen in quality audits.

Fabricators should setup key objectives for their assessment of good statistical practices. They could include: (1) determine your current state of compliance regulations; (2) determine impediments to compliances; (3) raise awareness of regulations; (4) measure improvements over time; (5) discover the best statistical practices in use throughout the company and share them with the rest of the company; and (6) provide advice on incorporating statistical tools into the quality improvement system.

## QUALITY SYSTEM REGULATION

Process validation (PV) and good manufacturing practice (GMP) may not at first have received the attention they deserved, but that situation changed in the medical device industry and also in other product industries where they wanted to follow strict processing procedures. The originator FDA (Food and Drug Administration) defines PV and GMP as a documented program which provides a high degree of assurance that a specific process will consistently produce a product meeting its predetermined specifications and quality attributes. Elements of validation are product specification, processing equipment, and process revalidation and documentation.

The GMP regulation became effective during 1978 [294]. As of October 7, 1996, GMP was revised incorporating many changes; it was renamed to quality system regulation (QSR) [155]. The GMP focused almost exclusively on production practices requiring very detailed manufacturing procedures and extremely very detailed documentation. The QSR major additional requirements are in the areas of design, management responsibility, purchasing, and servicing. It encompasses quality system requirements that apply to the entire life cycle of a device.

Quality system regulation was greatly influenced by the work of the Global Harmonization Task Force [294]. In slightly different wording, the regulation now incorporates all the requirements of ISO 9001, the fullest of the quality systems standards compiled by the International Organization of Standardization (ISO) (342). With the QSR's expanding scope, it brings its terminology into harmony with that of the ISO 9000 family of quality systems standards.

It includes requirements for quality planning, for a quality manual, and for an outline of the general quality procedure. All these new requirements were added in order to bring the regulation into harmony with the quality systems standards currently accepted in the member nations of the European Union (EU). Quality system regulation makes it clear that manufacturers need only comply with those parts of the regulation that apply to them [230].

So if you were manufacturing plastic (or any other material) products that went into the medical products such as the device industry, you are required to conformed to a very tight QSR. The plastic processor will find QSR enables him to review the detailed processing procedures required to set-up what might be said is to be the 'perfect' manufacturing operation.

## EXTRUSION VARIABLES

In extrusion, as in all other processes, an extensive theoretical analysis continues to be applied to facilitate understanding and to maximize the manufacturing operation. However, the real world must be understood and appreciated as well. The operator has to work within the many limitations of the plastic materials as well as the equipment that includes the extruder and all the auxiliary up-stream and down-stream equipment.

Variations in extruder performance that cause most of the problems in the extruder are rather extensive. These variations are principally caused during the melting process. They involve the following factors (that have been reviewed throughout this book): melt conditions with variations or nonuniformities in pressure, temperatures, insufficient melt capacity, solids conveying inefficiency, insufficient mixing capacity, screw and/or barrel wear, and screw speed, and others, such as plastic material flow in the hopper, unstable temperature at the barrel throat, insufficient pressure generated for proper pumping action, inadequate filtering screen, low diehead pressure, unsteady take-off equipment, etc.

If variations are excessive, all kinds of problems will develop. The interplay and interchange of process controls can help eliminate problems and aid the operations where variations exist (Chapter 6). The greatest degree of instability is due to improper screw design, or more often using the wrong screw. Proper instrumentation settings, particularly barrel heat, is important to diagnosis. For uniform or stable extrusion, it is important to check the drive system, the take-up device and other equipment periodically, and to compare the readings to their original settings.

An elaborate process control system can help, but it is best to improve stability in all facets of the extrusion line. Examples of instabilities and problem areas are: (1) nonuniform plastics flow into the hopper; (2) troublesome bridging that varies or stops plastic flow due to excessive barrel heat that melts and solidifies plastic in the hopper and the feed section of the screw; (3) variations in (a) barrel heat, (b) screw heat, (c) screw speed, (d) screw power drive, (e) die heat, (f) diehead pressure, and (g) take-up device; (4) insufficient melting and/or mixing capacity; (5) insufficient pressure-generating capacity; (6) wear or damage of the screw and/or barrel; and (7) melt fracture/sharkskin (Chapter 3).

One must also check if the extruder and the down-stream equipment are properly aligned. Proper alignment and isolation of the vibrators is essential for high quality, high-speed output. To support the alignment, proper instrumentation is vitally important for quick and accurate diagnosis [438].

A prerequisite for stable extrusion is a good extruder drive, a good temperature control system, an accurately controlled take-up device and, most important, a good screw design. The extruder should be equipped with some type of proportioning temperature control.

**Plastic material**

Even though equipment operations have or cause variabilities, the usual most uncontrollable variable in the process is the plastic material, blending plastics, as well as most additives and fillers used in the processing compounds. The degree of properly compounding by the plastic manufacturer or in-house by the fabricator/converter is important (Chapter 17).

Variations from one supplier to another of the same plastic material usually requires different settings on the equipment to maximize their performance. Even variations from delivered tank car to tank car or gaylord to gaylord within a lot from the same supplier can cause havoc. This situation is controllable by proper specification preparation and quality control from the material suppliers and/or in-house inspection upon the arrival of the material. If the materials are effected by aging (time in storage), it is important to follow up with another inspection in-house.

An important control concerns moisture content of the plastic materials. Throughout this century one can say that one of the biggest problems encountered by processors has been the contamination of plastics with moisture/water. Different plastics require different procedures to follow when drying plastics. Details on drying are in Chapter 3.

An example of a potential moisture problem in the plant concerns opening sealed bags. Their contents of a hygroscopic plastic have definitely dried by the supplier of the plastic. The fabricator only uses a few bags of the plastic. The remaining open bags are not properly sealed because they are expected to be used 'shortly.' However, in an area with high humidity, it does not take a long time for moisture to enter the bags and cause the plastics not to process properly.

**Melt flow**

As reviewed earlier, the target is to provide the necessary homogeneous melt by having the extruder to operate complete stable and working in equilibrium. In practice, even though with the developments that have occurred in the past and continue, this perfect stable situation is never achieved and there are variables that affect the output all along the extruder from the hopper to the die.

If the extrusion process is analyzed one can decide that two types of variables affect the quality and output rate. They can be identified as: (1) the variables of the extruder's design and manufacture; and (2) the operating or dynamic variables which control how the extruder is run.

To ensure the most efficient operation, there are parameters of design or intrinsic variables. These values are known or believed to be known. Once fixed, they generally cannot be altered without interrupting the process. They cover the geometry of the component parts which make up the extrusion system and include such important factors as the diameter and length of the screw, also the pitch and depth of the screw flights. Other variables in this category are the designs of the hopper, die, and barrel, together with the type, shape, and physical state of the plastic.

The operator or dynamic variables exist which can be altered at any time during the extrusion process. They can be identified as: (1) independent variables which are controllable that involve melt temperature, screw and barrel cooling/heating, and screw speed; and (2) dependent variables which result from the variables just listed, such as melt zone positions, homogeneity, output, and pressure.

Note that with extruders operating well below their output limit, it is not important that an absolute correct design be used. Where higher speeds are used, however, severe fluctuations in output do occur and both the choice of plastic and correct design of equipment are important factors.

Regarding the extruder's design or intrinsic variables, one orders the machine from the manufacturer that has designed an extruder to meet minimum variables and the customers requirements. Machine manufacturers have the ability to advance the state of the art to meet new and an endless list of new or past undiscovered variables. Thus, the extrusion hardware is not a fixed situation.

Regarding the extruder's operating or dynamic variables, the fabricator has control and capabilities that can be applied. An example is using the best screw design for processing a specific plastic, that may have come from the machine builder or screw manufacturing specialists. Information on the screw design and operation is reviewed in Chapter 4. Heat is developed, after start-up, from the frictional forces caused by the mixing and compression action of the screw. External heat and cooling action around the barrel varies in importance depending on product being produced.

In extrusion coating (Chapter 11) where very high operating temperatures are required, barrel heating is almost always required with the frictional heat in the melt. With film and profile extrusion, it is necessary to keep the temperature as low as is practicable. Screw design is carefully selected for a plastic so that the frictional heat developed during use is sufficient. A small amount of the external barrel heat is needed, if any.

**Process control**

Process controls (PCs) for the individual machines and the complete fabricating line can range from unsophisticated to extremely sophisticated devices. They can: (1) have closed loop control of temperature and pressure; (2) maintain preset parameters for the complete line; (3) monitor and correct machine operations; (4) constantly fine tune equipment; and (5) provide consistency and repeatability in the complete line.

Different degrees in PC exist, such as the open loop, monitor only, monitor and closed loop, and integrated machine and process control. Recognize that a PC is not a toy or a panacea. A PC demands a high level of expertise from the processor.

Based on the PC settings, different behaviors of the plastic will occur. Regardless of the type of controls available, the processor setting up equipment uses a systematic approach that should be outlined in the plant's operational manual. Once the machine is operating, the processor methodically makes usually one (major) change at a time and provides enough time to determine the result.

Adequate PC and its associated instrumentation are essential for product quality control. The goal in most cases is precise adherence to a control point. In other cases, maintaining the temperature within a comparatively narrow range is all that is necessary. For effortless controller tuning and lowest initial cost, the processor should select the simplest controller (temperature, pressure, time, melt flow rate, etc.) that will meet the desired results.

Once an extruder is running, it is important to ensure that conditions, such as the melt temperature, melt pressure, screw speed, and take-up devices, are consistent. If any of these conditions vary then the output will also vary and usually not meet performance requirements. Process variables, such as temperature and pressure, are measured by a transducer or sensor.

The output from the transducer is sent to a display unit where it can be displayed. A continuous display of temperature and pressure can be extremely useful and can lead to significant improvements in the line's productivity. For this reason, the display must be clear and easy to read and understand. So that information is not lost the equipment must not be over-damped; this is done to give steadier readings. Often the display output is only recorded if something is going wrong but a record of when it ran perfectly is very useful. The display may also have over and under alarms built in the system so that warning is given if the extrusion process is varying [434].

The temperature of the hot melt is measured by the melt temperature thermocouple. The location and depth of the thermocouple are important. It is better to use more thermocouples than not enough to cover from the hopper to the die. The thermocouple tip is usually approximately 6 mm (0.25 in) away from the barrel's inside wall. The thermocouples which feed the temperature control system should be located as deep into the barrel wall as safety will permit.

The melt temperature is kept constant because if it varies then the viscosity or ease of flow of the melt will also change, providing a variability in product performance. If melt temperature goes too high then the melt will degrade. Temperature variations can be smoothed out by using a static mixer (Chapter 2) between the screw tip and the die or gear pump. If the extruder is designed and operated correctly, the static mixer is not necessary, however it is often used just to ensure that no temperature variations will occur.

Melt pressure is measured, or sensed, by the melt pressure transducers which are usually located in the barrel and at the die end of the barrel and within the barrel, such as before and after a screen changer. They can be located down-stream of the screw to provide melt pressure conditions; they can spot where surging occurs. Such transducers are used because they definitely improve productivity and efficiency that can reduce or help eliminate variables. Any variation in pressure can cause large changes in the extruder's output.

To help eliminate extruder's pressure variations, gear pumps (Chapter 2) that stabilize the melt's pressure are located between the screw tip and the die. If a static mixer is also used, the usual line-up is from the screw tip, to the screen changer (Chapter 2), to the static mixer, to the gear pump, to the die. Several pressure transducers can be fitted around and/in these devices. They can automatically provide a signal if preset pressure limits are not being met; they can control automatic screen changers, and so on.

Screw speed (rotation) control is important to eliminate variables during the machine's operation. The rate at which the screw turns controls how much melt is pumped through the die. This pumping rate provides an important control factors such as the product's dimensions of thickness and width or diameter. In turn, down-stream controls are related to the screw speed, such as drawdown ratio, sizing/calibration equipment, and windup.

If the haul-off rate is kept constant then doubling the screw speed will almost double the thickness. To produce quality products requires accurate process control of screw speed to haul off/wrap-up speed. When using postforming equipment (Chapter 18) in-line, they also require adjustable and controllable settings to match the line speed. An example of where a line could have speed control variations is when a manufacturer of large products goes to producing very small products such as very small diameter tubing. The line requires a very sensitive and accurate speed control puller. If a nonuniform pull rate exists, variation in the product output is effected.

**Product**

The different extrusion processing lines reviewed each have situations where variations exist. They usually have common denominators regarding their problems and solutions, so consider reviewing them even though one is only interested in only one specific product. As an example, a certain chapter may review in depth a cooling system. This cooling system will be applicable in other chapters concerning other products. An example of specific product variables is with blown film. It can be difficult to describe the importance of the many variables which effect the film properties with any degree of precision. However, as reviewed in Chapter 7, process controls provide a means to control the variables that exist. Improvements continue to be made both on the plastic material being processed and the equipment used.

There are always advancements or new methods to control the pressure and/or temperature of the bubble, improve film cooling methods, ways to locate the freeze line, consistency of speed and tension in the handling and haul off system, and so on. What has been reviewed here has some applications to flat film and sheet lines.

There is continued progress in improving the variables that can develop on PE's film surface and optical properties. The two major types of optical irregularity are caused by: (1) melt flow phenomena effects on surface irregularities; and (2) crystallization behavior of the plastic. During the film-forming operation, there will be a change in the texture of the surface of the melt. This occurs because of a decrease in the depth of the extrusion and an increase in their length and width as the melt is drawn lengthwise and transversewise.

As long as the plastic remains molten, the overall magnitude of the defects will decrease under the influence of surface tension. Therefore, the freeze-line height is increased and the haze will diminish. At the same time that this action occurs, the opposing effect of the slower cooling times on the size of the crystallites will also change. Generally the blow-up ratio, freeze-line, die gap, and extrusion temperature should all be kept as high as possible.

## Processing intelligence

What is needed is to cut inefficiency, such as the variables, and in turn cut the costs associated with them. One approach that can overcome these difficulties is called intelligent processing of materials. This technology utilizes new sensors, expert systems, and process models that control processing conditions as materials are produced and processed without the need for human control or monitoring.

Sensors and expert systems are not new in themselves. What is novel is the manner in which they are tied together. In intelligent processing, new nondestructive evaluation sensors are used to monitor the development of a materials microstructure as it evolves during production in real time. These sensors can indicate whether the microstructure is developing properly. Poor microstructure will lead to defects in materials. In essence, the sensors are inspecting the material on-line before the product is produced.

The information these sensors gather is communicated, along with data from conventional sensors that monitor temperature, pressure, and other variables, to a computerized decision making system. This decision maker includes an expert system and a mathematical model of the process. The system then makes any changes necessary in the production process to ensure the material's structure is forming properly. These might include changing temperature or pressure, or altering other variables that will lead to a defect-free end product.

There are a number of benefits that can be derived from intelligent processing. There is, for instance, a marked improvement in overall product quality and a reduction in the number of rejected parts. And the automation concept that is behind intelligent processing is consistent with the broad, systematic approaches to planning and implementation being undertaken by industries to improve quality.

It is important to note that intelligent processing involves building-in quality rather than attempting to obtain it by inspecting a product after it is manufactured. Thus, industry can expect to reduce post-manufacturing inspection costs and time. Being able to change manufacturing processes or the types of material being produced is another potential benefit of the technique.

## Judgment and experience

The successful design and fabrication of good plastic products requires a combination of sound judgment and experience. Designing good products (profiles, etc.) requires a knowledge of plastics that includes their advantages and disadvantages, and also some familiarity with the processing methods [3].

Until the designer becomes familiar with processing, a fabricator must be taken into the designer's confidence early in the development stage and consulted frequently during those early days. The fabricator and the die designer should advise the product designer on plastic materials behavior and how to simplify the design to permit easier processability.

Although there is no limit theoretically to the shapes that can be created, practical considerations must be met. These relate not only to the part design, but also the die design, since there must be considered one entity in the total creation of a usable, economically feasible part.

### Problem and solution

Throughout this book problems have been presented to show what could happen when operating the extruder as well as the up-stream and down-stream equipment. In turn, these problems or variables are provided with solutions. The problem and solution guides are in the text and also presented in troubleshooting tables [190, 445].

In order to better understand potential problems or the variables, it is helpful to consider the relationships of machine capabilities, plastics processing variables, and product performance requirements [3]. It may be impossible to meet the product requirement because the equipment does not have the capability and/or the plastic does not have the capability. One can only get so much out of equipment and plastic. Companies that have to keep ahead of their competition have to consider new equipment when theirs is inadequate. Of course, this approach has been used for centuries even before plastic products were developed.

A distinction between machine conditions and processing variables must be made in order to avoid mistakes in using cause-effect relationships to their best advantage. As an example, Fig. 19.10 is to highlight a situation where it would normally be properly recognized and corrected by cleaning (assuming it could not be forced out, etc.) the die and ensuring that up-stream melt is not having a problem. However, for this review, examine it another way for the novice. It could be conceived as a die problem or someone else could assume it is a plastic problem.

To resolve variables or problems, a logical and systematic method of dealing with them is required. The method should use language that everyone understands. Terms and phrases are not to be ambiguous; they should not be prepared like a legal document where all kinds of definitions could exist just for one term or phrase. Unfortunately, certain terms and phrases may have different meanings to different people. Resolve this situation by clearly identifying them. It may be important to include what

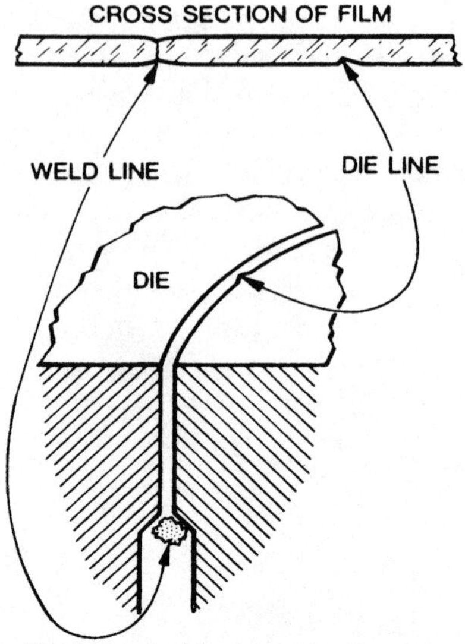


**Figure 19.10** Example of a contaminated die and a troubleshooting problem.

it does not mean to eliminate any misunderstandings. If it is necessary to include engineering equations or chemical formulas, it is important to explain them in terms understood by the nontechnical person (or even technical person that may misunderstand them).

When the specific problem(s) has been identified, record how to eliminate it. This type of information (as reviewed throughout this book) should be documented in the operating manual used in operating the production line. If applicable, include the information directly into the line's process control system.

Unfortunately at times one might be informed, particularly the operator, that Chisolm's Law is occurring. This law, which has been around a long time, states that if things appear to be going well, you must have overlooked something.

An important consideration is to analyze failures. You do not have to wear rose-colored glasses to view reverses as a route to eventual success. Put failures under the microscope of an objective critique. You may not want or need to schedule a full-scale inquest every time. But even a quick postmortem on a project that has foundered may keep you from botching another one.

# 20

# Summary

## INTRODUCTION

The extrusion plastic industry is comprised of mature practical and theoretical technology. Improved understanding and control of processes have significantly increased extruded product performances and reduced variabilities. These products can be characterized in many different ways, such as those with high impact and tear resistance properties, or flexible to high strength. They perform in all kinds of environments, are decorative to industrial structures, have short (throwaway) to very long service life. There are degradables and nondegradables. It is possible to process virgin with recycled plastics or recycled alone, to produce simple to complex shapes, including some which are difficult or impossible with other materials, to produce breathable film for use in horticulture, packaging, etc., and meet many more market requirements [1, 3, 14].

Extrusion technology continues to find practical solutions for problems to meet new performance requirements. Generally, the target is to further improve in the economy of the extrusion plant due to the endless demands of competition. In the past, a lack of technical data for TPs was a particular handicap. Gradually, complicated rheological data became available based on plastic performances during processing and these data helped advance the art of extruding plastics.

Plastics provide extruder processors with materials that are useful, meet product requirements, produce simple to complex shapes, and are economically beneficial. Processors must continually update their procedures and/or acquire additional knowledge on how to process plastics. New developments in this field are unlimited. Simultaneously, new developments continue to occur in equipment that continually simplify processability and process control.

However, it is important to recognize that identical machine models, including auxiliary equipment, built and delivered with consecutive serial numbers to the same site can perform so differently as to make some completely unacceptable by the customer, assuming they were installed properly.

## SUCCESS BY DESIGN

A skilled designer blends a knowledge of materials, an understanding of manufacturing processes, and imagination, with new and innovative designs. Recognizing the limits of design with traditional materials is the first step in exploring the possibilities for innovative design with plastics.

Some designers operate by creating only the stylish outer appearance, allowing basic engineers to work within that outside envelope. This approach is used very successfully in certain products, such as parts for furniture. There is also the combination of designing appearance with engineering so that the stylish product incorporates the best combination with ease of processing when using a specific plastic, and provides simplified assembly, capability of repair, streamlining quality control, and/or other conditions. The stylish envelope that eventually emerges will be a logical and aesthetic answer to the design challenge as summarized in Fig. 19.1.

The term 'design' has many connotations. Essentially, it is the process of devising a product that fulfills as completely as possible the total requirements of the user, and at the same time satisfies the needs of the fabricator in terms of cost-effectiveness (return on investment). The efficient use of the best available material and production process should be the goal of every design effort, including tool design. Product design is as much an art as a science. Design guidelines for plastics have existed for over a century [3, 43, 61]. Basically, design is the mechanism whereby a requirement is converted to a meaningful plan as summarized in Fig. 20.1.

Periodically, the reputation of plastics has been harmed a great deal by the fact that in many cases designers and engineers have, after deciding tentatively to try to introduce plastics, slavishly copied the metal part it is suppose to replace. Too much emphasis cannot be laid upon the general principle that if plastics are to be used to maximum advantage and with minimum risk of failure, it is essential for the designer inexperienced in working with plastics to do some homework, to become familiar and keep up to date with the plastic processes and materials (Fig. 20.2).

## COST

The production flexibility of the fabrication process, as reviewed throughout this book, is often the single most important economic factor in

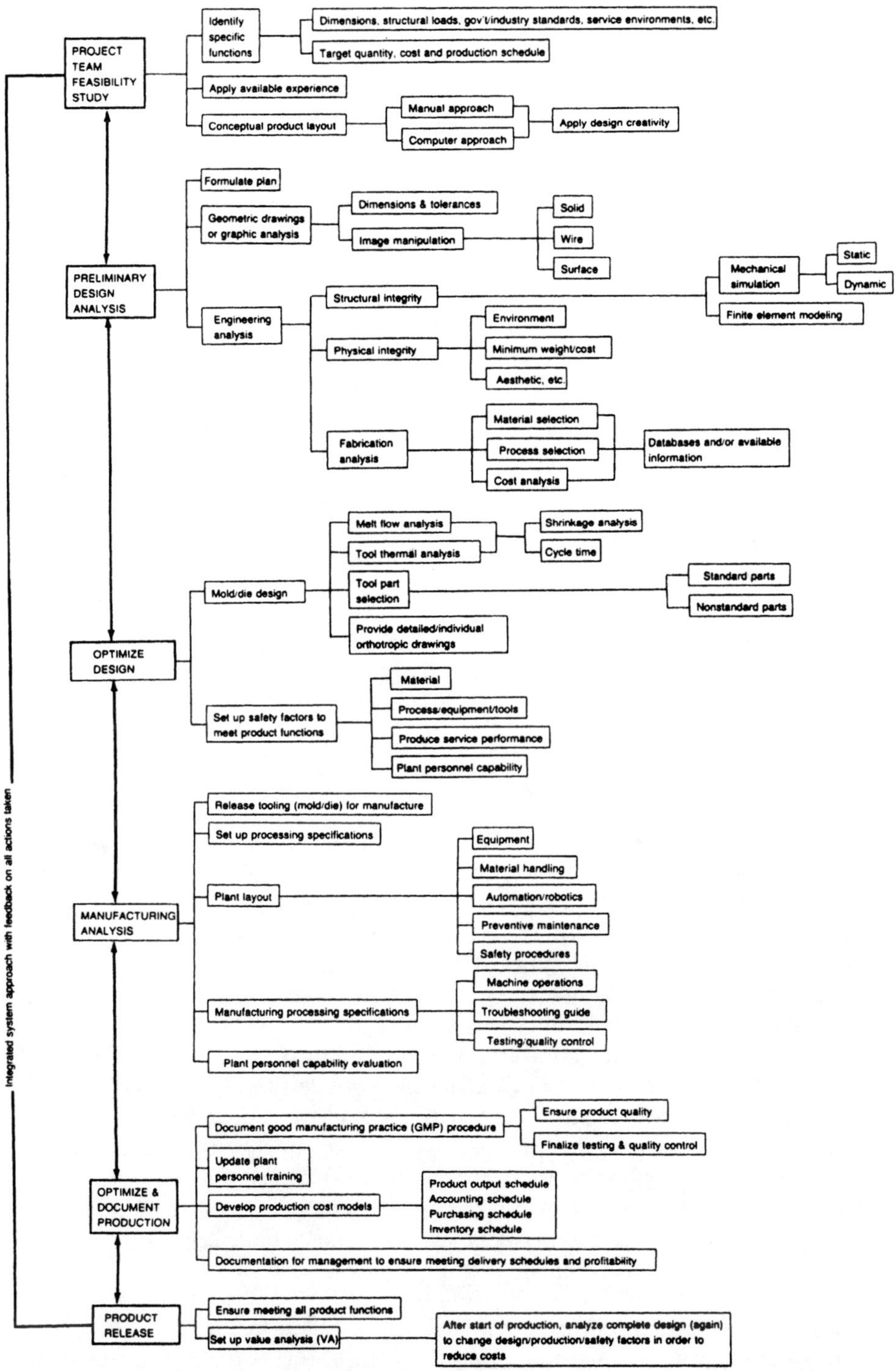


**Figure 20.1** Design flow diagram to produce products.

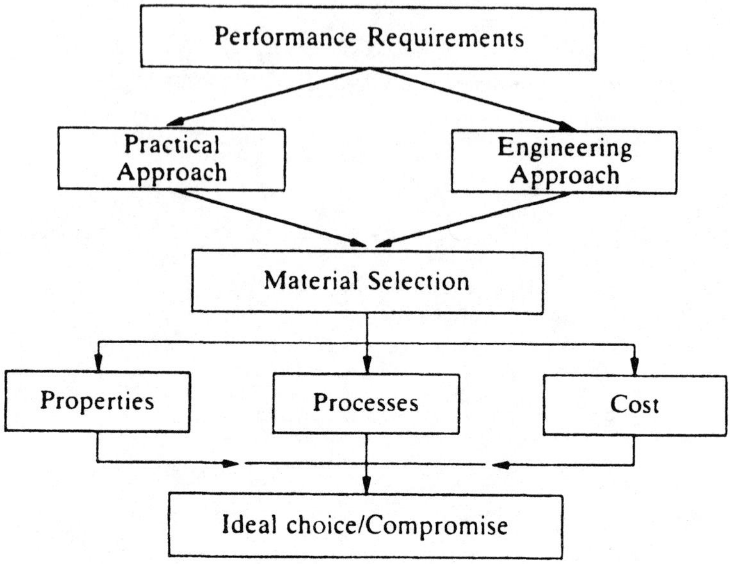


**Figure 20.2** Product manufacture: a simplified flow diagram.

producing a product. The product's size, shape, complexity, strength, orientation, etc., can be almost impossible to achieve. Thus, processing is the task of doing the 'impossible' at the lowest cost as summarized in Fig. 20.3.

Economics can be improved by targeting factors such as: (1) reduction in the use of material by minimizing tolerances; (2) improvement in product quality in terms of strength and/or other mechanical–physical characteristics; (3) reduction in setting-up times of start-up aids and automation systems; and (4) savings in electricity consumption by the optimization of the plasticizing and the use of efficient heating and cooling in the extruder.

The extruder and its components (gear pump, die, etc.) are costly. However, the other equipment in the line is usually more expensive and even a single piece of equipment can be more costly. As an example, up to one-third of the capital equipment cost in a film line goes for the winder [290]. Major costs may be incurred at the end of the extrusion line, on finishing, trimming, postforming, and very important the windup or stacking equipment. Sophisticated materials handling equipment becomes costly. Nevertheless, although these costs are high, there is normally a good return on investment provided you (or some one you trust) know what is required.

In a production line that has a relatively long run, the cost of equipment includes its financial amortization, and could be about 5%. Plastic material

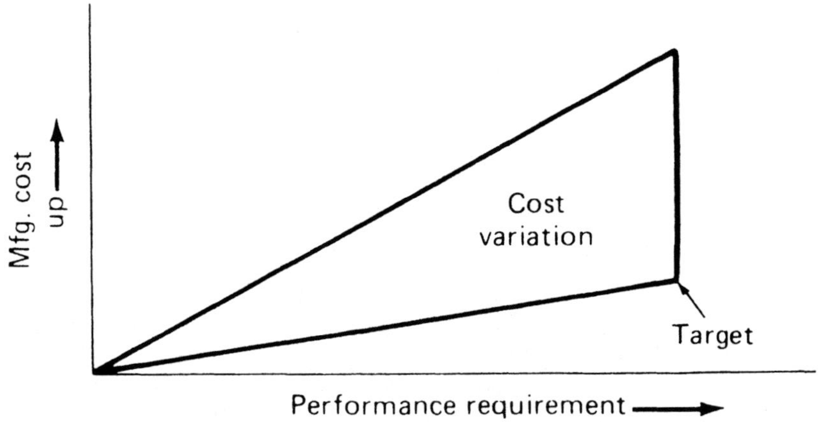


**Figure 20.3** Product cost: the influence of manufacturing costs and performance requirements.

cost could be as high 80% for high volume production. The other costs include power, water, labor, overhead and taxes. With precision, short-run costs could be equipment at 20–30%, and materials at 45–50%. So, as it is usually stated, do not buy equipment just because it cost less, since more profit can occur with the more expensive equipment. Of course, the reverse is possibly true. So, buyer has to know what is wanted when ordering. A line with 1000–4000 lb/h output would cost $0.5–1 million.

When you want to change a screw to see how your process is running, it seems expensive to take such action. Consequently, many people tend to run a poorly performing screw long after it should have been changed. Converting the cost of a screw into an equivalent volume of plastic or into a profit per day will determine payback. Assume a screw cost of $30 000–40 000 each, with an output at 1400 kg/h (3000 lb/h) of a $0.40/lb plastic. If you might waste 454 000 kg (1 000 000 lb) of plastic, then you have justification for a new screw. A new screw would represent 33 h of processing. So, it pays to replace the screw [187, E. Mount].

Whenever possible, observing the following practices will help to reduce costs: (1) strive for the simplest shape and form; (2) combine parts into single extrusions or use more than one die to extrude products/use multiple dieheads and openings; (3) make gradual changes in thickness to reduce frozen stress; (4) where bends occur, use maximum permissible radii; (5) purchase plastic material as economical as possible (Fig. 20.4); and (6) keep customers tolerance as liberal as possible, but once in production aim for tighter tolerances to save material costs and also probably reduce production costs.

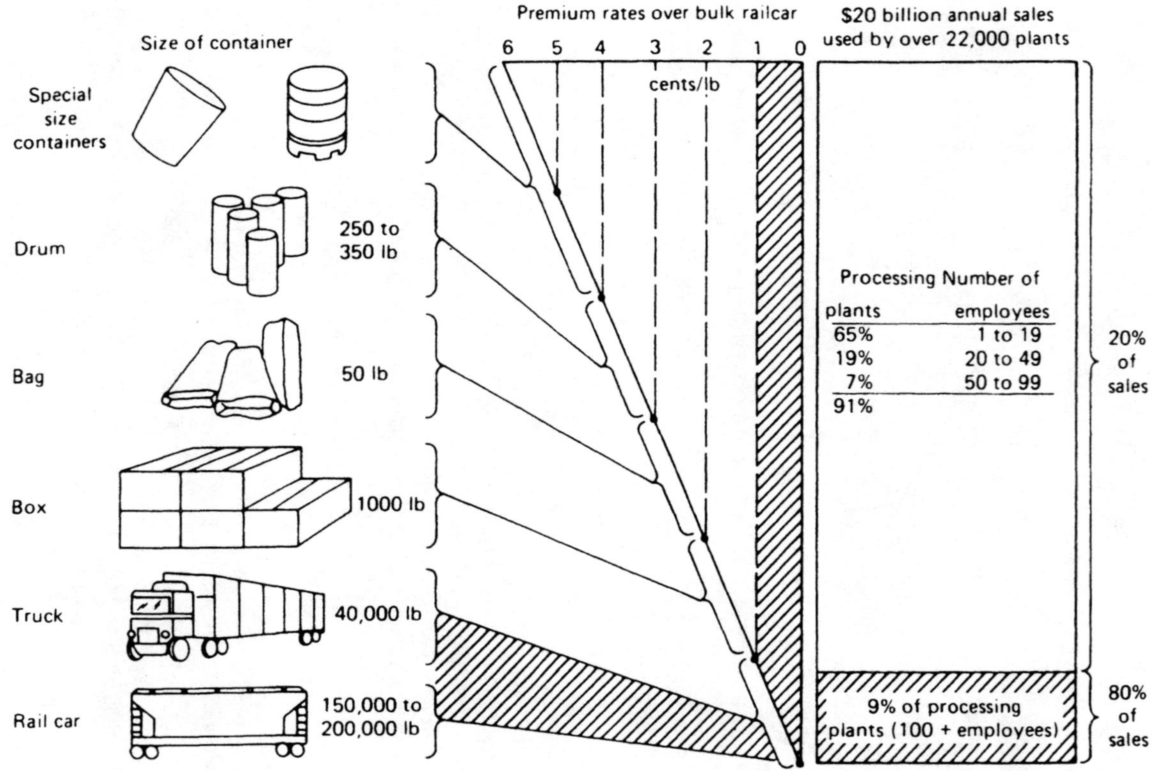


**Figure 20.4** Guide to plastic dollar purchases by all types of processing plant sizes and sizes of containers.

Minimizing costs is generally an overriding goal, whether a process is being selected for a new product or opportunities are being evaluated for replacing existing materials. The major elements of cost, as just reviewed, are equipment and material and those that could be called inefficiencies, such as scrap, repairs, waste, and machine down-time. Even though the scrap is recycled, it costs money to granulate, handle, and possible slow down the line.

Processing cost variations may be due to one or more of the following factors: (1) improper or unattainable performance requirements; (2) improper plastic selection; (3) improper in-line and off-line hardware and control selections; (4) improper selection of the complete line; (5) improper collection and/or handling at the end of the line; and (6) improper setup for testing, quality control, and troubleshooting.

### Estimating product cost

Estimating is a critical aspect of custom fabrication, yet it is often practiced with very little logic. It is shrouded in mystery and rarely discussed by processors. Indeed, it is considered among the dullest of topics.

In practice, it is considered excellent if one estimate in ten produces a successful bid. In other words, a 90% failure rate is considered excellent. No wonder estimating seems like some bizarre sacrificial rite. That does not include those estimates you just go through the motions of preparing because another company is bidding and you know you have no chance at all of landing the job. Or that company's supplier is overloaded or had an accident, so the customer needs 'temporary' help, and you know it; in which case, you provide a bid on the high side (to be safe)

But, what more directly represents the heart of your business than estimating? You are pulling together every facet of your operation, distilling it, putting numbers on it, and putting your company on the 'line.' You are stating that this is what we can do, and this is what we must charge to make a normal profit.

There are probably as many estimating techniques as there are estimators. Much contemporary estimating follows very vague procedures. The number of factors assembled to reach the appropriate figures is sometimes alarmingly small. Some estimators may not consider scrap, coloring, setup time with trial and error, and so on, to mention some of the more obvious omissions. Some estimates, that can work, are very simple creations, produced by determining the part weight, cost of plastic, and processing time, scribbling down some numbers, and adding a fudge factor (and possibly a little prayer).

Some companies do not even have their own standard forms whereby they could develop some useful history. Of course, what influences how one estimates generally relates to the fact that extruding is a highly competitive industry so, logically, the most time is spent preparing quotes where a payoff is a possibility.

### Direct and indirect cost

There are the operating quality costs of prevention and appraisal that are considered to be controllable quality costs. Also there are the internal and external failure costs. Figure 20.5 illustrates the relationship between the controllable and uncontrollable quality cost curves and the combined direct cost curve.

As the controllable cost of prevention and appraisal increases, the uncontrollable costs of internal and external failure decreases. At some point, the cost of prevention and appraising defective product exceeds the cost of correcting for the product failure. This point is the optimum operating quality cost.

In addition to the direct operating quality costs, the indirect quality costs and their effect on the total cost curve must be considered. Indirect quality costs can be divided into three categories: customer-incurred quality costs; customer-dissatisfaction quality costs; and loss-of-reputation costs.

These intangible, indirect quality costs are difficult to measure; however, they do effect the total quality cost curve. This influence is apparent when the indirect quality costs are added to the direct cost curve (Fig. 20.6). When the optimum point moves to the right, it indicates the

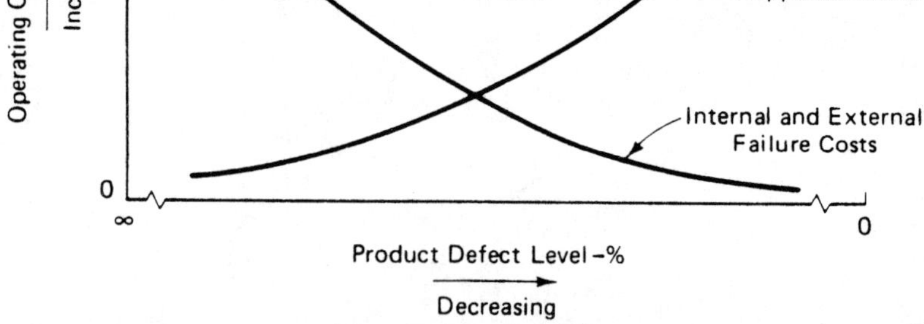


**Figure 20.5** Examples of direct operating quality costs.

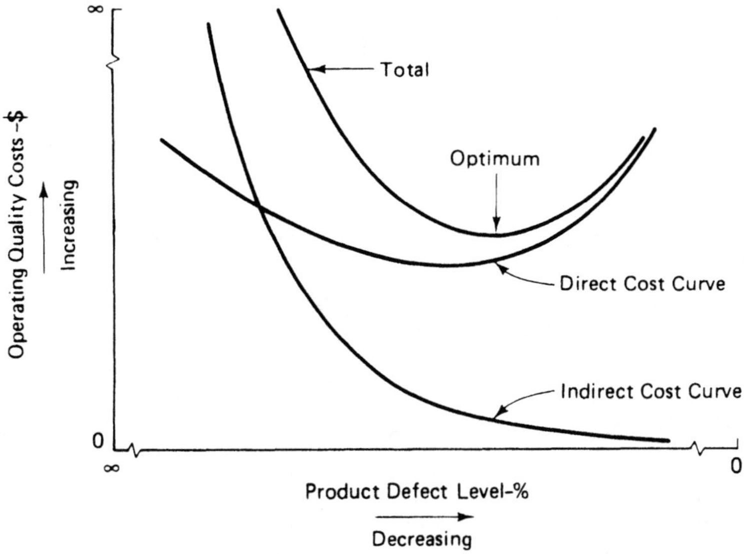


**Figure 20.6** Examples of direct and indirect quality costs.

need for a lower product defect level. A lower product defect level can be obtained by increasing the prevention and appraisal costs, which subsequently lowers the external failure costs. A lower external failure has a desirable influence on the direct costs.

The measurement of the actual indirect costs may be impossible. However, a knowledge that these costs exist and their relationship to the direct costs can aid in their control.

## Technical cost modeling

The adoption of any new technology for producing manufactured products is characterized by a wide range of processing, materials, and economic consequences. Although considerable talent can be brought to bear on the processing and engineering aspects, economic questions remain. Cost problems are particularly acute when the technology that will be employed is not fully understood as much of cost analysis is based on historical data, past experience, and individual accounting practices [2, 38].

Historically, technologies have been introduced on the shop floor incrementally, with their economic sequences measured on the shop floor incrementally, with their economic consequences measured directly. Although incorporating technical changes in the plant to test their viability may have been appropriate in the past, it is usually economically not feasible to explore today's wide range of alternatives in this fashion. Technical cost modeling (TCM) has been developed as a method for analyzing the economics of alternative manufacturing processes without the prohibitive economic burden of trial-and-error innovation and process utilization.

Technical cost modeling is an extension of conventional process modeling, with particular emphasis on capturing the cost implications of process variables and economic parameters. By coordinating cost estimates with processing knowledge, critical assumptions (processing rate, energy used, plastic consumed, melt behavior, etc.) can be made to interact in a consistent, logical, and accurate framework of economic analysis, producing cost estimates under a wide range of conditions.

For example, TCM can be used to determine the plastic process that is best for production without expensive expenditure of capital and time. Not only can TCM be used to establish direct comparisons between processes, but it can also determine the ultimate performance of a particular process, such extrusion, as well as identifying the limiting process steps and/or parameters.

Technical cost modeling uses an approach to cost estimating in which each of the elements that contribute to total cost is estimated individually. These individual estimates are derived from basic principles and the manufacturing process. Technical cost modeling reduces the complex problem of cost analysis to a series of simpler estimating problems, and rather than using intuition, it solves them by bringing processing expertise to bear.

In dividing cost into its contributing elements, the first distinction that can be made is that some cost elements depend upon the annual output of products, whereas others do not. For example, the cost contribution of the plastic is the same regardless of the number of items produced, unless the material price is discounted because of very high volume. On the other hand, the pre-price cost of tooling will vary with changes in production volume. These two elements, variable costs and fixed costs, create a natural division for preparing estimates.

Variable cost elements have values that depend on the number of pieces produced. For most plastics fabrication, the variable cost elements are principally material, direct labor, and energy.

Fixed costs are a function of the annual production volume. They are called fixed because they typically represent one-time capital investments (building, silo, extruders, etc.) or annual expenses unaffected by the number of parts produced (building rent, engineering support, administrative personnel, etc.). These costs tend to be distributed over the total number of products produced in a given period. For plastics fabrication, the fixed cost elements are principally the main extruders, auxiliary equipment, tooling, buildings, overhead labor, maintenance, and the cost of capital.

To demonstrate the use of such a comparative cost analysis, production of a panel was analyzed according to different processes as shown in Fig. 20.7 and under the following conditions: (1) panels measured 61 by 91 cm (24 by 36 in) with the wall thickness dictated by process and product requirements, so the weights of the panels differed; (2) production was at a level of 40 000 products per year; (3) plastics for all panels were of the same type, except that different grades had to be used (melt index, etc.) based on process requirement, so that cost changed; (4) each panel received one coat of paint, except that the structural foam also had a primer coating (to cover a relatively rough surface); and (5) costs were allocated as needed to those processes that required trimming and other secondary operations.

Technical cost modeling can keep cost data current, based on cost changes from day to day, region to region, and so on. Costs need to be collected on a regular basis and incorporated into the TCM, otherwise information will become out of date.

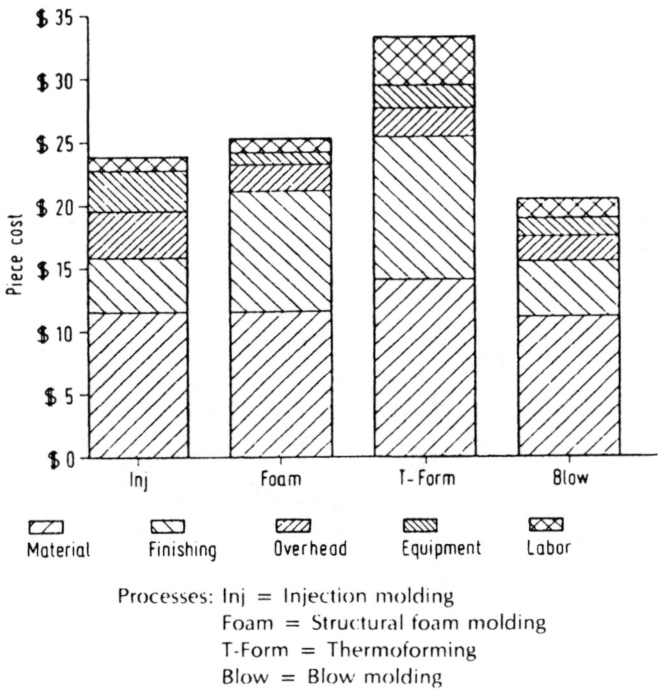


**Figure 20.7** Cost comparison of panel production using a technical cost-modeling program showing blow molding with the lowest piece cost.

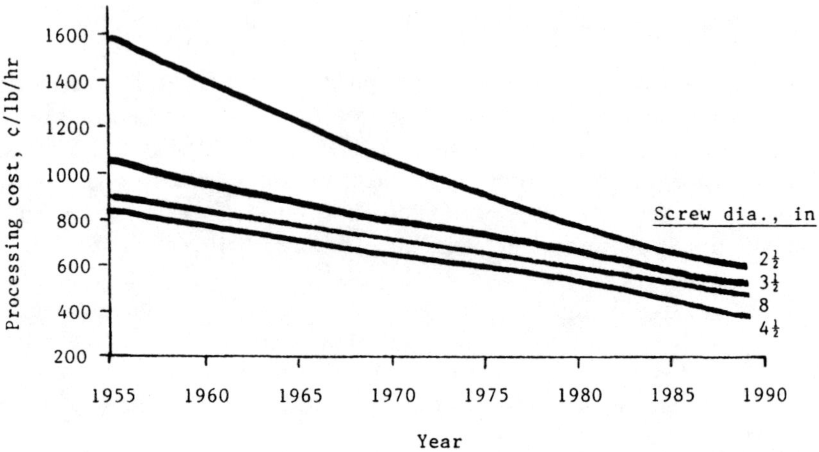


**Figure 20.8** Extrusion processing costs are steadily dropping based on output of different screws.

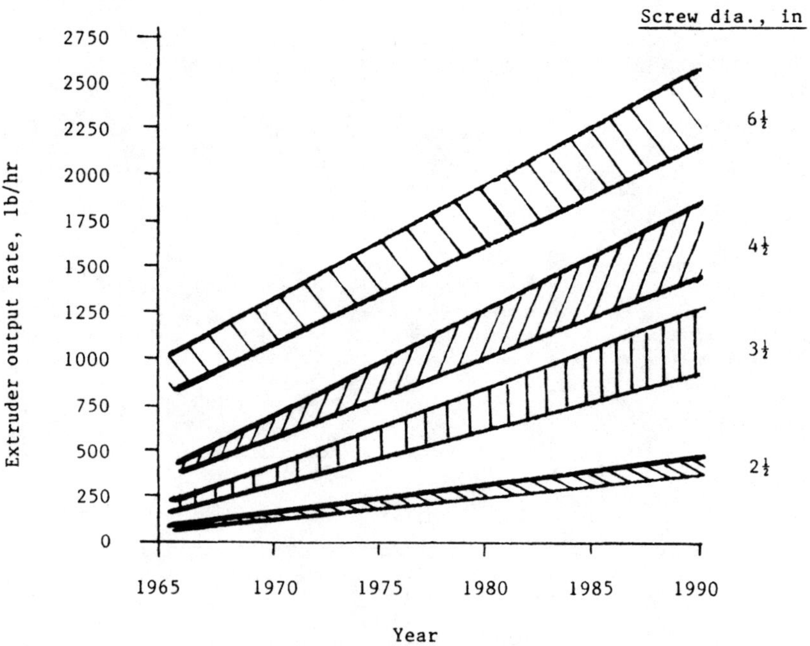


**Figure 20.9** Increasing outputs for extruders of various sizes versus time.

## PROCESSING IMPROVEMENT

New developments with equipment and plastic materials usually offer improvements in processing capabilities. These improvements do not occur at the rate computer changes occur but they definitely occur so keeping up to date on what is available is a full time job. As in the past, these improvements aid in meeting (asymptotic) goals of zero defects as well as to reduce costs. Figures 20.8 and 20.9 show the significant improvement in output from single screw extruders with similar action occurring with multi-screw extruders.

Although it is one of the oldest types of machinery for plastics processing, the single screw extruder has not yet reached the end of development. With changes in plastic materials, the machines have continually developed new processing requirements. Also, with improvements in machine

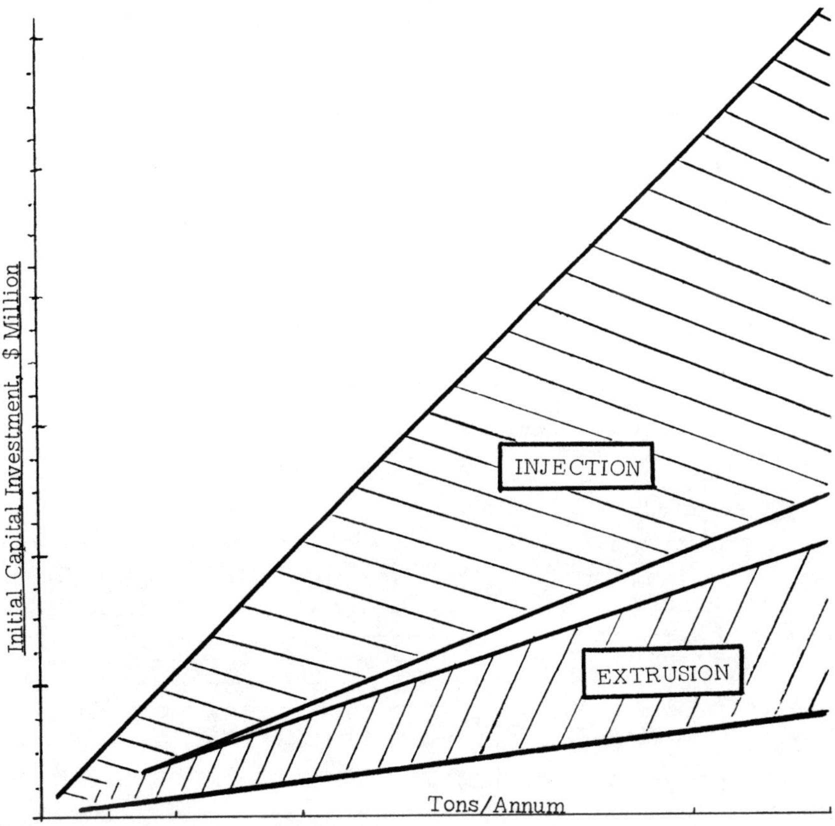


**Figure 20.10** Initial capital investment related to annual plastic tonnage throughput for the extrusion and injection molding processes.

operation, new plastic materials are required to take advantage of the changes. All these actions are taken and required to meet new market demands for products.

Output of the extruder lines (Fig. 20.10), once the major focus, remains paramount but qualities such as wear and versatility have become of equal importance. In addition with single screws, new developments also continue with the multiple screw extruders, particularly the twin screw that keep expanding compounding capabilities and processing the heat sensitive plastics.

In addition to the extruder improvements, up-stream to down-stream equipment is always being improved to permit better products. It seems that there is always a better way to provide for material handling from the supplier, to storage, and to the hopper. The very expensive windup equipment continues to simplify operations and provide improvements in controls. The different granulators provide better ways to obtain more useful recycled material sizewise with lower thermal degradation. And more can be said for other types of equipment, particular the controls on line.

## PROCESS TYPES

Table 20.1 provides a simplified guide to different plastic processes [1]. Another guide, Fig. 20.11, compares product size versus processes. Table 20.2 provides a very simplified guide comparing costs versus process type.

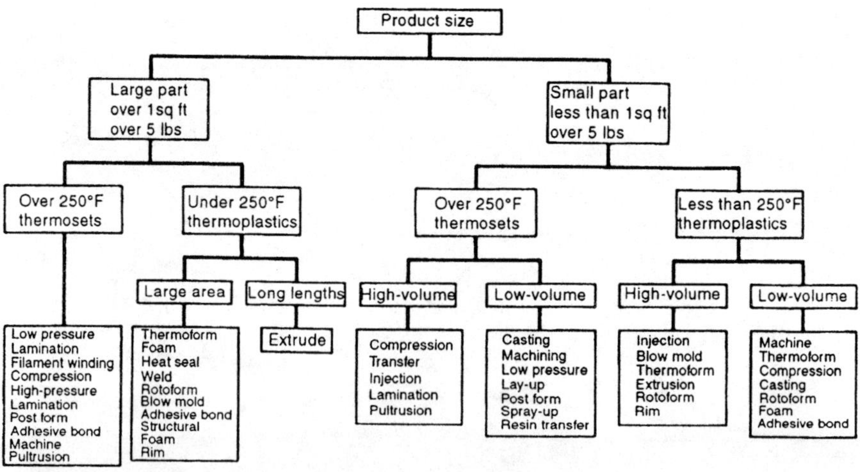


**Figure 20.11** Guide to product size versus process.

**Table 20.1** Plastic processing methods preliminary guide

| *Process* | *Description* | *Limitations* |
|---|---|---|
| Blow | An extruded parison tube of heated thermoplastic is positioned between two halves of an open split mold and expanded against the sides of the closed mold via air pressure. The mold is opened and the part ejected. Low tool and die costs, rapid production rates, and ability to mold fairly complex hollow shapes in one piece. | Generally limited to hollow or tubular parts; some versatile mold shapes, other than bottles and containers. |
| Calendering | Dough-consistent thermoplastic mass is formed into a sheet of uniform thickness by passing it through and over a series of heated or cooled rolls. Calenders are also utilized to apply plastic covering to the backs of other materials. Low cost, and sheet materials are virtually free of molded-in stresses. | Limited to sheet materials and very thin films are not possible. |
| Casting | Liquid plastic which is generally thermoset except for acrylics is poured into a mold without presssure, cured, and taken from the mold. Cast thermoplastic films are produced via building up the material (either in solution or hot-melt form) against a highly polished supporting surface. Low mold cost, capability to form large parts with thick cross sections, good surface finish, and convenient for low-volume production. | Limited to relatively simple shapes. Most thermoplastics are not suitable for this method. Except for cast films, method becomes uneconomical at high volume production rates. |

**Table 20.1** *Continued*

| *Process* | *Description* | *Limitations* |
|---|---|---|
| Centrifugal casting | Reinforcement is placed in mold and is rotated. Resin distributed through pipe; impregnates reinforcement through centrifugal action. Utilized for round objects, particularly pipe. | Limited to simple curvatures in single-axis rotation. Low production rates. |
| Encapsulation | Mixed compound is poured into open molds to surround and envelope components; cure may be at room temperature with heated postcure. Encapsulation generally includes several processes such as potting, embedding and conformal coating. | Low-volume process subject to inherent limitations on materials which can lead to product defect caused by exotherm, curing or molding conditions, low thermal conductivity, high thermal expansion, and internal stresses. |
| Extrusion | Widely used for continuous production of film, sheet, tube, and other profiles; also used in conjunction with blow molding. Thermoplastic or thermoset molding compound is fed from a hopper to a screw and barrel where it is heated to plasticity then forwarded, usually via a rotating screw, through a nozzle possessing the desired cross section. Production lines require input and takeoff equipment that can be complex. Low tool cost, numerous complex profile shapes possible, very rapid production rates, can apply coatings or jacketing to core materials (such as wire). | Usually limited to sections of uniform cross section. |

| | | |
|---|---|---|
| Filament winding | Excellent strength-to-weight. Continuous, reinforced filaments, usually glass, in the form of roving are saturated with resin and machine-wound onto mandrels having shape of desired finished part. Once winding is completed, part and mandrel are cured; mandrel can then be removed through porthole at end of wound part. High-strength reinforcements can be oriented precisely in direction where strength is required. Good uniformity of resin distribution in finished part; mainly circular objects such as pressure bottles, pipes, and rocket cases. | Limited to shapes of positive curvature; openings and holes can reduce strength if not properly designed into molding operations. |
| Injection molding | Very widely used. High automation of manufacturing is standard practice. Thermoplastic or thermoset is heated to plasticity in cylinder at controlled temperature, then forced under pressure through a nozzle into sprues, runners, gates, and cavities of mold. The resin undergoes solidification rapidly, the mold is opened, and the part ejected. Injection molding is growing in the making of glass-reinforced parts. High production runs, low labor costs, high reproducibility of complex details, and excellent surface finish. | High initial tool and die costs; not economically practical for small runs. |

**Table 20.1** *Continued*

| *Process* | *Description* | *Limitations* |
|---|---|---|
| Laminating | Material, usually in form of reinforcing cloth, paper, foil, metal, wood, glass fiber, plastic, etc., preimpregnated or coated with thermoset resin (sometimes a thermoplastic) is molded under pressure greater than 1000 psi (7 MPa) into sheet, rod, tube, or other simple shapes. Excellent dimensional stability of finished product; very economical in large production of parts. | High tool and die costs. Limited to simple shapes and cross sections. |
| Matched-die molding | A variation of the conventional compression molding this process employs two metal molds possessing a close-fitting, telescoping area to seal in the plastic compound being molded and to allow trim of the reinforcement. The mat or preform reinforcement is positioned in the mold and the mold is closed and heated under pressures of 150–400 psi (1–3 MPa). The mold is then opened and the part is removed. | Prevalent high mold and equipment costs. Parts often require extensive surface finishing. |
| Pultrusion | This process is similar to profile extrusion, but it does not provide flexibility and uniformity of product control, and automation. Used for continuous production of simple shapes (rods, tubes, and angles) principally incorporating fiberglass or other reinforcement. High output possible. | Close tolerance control requires diligence. Unidirectional strength usually the rule. |

| | | |
|---|---|---|
| Rotational molding | A predetermined amount of powdered or liquid thermoplastic or thermoset material is poured into mold; mold is closed, heated, and rotated in the axis of two planes until contents have fused to inner walls of mold; mold is then opened and part is removed. Low mold cost, large hollow parts in one piece can be produced, and molded parts are essentially isotropic in nature. | Limited to hollow parts; production rates are usually slow. |
| Slush molding | Powdered or liquid thermoplastic material is poured into a mold to capacity; mold is closed and heated for a predetermined time in order to achieve a specified buildup of partially cured material on mold walls; mold is opened and unpolymerized material is poured out; and semifused part is removed from mold and fully polymerized in oven. Low mold costs and economical for small production runs. | Limited to hollow parts; production rates are very slow; and limited choice of materials that can be processed. |
| Thermoforming | Heat-softened thermoplastic sheet is positioned over male or female mold; air is evacuated from between sheet and mold, forcing sheet to conform to contour of mold. Variations are vacuum snapback, plug assist, drape forming, etc. Tooling costs are generally low, large part production with thin sections possible, and often comes out economical for limited part production. | Limited to parts of simple configuration, high scrap, and limited number of materials from which to choose. |

**Table 20.1** *Continued*

| *Process* | *Description* | *Limitations* |
|---|---|---|
| Transfer molding | Related to compression and injection molding processes. Thermoset molding compound is fed from hopper into a transfer chamber where it is then heated to plasticity; it is then fed by a plunger through sprues, runners, and gates into a closed mold where it cures; mold is opened and part ejected. Good dimensional accuracy, rapid production rate, and very intricate parts can be produced. | High mold cost; high material loss in sprues and runners; size of parts is somewhat limited. |
| Wet-layup or contact molding | Several layers, consisting of a mixture of reinforcement (generally glass cloth) and thermosetting resin are positioned in mold and roller contoured to mold's shape; assembly is usually oven-cured without the application of pressure. In spray molding, a modification, resin systems and chopped fiber are sprayed simultaneously from a spray gun against the mold surface. Wet-layup parts are sometimes cured under pressure, using vacuum bag, pressure bag, or autoclave, and depending on the method employed, wet-layup can be called open molding, hand layup, sprayup, vacuum bag, pressure bag, or autoclave molding. Little equipment required, efficient, low cost, and suitable for low-volume production of parts. | Not economical for large-volume production; uniformity of resin distribution difficult to control; only one good surface; limited to simple shapes. |

**Table 20.2** Cost comparison guide of processes (Cost factor × Material cost = Purchased cost of product)

| | *Cost factor* | |
|---|---|---|
| *Process* | *Overall* | *Average* |
| Blow molding | $1^1/_{16}$–4 | $1^1/_8$–2 |
| Calendering | $1^1/_2$–5 | $2^1/_2$–$3^1/_2$ |
| Casting | $1^1/_2$–3 | 2–3 |
| Centrifugal casting | $1^1/_2$–4 | 2–4 |
| Coating | $1^1/_2$–5 | 2–4 |
| Cold pressure molding | $1^1/_2$–5 | 2–4 |
| Compression molding | $1^3/_8$–10 | $1^1/_2$–4 |
| Encapsulation | 2–8 | 3–4 |
| Extrusion forming | $1^1/_{16}$–5 | $1^1/_8$–2 |
| Filament winding | 5–10 | 6–8 |
| Injection molding | $1^1/_8$–3 | $1^3/_{16}$–2 |
| Laminating | 2–5 | 3–4 |
| Match-die molding | 2–5 | 3–4 |
| Pultrusion | 2–4 | 2–$3^1/_2$ |
| Rotational molding | $1^1/_4$–5 | $1^1/_2$–3 |
| Slush molding | $1^1/_2$–4 | 2–3 |
| Thermoforming | 2–10 | 3–5 |
| Transfer molding | $1^1/_2$–5 | $1^3/_4$–3 |
| Wet lay-up | $1^1/_2$–6 | 2–4 |

## PLANT CONTROL

It is seldom the case that a processing plant has only one processing machine, and if it has more, it is not what happens on the individual machine that determines profitability, but the performance of all machines. With many machines, it may become difficult to keep track of all the details (hundreds to thousands) that go into the plant's overall operation. It also becomes increasingly difficult for processors, quality control people, maintenance people, and others, always to be available when needed. Also, it may become difficult for a person to make a major decision when it is needed.

Modern central control and management systems have changed this situation. These systems have been called by different names, such as supervisory control, distributed control, CAD, CAM, CAE, CIM, and so on [4]. These different systems can provide monitoring and controlling of all operating parameters for every extruder as well as all up-stream through down-stream equipment. The systems receive inputs on all parameters and can issue instructions to each machine to ensure efficient and profitable operation. What is required is someone to set-up the settings and instructions for these control and management. Throughout this book, facts and guidelines have been provided as to what is needed and required for implementation.

For these systems to operate efficiently, talented people are needed to completely integrate the systems. These people must be available and must know what is required for all plant operations, particularly the extruders. System control is based on these requirements. In turn, these people must establish proper start-up procedures for all equipment following a methodology for 'threading' the line on start-up. They build limits into the system control, and interface them with control instructions that are best suited to keeping the machines' product outputs that meet performance requirements at the lowest processing cost. Proper training of people is required so they understand how best to operate and setup the equipment and to pay attention and set-up safety procedures.

As reviewed in Chapter 19, intelligent process control (IPC) is being used. It uses computer simulation to link a process, such as rate of extruding the plastic melt from the die, to real-time machine control. Various computer software programs are available [279].

## ENERGY

Cost savings via energy conservation can be considered from the viewpoints of machine operation, the plastic material, and the finished product. Extrusion machines are energy intensive. Thus it becomes very advantageous to reduce the energy requirements wherever it is possible starting with the purchase of any equipment in the extrusion line [188].

Energy conservation is one of many factors that should be considered in the selection of an automated materials-conveying system. Fortunately, in many cases, any steps taken to save energy will also save money. The traditional arguments favor the silo savings on plastic costs, labor savings through the elimination of handling bags and cartons, savings of costly warehouse space, and energy savings. For example, if a plant used a large quantity of plastics and did not use silos, then during the winter months bags and gaylords would be delivered repeatedly through 'open' delivery doors, and warm plant air would be lost.

With automatic delivery from silos, all plastic handling lines are kept as short as possible. There is no reason for lines to conform to the right angles of the walls; they should follow a straight line from the plastic's source to where it has to be delivered. There are graphs from handling systems suppliers that show the relationship between the length of conveyor lines and power requirements [2]. The graphs also show the horsepower (hp) required, based on different factors, such as the length and diameter of the delivery pipe, the position of the pipe, the type of plastic being conveyed, the size of the hopper at the machine, and the rate of flow deliverable.

As an example, with an average pellet size of 3.2 mm (0.125 in), 560 kg/m$^3$ (35 lb/ft$^3$) bulk density and a conveying vacuum of 300 mm (12 in) of mercury, PE can be moved in different ways. With a 19 kW (25 hp) vacuum hopper unit, a line will convey 8200 kg/h (18 000 lb/h) of PE if it is only 30 m (100 ft) long. Using the same power and a 137 m (450 ft) line, less PE will be moved. Equipment suppliers' data will show the power required to move a material in the fastest way with the least energy consumption. To convert energy consumption to electric-bill charges, one can use the following formulas where:

$$1\,\text{hp} = 0.745\,\text{kW}, \quad 1\,\text{kWh} = 3.6\,\text{MJ} = 3143\,\text{Btu}\ [2]$$

With a long pipeline, a 19 kW (25 hp) vacuum pump could be used. If the line could be shortened, a 7.5 kW (10 hp) unit would convey the same amount of plastic, resulting in power savings. One meter of vertical height in the line equals 2 m of horizontal distance in its effect on conveying rates. Bends in pipe add a considerable amount of equivalent footage.

If the lines cannot be kept short and relatively free of bends, then pressure drops will exceed 40 kPa (12 in Hg), requiring more power. Where the pressure drop is greater than 40 kPa (12 in Hg), which is the normal operating and limiting pressure for most vacuum systems, it is necessary to use a positive pressure system. All these type factors must be considered initially to obtain the most economic delivery system.

**Plastics and energy**

Numerous studies have shown: (1) plastics consume less energy to produce and fabricate products (Figs. 20.12 and 1.16) than other materials with glass being the major consumer of energy; (2) their use as a product reduces energy consumption; and (3) more energy can be produced when products are incinerated. In a cradle-to-grave analysis comparing different products made with different materials, plastic far outperforms other materials. The studies compared materials on basis such as energy consumption, air emissions, waterborne wastes, and solid waste products [4].

Opponents of plastics point to the fact that plastics are made from petroleum, a nonrenewable resource. Most of the plastics familiar to USA consumers are manufactured from ethylene. During the distillation of petroleum, the ethylene fraction comes off as a byproduct which many years ago was just burned since it was waste. This waste gas became the building block for most plastics. Reports show that in USA from 2–3% of all total annual petroleum production consumed is converted to ethylene.

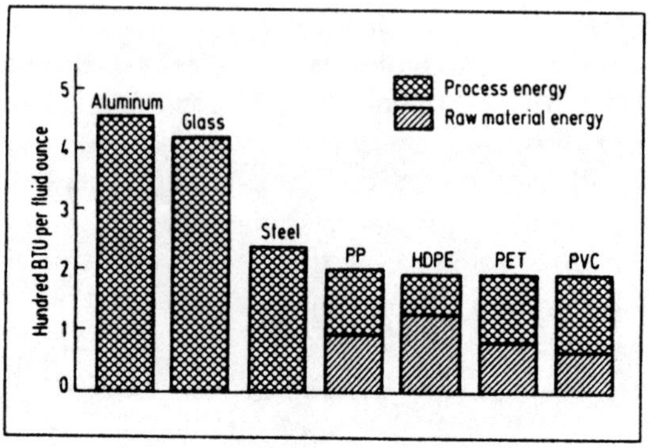


(a)

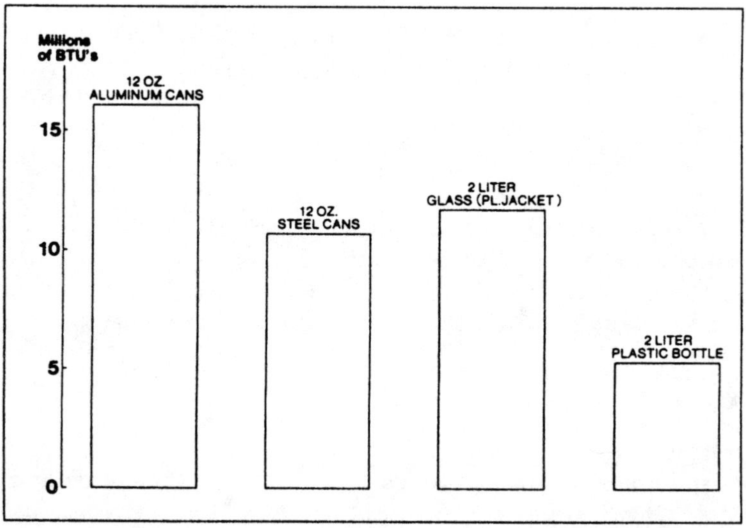


(b)

**Figure 20.12** (a) Total energy required to produce equivalent 0.7 kg (24 oz) container; (b) Total energy required to deliver 1000 liters of beverage.

Figure 20.13 provides a flow diagram going from energy sources to producing products fabricated by using different processes.

Table 20.3 gives examples of energy consumption and energy recovery that occurs for different packaging materials. In comparison with other materials, plastics have the lowest specific energy requirement for their manufacturer; also lowest recycling energy consumption. Figure 20.12 and Table 20.4 shows values for the energy required for finished products of various materials.

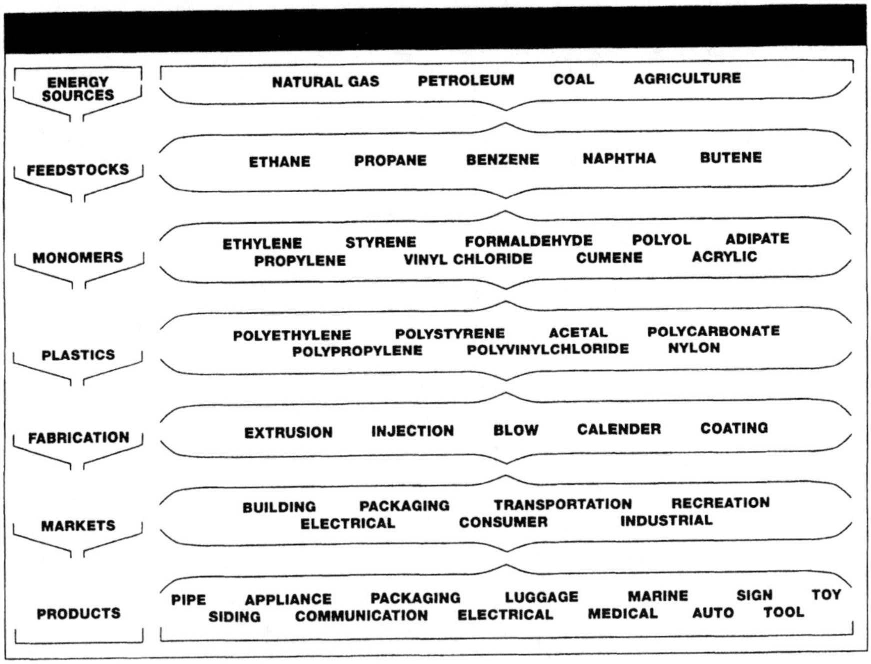


**Figure 20.13** Simplified flowchart of plastics from raw materials to products.

**Table 20.3** Examples of energy consumption of packaging materials

| *BTU/lb* | *Paper* | *Glass* | *Steel* | *Aluminum* | *Plastic* |
|---|---|---|---|---|---|
| BTU required to make originally | 20000 | 8700 | 22800 | 120000 | 35950 |
| BTU required to recycle into new container | 11500 | 6525 | 15960 | 19100 | 1000 |
| BTU recovered by burning | 8500 | 0 | 0 | 0 | 17875 |

Energy conservation occurs when plastics products are in use. With increasing use of plastics in automobiles, aircraft, and other means of transportation, weight is reduced. Where material substitutions occurred in place of steel, glass, etc., less energy is required to move the lighter vehicles. Plastics in the average automobile cuts fuel consumption by at least 5% and contributes to improved safety performance.

An interesting historical example is the small injection blow molded whiskey bottles that were substituted for glass blown bottles in commercial aircraft. In USA alone, over $500 \times 10^{12}$ Btu or the amount of energy equivalent to over $80 \times 10^{6}$ barrels of oil were saved. Plastics packaging

**Table 20.4** Guide for estimating specific energy requirements of different materials used for packaging and utensils

| *Materials* | *Energy requirement* E *(MJ/kg)* | *Recovered heat to be subtracted* H *(MJ/kg)* | *Net requirement E–H (theoretical)* *(MJ/kg)* | *(MJ/l)* | *Requirement relative to PE (approx.)* | *Specific energy requirement per functional unit (MJ/l of material)* |
|---|---|---|---|---|---|---|
| Polyethylene (PE) | 70–85 | 43 | 27–42 | 35 | 1.0 | 35 |
| Polystyrene (PS) | 80–90 | 40 | 40–50 | 47 | 1.0 | 47 |
| Polyvinylchoride (PVC) | 57–61 | 18 | 39–43 | 45 | 0.8 | 36 |
| Paper | 60–75 | 18 | 45–57 | 31 | 3.3 | 102 |
| Glass | 10 | – | 10 | 25 | 5–8 | 125–200 |
| Iron/steel | 20–25 | – | 20–25 | 172 | 0.3 | 52 |
| Aluminum | 115–140 | – | 115–140 | 344 | 0.3 | 103 |
| Tin-plate | 30 | – | 30 | 235 | 0.6 | 140 |

has significantly helped reduce food spoilage in the industrialized world to at least 2%. Developing countries have a spoilage rate of between 30–50%. It is important to recognize that plastics can be recycled so, if economics permit, industries will automatically consume the recycled material; and, of course, government mandates have required recycling plastics regardless of the cost to the public and the probable cost-inefficiency in the reprocessing of these operations.

Plastics are suitable for energy thermal reclamation or recycling because of their energy content or heat value is higher than materials such as coal. Often, the energy content of plastics is much the same as that of heating oil. For these reasons, it is rational to reuse plastic wastes that are not suitable for recycling, chemical thermal reclamation, etc. In waste incineration plants, energy is obtained in the form of community heat and electrical power. Practically all the plastics can be incinerated without problems and with a low (controllable and confining) emission of pollutants. For certain plastics, such as halogenated PVC, special care and systems can be used to permit incineration.

## PEOPLE

The recipe for productivity includes a list of ingredients that requires a company to have in-houe and/or keeping up to date with the outside world: (1) research and development; (2) new technologies; (3) update on equipment; (4) automated systems; (5) modern facilities; and (6) new plastic materials, to name but a few. However, as explained throughout this book, the one ingredient that ties the recipe together is people. None of the other factors has much impact without the appropriate people. Without these people employed in being responsible for the fabricating operation, you are not going to be operating efficiently no matter how large your capital expenditure becomes.

### Processing and patience

When making process changes, allow enough time to achieve a steady state before collecting data.

## PLASTIC GROWTH WITH EXTRUSION

Plastics (about $300 billion industry) are among the world's most widely used materials with thelargest segment at 38 wt% going through (continuous) extruders to produce products going into practically all markets. However, at least 65% of the total plastics going through all the processes (Table 20.1) are compounded in extruders. The US plastics manufacturing

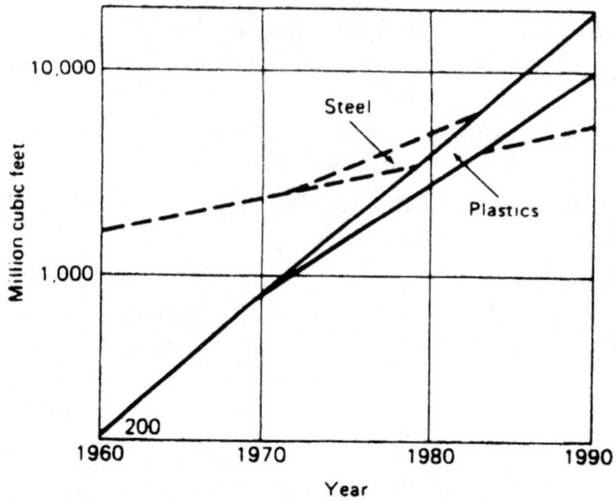


**Figure 20.14** World consumption of plastics by volume.

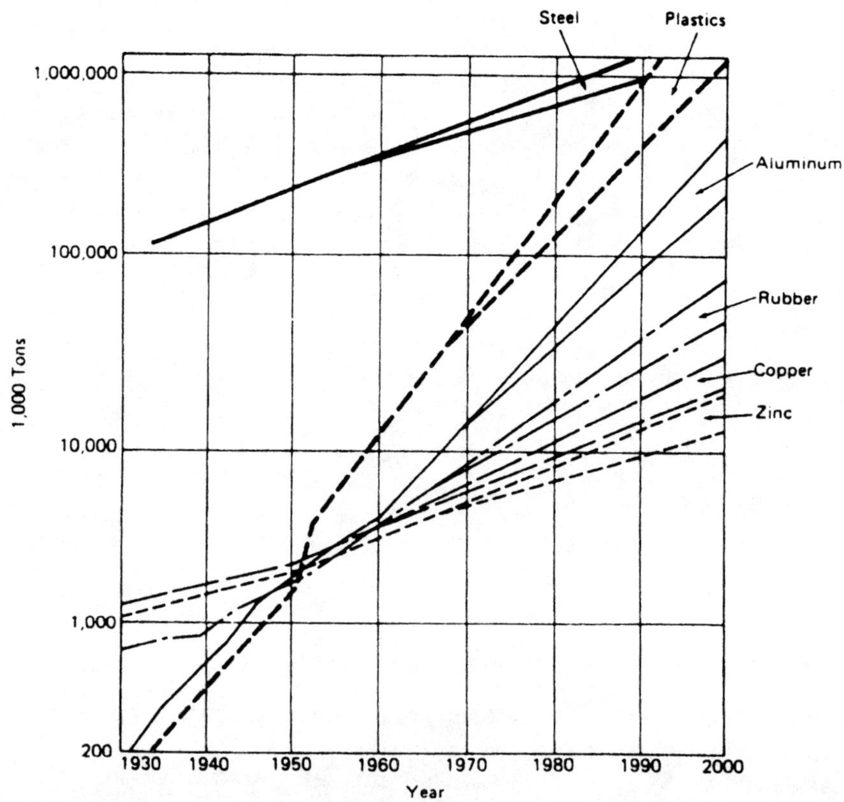


**Figure 20.15** World consumption of plastics by weight.

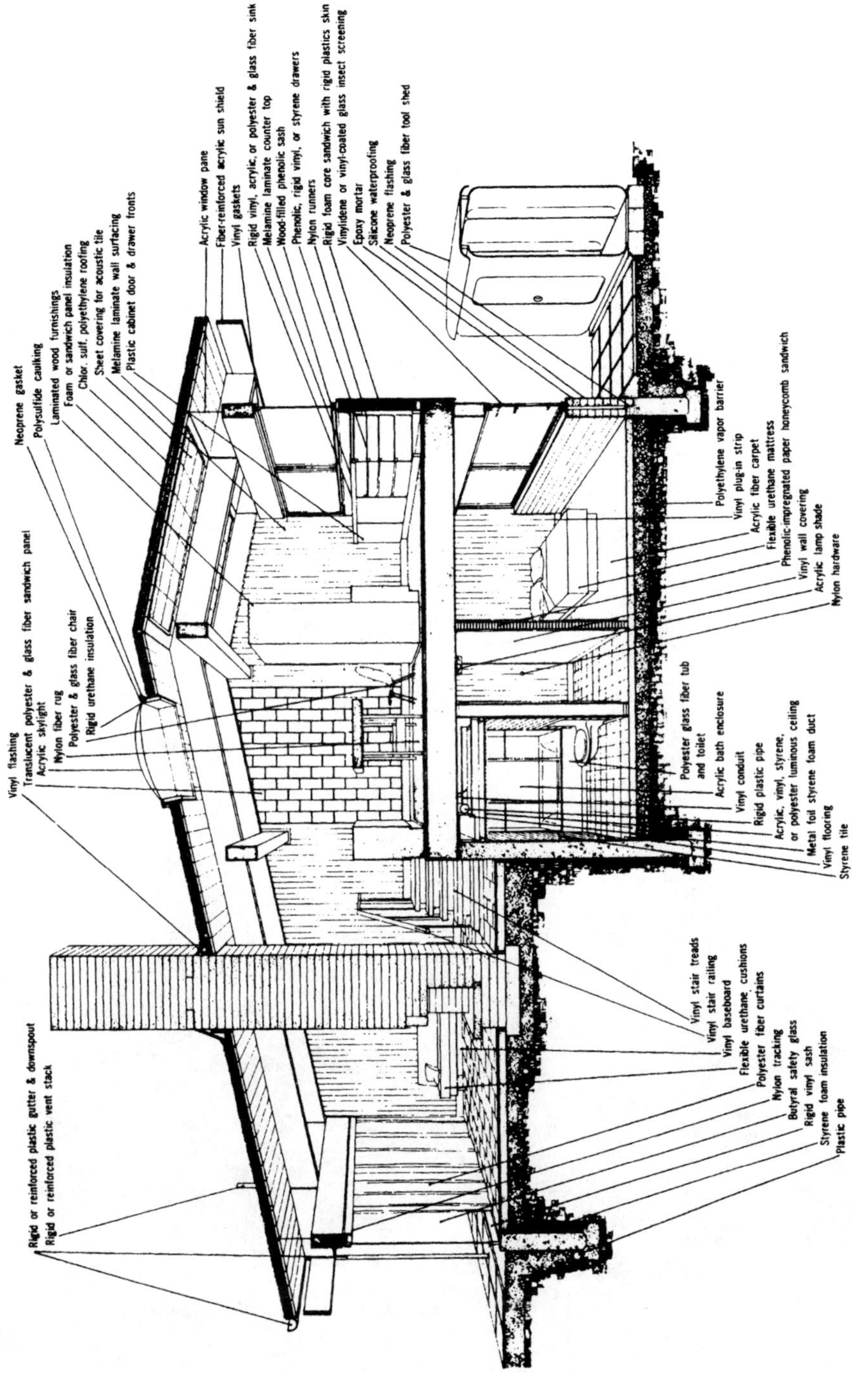


**Figure 20.16** The 1940s to 1950s concept of plastics in buildings.

industry is now in fourth place among all manufacturing; after motor vehicles, petroleum refining, and electronics compounding. During 1995 plastics growth moved to fourth place by passing meat products. Consumption by major markets are: packaging 30%, building/construction 20%, consumer and institutional 10%, electrical/electronic 5%, furniture/furnishing 4%, coating/adhesive/ink 3%, industrial machinery 1%, exports 15%, and others 12%. As Figure 20.14 shows, plastic materials surpassed steel on a volume basis in 1983. By about the turn of the twentieth century, plastics are expected to surpass steel on a weight basis (Fig. 20.15).

The data in Fig. 20.14 only represent up to possibly 10 vol% of all materials consumed worldwide. They do not include the two major and important categories, wood (building, paper, etc.) and construction material (concrete, stone, etc.). Wood and construction materials each represent perhaps about 45 vol% worldwide with plastics at 2.5 vol% of the grand total worldwide. It is probable that if wood had never been used in buildings and elsewhere, our regulations on fire, contamination, etc.,

**Figure 20.17** GE's plastics living environment house makes extensive use of extruded products.

would outlaw its use. Of course, it would be ridiculous to even imagine such a situation, but we have something like 'blue laws and regulations' that would cause this ridiculous situation to occur.

Plastic materials and products cover the entire spectrum of the world's economy, so that fortunes are not tied to any particular business segment. Plastic fabricators are in a position to benefit in a wide variety of markets. Figures 20.16 and 20.17 shows examples where growth is expected with plastics, even though the building market is plastics' second biggest market and represents less than 2% of all the plastics used. Extruded plastic products in buildings will continue to expand when basically the performance requirements to cost are right.

The GE living environmental house (Fig. 20.17) in Pittsfield, MA., USA, with 280 m$^2$ (3000 ft$^2$) floor space was opened to the public October 23, 1989. It features advanced design and building methods, processes, and materials to serve as a laboratory for the entire building industry. It explores the feasibility of widespread use of plastics in construction. The two interior floors reflect modern life-styles and the daily living environment of a family of four. The basement contains a 232 m$^2$ (2500 ft$^2$) business

**Figure 20.18** Location filming: plastics films could be fabricated directly onto farmland; this tractor is using two extruders.

center, offices, and a display area for prototype models, such as factories to produce highly automated housing systems.

A $2.3 million Tiffany Street Pier in the Bronx of New York City is one of largest single user of 'plastic lumber' that was extruded from recycled plastic into board-like shapes. It measures 125 m (410 ft) long by 15 m (49 ft) wide with a gazebo was built during 1995. In 1996 it was hit by

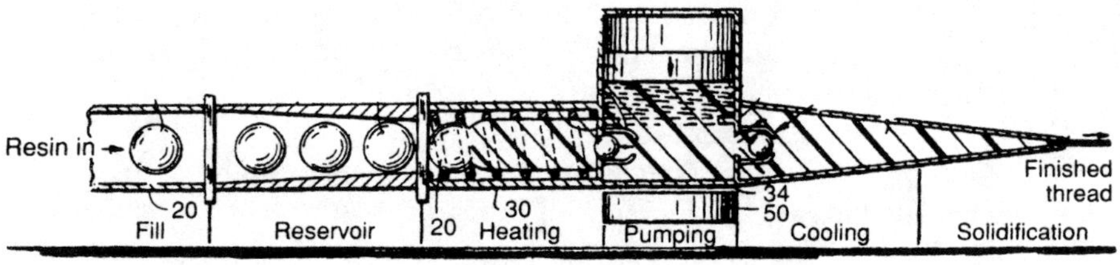


**Figure 20.19** Sewing machine that makes its own thread.

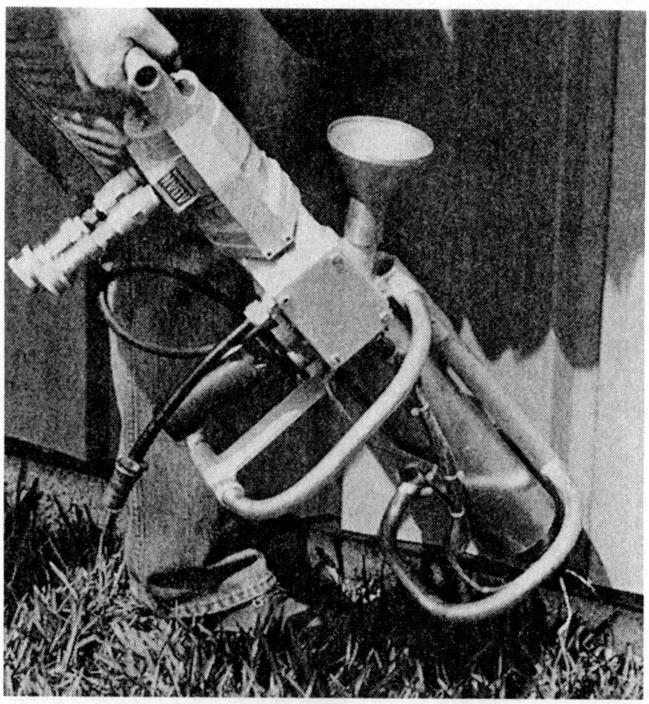

**Figure 20.20** Caulking is an example where an extruder can provide special services.

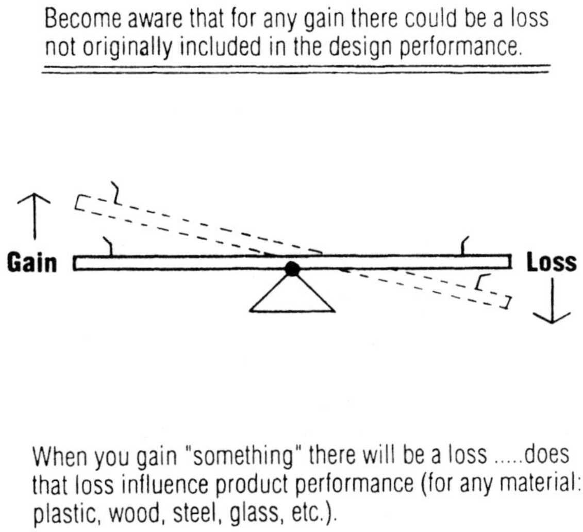


**Figure 20.21** Gains and losses: a judicious balance for design and processing.

lightning and, according to fire experts, it survived better than if it had been a traditional wooden pier even though one-third of the plastic was damaged. The damaged section was replaced with the extruded plastic lumber [130].

The plastics industry is a highly diversified economic activity that offers abundant opportunities for both employment and the entrepreneur. The prospect of growth in plastics is always very bright even with the expected ups and downs that occur in the business cycle. Effective exploitation of extrusion opportunities are really limitless such as those shown in Figs. 20.18 and 20.19.

Figure 20.20 shows an example where the process (extrusion) fits into the overall scheme to produce new applications. It is important is to recognize, as shown in Fig. 20.21, there can be a balance or a compromise between what could be either a gain or loss. Extrusion processing equipment provides an enormous scope for improving profitability by choosing the best available machines to manufacture the specified product at a profitable competitive cost [3, 103–105, 112, 116, 231, 357].

# Appendices

## METRIC CONVERSION CHARTS

| *US to metric* | | | *Metric to US* | | |
|---|---|---|---|---|---|
| *US* | *Metric* | *Multiply by* | *Metric* | *US* | *Multiply by* |
| | | **Density** | | | |
| $lb\,in^{-3}$ | $kg\,m^{-3}$ | 27 680 | $kg\,m^{-3}$ | $lb\,in^{-3}$ | 0.000 036 |
| $lb\,ft^{-3}$ | $g\,cm^{-3}$ | 0.0160 | $g\,cm^{-3}$ | $lb\,ft^{-3}$ | 62.43 |
| $lb\,ft^{-3}$ | $kg\,m^{-3}$ | 16.0185 | $kg\,m^{-3}$ | $lb\,ft^{-3}$ | 0.0624 |
| $lb\,in^{-3}$ | $g\,cm^{-3}$ | 27.68 | $g\,cm^{-3}$ | $lb\,in^{-3}$ | 0.036 13 |
| | | **Temperature** | | | |
| $°F^{-1}$ | $°C^{-1}$ | 1.8 | $°C^{-1}$ | $°F^{-1}$ | 0.556 |
| °F | °C | (°F − 32)/(1.8) | °C | °F | 1.8°C + 32 |
| °F | K | (°F + 459.67)/(1.8) | K | °F | 1.8K − 459.67 |
| | | **Pressure** | | | |
| psi | kPa | 6.8948 | kPa | psi | 0.145 |
| psi | MPa | 0.006 89 | MPa | psi | 145 |
| psi | GPa | 0.000 00689 | GPa | psi | 145 038 |
| psi | bar | 0.0689 | bar | psi | 14.51 |

| | | | | | |
|---|---|---|---|---|---|
| | | | **Energy and power** | | |
| ft lbf | J | 1.3558 | J | ft lbf | 0.7376 |
| in lbf | J | 0.113 | J | in lbf | 8.850 |
| ft lbf in$^{-1}$ | J m$^{-1}$ | 53.4 | J m$^{-1}$ | ft lbf in$^{-1}$ | 0.0187 |
| ft lbf in$^{-1}$ | J cm$^{-1}$ | 0.534 | J cm$^{-1}$ | ft lbf in$^{-1}$ | 1.87 |
| ft lbf in$^{-2}$ | kJ m$^{-2}$ | 2.103 | kJ m$^{-2}$ | ft lbf in$^{-2}$ | 0.4755 |
| kW | metric horsepower | 1.3596 | metric horsepower | kW | 0.7355 |
| US horsepower | kW | 0.7457 | kW | US horsepower | 1.3419 |
| Btu | J | 1055.1 | J | Btu | 0.00095 |
| Btu | Wh | 0.2931 | Wh | Btu | 3.412 |
| Btu in h$^{-1}$ ft$^{-2}$ °F$^{-1}$ | W m$^{-1}$ K$^{-1}$ | 0.1442 | W m$^{-1}$ K$^{-1}$ | Btu in h$^{-1}$ ft$^{-2}$ °F$^{-1}$ | 6.933 |
| Btu lb$^{-1}$ | kJ kg$^{-1}$ | 2.326 | kJ kg$^{-1}$ | Btu lb$^{-1}$ | 0.4299 |
| Btu lb$^{-1}$ °F$^{-1}$ | J kg$^{-1}$ °C$^{-1}$ | 4187 | J kg$^{-1}$ °C$^{-1}$ | Btu lb$^{-1}$ °F$^{-1}$ | 0.000239 |
| V mil$^{-1}$ | MV m$^{-1}$ | 0.0394 | MV m$^{-1}$ | V mil$^{-1}$ | 25.4 |
| | | | **Output** | | |
| lb min$^{-1}$ | g s$^{-1}$ | 7.560 | g s$^{-1}$ | lb min$^{-1}$ | 0.1323 |
| lb h$^{-1}$ | kg h$^{-1}$ | 0.4536 | kg h$^{-1}$ | lb h$^{-1}$ | 2.2046 |
| | | | **Velocity** | | |
| in min$^{-1}$ | cm s$^{-1}$ | 0.0423 | cm s$^{-1}$ | in min$^{-1}$ | 23.6220 |
| ft s$^{-1}$ | m s$^{-1}$ | 0.3048 | m s$^{-1}$ | ft s$^{-1}$ | 3.2808 |
| | | | **Viscosity** | | |
| P | Pa s | 0.1 | Pa s | poise | 10 |
| | | | **Length** | | |
| mil | millimeter | 0.0254 | millimeter | mil | 39.37 |
| inch | millimeter | 25.4 | millimeter | inch | 0.0394 |
| inch | centimeter | 2.54 | centimeter | inch | 0.3937 |
| foot | centimeter | 30.48 | centimeter | foot | 0.0328 |
| foot | meter | 0.3048 | meter | foot | 3.2808 |
| yard | meter | 0.9144 | meter | yard | 1.0936 |

| US to metric | | | Metric to US | | |
|---|---|---|---|---|---|
| *US* | *Metric* | *Multiply by* | *Metric* | *US* | *Multiply by* |
| | | **Area** | | | |
| $inch^2$ | $millimeter^2$ | 645.16 | $millimeter^2$ | $inch^2$ | 0.0016 |
| $inch^2$ | $centimeter^2$ | 6.4516 | $centimeter^2$ | $inch^2$ | 0.155 |
| $foot^2$ | $centimeter^2$ | 929.03 | $centimeter^2$ | $foot^2$ | 0.0011 |
| $foot^2$ | $meter^2$ | 0.0929 | $meter^2$ | $foot^2$ | 10.7639 |
| $yard^2$ | $meter^2$ | 0.8361 | $meter^2$ | $yard^2$ | 1.1960 |
| | | **Voume, capacity** | | | |
| $inch^3$ | $centimeter^3$ | 16.3861 | $centimeter^3$ | $inch^3$ | 0.061 |
| fluid ounce | $centimeter^3$ | $29.5735^3$ | $centimeter^3$ | fluid ounce | 0.0338 |
| quart (liquid) | $decimeter^3$ (liter) | 0.9464 | $decimeter^3$ (liter) | quart (liquid) | 1.0567 |
| gallon (US) | $decimeter^3$ (liter) | 3.7854 | $decimeter^3$ (liter) | gallon (US) | 0.2642 |
| gallon (US) | $meter^3$ | 0.0038 | $meter^3$ | gallon (US) | 264.17 |
| $foot^3$ | $decimeter^3$ | 28.3169 | $decimeter^3$ | $foot^3$ | 0.0353 |
| $foot^3$ | $meter^3$ | 0.0283 | $meter^3$ | $foot^3$ | 35.3147 |
| $yard^3$ | $meter^3$ | 0.7646 | $meter^3$ | $yard^3$ | 1.3079 |
| $in^3 lb^{-1}$ | $m^3 kg^{-1}$ | 0.000 036 | $m^3 kg^{-1}$ | $in^3 lb^{-1}$ | 27 680 |
| $ft^3 lb^{-1}$ | $m^3 kg^{-1}$ | 0.0624 | $m^3 kg^{-1}$ | $ft^3 lb^{-1}$ | 16.018 |
| | | **Mass** | | | |
| ounce (avdp.) | gram | 28.3495 | gram | ounce (avdp.) | 0.035 27 |
| pound | gram | 453.5924 | gram | pound | 0.0022 |
| pound | kilogram | 0.4536 | kilogram | pound | 2.2046 |
| pound | metric ton | 0.000 45 | metric ton | pound | 2204.6 |
| US ton (short) | metric ton | 0.9072 | metric ton | US ton (short) | 1.1023 |
| | | **Force** | | | |
| lbf | N | 4.448 | N | lbf | 0.225 |

Source: Institute for Science & Technology Metric Information Office.

| *Standard metric symbols* | | | | *Metric prefixes*[b] *Numerical value* | *Term* | *Symbol* |
|---|---|---|---|---|---|---|
| A | ampere | kg | Kilogram | 10 | deka | da |
| bar | bar | $dm^3$ | liter | $10^2$ | hecto | h |
| cd | candela | m | meter | $10^3$ | kilo | k |
| C | celsius[a] | N | newton | $10^6$ | mega | M |
| g | gram | Pa | pascal | $10^9$ | giga | G |
| h | hour | S | siemens | $10^{12}$ | tera | T |
| Hz | hertz | s | second | $10^{-1}$ | deci | d |
| J | joule | t | metric ton | $10^{-2}$ | centi | c |
| K | kelvin | V | volt | $10^{-3}$ | milli | m |
| | | W | watt | $10^{-6}$ | micro | μ |
| | | | | $10^{-9}$ | nano | n |
| | | | | $10^{-12}$ | pico | p |

[a] Formerly called centigrade.
[b] These prefixes may be used with all metric units.

## MATHEMATICAL SYMBOLS AND ABBREVIATIONS

| | | | |
|---|---|---|---|
| + | plus (addition) | $a'$, $a''$ | $a$-prime, $a$-second |
| − | minus (subtraction) | $a_1$, $a_2$ | $a$-sub one, $a$-sub two |
| ± ∓ | plus or minus, (minus or plus) | | |
| × | times, by (multiplication) | (), [], {} | parentheses, brackets, braces |
| ÷, / | divided by | | |
| : | is to (ratio) | ∠ ⊥ | angle, perpendicular to |
| :: | equals, as, so is | $a^2$, $a^3$ | $a$-square, $a$-cube |
| ∴ | therefore | $a^{-1}$, $a^{-2}$ | $1/a$, $1/a^2$ |
| = | equals | | |
| ∼ ≈ | approximately equals | $\sin^{-1} a$ | the angle whose sine is $a$ |
| > | greater than | π | pi = 3.141593+ |
| < | less than | | |
| ≧ | greater than or equals | | |
| ≦ | less than or equals | σ | summation of |
| ≠ | not equal to | ε, e | base of hyperbolic, natural or Napierian logs = 2.71828+ |
| ≐ | approaches | | |
| ∝ | varies as | Δ | difference |
| ∞ | infinity | g | acceleration due to gravity |
| ‖ | parallel to | | |
| $\sqrt{\ }$, $\sqrt[a]{\ }$ | square root, cube root | $E$ | coefficient of elasticity |
| □ | square | $v$ | velocity |
| ○ | circle | $f$ | coefficient of friction |
| ° | degrees (arc or thermometer) | $P$ | pressure of load |
| ′ | minutes or feet | hp | horsepower |
| ″ | seconds or inches | RPM | revolutions per minute |

## GREEK ALPHABET

| | | | | | | | |
|---|---|---|---|---|---|---|---|
| Α, α | alpha | Η, η | ēta | Ν, ν | nū | Τ, τ | tau |
| Β, β | bēta | Θ, θ ϑ | thēta | Ξ, φ | xi | Υ, υ | ūpsilon |
| Γ, γ | gamma | Ι, ι | iōta | Ο, ο | omicron | Φ, ϕ ψ | phi |
| Δ, δ | delta | Κ, κ | kappa | Π, π | pi | χ, φ | chi |
| Ε, ε | epsīlon | Λ, λ | lambda | Ρ, ϱ | rhō | Ψ, ψ | psi |
| Ζ, ζ | zēta | Μ, μ | mū | ∂, σ, ς | sigma | Ω, ω | ōmega |

## PROPERTIES OF WATER

Density of water is 62.38 lb ft$^{-3}$ or 0.0361 lb in$^{-3}$
Specific gravity of water is 1 g cm$^{-3}$
1 ft$^3$ of water at 3.9°C (39.1 °F) equals 0.4335 lb in$^{-2}$ (ft high)
1 in$^3$ of water weighs 0.576 oz
1 gallon of water contains 231 in$^3$ or 0.133 68 ft$^3$
1 gallon of water weighs 8.3356 lb at 17°C (62°F) (air-free, weighed in vacuum)
1 ft$^3$ of water equals 7.4805 gallons; a cylinder 7 in diameter and 6 in high contains 1 gallon of water
The maximum density of water is at 3.9°C (39.1°F)
The freezing point of water at sea level is 0°C (32°F)
The boiling point of water at sea level is 100°C (212°F)

## ANGLE CONVERSIONS

| | | |
|---|---|---|
| Degrees | $\times$ 60 | = minutes |
| | $\times$ 1.745 329 3 $\times 10^{-2}$ | = radians |
| Degrees/foot | $\times$ 5.726 145 $\times 10^{-4}$ | = radians/centimeter |
| Degrees/minute | $\times$ 2.908 8 $\times 10^{-4}$ | = radians/second |
| | $\times$ 4.629 629 $\times 10^{-5}$ | = revolutions/second |
| Degrees/second | $\times$ 1.745 329 3 $\times 10^{-2}$ | = radians/second |
| | $\times$ 0.166 | = revolutions/minute |
| | $\times$ 2.77 $\times 10^{-3}$ | = revolutions/second |
| Minutes | $\times$ 1.667 $\times 10^{-2}$ | = degrees |
| | $\times$ 2.908 8 $\times 10^{-4}$ | = radians |
| | $\times$ 60 | = seconds |
| Radians | $\times$ 0.159 154 | = circumferences |
| | $\times$ 57.295 77 | = degrees |
| | $\times$ 3.437 746 $\times 10^{3}$ | = minutes |
| Seconds | $\times$ 2.777 $\times 10^{-4}$ | = degrees |
| | $\times$ 1.667 $\times 10^{-2}$ | = minutes |
| | $\times$ 4.848 136 8 $\times 10^{-6}$ | = radians |
| Steradians | $\times$ 0.159 154 9 | = hemispheres |
| | $\times$ 7.957 74 $\times 10^{-2}$ | = spheres |
| | $\times$ 0.636 619 7 | = spherical right angles |

# References


1. Rosato, D.V. (1997) *Plastics Processing Data Handbook*, 2nd edn, Chapman & Hall.
2. Rosato, D.V. (1996) *Injection Molding Handbook*, 2nd edn, Chapman & Hall.
3. Rosato, D.V. (1991) *Designing with Plastics and Composites Handbook*, Chapman & Hall.
4. Rosato, D.V. (1993) *Rosato's Plastic Encyclopedia and Dictionary*, Hanser.
5. Rosato, D.V. (1997) Fundamentals of Processing, *SPE-IMD Newsletter*, No. 45.
6. Rosato, D.V. (1991) *Design Features that Influence Performance: Detractors/ constraints*, SPE-ANTEC, May.
7. Rosato, D.V. (1990) *Product Design – Plastics Selection Guide*, SPE-ANTEC, May.
8. Rosato, D.V. (1990) Materials selection and reinforced plastics, in *Concise Encyclopedia of Polymer Science Engineering* (ed. J.I. Kroschwitz), Wiley.
9. Rosato, D.V. *et al.* (1990) *Plastics Processing Data Handbook*, Chapman & Hall
10. Rosato, D.V. (1994) Plastics contribute to success and efficiency of today's major appliances, *SE-IMD Newsletter*, No. 37.
11. Rosato, D.V. (1997) Additives, plastics, in *Encyclopedia of Packaging Technology*, 2nd edn, Wiley.
12. Rosato, D.V. (1986–97) Manufacturing – designing with plastics, Rhode Island School of Design.
13. Rosato, D.V. (1994) *Designing with Plastics – Materials, Processes, Products*, Open University Seminars, Geneva, Switzerland, May.
14. Rosato, D.V. (1990) Current and future trends in the use of plastics for blow molding, *SME Tech. Paper MS90-198*, June.
15. Rosato, D.V. (1987) Plastics replaced aorta permits living normal long life, Newton-Wellesley Hospital, Mar.
16. Rosato, D.V. (1997) Simple sprue pickers and parts removal with on-robot secondary, in *Material and Part Handling in Manufacturing*, SME.
17. Rosato, D.V. (1988) *Injection Molding Technology: Economics and Markets*, SPE-ANTEC, May.
18. Rosato, D.V. (1989) *Injection Mold Design and Product Design*, China's Chemical Planning Institute, Beijing, China.

19. Rosato, D.V. (1989) Plastics and solid waste, Rhode Island School of Design, Oct.
20. Rosato, D.V. (1985–91) *Advanced Engineering Design, etc.*, Open University Seminars, Northhampton, UK.
21. Rosato, D.V. (1977) Uniform resin code, *Plastics World*, May.
22. Rosato, D.V. (1977) The DMC-12 car, *Plastics World*, Sep.
23. Rosato, D.V. (1960–70) (ed.) *Polymer Engineering and Technology*, including Epoxy technology (Bruins, P.F.), Man-made fibers (Mark, H.F.), etc., Wiley.
24. Rosato, D.V. (1970) Electrical wire and cable plastic coating – what's ahead, *Wire & Wire Products*, Mar.
25. Rosato, D.V. (1970) Insulation news, *Wire & Wire Products*, Nov.
26. Rosato, D.V. (1970) Wire and wire products – insulation news, *Wire & Wire Products*, Dec.
27. Rosato, D.V. (1987–98) Processing with plastics, Rhode Island School of Design course.
28. Rosato, D.V. (1983–86) *Reaction Injection Molding*, University of Massachusetts Lowell Seminars.
29. Rosato, D.V. *et al.* (1973) *Plastics Industry Safety Handbook*, Cahners.
30. Rosato, D.V. (1983) Advanced engineering design short course, *ASME Engineering Conference*.
31. Rosato, D.V. (1983) Designing with plastics, *MD & DI*, July 26–29.
32. Rosato, D.V. (1980–87) *Process Validation and Good Manufacturing Practices*, University of Massachusetts Lowell Seminars.
33. Rosato, D.V. (1974–86) *Fundamentals of Plastics*, University of Massachusetts Lowell Seminars.
34. Rosato, D.V. (1987) Role of additives in plastics: function of processing aids, *SPE-IMD Newsletter*, Nov.
35. Rosato, D.V. (1987) *Blow Molding Expanding Technologywise and Marketwise*, SPE-ANTEC, May.
36. Rosato, D.V. (1987) History-IM machine, *SPE-IMD Newsletter*, No. 15.
37. Rosato, D.V. (1974–86) *Quality Control and Statistical Control*, University of Massachusetts Lowell Seminars.
38. Rosato, D.V. (1989) *Blow Molding Handbook*, Hanser.
39. Rosato, D.V. (1987) *Extrusion: Technology, Markets, Economics*, University of Massachusetts Lowell Seminars.
40. Rosato, D.V. (1989) Choosing the blow molding machines to meet product demand, *Plastics Today*, Jan.
41. Rosato, D.V. (1980) Polymers, processes and properties of medical plastics-including applications, in *Synthetic Biomedical Polymers* (eds Szycher, M. and Robinson, R.T.), Technomic.
42. Rosato, D.V. (1982) An overview of composites, in *Handbook of Composites* (ed. Lubin, G.), Chapman & Hall.
43. Rosato, D.V. (1997) *Designing with Reinforced Plastic Composites*, Hanser.
44. Rosato, D.V. (1987–97) *Processing with Plastics*, Rhode Island School of Design course.
45. Rosato, D.V. (1997) Injection molding, in *Guide to Short and Medium Fiber Reinforced Plastics* (ed. Jones, R.F.), Hanser.

46. Rosato, D.V. (1987) Materials selection, in *Encyclopedia of Polymer Science and Engineering* (eds Mark, Bikales, Overberger, and Menges), Vol. 9, Wiley.
47. Rosato, D.V. (1988) Thermosets, in *Encyclopedia of Polymer Science and Engineering* (eds Mark, Bikales, Overberger, and Menges), Vol. 14, Wiley.
48. Rosato, D.V. (1996) *Injection Molding Higher Performance Reinforced Plastic, SPE-ANTE,* May.
49. Rosato, D.V. (1983–85) Advanced engineering design – short course, *ASME Engineering Conferences.*
50. Rosato, D.V. (1979) New approach to producing flexible PVC film, *Plastics World,* Jan.
51. Rosato, D.V. (1968) Heat resistance resins, *Plastics World,* Mar.
52. Rosato, D.V. (1967) Weighing out the aircraft market, *Plastics World.*
53. Rosato, D.V. (1966) The all-RP aircraft, *Plastics World,* Mar.
54. Rosato, D.V. (1966) Processing and structural integrity of filament winding, *Iowa State University Conference,* Apr.
55. Rosato, D.V. (1966) Film and sheet report, *Plastics World,* Nov., Dec.
56. Rosato, D.V. (1960) Plastics in missiles, *Plastics World,* Aug.
57. Rosato, D.V. (1994) Troubleshooting problems – solutions, *SPE-IMD Newsletter,* No. 37.
58. Rosato, D.V. (1983) Designing with plastics, *MD & DI,* July 26–29.
59. Rosato, D.V. *et al.* (1969) *Markets for Plastics,* Chapman & Hall.
60. Rosato, D.V. (1995) An injection molder's dream come true, *SPE-IMD Newsletter,* No. 40.
61. Rosato, D.V. and Schwartz, R.T. (1968) *Environmental Effects on Polymeric Materials, Vol. 1; Environment and Materials,* Vol. 2, Wiley.
62. Rosato, D.V. (1964) Radomes and antennas, *Plastics World,* Apr.
63. Rosato, D.V. (1964) *Filament Winding,* Wiley.
64. Rosato, D.V. (1969) *Filament Winding,* Wiley. *Hamotka ctekjiohntbio,* Russian translation, MaIIInhoctpoehn, Mockba.
65. Rosato, D.V. (1996) *Injection Molding Higher Performance Reinforced Plastic Composites,* SPE-ANTEC, May.
66. Rosato, D.V. (1974–86) *Fundamentals of Plastics,* University of Massachusetts Lowell Seminars.
67. Rosato, D.V. (1963) *Outer Space Parabolic Reflector Energy Converters,* SAMPE, June.
68. Rosato, D.V. (1944) Capt. Theoretical potential for polyethylene, US–British correspondence, USAF Materials Lab., WPAFB, OH., USA.
69. Rosato, D.V. (1944) Capt., All plastic military airplane successfully flight tested, Wright-Patterson AFB, Dayton, OH., USA.
70. Rosato, D.V. (1991) Materials selection, polymeric matrix composites, in *International Encyclopedia of Composites* (ed. Lee, S.M.), Vol. 3, VCH.
71. Rosato, D.V. (1974–89) Technical papers for seminars presented worldwide on 21 different subjects including different aspects on extrusion via University of Massachusetts Lowell, Tufts University, ASME, General Motors Inst., SPE, SPI, Hong Kong Production Center, Singapore Inst., China National Chem. Construction, Open University, etc.
72. Rosato, D.V. (1989) What to consider when picking an injection molding machine, *Plastics Today,* Jan.

73. Rosato, D.V. (1986) *Reinforced Plastics/composites*, University of Massachusetts Lowell Seminars.
74. Rosato, D.V. (1985) *Injection Molding Thermosets*, SPI Reinforced Plastics/ Composites Annual Meeting Seminar, Jan.
75. Rosato, D.V. (1962) Non-woven fibers in reinforced plastics, *Industrial Engineering Chemistry*, **54**(8).
76. Rosato, D.V. *et al.* (1980) The History of Reinforced Plastics, *ACS Annual Conference*, May.
77. Rosato, D.V. (1963) *Outer Space Parabolic Reflector Energy Converters*, SAMPE, June.
78. Rosato, D.V. (1959) *Asbestos*, Chapman & Hall.
79. Rosato, D.V. (1986) *Testing and Statistical Quality Control*, University of Massachusetts Lowell seminars.
80. Rosato, D.V. *et al.* (1969) Structural sandwich construction, in *Composite Engineering Materials* (ed. Dietz, A.G.), MIT Press.
81. Rosato, D.V. (1957) Nose cone of first US moon vanguard rocket is made in Manheim-US, *New Era-Lancaster PA Newspaper*, Nov. 30.
82. Rosato, D.V. (1967) 25 years of polyethylene, *Plastics World*, Jan.
83. Dorgham, M.A. and Rosato, D.V. (1986) *Designing With Plastic Composites*, Underscience Enterprises, Ltd., Switzerland.
84. DuBois, J.H. and Rosato, D.V. (1968) From laminates to composites, *Plastic World*, Apr.
85. Lubin, G. and Rosato, D.V. (1964) Applications of reinforced plastics, in *Proceedings of the 4th International Reinforced Plastics Conference*, British Plastics Federation, London, UK, Nov. 25–27.
86. Abbott, N.J. (1975) *Some Mechanical Properties of Kevlar and Other Heat Resistant Nonflammables*, AFML-TR-74-65.
87. Acierno, D. *et al.* (1996) *Rheology and Processing of Liquid Crystal Polymer*, Chapman & Hall.
88. Adolfsson, G. (1995) Why a tube? *Tube Topics*, **29**(4).
89. Advanced composites (1997) Compressed air, Sep.
90. Alger, M. (1997) *Polymer Science Dictionary*, 2nd edn, Chapman & Hall.
91. Al-Malaika, S. (1996) *Reactive Modifiers for Polymers*, Chapman & Hall.
92. Amon, M. (1997) *Measurement of the Curvature Defect in Films*, SPE-ANTEC, May.
93. Anderson, J. *et al.* (1997) *Troubleshooting Extrusion Problems*, SPE-ANTEC, May.
94. Ash, M. *et al.* (1995) *Gower Handbook of Plastic and Rubber Additives*, Gower Press.
95. Ashok, B.K. *et al.* (1995) PE cylindrical rods were produced by ram right angle extrusions at room temperatures, *J. Polym. Sci. B: Phys.*, **33**(1).
96. ASTM compilation of standard definitions (1995) Periodically updated.
97. ASTM Index-annual book update, ASTM standards, annual.
98. ASTM plastics handbooks and standards, Vols. 08.01, 08. 02, 08.03, and 08.04, yr. updated.
99. Avenas, J.F. (1990) *Modeling Polymer Processing*, Hanser.
100. Babriele, M.C. (1997) Multi-parison blow molding meets new volume needs, *Modern Plastics*, Feb.

101. Baer, E. (1964) *Engineering Design with Plastics*, Chapman & Hall.
102. Baer, E. *et al.* (1990) *High Performance Polymer*, Hanser.
103. Baker, M. (1997) European industry's role in a rapidly globalizing market, *MD & DI*, May.
104. Bakerjian, R. (1993) *Continuous Improvement*, Vol. 7, SME.
105. Baker, F. *et al.* (1997) Efficient integration boosts on-line rheometer benefits, *Modern Plastics*, May.
106. Barker, T.B. (1985) *Quality by Experimental Design*: Vol. 4, Quality and Reliability Series, Marcel Dekker.
107. Bang, D.-S. *et al.* (1997) *Modular Tangential Counter-rotating Twin Screw Extrusion: Non-Newtonian and Non-isothermal Simulation*, SPE-ANTEC, May.
108. Bashford, D.P. (1997) *Thermoplastics Directory and Data Book*, Chapman & Hall.
109. Battaglini, T.A. (1966) Current trends and future needs for plastics in computers, *SPI Annual Conference*, June.
110. Bayer ABS sheet extrusion guide (1992) *Bayer Bulletin No. 6407H.*
111. Beck, R.D. (1980) *Plastics Product Design*, Chapman & Hall.
112. Belofsky, H. (1995) *Plastics: Product Design and Process Engineering*, Hanser.
113. Berghaus, U. *et al.* (1995) Direct extrusion, *Kunststoffe Plast. Europe*, 2nd edn.
114. Berins M.L. (1994) *SPI Plastics Engineering Handbook*, 5th edn, Chapman & Hall.
115. Bernhardt, A. (1996) Evolution of machine controls and parameter settings, SPE-RETEC, Mar.
116. Bevis, M.J. (1996) Looking to the future – processing, *World Plastics & Rubber Technology*, No. 8.
117. Biklen, D. (1995) Planning the extrusion line, *SPE Extrusion Newsletter*, Dec.
118. Bird, A. *et al.* (1995) *Dynamics of Polymeric Liquids*, 2nd edn, Wiley.
119. Birley, A.W. *et al.* (1992) *Physics of Plastics*, Hanser.
120. Bode, W.W. (1997) *The Versatility of Grooved Feed Extrusion*, Battenfeld-Gloucester Engineering.
121. Bode, W.W. (1996) *Optimizing Coextruded Film Quality Using a Stationary Die and an Oscillating Hauloff*, Battenfeld-Gloucester Engineering.
122. Bode, W.W. (1989) *Proceedings of the Equipment for Processing HMWHDPE Film, Polymers, Laminations and Coatings Conference*, TAPPI.
123. Booth, A.F. (1997) The manufacturer's role in safe reuse, *MD & DI*, May.
124. Botros, M.G. (1996) *An Improved Purging Compound for Polyamides and Ethylene Vinyl Alcohol Copolymers*, SPE-ANTEC, May.
125. Boussinesq, M.J. (1868) *Journal de Mathematiques pures et appliquees*, **2**(13), 377–424.
126. Boyer, H.E. (1987) *Selection of Materials for Service Environments*, ASM.
127. Brady, G.S. and Clauser, H.R. (1985) *Materials Handbook*, McGraw-Hill.
128. Brandrup, S. *et al.* (1989) *Polymer Handbook*, 3rd edn, Wiley.
129. Braun, D. (1982) *Simple Methods for the Identification of Plastics*, Hanser.
130. Bregar, B. (1996) Lightning zaps, doesn't destroy New York plastic pier, *Plastics News*, Oct. 14.
131. Brostow, W. (1997) *Mechanical and Thermophysical Properties of Polymer Liquid Crystal*, Chapman & Hall.
132. Browstow, W. *et al.* (1986) *Failure of Plastics*, Hanser.

133. Buckleitner, E.L. (1995) *DuBois and Pebble's Plastics Mold Engineering Handbook*, Chapman & Hall.
134. Chanda, M. and Roy, S.K. (1987) *Plastics Technology Handbook*, Marcel Dekker.
135. Callari, J. (1996) Flow tuner, precise block, the latest in flat dies, *Plastics World*, Oct.
136. Callari, J. (1996) Compounding machinery suppliers in high-torque tug of war, *Plastics World*, Oct.
137. Callari, J. (1996) Line stretches speed envelope, *Plastics World*, May.
138. Callari, J. (1996) Robot extrusion gives TPE profiles new twist, *Plastics World*, Feb.
139. Campbell, G.A. *et al.* (1996) *Screw design and Newtonian Fluid Flow*, SPE-ANTEC, May.
140. Canby, T.Y. (1989) Reshaping our lives-advanced materials, *National Geographic*, Dec.
141. Carley, J.F. (1962) *Rheology and Die Design*, SPE-RETEC Newark, Nov.
142. Carton, D.R. (1991) *The Physics of Coextrusion*, EDI.
143. Cary, P. (1995) The asbestos panic attack, *News & World Report*, Feb. 20.
144. Chan, C.-M. (1995) Viscosity and the formation of die drool at the polymer-metal surfaces, *Int. Polym. Proc.*, **10**(3), Hong Kong University.
145. Chanda, M. *et al.* (1987) *Plastics Technology Handbook*, Marcel Dekker.
146. Chandrasekaran, M. *et al.* (1997) *Velocity Profiles in the Nip Region of a Co-rotating Twin-screw Extruder: Transitions Between Forward and Reverse Elements*, SPE-ANTEC, May.
147. Charrier, J.-M. (1991) *Polymeric Materials and Processing*, Hanser.
148. Chen, T. *et al.* (1995) On-line measurement of residence time distribution in a twin-screw extruder, *Int. Polym. Proc.*, **10**(1), McGill University.
149. Cheremisinoff, N.P. (1993) *An Introduction to Polymer Rheology and Processing*, CRC Press.
150. Christensen, R.M. (1971) *Theory of Viscoelasticity*, Academic Press.
151. Chrietiano, J.P. *et al.* (1997) *Investigation of Mixing Patterns in Co-rotating Fully Intermeshing Twin-screw Extruder Mixing Elements Using Dynamic Pressure Distributions*, SPE-ANTEC, May.
152. Chung, C.I. (1997) *A Scientific Approach to Screw Design*, SPE-ANTEC, May.
153. Chung, C.I. (1976) A new look at the mechanism of melting, *Plastics Today*, June.
154. Chung, C.I. *et al.* (1983) *A Novel Energy Efficient Extruder Screw Design*, SPE-ANTEC May.
155. *Code for Good Manufacturing Practice for Medical Devices* (1978) Federal Regulations 21 CFR 820 and 43 FR:31508, July 21.
156. Colvin, R. (1997) Processors seek to tame a new generation of blown film resins, *Modern Plastics*, May.
157. Coops, J. (1995) A new line based on cold feed extrusion for aerosol cup gaskets, *World Plastics & Rubbber Technology*, No. 7.
158. Cornell, W.E. (1996) How to make installations a breeze, *Plastics World*, Nov.
159. *Continuous Production of Square Mesh, Integral Plastic Net* (1968) Extruded British Plastics, Sep.
160. Cowell, T. (1997) *Spherilene Process LLDPE for Conventionally Blown Shrink Film*, SPE-ANTEC, May.

161. Cross, N. (1989) *Engineering Design Methods*, Wiley.
162. Davis, W.H. (1996) Heat transfer in extruders-an introduction, *SPE Extrusion Newsletter*, May.
163. Dealy, J.M. *et al.* (1990) *Melt Rheology and its Role in Plastic Processing*, Chapman & Hall.
164. Deanin, R.D. (1985–97) *Advances in Compounding*, University of Massachusetts Lowell seminars.
165. DeGaspari, J. (1996) Shake some energy savings out of your auxiliary equipment, *Plastics Today*, Oct.
166. DeGaspari, J. (1997) Six ways to make your plant safer, *Plastics Today*, Jan.
167. DeMatteo, A.J. (1968) *An Introduction to Extrusion*, SPI
168. Deming, E.W. (1986) *Out of Crisis*, MIT Center for Advanced Engineering Study, MIT.
169. Derezinski, S.J. (1997) *Calculating Surge Dampening in Melt Delivery Systems*, SPE-ANTEC, May.
170. Derezinski, S.J. (1996) *Calculating Power of Extruder Melt Sections*, SPE-ANTEC, May.
171. Designing with plastics: fundamentals (1989) Hoeschst Celanese Design manual TDM-1.
172. Domininghaus, H. (1993) *Plastics for Engineers*, Hanser.
173. Dooley, J. *et al.* (1995) Coextrusion layer thickness variations-effect of polymer viscoelasticity on layer uniformity, *SPE Extrusion Newsletter*, Sep.
174. Dooley, J. *et al.* (1996) Numerical simulation of viscoelastic polymer flow effects of secondary flows on multilayer coextrusion, *PE*, Apr.
175. Dostal, C.A. (1988) *Engineering Material Handbook*, Vol. 2, Engineering Plastics, ASM International.
176. Drewer, J.M. (1995) Everything transparent – extruding profiles, *Kunststoffe Plast Europe*, Sep.
177. Drozda, T. *et al.* (1983) *Machinery*, Vol. 1, SME.
178. Duff, J.M. *et al.* (1995) *Freehand Sketching for Engineering Design*, PWS.
179. DuPont (1995) Fluoropolymer bulletins.
180. Earle, J.H. (1991) *Drafting Technology*, 2nd edn, Addison-Wesley.
181. Edenbaum, J. (1992) *Plastics Additives and Modifiers Handbook*, Chapman & Hall.
182. Ehrig, R.J. (1992) *Plastics Recycling: Products and Processes*, Hanser.
183. Ellis, B. *et al.* (1997) *Polymers on CD-ROM*, Chapman & Hall.
184. *Engineering Plastics Handbook* (1988) Vol. 2: Engineering plastics, ASM International.
185. *Extruded Sheet Guideline* (1994) for standardized use of test method described in ASTM D 3029-82 for determining the relative impact resistance, SPI/SPD 0993-04, Apr.
186. *Extruder Polyethylene Sheet Guideline* (1995) SPI/SPD 0994-02, June 28.
187. *Extrusion* (1972–95) Collection of articles from extruder's consultant corners, SPE Extrusion Division.
188. Extrusion overcomes the energy challenges (1995) *Modern Plastics*, Apr.
189. *Extrusion Software Review* (1996) The extruder's technician, Spirex Corp., Youngstown, OH, USA.
190. Ezrin, M. (1996) *Plastics Failure Guide – Cause and Prevention*, Hanser.

191. Farnham, S.E. (1972) *A Guide to Thermoformed Plastic Packaging*, Cahners.
192. Feron, B. *et al.* (1996) *Optimized Cooling and Gauge Tolerances in Blown Film Extrusion*, SPE-ANTEC, May.
193. Finkel, R.W. *et al.* (1960) *IBM share program No. 989*, Rw spad.
194. *Fire Safety Aspects of Polymeric Materials* (1977) Vol. 4, National Material Advisory Board Publ. NMAB 318-1, National Academy of Science.
195. Fisher, E.G. (1964) *Extrusion of Plastics*, Iliffe.
196. Flanagan, J. (1992) *Tools for Optimizing the Cast Film Process*, Plastics and Rubber Institute.
197. Foston, A.C. *et al.* (1991) *Fundamentals of Computer Integrated Manufacturing*, Prentice-Hall.
198. Frederick, C.D. (1997) Blown PE film line has potential to raise margins on commodity resins, *Modern Plastics*, Feb.
199. Frederick, C.D. (1996) The voice of rotational molding, *World Plastics & Rubber Technology*, No. 8.
200. Freiherr, G. (1997) Product development making the most of the company team, *MD & DI*, May.
201. French, T.E. *et al.* (1986) *Engineering Drawing and Graphic Technology*, McGraw-Hill.
202. Gabriele, M.C. (1994) Coextrusion makes LCP competitive in barrier packaging, *Plastics Today*, July.
203. Gachter, R. *et al.* (1990) *Plastics Additives*, Hanser.
204. Geng, X. *et al.* (1997) *The Model of Solid Conveying in the Intermeshing Co-rotating Twin-screw Extruders with Three Sub-conveying Zones*, SPE-ANTEC, May.
205. Gifford, W.A. (1997) *A Three-dimensional Analysis of Coextrusion*, SPE-ANTEC, May.
206. Glanvill, A.B. (1971) *The Plastics Engineer's Data Book*, Machinery Publishing.
207. Garton, D.R. (1997) *The Physics of Coextrusion*, EDI, Chillewa Falls, WI., USA.
208. Gordon, J.M. (1992) *Total Quality Process Control for Injection Molding*, Hanser.
209. Griff, A.L. (1968) *Plastics Extrusion Technology*, 4th edn, Chapman & Hall.
210. Griffith, A.A. (1921) The phenomena of rupture in flow, *Philosophical Transactions of the Royal Society of London, Series A*, **221**, 163–98.
211. Griffith, A.A. (1924) The theory of rupture, in *Proceedings of the 1st International Congress on Applied Mechanics*, p. 55.
212. Griskey, R.G. (1995) *Polymer Process Engineering*, Chapman & Hall.
213. Gruenwald, G. (1993) *Plastics: How Structure Determines Properties*, Hanser.
214. Haddad, Y.M. (1995) *Viscoelasticity of Engineering Materials*, Chapman & Hall.
215. Hagen, P.R. (1995) *Process for Forming a Cured Article of Consistent Shape for an Extruded Hydrolyzable Siloxane Composition (TS silicone)*, US patent 5401781.
216. Hall, C.R. (1997) *Physical Methods of Materials Caracterization*, Chapman & Hall.
217. Harper, C.A. (1996) *Handbook of Plastic Elastomers and Composites*, 3rd edn, McGraw-Hill.
218. Harrel tubing lines optimized (1996) Harrel news release.
219. Harrington, H.J. (1997) *Benchmark 1.0: Innovative Software Applications*, ASTM.
220. Harrington, T.P. (1997) *Designing an Extruder Additive Injection Valve*, SPE-ANTEC, May.

221. Hayashi, T. *et al.* (1995) *Process for Producing TP Polyester Series Extruded Formed Material*, US patent 5389319.
222. Hensen, F. (1988) *Plastics Extrusion Technology*, Hanser.
223. Hertzberg, R.W. (1976) *Deformation and Structure Mechanics of Engineering Materials*, Wiley.
224. Hertzberg, R.W. (1980) *Fatigue of Engineering Plastics*, Academic Press.
225. Hesse, A. *et al.* (1995) Study of alcoholysis of EVA copolymers with menthanol in the extrusion process, *J. Macromol Sci. A: Pure Appl. Chem.*, **32**(7).
226. Higgins, R.A. (1997) *Materials for the Engineering Technician*, 3rd edn, Chapman & Hall.
227. Hilado, C.J. (1974) *Flammability Handbook of Plastics*, Technomic.
228. Hill, R. (1950) *Mathematical Theory of Plasticity*, Oxford Press.
229. Hilyard, N.C. (1994) *Low density cellular plastics*, Chapman & Hall.
230. Hooten, W.F. (1997) FDA makes quality the rule, *DM*, Jan.
231. Hormozi, N. (1997) *Plastics and Rubber Technology Handbook*, Forward.
232. Horns, J. *et al.* (1997) *The Influence of Using Fluoropolymer Processing Additives to Improve the Extrusion Characteristics of LDPE/LLDPE Resin Blends*, SPE-ANTEC, May.
233. Hulbert, W. (1994) Screw architecture and control systems influence extrusion quality, *Modern Plastics*, Nov.
234. Huneault, M.A. (1997) *Pressure Generation and Screw Design in Corotating Twin-screw Extruders*, SPE-ANTEC, May.
235. Hunkar, D.B. (1995) What you should know before buying equipment, *Plastics World*, Oct.
236. Hytrel engineering thermoplastic elastomer (1995) *DuPont Bulletin 201714C.*
237. Hyun, K.S. *et al.* (1997) *A New Model for Solids Conveying in Single-screw Plasticating Extruders*, SPE-ANTEC, May.
238. Hyun, K.S. *et al.* (1996) *Theoretical and Experimental Analysis of Solids Conveying in Single-screw Extruders*, SPE-ANTEC, May.
239. Hyun, K.S. *et al.* (1996) Elimination of a restriction at the entrance of the barrier flighted extruder screw-sections, *PE*, Apr.
240. Imthurn, A. (1997) *Lubrication of Hydrodynamic Bearings in Gear Pumps*, SPE-ANTEC, May.
241. *International Selector Plastics Digest* (annual) Data on 23 000 plastic materials and 250 manufacturers, ranked by properties, etc., Hard copy in two volumes, and CD-ROM, DATA. Business Publishing.
242. Ishibashi, J. *et al.* (1997) *Experimental Study of Extruding Behavior Against Material Property in the Filler Mixing by Intermeshing Co-rotating Twin-screw Extruder (II)*, SPE-ANTEC, May.
243. Jacob, C. *et al.* (1995) Inert gases as alternative blowing agents for extruded LDPE foam, *J. Cellular Plast.*, **31**(1).
244. Jacobi, H.R. (1963) *Screw Extrusion of Plastics*, Iliffe.
245. James River Corp. (1996) Ceramic rollers boost corona-treatment uptime, *Plastics Today*, Feb.
246. Johannaber, F. (1994) *Injection Molding Machines*, 3rd edn, Hanser.
247. Joo, Y.L. (1996) *Rheology and Processing of PC/*carbon Black (Twin-screw Extruder) Composites, SPE-ANTEC, May.

248. Joshi, S. *et al.* (1994) *Computer Control of Flexible Manufacturing*, Chapman & Hall.
249. Jovic, F. (1991) *Process Control Systems*, 2nd edn, Chapman & Hall.
250. Kabumoto, A. *et al.* (1995) *Method for Manufacturing (Extruded) Foamed Plastics of Saturated Polyester Using a Cyclic Tetramer as Foaming Agent*, US patent 5458832.
251. Kamykowski, G.W. *et al.* (1996) *The Effect of Layer Geometry on Tie Layer Adhesion of Coextruded Films*, SPE-ANTEC, May.
252. Karger-Kocsis, J. (1995) *Polypropylene: Structure, Blends, and Composites*, Chapman & Hall.
253. Kuye, B.H. (1997) *Powder Mixing*, Chapman & Hall.
254. Kayl, P.B. (1997) *Non-isothermal Viscoelastic Flow Simulation of a High Speed Wire Coating Process*, SPE-ANTEC, May.
255. Kennedy, P. (1993) *Flow Analysis*, Moldflow Pty Ltd.
256. Kiani, A. (1997) *Study of Flow and RTD in a ZSK Twin-screw Extruder*, SPE-ANTEC, May.
257. Kirkpatrick, J.M. (1996) *The AutoCAD Book*, 4th edn, Prentice-Hall.
258. Klinzing, G.E. (1997) *Pneumatic Conveying of Solids*, 2nd edn, Chapman & Hall.
259. Knights, M. (1995) Mastering micropellets: a processing primer, *Plastics Technology*, **41**(7).
260. Knittel, R. (1996) *A Review of the Film Extrusion Process*, SPE-ANTEC, May.
261. Krause, J.W. (1996) *The Applications and Benefits of Total Integrated Process Control*, SPE-ANTEC, May.
262. Knittel, R. (1996) Retrofitting a screw into an existing extruder, *SPE Extrusion Newsletter*, Dec.
263. Kroschwitz, J.I. (1990) *Concise Encyclopedia of Polymer Science and Engineering*, Wiley.
264. Kurtz, S.J. (1995) Relationship of stresses in blown film processes, *Int. Polym. Proc.*, **10**(2), Union Carbide.
265. Lahti, G.P. (1963) Calculation of pressure drops and outlets, *SPE Journal*, July.
266. Lachtermacher, M.G. *et al.* (1995) Reactive processing of LLDPEs in corotating intermeshing twin-screw extruder, *Journal Polym. Sci.*, **58**(11), University of Waterloo, Canada.
267. Lamont, P.R. *et al.* (1997) *Extrusion of Thin Gauge Polypropylene Sheet*, SPE-ANTEC, May.
268. Landis, H. (1993) *Environmental Toxicology and Risk Management*, STP 1179, ASTM.
269. Lang, U. *et al.* (1997) *Development of a Mathematical Model for the Calculation of the Pressure Drop in Extrusion Dies*, SPE-ANTEC, May.
270. Lashway, W.H. (1988) Melt-flow match is secret to PC/PP coextrusion Quality, *Plastics World*, Mar.
271. Latinen, G. (1962) Devolatilization of viscous polymer systems, *Adv. Chem. Ser. 41:34*.
272. Lauzon, M. (1997) Extrusion equipment will run the gamut, *Plastics News*, May 12.
273. Lawal, A. *et al.* (1995) Simulation of intensity of segregation distributions

using 3-D FEM analysis: application to corotating twin screw extrusion processing, *Journal Polym. Sci.*, B58(9), Stevens Inst of Tech.
274. Lee, N.C. (1990) *Plastic Blow Molding Handbook*, Chapman & Hall.
275. Lee, S.-T. (1995) *Expandable Composition and Process for Extruding TP Foams*, US patent 5462974.
276. Leingartner, L. (1995) Conical or parallel, *Kunststoffe Plast. Europe*, Sep.
277. Lengerich, K.-P. (1995) Film bubble stretching, *Kunststoffe*, Sep.
278. Levy, S. (1981) *Plastics Extrusion Technology Handbook*, Industrial Press.
279. Limburg, G.B. (1995) Production data acquisition for extrusion process, *Kunststoffe Plast. Europe*, Jan.
280. Lin, S.-C. (1993) *High Performance Thermosets*, Hanser.
281. Lindt, J.T. (1989) *A Comprehensive Model of Devolatilization in Twin-screw Extruders*, SPE-ANTEC, May.
282. Lobo, H. *et al.* (1996) Plastics analysis-improved characterization of polymer behavior and composition, *PE*, Nov.
283. Lubin, G. (1982) *Handbook of Composites*, Chapman & Hall.
284. Lucchesi, P.J. *et al.* (1985) Reducing draw resonanace of LLDPE film resins, *PE*, May.
285. Lu, J. *et al.* (1997) *The Effect of Die Lip Temperature on the Surface Characteristics of Polyethylenes*, SPE-ANTEC, May.
286. Luker, K. (1995) Extruder is designed with built-in surge control, *Modern Plastics*, Nov.
287. Lusignea, R. *et al.* (1997) *Extrusion of Oriented LCP Film*, SPE-ANTEC, May.
288. Mack, A.C. (1997) *Quality Management in an In-line Thermoforming Operation*, SPE-ANTEC, May.
289. Manas-Zloczower, I. (1994) *Mixing and Compounding of Polymers*, Hanser.
290. Mapleston, P. (1996) Winders manage tough materials, ultrafast speeds, *Modern Plastics*, Oct.
291. Marks, L.S. (1951) *Marks' Handbook of Mechanical Engineering*, McGraw-Hill.
292. Mead, D.W. (1995) Evolution of molecular weight distribution and linear viscoelastic rheological properties during the reactive extrusion of PP, *Journal Polym. Sci.*, **57**(2), University of Calif.
293. *Medical Devices; Current Good Manufacturing Practice (CGMP)* (1993) Regulations; proposed revisions; request for comments, 58 FR:61952-61986, Nov. 23.
294. *Medical Devices; Current Good Manufacturing Practice (CGMP)* (1996) Final rule; quality system regulation, Federal register, 61 FR:52602-52662, Oct. 7.
295. *Medical Devices; Working Draft of the Current Good Manufacturing Practice (CGMP)* (1995) Final rule; notice of availability, request for comments; public meeting, 60 FR:37856, July 24.
296. Michaeli, W. (1995) Conveying behavior of twin-screws, *Kunststoffe Plast. Europe*, Feb.
297. Michaeli, W. (1992) *Extrusion Dies for Plastics and Rubbers*, Hanser.
298. Miles, E. (1997) Simplex keeping ahead of the submarine cable revolution, *Kabelnews*, Union Carbide, Spring.
299. Miles, E. (1965) *Polymer Technology*, Chemical Publishing Co.
300. Miller, B. (1996) Rotating die paves way for LCP's benefits in tubing, *Plastics World*, July.

301. Miller, B. (1996) New compounding machines push productivity, flexibility envelope, *Plastics World*, Dec.
302. Mills, N. (1997) *Mechanical Design of Polymer Foam Products*, Chapman & Hall.
303. Mitchell, P. (1996) Plastic part manufacturing, *SME*, Vol. 8.
304. Mitsubishi water cooled sheet line yields high clarity PP (1996) *Modern Plastics*, Sep.
305. Molerus, O. (1997) *Heat Transfer in Fluidized Beds*, Chapman & Hall.
306. Moore, S. (1996) Blown film die eliminates spider lines, cuts back pressure, *Modern Plastics*, Oct.
307. Murao, T. *et al.* (1995) *Method of Blow or Extrusion Molding Polyoxymethylene Resin*, patent 5458839, Polyplastics Co. Ltd./Hoechst, Japan/Germany.
308. Myung-ho, K. *et al.* (1997) *Filling Length and Degree of Fill in Co-rotating Intermeshing Twin-screw Extruders*, SPE-ANTEC, May.
309. Naitove, M.H. (1996) Belt processor makes PP sheet run faster and clearer, *Plastics Today*, Apr.
310. Neaville, L.S. (1996) Chapparal buy provides ITW cast film niche, *Plastics News*, Sep. 30.
311. Newcome, J. (1990) Injection molding polycarbonate optical discs, *SPE-IMD Newsletter*, No. 23.
312. Nichols, W. (1983) *Melting in Counter-rotating Tangential Twin-screw Extruder*, SPE-ANTEC, May.
313. Nissel, F.R. (1990) *Coextrusion for Shelf Stable Food Packaging*, SPE-RETEC, Sep. 18–19.
314. Nissel, F.R. (1991), Coextrusion systems for high barrier containers, *NPE*, June 17–21.
315. Nissel, F.R. (1993) *Reminiscences About Sheet Extrusion – 1993*, Welex.
316. Nissel, F.R. (1996) V-PET = PET sheet by vented extrusion, in *Proceedings of the World Congress PET'96*, Zurich, Switzerland, Oct. 21.
317. Oehler, P.R. *et al.* (1995) Design-based material selection, *PE*, Jan.
318. Oertel, G. (1993) *Polyurethane Handbook*, Hanser.
319. Ogando, J. (1997) Open up window on your process, *Plastics Today*, Mar.
320. *Orientation of Plastics Sheeting Guide* (1995) SPI/SPD 0395-01, June 28.
321. Orth, F. (1989) Selection of material under dynamic loading, *Kunststoffe*, Aug.
322. Osswald, T.A. *et al.* (1995) *Materials Science of Polymers*, Hanser.
323. Pahl. M.H. *et al.* (1995) Multi-screw extruder; continuous high-performance devolatilization, *Kunststoffe Plast. Europe*, Nov.
324. *Permeability and Other Film Properties of Plastics and Elastomers* (1995) Plastics Design Library.
325. Pickel, H. *et al.* (1995) Electric and hydraulic-hybrid, *Kunststoffe*, Sep.
326. Pigorsch, E. (1997) *Advanced Materials*, 7th edn, Chapman & Hall.
327. *Pipe, Plastic and Competition* (1996) Freedonia Group.
328. Pittman, J.F. *et al.* (1995) *Mechanical Effects in Extrusion: Slit Dies*, University of Wales.
329. *Plastics Technology Data Bank*, Bill Publishing., annually updated.
330. *Plastiserve-Plastics Information Systems*, Data Services inc., annually updated.
331. Poslinski, A.J. (1997) Effects of small changes in the melt viscosity on the filling and packing stages of injection molding, *Journal Injection Molding Technology*, **1**(1).

332. Potente, H. *et al.* (1997) *Analysis and Modeling of the Residence Time Distribution in Intermeshing Co-rotating Twin-screw Extruders Based on Finite Element Simulations*, SPE-ANTEC, May.
333. Potente, H. *et al.* (1995) Bulk density of plastic pellets in a screw channel, *Int. Polym. Proc.*, **10**(1), Paderborn University, Germany.
334. Potter, K. (1997) *Resin Transfer Molding*, Chapman & Hall.
335. Powers, J.R. *et al.* (1997) *Extrusion of Syndiotactic Polystyrene*, SPE-ANTEC, May.
336. Powers, J.R. *et al.* (1996) *Coextrusion Layer Thickness Distribution Improvements via Equipment Design*, SPE-ANTEC, May.
337. Predohl, W. *et al.* (1982) Extrusion of flat film, thermoforming film sheets, *Industrial and Production Engineering*, Apr.
338. Priester, D.E. (1997) *Increasing Pelletizing Extruder Output with Fluoropolymer Alloys*, SPE-ANTEC, May.
339. *Product Safety Management Handbook* (1996) SPI.
340. *Prototyping Directory* (1996) 230 services worldwide, CAD/CAM Publishing, San Diego, CA., USA.
341. Pryweller, J. (1997) Echlin buys French tube extruder, molder, *Plastics News*, Feb. 17.
342. *Quality Systems* (1994) Model for quality assurance in design, development, production, installation, and servicing, ISO 9001, ISO Geneva.
343. Ramanathan, R. *et al.* (1996) Wave pattern instability in multilayer coextrusion – an experiment, *SPE Extrusion Newsletter*, Oct.
344. Rao, N.S. *et al.* (1997) *Advanced Film Winding System for Extrusion Lines*, SPE-ANTEC, May.
345. Rao, N.S. *et al.* (1991) *Design Formulas for Plastics Engineers*, Hanser.
346. Rao, N.S. *et al.* (1981) *Design Machines and Dies for Polymer Processing with Computer*, Hanser.
347. Rathji, W. *et al.* (1992) Five major myths about garbage, and why they're wrong, *Smithsonian*, July.
348. Rauwendaal, C. (1997) Screw design for foam processing, *Plastics World*, May 1997.
349. Rauwendaal, C. (1986) *Polymer Extrusion*, Hanser.
350. Rauwendaal, C. (1995) *Statistical Process Control in Extrusion*, Hanser.
351. Rauwendaal, C. (1995) Statistical process control in extrusion, *Plastics World*, Mar.
352. Rauwendaal, C. (1994) *Polymer Extrusion*, 3rd edn, Hanser.
353. Rauwendaal, C. (1991) *Mixing in Polymer Processing*, Marcel Dekker.
354. *Recycling and Recovery of Plastics – Global* (1997) APC.
355. *Recycling Plastic Products* (1994) APC.
356. Rees, C. (1997) *Product Design for Injection Molding*, Hanser.
357. *Reifenhauser News* (1995) Future prospects for extrusion technology, Issue 24, K-95.
358. Renstrom, R. (1997) Act to export use of LCP to catheter tubing, *Plastics News*, Mar. 3.
359. Reynolds, P. (1994) Closed-loop controls drive new film line, *Packaging World*, Sep.
360. Richardson, P.N. (1974) *Introduction to Extrusion*, 7th edn, SPE.

361. Richman, D.A. *et al.* (1997) FDA contends with regulation the internet, *MD & DI*, May.
362. Rincon, A. *et al.* (1997) *Transient Simulation of Coextrusion Flows in Coat-hanger Dies*, SPE-ANTEC, May.
363. Robert, D. (1996) In wire and cable, keep track of details, *Plastics World*, Nov.
364. Rubin, I.I. (1990) *Handbook of Plastic Materials and Technology*, Wiley.
365. Saechtling, H. (1995) *International Plastics Handbook*, Hanser.
366. Sagar, A.D. *et al.* (1995) Starch fragmentation during extrusion processing, *Polym.*, **36**(9).
367. Sangani, H. *et al.* (1997) Cold treatment extrusion of recycled commingled plastics, *PE*, Feb.
368. Satya, S. *et al.* (1995) *Process (Extrusion) for Coloring TPs*, US patent 5462709.
369. Schindler, B.M. (1983) Made in Japan (W. Edwards Deming), *ASTM Standardization News*, Feb.
370. Schenkel, G. (1963) *Plastics Extrusion Technology and Theory*, Hanser.
371. Schlichting, H. (1960) *Boundary Layer Theory*, McGraw-Hill.
372. Schmitt, S.W. (1990) Vulcanization method for Neoprene, *DuPont Report NP-460.1*.
373. Schott, N. (1997) *Guide to Mixing*, University of Massachusetts Lowell Lectures, 1997.
374. Schumacher, F. (1986) Temperature controlled flow regulation in flat sheet dies, *Industrial and Production Engineering*, Oct.
375. Schwartz, S.S. and Goodman, S. (1982) *Plastic Materials and Processes*, Chapman & Hall.
376. Schweizer, P. *et al.* (1997) *Liquid Film Coating*, Chapman & Hall.
377. Scott, G. *et al.* (1996) *Degradable Polymers – Principles and Applications*, Chapman & Hall.
378. Scycher, M. *et al.* (1980) *Synthetic Biomedical Polymers*, Technomic.
379. Secor, R. (1986) A mass transfer model for a twin screw extruder, *Polym. Engng Sci.*, 26:9.
380. Seymour, R.B. (1995) *Engineering Polymer Source Book*, McGraw-Hill.
381. Shaw, F.S. (1953) *Relaxation Methods*, Dover.
382. *Sheet*, 2nd edn (1997) SPI Sheet Producers Division.
383. Sherman, L.M. (1996) Stretch film maker sets speed record, *Plastics Today*, May.
384. Skeist, I. (1990) *Handbook of Adhesives*, 3rd edn, Chapman & Hall.
385. Skeist, I. (1966) *Plastics in Building*, Chapman & Hall.
386. Smith, D.J. (1995) *Achieving Quality Software*, Chapman & Hall.
387. Smith, D.J. (1991) *Quality Procedures for Hardware and Software*, Chapman & Hall.
388. Smith, D.J. (1990) Factors effecting rate of blown film production, in *Proceedings of the Polymers, Laminations and Coatings Conference*, TAPPI.
389. Song, K. *et al.* (1997) *Formation and Characterization of Cast, Tubular and Biaxially Stretched Poly(butylene terephthalate) Film*, SPE-ANTEC, May.
390. Sors, L. *et al.* (1994) *Plastic Mold and Dies*, Chapman & Hall.
391. Spalding, M.A. *et al.* (1997) *An Experimental Study of the Frictional and Viscous Forces Responsible for Solids Conveying and Melting in Single-screw Plasticating Extruders*, SPE-ANTEC, May.

392. Spirex, Colby (1996) *Plasticating Components Technology: Up to Date Version of Screw and Barrel Technology*, Spirex Corp., Youngstown, OH 44512, USA.
393. Stangle, G. (1997) *Modeling of Materials Processing*, Chapman & Hall.
394. Stasiek, J. (1995) State of the art in single-screw extruder plasticating systems, *Polimery*, **40**(4), Metalchem, Poland.
395. Stasiek, J. (1995) Twin screw extrusion of PVC pipes-effect on final compounding and properties of pipes, *Polimery*, **40**(9), Metalchem, Poland.
396. *Statistics Handbook – Global* (1997) British Plastics Federation.
397. Stevens, M.J. *et al.* (1996) *Extruder Principles and Operation*, 2nd edn, Chapman & Hall.
398. Steward, S. (1997) Operational efficiency is no longer just an option, *MD & DI*, May.
399. Steward, S. (1997) 3-D object printing improves designer communication, *MD & DI*, May.
400. Strub, R.A. (1952) Theory of screw extruders, *Proceedings of the 2nd Conference*, Ohio State University.
401. *Structural Plastics Selection Manual* (1984) No. 66, ASCE.
402. Sturrock, B. (1995) *Computerization and Networking of Materials Bases*, 4th edn, STP 1257, ASTM.
403. Szefi, R.J. (1997) Back to basics, *Automotive Body International*, Spring.
404. Tadmor, Z. *et al.* (1970) *Engineering Principles of Plasticating Extrusion*, Wiley.
405. Tadmor, Z. *et al.* (1979) *Principles of Polymer Processing*, Wiley.
406. Tanchoco, J. (1994) *Material Flow Systems in Manufacturing*, Chapman & Hall.
407. Thibault, F. *et al.* (1995) Modeling of single-screw extruders with mixing pins: the case of PVC, *Journal Vinyl Additive Technol.*, **1**(3), URPEI Polytechnic School, Canada.
408. Throne, J.L. (1987) *Thermoforming*, Hanser.
409. Throne, J.L. (1997) *Technology of Thermoforming*, Hanser.
410. Timoshenko, S. *et al.* (1961) *Theory of Elasticity*, McGraw-Hill.
411. Todd, D.B. (1997) *Practical Aspects of Processing in Intermeshing Twin-screw Extruders*, SPE-ANTEC, May.
412. Todd, D.B. (1994) Determining pressure drop in extrusion, *PC*, Sep.
413. Tomka, J.G. (1996) *Chemistry and Technology of Melt Spun Synthetic Fibers*, Chapman & Hall.
414. Tres, P.A. (1994) *Designing Plastic Parts for Assembly*, Hanser.
415. *Triple-pipe Extrusion* (1988) PM&E, May.
416. Tsai, H.Y. *et al.* (1995) The flow induced phase morphology of fluoroelastomer and PC blends during extrusion, *Polym. Engng. Sci.*, **35**(7), University of Akron, OH., USA.
417. Tucker, C.L. (1990) *Fundamentals of Computer Modeling for Polymer Processing*, Hanser.
418. Ulcej, J. (1997) Extrusion dies for flat sheet, film, and coatings, *EDI Bulletin*, EDI, Chillewa Falls, WI., USA.
419. Ulracki, L.A. (1990) *Polymer Rheology and Processing*, Chapman & Hall.
420. Vainio, T.P. *et al.* (1995) Screw optimization of a co-rotating twin-screw extruder binary immiscible blend, *Polym. Engng. Sci.*, **35**(3), Helsinki University of Tech., Finland.

421. Vernyi, B. (1996) Stretch, specialty films require tweaking, *Plastics News*, Oct. 7.
422. Vlcek, J. (1996) *Computer Simulation of Conventional and Barrier Screw Extruders*, SPE-ANTEC, May.
423. *W&P Processing Line* (1996) Brochure Vol. 5, No. 3, Werner & Pfleiderer.
424. Wadsworth, H.M. (1994) Standards for tools and techniques, *48th Annual Quality Congress*, ASQc.
425. Walters, R.B. (1991) *Hydraulic and Electro-hydraulic Control Systems*, Chapman & Hall.
426. Wang, H.P. (1996) Designing profile dies: CAE removes the guess work, *Plastics Today*, Feb.
427. Ward, I.M. (1987) *Developments in Oriented Polymers*, Chapman & Hall.
428. Warmedinger, M. *et al.* (1997) Quality management of geometric data in car body and interior trim development, *Auto. Body Int. Quarterly*, **2**(1), Winter.
429. Waterman, N.A. *et al.* (1997) *The Materials Selector*, 2nd edn, Chapman & Hall.
430. Waterman, N.A. *et al.* (1991) *The Materials Selector*, Chapman & Hall.
431. Weismantel, R. (1994) IM part geometry, quality factors and cycle times effect equipment choices, *Modern Plastics*, Nov.
432. Weizer, W. (1996) *Polypropylene in North America*, Freeddonia Group, Cleveland, OH., USA.
433. Wheeler, A. (1993) Good converting needs boost from quality film, Paper, *Film & Foil Converter*, May.
434. Whelan, T. *et al.* (1994) *The Dynisco Extrusion Processors Handbook*, Dynisco Co., MA., USA.
435. White, J.L. (1995) *Rubber Processing*, Hanser.
436. Wick, C. *et al.* (1984) *Forming*, Vol. 2, SME.
437. Wick, C. *et al.* (1987) *Quality Control and Assembly*, Vol. 4, SME.
438. Wigotsky, V. (1996) Extrusion, *PE*, Apr.
439. Wilson, A. *et al.* (1995) *Waterborne Coatings*, Chapman & Hall.
440. Wilson, G.M. (1997) *Interfacial Instability in Multilayer Extrusion Processes*, EDI, Chippewa Falls, WI., USA.
441. Wittfoht, A.M. (1981) *Plastics Technical Dictionary*, Part 1: Alphabetical Dictionary English–German, Hanser.
442. Wood, R. (1984) Coextrusion advances in cast film, sheet, *PM & E*, July.
443. Woodcock, C.R. *et al.* (1997) *Bulk Solids Handling*, Chapman & Hall.
444. Woodward, A.E. (1988) *Atlas of Polymer morphology*, Hanser.
445. Woon, H.S. *et al.* (1995) *Pulp-like Short Fibers from LCP*, US patent 5454910, Korean Institute of Science.
446. Wright, D. (1996) *Environmental Stress Cracking of Plastics*, SPE.
447. Wright, R.E. (1995) *Injection/transfer of Thermoset Plastics*, Hanser.
448. Xanthos, M. (1992) *Reactive Extrusion Principles and Practices*, Hanser.
449. Xie, Y. *et al.* (1997) *Experimental and Statistical Analysis of Residence Time Distribution in Corotating Twin-screw Extruder*, SPE-ANTEC, May.
450. Xing, K.C. *et al.* (1997) *Extrusion of Non-olefinic Polymers with a Fluoroelastomer Processing Aid*, SPE-ANTEC, May.
451. Yanovsky, Y.G. (1993) *Polymer Rheology*, Chapman & Hall.
452. Yoon, L.K. *et al.* (1995) Reactive extrusion of PP/natural rubber blends, *Journal Appl. Sci.*, **56**(2).

453. Yorkgitis, E. (1997) Adhesive bonding technology for composites and dissimilar materials, *Automotive Body Int. Quarterly*, Spring.
454. Zambetti, P.F. *et al.* (1995) Alpha tocopherol as an antioxidant for extrusion coating polymers, Hoffmann-LaRoche-Dow Chem., TAPPI, **78**(4).
455. Zambetti, P.F. (1995) Vitamin E: effective antioxidant for extrusion coating polymers, *Pack. Technol. Engng.*, **4**(7).
456. Zloczower, I. M.-Z. *et al.* (1997) *Influence of Design on Mixing Efficiency in the Kneading Disc Region of Co-rotating Twin-screw Extruders*, SPE-ANTEC, May.
457. Zozicki, W. *et al.* (1967) Extrusion die design, *Chem. Engineer*, **45**, 127.

# Index

Abbreviation 164–5, 735
Acceptable risk, *see* risk
Acrylic barrier 39
Adapter system 17, 82
Additive 168
Adiabatic extruder 55–6
Agitator, *see* Mixer, extensive
Agricultural film on site 729
Air
  blower 81
  blow molding, *see* Blow molding
  blown film, *see* Blown film
  contamination 15
  cooled 26, 82
  emission 42
  entrapment 97, 139
  temperature 196
Alignment 67
Aluminum
  energy 41–2
  property 169, 604–6
American National Safety Institute 39
American National Standards Institute 158
Amorphous plastic 165–7, 185
Amperage required, *see* Drive motor
Angle conversion 736
Applesauce 414
Astro turf 52–3
Automatic screen changer 86, 88
  *see also* Screen changer
Autothermal extruder 56
Auxiliary equipment
  basic 24, 98–116
  constrained 27–8
  in-line interface 99
  planning 118

Back pressure
  flow 9, 59
  flush 88
  screen back 83
  valve 83
Barrel
  alignment 67–70
    incorrect 69
  assembly 68
  basic 93–8
  bimetallic coating 76
  bore 68, 77
  cleaning 17
  concentricity 78
  construction 75–6, 79–80
  cooling 76
  dimension 77–80
  downsizing 79
  dry 16
  eccentric 7
  energy loss 40
  flange attachment 79
  groove 55, 139
  heat 22, 67, 76
  liner 76
  output rate 93
  parallelism 79
  pressure 11, 75
  rib 81–2
  size 201
  screw, *see* Screw
  sensor 96
  sleeve 80
  slot 55
  support 67
  temperature 12, 22–4, 205
  thermal barrier 96
  throat geometry 93
  vented, *see* Vented barrel
  water cooled feed throat 96, 127
  wobble 69, 138
  *see also* Borescoping; Feed
Barrel-screw, *see* Screw-barrel
Barrier
  property 28

Barrier (*Contd*)
  screw, *see* Screw
Battenfeld Gloucester
  blown coextruded film line
    adjustable height oscillating haul-off 4
    close-up Traversanip oscillating haul-off 342
    in-line grocery bag making system 310
    with flat slat collapsing frame 307
    with lay-flat tube winder 309
  cast coextruded film line 359–60
  extruder screen assembly with adapter 232
  motorized adjustable sizing basket 308
  radial fed die 337
  triple air-chamber ring with internal bubble cooler 320
  turret winder 354
  window profile line 529
Bearing, *see* Thrust bearing
Belt-driven extruder 71–2
Bernhardt, Ernest 52
Beverage container 722
Biklin, D. 129
Billings, M. 213
Bleedover 137
Blending 4
  clean 20
  error 45
  gravimetric 114
  manufacture 156
  volumetric 114
Blocking 35
Blow/injection molding with rotation 643–4
Blow molding
  accumulator
    cycle 563
    melt flow 577
    operation 561–2
  auxiliary equipment 553–4
  basic 552–4
  blow pin
    atomized gas nozzle 565
    entrance diameter 564
    hypodermic needle 564
    location 558, 561, 563–4
    orifice 564
  blow time 564–5
  characteristic
    extrusion 556–61
  clamping 566
  coextrusion 556–7
  continuous 556–61, 579
  control
    basic 574–7
    clamp 567
    die 578
    microprocessor 577
    parison thickness 575–6
    *see also* Control
  cooling
    chill water 568
    liquefied gas 565
    postcool 566
    type 565
  cost
    bottle 589
    parison control 575–6
    plastic 567
  cutter 558
  Dawn soap 575
  die
    basic 577–9
    nomenclature 576
    shaping 576
    spider 579
    type 559, 578
    *see also* Die
  extrusion
    basic 556–8
    die 578–9
    mold air venting 578
    orifice modulation 575–6
    parison
      basic 558
      cycle 560
      expanding 563
      mold 577, 580
      pinch 560
      pressure 563
      sag 577
      schematic 558, 561
      swell 556
      thickness control 575–6
    pinchoff 581–3
    shape, complex 580
  extrusion vs. injection 553, 570
  flash 557, 570
  freeze-off 563
  gas
    air 558, 563
    carbon dioxide 565
    discharge 565
    nitrogen 565
    pressure 554, 564
    temperature 565
    type 554
    volume 563
  injection
    basic 567–70
    blow mold with rotation 643–4
    cycle 570
    method 570

mold
air venting 578
type 577, 580
preform 567, 571
schematic 568, 569
three stage 569
injection molding 562
intermittent method 561
machine setting 575
manufacturing 156
melt
flow head 562, 579
hang-up 577
heat history 577
pressure 577
property 574
residence time 577
stress 563
mold
air venting 578
construction 580–1
extrusion BM 558, 577, 580
injection BM 568
shape, complex 580
stretch BM 571
temperature 583
operational mode 559–61
optimization 574–7
pinchoff schematic 582
plastic 554–6, 571, 578, 583
platen daylight 566
processing stage/station 568
product
air duct 580
aqua cycle wheel 586
auto panel 585
floating pontoons 585
geometry 567
hot water heater tank 584
lid 580, 586
panel, complex 587
performance 588
shape, complex 580
truck fascia 559
production 561
property vs. processing 122
quality 568
ram type 562
reciprocating screw 562
reheat 571
rotary 560
shape 588
shot size 562
scrap 555, 570
shrinkage
amorphous plastic 567
behavior 567
crystalline plastic 567
shuttle 561, 566
speed 566
stress 563
stretch
basic 570–4
cooling 574
draw ratio 572
extrusion BM 570
gas barrier 572
heat profile 573–4
injection BM 571
in-line 573
orientation 571–4
*see also* Orientation
plastic 572
shrinkage 572
strength 571–2
temperature 574
two-stage 573
temperature 554
thermal sensitivity 555
thermodynamic study 565
tolerance 567
troubleshooting 590–2
type 552
variable 575
Blown film
basic 305, 321
blow-up ratio 106, 312, 319
bubble
cooling 311, 318–20
shape 314
stability 317–18, 330
width 317
characteristic 315–20
coextruded 313, 335
*see also* Coextrusion
collapsible bubble 306
collapsing frame 343
consumption 305
control
basic 315–20
*see also* Control
computer 323
ultrasonic bubble thickness 316
cooling
air ring basic 309
dual-chamber air ring 311
internal bubble cooling 309, 311
triple-chamber air ring 320
uneven 321
cost 312–13, 323
design 233–4
die
basic 334–5
coextrusion 335–6

Blown film (*Contd*)
gap 317
radial fed 337
spiral 335
stacked 336
*see also* Die
feeding 325
*see also* Gravimetric weight control
frost/freeze line 317–18
grocery bag 310
haze 333
inflatable mandrel 315
lay-flat tube 309
line schematic 321
machine setting 326
melt 328, 335–6
*see also* Melt
optimization 324–6
orientation 106, 323–4
*see also* Orientation
output 308, 316, 326, 332
personnel 331
pinch roll 319
plant temperature 328
plastic 312–15, 323
pressure 11
product 333
randomization
advantage 339
basic 338–43
gauge band problem 339–40
location 340–1
melt distribution 339
oscillating haul-off 341–3
rotating extruder 340
rotating/oscillating die 338–9
rotating tower-winder 340
thickness 338
recycling 188
roll adjustment 322–3
*see also* Roll
roll transfer 322
safety 321
schematic 306
scrap rate 313
screw 214, 328
*see also* Screw
shut-down 325
sizing basket 308
start-up 320–3
tear strength 318, 320
thermal history 313
*see also* Thermal
thickness 313, 318, 325, 332
tolerance 313
trapped air 309
troubleshooting 344–8
variable 318, 325, 334
vs. flat film 357
winder 307, 322
*see also* Windup
wine glass shape 314
wrinkle 331
Blow up ratio 106
*see also* Orientation; Blown film
Borescoping 69
Bottle 39
energy conservation 41
Boussinesq, M. J. 270
Bramah, J. 50
Brass scraper 17
Breaker plate 84, 205
cleaning 18
screen pack, *see* Screen pack
spare 18
Bridging, *see* Melt; Screw
Building 669–70, 727–8
Burrs, *see* Safety
Butler, T. I. 148

Cable covering, *see* Wire covering
Cable jacketing precision 91
Calendering
bank mark 440
basic 418–20
blending 430
*see also* Blending
coating 424
compounding 430–3
contamination 434
control 437
*see also* Control
cost 422, 438, 440
credit card 428–9
energy 434
feeding 432–3
film 418
fluxing 432
gel 431
heat, critical 431, 433–4
nonwoven sheet 602
operation 422–7
optimizing 435
orientation 435
*see also* Orientation
plastic
basic 422, 430–2
polyvinyl chloride 422–5, 430
plate-out 431
preparation 419
product 437–8
recycling 434–4
roll
basic 435–7
configuration 424–5
corotating 418

covering 429–30
diameter 427
embossing 428
pressure 425–6
quick change 428
safety
basic 438
interlocking bar 439
schematic 419–20, 423–4
sheet 418
speed 425, 440
substrate covering 432
surface finish 427–8
temperature 426
thickness 427
troubleshooting 440
vs. extrusion 420–2
Cancer and Coca Cola 39
Capital investment 43, 706–11
Captive processor 43
Carothers, W. H. 595
Carousal molding 650
Carpet fiber stretching 105
*see also* Orientation
Cast film
die
design 235
pressure 11
*see* Die
speed 28
*see also* Flat film, control
Caterpillar drive 27
Caulking extruder 730
Chauvin, T. J. 45
Check list, safety 33
Chemical
reaction 64
reclamation 42
Chill roll film, *see* Flat film
Chisolm's Law 698
Chung, Chan I. 225
Cincinnati Milacron 6
Cleaning
basic 17
safe 33
shut-down 124
*see also* Purging
Cloren 260
Coating
adhesion 454, 460–2
air jet 455
basic 441–44
beading 454
blocking 463
control 463–4
*see also* Control
die, *see* Die
draw resonance 468
characteristic 449–56
control 448, 455
*see also* Control
cooling water nomogram 452
curtain flow 450
die 464–5
*see also* Die
dimensional stability 464
drying plastic 193
flame treatment 462
heat sealing 464
laminator 443
machine setting 458–9
melt index 458
melt sticking 451
method 445–6
neck-in 454, 457–8
nonporous 461
optimization 457
output 451
plastic 448
post polymerization 448
preheat 461
printing 462
processing 447
product 465
recycle 448
roll
chill 451
pressure 450
safety 465
schematic 442–3, 449–50
shut-down 456
slip surface 463
special 13–14
speed 28, 447
start-up 449
substrate 444, 447–8, 459–61
surface coverage 453–4
temperature 447, 456
thickness 451
trim & slit 453
troubleshooting 466–8
unwind 455
variable 441
weight 453
winder, turret 442–4
Coca Cola and cancer 39
Coefficient of friction 56, 60, 213
Coefficient of heat transfer 58
Coextrusion
adapter 82
advantages 152
basic 144–52
blown film 4
die
design 146, 151
swell 147

Coextrusion (*Contd*)
*see also* Die
feedblock dependent 148, 151
finite element analysis 150
instability 142
layer number 145
melt flow
characteristics 146
instability 146
non steady-state instability 147
output rate 152
plastic
behavior 149
compatibility 145
liquid crystal polymer film and sheet 151
type 145
viscoelasticity 148
processing window 149
safety 36
screw
channel geometry 150
drive motor 149
*see also* Drive motor
sheet line 144
steady-state instability 146
tie-layer 145
torque rating 149
Coiling device 25
Coining 642–3
Colby, P. T. 142
Colombo, Roberto 52
Color
fluctuation 137
masterbatch 328
Commodity plastics 168
Communication protocol 100
Complex 405–6
Compounding
additive
performance 611–12, 629
predispersed 611
type 609–14
adjustable 610
barrel, split 619
basic 608–9
batch 616
blending
basic 609, 615–16, 618–20
vs. coloring 621
calendering 432
characteristic 616
color 611, 621
concentrate 611
consumption 5, 608
contamination 628
continuous 617–18
control
basic 629
*see also* Control
microprocessor 619
quality 613
converters 608
cost 611
dicer
basic 627–8
dry-face 623–5
shape 627
underwater 623–5
die 629
*see also* Die
dimensional stability 612
drawdown ratio 624
equipment 608–9
feeding 618–20, 628
good manufacturing practice 629, 689
heat
resistance 612
sensitive 613
improper 630
manufacture 156, 610–11, 613, 623, 626
modular design 610
mixing
basic 609
concentric-screw 610
dispersive and distributive 615
*see also* Mixing
gravimetric 619
high-intensity 616
nomograph 614
single-screw 615, 617
*see also* Extruder
starve-fed 617, 629
twin-screw 610–11, 617–20
*see also* Twin-screw
volumetric 619
operating
mode 610–11
sequence 613
optimization 628–9
output 617, 628
pellet
basic 622–3
cost 721
die 623
dimensions and shapes 627
fines and longs 627
form 622
manufacturing 613
output 623
sticking 625
strand 623
twin-screw 626
plastic
basic 609–15
property 616

vinyl 611, 613
reactive extrusion 620, 622
reinforced 612
rheology 616
*see also* Rheology
shoe sole 613
speed 617
supplier 608
thermal conductivity 612
troubleshooting 629–30
variable 609
weight
additive 613
nomograph on mix 614
Compression/injection molding 642–3
Compression ratio, *see* Screw
Computer
blown film 323
calculation 44
capability 46, 693
die design 270–3
human mind 45–6
lock-out 36
modeling 215
panacea 45, 693
processor 285
safety 159
*see also* Process Control; Programmable controller; Finite element analysis; Software
Conair/Gatto auxiliary equipment 5, 499, 504–5, 513–14
Condensation, surface 20
Construction
machine 6
Consumption
film 14
plastic 5
sheet 14
Contamination
feed 18, 20, 49, 434
gel 141–2
purging 121–2
Continuous extruder 6
Control
accuracy 293, 296, 301
advantage 283
blending 300
blow molding 300
blown film 303
coating 302
coextrusion 285
color 293
compounding 300
conventional 289
equipment standard 296
extruder 297–9
feed 55
fiber 299
film 285, 301
flat film 298
flutter sensitivity 293
foam 299
fuzzy logic 292
gloss 293
gravimetric feed 303
haze 293
injection molding 300
intelligent processing 304
laminating 298
lock-out 36–7
melt pressure 290
fluctuation 292
transducer 293–4
microprocessor 151, 284–5
overheat 34
panacea 283
paper coating 298
performance 293
PID 292
pipe 303
problem 44–5
process
integration 301
requirement 287, 297–300
profile 297
programmable safety 159
quantification 296
resistance temperature detector 292
response loop 292
rod 297
sensor 294, 290–6
signal interpretation 292
sheet 298, 302
sophisticated 26
tape 298
tension 285
thermocouple 192
toy 283
tradeoff 287
transducer 290, 293–4
tube 297
type 291, 294
variable 284–5, 288–9, 296, 301
web gauging 295
wire coating 297
*see also* Computer; Process control
Convection heat loss 40
Conveyor 28
Cooling
air *see* Air
fluid 82
frozen strain 180
minimize 81
nonhomogeneous 180
on-off 129

Cooling (*Contd*)
  profile 22
Copper
  energy 41
  scrapper 17
Corona treatment, *see* Roll
Corrugated pipe 664–5
Cost
  auxiliary equipment 98
  basic 700–5
  blow molding 567, 575, 589
  blown film 312–3, 323
  calendering 422, 438, 440
  controller 693
  direct and indirect 706–7
  disadvantage 60–1
  energy 25, 64, 720–5
    *see also* Cost
  estimating 705
  extruder output 710–11
  garbage bags 150
  labor 8
  liquid crystal polymer 151
  lowest 50
  machine alignment 69
  maintenance 60–1
  manufacturing 117
  material in part 113
  orientation performance gains, *see* Orientation
  output estimating
  pipe 120, 495
  plastic-commodity and engineering 168
  process 710
  process efficiency 19
  product 44
  production 31
  profile 525
  reduction 26, 44–5
  screw output 215
  siding, house 119
  technical modeling 707–9
  type extruder 55
  utility 29
Counter-rotating screw, *see* Twin-screw
Cradle-to-grave analysis 42
Credit card 428
Crosslinking 162
Crouch, T. 74
Cryogenic cooling 26
Crystalline plastic
  initial heat 173
  property 165–7
  temperature vs. volume 185
Custom
  extruder 30
  processor 43
Cutter
  basic 25,
  capability 286
  in-line 27
  slitting film 352
Cylinder, *see* Barrel

Data, plastic property 604–6
Davis, W. M. 76
Davis-Standard 263, 302, 397, 480–1, 491
Death, premature 39
DeBoo, R. 516
Deckle 14
Decomposition/degradation 33
  coextrusion 150
  overheat 34
  rate 186
  temperature 173
Defect 178
Definitions plastic, polymer, resin 160
Density
  bulk factor 20, 63
Design
  basic 43–4
  equipment 157–8
  extruder 51, 61
  flow diagram 701
  fundamentals 674
  quality, *see* Testing; Quality control
  responsibility 43
  screw mechanism 61
  successful 46, 700
  thin part 31
  training program 142
  *see also* Finite element analysis
Dewatering, *see* Water
Dew point 195–6
Diced plastic, *see* Compounding
Die 228–39
  adapter 232
  adjustment 236, 248, 253, 260
  assembly 232
  Autoflex 248
  barrel coupler 229
  blocked 48
  blow molding
  blown film 233–4, 247
    coextrusion 260
    troubleshooting 279–80
  cast film 235
    troubleshooting 280
  chrome plate 19
  cleaning 18, 276
  coathanger 237, 250–1
  coating 238, 253
  coextrusion 256, 258–66
    iridescent effect 268–9
  construction 273–5
    coating 274

material 273
control 237, 252–3
cooling 242
crosshead 235, 238
dam 252
deckle rod 251–2
design
computer 270–3
geometry 228
uniform thickness 239
drawdown 241
efficiency 236
energy loss 40
extrudate 243
feedblock 252–3, 258
comparison 261
film 248, 259, 269
finite element analysis 255
gap 233, 235
gauge control 236
geometry 271
heat 236
heater 247
heat pipe 249
honing 19
land length
basic 241–4
balance 254
rule 245
layer control plate 262
leak 277
line 131
lip
clean 16
melt build-up 282
type 248
maintenance 275–6
manifold 250
matching 232
melt
balance 254
tuning 252–3
behavior 241
build-up 282
diagonal channel 242
equation 245–5
flow
basic 238
coefficient 272
model 272
orientation 241
pressure 238
residence time 264
*see also* Residence time
shear rate 242
temperature 12, 23, 246–9
transfer 264, 266
*see also* Melt
multiple 240
netting 266–8
operating 241, 271
orifice shape 242–6
pipe 235
troubleshooting 280–1
polishing 18
pressure
drop 246
melt 11, 228
screen pack 88
processing plastic 233
profile
basic 239, 244, 254
troubleshooting 280–1
pumping 16
re-assemble 19
removal 17
restrictor bar
basic 242
leak 278
safety 266
sensor 237
shape 243–5, 254, 273
sheet
basic 229, 232, 235–8
coextrusion 259–64
design 245–6
simplification 236
special 266–9
spiral 234
streamline 255–7
swell 147, 242
tear drop 246
term 228
thickness control 239
troubleshooting 276–82
tube 235
tuning device 252–3
type 229–31
vacuum box chamber 262, 265
variable
basic 239–40
geometry 271
melt influence 243–4
Diffusion process 64
Dimensional stability 25, 184
Direct-coupled extruder 71
Discharge, free 9
Double H Plastics Co. 3
Dow Chemical 217, 262, 268, 384–6, 661
Dow Corning 483
Down-stream, *see* Auxiliary equipment
Down-time 47
Draw down ratio, 106, 489–90
*see also* Orientation
Drive motor
amperage 123, 138

Drive motor (*Contd*)
energy 22
factor 72
horsepower 73
loss 40
overload 138
power
calculation 73
loss 24
speed 74, 117
torque, constant 73
*see also* Screw
type 70–5
variation 140
*see also* Screw drive
Drum
cooling 102
dry-bulb temperature 192
rotating 7
Drying 190–8
behavior 191
coating 193
equipment 114–15
hygroscopic plastic, *see* Hygroscopic plastic
moisture affects property 190
troubleshooting 194–7
DuPont 270

Economic, *see* Cost
Elastic interfacial melt instability 147
Elastomer
behavior 162
Electrical
power 17
safety 36–7
Eliminate myth 45
Empirical data 10
Energy 25
conservation 40–1, 721–5
consumption 42
content 42
cost saving 720–5
*see also* Cost
efficiency 26
flow to products 723
glass blown bottle 723
gear pump 92
heat capacity 182, 184–5
loss 40
manufacturing 117
plastics and 721
requirement 41
screw, *see* Screw
Engineer 43–4
Engineering plastics 168
English conversion 732–4
Equipment
quotation 158
rebuilding 157–8
safety, *see* Safety
selection 152–7
supplier 32
Estimating product cost 705
Ethylene 41
Expandable polystyrene
washed 64
Expander roll, *see* Roll
Experience
basic 10, 60
*see also* People
judgment 696
planning 118–21
processing 215
*see also* Training
Extrapolation process 64
Extrudate 12, 221
collection 121–2
heat removal 22
shape 243
*see also* Die
Extruder
adiabatic 55–6
advantage 699
alignment 67
components 65–6
compounding 608–30
foam 667–70
in-line 66, 118
machine similarity 699–70
line 54, 66
machine 5
operation
basic 1, 116–18
cost 710
output rate 29–30, 710
performance 96
planning 118–21
plastic growth 725–9
ram, *see* Ram extruder
reactive 620, 622
rotating extrusion-injection die/mold 647–9
rupture disc, *see* Safety
schematic 57, 71
line 4, 5
single-screw 56
size 24
twin-screw, *see* Twin-screw
type 6, 71
valve 83
vs. calender 420–2
vs. twin-screw 55, 60
Extrusion and plastics growth 725–9
Extrusion area diagram 19
Extrusion Dies Inc. 229, 248, 250, 252, 272–3
Extrusion volume diagram 19–20

Fabrication, *see* Processing
Facts and myths 44
FALLO 2, 50
FDA 28, 39
Feed
  air entrapment, *see* Air
  basic 93–8
  control 95–7
  crammer 21
  energy loss 40
  grooved feed screw 94
  inconsistent 45
  magnetic screen 98
  opening positions 95
  precise 60
  problem 94
  safety 35
  screw zone 58
  vacuum 97
  vibrating pad 97
  *see also* Barrel
Feedstock
  coefficient of friction 56
  energy 41
Festooning 368
Fiber
  basic 593–5
  bath
    cooling 603
    ripple 603
    tank 598
  breakage 600–1, 607
  calender bonded 602
  capillary rheometer 604
  characteristic 596
  control, *see* Control
  cutting/slitting 602
  definition 593
  denier 594
  diameter 593, 597
  die pressure 11
    *see also* Die
  draw ratio 604
  fibrillation 599
  filament 594
  film-to-fiber 599
  finishes 596
  gas heat 598, 601
  gear pump 597–8
    *see also* Gear pump
  length to weight 594
  manufacture 155
  melt
    blown 599
    filtration 597, 600–1
    pressure 597–8
    spinning 597
  monofilament 599–600, 603, 606
  nonwoven
    basic 601–2
    spinbonded 601–2
  optic sheathing precision 91
  optimization 603–4
  orientation 596–9, 603–4
    *see also* Orientation
  plastic
    acrylic 596
    arimid 595
    basic 595–6
    carbon 604
    nylon 595, 606
    polyester 596
    polyolefin 595, 599
    polypropylene 599
    spandex 596
    vinyl 596
    volumetric discharge 602–3
  product 599, 602–3, 606
  production rate 595–6
  property 604–6
  pulp-like olefin 599
  recycle 606
  roll, haul-off 598
  sewing machine 730
  shape 603
  slit-film 599
  solid-state extruded 604
  solvent 598
  speed 593, 597–8, 600
  spinneret 597, 602–3
  spinning
    basic 593, 599
    dry 598
    gel 599
    line 596
    melt 599
    slitting 599
    spurting 599
    wet 598
  spurted 599
  staple 594–5, 606
  strength
    maximum 603
    theoretical vs. actual 604–6
  stretching, *see* Orientation
  S-twist 600
  troubleshooting 606–7
  twist 599–600
  weight 594
  Z-twist 600
Filament, *see* Fiber
Film
  control, *see* Control
  breathable 699
  die, *see* Die
  original 52
  shrink 111–12
  stretching, *see* Roll; Orientation

Film (*Contd*)
  *see also* Blown film; Flat film; Tape
Film and sheet thickness 376
Filter plastic
  gear pump 601
    *see also* Gear pump
  unloading 29
Finch, C. 152
Finite element analysis
  basic 44
  coextrusion 150
  die 255
Fire 39, 729
Fish eye 62
Flame treatment, *see* Roll
Flaming torch 18
Flat film
  air knife 371
  annealing 351
  basic 349–51
  bead 349, 367
  characteristic 361
  chill roll 350
  coextruded 359–60, 371
  control
    basic 367–9
    line 369
    tension web 354
    *see also* Control
  die
    adjustment 351
    basic 370–2
    coathanger 349
    T-shape slot 349
    *see also* Die
  drawdown 350
  draw ratio 369
  draw resonance eliminator 370
  feed, starve 362
  festoon 368
  line schematic 350, 365
  machine setting 361
  manufacture 154, 350, 355–6
  melt
    beading 349
    flow 370–1
    neck-in 349–5
    shrink 349, 353
    viscosity 349
  orientation 363–7
    *see also* Orientation
  output rate 369
  plastic
    basic 358–9
    polytetrafluoroethylene 357
  post-crystallization 353
  problem 368
  product 356, 364
  puller 351
  roll
    corona treating 369
    engraving/embossing 353
    grooved 351
    *see also* Roll
  safety 369
  screw, *see* Screw
  shut-down 362
  skiving 357
  slitting 352
  start-up 362
  stretch 350–6
  tape 356, 364–6
  thermal transfer 371
  thin 355–6
  troubleshooting 372–5
  type 356
  variable 361, 371
  vs. blown film 357
  water quench 355
  winder 354
Flow, plastic, *see* Melt
Fluoropolymer 177
Flushing, *see* Purging
Fluxing 432
Foaming
  basic 663, 667–70
  bubble 49
  manufacture 155
Food additive 39
Foster-Miller 151, 645
Foundation 67
Fow, L. 539
Franklin Assoc. 42
Freeze-off 34
Frictional buildup 8, 10, 22
Fuel energy 41
Furnace 18

Gains and losses 731
Gasoline fuel tank 657
  *see also* Blow molding
Gauge uniformity 91–2
Gearbox
  alignment 67–8
  basic 8
  energy loss 40
  limitation 74–5
  melt pressure 74
  reducer 70
  twin-screw 74
  type gear 74
Gear pump
  basic 17, 89–92
  energy efficient 92
  fiber 597, 601
  location 89–90

patent 51
pipe line 120
positive displacement pump 90
pressure 89
schematic 90
twin-screw type 62
Volumetric efficiency 90–1
Gel
particle 62
problem 86, 141, 415, 431, 466, 590
spinning 599
Glass
blown bottle energy 723
energy 42
property 169, 604–6
recycling 42
Glass transition temperature 181, 185
Good manufacturing practice 689–90
Goodrich, B. F. 245
Granulating
beside press 5
plastic behavior 115–16
*see also* Recycle
Grass field 52–3
Gravimetric weight control
accuracy 114
feed 96–7
Greek alphabet 736
Gregory, Robert B. 601
Griff, A. L. 118
Grocery bag, *see* Blown film
Grooved feed screw, *see* Screw
Growth, plastic with extrusion 725
Guide
barrel construction 75–6
blown film yield 316
compounding 609
coating coverage 453
die land cleaning
drier 115
extruder
operation 116–18
settings 170–2
flat film performance 372, 375
maintenance checkup 129
manufacturing 32
products 152–7
operator safety 35
output estimating 143
pellet cost 721
plastic
changes during processing 187
drying 194–7
type and abbreviation 164–7
pressure effect on output 126
processing product 23
process selection 712–19
product size vs. process 712
profile plate die 546
purging agent 186
raw material to product 723
safety check list 33
screen
classification 87
screen to particle size 87
sensor 291
shut-down 124–5
start-up 121–4
surging problem 140–2
temperature change effect on output 125–6
training 142–4
troubleshooting 47–9, 130
*see also* Troubleshooting
tubing 509
web gauging 295
Gutta percha 50

Harrington, T. P. 75
Hazardous condition, *see* Safety
Health problem 39
Heat
capacity 182, 184
conductor, poor 82
energy loss 40
fluid 80
generation 60
history 91, 174
interrupted 48
method 22
overheat 84
pipe 249
profile 22, 126, 173–4
steam 81
*see also* Barrel heat
Heat transfer
basics 58
device 248
Herringbone gear 74
History 50
Hopkins, Dave R. 142
Hopper
clean 20
cool 24
vibrating pad 97
Horsepower 23
Housekeeping, *see* Cleaning; Safety
Hygroscopic plastic
drying 114, 190–8
residence time 114
*see also* Residence time
*see also* Vented barrel

Ice skating rink 53
Impurity, *see* Contamination
Incineration 42

Injection molding
  advantage 633
  basic 631–8, 637
  carousal molding 650
  clamp pressure 638
  component 632–3
  continuous extruder 640
  continuous molding 645–8
  control
    machine 635
    mold 636
    *see also* Control
  foam 667
  injection/blow molding with orientation 643
    *see also* Orientation
  injection-compression molding 642–3
  machine capability 634
  manufacturing 156–7
  melt
    pressure 637
  method
    one-stage 631–2
      noncontinuous extruder 638
    two-stage 631, 633
      continuous extruder 640
  mold
    basic 641
    cold runner 638
    hot runner 637
    wheel-shaped 646
  noncontinuous extruder 638
  performance maximized 633
  plastic 638, 645
  plasticizing 631
  processing cycle 631–2, 637
  product
    performance 634
    Velcro strip 645–8
  reciprocating screw action 640
  rotating
    mold 650
    wheel 648
  shot size 634
  start-up settings 170–2
  tolerance 646
  troubleshooting 639
  variable 634
Injection/blow molding with rotation 643–4
Injection-compression molding 642–3
In-line equipment, *see* Extruder; Auxiliary equipment
Inspection 226–7
Instrument
  energy loss 40
  *see* Sensor; Control
Intelligent processing 304, 696
Intensive mixing, *see* Mixing, intensive
Interchangeable screw section 61–2
Intermeshing screws, *see* Twin-screw
Investment, return on 31
ISO-9000, 290
Isothermal system 7

Judgment 696
  *see also* People

Kneading
  history 54–5
  pump 62
  screw 207
    *see also* Screw
  *see also* Mixer, extensive
Kruder, George 219

Lahti, G. P. 270
Laminar melt flow 66
Laminated sheet, *see* Sheet
Laminate manufacture 154
Lamination, *see* Coating
Latex, coagulated 64
Laws and regulations 39
L/D, *see* Screw
Leakage 9
Length-to-diameter, *see* Screw, L/D
Liquid crystal polymer film and sheet 151, 165
Liquid injection 138
Literature 54
Lock-out procedure 36–7
Losses and gains 730
Lubricant additive 96
Lumber 729

Machine
  alignment 67
  conditions 121
  differences 54
  foundation 67
  installation 67
  support 67
  variable 55
  ventilation 122
  *see also* Extruder; Operation
Maddock, Bruce 52
Madison Square Garden 53
Magnetic
  screen 98
  tape precision 91
Maintenance
  alignment 67
  basic 129–30
  cost 60–1
  die 275–6
  noise 129

preventative 130
see also Troubleshooting
Management 719
Manual 12, 122
Manufacturing, *see* Processing
Market share 31
Marshall and Williams 108
Material comparison 169
*see also* Plastic
Material handling
basic 28, 113–16
quantity 704
unloading rate 113
*see also* Rail car; Truck
Mathematical
analysis 10
human mind 45–6
symbols 735
Mechanical property, plastic 169, 182, 604–6
Medical tubing precision 91
Melt
behavior 123
*see also* Die
block 132
bridging 24, 35
*see also* Screw
counter pressure 59
decomposed 15
deformation effect 174
die, *see* Die
elastic interfacial instability 147
elasticity and strain behavior 175
exiting extruder 19
flow 12
direction 61
equation 245
flake 94
instability 146
property 176
rate 126
regrind 94
and rheology 174
fracture 147, 176–9
free discharge 59
instability 139
model, *see* Screw
orientation, *see* Orientation
problems, *see* Troubleshooting
processing temperatures 161
plug 15
premature 24
pressure 11
maximum 16
variation 59
pump, *see* Gear pump
quality 11
simulator, *see* Screw
temperature
change 125
false reading 136
processing 161
thermocouple 136
plastic 23
profile 12
residual stress 180
screw action 205
*see also* Screw
temperature 181
throttling 59
variable 45
zig-zag instability 147
Melt index 680, 682–6
Melting point 165
Memory unaccessed 37
Mesh size 85
Metallocene plastic 163, 233, 314, 359
Metering pump, *see* Gear pump
Metric conversion 732–4
Mitered cut 27
Mixing
distributive 64, 615
extensive 61, 63
external 61
homogenizing 60
intensive 61–2
internal 61
uniform, *see* Static mixer
*see also* Screw
Modulus of elasticity, plastic 169, 182, 604–6
Moisture
absorption 20
mechanics 191
barrier 28
migration mechanics 191
removal 114
Molding area diagram 20
Molding volume diagram 20
Molding with rotation 643–4
Molecular weight 165–8
Molecular weight distribution
basic 165–8
narrow 179
Monofilament, *see* Fiber
Monomer
removal 63
Motor, *see* Drive motor
Mount III, E. M. 102
Myths and facts 44–5

National Safety Council 38
Netting 663, 666
*see also* Die
Newtonian flow 44, 176, 179
Nip roll 36

Nissel, Frank R. 52, 144, 151
  *see also* Welex
Nitriding 76, 80
  *see also* Barrel construction
Noise 73
  trouble 129
Nonintermeshing screws, *see* Twin-screws
Nonisothermal condition 180
NonNewtonian flow 176, 179
Nonsteady state melt instability 147
Nylon fastener strip molding 645–8

Operation
  cost 706
  environment 158
  lock-out 36
  mode 47
  optimized 46
  over-heated 12
  people, *see* People
  product 23
  safe 33
  setting 11
  *see also* Processing; Extruder
Operator 12
  *see also* People, Training; Safety
Opponents of plastics 721–2
Orange peel 492, 590
Orientation
  basic 104–13
  biorientation 107, 109
  blown film, *see* Blown film
  cost 113
  fiber 104
  film 104
  frozen-in stress released 111
  injection/blow molding 643–4
  memory 111
  molding with rotation 643–4
  plastic performance 112
  processing conditions 107, 109
  rate of stretching 105
  roll stretching, *see* Roll stretching
  rules 111
  schematic 106, 108–10
  shrink film 111–12
  stretching degree 104–13
  tape 105, 110
  temperature of 107
  tentering frame 107–9
  terminology 106
  unidirectional 107
Original equipment manufacturer 37
Output direction 66
Output rate
  basic 29–30
    *see also* Screw
  dewatering 65
  estimator 143
  gear pump efficiency 90–1
  increase 31, 127
  interaction extruder to die 125
  reduction 48
  screen pack 85
  sheet line 7
  target 119
  temperature effect 30, 125
  uniform 123
Overload 138
Overview 1
Oxidation 15
Oxygen barrier 28
Ozone discharge, *see* Roll, corona treatment

Packaging
  energy consumption 722–5
  FDA 28
  film stretching, *see* Orientation
  shelf-stable food 151
Paper
  coating manufacture 155
  recycling 42
Patent 54
Patience and processing 47, 725
Pellet, *see* Compounding
People
  basic 47, 725
  computer 45
  experience 696
  productivity 46
Perfection
  material 46, 168
  people 46
  process 46
Performance prediction 43
Performance-to-cost 703
Personnel
  basic 10
  safety 33, 35
Petroleum consumption 41, 721–2
Phoenix Gummiwerke AG 51
Physical preperty, plastic 169, 182, 604–6
Pilot scale problem 64
Pipe
  basic 494–9
  calibration 497, 501, 510
  characteristic 503–14
  coextrusion 515
  collapsing 509
  control 494, *see* Control
  cooling 501
  corrugated 524, 664–5
  cost 120, 495, 525
  cutter 512–4
  definition tube or 494
  die

adjustment 522
basic 235, 520–2
calculation 516
counter-rotating 522
crosshead 520
design 521–2
floating plug 520
multihead 498, 516
pressure 11
streamline 520
*see also* Die
draw down control 496
feeding 495
free extrusion 496, 503, 505
guide 509
jacketing 516
line 3, 495
machine control 518
optimization 514–16
orientation 518–20, 523
*see also* Orientation
output rate 500
plastic
acetal 502–3
basic 501–3
crosslinked PE 502
liquid crystal polymer 522–3
plug expansion 519
product 496, 522–3, 524
puller 499
shrinkage 511
size/dimension 494, 498, 509
sizing
basic 495, 497, 500, 505–6
calibrating 519
differential pressure sizing 509
method 511
sophisticated 507
spacer plates 510
vacuum 504, 511
speed 517
squeeze-tube 524
start-up 517
take-away conveyor 505
thin wall 517
troubleshooting 525–7
variable 518
water lubrication 508, 510
water tank 504, 512
weld line 521
Planning
basic 28, 32
line 118–21
reasoning 30
Plant
clean 29
control 719–2
equipment selection 152–7
lighting 120
operation 28
operational manual 12
productivity influence 99
safety 32
water line problem 120
Plastics
abbreviations 164–5
consumption, *see* Consumption
decomposition 33
energy and 41, 721–5
from raw material to product 723
growth with extrusion 725–9
handle with care 113
melting 23
melt performance 12
orientation, 104
*see also* Orientation
oxidizing 15
petroleum consumption 41
poor heat conductor 82
powder 64
preheating 128
processing 160
property 168–72, 182–3, 604–6
*see also* Orientation
reaction 64
screw type 200
selection 702
stretching, *see* Orientation
type 162–7
virgin vs. regrind 116
Plasticator 199
history 54–5
plastic friction 56
purge *see* Purging
screw, *see* screw
Plastics and energy 721–5
Plastics Institute of America 142
Plate-out 372, 431
Plug
flow phenomena 45
freeze-off 34
pock mark 134
unmelted 15
Positive displacement pump, *see* Gear pump
Postforming 661–6
Predicting performance 43
Preheating plastic, *see* Plastic
Pressure
flow 9, 23
gear pump reduction 93
gradient 58
loss 59
output effect 126
screen pack reduction 86
vapor 191

Pressure (*Contd*)
  variation 60
Printing problem 133
Problem
  basic 697
  encountered 44, 50
  pilot scale 64
  software solution 142–4
  vs. solution 45
  *see also* Troubleshooting
Process control 37, 159, 283–6, 687
  *see also* Computer; Control
Process improvement 710–1
Processing
  analysis 1, 19, 32
  change 14
  condition 1
  explosive force 33
  extruder operation 116–17
  facts 44–5
  feature 46
  high stress (near melt elastic limit) 175
  intelligent 304, 696
  line reference base 22
  lock-out 36
  optimum 180–1
  orientation, *see* Orientation
  and patience 725
  plastic 160
    changing plastic 187
  rheology effect 174
  safety 38
    *see also* Safety
  temperature range 181
  time 47
  variable 688, 690–8
  viscoelasticity effect 174
  window 19, 149
  *see also* Operation; Extrusion volume diagram
Processor 42–3
Process selection 701–2, 712–19
Process validation 689–90
Product
  appearance 30
  caulking 730
  consistency 59
  estimating 142–4
  film extruded on site 729
  laboratory 43
  performance 702
  product size vs. process 712
  sewing machine thread 730
  type 42
  variation 59
  *see* the different processes
Production requirement
  equipment 25
    rebuilding 157
    selection 152–7
  fine tuning 121
  line 54
  lock-out 36
  pilot scale 64
  product estimating 142–4
Productivity
  communication protocol 100
  equipment control 99
  maximum 32
  people 46
  sacrifice 24
Profile
  basic 528–30
  building wall panel 531–2
  characteristic 533, 535
  coated 530, 535
  coextrusion 547
  control, *see* Control
  cooling 536
  decorative strip 547
  design product ideas 547–9
  die
    basic 542–6
    conical shape 543
    crosshead 535
    modular 543
    plate design 546
    spider 529
    *see also* Die
  dimensional tolerance 535
  dry-sleeve calibrator 538
  line 529, 535
  manufacture 153, 530
  mechanical strength advantage 540–1
  melt 11
  optimization 536
  plastic 532–3
  product 541
  robotics process 540
  rod 539
  rotated strip construction 541
  shape 528–32, 544
  shrink 533
  substrate 528, 535
  swelling 545
  troubleshooting 550–1
  vacuum sizing calibrator 538
  vs. pultrusion 528
  window frame 528, 537
Profitability 3, 730
Programmable controller 37
  *see also* Computer; Process Control
Property improvement 104
  *see also* Orientation

Property of plastic 604–6
Proprietary processor 43
Proteus feedblock die 252–3
  *see also* Die
Psychrometric chart 192
Pulling device 25
Pulses 91
Pumping action, *see* Screw
Purging
  agent 186–8
  plastics processed 13
  material handling 113
    *see also* Material handling
  quality 123
  start-up 121–2
Pyrometer 134

Quality control
  acceptable quality level 687
  additive 673–5
  basic 686–9
  Chisolm's Law 698
  cost 693, 706
  customer satisfaction 673
  density and specific gravity 677
  good manufacturing practice 689
  intelligent processing 696
  judgment and experience 696–7
  on-lie 687
  planning 689
  problem and solution 697
    *see also* Troubleshooting
  process control 693–5
  processing defect 673
  product 54, 123
  rule 676
  statistical process control 687–8
    *see also* Control
  system regulation 689
  testing, *see* Testing
  variable 688–9, 690–3, 695
Quality assurance 674–5, 686

Rail car unloading 4, 28–9
  *see also* Material handling
Ram extruder 6, 50, 483
Rauwendaal, C. 126
Reactive extrusion 620, 622
Reclamation 42
Recycle 188–90
  beverage bottle 189
  calendering 434
  energy 41
  lumber 729
  screen pack 88
  thermal reclamation 42
  *see also* Granulating
References 737–53
Refrigerator liner 43
Reifenhauser 319, 500, 601–2
Rensselaer Polytechnic Inst. 225
Requirement 31
Residence time
  die 264
  hygroscopic plastic 114
  short 65
  thermoplastic 174
Return on investment 31
Rheology 174
Risk
  acceptable 39
  zero 39
Rod manufacture 153
Roll
  adjustable 103, 322–3
  alignment 101
  caterpillar 25
  coated web 102
  cooling 102
  corona treatment 103–4
  dancer 101
  expander 102
  film stretching 102
  flag 14
  flame treatment 103
  grooved 351
  idler 102
  lay-on force 101
  limitation 103
  nip 36, 103
  stretching 102
    *see also* Orientation
  takeoff 25
  tension control 101
  transfer 28
  types 100–4
  winder 101
  winding strain 102
Roll cast film, *see* Flat film
Roman extruder 40
Rope stretching, *see* Orientation
Rotating molding 645–9, 650
Rubber behavior 162
Rupture disc, *see* Safety

Sack stretching, *see* Orientation
Safety
  automobile 39
  blown film 321
  bolt 84
  calender 438–9
  check list 33
  clothing 15
  device 33, 36

Safety (*Contd*)
  equipment 158–9
  freeze-off 34
  industry 38
  interlock 37
  lock-out procedure 36–7, 158
  medical attention 35
  operator 35
  plant 32
  potential problem 12, 121
  process 38
  programmable controller 159
  purged start-up 121–2
  rupture disc 36, 73
  shear pin 36
  warning signs 158
Salt bath 481–2
Sansone, L. F. 64
Scale-up 44
Schematic, single-screw extruder 57
Science fiction, *see* Myths
Scrap, clean 20
Scrapless forming 660–1
Screen changer 86
  *see also* Screen pack
Screen pack 84–9
  adapter 82
  backflush 88
  breaker plate 84
    *see also* Breaker plate
  filter media 88
  output 85
  mesh size 85–8
  schematic 85
  screw flight change 89
  shut-down 124
  troubleshooting 135
Screw 199–208
  adapter to die, *see* Adapter
  auger action 205
  barrel, *see* Barrel
  barrier 219
  bridging 138
    *see also* Melt
  channel
    melt shear 177
    problem 94
    shallow 63
  cleaning 17
  component 199
  compression ratio 209–11
  compression zone, *see* Screw, transition zone
  compressor 20
  computer modeling 215
  concentricity 203
  conical 6
  conveying 20, 207
  cooling 213
  C/R, *see* Screw, compression ratio
  design 10, 208–10
  dimension 202–4
  double wave 219
  drive, *see* Drive motor
  efficiency 212
  energy required 212
  feed zone 200, 206
  freeze test 221–2
  friction action 213
  general purpose 201
  geometry 22, 207
  grooved feed 94
  heat transfer 207
  hypothetical data 209–10
  inspection 226–7
  interchangeable section 61–2
  kneading 207
  L/D
    basic 75, 200, 209
    power requirement 81
  manufacturing type 152–7
  melting
    action 205, 220–1
    capacity 208
    inspect 124
    model 206, 209, 219
    simulation 223, 225
  metering zone 200, 206
    depth ratio 223
  misconception 210
  mixing 216–18
    action 208–9
    Davis-Standard 217
    Dulmage 217
    Maddock 217–18
    pin 218
    Pulsar 218
    type 217
  multiple 207
    *see also* Twin-screw
  nomenclature 200
  output
    basic 210–2
      *see also* Output
    cost 215, 710
    estimator 143
    formula 210–11
    loss 212
    rate 202, 204–5, 710
    thermoplastic 214
  parabolic 6
  parallelism 204
  performance 214, 216, 224, 225–6
  pitch 207
  plastic 200
  plasticator, 208

*see also* Plasticator
processing variable 224
pumping
basic 8, 56
force 216
process 199
rate 60
ratio 223
stability 59
rotation 22
screen pack influence 89
secretive 208
single 56
schematic 57
single vs. twin 55
size 201
speed 14, 70
speed and torque 24
starve-fed 65, 141
static mixer effect 93
surface 204
surging 91, 213
*see also* Surging
temperature 205
tip 202
tolerance 202–4
torque 8, 212
*see also* Drive motor
transition zone 200, 206
twin-screw, *see* Twin-screw
two-stage
basic 217
nomenclature 222
type 202–4
undercut 9, 202
vapor 201
variable 55
vented
basic 221–4
double 223
non vented 222
two-stage 222
*see also* Vented barrel
wear 226
wobble 138
zone 199–201
Screw/barrel
centerline alignment 67
clearance 9, 58, 202
Screwless 7
Secondary operation 116
equipment 25
planning 118
Sensor
barrel 96
control 290–6
*see also* Control
Shape 11
down sizing 26
fixture 26
vacuum box 26
Shear
behavior 176
rate 61, 118
screw channel 177
stress vs. shear rate 178
Shear pin 36
Sheet
air knife 380, 400
alignment 386–7
antistatic bath 401, 404
basic 376–83
coextruded 396–400
control 402–4, *see* Control
cooling 377
cutter 403, 405–6
die
basic 409–11
coathanger 409
design 236
pressure 11
*see also* Die
equipment 81, 378–82
heat glazier
laminated
basic 396–400
capping 399
substrate cover 398
line
component 377–82, 388–96
schematic 377, 380,385, 389, 392, 398, 399, 404
safety 33
manufacturing 154
melt
block 413
profile 411
*see also* Melt
method 378
optical 399
optimization 388
orientation 387, *see* Orientation
output rate 384
plastic
ABS 383–6
basic 383
polyethylene terephthalate 386–7
polystyrene, glossy high-impact 151
product 396, 411–12
residual memory 383
roll
chain drive 391
double-shell 391
embossed 388, 396
polishing 389, 393
pull 400

Sheet (*Contd*)
size 389
stack arrangement 377, 379–82, 388–96
sticking 393
temperature 377, 380, 384, 387, 389, 391–2
scrap reclaiming 398
take-off rolls 379
tape 402
temperature 384–6, 387
thickness gauge 405
trim and slit 400–2
troubleshooting 412–17
vented extruder 379
winder 406–8
Sheet and film thickness 376
Shrink film 111–2
Shut-down 15, 124–5
flat film 362
maintenance 129
prevent 24
safe 33
Side feeder, *see* Feed
Siding, house in-line 119
Silicone grease 17
Silver streak 194
Silo
clean 20
controller 114
material transfer 28–9
protection 114
Single-screw, *see* Extruder
Single-screw extruder
basic 56, *see* Extruder
vs. twin-screw 60
*see also* Twin-screw
Sizing
fixture 25
plate 26
Skiving film 357, 483
Slide-plate screen changer 86
*see also* Screen changer
Slot cast film, *see* Flat film
Society of Plastics Industry
barrel construction 75, 76
communication protocol 100
safety 38–9, 158
screw dimension 202
Software
die design 270–3
operation 45
training 142–4
*see also* Computer
Solid-phase forming 660–1
Solids flow/liquid flow 97
*see also* Feed; Melt flow
Solution 45
Solvent removal 63
Specific heat 182, 184
Spiral die, *see* Die
Spirex 142, 218
Stamping/injection molding 642–3
Start-up
basic 121–4
blown film 320–3
common factor 14
extrudate 26
flat film 362
freeze-off 34
operation mode 47
retread 15
safe 33
static mixer reduce damage 93
stringing 12
temperature 13
threading 12
Starve-fed 65
*see also* Screw
Static mixer
adapter 17, 82
basic 92–3
L/D effect 93
location 92
melt uniformity 92
Statistics
accidents 38
melt distribution 61
process control 687–8
Steady-state melt instability 146
Steel
energy 41
property 169, 604–6
recycling 42
Strength of plastic 169, 182, 604–6
Stress
analysis 44
streak 137
Stretching plastic, *see* Orientation
Success by design 700–1
*see also* Design
Summary 699
Superex Polymer 151, 522–3
Surging 9, 91, 131, 139, 415, 492
coextrusion 146
*see also* Screw

$T_g$, *see* Glass transition temperature
Takeoff 25
Tape
manufacture 155, 356
*see also* Flat film
oriented 105, 110
*see also* Orientation
Technical capability 31
Technical cost modeling 707–9
Techware Design 142

Temperature
  dry-bulb 192
  false reading 136
  melting point 181
  output effect 30, 126
  overheat 34
  plastic decomposition 173
  rise 48
  variable 45
Tensile strength 169, 604–6
Tension control 25
Tentering frame, *see* Orientation
Term
  plastic, polymer, resin 160
  tool, die, mold 228
Testing
  basic 675–7
  Chisolm's law 698
  computer 44
  condition 676
  density and specific gravity 677, 681
  experience 696–7
  failure type 676
  field 43
  intelligent processing 696
  judgment 696–7
  melt
    flow 692–3
    index 680–6
    pressure 694
    rheometer 682–6
  moisture 691
  molecular weight distribution 682
  people 676
  plastic 674–5
  process control 693–4
    *see also* Control
  processing intelligent 696
  product 44, 695–6
  quality, *see* Quality
  regulation 679–80
  rule 676
  service life 676
  specification and standard 677, 679
  stability 690
  test method 678
  troubleshooting 697–8
  type 677
  variable
    control 693–5
    extrusion 690–1
    melt flow
    plastic 691
    product 695–6
    temperature 694
Theoretical analysis 10
Theoretical vs. actual plastic property 169, 604–6
Thermal
  condition 22
  conductivity 58, 182, 184
  diffusivity 182, 184–6
  pin 249
  property 180, 182
  reclamation 42
  stability 184
Thermodynamic analysis 10, 30
Thermoforming
  advantage 650
  basic 650–2
  building 669–70
  coextruded 656
  corrugated pipe 664–5
  cost 652
  die, *see* Die
  equipment 651–7
  extrusion line 652
  film/sheet
    precision 91
  foaming
    basic 663, 667–70
    film 671
    troubleshooting 671–2
  fuel tank 657
  mold 656
  netting 663, 666
  output 652
  postforming 661–3
  processor 43
  product 654, 656, 657
  rotary
    carousel 655
    drum 652–4, 657
  scrapless 660–1
  shape 650–4, 656–6, 660–9
  shuttle type 655
  thickness 656
  troubleshooting 658–9
  *see also* Orientation
Thermoplastic
  behavior 162
  consumption 5
  melt profile 163
  screw 199
  window 19
Thermoset
  behavior 162
  consumption 5, 80
  melt profile 163
  screw 199, 201
Theysohn Maschinenbau GmbH 619
Thrust bearing 75
Tin recycling 42
Tolerance
  correct/required 174
  precision 91

Tooling
  change 33
  damage 89
Torque winder 28
  *see also* Screw
Total indicator reading 100
Toxic 39
Training
  basic 12, 142
  commonsense 34
  productivity 46
  *see also* Experience
Transducer, *see* Control
Trial and error 215
Troester, P. 51
Troubleshooting
  air entrapment 139
  basic 130–42, 697
  blow molding 590–2
  blown film 344–8
  calender 440
  checklist 47–9, 130
  coating 466–8
  coextrusion instability 142
  compounding 629–30
  die 276–82, 279–81
  drying plastic 194–7
  fiber 606–7
  flat film 372–5
  foam film 671
  gel 141–2
  pipe 525
  profile 525, 550–1
  screen pack 135
  sheet 412–7
  software 142–4
  surging 139–41
  thermoforming 658–9
  tubing 526–7
  wire and cable 492–3
  *see also* Maintenance
Truck
  forklift 119
  unloading 28
  *see also* Material handling
Tube
  definition pipe or 494
  die 235
  manufacture 153
  *see also* Pipe
Tubular film manufacture 155
Twin-screw
  compounding 608–23
  developed 52
  extruder 60
  horsepower 65
  maintenance 129
  melting action 207
    *see also* Screw
  monopoly 63
  starve-fed 65
  tapered 63
  torque 65
  variable 55
  vs. single-screw 55, 60
  *see also* Screw

Unload plastics, *see* Material handling; Rail car; Truck
Unorientation, *see* Oriented
Up-stream, *see* Auxiliary equipment
USA equipment 5
Used equipment 157–8
Utility 29

Vacuum box 26
Valve 83
Vapor pressure 191
Variable
  pressure 23
  temperature 24
Velcro strip injection molding 645–8
Velocity gradient 62
Vented barrel
  basic 36
  Bernhardt, Ernest 52
  hygroscopic plastic 114
  *see also* Screw
Viscoelasticity
  behavior 161, 174
  coextrusion channel 148
Viscosity
  behavior 168, 179
  friction 56
  gear pump 89
  measurement 118
  temperature influence 24, 125
Volumetric blending, *see* Blending
Vulcanization 479

Warehouse option 120
Warm-up 11
  *see also* Start-up
Waste incineration 42
Water
  cooling 26
  pollution 39
  problem 120
  property 736
  removal/extracted 63–4, 114
  waste 42
Wear, *see* Screw
Web guide 4
Welex
  clamping system
    bolted 84

swinging gate 83
coextruder 144
compounding 617
extruder
air knife 380
belt-driven 72
double-vented 223
feed section 94
in-line 66
vented 379
feed crammer 21
microprocessor 285
pipe/profile line 3
profile and tube 497–8
screen
changer 17
hopper 98
sheet die 229
sheet line 81, 285, 378–9
anti-static bath 404
jumbo roll winder 408
razor edge trimmer 401
rotary slitter 413
stack roll positions 379, 381–2, 394–5
take-off 3-roll 91
winding into roll stock 407
single-screw extruder schematic 56–7
thermoforming
in-line rotary drum 654
product 654
sheet line 652–3
sheet line reduced 653
tubing 496
*see also* Nissel, Frank R.
Werner and Pfleider 62, 610–11, 618, 620, 626
Whisker, *see* Fiber
Winder
speed 28
tension 14
turret 354
Windmoeller and Hoelscher 341
Window, *see* Processing
Wire braiding 120
Wire covering
adjustment 484, 490
basic 469–71
capstan 470, 475
catenary continuous vulcanization 479
characteristic 474–8
coextruded 473, 491
compounding 474
concentricity 484
control 477
*see also* Control
cooling 476
definition 471
diameter measurement 477
die
basic 486–91
crosshead 470, 475
high pressure 487
low pressure 488
pressure 11, 487
sleeve/tube 488
tapering 489
*see also* Die
drawdown ratio 488–90
draw ratio balance 489–90
eccentricity 477
liquid salt continuous vulcanization 481–2
machine setting 485
manufacture 153
melt
shear 470
temperature 485
moisture initiated crosslinking 483
multi 491
optimization 48–6
output rate 470, 473
plastic
basic 471–4
crosslinked PE 478, 491
polytetrafluoroethylene 483, 491
preheating 475, 490–1
radiation curing 481
ram extruder 483
roll
dual unwind 475
windup 470
schematic 470, 479–84, 487–91
sheaths and protective 474
size and shape 469
skiving 483
special line 478–84
speed 473, 475–6
start-up 483–4
steam continuous vulcanization 479
thermoset gas curing 481
troubleshooting 492–3
variable 485
Wood pier 729
Worldwide plastics growth 726

Yarn, *see* Fiber
Yogurt cup 151

Zero risk, *see* Risk